OT 45
Operator Theory: Advances and Applications
Vol. 45

Editor:
I. Gohberg
Tel Aviv University
Ramat Aviv, Israel

Editorial Office:
School of Mathematical Sciences
Tel Aviv University
Ramat Aviv, Israel

Birkhäuser Verlag
Basel · Boston · Berlin

Joseph A. Ball
Israel Gohberg
Leiba Rodman

Interpolation of Rational Matrix Functions

1990

Birkhäuser Verlag
Basel · Boston · Berlin

Authors' addresses:

Prof. Joseph A. Ball
Department of Mathematics
Virginia Polytechnic Institute
and State University
Blacksburg, Virginia 24061
USA

Prof. Leiba Rodman
Department of Mathematics
The College of William and Mary
Williamsburg, Virginia 23187-8795
USA

Prof. Israel Gohberg
School of Mathematical Science
Raymond and Beverly Sackler Faculty
of Exact Sciences
Tel-Aviv University
Ramat Aviv 69978 Tel-Aviv
Israel

Deutsche Bibliothek Cataloguing-in-Publication Data

Ball, Joseph A.:
Interpolation of rational matrix functions / Joseph A. Ball ;
Israel Gohberg ; Leiba Rodman. – Basel ; Boston ; Berlin :
Birkhäuser, 1990
   (Operator theory ; Vol. 45)
   ISBN 3-7643-2476-7 (Berlin ...)
   ISBN 0-8176-2476-7 (Boston)
NE: Gochberg, Izrail':; Rodman, Leiba:; GT

© 1990 Birkhäuser Verlag Basel
Printed in Germany on acid-free paper
ISBN 3-7643-2476-7
ISBN 0-8176-2476-7

To my mother

ANGELA BALL

and in memory of my father

W. HOWARD BALL

J.A.B.

In memory of my parents

HAYA CLARA and TSUDIC GOHBERG

I.G.

To my parents

HAYA and ZALMAN RODMAN

L.R.

PREFACE

This book aims to present the theory of interpolation for rational matrix functions as a recently matured independent mathematical subject with its own problems, methods and applications. The authors decided to start working on this book during the regional CBMS conference in Lincoln, Nebraska organized by F. Gilfeather and D. Larson. The principal lecturer, J. William Helton, presented ten lectures on operator and systems theory and the interplay between them. The conference was very stimulating and helped us to decide that the time was ripe for a book on interpolation for matrix valued functions (both rational and non-rational). When the work started and the first partial draft of the book was ready it became clear that the topic is vast and that the rational case by itself with its applications is already enough material for an interesting book. In the process of writing the book, methods for the rational case were developed and refined. As a result we are now able to present the rational case as an independent theory. After two years a major part of the first draft was prepared. Then a long period of revising the original draft and introducing recently acquired results and methods followed. There followed a period of polishing and of 25 chapters and the appendix commuting at various times somewhere between Williamsburg, Blacksburg, Tel Aviv, College Park and Amsterdam (sometimes with one or two of the authors). It took us four years to complete the whole project. Much of this time was spent on filling in the gaps and missing connections between different parts of the theory. If the reader finds a unified and appealing basic theory as a result of all these efforts, the authors will be gratified.

We are aware that even for rational matrix functions many developed topics of interpolation are not included in this book. These include the theory of singular or degenerate cases, problems with two or more blocks, interpolation with symmetries, Schur-like methods, methods based on commutant lifting theorems, methods connected with reproducing kernel Hilbert spaces and others. Remarks on and references to these matters can be found in the Notes for Part V. We hope to return to some of these themes in future work.

The authors would like to thank their home institutions, Virginia Tech, Tel Aviv University, and the College of William and Mary for supporting this project. It is also a pleasure to thank the Vrije Universiteit in Amsterdam, the University of Maryland in College Park, and the AMS 1988 Summer Institute on Operator Theory/Operator Algebras at the University of New Hampshire in Durham for their hospitality during various extended meetings of the authors. Finally we wish to thank the National Science Foundation, the Air Force Office of Scientific Research, and the Lilly and Nathan Silver Chair in Mathematical Analysis and Operator Theory in Tel Aviv University for partial support.

It is a pleasure to thank also our colleagues Daniel Alpay, Harm Bart, Nir Cohen, Harry Dym, Ciprian Foias, Rien Kaashoek, Peter Lancaster, Leonid Lerer, André Ran, Marvin Rosenblum and Jim Rovnyak for their advice and interest. We would also like to thank Jeongook Kang for proof-reading a part of the manuscript.

CONTENTS

PART III   SUBSPACE INTERPOLATION PROBLEMS

PART IV   NONHOMOGENEOUS INTERPOLATION PROBLEMS

INTRODUCTION

This book develops systematically a self-contained theory of interpolation for rational matrix functions and includes samples of applications to modern systems theory and control. Both homogeneous and nonhomogeneous interpolation problems, with and without additional constraints are treated. In each case conditions for the existence of a solution and explicit formulas and procedures are presented.

As is well known any proper rational matrix function $W(z)$ can be expressed in terms of four complex matrices A, B, C, D as

$$W(z) = D + C(zI - A)^{-1}B.$$

This representation is called a realization of the function $W(z)$. Realizations allow one to reduce interpolation problems to problems in matrix theory. One of the main features of the book is that the whole theory is developed in terms of the matrices A, B, C, D within the framework of finite matrices and finite dimensional linear spaces. The idea of using realization as a working tool for dealing with rational matrix functions, which comes from systems theory, underlies the unifying approach of the book. Systems theory also provides the motivation and applications for many of the interpolation problems considered in the book. The choice of the class of rational matrix functions as well as the demand for solutions to be presented in an explicit realized form is also dictated by systems theory. The last of the six parts of the book is devoted explicitly to applications in systems theory.

We would next like to describe some of the problems solved in the book in the most transparent scalar formulations. The first interpolation problem is to find a rational function with given zeros and poles (counted with multiplicities). We call this problem homogeneous because any nonzero scalar multiple of one solution is again a solution. A second problem is the nonhomogeneous interpolation problem (which is usually attributed to Lagrange or Lagrange-Sylvester and also to Newton) of finding a rational function with specified values at certain specified points. The next problem incorporates a metric constraint; the object is to find a rational function which is analytic and with modulus at most one on the unit disk which takes certain specified values at certain points inside the unit disk; this is the famous problem of Nevanlinna and Pick. The well-known problem of Nehari is that of finding a rational function with values of modulus at most one on the unit circle having prescribed Fourier coefficients with negative indices. The problem of Caratheodory-Toeplitz is that of finding a rational function with the first n Taylor coefficients at the origin prescribed, which maps the unit disk into the right half plane. These last three problems are all nonhomogeneous with metric constraints. We also consider the problem of finding a rational function having prescribed zeros inside the disk and having values of modulus one on the unit circle; this is a homogeneous interpolation problem with an additional metric constraint. Problems with other types of supplementary constraints will also be considered.

It is perhaps well to emphasize that we understand interpolation problems in a broad sense. For instance, we consider the problem of finding a common multiple of a collection of polynomials as a homogeneous interpolation problem and the problem of finding a matrix polynomial with given remainders after division by various other given polynomials as a nonhomogeneous interpolation problem.

Matrix versions of all problems described above are treated in the book. In general a scalar interpolation problem often leads to many different possible generalizations for matrix valued functions. For instance, the condition for a scalar function P that

$$(1) \qquad\qquad\qquad P(z_0) = w$$

can be generalized to the matrix-valued case in the following ways:

$$(2) \qquad\qquad\qquad P(z_0)u = v$$
$$(3) \qquad\qquad\qquad xP(z_0) = y$$
$$(4) \qquad\qquad\qquad xP(z_0)u = \rho$$
$$(5) \qquad\qquad\qquad P(z_0) = M$$

where u and v are given column vectors and x and y are given row vectors, M is a given matrix, $\rho \in \mathbb{C}$, and $P(z)$ is the unknown matrix valued function. Depending on the choice of the generalization (2)–(5) of condition (1), one can state several distinct matrix interpolation problems. Following the suggestion of M.G. Krein (see Fedchina [1975]), we refer to conditions of type (2) and (3) as tangential interpolation conditions. For example, (2) is called a right tangential interpolation condition since it specifies the value of $P(z_0)$ along a direction on the right side, and hence similarly we refer to (3) as a left tangential interpolation condition and to condition (4) as a bitangential (or two-sided) condition. Another source of difficulty in the generalizations for matrix valued functions is that in contrast with the scalar case a rational matrix valued function and its inverse both may have a pole at the same point; for example $W(z) = \begin{pmatrix} 1 & z^{-1} \\ 0 & 1 \end{pmatrix}$ and $W(z)^{-1} = \begin{pmatrix} 1 & -z^{-1} \\ 0 & 1 \end{pmatrix}$ both have an entry with a pole at 0.

As a simple representative of the results in this book, we describe the left tangential Nevanlinna-Pick interpolation problem over the right half plane $\Pi^+ = \{z : \operatorname{Re} z > 0\}$. The data for this problem consist of a set of given nonzero $1 \times M$ row vectors $x_1, \ldots, x_n$ and $1 \times N$ row vectors $y_1, \ldots, y_n$ and a set of distinct points $z_1, \ldots, z_n$ in Π^+. The problem is to describe all rational $M \times N$ matrix functions $F(z)$ such that

$$(\text{i}) \qquad\qquad\qquad F \text{ is analytic on } \Pi^+$$
$$(\text{ii}) \qquad\qquad\qquad \sup_{z \in \Pi^+} \|F(z)\| < 1$$

and

$$(\text{iii}) \qquad\qquad x_j F(z_j) = y_j, \qquad 1 \le j \le n.$$

The complete solution of this problem is:

THEOREM I. *The left tangential Nevanlinna-Pick interpolation problem over* Π^+ *with given data* $\{x_1, \ldots, x_n; y_1, \ldots, y_n; z_1, \ldots, z_n\}$ *is solvable if and only if the* $n \times n$ *matrix*

$$\Lambda = \left[\frac{x_i x_j^* - y_i y_j^*}{\overline{z}_j + z_i} \right]_{1 \le i,j \le n}$$

is positive definite. In this case one particular solution is given by

$$F(z) = [X^*(zI + A_\zeta^*)^{-1}\Lambda^{-1}Y][I + Y^*(zI + A_\zeta - \Lambda^{-1}YY^*)^{-1}\Lambda^{-1}Y]$$

where

$$X = \begin{bmatrix} x_1 \\ \vdots \\ x_n \end{bmatrix}, \qquad Y = \begin{bmatrix} y_1 \\ \vdots \\ y_n \end{bmatrix}$$

and

$$A_\zeta = \begin{bmatrix} z_1 & & 0 \\ & \ddots & \\ 0 & & z_n \end{bmatrix}.$$

The general solution is given by

$$F(z) = \left(\Theta_{11}(z)G(z) + \Theta_{12}(z) \right) \left(\Theta_{21}(z)G(z) + \Theta_{22}(z) \right)^{-1}$$

where

$$\Theta_{11}(z) = I_M - X^*(zI + A_\zeta^*)^{-1}\Lambda^{-1}X$$
$$\Theta_{12}(z) = X^*(zI + A_\zeta^*)^{-1}\Lambda^{-1}Y$$
$$\Theta_{21}(z) = Y^*(zI + A_\zeta^*)^{-1}\Lambda^{-1}X$$
$$\Theta_{22}(z) = I_N - Y^*(zI + A_\zeta^*)^{-1}\Lambda^{-1}Y.$$

More complicated problems of Nevanlinna-Pick type where the interpolation conditions of the type (1) are replaced by conditions of the form

(6) $$\frac{1}{2\pi i} \int_\gamma F(z)C_-(zI - A_\pi)^{-1} = C_+$$

(7) $$\frac{1}{2\pi i} \int_\gamma (zI - A_\zeta)^{-1}B_+F(z) = -B_-$$

(8) $$\frac{1}{2\pi i} \int_\gamma (zI - A_\zeta)^{-1}B_+F(z)C_-(zI - A_\zeta)^{-1} = \Gamma$$

where C_+, C_-, A_π, A_ζ, B_+, B_-, Γ are given matrices of compatible sizes and γ is a suitable simple closed contour are also treated in the book as are problems of the Nevanlinna-Pick-Takagi type, where the solutions are allowed to have a prescribed number of poles in the unit disk.

The interpolation problems considered in this book have a rich history. The results for the scalar case were obtained many decades ago (some were even resolved in the previous century). Especially influential during this first classical period were such mathematicians as Lagrange, Sylvester, Hermite, Schur, Caratheodory, Toeplitz, Nevanlinna, Pick, Takagi, Loewner, and Nehari. Most of this work in the classical function theory style took place before World War II. A new phase began with the work of Akhiezer and Krein in the 1930's who introduced functional analysis and operator theoretic methods to interpolation problems. In the early 1950's a new period started where interpolation problems for matrix functions were considered. Important contributions here were made by Potapov and his associates, Sz.-Nagy-Koranyi, Adamian-Arov-Krein, Krein, Nudelman, Sarason, Sz.-Nagy-Foias, Helton and Rosenblum-Rovnyak. The area of interpolation problems in general has been and continues to be very active. In fact M. Heins [1982] has collected a bibliography with some 350 entries on interpolation up to the year 1982.

In the last fifteen years or so research activities in interpolation theory started to flourish again and the list of important contributors is quite long. We shall quote and describe in more detail the contributions which are related to the developments and results described in this book in the Notes for each part. It is remarkable that the theory of interpolation which developed over close to 100 years as a purely mathematical quest turned out to have great importance for practical problems. In recent years the number of these important problems has increased considerably. A sample of some such recent engineering applications in also included.

Part I of the book sets up the machinery and terminology needed for dealing with rational matrix functions and matrix polynomials and includes the basic results about rational matrix functions and results on the basic homogeneous interpolation problems which are used later. Part II considers homogeneous interpolation problems related in a natural way to the problems considered in Part I but which are not needed or used in the sequel; this Part is included for completeness. In Part III we set up the theory of null-pole subspaces and solve the corresponding homogeneous interpolation problem. These results present a finite dimensional analogue of the Ball-Helton theory; in this form it appears here for the first time. Only some results from Chapters 12 and 13 from this Part are used later in the book. Part IV contains a matrix function version of Lagrange and Lagrange-Sylvester interpolation and their generalizations. Also treated in this Part as another nonhomogeneous interpolation problem is the partial realization problem from systems theory. Part V treats the matrix generalizations of the classical problems of Nevanlinna-Pick, Schur-Takagi, Nehari, boundary Nevanlinna-Pick and Caratheodory-Toeplitz; here the results from Parts I, III, IV are used in an essential way. Finally Part VI presents three applications from modern systems theory, namely, sensitivity minimization, model reduction and robust stabilization. These applications are all presented in a self-contained form which include the engineering motivation and does not assume any previous knowledge of systems theory.

PART I

HOMOGENEOUS INTERPOLATION PROBLEMS
WITH STANDARD DATA

The main content of this part is the generalization for matrix valued functions of the following problems and formulas for scalar polynomials and scalar rational functions. First is the problem of the construction of a polynomial if its zeros $z_1, \ldots, z_m$ and their multiplicities $k_1, \ldots, k_m$ are given. In the scalar case the polynomial is obtained via the formula

$$(1) \qquad p(z) = c(z - z_1)^{k_1}(z - z_2)^{k_2} \cdots (z - z_m)^{k_m}$$

where c is a nonzero constant. The next problem in the scalar case is to find the rational function if its zeros $z_1, z_2, \ldots, z_m$ and poles $w_1, w_2, \ldots, w_p$ and their respective multiplicities $k_1, k_2, \ldots, k_m$ and $n_1, n_2, \ldots, n_p$ are given. In this case the condition $z_j \neq w_k$ for all j and k $(1 \leq j \leq m, 1 \leq k \leq p)$ has to be fulfilled for a solution to exist. If this condition holds, then a solution is given by the formula

$$(2) \qquad r(z) = c \frac{(z - z_1)^{k_1}(z - z_2)^{k_2} \cdots (z - z_m)^{k_m}}{(z - w_1)^{n_1}(z - w_2)^{n_2} \cdots (z - w_p)^{n_p}}$$

where c is a nonzero complex number. These problems we call homogeneous because any solution remains a solution after we multiply by a nonzero complex constant. The other problems which we are planning to generalize in this part concern construction of rational functions with values of modulus one on the unit circle or imaginary axis given all their zeros or all their poles in the complex plane $\mathbb{C}$, or all zeros and poles inside the unit disk or right half plane (including multiplicities). For example, given the zeros $z_1, z_2, \ldots, z_m$ and the poles $w_1, w_2, \ldots, w_p$ inside the unit disk with multiplicities $k_1, k_2, \ldots, k_m$ and $n_1, n_2, \ldots, n_p$, respectively, then a necessary and sufficient condition for a rational function $u(z)$ having these zeros and poles inside the unit disk and such that $|u(z)| = 1$ for $|z| = 1$ is that $z_j \neq w_k$ for all j and k; in this case the solution is given by

$$(3) \qquad u(z) = c \left(\frac{z - z_1}{1 - z \overline{z}_1} \right)^{k_1} \cdots \left(\frac{z - z_m}{1 - z \overline{z}_m} \right)^{k_m} \left(\frac{1 - z \overline{w}_1}{z - w_1} \right)^{n_1} \cdots \left(\frac{1 - z \overline{w}_p}{z - w_p} \right)^{n_p}$$

where c is a complex number of modulus one. These problems we also consider as homogeneous interpolation problems since multiplication by constants of modulus one transforms one solution to another solution of this problem.

The general problem stated above of generalization of interpolation problems to matrix valued functions consists of two parts. First is the generalization of the concept of null and pole structure to the matrix valued case. Second is the proper generalization of the interpolation problems and formulas to the matrix valued case. A new feature is that a rational matrix-valued function may have a zero and a pole at the same point; this complicates the analysis considerably.

Generalization of the formulas (1), (2) and (3) presented here is far from being straightforward; it is based on the theory of realization for rational matrix functions from systems theory (the necessary portion of which is also explained in this part). We would like to remark that recovering the formulas (1), (2) and (3) for the scalar case from the more general matrix formulas is not easy; we will do this calculation for two examples.

This part consists of seven chapters. In Chapters 1 and 3 is introduced the null and pole structure. In Chapters 2 and 4 the interpolation problems are solved and the formulas analogous to (1) and (2) are obtained. The case of singularities at infinity for matrix functions requires a special treatment; this is done in Chapter 5. Chapters 6 and 7 contain generalizations related to formula (3). Here instead of functions having values of modulus 1 we consider the much more general case of rational matrix valued functions with J-unitary values on the unit circle or imaginary axis. The results of this part are essentially used in Part V. Many of the results of this part are basic for the rest of the book.

CHAPTER 1

NULL STRUCTURE FOR ANALYTIC MATRIX FUNCTIONS

For scalar analytic functions null structure is described by the zeros together with their multiplicities. In this chapter we develop the appropriate generalization of this structure for analytic matrix functions. It is now described in various equivalent forms, namely: null functions, null chains, null pairs, singular parts, null subspaces and null triples. The connections among them form the main content of this chapter. These connections allow greater flexibility in the applications in later chapters.

1.1 LOCAL NULL STRUCTURE OF ANALYTIC MATRIX FUNCTIONS

In this section we shall give the basic definitions of zero or null structure at a point for a matrix analytic function.

Let $A(z)$ be an $n \times n$ matrix function analytic in a neighborhood of the point $z_0 \in \mathbb{C}$ and with $\det A(z)$ not vanishing identically. Such a function we call a *regular analytic matrix function*. We say that the $\mathbb{C}^n$-valued function $\varphi(z)$ is a *right null function for $A(z)$ at the point z_0 of order k* if $\varphi(z)$ is analytic at z_0 with $\varphi(z_0) \neq 0$ and the analytic vector function $A(z)\varphi(z)$ has a zero of order k at z_0:

$$A(z)\varphi(z) = O(z - z_0)^k.$$

It is easy to see that if $\varphi^{(1)}(z), \ldots, \varphi^{(m)}(z)$ are right null functions for $A(z)$ at z_0 and $c_1, \ldots, c_m$ are complex numbers, then the linear combination $\varphi(z) = c_1\varphi^{(1)}(z) + \cdots + c_m\varphi^{(m)}(z)$ is again a right null function as long as $\varphi(z_0) \neq 0$; if no $c_j = 0$, then the order of $\varphi(z)$ as a null function is at least the minimum of the orders of the null functions $\varphi^{(1)}(z), \ldots, \varphi^{(m)}(z)$. If φ is a right null function and $s(z)$ is an arbitrary scalar analytic function with $s(z_0) \neq 0$, then $\widetilde{\varphi}(z) = s(z)\varphi(z)$ is also a right null function of the same order.

EXAMPLE 1.1.1. Let $A(z)$ have the form $A(z) = zI - A$ where A is the $n \times n$ Jordan block with eigenvalue z_0:

$$A = \begin{bmatrix} z_0 & 1 & 0 & \cdots & 0 & 0 \\ 0 & z_0 & 1 & \cdots & 0 & 0 \\ \vdots & \vdots & \vdots & \cdots & \vdots & \vdots \\ & & & & z_0 & 1 \\ 0 & 0 & 0 & \cdots & 0 & z_0 \end{bmatrix}.$$

We analyze the form of a right null function $\varphi(z) = \sum_{j=0}^{\infty} \varphi_j(z - z_0)^j$ of $A(z)$ at z_0. Write $A(z)$ in the form $A(z) = (z_0 I - A) + (z - z_0)I$ and multiply out power series in

$(z - z_0)$ to get

$$A(z)\varphi(z) = \sum_{j=0}^{\infty}[(z_0 I - A)\varphi_j + \varphi_{j-1}](z - z_0)^j$$

(where we set $\varphi_{-1} = 0$). Thus, for $j = 0$ we must have

$$(z_0 I - A)\varphi_0 = 0$$

so $\varphi_0 = c_1 e_1$ for a nonzero constant c_1, where $e_1, e_2, \ldots, e_n$ are the standard basis vectors for $\mathbf{C}^n$. For φ to be a right null function of order at least two, we must have in addition

$$(z_0 I - A)\varphi_1 + \varphi_0 = 0.$$

This then forces $\varphi_1 = c_1 e_2 + c_2 e_1$ for some constant c_2. To achieve a right null function of maximum possible order we choose

$$\varphi_j = c_1 e_{j+1} + c_2 e_j + \cdots + c_{j+1} e_1 \quad \text{for} \quad 0 \leq j \leq n-1.$$

Here $c_1, \ldots, c_n$ are constant complex numbers with $c_1 \neq 0$. The resulting vector function

$$(1.1.1) \quad \varphi(z) = c_1 e_1 + (z-z_0)(c_1 e_2 + c_2 e_1) + \cdots + (z-z_0)^{n-1}(c_1 e_n + \cdots + c_n e_1) + O(z-z_0)^n$$

is a right null function at z_0 of order n. As $\varphi_{n-1} = c_1 e_n + \cdots + c_n e_1$ is not in the image of $z_0 I - A$, there is no way to define φ_n as to make $\varphi(z)$ a right null function at z_0 of order larger than n. Thus n is the largest possible order for a right null function $\varphi(z)$, and we have shown that every right null function of order n has the form (1.1.1). Note that for this example the Taylor coefficients $\varphi_0, \varphi_1, \ldots, \varphi_{n-1}$ for the right null function φ coincide with a Jordan chain of vectors for the matrix A. This gives a connection between right null functions for analytic matrix functions and the theory of Jordan canonical form for finite matrices. $\square$

To organize the collection of all right null functions we now define the notion of a *canonical set of right null functions*. We shall see later (see Corollary 1.1.6) that the set of possible orders for a right null function at z_0 is bounded. Let $\varphi^{(1)}$ be a right null function having this maximum order. If $\varphi^{(1)}(z_0)$ already spans $\operatorname{Ker} A(z_0)$, then the single right null function $\{\varphi^{(1)}(z)\}$ forms a canonical set of right null functions. Otherwise we choose a direct complement $\mathcal{K}_1$ of $\operatorname{Span}\{\varphi^{(1)}(z_0)\}$ in $\operatorname{Ker} A(z_0)$ and let $\varphi^{(2)}(z)$ be a right null function of maximum order such that $\varphi^{(2)}(z_0) \in \mathcal{K}_1$. If $\operatorname{Ker} A(z_0)$ is spanned by $\varphi^{(1)}(z_0)$ and $\varphi^{(2)}(z_0)$, stop. Otherwise, choose a direct complement $\mathcal{K}_2$ of $\operatorname{Span}\{\varphi^{(1)}(z_0), \varphi^{(2)}(z_0)\}$ in $\operatorname{Ker} A(z_0)$, and let $\varphi^{(3)}(z)$ be a right null function of maximum order such that $\varphi^{(3)}(z_0) \in \mathcal{K}_2$. Continuing in this way we construct right null functions $\varphi^{(1)}(z), \varphi^{(2)}(z), \ldots, \varphi^{(m)}(z)$ with nonincreasing orders $k_1 \geq k_2 \geq \cdots \geq k_m$ such that $\{\varphi^{(1)}(z_0), \varphi^{(2)}(z_0), \ldots, \varphi^{(m)}(z_0)\}$ is a basis for $\operatorname{Ker} A(z_0)$. A collection of right null functions $\{\varphi^{(1)}(z), \varphi^{(2)}(z), \ldots, \varphi^{(m)}(z)\}$ with these properties we call a *canonical set of right null functions for $A(z)$ at z_0*[1].

---

[1] The set of all analytic vector functions $\varphi(z)$ such that $A(z)\varphi(z)$ has a zero at z_0 of order more

EXAMPLE 1.1.2. We return to Example 1.1.1 to compute a canonical set of right null functions for this example. As $\dim \operatorname{Ker} A(z_0) = 1$, it is clear that any canonical set of right null functions consists of a single right null function of maximum order. Thus any canonical set has the form

$$\{\varphi(z) = c_1 e_1 + (z - z_0)(c_1 e_2 + c_2 e_1) + \cdots + (z - z_0)^{n-1}(c_1 e_n + \cdots + c_n e_1) + (z - z_0)^n \psi(z)\},$$

where $c_1, \ldots, c_n \in \mathbf{C}$ with $c_1 \neq 0$ and the vector function $\psi(z)$ analytic at z_0 are arbitrary but fixed. $\square$

EXAMPLE 1.1.3. We suppose that $A(z)$ has a block diagonal form

$$A(z) = \operatorname{diag}(A_j(z))_{j=1}^m$$

where (say) each $A_j(z)$ has size $n_j \times n_j$. Then a right null function $\varphi(z)$ for $A(z)$ can be partitioned conformally with respect to the block diagonal structure of $A(z)$ as $\varphi(z) = \operatorname{col}(\varphi^{(j)}(z))_{j=1}^m$ where $\varphi^{(j)}(z)$ is $\mathbf{C}^{n_j}$-valued. We first observe that if $\varphi^{(k)}(z)$ is a null function for $A_k(z)$ at z_0, then $\varphi(z) = \operatorname{col}(\delta_{kj} \varphi^{(k)}(z))_{j=1}^m$ ($\delta_{kj} = 1$ for $k = j$ and 0 otherwise) is a right null function for $A(z)$ of the same order. Conversely, if $\varphi(z) = \operatorname{col}(\varphi^{(j)}(z))_{j=1}^m$ is a right null function for $A(z)$ of order (say) k, then $\varphi^{(j)}(z)$ is a right null function for $A_j(z)$ of order at least k for each j such that $\varphi^{(j)}(z_0) \neq 0$. Thus the maximum order for a right null function for $A(z)$ is equal to the maximum orders of the right null functions for the $A_j(z)$'s. In this way it is possible to construct a canonical set of right null functions for $A(z)$ from canonical sets of right null functions for all the blocks $A_j(z)$. So by using the Jordan canonical form of A and appealing to Example 1.1.2, we can write out a canonical set of right null functions for any linear matrix function of the type $zI - A$. $\square$

EXAMPLE 1.1.4. Here we consider the familiar scalar case

$$a(z) = (z - z_0)^k a_k + (z - z_0)^{k+1} a_{k+1} + \cdots$$

where $a_k \neq 0$, $k \geq 0$. Then $a(z)$ has a null function at z_0 if and only if $k > 0$. In this case, *any* scalar function $\varphi(z)$ analytic at z_0 with $\varphi(z_0) \neq 0$ is a null function of precise order k. $\square$

---

than the order of the zero of φ at z_0, together with zero, forms a submodule of the module of all analytic vector functions over the ring of scalar functions analytic at z_0. This submodule, which we might call the right *null module* for $A(z)$ at z_0, consists of all right null functions together with functions φ satisfying all the requirements to be a right null function except $\varphi(z_0) \neq 0$. Then it can be shown (cf. Proposition 1.2.2) that a canonical set of right null functions is just a module basis for the right null module. It is well known (see, e.g., Section XIII.3 in Lang [1965]) that given a canonical set $\varphi^{(1)}, \ldots, \varphi^{(m)}$ of right null functions, any other set $\psi^{(1)}, \ldots, \psi^{(p)}$ of right null functions of $A(z)$ at z_0 is canonical if and only if $p = m$ and

$$\operatorname{row}(\psi^{(i)})_{i=1}^m = \operatorname{row}(\varphi^{(i)})_{i=1}^m \cdot Q,$$

where Q is $m \times m$ matrix whose entries are analytic functions in a neighborhood of z_0, and $\det Q(z_0) \neq 0$.

We next examine the behavior of right null functions for a matrix function $A(z)$ under multiplication on the left and right by invertible analytic matrix functions.

LEMMA 1.1.1. *Suppose* A_1, E, F *are* $n \times n$ *matrix functions analytic at* z_0 *with* $\det E(z_0) \neq 0$, $\det F(z_0) \neq 0$, *and set* $A_2(z) = E(z)A_1(z)F(z)$. *Then* $\varphi(z)$ *is a right null function for* $A_2(z)$ *at* z_0 *if and only if* $F(z)\varphi(z)$ *is a right null function for* $A_1(z)$ *of the same order. Moreover, if* $\{\varphi^{(1)}(z), \ldots, \varphi^{(m)}(z)\}$ *is a canonical set of right null functions for* $A_2(z)$, *then* $\{F(z)\varphi^{(1)}(z), \ldots, F(z)\varphi^{(m)}(z)\}$ *is a canonical set of right null functions for* $A_1(z)$.

PROOF. If $\varphi(z_0) \neq 0$ then $F(z_0)\varphi(z_0) \neq 0$ since $\det F(z_0) \neq 0$. If $A_2(z)\varphi(z)$ has a zero of order k at z_0, then the same holds for $A_1(z)F(z)\varphi(z) = E(z)^{-1}A_2(z)\varphi(z)$ since $E(z)^{-1}$ is analytic at z_0 with $\det E(z_0)^{-1} \neq 0$. This essentially proves the first assumption, and the second assertion is an easy consequence, again by using $\det F(z_0) \neq 0$. $\square$

Our next goal is to compute a canonical set of right null functions at a point z_0 for a general analytic matrix function $A(z)$. As we shall see, the construction reduces to piecing together the scalar case (Example 1.1.4), the behavior under direct sums (Example 1.1.3) and the behavior under multiplication (Lemma 1.1.1). The main tool for achieving this reduction is the local Smith form for an analytic matrix function.

THEOREM 1.1.2. *Let* $A(z)$ *be a matrix function analytic at* z_0 *with determinant not vanishing identically. Then* $A(z)$ *admits the following representation*

$$(1.1.2) \qquad A(z) = E(z) \operatorname{diag}\big((z - z_0)^{\kappa_1}, \ldots, (z - z_0)^{\kappa_n}\big) F(z),$$

where $E(z)$ *and* $F(z)$ *are matrix functions which are analytic and invertible in a neighborhood of* z_0, *and* $\kappa_1 \geq \cdots \geq \kappa_n \geq 0$ *are integers.*

PROOF. The representation (1.1.2), as well as the diagonal matrix $\operatorname{diag}\big((z - z_0)^{\kappa_1}, \ldots, (z - z_0)^{\kappa_n}\big)$, will be called the local Smith form of $A(z)$ at z_0. We shall follow the arguments from the proof of the Smith form for matrix polynomials (see, e.g., Gohberg-Lancaster-Rodman [1982]). Let $A(z) = [a_{ij}(z)]_{i,j=1}^n$, and for every $a_{ij}(z)$, let m_{ij} be the multiplicity of z_0 as a zero of $a_{ij}(z)$ (we set $m_{ij} = \infty$ if $a_{ij} \equiv 0$ and $m_{ij} = 0$ if $a_{ij}(z_0) \neq 0$). Let $p = \min m_{ij}$. Since $\det A(z) \not\equiv 0$, we have $p < \infty$. By permuting rows and columns of $A(z)$, if necessary, we can assume that $m_{11} = p$. Multiply the first row in $A(z)$ by $\frac{(z-z_0)^p}{a_{11}(z)}$ to obtain a matrix $A_1(z)$ where the $(1,1)$ entry is $(z - z_0)^p$. Write $A_1(z) = [b_{ij}(z)]_{i,j=1}^n$; then

$$b_{ij}(z) = (z - z_0)^p c_{ij}(z)$$

for some analytic functions $c_{ij}(z)$. Subtracting the first row (resp. first column) of $A_1(z)$ multiplied by $c_{j1}(z)$ (resp. by $c_{ij}(z)$) from the j-th row (resp. j-th column), we obtain

$$A_2(z) = \begin{bmatrix} (z - z_0)^p & 0 & \cdots & 0 \\ 0 & & & \\ \vdots & & A_3(z) & \\ 0 & & & \end{bmatrix}$$

where $A_3(z)$ is $(n-1) \times (n-1)$ analytic matrix function. Apply the above procedure to $A_3(z)$ (in place of $A(z)$), and so on. Eventually the formula (1.1.2) follows. □

The proof of Theorem 1.1.2 shows that $F(z)$ can be chosen to be unimodular (i.e. with constant non-zero determinant) matrix polynomial. By considering $A(z)^T$ in place of $A(z)$, one obtains formula (1.1.2) with $E(z)$ a unimodular matrix polynomial.

From the Binet-Cauchy formula for the minors of a product of matrices it is clear that the numbers $\kappa_1, \ldots, \kappa_n$ from (1.1.2) are uniquely determined by $A(z)$ and z_0. They are called the *partial multiplicities* of $A(z)$ at z_0.

Using the local Smith form it is easy to construct a canonical set of right null functions for a general regular analytic matrix function.

THEOREM 1.1.3. *Suppose $A(z)$ is a regular analytic matrix function with local Smith form at z_0*

$$A(z) = E(z) \operatorname{diag}\left((z-z_0)^{\kappa_1}, \ldots, (z-z_0)^{\kappa_n}\right) F(z).$$

Let $\kappa_1 \geq \kappa_2 \geq \cdots \geq \kappa_j > 0$ (where $j \leq n$) be the positive partial multiplicities of $A(z)$. Then $\varphi^{(i)}(z) = F(z)^{-1} e_i$ is a right null function for $A(z)$ at z_0 of order κ_i for $1 \leq i \leq j$ and $\{F(z)^{-1} e_1, F(z)^{-1} e_2, \ldots, F(z)^{-1} e_j\}$ is a canonical set of right null functions for $A(z)$ at z_0.

PROOF. It is elementary to see (by combining Examples 1.1.3 and 1.1.4) that e_i is a right null function for $\operatorname{diag}\left((z-z_0)^{\kappa_1}, \ldots, (z-z_0)^{\kappa_j}, 1, \ldots, 1\right)$ of order κ_i and that $\{e_1, e_2, \ldots, e_j\}$ is a canonical set of right null function. In particular, it is easy to see for the diagonal case that the order of a right null function is finite and bounded by κ_1. Now Theorem 1.1.3 follows immediately from the multiplication lemma (Lemma 1.1.1). □

Theorem 1.1.3 establishes that the positive partial multiplicities in the local Smith form for $A(z)$ at z_0 coincide with the orders of the right null functions in at least one canonical set of right null functions for $A(z)$ at z_0. It is not immediately obvious that these are the orders of the right null functions in any canonical set of right null functions; this gap will be taken care of later (in Corollary 1.1.6).

Consider now the right null functions of the local Smith form. Let

(1.1.3) $A(z) = \operatorname{diag}\left((z-z_0)^{\kappa_1}, \ldots, (z-z_0)^{\kappa_r}, 1, \ldots, 1\right),$

where $\kappa_1 \geq \cdots \geq \kappa_r$ are positive integers. Let $k_1 > k_2 > \cdots > k_m > 0$ be the distinct partial multiplicities, each with multiplicity $n_1, n_2, \ldots, n_m$ respectively (thus $k_1 = \kappa_1 = \cdots = \kappa_{n_1} > k_2 = \kappa_{n_1+1} = \cdots = \kappa_{n_1+n_2} > \cdots > k_m = \kappa_{n_1+\cdots+n_{m-1}+1} = \cdots = \kappa_{n_1+\cdots+n_m} > 0 = \kappa_{n_1+\cdots+n_m+1} = \cdots = \kappa_n$ where $n_1 + \cdots + n_m = r$). We wish to describe the form of an arbitrary right null function for $A(z)$. Thus let $\{\varphi^{(1)}(z), \varphi^{(2)}(z), \cdots, \varphi^{(r)}(z)\}$ be a canonical set of right null functions for $A(z)$ with Taylor expansions

(1.1.4) $$\varphi^{(j)}(z) = \sum_{\alpha=0}^{\infty} \varphi_\alpha^{(j)} (z-z_0)^\alpha$$

at z_0. By definition, $\varphi^{(1)}(z_0) \neq 0$ and $A(z)\varphi^{(1)}(z)$ has a zero at z_0 of maximum possible order subject to this constraint. By the diagonal form of $A(z)$ it is easy to check that this forces $0 \neq \varphi^{(1)}(z_0) \in \mathrm{Span}\{e_1, e_2, \ldots, e_{n_1}\}$ and more generally $\varphi_\alpha^{(1)} \in \mathrm{Span}\{e_1, \ldots, e_{n_1}, e_{n_1+1}, \ldots, e_{n_1+n_2+\cdots+n_j}\}$ if $k_j + \alpha \geq \kappa_1$, and then the order is precisely κ_1. Conversely, any such $\varphi^{(1)}(z)$ is a right null function of the maximum possible order $\kappa_1 = k_1$, and may serve as the first right null function in a canonical set.

In general we claim:

THEOREM 1.1.4. *Let $A(z)$ be given by (1.1.3) with $\kappa_1 \geq \kappa_2 \geq \cdots \geq \kappa_r > 0$ and let $\{\varphi^{(1)}, \ldots, \varphi^{(K)}\}$ be vector valued functions analytic at z_0 with Taylor expansions (1.1.4). Then the set $\{\varphi^{(1)}, \ldots, \varphi^{(K)}\}$ satisfies*

(i) for $1 \leq i \leq K$, $\varphi^{(i)}(z)$ is a right null function for $A(z)$ at z_0 of maximal possible order such that $\varphi_0^{(i)}$ is not in $\mathrm{Span}\{\varphi_0^{(1)}, \ldots, \varphi_0^{(i-1)}\}$ if and only if

(ii) $\{\varphi_0^{(1)}, \ldots, \varphi_0^{(K)}\}$ is linearly independent

and

(iii) for $1 \leq i \leq K$, $\varphi_\alpha^{(i)} \in \mathrm{Span}\{e_1, \ldots, e_{n_1}, \ldots, e_{n_1+n_2+\cdots+n_j}\}$ where $j = \max\{m: k_m + \alpha \geq \kappa_i\}$, and in this case $\varphi^{(i)}$ is a right null function of precise order κ_i.

PROOF. The proof is by induction on K. The case $K = 1$ was verified above. Assume that the statement in the theorem holds for $i \leq K$. We wish to prove that it holds for $i = K + 1$.

Case 1. $K = n_1 + n_2 + \cdots + n_\beta$. By condition (ii), the set $\{\varphi_0^{(i)}: 1 \leq i \leq K\}$ is linearly independent. Since $\kappa_1 \geq \kappa_2 \geq \cdots \geq \kappa_K$, a particular consequence of (iii) is that

$$\varphi_0^{(i)} \in \mathrm{Span}\{e_1, \ldots, e_{n_1}, \ldots, e_{n_1+n_2+\cdots+n_j}\}$$

for $1 \leq i \leq K$ if $j = \max\{m: k_m \geq \kappa_K\}$. Since $K = n_1 + n_2 + \cdots + n_\beta$, we have $\kappa_K = k_\beta$ and hence $j = \beta$. We have now established

$$\varphi_0^{(i)} \in \mathrm{Span}\{e_1, \ldots, e_{n_1}, \ldots, e_{n_1+n_2+\cdots+n_\beta}\}$$

for $1 \leq i \leq K$. By dimension count

$$\mathrm{Span}\{\varphi_0^{(i)}: 1 \leq i \leq K\} = \mathrm{Span}\{e_i: 1 \leq i \leq K\}.$$

By (ii) we must therefore choose $\varphi^{(K+1)}$ so that $\varphi_0^{(K+1)}$ is linearly independent of $\mathrm{Span}\{e_i: 1 \leq i \leq K\}$. By the diagonal form of $A(z)$ we read off that the maximal possible order for a right null function $\varphi^{(K+1)}$ for $A(z)$ subject to this linear independence constraint is $k_{\beta+1}$, and this is achieved as long as

$$\varphi_\alpha^{(K+1)} \in \mathrm{Span}\{e_1, \ldots, e_{n_1}, \ldots, e_{n_1+n_2+\cdots+n_j}\}$$

where

$$j = \max\{m: k_m + \alpha \geq \kappa_{K+1} = k_{\beta+1}\}.$$

Conversely, any such $\varphi^{(K+1)}(z)$ may serve as the $(K+1)$-st function in a set of $K+1$ null functions for $A(z)$ satisfying (i).

Case 2. $n_1 + \cdots + n_\beta < K < n_1 + \cdots + n_{\beta+1}$. In this case also $\varphi_0^{(i)} \in$ Span$\{e_i: 1 \le i \le n_j\}$ for $1 \le i \le k$ where $j = \max\{m: k_m \ge \kappa_K\}$. For this case $\kappa_K = k_{\beta+1}$, so $j = \beta + 1$. In this case by dimension count Span$\{\varphi_0^{(i)}: 1 \le i \le K\}$ does not fill out Span$\{e_i: 1 \le i \le n_1 + \cdots + n_{\beta+1}\}$. The maximal possible order for a right null function $\varphi^{(K+1)}$, subject to the restriction that $\varphi_0^{(K+1)}$ is linearly independent of Span$\{\varphi_0^{(i)}: 1 \le i \le K\}$ is therefore $k_{\beta+1} = \kappa_{K+1}$, and this is achieved as long as $\varphi_0^{(K+1)}$ is chosen linearly independent of Span$\{\varphi_0^{(i)}: 1 \le i \le K\}$ but still in Span$\{e_i: 1 \le i \le n_1 + \cdots + n_{\beta+1}\}$, and also $\varphi_\alpha^{(K+1)} \in$ Span$\{e_i: 1 \le i \le n_1 + \cdots + n_j\}$ where $j = \max\{m: k_m + \alpha \ge k_{\beta+1} = \kappa_{K+1}\}$. Conversely, any right null function $\varphi^{(K+1)}$ of maximal possible order $k_{\beta+1}$ such that $\varphi_0^{(K+1)}$ is linearly independent of Span$\{\varphi_0^{(i)}: 1 \le i \le K\}$ must arise in this way. $\square$

Theorem 1.1.4 implies easily the description of an arbitrary canonical set of right null functions for $A(z)$.

THEOREM 1.1.5. *For $A(z)$ given by (1.1.3) with $\kappa_1 \ge \kappa_2 \ge \cdots \ge \kappa_r > 0$, the set of vector functions $\{\varphi^{(1)}, \ldots, \varphi^{(r)}\}$ with Taylor expansions (1.1.4) is a canonical set of right null functions for $A(z)$ at z_0 if and only if $\hat{r} = r$, $\{\varphi_0^{(1)}, \ldots, \varphi_0^{(r)}\}$ is linearly independent, and*

$$\varphi_\alpha^{(i)} \in \text{Span}\{e_1, \ldots, e_{n_1}, \ldots, e_{n_1+n_2+\cdots+n_j}\}$$

where $j = \max\{m: k_m + \alpha \ge \kappa_i\}$, and in this case $\varphi^{(i)}$ is a right null function of precise order κ_i.

PROOF. Note that the inductive construction in the proof of Theorem 1.1.4 continues until $K = n_1 + n_2 + \cdots + n_m = r$. If $\varphi(z_0) \notin \text{Span}\{e_1, e_2, \ldots, e_r\}$, then $A(z_0)\varphi(z_0) \ne 0$; thus the set of right null functions $\{\varphi^{(i)}(z): 1 \le i \le r\}$ cannot be enlarged to a larger set and still satisfy (i). Thus the number of right null functions in a canonical set is precisely r and all canonical sets are of the form in the theorem. $\square$

We return now to the general case of a regular $n \times n$ matrix function $A(z)$ which is analytic in a neighborhood of z_0.

COROLLARY 1.1.6. *If $\{\varphi^{(1)}, \ldots, \varphi^{(j)}\}$ is any canonical set of right null functions for $A(z)$ at the point z_0, then the orders of $\varphi^{(1)}, \ldots, \varphi^{(j)}$ coincide with the positive partial multiplicities $\kappa_1 \ge \kappa_2 \ge \cdots \ge \kappa_j > 0$ in the local Smith form for $A(z)$ at z_0.*

PROOF. By using the local Smith form and Lemma 1.1.1, it suffices to consider the case where $A(z) = \text{diag}((z - z_0)^{\kappa_1}, \ldots, (z - z_0)^{\kappa_j}, 1, \ldots, 1)$. In this case the corollary follows from Theorem 1.1.5. $\square$

We next record how the partial multiplicities of $A(z)$ are related to the order of the zero of the scalar function $\det A(z)$ at a point z_0.

COROLLARY 1.1.7. *Suppose $A(z)$ is a regular analytic matrix function with partial multiplicities $\kappa_1 \geq \kappa_2 \geq \cdots \geq \kappa_n$ at the point z_0. Then the sum $\kappa_1 + \kappa_2 + \cdots + \kappa_n$ of the partial multiplicities is equal to the order of z_0 as a zero of the scalar function $\det A(z)$.*

PROOF. This is an immediate consequence of the local Smith form and elementary properties of the determinant. $\square$

All of the above theory can be repeated with left null functions replacing right null functions. By a *left null function* $\varphi^*(z) = \varphi_0^* + \varphi_1^*(z - z_0) + \cdots$ for the regular analytic matrix function $A(z)$ at the point z_0, we mean a row vector function $\varphi^*(z)$ analytic at z_0 with $\varphi^*(z_0) \neq 0$ for which the analytic row vector function $\varphi^*(z)A(z)$ has value 0 at z_0; the order of the zero of $\varphi^*(z)A(z)$ at z_0 is then the order of $\varphi^*(z)$ as a left null function. Note that the row vector function $\varphi^*(z)$ is a left null function for $A(z)$ at z_0 if and only if $\varphi(z) = \varphi^*(z)^T$ is right null function for $A(z)^T$ at z_0 of the same order. In this way all the examples and theorems of this section can be adapted to the setting of left null functions, or alternatively, one could develop all the results from first principles using parallel arguments. A property of the local Smith form is that the partial multiplicities for $A(z)$ are the same as the partial multiplicities for $A(z)^T$; indeed, if $A(z)$ has local Smith form $A(z) = E(z) \operatorname{diag}\big((z - z_0)^{\kappa_1}, \ldots, (z - z_0)^{\kappa_n}\big) F(z)$, then $A(z)^T$ has local Smith form $A(z)^T = F(z)^T \operatorname{diag}\big((z - z_0)^{\kappa_1}, \ldots, (z - z_0)^{\kappa_n}\big) E(z)^T$. From this it follows immediately that the orders of the left null functions in a canonical set of left null functions for a given matrix function $A(z)$ are the same as the orders of the right null functions in a canonical set of right null functions for $A(z)$ at z_0.

We leave it as an exercise to the reader to characterize the form of an arbitrary left null function for the matrix function $zI - A$ where A is a general square matrix in Jordan canonical form.

1.2 NULL CHAINS AND NULL PAIRS

In this section we introduce some alternative but equivalent language to that of the last section for the discussion of null structure for an analytic matrix function.

Let $A(z)$ be a regular matrix function analytic at z_0. Recall that the analytic vector function $\varphi(z)$ is a right null function for $A(z)$ at z_0 of order r if $\varphi(z_0) \neq 0$ and the analytic vector function $A(z)\varphi(z)$ has a zero at z_0 of order r. If we express $A(z)$ and $\varphi(z)$ in terms of their Taylor series expansion at z_0

$$A(z) = \sum_{i=0}^{\infty} \frac{1}{i!} A^{[i]}(z_0)(z - z_0)^i$$

$$\varphi(z) = \sum_{j=0}^{\infty} \frac{1}{j!} \varphi^{[j]}(z_0)(z - z_0)^j$$

then $A(z)\varphi(z)$ has Taylor series expansion

$$A(z)\varphi(z) = \sum_{j=0}^{\infty} \left[\sum_{i=0}^{j} \frac{1}{i!} A^{[i]}(z_0) \frac{1}{(j - i)!} \varphi^{[j-i]}(z_0) \right] (z - z_0)^j.$$

This function has a zero of order r at z_0 if and only if its first r Taylor coefficients vanish, i.e.

$$(1.2.1) \qquad \sum_{i=0}^{j} \frac{1}{i!} A^{[i]}(z_0) \frac{1}{(j-i)!} \varphi^{[j-i]}(z_0) = 0, \qquad 0 \le j \le r-1.$$

In particular, we see that the Taylor coefficients with index $j > r$ of $\varphi(z)$ are irrelevant. This suggests that we define a chain of column vectors $(x_0, x_1, \ldots, x_{r-1})$ to be a *right null chain of order* r for $A(z)$ at z_0 if $x_i = \frac{1}{i!} \varphi^{[i]}(z_0)$ for $0 \le i \le r-1$, where $\varphi(z)$ is a right null function for $A(z)$ of order r at z_0. By (1.2.1) we see that then the chain of column vectors $(x_0, x_1, \ldots, x_{r-1})$ is a right null chain of order r for $A(z)$ at z_0 if and only if $x_0 \ne 0$ and

$$(1.2.2) \qquad \sum_{i=0}^{j} \frac{1}{i!} A^{[i]}(z_0) x_{j-i} = 0, \qquad j = 0, \ldots, r-1.$$

Note that in particular $x_0 = \varphi(z_0)$ satisfies $A(z_0) x_0 = 0$; x_0 we call an *eigenvector* for $A(z)$ at z_0 and the remaining vectors $x_1, \ldots, x_{r-1}$ *generalized eigenvectors*. In particular, there is an eigenvector of $A(z)$ at z_0 if and only if $\det A(z_0) = 0$ i.e. z_0 is a zero of $A(z)$. The length r of a right null chain $(x_0, x_1, \ldots, x_{r-1})$ is by definition no more than the order of the corresponding right null function $\varphi(z) = \sum_{i=1}^{r-1} (z - z_0)^i x_i$. We define the *rank* of the eigenvector x_0 to be the maximum of the orders of all right null functions φ having $\varphi(z_0) = x_0$.

The structure of the set of eigenvectors with respect to their ranks is captured in the following theorem.

THEOREM 1.2.1. *Let $A(z)$ be a regular analytic matrix function in a neighborhood of z_0, and assume that z_0 is a zero of $A(z)$.*

Denote by $\mathcal{X}_k \subset \mathbf{C}^n$ $(k = 1, 2, \ldots)$ a maximal subspace in $\mathbf{C}^n$ with the property that all its nonzero elements are eigenvectors of $A(z)$ of rank k corresponding to z_0. Then

(i) there exists a finite set $\{k_1, \ldots, k_p\} \subset \{1, 2, \ldots\}$ such that $\mathcal{X}_k \ne \{0\}$ if and only if $k \in \{k_1, \ldots, k_p\}$;

(ii) the subspace $\operatorname{Ker} A(z_0) \subset \mathbf{C}^n$ is decomposed into a direct sum:

$$(1.2.3) \qquad \operatorname{Ker} A(z_0) = \mathcal{X}_{k_1} \dotplus \cdots \dotplus \mathcal{X}_{k_p};$$

(iii) if $x = \sum_{j=1}^{p} x_j \in \operatorname{Ker} A(z_0)$, where $x_j \in \mathcal{X}_{k_j}$, then the rank of x is equal to $\min\{k_j \mid x_j \ne 0\}$.

This theorem follows immediately from Theorem 1.1.5 upon translating to the language of right null pairs.

By a *canonical set of right null chains* we mean an ordered set of null chains

$$(1.2.4) \qquad x_0^{(1)}, \ldots, x_{r_1-1}^{(1)}; x_0^{(2)}, \ldots, x_{r_2-1}^{(2)}; \ldots; x_0^{(k)}, \ldots, x_{r_k-1}^{(k)}$$

such that the corresponding ordered set of right null functions

$$\left\{ \varphi^{(j)}(z) = \sum_{i=0}^{r_j-1} (z - z_0)^i x_i^{(j)} : j = 1, \ldots, k \right\}$$

is a canonical set of right null functions in the sense of the previous section. Here $r_j = \operatorname{rank} x_0^{(j)}$, $j = 1, \ldots, k$; $k = \dim \operatorname{Ker} A(z_0)$.

The next proposition shows that a canonical system plays the role of a basis in the set of all null chains of $A(z)$ corresponding to a given eigenvalue z_0.

Let μ be the length of the longest possible null chain of $A(z)$ corresponding to z_0. It will be convenient to introduce into consideration the subspace $\mathcal{N} \subset \mathbb{C}^{n\mu}$ consisting of all sequences $(y_0, \ldots, y_{\mu-1})$ of n-dimensional vectors such that

(1.2.5)
$$\begin{bmatrix} A(z_0) & 0 & \cdots & 0 \\ A'(z_0) & A(z_0) & \cdots & 0 \\ \vdots & \vdots & \cdots & \vdots \\ \frac{1}{(\mu-1)!}A^{[\mu-1]}(z_0) & \frac{1}{(\mu-2)!}A^{[\mu-2]}(z_0) & \cdots & A(z_0) \end{bmatrix} \begin{bmatrix} y_0 \\ y_1 \\ \vdots \\ y_{\mu-1} \end{bmatrix} = 0.$$

Note that $\mathcal{N}$ consists of null chains for $A(z)$ corresponding to z_0, after we drop first zero vectors (if any) in the sequence $(y_0, \ldots, y_{\mu-1}) \in \mathcal{N}$.

PROPOSITION 1.2.2. *Let*

(1.2.6)
$$\varphi_{i0}, \ldots, \varphi_{i,\mu_i-1}, \qquad i = 1, \ldots, s$$

be a set of null chains of $A(z)$ corresponding to z_0. Then the following conditions are equivalent:

(i) *the set (1.2.6) is canonical;*

(ii) *the eigenvectors $\varphi_{10}, \ldots, \varphi_{s0}$ are linearly independent and $\sum_{i=1}^{s} \mu_i = \sigma$, the multiplicity of z_0 as a zero of $\det A(z)$;*

(iii) *the set of sequences*

(1.2.7)
$$\gamma_{ij} = (0, \ldots, 0, \varphi_{i0}, \ldots, \varphi_{ij})^T, \qquad j = 0, \ldots, \mu_i - 1, \quad i = 1, \ldots, s,$$

where the number of zero vectors preceding φ_{i0} in γ_{ij} is $\mu - (j + 1)$, form a basis in $\mathcal{N}$.

PROOF. We shall use the reduction to local Smith form

$$A(z) = E(z) \operatorname{diag}\left((z - z_0)^{\nu_i}\right)_{i=1}^{n} F(z),$$

(where $\nu_1 \geq \cdots \geq \nu_n \geq 0$). For brevity, write A for the matrix appearing in the left-hand side of (1.2.5). Similarly, E, F and D will denote the corresponding matrices formed from the analytic matrix function $E(z)$, $F(z)$ and $D(z) = \operatorname{diag}\left((z - z_0)^{\nu_i}\right)_{i=1}^{n}$ at z_0, respectively. One verifies by direct computations that

$$A = EDF,$$

and consequently

(1.2.8) $$\tilde{\mathcal{N}} = F\mathcal{N},$$

where the subspace $\tilde{\mathcal{N}} \subset \mathbf{C}^{n\mu}$ is defined by the formula (1.2.5) with $A(z)$ replaced by $D(z)$. (Note that according to Corollary 1.1.6 the length of the longest right null chain at z_0 for $A(z)$ and $D(z)$ is the same μ.) Formula (1.2.8) allows the reduction to the case when

(1.2.9) $$A(z) = \operatorname{diag}\left((z - z_0)^{\nu_i}\right)_{i=1}^{n}.$$

Thus it remains to prove Proposition 1.2.2 for $A(z)$ given by (1.2.9). The part (i) $\Rightarrow$ (ii) follows from the definition of a canonical set and Corollary 1.1.7. Let us prove (ii) $\Rightarrow$ (iii). Let the chains (1.2.6) be given such that $\varphi_{10}, \ldots, \varphi_{s0}$ are linearly independent and $\sum_{i=1}^{s} \mu_i = \sigma$. From the linear independence of $\varphi_{10}, \ldots, \varphi_{s0}$ it follows that the vectors

(1.2.10) $$\gamma_{ij} = (0, \ldots, 0, \varphi_{i0}, \ldots, \varphi_{ij}), \qquad j = 0, \ldots, \mu_i - 1, \quad i = 1, \ldots, s$$

are linearly independent in $\mathcal{N}$. On the other hand, it is easy to see that

(1.2.11) $$\dim \mathcal{N} = \sum_{j=1}^{n} \nu_i,$$

which coincides with the multiplicity of z_0 as a zero of $L(z)$, i.e., with $\sigma = \sum_{i=1}^{s} \mu_i$. It follows that the sequences (1.2.9) form a basis in $\mathcal{N}$, and (iii) is proved.

It remains to prove that (iii) $\Rightarrow$ (i). Without loss of generality we can suppose that μ_i are arranged in the nonincreasing order: $\mu_1 \geq \cdots \geq \mu_s$. Suppose that $\nu_r > \nu_{r+1} = \cdots = \nu_n = 0$ (where ν_i are taken from (1.2.9)). Then necessarily $s = r$ and $\mu_i = \nu_i$ for $i = 1, \ldots, s$. Indeed, $s = r$ follows from the fact that the dimension of the subspace spanned by all the vectors of the form $(0, \ldots, 0, x)^T$, which belong to $\mathcal{N}$, is just r. Further, from Theorem 1.2.1 it follows that the dimension of the subspace spanned by all the vectors of the form $(0, \ldots, x_0, \ldots, x_{j-1})^T$, which belong to $\mathcal{N}$ is just

(1.2.12) $$\sum_{p \geq j} j \cdot |\{k \mid \nu_k = p\}| + \sum_{p=1}^{j-1} p \cdot |\{k \mid \nu_k = p\}|, \qquad j = 1, \ldots, \mu,$$

where $|A|$ denotes the number of elements in a finite set A, and μ is the length of the longest null chain of $A(z)$ corresponding to z_0. On the other hand, from (iii) it follows that this dimension is

(1.2.13) $$\sum_{p \geq j} j \cdot |\{k \mid \mu_k = p\}| + \sum_{p=1}^{j-1} p \cdot |\{k \mid \mu_k = p\}|, \qquad j = 1, \ldots, \mu.$$

Comparing (1.2.12) and (1.2.13), we obtain eventually that $\mu_i = \nu_i$. Now (i) follows from the definition of a canonical set of null chains. $\square$

We will need later on also the following property of a canonical set of null chains.

LEMMA 1.2.3. *Let $A(z)$ be an analytic matrix function with a zero z_0, and let*

$$x_0^{(1)}, \ldots, x_{r_1-1}^{(1)}; \cdots; x_0^{(k)}, \ldots, x_{r_k-1}^{(k)}$$

be a canonical set of right null chains of $A(z)$ at z_0 (here $r_1 \geq r_2 \geq \cdots \geq r_k > 0$). Then for every right null chain $y_0, \ldots, y_r$ of $A(z)$ at z_0 there exist complex numbers α_{ij} ($0 \leq i \leq r$; $1 \leq j \leq p$; where p is the maximal integer for which $r_p \geq r$) such that

$$y_0 = \sum_{j=0}^{p} \alpha_{0j} x_0^{(j)};$$

$$y_1 = \sum_{j=0}^{p} \alpha_{0j} x_1^{(j)} + \sum_{j=0}^{p} \alpha_{1j} x_0^{(j)};$$

$$\vdots$$

$$y_r = \sum_{j=0}^{p} \alpha_{0j} x_r^{(j)} + \sum_{j=0}^{p} \alpha_{1j} x_{r-1}^{(j)} + \cdots + \sum_{j=0}^{p} \alpha_{rj} x_0^{(j)}.$$

Indeed, by Proposition 1.2.2 the set of sequences

(1.2.14) $(0, \ldots, 0, x_{i0}, \ldots, x_{ij})^T, \qquad j = 0, \ldots, r_i - 1, \ i = 1, \ldots, k$

forms a basis in $\mathcal{N}$. Upon expressing the sequences

$$(0, \ldots, 0, y_0, \ldots, y_j)^T, \qquad j = 0, \ldots, r-1$$

as linear combinations of (1.2.14), the conclusion of the lemma follows.

We return now to a canonical set (1.2.4) of right null chains of $A(z)$ at z_0. We write such a canonical set of null chains in matrix form:

$$X = [x_0^{(1)} \cdots x_{r_1-1}^{(1)} x_0^{(2)} \cdots x_{r_2-1}^{(2)} \cdots x_0^{(k)} \cdots x_{r_k-1}^{(k)}];$$

$$J = \operatorname{diag}(J_1, \ldots, J_k),$$

where J_i is the Jordan block of size $r_i \times r_i$ with eigenvalue z_0; thus J is a Jordan matrix of size $\sum_{i=1}^{k} r_i \times \sum_{i=1}^{k} r_i$ with spectrum $\sigma(J) = \{z_0\}$. The pair (X, J) constructed in this way is called a *right null pair* of $A(z)$ corresponding to z_0. The point here is that the matrix X carries only the ordered list of vectors in a canonical set of right null chains; information about where one chain ends and the next begins has been lost. At this stage the addition of the Jordan matrix J is a matricial bookkeeping device by which we recover the information about where one null chain ends and the next begins in the list of columns of the matrix X. Eventually we shall consider several zeros $z_1, z_2, \ldots, z_m$ of an analytic matrix function simultaneously rather than such a single zero z_0. For this reason it is convenient at this stage to define J to have spectrum $\sigma(J) = \{z_0\}$ when (X, J) is a null pair for $A(z)$ at the point z_0. In any case it is clear that a right null pair (X, J) for $A(z)$ contains the same information as a canonical set of right null chains.

To give greater flexibility in the manipulation of null pairs it is sometimes more convenient to use a pair of matrices (Z, T), where T is not necessarily in the Jordan canonical form, as a right null pair for $A(z)$ at z_0. We extend now the definition of a right null pair for $A(z)$ at z_0 as follows: the pair (Z, T) is called a *right null pair* for $A(z)$ at z_0 if it is *similar* to a right null pair (X, J) constructed as above using a canonical set of null chains, i.e. for some square and invertible matrix S the equalities $Z = XS$, $T = S^{-1}JS$ hold.

When we study interpolation problems later in this book, one question of interest will be a characterization of which pairs of matrices (Z, T) can arise as the right null pair of some regular analytic matrix function (or roughly equivalently, which sequences of chains of vectors $\{x_0^{(1)}, x_1^{(1)}, \ldots, x_{r_1-1}^{(1)}; x_0^{(2)}, \ldots, x_{r_2-1}^{(2)}; \ldots; x_0^{(k)}, \ldots, x_{r_k-1}^{(k)}\}$ can arise as the canonical set of null chains at a point z_0 for some regular analytic matrix function). The following gives a necessary condition which, as we shall see later, turns out also to be sufficient.

THEOREM 1.2.4. *If (Z, T) is a right null pair for a regular analytic matrix function $A(z)$ at a point z_0, where Z is of size $n \times p$ and T is a square matrix of size $p \times p$, then*

(i) *the order of z_0 as a zero of $\det A(z)$ is p and*

(ii) $\operatorname{rank} \operatorname{col}(ZT^i)_{i=1}^{\ell-1} = p$ *for some integer $\ell \geq 1$.*

PROOF. By definition of a right null pair, the size p of T equals the sum of the lengths of the right null chains in a canonical set, i.e. the sum of the orders of the right null functions in a canonical set of right null functions. By Corollary 1.1.7 this in turn is the sum of the partial multiplicities in the local Smith form for $A(z)$ at z_0 which is clearly the order of z_0 as a zero of $\det A(z)$. This establishes (i).

For (ii), first note that the pair (Z, T) satisfies (ii) if and only if the pair $(ZS, S^{-1}TS)$ satisfies (ii), where S is any $p \times p$ matrix with $\det S \neq 0$. Thus we may assume that $(Z, T) = (X, J)$ is the null pair constructed from a canonical set of right null chains as done above. It is elementary to see that a vector $v \in \mathbb{C}^p$ is in $\operatorname{Ker} \operatorname{col}(XJ^i)_{i=0}^{\ell-1}$ if and only if v is in $\operatorname{Ker} \operatorname{col}(X(J - z_0 I)^i)_{i=0}^{\ell-1}$, and thus $\operatorname{rank} \operatorname{col}(XJ^i)_{i=0}^{\ell-1} = \operatorname{rank} \operatorname{col}(X(J - z_0 I)^i)_{i=0}^{\ell-1}$. As $J - z_0 I$ is the direct sum of the shift matrices

$$\begin{bmatrix} 0 & 1 & & & \\ & 0 & 1 & & \\ & & \ddots & \ddots & 1 \\ & & & & 0 \end{bmatrix},$$

we see that $X(J - z_0 I)^i$ has the form

$$X(J - z_0 I)^i = [\underbrace{0, \ldots, 0, x_0^{(1)}, \ldots, x_{r_1-i-1}^{(1)}}_{s_1}; \underbrace{0, \ldots, 0, x_0^{(2)}, \ldots, x_{r_2-i-1}^{(2)}}_{s_2}; \ldots;$$

$$\underbrace{0, \ldots, 0, x_0^{(k)}, \ldots, x_{r_k-i-1}^{(k)}}_{s_k}],$$

where $s_j = \min(r_j, i)$. Using the linear independence of the leading vectors $\{x_0^{(1)}, x_0^{(2)}, \ldots, x_0^{(k)}\}$ in the chains, one can now deduce, assuming $r_1 = \max\{r_j : 1 \leq j \leq k\}$, that $\mathrm{col}(X(J - z_0 I)^i)_{i=1}^{r_1-1}$ has $r_1 + r_2 + \cdots + r_k = p$ linearly independent columns, and thus has full rank p. $\square$

Theorem 1.2.4 suggests the following definition. We shall say that the pair of matrices (Z, T), where Z has size $n \times p$ and T has square size $p \times p$, is a *null kernel pair* if $\bigcap_{i=0}^{\infty} \mathrm{Ker}(ZT^i) = \{0\}$, or equivalently,

$$\mathrm{rank} \, \mathrm{col}(ZT^i)_{i=1}^{\ell-1} = p$$

for some integer $\ell \geq 1$. The size p of the square matrix T we shall call the *order* of the pair. Thus Theorem 1.2.4 says that a right null pair (Z, T) for an analytic matrix function $A(z)$ at a point z_0 is a null kernel pair of order equal to the order of z_0 as a zero of $\det A(z)$.

Our next goal is to show that a right null pair for $A(z)$ at z_0 is uniquely determined up to similarity. To this end we first note the following.

PROPOSITION 1.2.5. *Let $A(z)$ be a $n \times n$ matrix function regular and analytic at the point z_0. Let $E(z)$ and $F(z)$ be $n \times n$ matrix functions analytic at z_0 such that $\det E(z_0) \neq 0$ and $\det F(z_0) \neq 0$. Then $y_0, \ldots, y_k$ is a right null chain for $E(z)A(z)F(z)$ at z_0 if and only if the vectors*

$$(1.2.15) \qquad z_j = \sum_{i=0}^{j} \frac{1}{i!} F^{[i]}(z_0) y_{j-i}, \qquad j = 0, \ldots, k$$

form a right null chain for $A(z)$ at z_0.

PROOF. By definition of right null chain, $y_0, \ldots, y_k$ is a right null chain for $E(z)A(z)F(z)$ if and only if $\varphi(z) = y_0 + (z - z_0)y_1 + \cdots + (z - z_0)^k y_k$ is a right null function for $E(z)A(z)F(z)$ of order $k + 1$. By the multiplication lemma (Lemma 1.1.1), $\psi(z)$ is a right null function for $A(z)$ if and only if $\psi(z) = F(z)\varphi(z)$ for a right null function $\varphi(z)$ for $E(z)A(z)F(z)$ of the same order at z_0. Computing the first $k + 1$ Taylor coefficients of $\psi(z)$ at z_0 then implies that any right null chain for $A(z)$ has the form (1.2.15) as asserted. $\square$

For the next theorem and in general, given (column) vectors $x_r, x_{r+1}, \ldots, x_s$,

we denote by $\mathrm{col}(x_r, x_{r+1}, \ldots, x_s)$ the column vector $\begin{bmatrix} x_r \\ x_{r+1} \\ \vdots \\ x_s \end{bmatrix}$.

THEOREM 1.2.6. *Let (Z_1, T_1) and (Z_2, T_2) be right null pair of $A(z)$ at z_0. Then*

$$Z_1 = Z_2 S, \qquad T_1 = S^{-1} T_2 S$$

for a unique invertible matrix S (in particular, (Z_1, T_1) and (Z_2, T_2) are similar).

PROOF. By definition of null pair and by Corollary 1.1.6 we can assume that $T_1 = T_2 = T$ is a Jordan matrix, and Z_1 (resp. Z_2) form a canonical set of null chains of $A(z)$ corresponding to z_0. Let us show that

(1.2.16)
$$\operatorname{Im} \operatorname{col}\big(Z_j (T - z_0 I)^i\big)_{i=0}^{\ell-1} = \operatorname{Span}\{\operatorname{col}(x_k, x_{k-1}, \ldots, x_0, 0, \ldots, 0) \mid x_0, \ldots, x_k$$
is a right null chain for $A(z)$ at $z_0\}, j = 1, 2$.

Here ℓ is such that the columns of both matrices $\operatorname{col}(Z_1 T^i)_{i=0}^{\ell-1}$ and $\operatorname{col}(Z_2 T^i)_{i=0}^{\ell-1}$ are linearly independent. Indeed, equality (1.2.16) can be checked easily for the case that $A(z) = \operatorname{diag}\big((z - z_0)^{\kappa_1}, \ldots, (z - z_0)^{\kappa_n}\big)$, where $\kappa_1 \geq \cdots \geq \kappa_n$ are non-negative integers. Consider now the general case, and let $A(z) = E(z) \cdot \operatorname{diag}\big((z - z_0)^{\kappa_1}, \ldots, (z - z_0)^{\kappa_n}\big) F(z)$ be a local Smith form of $A(z)$. Write

(1.2.17)
$$Z_1 = [x_0^{(1)} \cdots x_{r_1-1}^{(1)} x_0^{(2)} \cdots x_{r_2-1}^{(2)} \cdots x_0^{(k)} \cdots x_{r_k-1}^{(k)}];$$

$$T = \operatorname{diag}(J_1, \ldots, J_k),$$

where J_i is the Jordan block of size $r_i \times r_i$ with eigenvalue z_0. Put

(1.2.18)
$$y_r^{(i)} = \sum_{j=0}^{r} \frac{1}{j!} F^{[j]}(z_0) x_{r-j}^{(i)};$$

let $\widetilde{Z}_1$ be the matrix obtained from Z_1 by replacing $x_r^{(i)}$ by $y_r^{(i)}$. Then (cf. Proposition 1.2.5) $(\widetilde{Z}_1, T)$ is a right null pair for $D(z) = \operatorname{diag}\big((z - z_0)^{\kappa_1}, \ldots, (z - z_0)^{\kappa_n}\big)$ at z_0. Since (1.2.16) is already proved for diagonal matrix functions,

(1.2.19)
$$\operatorname{Im} \operatorname{col}\big(\widetilde{Z}_1 (T - z_0 I)^i\big)_{i=0}^{\ell-1} = \operatorname{Span}\{\operatorname{col}(y_k, y_{k-1}, \ldots, y_0, 0, \ldots, 0) \mid y_0, \ldots, y_k$$
is a null chain of $D(z)$ corresponding to $z_0\}$.

Applying the matrix

$$\begin{bmatrix} F^{-1}(z_0) & \big(F^{-1}(z_0)\big)^{[1]} & \cdots & \big(F^{-1}(z_0)\big)^{[\ell-1]} \\ 0 & F^{-1}(z_0) & \cdots & \big(F^{-1}(z_0)\big)^{[\ell-2]} \\ \vdots & \vdots & & \vdots \\ 0 & 0 & \cdots & F^{-1}(z_0) \end{bmatrix}$$

to both sides of (1.2.19), we obtain the equality (1.2.16) for $j = 1$ (and analogously for $j = 2$).

In particular, $\operatorname{Im}\operatorname{col}\big(Z_1(T-z_0I)^j\big)_{j=0}^{\ell-1} = \operatorname{Im}\operatorname{col}\big(Z_2(T-z_0I)^j\big)_{j=0}^{\ell-1}$. So there exists a $p \times p$ matrix S (where p is the size of T) such that

$$(1.2.20) \qquad \operatorname{col}\big(Z_1(T-z_0I)^j\big)_{j=0}^{\ell-1}S = \operatorname{col}\big(Z_2(T-z_0I)^j\big)_{j=0}^{\ell-1}.$$

It is easily seen that such a matrix S is invertible and uniquely determined. The first block row in (1.2.20) implies that $Z_1S = Z_2$. To prove that $TS = ST$ let us consider the matrices $Q_i \overset{\text{def}}{=} \operatorname{col}\big(Z_i(T-z_0I)^j\big)_{j=0}^{\ell}$, $i = 1, 2$. As above, we have $Q_1\widetilde{S} = Q_2$ for some $p \times p$ matrix $\widetilde{S}$. By uniqueness of S we obtain that in fact $\widetilde{S} = S$. Then

$$\operatorname{col}\big(Z_1(T-z_0I)^j\big)_{j=0}^{\ell-1} \cdot (T-z_0I)S = \operatorname{col}\big(Z_2(T-z_0I)^j\big)_{j=0}^{\ell-1}(T-z_0I)$$
$$= \operatorname{col}\big(Z_1(T-z_0I)^j\big)_{j=0}^{\ell-1}S(T-z_0I).$$

Since the columns of $\operatorname{col}\big(Z_1(T-z_0I)^j\big)_{j=0}^{\ell-1}$ are linearly independent, the equality $(T-z_0I)S = S(T-z_0I)$, or $TS = ST$, follows. So $Z_1S = Z_2$ and $TS = ST$, i.e. the pairs (Z_1, T) and (Z_2, T) are similar. $\square$

The following proposition gives a useful description of right null pairs.

PROPOSITION 1.2.7. *Let (Z, T) be a pair of matrices of sizes $n \times p$ and $p \times p$ respectively. The following conditions are necessary and sufficient in order that (Z, T) be a right null pair for the analytic matrix function $A(z) = \sum_{j=0}^{\infty} A_j(z - z_0)^j$ at the point z_0:*

(i) $\sigma(T) = \{z_0\}$.

(ii) $p = $ *the order of z_0 as a zero of $\det A(z)$.*

(iii) $\operatorname{rank}\operatorname{col}(ZT^i)_{i=0}^{\ell-1} = p$ *for some integer $\ell \geq 1$.*

(iv) $\sum_{j=0}^{\infty} A_j Z(T - z_0I)^j = 0$.

(*Note that only a finite number of terms in the left hand side of* (iv) *is different from zero since $\sigma(T) = \{z_0\}$.*)

PROOF. Necessity of (i), (ii) and (iii) have already been observed ((i) is by definition, (ii) and (iii) by Theorem 1.2.4). To prove (iv) we may assume that $(Z, T) = (X, J)$ is the right null pair constructed from a canonical set of right null chains for $A(z)$ at z_0. Let $X_i = [x_0^{(i)} x_1^{(i)} \cdots x_{r_i-1}^{(i)}]$, $i = 1, \ldots, k$. Evidently we need only check

$$(1.2.21) \qquad \sum_{j=0}^{\infty} A_j X_i(J_i - z_0I)^j = 0; \qquad i = 1, \ldots, k.$$

But the α-th column $(\alpha = 1, \ldots, r_i)$ in (1.2.21) is $A_0 x_{\alpha-1}^{(i)} + A_1 x_{\alpha-2}^{(i)} + \cdots + A_{\alpha-1} x_0^{(i)}$, which is zero in view of the definition of a right null chain.

Conversely, suppose (Z, T) satisfies the conditions (i)–(iv). We can assume that

$$(1.2.22) \qquad T = J = \operatorname{diag}(J_1, \ldots, J_k)$$

is in the Jordan normal form with Jordan blocks $J_1, \ldots, J_k$. Let $Z = [Z_1 Z_2 \cdots Z_k]$ be the partition of Z consistent with (1.2.22). As we have seen in the preceding paragraph, equality (iv) implies that the columns of Z_i form a null chain of $A(z)$ corresponding to z_0, $i = 1, \ldots, k$. On the other hand, the condition (iii) implies that the eigenvectors in Z (i.e. first columns in each block Z_i, $i = 1, \ldots, k$) are linearly independent. Using this fact as well as the assumption (ii) that the order of (Z, J) is p, we deduce by reduction to a local Smith form, that the numbers of columns in $Z_1, \ldots, Z_k$ coincide with the nonzero partial multiplicities of $A(z)$ at z_0. By the definition of a right null pair it is clear that (Z, J) is a right null pair for $A(z)$ at z_0. $\square$

Analogously one defines a left null pair for the regular analytic matrix function $A(z)$ at z_0 as follows. Row vectors $y_0, \ldots, y_{r-1}$ form a *left null chain* of $A(z)$ corresponding to z_0, if $y_0 \neq 0$ and

$$\sum_{i=0}^{j} \frac{1}{i!} y_{j-i} A^{[i]}(z_0) = 0, \qquad j = 1, \ldots, r-1.$$

Thus, $y_0, \ldots, y_{r-1}$ is a left null chain for $A(z)$ if and only if the transposed vectors $y_0^T, \ldots, y_{r-1}^T$ form a right null chain of the transposed function $A(z)^T$ (corresponding to the same z_0). The construction of a canonical set of left null chains follows the pattern described above. Observe that lengths of the left null chains in a canonical set coincide with the non-zero partial multiplicities of $A(z)$ at z_0.

Now let

$$\psi_0^{(1)}, \ldots, \psi_{s_1-1}^{(1)}; \psi_0^{(2)}, \ldots, \psi_{s_2-1}^{(2)}; \cdots; \psi_0^{(k)}, \ldots, \psi_{s_k-1}^{(k)}$$

be a canonical set of left null chains for $A(z)$ at z_0; $\psi_m^{(j)}$ are n-dimensional row vectors. Form the matrices

$$(1.2.23) \qquad\qquad K(z_0) = \operatorname{diag}(J_1^T, \ldots, J_k^T),$$

where J_i is the $s_i \times s_i$ Jordan block with eigenvalue z_0,

$$(1.2.24) \qquad\qquad Y(z_0) = \operatorname{col}\left(\begin{bmatrix} \psi_0^{(i)} \\ \psi_1^{(i)} \\ \vdots \\ \psi_{s_i-1}^{(i)} \end{bmatrix} \right)_{i=1}^{k}.$$

Thus, $K(z_0)$ is of size $\mu \times \mu$, where μ is the multiplicity of z_0 as a zero of $\det A(z)$, and $Y(z_0)$ is of size $\mu \times n$. The pair $\big(K(z_0), Y(z_0)\big)$ will be called *left null pair* for $L(z)$ at z_0. Again, as it was done for right null pairs, we shall extend this definition to a larger class of pairs by similarity. So, let (S_1, Y_1) and (S_2, Y_2) be two pairs of matrices, where the size of Y_1 and Y_2 is $\mu \times n$ and that of S_1 and S_2 is $\mu \times \mu$. These pairs are called *similar* if for some invertible $\mu \times \mu$ matrix T the equalities

$$Y_1 = TY_2, \qquad S_1 = TS_2T^{-1}$$

hold. A pair of matrices (S, Y), where S (resp. Y) has size $\mu \times \mu$ (resp. $\mu \times n$) will be called a *left null pair* of $A(z_0)$ at z_0 if it is similar to a pair $(K(z_0), Y(z_0))$ given by (1.2.23), (1.2.24). Clearly, in this case μ must be equal to the multiplicity of z_0 as a zero of $\det A(z)$. Observe that (S, Y) is a left null pair of $A(z)$ at z_0 if and only if (Y^T, S^T) is right null pair of $A^T(z)$ (corresponding to the same z_0). This observation allows us to reproduce easily the results of this section for left null pairs. In particular, a left null pair (S, Y) of $A(z)$ at z_0 is always a *full range pair*, i.e.

$$\sum_{i=0}^{\infty} \mathrm{Im}(S^i Y) = \mathbf{C}^p,$$

where $p \times p$ is the size of S.

1.3 SINGULAR PARTS AND NULL SUBSPACES

In this section we shall introduce and study other ways to characterize the structure of a zero of an analytic matrix function. One way to describe this structure is by means of right and left null chains. Other ways are the singular part of the Laurent expansion of the inverse null function and the null subspace to be defined in this section.

Let $A(z)$ be regular analytic $n \times n$ matrix function with a zero at z_0. Then $A(z)^{-1}$ exists in a neighborhood of z_0 except for $z = z_0$. So $A(z)^{-1}$ has a Laurent series expansion

$$A(z)^{-1} = \sum_{\alpha=-\nu}^{\infty} (z - z_0)^{\alpha} K_{\nu+1+\alpha},$$

where $K_{\nu+1+\alpha}$ are $n \times n$ matrices. The singular part of this series will be denoted $SP(A^{-1}(z_0))$:

(1.3.1) $SP(A^{-1}(z_0)) = (z - z_0)^{-\nu} K_1 + (z - z_0)^{-\nu+1} K_2 + \cdots + (z - z_0)^{-1} K_\nu.$

Define $mn \times mn$ matrices $S_m = S_m(z_0; A)$ $(m = 1, \ldots, \nu)$:

(1.3.2)
$$S_m = \begin{bmatrix} K_1 & 0 & \cdots & 0 & 0 \\ K_2 & K_1 & \cdots & 0 & 0 \\ \vdots & \vdots & & \vdots & \vdots \\ K_{m-1} & K_{m-2} & \cdots & K_1 & 0 \\ K_m & K_{m-1} & \cdots & K_2 & K_1 \end{bmatrix}.$$

It is convenient to introduce the following notation: Given $n \times n$ matrices $A_1, A_2, \ldots, A_p$, we shall denote by $\Delta(A_1, \ldots, A_p)$ the block triangular $np \times np$ matrix

$$\begin{bmatrix} A_1 & 0 & \cdots & 0 & 0 \\ A_2 & A_1 & \cdots & 0 & 0 \\ \vdots & \vdots & & \vdots & \vdots \\ A_{p-1} & A_{p-2} & \cdots & A_1 & 0 \\ A_p & A_{p-1} & \cdots & A_2 & A_1 \end{bmatrix}.$$

For example, $S_m = \Delta(K_1, K_2, \ldots, K_m)$.

Recall that for given (column) vectors $x_r, \ldots, x_s$, we denote by $\mathrm{col}(x_r, x_{r+1}, \ldots, x_s)$ the column vector

$$\begin{bmatrix} x_r \\ x_{r+1} \\ \vdots \\ x_s \end{bmatrix}.$$

PROPOSITION 1.3.1. *Suppose* $A(z)$ *is a matrix function, regular and analytic at* z_0*, with the order of the pole of* $A^{-1}(z)$ *at* z_0 *equal to* ν *(i.e.* ν *is the smallest nonnegative integer such that* $(z - z_0)^\nu A(z)^{-1}$ *is analytic at* z_0*). Then*

$$\mathrm{Im}\, S_\nu = \{\mathrm{col}(0, \ldots, 0, x_0, x_1, \ldots, x_\ell) \colon (x_0, \ldots, x_\ell)$$
$$\text{is a right null chain for } A(z) \text{ at } z_0\},$$

where $S_\nu = S_\nu(z_0; A)$.

PROOF. Use the local Smith form for $A(z)$ at z_0

$$A(z) = E(z)D(z)F(z)$$

where $D(z) = \mathrm{diag}((z - z_0)^{\kappa_1}, \ldots, (z - z_0)^{\kappa_n})$ and $E(z)$, $F(z)$ are analytic matrix functions such that $E(z_0)$ and $F(z_0)$ are nonsingular. The idea is to reduce the proof of Proposition 1.3.1 to the diagonal case where it can be checked easily.

Let

$$SP(D^{-1}(z_0)) = (z - z_0)^{-\nu} L_1 + (z - z_0)^{-\nu+1} L_2 + \cdots + (z - z_0)^{-1} L_\nu.$$

It is easy to check that

$$S_m = \Delta(\widetilde{F}_0, \widetilde{F}_1, \ldots, \widetilde{F}_{m-1}) \cdot \Delta(L_1, L_2, \ldots, L_m) \cdot \Delta(\widetilde{E}_0, \widetilde{E}_1, \ldots, \widetilde{E}_{m-1}),$$
$$m = 1, 2, \ldots, \nu,$$

where

$$\widetilde{F}_j = \frac{1}{j!}(F^{-1})^{[j]}(z_0), \quad \widetilde{E}_j = \frac{1}{j!}(E^{-1})^{[j]}(z_0), \qquad j = 0, 1, \ldots.$$

Let now

$$\mathrm{col}(0, \ldots, 0, x_0, x_1, \ldots, x_\ell) \in \mathrm{Im}\, S_\nu$$
$$= \mathrm{Im}[\Delta(\widetilde{F}_0, \widetilde{F}_1, \ldots, \widetilde{F}_{\nu-1}) \cdot \Delta(L_1, L_2, \ldots, L_\nu)]$$

so

$$\mathrm{col}(0, \ldots, 0, x_0, x_1, \ldots, x_\ell) = \Delta(\widetilde{F}_0, \widetilde{F}_1, \ldots, \widetilde{F}_{\nu-1}) \cdot \Delta(L_0, L_1, \ldots, L_\nu) \mathrm{col}(z_1, \ldots, z_\nu)$$

for some n-dimensional vectors $z_1, \ldots, z_\nu$. Multiplying this equality from the left by $\Delta(F_0, F_1, \ldots, F_{\nu-1})$, where $F_j = \frac{1}{j!}F^{[j]}(z_0)$, we obtain that $\mathrm{col}(0, \ldots, 0, y_0, \ldots, y_\ell) =$

$\mathrm{col}(0, \ldots, 0, F_0 x_0, F_1 x_0 + F_0 x_1, \ldots, F_\ell x_0 + F_{\ell-1} x_1 + \cdots + F_0 x_\ell) \in \mathrm{Im}\,\Delta(L_0, L_1, \ldots, L_\nu)$, and conversely any vector $\mathrm{col}(0, \ldots, 0, y_0, \ldots, y_\ell)$ in $\mathrm{Im}\,\Delta(L_0, L_1, \ldots, L_\nu)$ arises in this way from a vector $\mathrm{col}(0, \ldots, 0, x_0, x_1, \ldots, x_\ell)$ in $\mathrm{Im}\,S_\nu$. By Lemma 1.1.1 and the definition of null chains, we know that $(x_0, x_1, \ldots, x_\ell)$ is a null chain for $A(z)$ at z_0 if and only if $(y_0, y_1, \ldots, y_\ell)$ is a null chain for $D(z)$ at z_0. Thus it suffices to verify Proposition 1.3.1 for the diagonal case.

For the diagonal case we have

$$D(z)^{-1} = L_1(z - z_0)^{-\nu} + L_2(z - z_0)^{-\nu+1} + \cdots + L_\nu(z - z_0)^{-1}$$
$$= \mathrm{diag}\big((z - z_0)^{-\kappa_1}, \ldots, (z - z_0)^{-\kappa_n}\big).$$

Assuming $x_1 \geq \cdots \geq x_n$, it follows that

$$L_j = P_{m(j)} - P_{m(j-1)}$$

where

(1.3.3) $$m(j) = \max\{j \colon \kappa_j \geq \nu + 1 - j\},$$

and P_k is the projection onto the first k standard basis vectors in $\mathbf{C}^n$. Thus one can verify

$$\mathrm{Im}\,\Delta(L_1, L_2, \ldots, L_\nu) = \{\mathrm{col}(y_1, \ldots, y_\nu) \colon y_j^{(i)} = 0 \text{ if } i > m(j)\}$$

where $y_j^{(i)}$ is the i-th component of the vector y_j in $\mathbf{C}^n$. By definition of $m(j)$, the condition $i > m(j)$ is equivalent to $\kappa_i \leq \nu - j$. Thus we have established

$$\mathrm{Im}\,\Delta(L_1, \ldots, L_\nu) = \{\mathrm{col}(y_1, \ldots, y_n) \colon y_j^{(i)} = 0 \text{ if } j \leq \nu - \kappa_i\}.$$

On the other hand $x_0, x_1, \ldots, x_\ell$ (where $\ell \leq \nu - 1$ and $x_0 \neq 0$) is a right null chain for $D(z)$ at z_0 if and only if $D(z) \cdot (z - z_0)^{\nu-\ell-1} \sum_{j=0}^{\ell} x_j(z - z_0)^j$ has a zero of order ν at z_0. If we set $\mathrm{col}(y_1, y_2, \ldots, y_\nu) = \mathrm{col}(0, \ldots, 0, x_0, x_1, \ldots, x_\ell)$, by the simple diagonal form of $D(z)$ we see that this also is equivalent to $y_j^{(i)} = 0$ if $j \leq \nu - \kappa_i$. This establishes Proposition 1.3.1 for the diagonal case, and hence also for the general case. $\square$

As a corollary to Proposition 1.3.1 we read off the following. Recall the definition of rank x, where $x \in \mathrm{Ker}\,A(z_0)$, given in Section 1.2.

COROLLARY 1.3.2. *With notation as in Proposition 1.3.1, we have*

(i) $\mathrm{Ker}\,A(z_0) = \{x \in \mathbf{C}^n \colon \mathrm{col}(0, 0, \ldots, x) \in \mathrm{Im}\,S_\nu\}$,

(ii) $\mathrm{rank}\,x = \max\{r \colon \mathrm{col}(0, 0, \ldots, x) \in \mathrm{Im}\,S_{\nu-r+1}\}$ *for every* $x \in \mathrm{Ker}\,A(z_0) \backslash \{0\}$.

PROOF. $\mathrm{Ker}\,A(z_0) \backslash \{0\}$ is the set of leading vectors for all possible right null chains for $A(z)$ at z_0; thus (i) is an immediate consequence of the characterization of $\mathrm{Im}\,S_\nu$ in Proposition 1.3.1. Statement (ii) also follows easily from this characterization of $\mathrm{Im}\,S_\nu$. $\square$

We also observe that one can build right null chains from S_ν:

PROPOSITION 1.3.3.   *If* $x \in \operatorname{Ker} A(z_0)\backslash\{0\}$, $\operatorname{rank} x = r$ *and* $S_{\nu-r+1}\operatorname{col}(y_1, y_2, \ldots, y_{\nu-r+1}) = \operatorname{col}(0, 0, \ldots, x)$, *then the vectors*

$$x_0 = x, \qquad x_1 = K_2 y_{\nu-r+1} + K_3 y_{\nu-r} + \cdots + K_{\nu-r+2} y_1,$$
$$\cdots, x_{r-1} = K_r y_{\nu-r+1} + K_{r+1} y_{\nu-r} + \cdots + K_\nu y_1$$

form a maximal right null chain for $A(z)$ *at* z_0.

For the proof note that if $x_0, x_1, \ldots, x_{r-1}$ are defined as in Proposition 1.3.3, then

$$\operatorname{col}(0, \ldots, 0, x_0, x_1, \ldots, x_{r-1}) = S_\nu \operatorname{col}(y_1, y_2, \ldots, y_{\nu-r+1}, 0, \ldots, 0),$$

and it remains to use Proposition 1.3.1. □

The left version of Propositions 1.3.1, 1.3.3 and Corollary 1.3.2 are as follows. Here, if X is a $p \times q$ matrix, $\operatorname{Im} \ell X$ denotes its image as a right multiplication operator from the vector space of row matrices $M_{1\times p}$ to the vector space of row matrices $M_{1\times q}$. Also, given $SPA^{-1}(z_0)$ as in formula (1.3.1), define:

$$\widetilde{S}_m = \begin{bmatrix} K_1 & K_2 & \cdots & K_m \\ 0 & K_1 & \cdots & K_{m-1} \\ \vdots & \vdots & & \vdots \\ 0 & 0 & \cdots & K_1 \end{bmatrix}, \qquad m = 1, \ldots, \nu.$$

PROPOSITION 1.3.4. *If* $A(z)$ *is a matrix function, regular and analytic at* z_0 *and* ν *is the order of the pole of* $A^{-1}(z)$ *at* z_0, *then*

$$\operatorname{Im}_\ell \widetilde{S}_\nu = \{[0, \ldots, 0, x_0, x_1, \ldots, x_\ell]: \text{ the row vectors}$$
$$x_0, x_1, \ldots, x_\ell \text{ form a left null chain for } A(z) \text{ at } z_0\}.$$

Moreover,

(i) x *is a left eigenvector for* $A(z)$ *at* z_0 *if and only if* $[0, \ldots, 0, x] \in \operatorname{Im}_\ell \widetilde{S}_\nu$. *In this case*

(ii) $\operatorname{rank} x = \max\{r: [0, 0, \ldots, x] \in \operatorname{Im}_\ell \widetilde{S}_{\nu-r+1}\}$.

If $\operatorname{rank} x = r$ *and* $[0, \ldots, 0, x] = [y_1, y_2, \ldots, y_{\nu-r+1}]\widetilde{S}_{\nu-r+1}$, *then the row vectors* $x_0 = x$, $x_1 = y_{\nu-r+1}K_2 + y_{\nu-r}K_3 + \cdots + y_1 K_{\nu-r+2}; \ldots; x_{r-1} = y_{\nu-r+1}K_r + y_{\nu-r}K_{r+1} + \cdots + y_1 K_\nu$ *form a left null chain for* $A(z)$ *at* z_0.

Observe that a right null pair (Z, T) for an analytic matrix function $A(z)$ at z_0 has a disadvantage as a local characteristic for $A(z)$; namely, it is not uniquely determined by $A(z)$ (and z_0): any pair similar to (Z, T) is also a right null pair for $A(z)$ at z_0. There is a local characteristic for $A(z)$ which is free of this disadvantage and at the same time can be constructed from a right null pair. Namely, the *right null subspace* $R_\nu = R_\nu(z_0; A) \subset \mathbb{C}^{n\nu}$ defined by the equality

$$R_\nu = \operatorname{Im} \operatorname{col}\big(Z(T - z_0 I)^{\nu-i}\big)_{i=1}^\nu$$

where ν is the order of z_0 as a pole of $A(z)^{-1}$, is such a local characteristic of $A(z)$ at z_0. Indeed it is easy to see (for instance, by using the local Smith form of $A(z)$) that ν coincides with the maximal length of null chains for $A(z)$ at z_0; then Theorem 1.2.6 implies that R_ν does not depend on the particular choice of the null pair (Z, T). We remark that in all the considerations and results below one can replace ν in the definition of R_ν by any fixed integer greater than or equal to the order of z_0 as a pole of $A(z)^{-1}$.

The following proposition gives an equivalent definition of the right null space in the language of right null functions.

PROPOSITION 1.3.5. *The right null subspace R_ν for the analytic matrix function $A(z)$ at the point z_0 can equivalently be defined as*

$$R_\nu = \{\mathrm{col}(\varphi_0, \varphi_1, \ldots, \varphi_{\nu-1}): A(z)\varphi(z) \text{ has a zero of order}$$

$$\text{at least } \nu \text{ at } z_0, \text{ where } \varphi(z) = \sum_{i=0}^{\nu-1} (z - z_0)^i \varphi_i + 0(z - z_0)^\nu \}.$$

PROOF. It is easy to see by multiplying out Taylor series that the analytic vector function $\varphi(z) = \sum_{i=0}^{\infty}(z - z_0)^i \varphi_i$ is such that $A(z)\varphi(z)$ has a zero of order at least ν at z_0 if and only if the vector $\mathrm{col}(\varphi_i)_{i=0}^{\nu-1}$ is in the kernel of the matrix $G_\nu = \Delta(A(z_0), A'(z_0), \ldots, \frac{1}{(\nu-1)!}A^{[\nu-1]}(z_0))$. The proposition therefore follows immediately from the following general lemma.

LEMMA 1.3.6. *Let $A(z)$ be a regular analytic matrix function with null pair (Z, T) at z_0. Then*

$$(1.3.4) \qquad \mathrm{Im\,col}\big(Z(T - z_0 I)^{\nu-m}\big)_{m=1}^{\nu} = \mathrm{Ker}\, G_\nu = \mathrm{Im}\, S_\nu$$

where $S_\nu = S_\nu(z_0; A)$, $G_\nu = \Delta(A(z_0), A'(z_0), \ldots, (\frac{1}{(\nu-1)!})A^{[\nu-1]}(z_0))$, and ν is the order of z_0 as a pole of $A^{-1}(z_0)$.

PROOF. Recall that ν coincides with the maximal length of the null chains for $A(z)$ at z_0. By definition, $x_0, \ldots, x_k$ is a null chain for $A(z)$ at z_0 if and only if

$$\mathrm{col}(0, \ldots, 0, x_0, \ldots, x_k) \in \mathrm{Ker}\, G_\nu.$$

Now use equality (1.2.16) in the proof of Theorem 1.2.6 to deduce the left equality in (1.3.4). The right equality in (1.3.4) follows from the above characterization of $\mathrm{Ker}\, G_\nu$ together with Proposition 1.3.1. $\square$

The following theorem provides a characteristic property of the (right) null subspace R_ν. Let $U: \mathbf{C}^{n\nu} \to \mathbf{C}^{n\nu}$ be the shift operator: $U(\mathrm{col}(x_1, x_2, \ldots, x_\nu)) = \mathrm{col}(0, x_1, x_2, \ldots, x_{\nu-1})$; $x_i \in \mathbf{C}^n$.

THEOREM 1.3.7. *The right null subspace of an analytic matrix function at a point z_0 is invariant for U.*

This theorem will be proved together with the next result, which gives explicit formulas for the right null subspaces in terms of the singular part of the Laurent expansion in a neighborhood of z_0, as well as formulas for a right null pair in terms of the right null subspace.

THEOREM 1.3.8. *The right null subspace $R_\nu \subset \mathbf{C}^{n\nu}$ of an analytic matrix function $A(z)$ at z_0 determines uniquely (up to similarity) a right null pair (Z, T) of $A(z)$ at z_0, according to the following formulas:*

$$(1.3.5) \qquad Z = P_{\nu|R_\nu}, \qquad T = (U + \lambda_0 I)_{|R_\nu}$$

where $P_\nu: \mathbf{C}^{n\nu} \to \mathbf{C}^{n\nu}$ projects onto the last n coordinates, and $U: \mathbf{C}^{n\nu} \to \mathbf{C}^{n\nu}$ is the shift operator. Conversely, if $SP(A^{-1}(z_0))$ is given, then the right null subspace is determined uniquely by the formula

$$(1.3.6) \qquad R_\nu = \operatorname{Im} S_\nu,$$

where $S_\nu = S_\nu(z_0; A)$ is defined by (1.3.2).

PROOF OF THEOREMS 1.3.7 and 1.3.8. If $\varphi(z)$ is a vector function analytic at z_0 such that $A(z)\varphi(z)$ has a zero of order at least ν at z_0, then $\widetilde{\varphi}(z) = (z - z_0)\varphi(z)$ is certainly another such function; this combined with the characterization of R_ν in Proposition 1.3.5 immediately implies that R_ν is U-invariant.

The formula (1.3.5) follows from the identity

$$\begin{bmatrix} 0 & 0 & \cdots & 0 \\ I & 0 & \cdots & 0 \\ 0 & I & \cdots & 0 \\ \vdots & \vdots & & \vdots \\ 0 & 0 & \cdots & I \end{bmatrix} \begin{bmatrix} Z(T - z_0 I)^{\nu-1} \\ Z(T - z_0 I)^{\nu-2} \\ \vdots \\ Z \end{bmatrix} = \begin{bmatrix} Z(T - z_0 I)^{\nu} \\ Z(T - z_0 I)^{\nu-1} \\ \vdots \\ Z(T - z_0 I) \end{bmatrix}$$

where we take into account that $(T - z_0 I)^\nu = 0$. Finally, (1.3.6) follows from Lemma 1.3.6. $\square$

Analogously to the notion of the right null subspace one can introduce the notion of the left null subspace. Namely, for an analytic $n \times n$ matrix function $A(z)$ at z_0 with left null pair (T, Y), define the *left null subspace* $R_\nu^{(\ell)}$ as the subspace of all $n\nu$-dimensional row vectors of the form

$$\varphi[Y, (T - z_0 I)Y, \ldots, (T - z_0 I)^{\nu-1} Y],$$

where φ is an arbitrary p-dimensional row vector and $p \times p$ is the size of T. (As before, ν is the order of z_0 as a pole of $A(z)^{-1}$.) All the results concerning the left null subspaces can be deduced from the corresponding results on the right null subspaces, using the simple observation that φ belongs to the left null subspace of $A(z)$ at z_0 if and only if φ^T belongs to the right null subspace for $A(z)^T$. We leave the statements and deduction of these results to the reader.

1.4 NULL TRIPLES

Let $A(z)$ be an $n \times n$ analytic matrix function (as usual we assume $\det A(z) \not\equiv 0$), and let z_0 be a zero of $\det A(z)$. The singular part of the Laurent series of $A(z)^{-1}$ in a neighborhood of z_0 (denoted, as before, $SP(A^{-1}(z_0))$) is an important local characteristic of zeros of $A(z)$ at z_0, and its connections with right and left null chains were explored in Section 1.3. In this section we express the relationships between $SP(A^{-1}(z_0))$ and the null chains of $A(z)$ at z_0 using the more convenient language of null triples.

A triple of matrices (Z, T, Y) of sizes $n \times p$, $p \times p$, and $p \times n$, respectively, will be called a *null triple* for $A(z)$ at z_0 if the matrix function

$$A(z)^{-1} - Z(zI - T)^{-1}Y$$

is analytic in a neighborhood of z_0, $\sigma(T) = \{z_0\}$ and the pairs (Z, T) and (T, Y) are null kernel and full range pairs, respectively, i.e. for q large enough we have

$$(1.4.1) \qquad \operatorname{rank} \operatorname{col}(ZT^i)_{i=0}^{q-1} = \operatorname{rank}[Y, TY, \ldots, T^{q-1}Y] = p.$$

(Here p depends on $A(z)$ and z_0.) Leaving aside for the time being the problem of existence of null triples (we shall see later in Chapter 2 that they always exist), let us point out that the null triple of $A(z)$ at z_0 is uniquely determined up to a naturally defined similarity. This follows from the following proposition:

PROPOSITION 1.4.1. *Let (Z_1, T_1, Y_1) and (Z_2, T_2, Y_2) be triples of matrices with the following properties:*

(i) *the sizes of Z_i, T_i, Y_i are $n \times m_i$, $m_i \times m_i$, $m_i \times n$, respectively, for $i = 1, 2$;*

(ii) *(Z_i, T_i) are null kernel pairs, and (T_i, Y_i) are full range pairs;*

(iii) *the matrix function $Z_1(zI - T_1)^{-1}Y_1 - Z_2(zI - T_2)^{-1}Y_2$ is analytic at z_0;*

(iv) *$\sigma(T_1) = \sigma(T_2) = \{z_0\}$.*

Then $m_1 = m_2$, and (Z_1, T_1, Y_1) and (Z_2, T_2, Y_2) are similar, i.e. there exists an invertible matrix S such that $Z_1 = Z_2 S$, $T_1 = S^{-1}T_2 S$, $Y_1 = S^{-1}Y_2$.

PROOF. Let q be such that

$$\operatorname{rank}[Y_i, T_i Y_i, \ldots, T_i^{q-1}Y_i] = \operatorname{rank} \operatorname{col}(Z_i T_i^j)_{j=0}^{q-1} = m_i$$

for $i = 1, 2$. From condition (iii) it follows that

$$(1.4.2) \qquad \operatorname{col}(Z_1 T_1^k)_{k=0}^{q-1} \cdot [Y_1, \ldots, T_1^{q-1}Y_1] = \operatorname{col}(Z_2 T_2^k)_{k=0}^{q-1} \cdot [Y_2, \ldots, T_2^{q-1}Y_2]$$

and

$$(1.4.3) \qquad \operatorname{col}(Z_1 T_1^k)_{k=0}^{q-1} \cdot T_1[Y_1, \ldots, T_1^{q-1}Y_1] = \operatorname{col}(Z_2 T_2^k)_{k=0}^{q} \cdot T_2[Y_2, \ldots, T_2^{q-1}Y_2].$$

Let U be some left inverse of $\operatorname{col}(Z_1 T_1^k)_{k=0}^{q-1}$, and let V be some right inverse of $[Y_1, \ldots, T_1^{q-1}Y_1]$. Let $S = [Y_2, \ldots, T_2^{q-1}Y_2]V$ be an $m_2 \times m_1$ matrix. Then (1.4.2) implies that $I = U \operatorname{col}(Z_2 T_2^k)_{k=0}^{q-1} \cdot S$, i.e. S is left invertible with a left inverse

$S^{-1} = U \cdot \mathrm{col}(Z_2 T_2^k)_{k=0}^{q-1}$. In particular, $m_2 \geq m_1$; analogously one obtains $m_1 \geq m_2$. So $m_1 = m_2$ and S is square and (two-sided) invertible. Now (1.4.3) gives $T_1 = S^{-1} T_2 S$. Equating the upper block row in both sides of (1.4.2) leads to $Z_1 = Z_2 S$, and equating the left block column in (1.4.2) leads to $Y_1 = S^{-1} Y_2$. $\square$

Note that conditions (iii) and (iv) of Proposition 1.4.1 yield in fact the equality $Z_1(zI - T_1)^{-1} Y_1 = Z_2(zI - T_2)^{-1} Y_2$. Indeed, the matrix function $F(z) = Z_1(zI - T_1)^{-1} Y_1 - Z_2(zI - T_2)^{-1} Y_2$ is analytic in $\mathbb{C}$ and $F(\infty) = 0$. By Liouville's theorem, $F(z) \equiv 0$.

We record here a more general statement that will be used later.

PROPOSITION 1.4.2. *Let* (Z_i, T_i, Y_i), $i = 1, 2$ *be two triples of matrices such that the properties* (i) *and* (ii) *of Proposition 1.4.1, as well as the following property is satisfied:*

(v) *the matrix function* $Z_1(zI - T_1)^{-1} Y_1 - Z_2(zI - T_2)^{-1} Y_2$ *is analytic on some open set that contains all the eigenvalues of* T_1 *and* T_2.

Then (Z_1, T_1, Y_1) *and* (Z_2, T_2, Y_2) *are similar.*

The proof can be done in the same way as the proof of Proposition 1.4.1, using the fact that

$$Z_1(zI - T_1)^{-1} Y_1 - Z_2(zI - T_2)^{-1} Y_2 \equiv 0$$

by Liouville's theorem, and hence equality (1.4.2) holds.

To prove the existence of a null triple, we find first a triple of matrices (Z, T, Y) such that $\sigma(T) = \{z_0\}$ and the function

$$A(z)^{-1} - Z(zI - T)^{-1}$$

is analytic in a neighborhood of z_0 (we do not require that (1.4.1) holds). Indeed, write

$$SP(A^{-1}(z_0)) = \sum_{k=1}^{p} (z - z_0)^{-k} K_k,$$

then

$$A(z)^{-1} - \begin{bmatrix} I & 0 & \cdots & 0 \end{bmatrix} \left(zI - \begin{bmatrix} z_0 I & I & & 0 \\ & z_0 I & I & \\ & & \ddots & I \\ 0 & & & \ddots & z_0 I \end{bmatrix} \right)^{-1} \begin{bmatrix} K_1 \\ \vdots \\ K_p \end{bmatrix}$$

is analytic in a neighborhood of z_0.

Now using the reduction of the triple

$$\left(\begin{bmatrix} I & 0 & \cdots & 0 \end{bmatrix}, \begin{bmatrix} z_0 I & I & & 0 \\ & z_0 I & I & \\ & & \ddots & I \\ 0 & & & \ddots & z_0 I \end{bmatrix}, \begin{bmatrix} K_1 \\ \vdots \\ K_p \end{bmatrix} \right)$$

(see Section 4.1, especially Theorem 4.1.2, to come), we find a null triple for $A(z)$ at z_0.

A different approach for constructing a null triple for $A(z)$ at z_0 (based on the theory of matrix polynomials) will be given in Chapter 2.

Next, we study the connections between null triples and right and left null pairs. First of all observe that if ν is the order of z_0 as a pole of $A(z)^{-1}$ (i.e. ν is the smallest nonnegative integer for which $(z - z_0)^\nu A(z)^{-1}$ is analytic at z_0) then a right null pair (Z, T) for $A(z)$ at z_0 satisfies

$$\bigcap_{i=0}^{\nu-1} \operatorname{Ker}(ZT^i) = \{0\}.$$

Indeed, because of the equality

$$\begin{bmatrix} I & 0 & 0 & \cdots & 0 \\ -z_0\binom{1}{0}I & \binom{1}{1}I & 0 & \cdots & 0 \\ \vdots & \vdots & \vdots & & \vdots \\ (-z_0)^{q-1}\binom{q-1}{0}I & (-z_0)^{q-2}\binom{q-1}{1}I & (-z_0)^{q-3}\binom{q-1}{2}I & \cdots & \binom{q-1}{q-1}I \end{bmatrix}$$
$$\times \begin{bmatrix} Z \\ ZT \\ \vdots \\ ZT^{q-1} \end{bmatrix} = \begin{bmatrix} Z \\ Z(T - z_0 I) \\ \vdots \\ Z(T - z_0 I)^{q-1} \end{bmatrix}$$

$(q = 1, 2, \ldots)$ it suffices to show that

$$\bigcap_{i=0}^{\nu-1} \operatorname{Ker}\big(Z(T - z_0 I)^i\big) = \{0\}.$$

It is easy to see (for example, by reduction to the local Smith form) that $A(z)$ cannot have a right null chain at z_0 with more than ν vectors in it. Hence $(T - z_0 I)^i = 0$ for $i \geq \nu$, and consequently,

$$\bigcap_{i=0}^{\nu-1} \operatorname{Ker}\big(Z(T - z_0 I)^i\big) = \bigcap_{i=0}^{\infty} \operatorname{Ker}\big(Z(T - z_0 I)^i\big) = \{0\},$$

where the last equality follows from the null kernel property of the pair $(Z, T - z_0 I)$ (which in turn is a consequence of the null kernel property of (Z, T)). Analogously for a left null pair (T, Y) for $A(z)$ at z_0 one proves that the rows of

$$[Y, TY, \ldots, T^{\nu-1}Y]$$

are linearly independent.

The following result gives the precise procedure to build a null triple from given right and left null pairs.

THEOREM 1.4.3. *Let (Z_1, T_1) and (T_1, Y_1) be right and left null pairs, respectively, of the analytic matrix function $A(z)$ corresponding to z_0. Let*

$$
(1.4.4) \qquad
G =
\begin{bmatrix}
Z_1 \\
Z_1(T_1 - z_0 I) \\
\vdots \\
Z_1(T_1 - z_0 I)^{\nu-1}
\end{bmatrix}^{-1}
\begin{bmatrix}
K_1 & K_2 & \cdots & K_{\nu-1} & K_\nu \\
K_2 & K_3 & \cdots & K_\nu & 0 \\
\vdots & \vdots & & \vdots & \vdots \\
K_\nu & 0 & \cdots & 0 & 0
\end{bmatrix}
$$
$$
\times [Y_1, (T_1 - z_0 I)Y_1, \ldots, (T_1 - z_0 I)^{\nu-1}Y_1]^{-1},
$$

where

$$
SP\big(A^{-1}(z_0)\big) = \sum_{j=1}^{\nu} (z - z_0)^{-j} K_j
$$

and the inverse matrices are understood as some one-sided inverses from the appropriate side. (Observe that by the preceding discussion the matrix $\operatorname{col}\big(Z_1(T_1 - z_0 I)^i\big)_{i=0}^{\nu-1}$ is left invertible while

$$
[Y_1, (T_1 - z_0 I)Y_1, \ldots, (T_1 - z_0 I)^{\nu-1}Y_1]
$$

is right invertible.) Then G is invertible, and

$$
(1.4.5) \qquad SP\big(A^{-1}(z_0)\big) = Z_1(zI - T_1)^{-1}GY_1 = Z_1 G(zI - T_1)^{-1}Y_1
$$

and

$$
(1.4.6) \qquad GT_1 = T_1 G.
$$

The matrix G is the unique $p \times p$ matrix (p is the size of T_1) which satisfies (1.4.5) and (1.4.6).

In particular, $(Z_1 G, T_1, Y_1)$ is a null triple for $M(z)$ at z_0.

The proof of Theorem 1.4.3 is based on the matrix polynomial approach, and hence will be given in Chapter 2.

Using the matrix G given by (1.4.4) one can describe all possible singular parts of the inverse of an analytic matrix function with given right and left null pairs as follows.

THEOREM 1.4.4. *Let (Z, T) and (T, Y) be a null kernel pair and a full range pair of matrices respectively such that $\sigma(T) = \{z_0\}$. Then all possible singular parts at z_0 of the inverse of an analytic matrix function $A(z)$ having (Z, T) and (T, Y) as right and left null pairs corresponding to z_0, are given by the formula*

$$
(1.4.7) \qquad SP\big(A^{-1}(z_0)\big) = ZG(zI - T)^{-1}Y \qquad (= Z(zI - T)^{-1}GY),
$$

where G is any invertible matrix commuting with T. The correspondence, given by (1.4.7), between the set of all singular parts $SP\big(A^{-1}(z_0)\big)$ and the set of all invertible matrices commuting with T, is one-to-one.

PROOF. If $A(z)$ is an analytic matrix function with right and left null pairs (Z, T) and (T, Y) respectively corresponding to z_0, then by Theorem 1.4.3 there is a unique invertible G commuting with T for which (1.4.7) holds. Conversely, given invertible G such that $GT = TG$, we show that there is a matrix function $A(z)$ analytic in a neighborhood of z_0 such that

(1.4.8)
$$SP(A^{-1}(z_0)) = ZG(zI - T)^{-1}Y.$$

(Actually, one can choose $A(z)$ to be a rational matrix function.) Write

$$SP(ZG(z - T)^{-1}Y) = \sum_{j=1}^{k} (z - z_0)^{-j} M_j,$$

and let

$$M_j = [\alpha_{pq}^{(j)}]_{p,q=1}^{n}, \qquad j = 1, \ldots, k.$$

Define scalar rational functions by

$$b_{pq}(z) = \sum_{j=1}^{k} (z - z_0)^{-j} \alpha_{pq}^{(j)}, \qquad p, q = 1, \ldots, n.$$

For a suitable $\lambda_0 \in \mathbb{C}$ the matrix function

$$B(z) = \lambda_0 I + [b_{pq}(z)]_{p,q=1}^{n}$$

has the property that $\det B(z) \not\equiv 0$. Let

$$B(z) = E(z) \operatorname{diag}\big((z - z_0)^{\kappa_1}, \ldots, (z - z_0)^{\kappa_n}\big) F(z)$$

be a local Smith form for $B(z)$, so $\kappa_1 \geq \cdots \geq \kappa_n$ are integers and $E(z)$ and $F(z)$ are matrix functions analytic and invertible at z_0 (the existence of the local Smith form for $B(z)$ follows easily from Theorem 1.1.2). Then one can take

$$A(z) = \big(E(z) \operatorname{diag}((z - z_0)^{\mu_1}, \ldots, (z - z_0)^{\mu_n}) F(z)\big)^{-1},$$

where $\mu_i = \min(0, \kappa_i)$, to satisfy (1.4.8).

So (ZG, T, Y) is a null triple for $A(z)$ at z_0. On the other hand, comparing with the null triple for $A(z)$ at z_0 built in Theorem 1.4.3 from right and left null pairs (Z_1, T_1) and (T_1, Y_1) of $A(z)$ at z_0, and using the uniqueness of null triples up to similarity, we conclude that (ZG, T) and (T, Y) are right and left null pairs for $A(z)$ at z_0, respectively. As $GT = TG$, the pair (Z, T) is also a right null pair for $A(z)$ at z_0. This completes the proof of Theorem 1.4.4. $\square$

1.5 NULL CHARACTERISTICS WITH RESPECT TO SEVERAL ZEROS

Many notions, as well as results, describing local properties of an analytic matrix function $A(z)$ at z_0 can be carried out naturally for the case of several (but a finite number of) zeros of $\det A(z)$. In this section we shall outline some notions and results concerning this more general situation. As always in this book we assume throughout that $A(z)$ is regular: $\det A(z) \not\equiv 0$.

Fix a compact subset $Q \subset \Omega$, where Ω is the domain of analyticity of $A(z)$. Let $z_1, z_2, \ldots, z_k$ be all the different zeros of $\det A(z)$ lying in Q (their number is finite), and let $(Z_1, T_1), \ldots, (Z_k, T_k)$ be right null pairs of $A(z)$ corresponding to $z_1, \ldots, z_k$, respectively. A pair of matrices (Z, T) which is similar to $([Z_1 Z_2 \cdots Z_k], \operatorname{diag}(T_1, T_2, \ldots, T_k))$ is called a right null pair of $A(z)$ relative to Q. So the size of the matrix T in a null pair (Z, T) of $A(z)$ relative to Q is just the number of zeros (counting multiplicities) of $\det A(z)$ inside Q.

THEOREM 1.5.1. *The null pair (Z, T) of $A(z)$ relative to Q is a null kernel pair and is unique up to similarity.*

PROOF. The uniqueness up to similarity follows easily from the uniqueness up to similarity of each right null pair (Z_i, T_i) corresponding to z_i $(i = 1, \ldots, k)$.

We prove now that

$$(1.5.1) \qquad \bigcap_{j \geq 0} \operatorname{Ker}[Z_1 T_1^j, Z_2 T_2^j, \ldots, Z_k T_k^j] = \{0\},$$

where (Z_i, T_i) is a right null pair of $A(z)$ corresponding to z_i. Let $x = \operatorname{col}(x_1, \ldots, x_k)$ belong to the left-hand side of (1.5.1), i.e.

$$(1.5.2) \qquad \sum_{p=1}^{k} Z_p T_p^j x_p = 0 \quad \text{for} \quad j = 0, 1, \ldots .$$

As the spectra $\sigma(T_1), \ldots, \sigma(T_k)$ are mutually disjoint, there is a scalar polynomial $q(z)$ (Lagrange interpolating polynomial) such that $q(T_1) = I$ and $q(T_p) = 0$ for $p \neq 1$. Then for $j = 0, 1, \ldots$ we obtain using (1.5.2):

$$Z_1 T_1^j x_1 = Z_1 T_1^j q(T_1) x_1 = -\sum_{p=2}^{k} Z_p T_p^j q(T_p) x_p = 0.$$

As (Z_1, T_1) is a null kernel pair, we conclude that $x_1 = 0$. Analogously one proves $x_p = 0$ for $p \geq 2$, and (1.5.1) follows. □

The analogous construction for left null pairs leads to the notion of a left null pair (T, Y) of $A(z)$ relative to Q.

To study the properties of null pairs relative to Q it is convenient to use the following generalization of a local Smith form for $A(z)$:

THEOREM 1.5.2. *Let $Q \subset \Omega$ be a compact set. Then $A(z)$ admits a representation*

$$A(z) = E(z) \operatorname{diag}(\varphi_1(z), \ldots, \varphi_n(z)) F(z),$$

where $F(z)$ is a unimodular matrix polynomial, $E(z)$ is an analytic matrix function invertible for $z \in Q$, and $\varphi_1(z), \ldots, \varphi_n(z)$ are scalar polynomials with zeros in Q such that $\varphi_i(z)$ is divisible by $\varphi_{i+1}(z)$, $i = 1, \ldots, n-1$.

The proof follows the same line of argument as the proof of Theorem 1.1.2 (instead of the factors of type $(z - z_0)^p$ one has to consider scalar polynomial factors $(z - z_1)^{p_1} \cdots (z - z_0)^{p_k}$, where $\{z_1, \ldots, z_k\}$ are all distinct zeros of $\det A(z)$ in Q). We omit the details.

Of course, a result analogous to Theorem 1.5.2 holds with $E(z)$ (but generally not $F(z)$) required to be a unimodular matrix polynomial.

We describe now null triples of an $n \times n$ analytic function $A(z)$ relative to a compact set $Q \subset \Omega$ (as before, $\Omega \subset \mathbb{C}$ is the domain of analyticity of $A(z)$, and $A(z)$ is assumed to be regular). To avoid vacuousness, we assume that there is at least one zero of $\det A(z)$ in Q. A triple (Z, T, Y) of matrices of sizes $n \times p$, $p \times p$, $p \times n$, respectively, is called a *null triple* of $A(z)$ relative to Q if the matrix function $A(z)^{-1} - Z(zI - T)^{-1}Y$ is analytic in a neighborhood of Q, (Z, T) and (T, Y) are null kernel pair and full range pair, respectively, and $\sigma(T) \subset Q$. By Proposition 1.4.2 a null triple of $A(z)$ relative to Q is unique up to similarity. The existence of a null triple relative to Q follows from the existence of a null triple of $A(z)$ at each point. Indeed, let $z_1, \ldots, z_r$ be all the distinct zeros of $\det A(z)$ in Q, and let (Z_i, T_i, Y_i) be a null triple of $A(z)$ at z_i, $i = 1, \ldots, p$. Then

$$\left(\begin{bmatrix} Z_1 & Z_2 & \cdots & Z_r \end{bmatrix}, T_1 \oplus \cdots \oplus T_r, \operatorname{col}(Y_i)_{i=1}^r \right)$$

is a null triple of $A(z)$ with respect to Q. Also, as in the case of a null triple at each point, by using Theorem 1.4.3 one deduces that if (Z, T, Y) is a null triple of $A(z)$ relative to Q, then (Z, T) and (T, Y) are right and left null pairs, respectively, of $A(z)$ relative to Q.

We shall need the following properties of null triples.

THEOREM 1.5.3. *Let $A(z)$ be a regular $n \times n$ analytic matrix function in the domain $\Omega \subset \mathbb{C}$, and let $Q \subset \Omega$ be a compact set. A triple of matrices (Z, T, Y) of sizes $n \times p$, $p \times p$, $p \times n$, respectively, is a null triple of $A(z)$ relative to Q if and only if the following conditions hold:*

(i) *$\sigma(T) \subset Q$;*

(ii) *the matrix function $A(z)^{-1} - Z(zI - T)^{-1}Y$ has an analytic extension to a neighborhood of Q;*

(iii) *the matrix function $A(z)Z(zI - T)^{-1}$ is analytic in a neighborhood of Q;*

(iv) *(Z, T) is a null kernel pair.*

PROOF. Assume (Z, T, Y) is a null triple of $A(z)$ relative to Q. Then (i), (ii), (iv) obviously hold. To verify (iii), observe that (Z, T) is a right null pair of $A(z)$

relative to Q. Passing, if necessary, to a similar pair, we can assume that

$$Z = [Z_1 \cdots Z_r], \qquad T = \mathrm{diag}(T_1, \ldots, T_r),$$

where (Z_j, T_j) is a right null pair of $A(z)$ corresponding to z_j, and $z_1, \ldots, z_r$ are all the distinct zeros of $\det A(z)$ in Q. It is sufficient to prove that $A(z)Z_j(zI - T_j)^{-1}$ is analytic at z_j. To this end write

$$Z_j(zI - T_j)^{-1} = Z_j\big((z - z_j)I - (T_j - z_jI)\big)^{-1}$$
$$= \sum_{\alpha=0}^{\infty} Z_j(T_j - z_jI)^{\alpha}(z - z_j)^{-\alpha-1}$$

and use Proposition 1.2.7(iv).

      We now prove the converse statement. So let (Z, T, Y) be a triple satisfying the conditions (i)–(iv). We have to prove that (Z, T, Y) is a null triple for $A(z)$ relative to Q. Without loss of generality we can assume that there is only one point, call it z_0, in Q where $A(z)$ has a zero. Then $\sigma(T) = \{z_0\}$. Indeed, suppose a point $z_1 \in \Omega \backslash \{z_0\}$ is an eigenvalue of T. Applying a similarity, if necessary, to the pair (Z, T), without loss of generality it can be assumed that $T = T_1 \oplus T_0$, where $\sigma(T_1) = \{z_1\}$ and $z_1 \notin \sigma(T_0)$. Denote by Z_1 the part of Z that corresponds to T_1; the condition (iii), together with the invertibility of $A(z_1)$, implies that $Z_1(zI - T_1)^{-1}$ is analytic at z_1. As $\sigma(T_1) = \{z_1\}$, by Liouville's theorem we obtain that $Z_1(zI - T_1)^{-1} \equiv 0$ for all $z \in \mathbb{C}$. So $Z_1 T_1^j = 0$ for $j = 0, 1, \ldots$, which contradicts the null kernel property of (Z, T).

      Thus, $\sigma(T) = \{z_0\}$, and we can assume that T is in the Jordan form:

(1.5.3) $$T = J_1 \oplus \cdots \oplus J_p,$$

where J_i is $\alpha_i \times \alpha_i$ Jordan block with eigenvalue z_0. Partition Z conformally with (1.5.3):

$$Z = [Z_1 Z_2 \cdots Z_p].$$

Using (iii), one checks (cf. the proof of Proposition 1.2.7) that the column vectors in each Z_i form a right null chain of $A(z)$ corresponding to z_0, and condition (iv) ensures that the eigenvectors in these chains are linearly independent. It follows (using the reduction to the local Smith form of $A(z)$ at z_0) that the size of T cannot exceed the order of z_0 as a zero of $\det A(z)$.

      At this point we use some basic results on realizations and their minimality presented in Section 4.1. Let $(\widetilde{Z}, \widetilde{T}, \widetilde{Y})$ be a null triple of $A(z)$ at z_0. Then

$$W(z) \stackrel{\mathrm{def}}{=} \widetilde{Z}(zI - \widetilde{T})^{-1}\widetilde{Y}$$

is a minimal realization (Theorem 4.1.5). But the size of $\widetilde{T}$ is precisely equal to the order α of z_0 as a zero of $\det A(z)$. As

(1.5.4) $$W(z) = Z(zI - T)^{-1}Y,$$

and the size of T does not exceed α (see the previous paragraph), we conclude that T must be of size $\alpha \times \alpha$, and the realization (1.5.4) is minimal as well. But any two minimal realizations of $W(z)$ are similar (Theorem 4.1.4), so (Z, T, Y) is similar to $(\widetilde{Z}, \widetilde{T}, \widetilde{Y})$, and we are done. $\square$

A dual version of Theorem 1.5.3 is also available. Namely, the conclusions of Theorem 1.5.3 hold true if (iii) and (iv) are replaced by the following conditions:

(v) the matrix function $(zI - T)^{-1}YA(z)$ is analytic in a neighborhood of Q;

(vi) (T, Y) is a full range pair.

The proof of this statement is left to the reader.

The property of null triples described in Theorem 1.5.3(iii) allows us to prove easily the following additional characterization of right and left pairs, which will be used later on.

THEOREM 1.5.4. *Let (Z, T) be a right null pair for a regular $n \times n$ analytic matrix function $A(z)$, with respect to a compact set Q in the domain of definition of $A(z)$. Then the function $A(z)Z(zI - T)^{-1}$ is analytic on Q. Conversely, if (Z, T) is a null kernel pair with $\sigma(T) \subset Q$ and with the size of T equal to the number of zeros of $\det A(z)$ in Q (counted with multiplicities), and if $A(z)Z(zI - T)^{-1}$ is analytic in Q, then (Z, T) is a right null pair of $A(z)$ with respect to Q.*

PROOF. Given a right null pair (Z, T) of $A(z)$ with respect to Q, let Y be the unique matrix (by Theorem 1.4.3) such that (Z, T, Y) is a null triple of $A(z)$ relative to Q. Now apply Theorem 1.5.3(iii). For the converse statement, we can assume without loss of generality (by passing to a similar pair if necessary) that

$$Z = [\, Z_1 \quad Z_2 \quad \cdots \quad Z_k \,]$$
$$T = T_1 \oplus T_2 \oplus \cdots \oplus T_k,$$

where $\sigma(T_j) = \{z_j\}$ and $z_1, \ldots, z_k$ are distinct points in Q. The analyticity of $A(z)Z(zI - T)^{-1}$ shows that each z_j must be a zero of $A(z)$. Next, we show that for $j = 1, \ldots, k$, the size of T_j coincides with the multiplicity of z_j as a zero of $\det A(z)$. Suppose not; then, because the size of T is equal to the number of zeros of $\det A(z)$ in Q, for some j_0 the size of T_{j_0} is bigger than the multiplicity of z_{j_0} as a zero of $\det A(z)$. Say, $j_0 = 1$, and without loss of generality assume $z_1 = 0$. Further, we can take T_1 in the Jordan form:

$$T_1 = J_{p_1} \oplus \cdots \oplus J_{p_r},$$

where J_{p_j} is the $p_j \times p_j$ Jordan block with eigenvalue 0, and $p_1 + \cdots + p_r > q$, the multiplicity of $z_1 = 0$ as a zero of $\det A(z)$. Partition Z_1 conformally:

$$Z_1 = [\, Z_{11} \quad Z_{12} \quad \cdot \quad Z_{1r} \,],$$

where $Z_{1j} = [x_0^{(j)} \; x_1^{(j)} \; \cdots \; x_{p_j-1}^{(j)}]$ has p_j columns. As in the proof of Proposition 1.2.7, we see that the columns of Z_{1j} form a right null chain of $A(z)$ at 0, for $j = 1, \ldots, r$. Also, that the vectors $x_0^{(1)}, \ldots, x_0^{(r)}$ are linearly independent. Passing to a local Smith form of $A(z)$ at $z_1 = 0$ and using Theorem 1.1.3, we find a set of right null functions of $\text{diag}(z^{\kappa_1}, z^{\kappa_2}, \ldots, z^{\kappa_n})$ at 0 with linearly independent eigenvectors ($=$ the values of the right null functions at 0) and whose sum of the orders exceeds $\kappa_1 + \kappa_2 + \cdots + \kappa_n$. However, this contradicts Theorem 1.1.4. So we have established that the size of T_j coincides with the multiplicity of z_j as a zero of $\det A(z)$, for $j = 1, \ldots, k$. Now the proof of Theorem 1.5.4 is completed by applying Proposition 1.2.7. $\square$

Analogous statement holds for the left null pair (T, Y) of $A(z)$; in this case, the function $(zI - T)^{-1} Y A(z)$ plays the role of $A(z) Z (zI - T)^{-1}$ in Theorem 1.5.4.

CHAPTER 2
NULL STRUCTURE AND INTERPOLATION PROBLEMS
FOR MATRIX POLYNOMIALS

In this chapter we consider matrix polynomials, i.e. matrix functions $L(z)$ of the form $L(z) = \sum_{i=0}^{\ell} A_i z^i$, where $A_0, A_1, \ldots, A_\ell$ are $n \times n$ matrices. We will assume throughout also that $L(z)$ is regular, i.e. $\det L(z)$ does not vanish identically. One of our aims is to specify for matrix polynomials in a detailed way the theory developed in the first chapter for analytic matrix functions. We also solve here the first interpolation problem in this book, namely, how to construct a matrix polynomial with given null structure. Explicit formulas for the solution are obtained.

2.1 LOCAL DATA

The definitions and properties of null functions, null chains, canonical set of null functions or null chains, null pairs and subspaces given in the preceding chapter apply, in particular, for matrix polynomials. In this section we shall emphasize the properties of these local data peculiar to the polynomial case.

PROPOSITION 2.1.1. *The vectors* $x_0, \ldots, x_{k-1}$ *form a right null chain for the matrix polynomial* $L(z) = \sum_{j=0}^{\ell} A_j z^j$ *corresponding to* z_0 *if and only if* $x_0 \neq 0$ *and*

$$(2.1.1) \qquad A_0 X_0 + A_1 X_0 J_0 + \cdots + A_{\ell-1} X_0 J_0^{\ell-1} + A_\ell X_0 J_0^\ell = 0,$$

where

$$X_0 = [x_0 \cdots x_{k-1}]$$

is an $n \times k$ *matrix, and* J_0 *is the Jordan block of size* $k \times k$ *with* z_0 *on the main diagonal.*

For the proof rewrite formula (1.2.21) from Proposition 1.2.7 appropriately.

THEOREM 2.1.2. *Let* (X, T) *be a pair of matrices, where* X *is an* $n \times n$ *matrix and* T *is a* $\mu \times \mu$ *matrix with unique eigenvalue* z_0. *Then the following conditions are necessary and sufficient in order that* (X, T) *be a right null pair for* $L(z) = \sum_{j=0}^{\ell} z^j A_j$ *corresponding to* z_0:

(i) $\det L(z)$ *has a zero* z_0 *of multiplicity* μ;

(ii) $\operatorname{rank} \operatorname{col}(XT^j)_{j=0}^{p-1} = \mu$ *for sufficiently large* p;

(iii) $A_\ell XT^\ell + A_{\ell-1} XT^{\ell-1} + \cdots + A_0 X = 0$.

We shall see later that actually (ii) holds for $p = \ell$.

PROOF. If (X, T) is a right null pair for $L(z)$ at z_0 then (i) follows from Corollary 1.1.7, (ii) from Theorem 1.2.4 and (iii) from Proposition 2.1.1. Conversely, if (X, T) satisfies (i), (ii) and (iii) then (X, T) is a right null pair for $L(z)$ at z_0 by the sufficiency direction in Proposition 1.2.7. □

We may define right and left null pairs relative to a given set Ω in the complex plane for a matrix polynomial $L(z)$ just as was done in Section 1.5 for the case of analytic matrix functions. As was mentioned in Section 1.5 any two right null pairs of $L(z)$ relative to a given subset Ω are similar. We also have the following analogue of Theorem 2.1.2.

THEOREM 2.1.3. *Let (Z, T) be a pair of matrices where Z is of size $n \times \mu$ and T is of size $\mu \times \mu$. The following conditions are necessary and sufficient in order that (Z, T) is a right null pair of the matrix polynomial $L(z) = \sum_{j=0}^{\ell} z^j A_j$ relative to Ω.*

(i) *the eigenvalues of T coincide with the zeros of $\det L(z)$ which are in Ω;*

(ii) *the sum of the multiplicities of the zeros of $\det L(z)$ which are in Ω coincides with μ;*

(iii) *(Z, T) is a null kernel pair;*

(iv) *$A_\ell ZT^\ell + A_{\ell-1} ZT^{\ell-1} + \cdots + A_0 Z = 0$.*

PROOF. Without loss of generality we may assume that the right null pair (Z, T) for $L(z)$ relative to Ω has the form

$$(2.1.2) \qquad (Z, T) = ([Z_1, Z_2, \ldots, Z_r], T_1 \oplus T_2 \oplus \cdots \oplus T_r)$$

where $(Z_1, T_1), \ldots, (Z_r, T_r)$ are right null pairs for $L(z)$ at the points $z_1, \ldots, z_r$ respectively and where $z_1, \ldots, z_r$ are the distinct zeros of $\det L(z)$ inside Ω. Then the necessity of conditions (i) through (iv) is an immediate consequence of Theorem 2.1.2. (The verification that (Z, T) is a null kernel pair if each (Z_i, T_i) is can be done as in the proof of Theorem 1.5.1.)

For the converse by using a similarity we may assume that (Z, T) has the form (2.1.2) where the spectra of $T_1, \ldots, T_r$ are pairwise disjoint and each spectrum $\sigma(T_j)$ consists of a single point z_j. Since (Z, T) satisfies (iv), we see that each (Z_j, T_j) satisfies (iv) individually. Expanding $L(z)$ in powers of $(z - z_j)$ rather than of z, we see that (iv) implies that each (Z_j, T_j) satisfies (iv) in Proposition 1.2.7 for the point z_j. From this we get that the columns of Z_j form a set of right null chains for $L(z)$. By the null kernel property of (Z, T) from (ii), each (Z_j, T_j) must also be a null kernel pair, so the leading vectors from these right null chains form a linearly independent set. By a counting argument using (i), we see next that the columns of each Z_j must form a canonical set of right null chains at z_j for each j and that $z_1, \ldots, z_r$ must exhaust all the zeros of $\det L(z)$ in Ω. We have thus shown that indeed (Z, T) is a right null pair for $L(z)$ relative to Ω. $\square$

A pair (S, Y) which is similar to $(S_1 \oplus \cdots \oplus S_r, \mathrm{col}(Y_i)_{i=1}^r)$, where (S_j, Y_j) is a left null pair of $L(z)$ at the zero z_j is called *left null pair* of $L(z)$ relative to Ω. A result analogous to Theorem 2.1.3 holds for left null pairs:

THEOREM 2.1.4. *Let (S, Y) be a pair of matrices where S is $\mu \times \mu$ and Y is $\mu \times n$. The following conditions are necessary and sufficient in order that (S, Y) is a left null pair of the matrix polynomial $L(z) = \sum_{j=0}^{\ell} z^j A_j$ with respect to Ω:*

(i) *the eigenvalues of S coincide with the zeros of $\det L(z)$ which are in Ω;*

(ii) *the sum of the multiplicities of the zeros of* $\det L(z)$ *which are in* Ω *coincides with* μ;

(iii) (S, Y) *is a full range pair*;

(iv) $S^\ell Y A_\ell + S^{\ell-1} Y A_{\ell-1} + \cdots + Y A_0 = 0.$

As we shall see later, if (i)–(iv) holds, then (iii) holds with $p = \ell$, the degree of $L(z)$. The same observation applies to Theorem 2.1.3.

2.2 LINEARIZATION AND MONIC MATRIX POLYNOMIALS

In this section we prepare the necessary background for the solution of the basic interpolation problem for matrix polynomials, i.e. construction of matrix polynomials with prescribed null pairs. The solution to this problem will be given in the next two sections.

Let $L(z) = \sum_{j=0}^{\ell} z^j A_j$ be an $n \times n$ matrix polynomial (as always in this chapter we assume that $\det L(z) \not\equiv 0$). Consider the equation

$$(2.2.1) \qquad\qquad L(z)x = 0.$$

This equation is nonlinear in z. However, by putting

$$(2.2.2) \qquad\qquad x_0 = x; x_1 = zx_0; \ldots; x_{\ell-1} = zx_{\ell-2}$$

the equation (2.2.1) reduces to

$$(2.2.3) \qquad zA_\ell x_{\ell-1} + A_{\ell-1}x_{\ell-1} + A_{\ell-2}x_{\ell-2} + \cdots + A_0 x_0 = 0,$$

which is linear in z. (This is the standard method of reduction of n-th order linear differential equation to a system of first order linear differential equations.) The system of equations (2.2.2), (2.2.3) can also be written in the form

$$C_L(z) \begin{bmatrix} x_0 \\ x_1 \\ \vdots \\ x_{\ell-1} \end{bmatrix} = 0,$$

where

$$C_L(z) = \begin{bmatrix} I & 0 & \cdots & & 0 \\ 0 & I & \cdots & & 0 \\ \vdots & \vdots & & \vdots & \vdots \\ 0 & 0 & \cdots & I & 0 \\ 0 & 0 & \cdots & 0 & A_\ell \end{bmatrix} z + \begin{bmatrix} 0 & -I & 0 & \cdots & 0 \\ 0 & 0 & -I & \cdots & 0 \\ \vdots & \vdots & \vdots & \vdots & \vdots \\ 0 & 0 & & \cdots & -I \\ A_0 & A_1 & & \cdots & A_{\ell-1} \end{bmatrix}.$$

is a linear $n\ell \times n\ell$ matrix polynomial called the *companion polynomial* of $L(z)$.

This idea of a linearization is formalized as follows. Let $L(z) = \sum_{i=0}^{\ell} A_i z^i$ be $n \times n$ matrix polynomial. An $n\ell \times n\ell$ linear matrix polynomial $S_0 + S_1 z$ is called a *linearization* of $L(z)$ if

$$\begin{bmatrix} L(z) & 0 \\ 0 & I_{n(\ell-1)} \end{bmatrix} = E(z)(S_0 + S_1 z)F(z)$$

for some unimodular (i.e. with constant non-zero determinants) $n\ell \times n\ell$ matrix polynomials $E(z)$ and $F(z)$.

It turns out that the companion polynomial $C_L(z)$ is a linearization of $L(z)$. Indeed, define a sequence of polynomials $E_i(z)$, $i = 1, \ldots, \ell$, by induction: $E_\ell(z) = A_\ell$, $E_{i-1}(z) = A_{i-1} + zE_i(z)$, $i = \ell, \ldots, 2$. It is easy to see that

(2.2.4)
$$\begin{bmatrix} E_1(z) & \cdots & & E_{\ell-1}(z) & I \\ -I & 0 & \cdots & & 0 \\ 0 & -I & \cdots & & 0 \\ \vdots & \vdots & & \vdots & \vdots \\ 0 & 0 & \cdots & -I & 0 \end{bmatrix} C_L(z) =$$

$$= \begin{bmatrix} L(z) & 0 \\ 0 & I \end{bmatrix} \begin{bmatrix} I & 0 & 0 & \cdots & & 0 \\ -zI & I & 0 & \cdots & & 0 \\ 0 & -zI & I & \cdots & & 0 \\ \vdots & \vdots & \vdots & & \vdots & \vdots \\ 0 & 0 & 0 & \cdots & -zI & I \end{bmatrix},$$

so indeed $C_L(z)$ is a linearization of $L(z)$. In particular, every regular matrix polynomial has a linearization.

Of particular importance are matrix polynomials with leading coefficient I — *monic matrix polynomials*. For such a polynomial $L(z)$ the companion polynomial is $zI - C_L$, where

$$C_L = \begin{bmatrix} 0 & I & 0 & \cdots & 0 \\ 0 & 0 & I & \cdots & 0 \\ \vdots & & & \ddots & \\ 0 & 0 & 0 & \cdots & I \\ -A_0 & -A_1 & \cdots & & -A_{\ell-1} \end{bmatrix}$$

is the *companion matrix* of $L(z)$. Using the linearization, we prove the following important property of right and left null pairs for monic matrix polynomials.

THEOREM 2.2.1. *Let $L(z)$ be a monic matrix polynomial of degree ℓ with right null pair (Z, T) and left null pair (S, Y), relative to the whole complex plane $\mathbb{C}$. Then T and S are $n\ell \times n\ell$ matrices, and the matrices*

$$\mathrm{col}(ZT^i)_{i=0}^{\ell-1}, \qquad [Y, SY, \ldots, S^{\ell-1}Y]$$

are invertible.

Note a particular consequence of Theorem 2.2.1 is that the sum of the lengths of a canonical set of right null chains associated with all the zeros in $\mathbb{C}$ for a monic $n \times n$

matrix polynomial of degree ℓ in $n\ell$; in this sense Theorem 2.2.1 can be considered as a matrix analogue of the fundamental theorem of algebra.

PROOF. Let C_L be the companion matrix for $L(z)$. Then the equality

$$(2.2.5) \qquad C_L \cdot \text{col}(ZT^i)_{i=0}^{\ell-1} = \text{col}(ZT^i)_{i=0}^{\ell-1}T$$

holds. Indeed, the equality in all but the last block row in (2.2.5) is evident; the equality in the last block row follows from Theorem 2.1.3(iv). Equality (2.2.5) shows that if T is in Jordan canonical form, then the columns of $\text{col}(ZT^i)_{i=0}^{\ell-1}$ form right null chains for the linear matrix polynomial $zI - C_L$.

Let us show that these right null chains of $zI - C_L$ form a canonical set for each zero z_0 of $zI - C_L$. First note that in view of equality (2.2.4) we have $\sigma(L(z)) = \sigma(zI - C_L)$, and the multiplicity of an eigenvalue z_0 is the same, when z_0 is regarded as a zero of $L(z)$ or of $zI - C_L$. Thus, the number of vectors in the columns of $\text{col}(ZT^i)_{i=0}^{\ell-1}$ corresponding to some $z_0 \in \sigma(zI - C_L)$ coincides with the multiplicity of z_0 as a zero of $\det(zI - C_L)$. Assuming that T is in the Jordan form, let $x_1, \ldots, x_k$ be the eigenvectors of $L(z)$ corresponding to z_0 and which appear as columns in the matrix Z; then $\text{col}(z_0 x_1)_{i=0}^{\ell-1}, \ldots, \text{col}(z_0^i x_k)_{i=0}^{\ell-1}$ will be the corresponding columns in $\text{col}(ZT^i)_{i=0}^{\ell-1}$. By Proposition 1.2.5, in order to prove that the columns of $\text{col}(ZT^i)_{i=0}^{\ell-1}$ form a canonical set of right null chains of $zI - C_L$ corresponding to z_0 we have only to check that the vectors

$$\text{col}(z_0^i x_j)_{i=0}^{\ell-1} \in \mathbf{C}^{n\ell}, \qquad j = 1, \ldots, k$$

are linearly independent. But this is evident, since by the construction of a right null pair for $L(z)$ it is clear that the vectors $x_1, \ldots, x_k$, are linearly independent.

So the columns of $\text{col}(ZT^i)_{i=0}^{\ell-1}$ form a canonical set of right null chains for $zI - C_L$. This means that these form a basis in $\mathbf{C}^{n\ell}$ in which C_L is represented by its Jordan form. In particular, $\text{col}(ZT^i)_{i=0}^{\ell-1}$ is invertible.

The statement about $[Y, SY, \ldots, S^{\ell-1}Y]$ follows from the already proved part of the theorem taking into account that (Y^T, S^T) is a right null pair for the monic matrix polynomial $L(z)^T$. $\square$

Another consequence of the linearization formula (2.2.4) is that

$$(2.2.6) \qquad ([\, I \quad 0 \quad \cdots \quad 0\,], C_L)$$

is a right null pair of a monic matrix polynomial $L(z) = \sum_{j=0}^{\ell} z^j A_j$, $A_\ell = I$, while

$$(2.2.7) \qquad \left(C_L, \begin{bmatrix} 0 \\ 0 \\ \vdots \\ 0 \\ I \end{bmatrix} \right)$$

is a left null pair of $L(z)$. Indeed, we shall verify that the conditions (i)–(iv) of Theorem 2.1.3 hold for $Z = [I \quad 0 \quad \cdots \quad 0]$, $T = C_L$. Formula (2.2.4) shows that (i) is true.

By taking determinants in (2.2.4) we see that $\det L(z)$ has exactly $n\ell$ zeros (counting multiplicities). So (ii) holds. The property (iii) holds by Theorem 2.2.1. Finally, a direct computation shows that the first block row of C_L^j is $[\,0 \;\; \cdots \;\; 0 \; I \; 0 \;\; \cdots \;\; 0\,]$ with I on the $(j+1)$-th space for $j = 1, \ldots, \ell - 1$, and the first block row of C_L^ℓ is

$$[-A_0, -A_1, \ldots, -A_{\ell-1}].$$

Using this observation, one obtains (where $Z = [\,I \; 0 \;\; \cdots \;\; 0\,]$)

$$\sum_{j=0}^{\ell-1} A_j Z C_L^j + Z C_L^\ell = \sum_{j=0}^{\ell-1} A_j [\,0 \;\; \cdots \;\; 0 \; I \; 0 \;\; \cdots \;\; 0\,] + [-A_0, -A_1, \ldots, -A_{\ell-1}] = 0,$$

and (iv) follows.

We verify now that (2.2.7) is a left null pair for $L(z)$ relative to C. We shall use Theorem 2.1.4, where the only non-trivial verification is that of the statement (ii) (with

$$S = C_L, \qquad Y = \mathrm{col}(\delta_{i\ell} I)_{i=1}^\ell.)$$

First observe (by application of the already proved fact that

$$([\,I \; 0 \;\; \cdots \;\; 0\,], C_L^T)$$

is the right null pair for the matrix polynomial $L^T(z) = \sum_{j=0}^\ell z^j A_j^T$) that the pair

$$(2.2.8) \qquad \tilde{C} = \begin{bmatrix} 0 & 0 & \cdots & & -A_0 \\ I & 0 & \cdots & & -A_1 \\ 0 & I & & & \vdots \\ \vdots & \vdots & \ddots & & \\ 0 & 0 & \cdots & I & -A_{\ell-1} \end{bmatrix}, \begin{bmatrix} I \\ 0 \\ \vdots \\ 0 \end{bmatrix}$$

is a left null pair for $L(z)$ relative to C. Define the invertible $n\ell \times n\ell$ matrix

$$B = \begin{bmatrix} A_1 & A_2 & \cdots & A_{\ell-1} & I \\ A_2 & & & I & 0 \\ & & & 0 & \\ \vdots & & & \vdots & \vdots \\ A_{\ell-1} & I & & 0 & 0 \\ I & 0 & \cdots & 0 & 0 \end{bmatrix}$$

and by straightforward multiplication verify that

$$B C_L = \tilde{C} B,$$

or

$$\tilde{C} = B C_L B^{-1}.$$

Further, $\operatorname{col}(\delta_{1i}I)_{i=1}^{\ell} = B \operatorname{col}(\delta_{\ell i}I)_{i=1}^{\ell}$. Consequently, the pairs (2.2.8) and (2.2.7) are similar, and (2.2.8) is a left null pair of $L(z)$ (relative to $\mathbb{C}$), the same is true for (2.2.7).

From the preceding discussion we obtain the following important fact:

PROPOSITION 2.2.2. *Let $y_1, \ldots, y_{n\ell}$ be a Jordan basis for C_L, and let $P: \mathbb{C}^{n\ell} \to \mathbb{C}^n$ be the linear transformation defined by $P \operatorname{col}(x_i)_{i=1}^{n\ell} = \operatorname{col}(x_i)_{i=1}^{n}$. Then the vectors $Py_1, \ldots, Py_{n\ell}$ form canonical sets of right null chains corresponding to all the zeros of $L(z)$.*

PROOF. Define

$$S = [y_1 y_2 \cdots y_{n\ell}].$$

Then S is invertible and

$$S^{-1}C_L S = J$$

is in the Jordan form. Hence ($[\; I \quad 0 \quad \cdots \quad 0 \;]S, J$) is again a right null pair of $L(z)$ (relative to $\mathbb{C}$). Our assertion now follows by the definitions of a right null pair. □

2.3 INTERPOLATION THEOREMS: MONIC MATRIX POLYNOMIALS WITH GIVEN NULL PAIRS

In this section we shall consider the simplest and basic interpolation problem for monic matrix polynomials. It turns out that the invertibility properties described in Theorem 2.2.1 are characteristic of null pairs for monic matrix polynomials.

THEOREM 2.3.1. (a) *Let (Z, T) be a pair of matrices of sizes $n \times n\ell$ and $n\ell \times n\ell$, respectively, such that $\operatorname{col}(ZT^i)_{i=0}^{\ell-1}$ is invertible. Then there is a unique monic $n \times n$ matrix polynomial $L(z)$ of degree ℓ whose right null pair (relative to $\mathbb{C}$) is (Z, T). This polynomial is given by the formula*

(2.3.1) $$L(z) = z^{\ell}I - ZT^{\ell}(V_1 + V_2 z + \cdots + V_{\ell}z^{\ell-1}),$$

where $V_1, \ldots, V_{\ell}$ are $n\ell \times n$ matrices defined by

$$[\; V_1 \quad V_2 \quad \cdots \quad V_{\ell} \;] = [\operatorname{col}(ZT^i)_{i=0}^{\ell-1}]^{-1}.$$

(b) *Let (S, Y) be a pair of matrices of sizes $n\ell \times n\ell$ and $n\ell \times n$, respectively, such that*

$$[Y, SY, \ldots, S^{\ell-1}Y]$$

is invertible. Then there is a unique monic $n \times n$ matrix polynomial $M(z)$ of degree ℓ whose left null pair (relative to $\mathbb{C}$) is (S, Y). This polynomial is given by the formula

(2.3.2) $$M(z) = z^{\ell}I - (W_1 + W_2 z + \cdots + W_{\ell}z^{\ell-1})S^{\ell}Y,$$

where $W_1, \ldots, W_{\ell}$ are $n \times n\ell$ matrices such that

$$\operatorname{col}(W_i)_{i=1}^{\ell} = [Y, SY, \ldots, S^{\ell-1}Y]^{-1}.$$

PROOF. Again, we shall prove the part (a) only (part (b) follows by applying (a) to the pair (Y^T, S^T)).

Let C_L be the companion matrix for $L(z)$ given by (2.3.1). One verifies that

$$C_L \operatorname{col}(ZT^i)_{i=0}^{\ell-1} = \operatorname{col}(ZT^i)_{i=0}^{\ell-1} T,$$

and, obviously,

$$[\, I \quad 0 \quad \cdots \quad 0 \,] \operatorname{col}(ZT^i)_{i=0}^{\ell-1} = Z.$$

These equalities show that the pair (Z, T) is similar to $([\, I \quad 0 \quad \cdots \quad 0 \,], C_L)$. Since the latter pair is a right null pair of $L(z)$ relative to $\mathbb{C}$ (see Section 2.2), the same is true also for (Z, T). The uniqueness of a monic matrix polynomial $L(z)$ with given right null pair (Z, T) follows from the formula (2.2.5). $\square$

We note that for monic matrix polynomials right and left pairs determine each other. Namely, if (Z, T) is a right null pair (relative to $\mathbb{C}$) of a monic matrix polynomial $L(z)$ of degree ℓ, then (T, Y), where $Y = \left(\operatorname{col}(ZT^i)_{i=0}^{\ell-1}\right)^{-1} \operatorname{col}(\delta_{i\ell} I)_{i=1}^{\ell}$, is a left null pair of $L(z)$. Indeed, by passing to a similar right null pair, we can assume that

$$Z = [\, I \quad 0 \quad \cdots \quad 0 \,], \qquad T = C_L.$$

Then $\operatorname{col}(ZT^i)_{i=0}^{\ell-1} = I_{n\ell}$ and consequently $Y = \operatorname{col}(\delta_{i\ell} I)_{i=1}^{\ell}$. As we have proved in Section 2.2, the pair (C_L, Y) is indeed a left null pair for $L(z)$. Analogously one proves that if (T, Y) is a left null pair for $L(z)$, then (Z, T), where

$$Z = [\, 0 \quad \cdots \quad 0 \quad I \,][Y, TY, \ldots, T^{\ell-1}Y]^{-1}$$

is a right null pair for $L(z)$.

Another important form of the monic matrix polynomial with given right null pair is in terms of its inverse:

THEOREM 2.3.2. *Let $L(z)$ be a monic matrix polynomial of degree ℓ, and let (Z, T) be its right null pair (relative to $\mathbb{C}$). Then for every $z \notin \sigma(T)$ we have*

(2.3.3) $$L(z)^{-1} = Z(zI - T)^{-1} Y,$$

where

$$Y = \begin{bmatrix} Z \\ ZT \\ \vdots \\ ZT^{\ell-1} \end{bmatrix}^{-1} \begin{bmatrix} 0 \\ \vdots \\ 0 \\ I \end{bmatrix}.$$

Observe that the matrix $\operatorname{col}(ZT^i)_{i=0}^{\ell-1}$ is invertible (see Theorem 2.2.1).

PROOF. Since the theorem does not depend on the choice of the right null pair, (Z, T), we can take

$$Z = [\, I \quad 0 \quad \cdots \quad 0 \,]; \qquad T = \begin{bmatrix} 0 & I & 0 & \cdots & 0 \\ 0 & 0 & I & \cdots & 0 \\ & & & & \vdots \\ \vdots & \vdots & & \ddots & I \\ -A_0 & -A_1 & & \cdots & -A_{\ell-1} \end{bmatrix},$$

where $L(z) = z^\ell I + \sum_{j=0}^{\ell-1} z^j A_j$. Then $Y = \text{col}(\delta_{i\ell} I)_{i=1}^\ell$. Now use the formula (2.2.4), which implies

$$
\begin{bmatrix} L(z)^{-1} & 0 \\ 0 & I_{n(\ell-1)} \end{bmatrix} =
\begin{bmatrix}
I & 0 & 0 & \cdots & 0 & 0 \\
-zI & I & 0 & \cdots & 0 & 0 \\
0 & -zI & I & \cdots & \vdots & \vdots \\
& \vdots & & & I & 0 \\
0 & 0 & 0 & \cdots & -zI & I
\end{bmatrix} \times
$$

$$
\times (zI - T)^{-1}
\begin{bmatrix} 0 \\ 0 \\ \vdots \\ 0 \\ I \end{bmatrix} * \ .
$$

(The star denotes entries of no immediate interest.) Premultiplying by $[\, I \ 0 \ \cdots \ 0 \,]$ and postmultiplying by $\text{col}(\delta_{i1} I)_{i=1}^\ell$, we obtain the theorem.   □

The form (2.3.3) is called the *resolvent form* for $L(z)$. One can analogously obtain the resolvent form for $L(z)$ starting with a left null pair (T, Y) of $L(z)$:

$$
L(z)^{-1} = Z(zI - T)^{-1} Y,
$$

where $Z = [\, 0 \ \cdots \ 0 \ I \,][Y, TY, \ldots, T^{\ell-1} Y]^{-1}$.

We conclude this section with a scalar example.

EXAMPLE 2.3.1.   Consider the simplest interpolation problem for scalar polynomials: given m distinct (complex) numbers $z_1, \ldots, z_m$ find a monic polynomial of degree m with zeros $z_1, \ldots, z_m$. The answer is, of course, $(z - z_1) \cdots (z - z_m)$. We show that by using formula (2.3.1) one obtains the same polynomial (as it should be). To this end we need a formula for the inverse of a Vandermonde matrix. Given complex numbers $z_1, \ldots, z_m$, define the Vandermonde matrix

$$
V(z_1, \ldots, z_m) =
\begin{bmatrix}
1 & 1 & \cdots & 1 \\
z_1 & z_2 & \cdots & z_m \\
\vdots & \vdots & & \vdots \\
z_1^{m-1} & z_2^{m-1} & \cdots & z_m^{m-1}
\end{bmatrix} .
$$

It is well-known and can be easily proved by induction on m (see, e.g., Davis [1975]) that

$$
(2.3.4) \qquad \det V(z_1, \ldots, z_m) = \prod_{1 \le i < j \le m} (z_j - z_i).
$$

Introduce also the $(m-1) \times (m-1)$ matrices

$$
W(z_1, \ldots, z_{m-1}; s) =
\begin{bmatrix}
1 & 1 & \cdots & 1 \\
z_1 & z_2 & \cdots & z_{m-1} \\
\vdots & \vdots & & \vdots \\
z_1^{s-1} & z_2^{s-1} & \cdots & z_{m-1}^{s-1} \\
z_1^{s+1} & z_2^{s+1} & \cdots & z_{m-1}^{s+1} \\
\vdots & \vdots & & \vdots \\
z_1^{m-1} & z_2^{m-1} & \cdots & z_{m-1}^{m-1}
\end{bmatrix}
$$

$(1 \le s \le m - 2)$. The determinant of $W(z_1, \ldots, z_{m-1}; s)$ can be computed as follows:
Write

(2.3.5) $$\det V(z_1, \ldots, z_{m-1}; z) = \prod_{1 \le i < j \le m-1} (z_j - z_i) \prod_{1 \le k \le m-1} (z - z_k),$$

where z is a complex variable; on the other hand, expand $\det V(z_1, \ldots, z_{m-1}, z)$ along
the last column and get

(2.3.6)
$$\begin{aligned}
\det V(z_1, \ldots, z_{m-1}, z) = {}& z^{m-1} \det V(z_1, \ldots, z_{m-1}) \\
& - z^{m-2} \det W(z_1, \ldots, z_{m-1}; m-2) + \cdots \\
& + (-1)^m z \det W(z_1, \ldots, z_{m-1}; 1) \\
& + (-1)^{m+1} z_1 \cdots z_{m-1} \det V(z_1, \ldots, z_{m-1}).
\end{aligned}$$

Comparing the coefficients of z^p in (2.3.5) and (2.3.6) one obtains a formula for
$W(z_1, \ldots, z_{m-1}; p)$. As a by-product we record the formula
(2.3.7)
$$\prod_{1 \le i < j \le m-1} (z_j - z_i) \prod_{1 \le k \le m-1} (z - z_k) = z^{m-1} \det V(z_1, \ldots, z_{m-1})$$

$$\begin{aligned}
& - z^{m-2} \det W(z_1, \ldots, z_{m-1}; m-2) + \cdots \\
& + (-1)^m z \det W(z_1, \ldots, z_{m-1}; 1) \\
& + (-1)^{m+1} z_1 \cdots z_{m-1} \det V(z_1, \ldots, z_{m-1})
\end{aligned}$$

that will be used later.

Return now to the interpolation problem at hand. This problem can be
recast in the terminology of right null pairs; namely, given m distinct complex numbers
$z_1, \ldots, z_m$ find a polynomial with right null pair (relative to $\mathbf{C}$)

$$Z = [\, 1 \quad 1 \quad \cdots \quad 1\,]; \qquad T = \begin{bmatrix} z_1 & 0 & \cdots & 0 \\ 0 & z_2 & \cdots & 0 \\ \vdots & \vdots & & \vdots \\ 0 & 0 & \cdots & z_m \end{bmatrix}.$$

We see immediately that

$$\operatorname{col}(ZT^i)_{i=0}^{m-1} = V(z_1, \ldots, z_m),$$

which is invertible in view of the formula (2.3.4). So the required polynomial $L(z)$ is
given by (2.3.1):

(2.3.8) $$L(z) = z^m - [\, z_1^m \quad z_2^m \quad \cdots \quad z_m^m \,](V_1 + V_2 z + \cdots + V_m z^{m-1}),$$

where $V_1, V_2, \ldots, V_m$ are the columns of $V(z_1, \ldots, z_m)^{-1}$. The cofactor formula for the
inverse of $V(z_1, \ldots, z_m)$ gives

$$V(z_1, z_2, \ldots, z_m)^{-1} = \prod_{1 \le i < j \le m} (z_j - z_i)^{-1}$$

(2.3.9)
$$\begin{bmatrix}
\alpha_1 & -\beta_{11} & \cdots & (-1)^m \beta_{1,m-2} & (-1)^{m+1} \gamma_1 \\
-\alpha_2 & \beta_{21} & \cdots & (-1)^{m-1} \beta_{2,m-2} & (-1)^m \gamma_2 \\
\vdots & \vdots & \vdots & \vdots & \vdots \\
(-1)^{m-1} \alpha_m & (-1)^m \beta_{m1} & \cdots & -\beta_{m,m-2} & \gamma_m
\end{bmatrix},$$

where

$$\alpha_j = z_1 \cdots z_{j-1} z_{j+1} \cdots z_m \det V(z_1, \ldots, z_{j-1}, z_{j+1}, \ldots, z_m),$$
$$\beta_{js} = \det W(z_1, \ldots, z_{j-1}, z_{j+1}, \ldots, z_m; s)$$

and

$$\gamma_j = \det V(z_1, \ldots, z_{j-1}, z_{j+1}, \ldots, z_m).$$

Using formula (2.3.9) combined with (2.3.7) we compute

(2.3.10)
$$V_1 + V_2 z + \cdots + V_m z^{m-1} = \left[\prod_{1 \le i < j \le m} (z_j - z_i) \right]^{-1}$$
$$\times \begin{bmatrix} (-1)^{m+1} \det V(z_2, \ldots, z_m)(z - z_2) \cdots (z - z_m) \\ (-1)^m \det V(z_1, z_3, \ldots, z_m)(z - z_1)(z - z_3) \cdots (z - z_m) \\ \vdots \\ \det V(z_1, \ldots, z_{m-1})(z - z_1)(z - z_2) \cdots (z - z_{m-1}) \end{bmatrix}.$$

One can verify this equality by evaluating both sides at $z = z_1, z = z_2, \ldots, z = z_m$. Further, the right hand side of (2.3.10) is equal to

$$\begin{bmatrix} (-1)^{m+1} \prod_{j=2}^{m}[(z_j - z_1)^{-1}(z - z_j)] \\ (-1)^{m+1} \prod_{j \ne 2}[(z_j - z_2)^{-1}(z - z_j)] \\ \vdots \\ (-1)^{m+1} \prod_{j=1}^{m-1}[(z_j - z_m)^{-1}(z - z_j)] \end{bmatrix}$$
$$= \prod_{j=1}^{m} (z - z_j) \begin{bmatrix} (z - z_1)^{-1} \prod_{j=2}^{m}(z_1 - z_j)^{-1} \\ (z - z_2)^{-1} \prod_{j \ne 2}(z_2 - z_j)^{-1} \\ \vdots \\ (z - z_m)^{-1} \prod_{j=1}^{m-1}(z_m - z_j)^{-1} \end{bmatrix}.$$

Finally, substitute into the formula (2.3.8):

$$L(z) = z^m - \prod_{j=1}^{m}(z - z_j) \sum_{k=1}^{m} \left(z_k^m (z - z_k)^{-1} \prod_{i \ne k} (z_k - z_i)^{-1} \right)$$
$$= \prod_{j=1}^{m}(z - z_j) \left\{ z^m \prod_{j=1}^{m}(z - z_j)^{-1} - \sum_{k=1}^{m} \left(z_k^m (z - z_k)^{-1} \prod_{i \ne k}(z_k - z_i)^{-1} \right) \right\}.$$

The rational function in the braces is analytic at each point z_k (because its residue at each z_k is zero) and has value 1 at infinity. Consequently, this function is identically 1, and

$$L(z) = \prod_{j=1}^{m}(z - z_j),$$

as expected.

2.4 INTERPOLATION OF MONIC MATRIX POLYNOMIALS WITH INCOMPLETE DATA

Let now (Z, T) be a null kernel pair of matrices of sizes $n \times p$ and $p \times p$, respectively, so

(2.4.1)
$$\bigcap_{j=0}^{q-1} \operatorname{Ker}(ZT^j) = \{0\}$$

for some positive integer q. Consider the problem of existence of a monic matrix polynomial $L(z)$ such that (Z, T) is a right null pair of $L(z)$ relative to Ω, where Ω is a prescribed domain in the complex plane (different from $\mathbf{C}$ itself) which contains the spectrum of T. By Theorem 2.2.1 in general such an $L(z)$ whose right null pair on the whole complex plane is (Z, T) would not exist. However, if we allow the next best thing, namely, that $L(z)$ has just one spectral point outside Ω, then the answer is in the affirmative:

THEOREM 2.4.1. *Let (Z, T) be a pair of matrices of sizes $n \times p$ and $p \times p$, respectively satisfying the property (2.4.1), and let $\Omega \neq \mathbf{C}$ be a domain in the complex plane which contains the spectrum of T. Then, for every $z_0 \in \mathbf{C} \backslash \Omega$, there exists a monic matrix polynomial $L(z)$ of degree q whose set of zeros is contained in $\sigma(T) \cup \{z_0\}$ and whose right null pair relative to Ω is (Z, T).*

The proof of this theorem will show that actually one can characterize all possible Jordan structures of the monic matrix polynomials $L(z)$ satisfying the condition of Theorem 2.4.1. It will be also clear that the set of zeros of $L(z)$ coincides with $\sigma(T) \cup \{z_0\}$, unless the matrix $\operatorname{col}(ZT^i)_{i=0}^{q-1}$ is actually invertible, in which case there is a (unique) monic matrix polynomial of degree q with right null pair (Z, T) (with respect to $\mathbf{C}$). So in this case the set of zeros of $L(z)$ coincides with $\sigma(T)$.

The proof of Theorem 2.4.1 is based on the following result.

THEOREM 2.4.2. *Let (Z, T) and z_0 be as in Theorem 2.4.1. Let $\widetilde{T}$ be a matrix of size $(nq - p) \times (nq - p)$ with $\sigma(\widetilde{T}) = \{z_0\}$. Then there exists an $n \times (nq - p)$ matrix $\widetilde{Z}$ such that the square matrix*

(2.4.2)
$$\begin{bmatrix} Z & \widetilde{Z} \\ ZT & \widetilde{Z}\widetilde{T} \\ \vdots & \vdots \\ ZT^{q-1} & \widetilde{Z}\widetilde{T}^{q-1} \end{bmatrix}$$

is invertible if and only if the sizes $p_0 \geq \cdots \geq p_\nu$ of Jordan blocks in the Jordan form of $\widetilde{T}$ satisfy the inequalities

(2.4.3)
$$\sum_{j=0}^{i} p_j \geq \sum_{j=0}^{i} s_j \quad \text{for} \quad i = 0, 1, \ldots, \nu,$$

where s_k is the number of the integers $n + r_{q-j-1} - r_{q-j}$ $(j = 0, \ldots, q-1)$ larger than k, and the integers $\{r_j\}_{j=0}^{q}$ are defined by

$$r_0 = 0; \quad r_j = \operatorname{rank} \operatorname{col}(ZT^k)_{k=0}^{j-1}, \quad j \geq 1.$$

We will not prove Theorem 2.4.2 here but refer the reader for the proof to Gohberg, Rodman [1979] (Theorem 3.1).

The proof of Theorem 2.4.1 is now readily obtained. Namely, given (Z, T) and z_0 as in Theorem 2.4.1, choose a sequence of positive integers $p_0 \geq \cdots \geq p_\nu$ that sum up to $nq - p$ and satisfy inequalities (2.4.3) (one obvious choice is $\nu = 0$ and $p_0 = nq - p$; another obvious choice is $p_i = s_i$ ($i = 0, \ldots, \nu$), where ν is the largest nonnegative integer for which s_ν is positive). By Theorem 2.4.2 there exists a pair of matrices $(\widetilde{Z}, \widetilde{T})$ such that $\sigma(\widetilde{T}) = \{z_0\}$, the Jordan blocks in the Jordan normal form of $\widetilde{T}$ are of sizes $p_0, \ldots, p_\nu$, and the matrix (2.4.2) is invertible. By Theorem 2.3.1 there is a monic matrix polynomial $L(z)$ with right null pair (relative to **C**) ($[\, Z \;\; \widetilde{Z} \,], T \oplus \widetilde{T}$). The matrix polynomial $L(z)$ clearly satisfies the requirements of Theorem 2.4.1. $\square$

We present now another construction of a monic matrix polynomial with incomplete right null pair. This construction is based on the notion of special left inverse, and the following lemma plays an important role.

For given subspaces $\mathcal{U}_1, \ldots, \mathcal{U}_\ell \subset \mathbf{C}^n$ we denote by $\mathrm{col}(\mathcal{U}_i)_{i=1}^\ell$ the subspace in $\mathbf{C}^{n\ell}$ consisting of all ordered ℓ-tuples $(x_1, \ldots, x_\ell)$ of n-dimensional vectors $x_1, \ldots, x_\ell$ such that $x_1 \in \mathcal{U}_1, x_2 \in \mathcal{U}_2, \ldots, x_\ell \in \mathcal{U}_\ell$.

LEMMA 2.4.3. *Let (Z, T) be a null kernel pair such that $\bigcap_{i=0}^{\ell-1} \mathrm{Ker}(ZT^i) = \{0\}$ with an invertible matrix T. Then there exists a sequence of subspaces $\mathbf{C}^n \supset \mathcal{W}_1 \supset \cdots \supset \mathcal{W}_\ell$ such that*

$$(2.4.4) \qquad \mathrm{col}(\mathcal{W}_j)_{j=k+1}^\ell \dotplus \mathrm{Im}\,\mathrm{col}(ZT^{j-1})_{j=k+1}^\ell = \mathbf{C}^{n(\ell-k)}$$

for $k = 0, \ldots, \ell - 1$.

PROOF. Let $\mathcal{W}_\ell$ be a direct complement to $\mathrm{Im}\,ZT^{\ell-1} = \mathrm{Im}\,Z$ (this equality follows from the invertibility of T) in $\mathbf{C}^n$. Then (2.4.4) holds for $k = \ell - 1$. Suppose that $\mathcal{W}_{i+1} \supset \cdots \supset \mathcal{W}_\ell$ are already constructed so that (2.4.4) holds for $k = i, \ldots, \ell - 1$. It is then easy to check that $\mathrm{col}(\mathcal{Z}_j)_{j=i}^\ell \cap \mathrm{Im}\,\mathrm{col}(ZT^{j-1})_{j=i}^\ell = \{0\}$, where $\mathcal{Z}_k = \mathcal{W}_k$ for $k = i+1, \ldots, \ell$, and $\mathcal{Z}_i = \mathcal{W}_{i+1}$. Hence, the sum

$$\mathcal{J} = \mathrm{col}(\mathcal{Z}_j)_{j=i}^\ell \dotplus \mathrm{Im}\,\mathrm{col}(ZT^{j-1})_{j=i}^\ell$$

is a direct sum. Let $\mathcal{I}$ be a direct complement of $\mathrm{Ker}\,A$ in $\mathbf{C}^{n(\ell-i)}$, where $A = \mathrm{col}(ZT^{j-1})_{j=i+1}^\ell$. Let A^I be the generalized inverse of A such that $\mathrm{Im}\,A^I = \mathcal{I}$ and $\mathrm{Ker}\,A^I = \mathrm{col}(\mathcal{W}_j)_{j=i+1}^\ell$. Here and everywhere in the book a $p \times q$ matrix B is called a *generalized inverse* of a $q \times p$ matrix A if the equalities $BAB = B$, $ABA = A$ hold. Let P be the projection on $\mathrm{Ker}\,A$ along $\mathcal{I}$. One verifies easily that $A^I A = I - P$. We shall prove the equality

$$(2.4.5) \qquad \mathcal{J} = \left\{ \begin{bmatrix} y + ZT^{i-1}Pz + ZT^{i-1}A^I x \\ x \end{bmatrix} \middle| y \in \mathcal{W}_{i+1}, x \in \mathbf{C}^{n(\ell-i)}, z \in \mathbf{C}^r \right\},$$

where r is the size of T.

Indeed, if $x \in \mathbf{C}^{n(\ell-i)}$ then by the induction hypothesis $x = Az_1 + x_1$ where $z_1 \in \mathcal{I}$ and $x_1 \in \mathrm{col}(W_j)_{j=i+1}^{\ell}$. Hence

$$\left[\begin{array}{c} ZT^{i-1}A^I x \\ x \end{array} \right] = \left[\begin{array}{c} ZT^{i-1}A^I(Az_1 + x_1) \\ Az_1 + x_1 \end{array} \right] = \left[\begin{array}{c} ZT^{i-1}z_1 \\ Az_1 \end{array} \right] + \left[\begin{array}{c} 0 \\ x_1 \end{array} \right].$$

From the definition of $\mathcal{J}$ it follows that

$$\left[\begin{array}{c} ZT^{i-1}A^I x \\ x \end{array} \right] \in \mathcal{J}.$$

For any $z \in \mathbf{C}^r$, also

$$\left[\begin{array}{c} ZT^{i-1}Pz \\ 0 \end{array} \right] = \left[\begin{array}{c} ZT^{i-1}Pz \\ APz \end{array} \right] \in \mathcal{J},$$

and clearly

$$\left[\begin{array}{c} y \\ 0 \end{array} \right] \in \mathcal{J}$$

for any $y \in \mathcal{W}_{i+1}$. The inclusion $\supset$ in (2.4.5) thus follows. To check the converse inclusion, take $y \in \mathcal{W}_{i+1}, x_1 \in \mathrm{col}(W_j)_{j=i+1}^{\ell}$, and $z \in \mathbf{C}^r$. Then

$$\left[\begin{array}{c} y \\ x_1 \end{array} \right] + \left[\begin{array}{c} ZT^{i-1} \\ A \end{array} \right] z = \left[\begin{array}{c} y \\ x_1 \end{array} \right] + \left[\begin{array}{c} ZT^{i-1} \\ A \end{array} \right][Pz + (I - P)z]$$

$$= \left[\begin{array}{c} y \\ x_1 \end{array} \right] + \left[\begin{array}{c} ZT^{i-1}Pz \\ 0 \end{array} \right] + \left[\begin{array}{c} ZT^{i-1}(I - P)z \\ A(I - P)z \end{array} \right]$$

$$= \left[\begin{array}{c} y \\ 0 \end{array} \right] + \left[\begin{array}{c} ZT^{i-1}Pz \\ 0 \end{array} \right] + \left[\begin{array}{c} ZT^{i-1}A^I(x_1 + A(I - P)z) \\ x_1 + A(I - P)z \end{array} \right]$$

(the last equality follows from $A^I x_1 = 0$ and $I - P = A^I A$), and (2.4.5) is proved.

Now, let $\mathcal{Y}$ be a direct complement of $\mathcal{W}_{i+1} + \mathrm{Im}\, ZT^{i-1}P$ in $\mathbf{C}^n$. Then from (2.4.5) we obtain

$$\mathcal{J} \dotplus \left[\begin{array}{c} \mathcal{Y} \\ 0 \end{array} \right] = \mathbf{C}^{n(\ell-i+1)},$$

and so we can put $\mathcal{W}_i = \mathcal{W}_{i+1} + \mathcal{Y}$. $\square$

We note some remarks concerning the subspaces $\mathcal{W}_j$ constructed in this lemma.

(1) The subspaces $\mathcal{W}_j$ are not uniquely determined, but the dimensions $\nu_j = \dim \mathcal{W}_j, j = 1, \ldots, \ell$ are determined uniquely by the pair (Z, T): namely,

$$(2.4.6) \quad \nu_{k+1} = n + \mathrm{rank}\, \mathrm{col}(ZT^{j-1})_{j=k+1}^{\ell} - \mathrm{rank}\, \mathrm{col}(ZT^{j-1})_{j=k}^{\ell}, \qquad k = 0, \ldots, \ell - 1.$$

(2) Equality (2.4.4) for $k = 0$ means

$$\mathrm{col}(W_j)_{j=1}^{\ell} \dotplus \mathrm{Im}\, \mathrm{col}(ZT^{j-1})_{j=1}^{\ell} = \mathbf{C}^{n\ell},$$

and consequently, there exists an $r \times n\ell$ matrix V such that

(2.4.7) $V \cdot \mathrm{col}(ZT^{j-1})_{j=1}^{\ell} = I, \qquad V(\mathrm{col}(\mathcal{W}_j)_{j=1}^{\ell}) = \{0\}.$

The left inverse V of $\mathrm{col}(ZT^{j-1})_{j=1}^{\ell}$ with this property will be called the *special left inverse* (of course, it depends on the choice of $\mathcal{W}_j$).

Given a special left inverse $V = [\, V_1 \ V_2 \ \cdots \ V_\ell \,]$ (so V_i are $r \times n$ matrices), we can form the monic matrix polynomial

(2.4.8) $L(z) = z^\ell I - ZT^\ell(V_1 + V_2 z + \cdots + V_\ell z^{\ell-1}).$

The polynomial $L(z)$ will be called the *special* matrix polynomial associated with the pair (Z, T).

THEOREM 2.4.4. *Let $L(z)$ be a special monic matrix polynomial associated with a null kernel pair. Then $z_0 = 0$ is a zero of $\det L(z)$, and the partial multiplicities $\kappa_1 \geq \kappa_2 \geq \cdots \geq \kappa_n \geq 0$ of $L(z)$ at 0 are defined by*

(2.4.9) $\kappa_i = \{\text{the number of indices } j, 1 \leq j \leq \ell, \text{ such that } \nu_j \geq i\}$

where ν_j are given by (2.4.6).

PROOF. Let $\mathcal{W}_1 \supset \cdots \supset \mathcal{W}_\ell$ be the sequences of subspaces from Lemma 2.4.3. Let

$$V = [\, V_1 \quad \cdots \quad V_\ell \,]$$

be a special left inverse of $\mathrm{col}(ZT^i)_{i=0}^{\ell-1}$ such that (2.4.7) and (2.4.8) hold. From (2.4.7) it follows that

$$\mathrm{Ker} \, V_j \supset \mathcal{W}_j, \qquad j = 1, \ldots, \ell.$$

It is easy to see that $x \in \mathcal{W}_i \backslash \mathcal{W}_{i+1}$ ($i = 1, \ldots, \ell; \ \mathcal{W}_{\ell+1} = \{0\}$) is an eigenvector of $L(z)$ generating a right null chain $x, 0, \ldots, 0$ ($i - 1$ zeros). Taking a basis in each $\mathcal{W}_i$ modulo $\mathcal{W}_{i+1}$, we obtain in this way a canonical set of right null chains of $L(z)$ at zero. Now Theorem 2.4.4 follows from Corollary 1.1.6. □

Let $L(z)$ and (Z, T) be as in Theorem 2.4.4. It is easy to see that

$$\nu_1 + \cdots + \nu_\ell = n\ell - r,$$

where ν_k are given by (2.4.6). On the other hand, (2.4.9) implies that

$$\kappa_1 + \kappa_2 + \cdots + \kappa_n = \nu_1 + \cdots + \nu_\ell.$$

We see by Theorem 2.4.4 that the multiplicity of $z_0 = 0$ as a zero of $\det L(z)$ is precisely $n\ell - r$. Combining this observation with the formula (2.2.4) we see that $z_0 = 0$ is an eigenvalue of the companion matrix of $L(z)$ of algebraic multiplicity $n\ell - r$.

On the other hand, let $([\, I \ 0 \ \cdots \ 0 \,], C_L)$ be the right null pair of $L(z)$ (relative to the whole complex plane) with the companion matrix C_L. One easily verifies that

(2.4.10)
$$[\, I \quad 0 \quad \cdots \quad 0 \,] \mathrm{col}(ZT^i)_{i=0}^{\ell-1} = Z$$
$$C_L \, \mathrm{col}(ZT^i)_{i=0}^{\ell-1} = \mathrm{col}(ZT^i)_{i=0}^{\ell-1} T.$$

These equalities imply, in particular, that the subspace $\mathcal{M} \overset{\text{def}}{=} \operatorname{Im}\operatorname{col}(ZT^i)_{i=0}^{\ell-1}$ is C_L-invariant and the restriction $C_L\,|_{\mathcal{M}}$ is similar to T. As $\dim\mathcal{M} = r$, a dimension count together with the observation made in the preceding paragraph shows that $\mathcal{M}$ is the spectral C_L-invariant subspace corresponding to the non-zero eigenvalues of C_L. Now the equalities (2.4.10) show that (Z,T) is a right null pair of $L(z)$ relative to $\mathbb{C}\backslash\{0\}$. We have proved

THEOREM 2.4.5. *Let (Z,T) be a pair of matrices with $\bigcap_{i=0}^{\ell-1}\operatorname{Ker}(ZT^i) = \{0\}$ and invertible T. Then (Z,T) is a right null pair relative to $\mathbb{C}\backslash\{0\}$ for every special monic matrix polynomial associated with (Z,T).*

Thus, the special matrix polynomials solve the interpolation problem for monic matrix polynomials with incomplete right null pair.

It is not hard to generalize Theorem 2.4.5 in such a way as to remove the invertibility condition on T. Namely, let $\alpha \in \mathbb{C}$ be any number outside the spectrum of T, and consider the pair $(Z, -\alpha I + T)$ in place of (Z,T). Now the pair $(Z, T - \alpha I)$ is also a null kernel pair because of the identity

$$
\begin{bmatrix}
I & 0 & 0 & \cdots & 0 \\
\alpha\binom{1}{0}I & \binom{1}{1}I & 0 & \cdots & 0 \\
\vdots & & \vdots & & \vdots \\
\alpha^{\ell-1}\binom{\ell-1}{0}I & \alpha^{\ell-2}\binom{\ell-1}{1}I & \cdots & & \binom{\ell-1}{\ell-1}I
\end{bmatrix}
\begin{bmatrix}
Z \\
Z(T - \alpha I) \\
\vdots \\
Z(T - \alpha I)^{\ell-1}
\end{bmatrix}
=
\begin{bmatrix}
Z \\
ZT \\
\vdots \\
ZT^{\ell-1}
\end{bmatrix}.
$$

So we can apply Theorem 2.4.5 for the pair $(Z, -\alpha I + T)$ producing a monic matrix polynomial $L_\alpha(z)$ whose right null pair relative to $\mathbb{C}\backslash\{0\}$ is $(Z, T - \alpha I)$. Then the monic matrix polynomial $L(z) = L_\alpha(z + \alpha)$ has right null pair (Z,T) relative to $\mathbb{C}\backslash\{\alpha\}$.

2.5 NON-MONIC MATRIX POLYNOMIALS WITH GIVEN NULL PAIRS

We reduce the problem of existence of non-monic matrix polynomials with a given null pair (relative to $\mathbb{C}$) to the case of monic polynomials which has already been considered in the previous section; another approach will be presented in Section 5.7.

Let $L(z)$ be a regular (i.e. with determinant not identically zero) $n \times n$ matrix polynomial and let (Z,T) be a right null pair of $L(z)$ relative to the whole complex plane. If ℓ is any integer greater than or equal to the degree of $L(z)$ (i.e. $L^{[\ell+1]}(z) \equiv 0$), then $\widetilde{L}(z) = z^\ell L(z^{-1})$ is again a regular matrix polynomial. Let (Z_s, T_s) be a right null pair of $\widetilde{L}(z)$ at 0 (if $\det\widetilde{L}(0) \neq 0$, then (Z_s, T_s) is an empty pair of matrices). The pair (Z_s, T_s) will be called a *supplementary right null pair* of $L(z)$. Observe that (assuming (Z_s, T_s) is not empty) $\sigma(T_s) = \{0\}$.

Using Proposition 1.2.7, we obtain the following characterization of right null pairs and supplementary right null pairs of $L(z)$:

PROPOSITION 2.5.1. (a) *Let there be given a pair of matrices (Q,T), where Q is of size $n \times \nu$ and T is a matrix of size $\nu \times \nu$. Then (Q,T) is a right null pair (relative to $\mathbb{C}$) of a matrix polynomial $L(z) = \sum_{j=0}^m z^j L_j$ if and only if the following conditions hold:*

(i) ν is the degree of det $L(z)$;

(ii) (Q, T) is a null kernel pair;

(iii) $L_0 Q + L_1 Q T + \cdots + L_m Q T^m = 0$.

(b) Let there be given a pair of matrices (Q', T'), where Q' is of size $n \times \mu$ and T' is a nilpotent of size $\mu \times \mu$. Then (Q', T') is a supplementary null pair of a matrix polynomial $L(z) = \sum_{j=0}^{m} z^j L_j$ of degree m if and only if the following conditions hold:

(i) $\mu + \text{degree}(\det L(z)) = nm$;

(ii) (Q', T') is a null kernel pair;

(iii) $\sum_{k=0}^{m} L_{m-k} Q' T'^k = 0$.

As we shall see later (Theorem 2.5.4) we actually have

$$(2.5.1) \qquad \text{rank} \, \text{col}(Q T^i)_{i=0}^{k-1} = \nu$$

for every $k \geq m$ in part (a) and

$$(2.5.2) \qquad \text{rank} \, \text{col}(Q' T'^i)_{i=0}^{k-1} = \mu$$

for every $k \geq m$ in part (b).

Next, we study the behaviour of null pairs under the linear change of variable $z \to z + \alpha$.

PROPOSITION 2.5.2. Let $(\widehat{Q}, \widehat{T})$ and $(\widehat{Q}_s, \widehat{T}_s)$ be a right null pair (relative to $\mathbb{C}$) and a supplementary right null pair of the matrix polynomial $\widehat{L}(z) = L(z + \alpha)$, where $\alpha \in \mathbb{C}$ is fixed. Then $(\widehat{Q}, \widehat{T} + \alpha I)$ is a right null pair for $L(z)$, and $(\widehat{Q}_s, \widehat{T}_s(I + \alpha \widehat{T}_s)^{-1})$ is a supplementary right null pair for $L(z)$.

PROOF. The assertion concerning right null pair of $L(z)$ follows easily from Proposition 2.5.1.

We prove the statement concerning $(\widehat{Q}_s, \widehat{T}_s(I + \alpha \widehat{T}_s)^{-1})$. Note first that the coefficients $\widehat{L}_j$ $(j = 0, 1, \ldots, m)$ of the polynomial $\widehat{L}(z) = \sum_{j=0}^{m} z^j \widehat{L}_j$ are given by the formulas

$$(2.5.3) \qquad \widehat{L}_k = \sum_{j=k}^{m} \binom{j}{j-k} \alpha^{j-k} L_j, \qquad k = 0, \ldots, m,$$

where $L(z) = \sum_{j=0}^{m} z^j L_j$. Since $(\widehat{Q}_s, \widehat{T}_s)$ is a supplementary right null pair of $\widehat{L}(z)$, we have by Proposition 2.5.1:

$$(2.5.4) \qquad \sum_{k=0}^{m} \widehat{L}_k \widehat{Q}_s \widehat{T}_s^{m-k} = 0.$$

From (2.5.3) and (2.5.4) it is not hard to deduce that the matrices $Q_s = \widehat{Q}_s$, $T_s = (I + \alpha \widehat{T}_s)^{-1} \widehat{T}_s$ satisfy the equality

$$\sum_{j=0}^{m} L_j Q_s T_s^{m-j} = 0.$$

In view of Proposition 2.5.1 it remains to show that (Q_s, T_s) is a null kernel pair. It is easy to see (using the property that $\sigma(T_s) = \{0\}$) that (Q_s, T_s) is a null kernel pair if and only if $Q_s x \neq 0$ for every $x \in \operatorname{Ker} T_s \backslash \{0\}$. But since $(\widehat{Q}_s, \widehat{T}_s)$ is a null kernel pair by Proposition 2.5.1 we obtain that $\widehat{Q}_s x \neq 0$ for every x, $x \in \operatorname{Ker} \widehat{T}_s \backslash \{0\}$. Since $\operatorname{Ker} T_s = \operatorname{Ker} \widehat{T}_s$, we are done. $\square$

A complete description of regular matrix polynomials with prescribed right null pair and supplementary right null pair is given by the following theorem.

THEOREM 2.5.3. *Let* (Q, T) *and* (Q_s, T_s) *be pairs of matrices such that* $\sigma(T_s) = \{0\}$ *and the sizes of* Q, T, Q_s, T_s *are* $n \times p$, $p \times p$, $n \times q$, $q \times q$, *respectively.*

Then there exists a matrix polynomial $L(z)$ *of degree* $\leq m$ *such that* (Q, T) *is its right null pair (relative to* $\mathbb{C}$*) and* (Q_s, T_s) *is its supplementary right null pair if and only if the matrix*

$$\operatorname{col}([QT^{j-1}, Q_s T_s^{m-j}])_{j=1}^m$$

is square and invertible. If this condition is satisfied, then every solution $L_0, \ldots, L_m$ *of the equation*

$$(2.5.5) \qquad [L_0 \cdots L_m] \operatorname{col}([QT^{j-1}, Q_s T_s^{m+1-j}])_{j=1}^{m+1} = 0$$

such that $\operatorname{rank}[L_0 \cdots L_m] = n$ *determines a matrix polynomial* $L(z) = \sum_{j=0}^m z^j L_j$ *with the required properties. Conversely, the coefficients of every polynomial* $L(z)$ *of degree* $\leq m$ *and with the right null pair* (Q, T) *and the supplementary right null pair* (Q_s, T_s) *are determined by a solution* $L_0, \ldots, L_m$ *of (2.5.5) such that* $\operatorname{rank}[L_0 \cdots L_m] = n$. *The polynomial* $L(z)$ *is unique up to multiplication from the left by a constant invertible matrix.*

PROOF. Let us prove the first assertion of the theorem. Consider first the case the T is invertible. Suppose that $L(z)$ exists. We shall prove that $([\,Q \quad Q_s\,], T^{-1} \oplus T_s)$ is a right null pair (relative to $\mathbb{C}$) of the monic matrix polynomial $\widetilde{L}(z) = z^m (L(0))^{-1} L(z^{-1})$.

Indeed, applying a similarity transformation, if necessary, we can assume that T is in the Jordan normal form, so Q consists of right null chains for $L(z)$.

Let $B = [\,x_0 \quad x_1 \quad \cdots \quad x_r\,]$ be a right null chain for $L(z)$ taken from Q and corresponding to the eigenvalue $z_0 \neq 0$. Let K be the Jordan block of size $(r+1) \times (r+1)$ with eigenvalue z_0. By Proposition 2.5.1 we have

$$(2.5.6) \qquad \sum_{j=0}^m A_j B K^j = 0,$$

where A_j are the coefficients of $L(z)$. Since $z_0 \neq 0$, K is nonsingular and so

$$(2.5.7) \qquad \sum_{j=0}^m A_j B K^{j-m} = \sum_{j=0}^m A_{m-j} B (K^{-1})^j = 0.$$

Let K_0 be the matrix obtained from K by replacing z_0 by z_0^{-1}. Then K_0 and K^{-1} are similar:

$$(2.5.8) \qquad\qquad K^{-1} = M_r(z_0) K_0 \big(M_r(z_0) \big)^{-1},$$

where

$$M_r(z_0) = \left[(-1)^j \binom{j-1}{i-1} z_0^{j+i} \right]_{i,j=0}^r$$

and it is assumed that $\binom{-1}{-1} = 1$ and $\binom{p}{q} = 0$ for $q > p$ or $q = -1$ and $p > -1$. The equality (2.5.8) or, what is the same, $K M_r(z_0) K_0 = M_r(z_0)$, can be verified by multiplication. Insert in (2.5.7) the expression for K^{-1}; it follows from Proposition 2.5.1 that the columns of $B M_r(z_0)$ form a right null chain of $\widetilde{L}(z)$ corresponding to z_0^{-1}.

Let k be the number of (right) null chains in Q corresponding to z_0, and let B_i ($i = 1, \ldots, k$) be the matrix whose columns form the i-th null chain corresponding to z_0. We perform the above transformation for every B_i. It then follows that the columns of $B_1 M_{r_1}(z_0), \cdots, B_k M_{r_k}(z_0)$ form null chains of $\widetilde{L}(z)$ corresponding to z_0^{-1}. Let us check that these null chains of $\widetilde{L}(z)$ form a canonical set of null chains corresponding to z_0^{-1}. By Proposition 1.2.2, it is enough to check that the eigenvectors $b_1, \ldots, b_k$ in $B_1 M_{r_1}(z_0), \ldots, B_k M_{r_k}(z_0)$, respectively, are linearly independent. But from the structure of $M_{r_i}(z_0)$ it is clear that $b_1, \ldots, b_k$ are also the eigenvectors in $B_1, \ldots, B_k$, respectively, of $L(z)$ corresponding to z_0, and therefore they are linearly independent again by Proposition 1.2.2.

Finally let $K(r_i)$ be the $(r_i + 1) \times (r_i + 1)$ Jordan block with eigenvalue z_0 and let $K_0(r_i)$ be the corresponding object with z_0^{-1} in place of z_0. If we set $B = [B_1, \ldots, B_k]$ and $\widetilde{K}_0 = \mathrm{diag}\big(K_0(r_1), \ldots, K_0(r_k) \big)$, the above analysis shows that $\big(\widetilde{B} \cdot \mathrm{diag}\big(M_{r_1}(z_0), \ldots, M_r(z_0) \big), \widetilde{K}_0 \big)$ is a right null pair for $\widetilde{L}(z)$ for the point z_0^{-1}. But this full range pair is similar to the pair $(\widetilde{B}, \widetilde{K}^{-1})$, where $\widetilde{K} = \mathrm{diag}\big(K(r_1), \ldots, K(r_k) \big)$. As z_0 is an arbitrary nonzero point in $\mathbb{C}$, this establishes that (Q, T^{-1}) is a right null pair for $\widetilde{L}(z)$ over $\mathbb{C} \backslash (0)$, and hence that $([\, Q \quad Q_s\,], T^{-1} \oplus T_s)$ is a right null pair for $\widetilde{L}(z)$ over $\mathbb{C}$. We now conclude that

$$\mathrm{col}([Q T^{j-1}, Q_s T_s^{m-j}])_{j=1}^m$$

is square and invertible by Theorem 2.2.1.

Conversely, if $\mathrm{col}([Q T^{j-1}, Q_s T_s^{m-j}])_{j=1}^m$ is square and invertible, then put $L(z) = z^m \widetilde{L}(z^{-1})$, where $\widetilde{L}(z)$ is monic polynomial of degree m with right null pair $([\, Q \quad Q_s\,], T^{-1} \oplus T_s)$. (The polynomial $\widetilde{L}(z)$ is given by the formula of Theorem 2.3.1:

$$\widetilde{L}(z) = z^m I - [\, Q \quad Q_s\,](T^{-m} \oplus T_s^m)(V_0 + V_1 z + \cdots + V_{m-1} z^{m-1}),$$

where $[V_0 \cdots V_{m-1}] = \{\mathrm{col}([Q T^{1-j}, Q_s T_s^j])_{j=1}^m\}^{-1}.$)

Consider now the case that T is not invertible. Fix some $\alpha \in \mathbb{C} \backslash \sigma(T)$, and denote

$$\widehat{Q} = Q; \quad \widehat{T} = T - \alpha I; \quad \widehat{Q}_s = Q_s; \quad \widehat{T}_s = T_s(I - \alpha T_s)^{-1}.$$

It is easy to check that

(2.5.9)
$$((\tbinom{j}{k})(-\alpha)^{j-k}I)_{j,k=0}^{m-1} \cdot \text{col}([QT^{j-1}, Q_sT_s^{m-j}])_{j=1}^m =$$
$$= \text{col}([\widehat{Q}\widehat{T}^{j-1}, \widehat{Q}_s\widehat{T}_s^{m-j}])_{j=1}^m \cdot (I \oplus (I - \alpha T_s)^{m-1})$$

consequently, $\text{col}([QT^{j-1}, Q_sT_s^{m-j}])_{j=1}^m$ is invertible if and only if $\text{col}([\widehat{Q}\widehat{T}^{j-1}, \widehat{Q}_s\widehat{T}_s^{m-j}])_{j=1}^m$ is invertible. But the invertibility of the latter (in view of the already proved part of the theorem) is equivalent to existence of a matrix polynomial $\widehat{L}(z)$ with right null pair $(\widehat{Q}, \widehat{T})$ (over the whole complex plane) and the supplementary right null pair $(\widehat{Q}_s, \widehat{T}_s)$. Then by Proposition 2.5.2 (Q, T) is a right null pair of $L(z) = \widehat{L}(z + \alpha)$, and (Q_s, T_s) is a supplementary right null pair for $L(z)$. Equality (2.5.5) follows from Proposition 2.5.1. Since $\det L(z) \not\equiv 0$, we obtain that $\text{rank}[L_0 L_1 \cdots L_m] = n$. So indeed the coefficients of $L(z)$ form a full-rank solution of (2.5.5).

We claim that the matrix
$$\text{col}([QT^{j-1}, Q_sT_s^{m+1-j}])_{j=1}^{m+1}$$
has linearly independent columns. Indeed, if T is invertible, then this follows from the invertibility of $\text{col}([QT^{j-1}, Q_sT_s^{m-j}])_{j=1}^m$. If T is not invertible, use (2.5.9) and the equality which is obtained from (2.5.9) by replacing m by $m+1$ to establish this fact. Now it is clear that any two full-rank solutions of (2.5.5) differ only by multiplication from the left by an invertible $n \times n$ matrix; so any such solution determines a matrix polynomial $L(z)$ of degree $\leq m$ with the required properties.

Finally, the uniqueness of $L(z)$ follows from the description of $L(z)$ in terms of the full-rank solutions of (2.5.5), as proved above. $\square$

If we are looking for a (regular) matrix polynomial with prescribed right null pair only, then we have the following result.

THEOREM 2.5.4. *Let (Z, T) be a pair of matrices of sizes $n \times p$, $p \times p$, respectively, and such that*

(2.5.10)
$$\bigcap_{j=0}^{q-1} \text{Ker}(ZT^j) = \{0\}$$

for some $q > 0$. Then there exists a regular matrix polynomial $L(z)$ of degree $\leq q$ for which (Z, T) is a right null pair relative to $\mathbb{C}$.

Conversely, if (Z, T) is a right null pair (relative to $\mathbb{C}$) of a matrix polynomial of degree $\leq q$, then (2.5.10) is satisfied.

PROOF. We can assume that T is invertible (otherwise replace T by $T + \alpha I$ for a suitable $\alpha \in \mathbb{C}$ and replace $L(z)$ by $L(z - \alpha)$, using Proposition 2.5.2). Then

$$\bigcap_{j=0}^{q-1} \text{Ker}(ZT^{-j}) = \{0\}.$$

By Theorem 2.4.2 there are matrices Z_s and T_s with $\sigma(T_s) = \{0\}$ such that the matrix

$$\operatorname{col}([ZT^{-j}, Z_s T_s^j])_{j=0}^{q-1}$$

is square and invertible. So

$$\operatorname{col}([ZT^j, Z_s T_s^{q-1-j}])_{j=0}^{q-1}$$

is invertible as well, and now appeal to Theorem 2.5.3.

To prove the converse statement, let (Z, T) be a right null pair of a matrix polynomial $L(z)$ of degree $\leq q$, and again without loss of generality we can assume that T is invertible, and hence so is $L(0)$. As shown in the proof of Theorem 2.5.3, the pair $([\ Z\quad Z_s\ ], T^{-1} \oplus T_s)$, for suitable Z_s and T_s, is a right null pair of the monic matrix polynomial $z^q (L(0))^{-1} L(z^{-1})$ of degree q. By Theorem 2.2.1, the matrix

$$\operatorname{col}([ZT^{-j}, Z_s T_s^j])_{j=0}^{q-1}$$

is invertible, and (2.5.10) follows. □

We shall see later that the matrix polynomial $L(z)$ from Theorem 2.5.4 is defined uniquely up to multiplication from the left by a matrix polynomial with constant nonzero determinant.

We remark that all the results of Sections 2.4 and 2.5 admit counterparts concerning left null pairs, the statements and proofs of which are left to the reader. they can be obtained by applying the results of Sections 2.4 and 2.5 to the matrix polynomials with transposed coefficients.

2.6 PAIRS AND TRIPLES

Recall that (by the general definition given in Section 1.4) a triple of matrices (X, T, Y) is called a null triple for a matrix polynomial $L(z)$ with respect to $z_0 \in \mathbf{C}$ if

$$L(z)^{-1} - X(zI - T)^{-1} Y$$

is analytic in a neighborhood of z_0, $\sigma(T) = \{z_0\}$ and the pairs (X, T) and (T, Y) are null kernel and full range pairs, respectively.

In this section we make the connections between left and right null pairs for a matrix polynomial and its null triple. We start with monic polynomials.

Theorem 2.3.2 allows one to obtain easily a null triple for the monic matrix polynomial $L(z)$ at z_0. Indeed, let Z, T, Y be as in Theorem 2.3.2. Passing to a similar right null pair, we can assume without loss of generality that

$$T = \begin{bmatrix} T_1 & 0 \\ 0 & T_2 \end{bmatrix},$$

where $\sigma(T_1) = \{z_0\}$ and $z_0 I - T_2$ is invertible. Partition accordingly $Z = [\ Z_1\quad Z_2\ ]$ and $Y = \begin{bmatrix} Y_1 \\ Y_2 \end{bmatrix}$. Then

$$L(z)^{-1} = Z_1 (zI - T_1)^{-1} Y_1 + Z_2 (zI - T_2)^{-1} Y_2,$$

and since the matrices

$$\mathrm{col}(ZT^i)_{i=0}^{\ell-1}, \qquad [Y, TY, \ldots, T^{\ell-1}Y]$$

are invertible, it follows that (Z_1, T_1) and (T_1, Y_1) are null kernel and full range pairs, respectively. So (Z_1, T_1, Y_1) is a null triple of $L(z)$ at z_0. Observe that (Z_1, T_1) is a right null pair of $L(z)$ at z_0 (by construction). Passing to the null triple (Y_1^T, T_1^T, Z_1^T) at z_0 of the transposed matrix polynomial $L(z_1)^T$, we see that (T_1, Y_1) is a left null pair of $L(z)$ at z_0. In view of Proposition 1.4.1 this property does not depend on the choice of a null triple of $L(z)$ at z_0. The following result shows that this property extends to non-monic matrix polynomials as well:

THEOREM 2.6.1. *Let* $M(z)$ *be a (regular) matrix polynomial, and let* (Z, T, Y) *be a null triple of* $M(z)$ *at* z_0. *Then* (Z, T) *and* (T, Y) *are right and left null pairs, respectively, of* $M(z)$ *at* z_0.

PROOF. Without loss of generality we may assume that $z_0 \neq 0$ and $\det M(0) \neq 0$ (otherwise shift the argument $z \to z + \alpha$). Let q be the degree of $M(z)$, and let

$$(2.6.1) \qquad \widetilde{M}(z) = z^q (M(0))^{-1} M(z)^{-1}$$

be a monic matrix polynomial of degree q. Let $(\widetilde{Z}, \widetilde{T})$ be a right null pair of $\widetilde{M}(z)$ (relative to $\mathbb{C}$), and let

$$(2.6.2) \qquad \widetilde{Y} = \left(\mathrm{col}(\widetilde{Z}\widetilde{T}^i)_{i=0}^{q-1}\right)^{-1} \mathrm{col}(\delta_{iq} I)_{i=1}^q.$$

Then (Theorem 2.3.2)

$$(2.6.3) \qquad \widetilde{M}^{-1}(z) = \widetilde{Z}(zI - \widetilde{T})^{-1}\widetilde{Y} \quad \text{for} \quad z \notin \sigma(\widetilde{T}).$$

Using (2.6.3) and the property

$$\mathrm{col}(\widetilde{Z}\widetilde{T}^i)_{i=0}^{q-1} \cdot \widetilde{Y} = \mathrm{col}(0, 0, \ldots, I)$$

we deduce that

$$(2.6.4) \qquad z^{q-1}\widetilde{M}^{-1}(z) = \widetilde{Z}(zI - \widetilde{T})^{-1}\widetilde{T}^{q-1}\widetilde{Y}.$$

Without loss of generality we may (and will) assume that $\widetilde{T} = \widetilde{T}_1 \oplus \widetilde{T}_2$, where $\sigma(\widetilde{T}_1) = \{z_0^{-1}\}$; $\sigma(\widetilde{T}_2) \cap \{z_0^{-1}\} = \emptyset$. Let $\widetilde{Z} = [\widetilde{Z}_1, \widetilde{Z}_2]$, $\widetilde{Y} = \mathrm{col}(\widetilde{Y}_1, \widetilde{Y}_2)$ be the corresponding partitions of $\widetilde{Z}$ and $\widetilde{Y}$. Substitution of (2.6.1) into (2.6.4) leads to the equality

$$(2.6.5) \quad M(z)^{-1} = -\widetilde{Z}_1(zI - \widetilde{T}_1^{-1})^{-1}\widetilde{T}_1^{q-2}\widetilde{Y}_1 M(0)^{-1} + \widetilde{Z}_2(I - z\widetilde{T}_2)^{-1}\widetilde{T}_2^{q-1}\widetilde{Y}_2 M(0)^{-1}.$$

Put $Y_1' = -\widetilde{T}_1^{q-2}\widetilde{Y}_1 M(0)^{-1}$. Since the matrices $\mathrm{col}(\widetilde{Z}\widetilde{T}^i)_{i=0}^{q-1}$ and $[\widetilde{Y}, \widetilde{T}\widetilde{Y}, \ldots, \widetilde{T}^{q-1}\widetilde{Y}]$ are invertible (Theorem 2.2.1), the pairs $(\widetilde{Z}_1, \widetilde{T}_1^{-1})$ and $(\widetilde{T}_1^{-1}, Y_1')$ are null kernel and full range pairs, respectively. Comparing the definition of a null triple (which implies that

$$M(z)^{-1} - Z(zI - T)^{-1}Y$$

is analytic at z_0) and (2.6.5) and bearing in mind that $z_0^{-1} \notin \sigma(\widetilde{T}_2)$, we conclude that the matrix function

$$Z(zI - T)^{-1}Y - \widetilde{Z}_1(zI - \widetilde{T}_1^{-1})^{-1}Y_1'$$

is analytic at z_0. By Proposition 1.4.1 there exists an invertible matrix S such that

$$(2.6.6) \qquad Z = \widetilde{Z}_1 S, \quad T = S^{-1}\widetilde{T}_1^{-1}S, \quad Y = S^{-1}Y_1'.$$

It follows from Proposition 2.5.1 that $(\widetilde{Z}_1, \widetilde{T}_1^{-1})$ is a right null pair of $M(z)$ corresponding to z_0; by (2.6.6), the same is true for (Z, T). Applying the result just obtained to the transposed matrix polynomial $M^T(z)$ and the triple (Y^T, T^T, Z^T) one sees that (T, Y) is a left null pair of $M(z)$ at z_0. $\square$

The following theorem may be regarded as a partial converse of Theorem 2.6.1. In particular, it implies existence of null triple (Z, T, Y). The result is a polynomial version of Theorem 1.4.3.

We denote (as in Chapter 1) by $SP(M^{-1}(z_0))$ the singular part of the Laurent expansion of $M^{-1}(z)$ in a neighborhood of z_0 (here $M(z)$ is a matrix polynomial). Observe (cf. the discussion preceding Theorem 1.4.3) that a right null pair (Z, T) of $M(z)$ at z_0 has the property that $\mathrm{col}(Z(T - z_0I)^i)_{i=0}^{\nu-1}$ is left invertible, where ν is the order of z_0 as a pole of $M(z)^{-1}$. Analogously, for a left null pair (T, Y) of $M(z)$ at z_0, the matrix

$$[Y, (T - z_0I)Y, \dots, (T - z_0I)^{\nu-1}Y]$$

is right invertible.

THEOREM 2.6.2. *Let (Z_1, T_1) and (T_1, Y_1) be right and left null pairs, respectively, of a matrix polynomial $M(z)$ at z_0. Let*

$$(2.6.7) \qquad G = \begin{bmatrix} Z_1 \\ Z_1(T_1 - z_0I) \\ \vdots \\ Z_1(T_1 - z_0I)^{\nu-1} \end{bmatrix}^{-1} \begin{bmatrix} K_1 & K_2 & \cdots & K_{\nu-1} & K_\nu \\ K_2 & K_3 & \cdots & K_\nu & 0 \\ \vdots & \vdots & & \vdots & \vdots \\ K_\nu & 0 & \cdots & 0 & 0 \end{bmatrix}$$
$$\times [Y_1, (T_1 - z_0I)Y_1, \dots, (T_1 - z_0I)^{\nu-1}Y_1]^{-1},$$

where

$$SP(M^{-1}(z_0)) = \sum_{j=1}^{\nu}(z - z_0)^{-j}K_j$$

and inverse matrices are understood as some one-sided inverses from the appropriate side. Then G is invertible, and

$$(2.6.8) \qquad SP(M^{-1}(z_0)) = Z_1(zI - T_1)^{-1}GY_1 = Z_1G(zI - T_1)^{-1}Y_1$$

and

$$(2.6.9) \qquad\qquad\qquad GT_1 = T_1G.$$

The matrix G is the unique $p \times p$ matrix (p is the size of T_1) which satisfies (2.6.8) and (2.6.9). In particular, (Z_1G, T_1, Y_1) is a null triple for $M(z)$ at z_0.

PROOF. It will be assumed $z_0 \neq 0$ and $\det M(0) \neq 0$. We shall use the notation introduced in the proof of Theorem 2.6.1.

Equality (2.6.5) ensures that

$$SP(M^{-1}(z_0)) = \tilde{Z}_1(zI - \tilde{T}_1^{-1})^{-1}Y_1'.$$

Since $(\tilde{Z}_1, \tilde{T}_1^{-1})$ (resp. $(\tilde{T}_1^{-1}, Y_1')$) is a right (resp. left) null pair of $M(z)$ corresponding to z_0 (see the proof of Theorem 2.6.1), in view of their uniqueness up to similarity (Theorem 1.2.6) there exist invertible matrices H_1 and H_2 such that

$$Z_1 = \tilde{Z}_1 H_1; \quad T_1 = H_1^{-1}\tilde{T}_1^{-1}H_1; \quad Y_1 = H_2 Y_1'; \quad T_1 = H_2\tilde{T}_1^{-1}H_2^{-1}.$$

Put $G_0 = (H_2H_1)^{-1}$; then

(2.6.10) $$SP(M^{-1}(z_0)) = Z_1(zI - T_1)^{-1}G_0Y_1 = Z_1G_0(zI - T_1^{-1})Y_1$$

and

(2.6.11) $$G_0T_1 = T_1G_0.$$

Equalities (2.6.10) and (2.6.11) imply that in fact $G_0 = G$ (where G is defined by (2.6.7)). Indeed,

$$(zI - T_1)^{-1} = \sum_{i=1}^{\infty}(z - z_0)^{-i}(T_1 - z_0I)^{i-1};$$

so (2.6.10) gives $Z_1(T_1 - z_0I)^{i-1}G_0Y_1 = Z_1G_0(T_1 - z_0I)^{i-1}Y_1 = K_i$, $i = 1, \ldots, \nu$. Using (2.6.11), rewrite these equalities in the form

$$\operatorname{col}(Z_1(T_1 - z_0I)^i)_{i=0}^{\nu-1} \cdot G_0[Y_1, (T_1 - z_0I)Y_1, \ldots, (T_1 - z_0I)^{\nu-1}Y_1]$$

$$= \begin{bmatrix} K_1 & K_2 & \cdots & K_{\nu-1} & K_\nu \\ K_2 & K_3 & \cdots & K_\nu & 0 \\ \vdots & \vdots & & \vdots & \vdots \\ K_\nu & 0 & \cdots & 0 & 0 \end{bmatrix},$$

and $G_0 = G$ by (2.6.7). This proves (2.6.8) and (2.6.9).

Let us prove the uniqueness of G. Assume

(2.6.12) $$Z_1(zI - T_1)^{-1}GY_1 = Z_1(zI - T_1)^{-1}\tilde{G}Y_1$$

and

(2.6.13) $$GT_1 = T_1G; \qquad \tilde{G}T_1 = T_1\tilde{G}$$

for invertible matrices G and $\tilde{G}$. Equality (2.6.12) implies $Z_1T_1^rGY_1 = Z_1T_1^r\tilde{G}Y_1$, $r = 0, 1, \ldots$, or $\operatorname{col}(Z_1T_1^i)_{i=0}^r GY_1 = \operatorname{col}(Z_1T_1^i)_{i=0}^r \tilde{G}Y_1$. Since (Z_1, T_1) is a null kernel pair, for some r the columns of $\operatorname{col}(Z_1T_1^i)_{i=0}^r$ are linearly independent; consequently, $GY_1 = \tilde{G}Y_1$. Combining this equality with (2.6.13) we obtain

$$G[Y_1, T_1Y_1, \ldots, T_1^rY_1] = \tilde{G}[Y_1, T_1Y_1, \ldots, T_1^rY_1], \qquad r = 0, 1, \ldots .$$

For r large enough, the rows of $[Y_1, T_1Y_1, \ldots, T_1^rY_1]$ are linearly independent, and the equality $G = \tilde{G}$ follows. $\square$

2.7 PROOF OF THEOREM 1.4.3

We give here a proof of Theorem 1.4.3 (which was promised in Chapter 1). This theorem is actually an extension of Theorem 2.6.2 to the framework of analytic matrix functions, and the proof proceeds by reduction to the polynomial case. Let $A(z)$ be an analytic matrix function with zero z_0, and let

$$A(z) = E(z)\,\mathrm{diag}\big((z - z_0)^{\alpha_1}, \ldots, (z - z_0)^{\alpha_n}\big) F(z)$$

be a local Smith form of $A(z)$. So $\alpha_1 \geq \cdots \geq \alpha_n \geq 0$ are integers, and $E(z)$ and $F(z)$ are analytic matrix functions invertible at z_0. Write the Taylor series

$$E(z) = \sum_{j=0}^{\infty} (z - z_0)^j E_j; \qquad F(z) = \sum_{j=0}^{\infty} (z - z_0)^j F_j$$

(here E_0 and F_0 are invertible matrices). For $p = 1, 2, \ldots$ define

$$M_p(z) = \left(\sum_{j=0}^{p} (z - z_0)^j E_j \right) \mathrm{diag}\big((z - z_0)^{\alpha_1}, \ldots, (z - z_0)^{\alpha_n}\big) \left(\sum_{j=0}^{p} (z - z_0)^j F_j \right).$$

It is easy to see that

$$SP\big(M_p^{-1}(z_0)\big) = SP\big(A^{-1}(z_0)\big)$$

for p large enough. Also, Theorem 1.1.3 shows that for p large enough $A(z)$ and $M_p(z)$ have precisely the same right and left null pairs corresponding to z_0. As $M_p(z)$ is a polynomial, Theorem 2.6.2 is applicable, and we obtain the result of Theorem 1.4.3 by taking p to be sufficiently large.

CHAPTER 3

LOCAL DATA FOR MEROMORPHIC MATRIX FUNCTIONS

In this chapter we develop further the results from Chapter 1 for meromorphic matrix valued functions. In addition to null structure, pole structure is introduced and studied. In contrast to the scalar case, a new feature appears in the matrix case, namely, a meromorphic matrix function can have a zero and a pole at the same point. This chapter contains the analysis of null and pole chains, null and pole pairs, null and pole triples and singular parts, as well as connections among them.

3.1 NULL FUNCTIONS AND POLE FUNCTIONS

In this section we extend the notions of local null structure for a matrix function from the analytic to the meromorphic case and elucidate some notions of local pole structure for the meromorphic case.

Suppose $A(z)$ is an $n \times n$ matrix function which is meromorphic at the point $z_0 \in \mathbb{C}$ (i.e. the matrix entries $a_{ij}(z)$ of $A(z)$ are meromorphic at z_0) and for which $\det A(z)$ does not vanish identically at points of analyticity for $A(z)$; such a function we call a *regular* meromorphic matrix function. We say that the analytic $\mathbb{C}^n$-valued function $\varphi(z)$ is a *right null function* for $A(z)$ at z_0 of order $k > 0$ if $\varphi(z_0) \neq 0$ and if $A(z)\varphi(z)$ is analytic at z_0 with a zero of order k at z_0. If $A(z)$ is analytic at z_0, the notion is no different from that in Section 1.1. In the matrix case it is possible for a point z_0 to be both a zero and a pole. We illustrate with several examples.

EXAMPLE 3.1.1. Let $A(z) = \begin{bmatrix} 1 & z^{-1} \\ 0 & 1 \end{bmatrix}$ and suppose $\varphi(z)$ is a right null function for $A(z)$ at $z_0 = 0$. Write $\varphi(z)$ in component form $\begin{bmatrix} \varphi^{(1)}(z) \\ \varphi^{(2)}(z) \end{bmatrix}$. Then for φ to be a null function or order k we must have $\varphi(z_0) \neq 0$ and $A(z)\varphi(z) = O(|z|^k)$ for z near zero, i.e.

$$\varphi^{(1)}(z) + z^{-1}\varphi^{(2)}(z) = O(|z|^k)$$
$$\varphi^{(2)}(z) = O(|z|^k).$$

If $\varphi_j = \begin{bmatrix} \varphi_j^{(1)} \\ \varphi_j^{(2)} \end{bmatrix}$ is the j-th Taylor coefficient of $\varphi(z)$ at 0, this says

(3.1.1) $$\varphi_0^{(2)} = 0 \text{ and } \varphi_j^{(1)} + \varphi_{j+1}^{(2)} = 0 \quad \text{for} \quad j \leq k - 1$$

and

(3.1.2) $$\varphi_j^{(2)} = 0 \quad \text{for} \quad j \leq k - 1.$$

If $k > 1$ then (3.1.2) forces $\varphi_0^{(2)} = \varphi_1^{(2)} = 0$ whence from (3.1.1) we get $\varphi_0^{(1)} = -\varphi_1^{(2)} = 0$.
Thus $\varphi(0) = \begin{bmatrix} \varphi_0^{(1)} \\ \varphi_0^{(2)} \end{bmatrix} = \begin{bmatrix} 0 \\ 0 \end{bmatrix}$, a contradiction. On the other hand, for $k = 1$, the
general solution is $\varphi_0^{(2)} = 0$, $\varphi_1^{(2)} = -\varphi_0^{(1)} =$ arbitrary, $\varphi_1^{(1)} =$ arbitrary, $\varphi_j^{(1)}$ and $\varphi_j^{(2)}$
arbitrary for $j > 2$. Thus the general right null function for this $A(z)$ at $z = 0$ is

$$\varphi(z) = \begin{bmatrix} c \\ 0 \end{bmatrix} + z \begin{bmatrix} d \\ -c \end{bmatrix} + O(|z|^2)$$

where d and c are arbitrary complex numbers such that $c \neq 0$. Note that the second
Taylor coefficient $\begin{bmatrix} d \\ -c \end{bmatrix}$ cannot be chosen independently of the first Taylor coefficient
$\begin{bmatrix} c \\ 0 \end{bmatrix}$, unlike the situation for the analytic case. Also $\det A(z) \equiv 1$, so zeros of $\det A(z)$
do not coincide with the zeros of $A(z)$ for the matrix case. $\square$

EXAMPLE 3.1.2. Let us consider the scalar case

$$a(z) = a_k(z - z_0)^k + a_{k+1}(z - z_0)^{k+1} + \cdots .$$

Then a has a pole at z_0 if and only if $k < 0$. In this case $a(z)\varphi(z)$ is not analytic at z_0
whenever $\varphi(z)$ is a scalar analytic function with $\varphi(z_0) \neq 0$, and hence z_0 cannot be a
zero for $a(z)$ unless a is analytic at z_0. $\square$

As in the analytic case, it is easy to see that if $\varphi^{(1)}(z), \cdots, \varphi^{(m)}(z)$ are right
null functions for $A(z)$ at z_0 and $c_1, \ldots, c_m$ are complex numbers, then the linear combination $\varphi(z) = c_1\varphi^{(1)}(z) + \cdots + c_m\varphi^{(m)}(z)$ is again a right null function as long as
$\varphi(z_0) \neq 0$; if no $c_j = 0$, then the order of $\varphi(z)$ as a null function is at least the minimum
of the orders of the null functions $\varphi^{(1)}(z), \ldots, \varphi^{(m)}(z)$. Also as in the analytic case, if φ
is a right null function and $s(z)$ is an arbitrary scalar analytic function with $s(z_0) \neq 0$,
then $\tilde{\varphi}(z) = s(z)\varphi(z)$ is also a right null function of the same order.

If $0 \neq x$ is a column vector such that $x = \varphi(z_0)$ for some right null function
for the meromorphic matrix function $A(z)$ at z_0, we say that x is (right) *eigenvector* for
$A(z)$ at z_0. The set of all eigenvectors for $A(z)$ at z_0 together with 0 is a linear subspace
of $\mathbf{C}^n$ which we call the (*right*) *eigenspace* (denoted $\mathrm{Ker}(A; z_0)$) for $A(z)$ at z_0; note that
we cannot identify the eigenspace as $\mathrm{Ker}\, A(z_0)$ for the meromorphic case, since $A(z_0)$ is
not defined as a finite matrix if A has a pole at z_0. For example, from Example 3.1.1
we see that the right eigenspace for the matrix function $A(z) = \begin{bmatrix} 1 & \frac{1}{z} \\ 0 & 1 \end{bmatrix}$ at the point
$z_0 = 0$ is $\left\{ \begin{bmatrix} c \\ 0 \end{bmatrix} : c \in \mathbf{C} \right\}$.

In general we define a *canonical set* of *right null functions* for a meromorphic
matrix function $A(z)$ at a point z_0 in the same way as was done in Section 1.1 for the
analytic case. Namely, let $\varphi^{(1)}$ be a right null function for $A(z)$ at z_0 having the maximal
possible order (we shall see later (Theorem 3.1.2) that the set of possible orders for a
right null function for $A(z)$ at z_0 is bounded). If the eigenvector $\varphi^{(1)}(z_0)$ already spans

the right eigenspace of $A(z)$ at z_0, then the single right null function $\{\varphi^{(1)}(z)\}$ forms a canonical set of right null functions. Otherwise we choose a direct complement $\mathcal{K}_1$ of $\text{Span}\{\varphi^{(1)}(z_0)\}$ in the right eigenspace and let $\varphi^{(2)}(z)$ be a right null function of maximum order such that $\varphi^{(2)}(z_0) \in \mathcal{K}_1$. Continuing in this way we construct right null functions $\varphi^{(1)}(z), \varphi^{(2)}(z), \ldots, \varphi^{(m)}(z)$ with nonincreasing orders $k_1 \geq k_2 \geq \cdots \geq k_m$ such that $\{\varphi^{(1)}(z_0), \varphi^{(2)}(z_0), \ldots, \varphi^{(m)}(z_0)\}$ is a basis for the eigenspace. A collection of right null functions $\{\varphi^{(1)}(z), \varphi^{(2)}(z), \ldots, \varphi^{(m)}(z)\}$ with these properties we call a *canonical set of right null functions* for $A(z)$ at z_0.

For instance, in Example 3.1.1 the function

$$\varphi(z) = \begin{bmatrix} 1 \\ 0 \end{bmatrix} + z \begin{bmatrix} 0 \\ -1 \end{bmatrix}$$

by itself forms a canonical set of right null functions of $A(z)$ at $z_0 = 0$. More generally, any single function $\varphi(z)$ of the form

$$\varphi(z) = \begin{bmatrix} c \\ 0 \end{bmatrix} + z \begin{bmatrix} d \\ -c \end{bmatrix} + O(|z - z_0|^2),$$

where $0 \neq c$ and d are arbitrary but fixed, forms a canonical set. In the scalar case the function $\varphi(z) = 1$ forms a canonical set of null functions at each zero z_j of $A(z)$.

There is a completely analogous theory for the local pole structure for a meromorphic matrix function at a point z_0. We say that the analytic vector function $\psi(z)$ with $\psi(z_0) \neq 0$ is a *right pole function* for the meromorphic matrix function $A(z)$ at the point z_0 of order $k > 0$ if there is an analytic vector function $\varphi(z)$ such that $\varphi(z_0) \neq 0$ and $A(z)\varphi(z) = (z - z_0)^{-k}\psi(z)$. We shall always assume $A(z)$ is regular, so $A^{-1}(z)$ is meromorphic at z_0 whenever $A(z)$ is. Then if $\varphi(z)$ and $\psi(z)$ are related as above, upon premultiplying by $(z - z_0)^k A^{-1}(z)$ we get

$$A^{-1}(z)\psi(z) = (z - z_0)^k \varphi(z) = O(|z - z_0|^k).$$

That is, $\psi(z)$ is a right pole function for $A(z)$ at z_0 of order k if and only if $\psi(z)$ is a right null function for $A^{-1}(z)$ at z_0 of order k. Thus the analysis of right pole functions for a given meromorphic matrix function reduces to the study of right null functions for the inverse function. In particular, we may also speak of a *canonical set of right pole functions* for a given meromorphic matrix function at a point z_0. We illustrate this notion with some examples.

EXAMPLE 3.1.3. As in Example 3.1.1, let $A(z) = \begin{bmatrix} 1 & \frac{1}{z} \\ 0 & 1 \end{bmatrix}$. Then if

$$\varphi(z) = \begin{bmatrix} \varphi_0^{(1)} \\ \varphi_0^{(2)} \end{bmatrix} + O(|z|), \text{ we see that near } 0,$$

$$A(z)\varphi(z) = z^{-1} \begin{bmatrix} \varphi_0^{(2)} \\ 0 \end{bmatrix} + \begin{bmatrix} \varphi_0^{(1)} + \varphi_1^{(2)} \\ \varphi_0^{(2)} \end{bmatrix} + O(|z|).$$

Thus a pole function for $A(z)$ at 0 can have order at most 1, in which case

$$(3.1.3) \qquad \psi(z) = \begin{bmatrix} c \\ 0 \end{bmatrix} + z \begin{bmatrix} d \\ c \end{bmatrix} + O(|z|^2)$$

where $0 \neq c$ and d are any complex numbers, is a pole function for $A(z)$ of order 1. Alternatively, we may note that $A^{-1}(z) = \begin{bmatrix} 1 & -z^{-1} \\ 0 & 1 \end{bmatrix}$ and calculate right null functions for $A^{-1}(z)$ as in Example 3.1.1. Then any canonical set of right pole functions for $A(z)$ consists of a single function of the form (3.1.3) with $c \neq 0$. $\square$

EXAMPLE 3.1.4. For the scalar case

$$a(z) = a_k(z - z_0)^k + a_{k-1}(z - z_0)^{k+1} + \cdots$$

where $k < 0$ and $a_k \neq 0$, any analytic vector function $\psi(z)$ with $\psi(z_0) \neq 0$ is a pole function of order $-k$. $\square$

EXAMPLE 3.1.5. We suppose that $A(z)$ has a block diagonal form

$$A(z) = \operatorname{diag}(A_j(z))_{j=1}^m$$

where (say) each $A_j(z)$ has size $n_j \times n_j$. Then a canonical set of right null functions and of right pole functions for $A(z)$ can be determined from a knowledge of the corresponding objects for each of the $A_j(z)$'s, much as in Example 1.1.3. $\square$

EXAMPLE 3.1.6. Suppose that A_1 is a meromorphic matrix function at z_0 and E and F are analytic matrix functions such that $\det E(z_0) \neq 0$ and $\det F(z_0) \neq 0$, and $A_2(z) = E(z)A_1(z)F(z)$. Then $\varphi(z)$ is a right null function of order k for $A_2(z)$ at z_0 if and only if $F(z)\varphi(z)$ is a right null function of order k for $A_1(z)$ at z_0; this follows exactly as in the analytic case (Lemma 1.1.1). Since a right pole function for a meromorphic matrix function is just a right null function for the inverse, it follows that $\psi(z)$ is a right pole function of order k for $A_1(z)$ at z_0 if and only if $E(z)\psi(z)$ is a right pole function for $A_2(z)$ at z_0. $\square$

EXAMPLE 3.1.7. Suppose that $A(z) = \operatorname{diag}((z-z_0)^{\kappa_1}, \ldots, (z-z_0)^{\kappa_n})$ where $\kappa_1 \geq \kappa_2 \geq \cdots \geq \kappa_n$ are integers. Then the vector function $\varphi(z) = \operatorname{col}(\varphi^{(i)}(z))_{i=1}^n$ is a right null function for $A(z)$ of order k if and only if $\varphi(z_0) \neq 0$ and $\varphi^{(i)}(z)$ has a zero at z_0 of order at least $k - \kappa_i$ for $1 \leq i \leq n$. When $k > \kappa_1$, the second condition is incompatible with the first; thus κ_1 is the maximum possible order of a null function (assuming $\kappa_1 > 0$). If some κ_i's are negative, then the second condition places restrictions on some Taylor coefficients $\frac{1}{j!}\varphi^{[j]}(z)$ of φ at z_0 for $j > k$. Let $k_1 > k_2 > \cdots > k_r > 0 > k_{m+1-s} > k_{m+2-s} > \cdots > k_m$ be the distinct nonzero partial multiplicities, each with multiplicity $n_1, n_2, \ldots, n_r, n_{m+1-s}, \cdots, n_m$, respectively (so $k_1 = \kappa_1 = \cdots = \kappa_{n_1} > k_2 = \kappa_{n_1+1} = \cdots = \kappa_{n_2} > \cdots > k_r = \kappa_{n_{r-1}+1} = \cdots = \kappa_{n_r} > 0 > k_{m+1-s} = \kappa_{n-n_s+1} = \cdots - n_m+1 = \cdots = \kappa_{n-n_s+1-\cdots-n_m} > \cdots > k_m = \kappa_{n+1-n_m} = \kappa_{n+2-n_m} = \cdots = \kappa_n$). Then the form of an arbitrary canonical set of right null functions can be described as in the analytic case (see the proof of Theorem 1.1.5). Indeed, $\{\varphi^{(1)}, \varphi^{(2)}, \cdots, \varphi^{(r)}\}$ is a canonical set of right null functions for $A(z)$ at z_0 with Taylor expansions $\varphi^{(i)}(z) = \sum_{\alpha=0}^{\infty} \varphi_\alpha^{(i)}(z - z_0)^\alpha$, if and only if $\hat{r} = r$, the set $\{\varphi_0^{(i)}: 1 \leq i \leq r\}$ is linearly independent and $\varphi_\alpha^{(i)} \in$ Span$\{e_1, \ldots, e_{n_1 + \cdots + n_j}\}$ whenever $k_j + \alpha < \kappa_i$, and in this case $\varphi^{(i)}(z)$ is a right null function of order κ_i. Note that if some κ_j's are negative, then the Taylor coefficients $\varphi_\alpha^{(i)}$ are subject to restrictions even when $\alpha > \kappa_i$.

A canonical set of right pole functions can be given a similar description, since such a set is just a canonical set of right null functions for $A^{-1}(z)$ at z_0. The orders of the right pole functions in a canonical set of right pole functions are the negatives of the negative partial multiplicities of $A(z)$ at z_0. We leave the details to the reader. □

To compute a canonical set of right null functions and of right pole functions for a meromorphic matrix function $A(z)$, we need the extension of the local Smith form to the meromorphic case (called the *local Smith-McMillan form*).

THEOREM 3.1.1. *Let $A(z)$ be a matrix function meromorphic at z_0 with determinant not vanishing identically at points of analyticity. Then $A(z)$ admits the representation*

$$A(z) = E(z) \operatorname{diag}\left((z - z_0)^{\kappa_1}, \ldots, (z - z_0)^{\kappa_n}\right) F(z)$$

where $E(z)$ and $F(z)$ are matrix functions which are analytic and invertible in a neighborhood of z_0 and where $\kappa_1 \geq \kappa_2 \geq \cdots \geq \kappa_n$ are integers (positive, negative or zero).

PROOF. The theorem follows easily from the local Smith form for the analytic matrix function $(z - z_0)^p A(z)$ (Theorem 1.1.2), where p is the order of the pole z_0 of $A(z)$. □

The indices $\kappa_1 \geq \kappa_2 \geq \cdots \geq \kappa_n$ we again refer to as the *partial multiplicities* for $A(z)$ at z_0.

Using the local Smith-McMillan form, it is easy to compute a canonical set of right null functions and pole functions for a meromorphic matrix function at a point z_0.

THEOREM 3.1.2. *Suppose $A(z)$ is a (regular) meromorphic matrix function with local Smith-McMillan form at z_0*

$$A(z) = E(z) \operatorname{diag}\left((z - z_0)^{\kappa_1}, \ldots, (z - z_0)^{\kappa_n}\right) F(z)$$

as in Theorem 3.1.1. Let $\kappa_1 \geq \kappa_2 \geq \cdots \geq \kappa_j > 0$ be the positive partial multiplicities of $A(z)$ and let $0 > \kappa_{n-r+1} \geq \kappa_{n-r+2} \geq \cdots \geq \kappa_n$ be the negative partial multiplicities. Then:

(i) *$F(z)^{-1} e_\ell$ is a right null function for $A(z)$ at z_0 of order κ_ℓ for $1 \leq \ell \leq j$*

and

$$\{F(z)^{-1} e_1, F(z)^{-1} e_2, \ldots, F(z)^{-1} e_j\}$$

is a canonical set of right null functions for $A(z)$ at z_0.

(ii) *$E(z) e_\ell$ is a right pole function for $A(z)$ at z_0 of order $-\kappa_\ell$ for $n-r+1 \leq \ell \leq n$, and*

$$\{E(z) e_n, E(z) e_{n-1}, \ldots, E(z) e_{n-r+1}\}$$

is a canonical set of right pole functions for $A(z)$ at z_0.

(iii) *If $\{\varphi^{(1)}, \ldots, \varphi^{(p)}\}$ is any other canonical set of right null functions for $A(z)$ at z_0, then $p = j$ and $\varphi^{(\ell)}$ has order κ_ℓ for $1 \le \ell \le j$.*

(iv) *If $\{\psi^{(1)}, \ldots, \psi^{(q)}\}$ is any other canonical set of right pole functions, then $q = r$ and $\psi^{(\ell)}$ has order $\kappa_{n+1-\ell}$ for $1 \le \ell \le r$.*

PROOF. By using the result of Example 3.1.6 it suffices to consider the diagonal case. The diagonal case in turn is worked out in Example 3.1.7. □

All of the above developments can be repeated with left null functions in place of right null functions and left pole functions in place of right null functions. Explicitly, a *left null function* for the regular meromorphic matrix function $A(z)$ at the point z_0 is a row vector function $\varphi^*(z)$ analytic at z_0 with $\varphi^*(z_0) \ne 0$ for which $\varphi^*(z)A(z)$ has analytic continuation to z_0 which is zero at z_0; the order k of this zero defines the order $\varphi^*(z)$ as a left null function. Similarly, a *left pole function of order k* is a row vector function $\psi^*(z)$ analytic at z_0 with $\psi^*(z_0) \ne 0$ such that $(z - z_0)^{-k}\psi^*(z) = \varphi^*(z)A(z)$ for some analytic row vector function $\varphi^*(z)$. Equivalently, $\psi^*(z)$ is a right null function for $A^{-1}(z)$ of order k at z_0. Then one can define the notions of *canonical set of left null functions* and a *canonical set of left pole functions*. Using the local Smith-McMillan form, one can show that the orders of the right null functions in a right canonical set are the same as the positive partial multiplicities of $A(z)$, and thus are the same as the orders of the left null functions in a left canonical set, and similarly for pole functions.

3.2 NULL AND POLE CHAINS, NULL AND POLE PAIRS

In this section we introduce an alternative language for dealing with the local zero and pole structure of a meromorphic matrix function.

Suppose $A(z)$ is a (regular) meromorphic matrix function with Laurent expansion in a neighborhood of z_0 $A(z) = \sum_{j=-q}^{\infty}(z - z_0)^j A_j$. Here it is assumed that $q \ge 0$. Suppose that the analytic vector function $\varphi(z)$ with Taylor series expansion $\varphi(z) = \sum_{j=0}^{\infty} \frac{1}{j!}\varphi^{[j]}(z_0)(z - z_0)^j$ is a right null function for $A(z)$ at z_0. Thus $A(z)\varphi(z)$ has analytic continuation to z_0 which has a zero of order k at z_0, where $A(z)\varphi(z)$ has Laurent expansion

$$A(z)\varphi(z) = \sum_{i=0}^{\infty}\left[\sum_{\ell=0}^{i} A_{-q+\ell}\frac{1}{(i-\ell)!}\varphi^{[i-\ell]}(z_0)\right](z - z_0)^{i-q}.$$

Thus, if φ is a right null function of order k for $A(z)$ at z_0 we have

(3.2.1) $A_{-q}\frac{1}{i!}\varphi^{[i]}(z_0) + \cdots + A_{-q+i}\varphi(z_0) = 0, \qquad i = 0, \ldots, q + k - 1.$

Let us now say that an ordered set of column vectors $(x_0, \ldots, x_{k-1})$ $(k \ge 0)$ in $\mathbf{C}^n$ is a *right null chain* for the meromorphic matrix function $A(z)$ at z_0 if $x \ne 0$ and there exists a right null function $\varphi(z)$ for $A(z)$ at z_0 of order $\ge k$ having its first k Taylor coefficients satisfy

$$\frac{1}{i!}\varphi^{[i]}(z_0) = x_i, \qquad 0 \le i \le k - 1.$$

From (3.2.1) we see that equivalently, $(x_0, \ldots, x_{k-1})$ is a right null chain for $A(z)$ at z_0 if $x \neq 0$ and there exist vectors $x_k, \ldots, x_{q+k-1}$ in $\mathbb{C}^n$ such that

(3.2.2) $$A_{-q}x_i + \cdots + A_{-q+i}x_0 = 0, \qquad i = 0, \ldots, q + k - 1.$$

It is sometimes convenient to write equalities (3.2.2) in the form

(3.2.3)
$$\begin{bmatrix} A_{-q} & 0 & \cdots & 0 \\ A_{-q+1} & A_{-q} & \cdots & 0 \\ \vdots & \vdots & \ddots & \vdots \\ A_{k-1} & A_{k-2} & \cdots & A_{-q} \end{bmatrix} \begin{bmatrix} x_0 \\ x_1 \\ \vdots \\ x_{q+k-1} \end{bmatrix} = 0.$$

The number k is called the *length* of the chain. The leading vector x_0 in the chain is the value $\varphi(z_0)$ of the associated right null function for $A(z)$ at z_0, i.e. a right eigenvector for $A(z)$ at z_0. Given a right eigenvector x_0 for $A(z)$ at z_0, of course there may in general be many right null chains for $A(z)$ at z_0 which have x_0 as a first vector.

Note that condition (3.2.2) for a right null chain does not depend on the choice of q, i.e. writing

(3.2.4) $$A(z) = \sum_{j=-r}^{\infty} (z - z_0)^j A_j,$$

with $r > q$ and $A_{-r} = \cdots = A_{-q+1} = 0$, we obtain precisely the same right null chains by using (3.2.4). Note that if $(x_0, x_1, \ldots, x_{k-1})$ is a right null chain then there exist $x_k, \ldots, x_{q+k-1}$ such that

(3.2.5) $$\varphi(z) = \sum_{i=0}^{q+k-1} (z - z_0)^i x_i$$

is a right null function for $A(z)$ at z_0; unlike the analytic case, there may exist other choices of $x_k, x_{k+1}, \ldots, x_{q+k-1}$ for which (3.2.5) is not a right null function of order k.

We emphasize that right null functions are not determined generally by the right null chains (as we know is the case for analytic functions). Indeed, if $A(z)$ has a pole and a zero at z_0, a number of Taylor coefficients (beyond those in the right null chain) appear in the equations that define a right null function, and hence generally can not be arbitrarily chosen.

By a *canonical set of right null chains* for $A(z)$ at z_0 we mean an ordered set of vectors

(3.2.6) $$x_0^{(1)}, \ldots, x_{r_1-1}^{(1)}; x_0^{(2)}, \ldots, x_{r_2-1}^{(2)}; \cdots; x_0^{(p)}, \ldots, x_{r_p-1}^{(p)}$$

such that the corresponding set of right null functions

$$\left\{ \varphi^{(j)}(z) = \sum_{i=0}^{r_j+q-1} (z - z_0)^i x_i^{(j)} : j = 1, \ldots, p \right\}$$

is a canonical set of right null functions for $A(z)$ at z_0.

The information contained in (3.2.6) can be conveniently put into a pair of matrices, much in the same way it was done in Section 1.2.

Let X_i be the $m \times r_i$ matrix whose j-th column is $x_{j-1}^{(i)}$, and let J_i be the $r_i \times r_i$ Jordan block with z_0 on the main diagonal. Put

$$X_0 = [X_1, \ldots, X_p], \qquad J_0 = J_1 \oplus \cdots \oplus J_p.$$

So J_0 is Jordan matrix with the single eigenvalue z_0. Note that

$$\dim \operatorname{Ker}(z_0 I - J_0) = \dim \operatorname{Ker}(A; z_0).$$

The pair (X_0, J_0) is called a *right null pair* of A corresponding to z_0. Note that X_0 is the list of vectors in a canonical set of right null chains, organized as a matrix, and the Jordan matrix J_0 can be thought of as a bookkeeping device, which encodes where one chain ends and the next begins in X_0 and which point z_0 we are considering. Also, any pair (Z, T) similar to (X_0, J_0) (i.e. such that $Z = X_0 S$, $T = S^{-1} J_0 S$ for some invertible matrix S) will be called the *right null pair* of $A(z)$ at z_0.

Recall in Section 1.2 we defined a pair of matrices (Z, T), where Z has size $n \times p$ and T has square size $p \times p$, to be a *null kernel pair* if

$$\operatorname{rank} \operatorname{col}(ZT^i)_{i=0}^{\ell-1} = p$$

for some integer $\ell \geq 1$. The size p of T is said to be the order of the pair (Z, T). By using a similarity transformation to reduce to the case where the right null pair (Z, T) is one constructed from a canonical set of right null chains of $A(z)$ at z_0, it is easy to establish the following analogue of Theorem 1.2.4.

THEOREM 3.2.1. *If (Z, T) is a right null pair for a meromorphic matrix function $A(z)$ at z_0, then (Z, T) is a null kernel pair with order equal to the sum of the positive Smith-McMillan partial multiplicities for $A(z)$ at z_0.*

Just as in the analytic case, the right null pair at a point is uniquely determined up to similarity.

THEOREM 3.2.2. *The right null pair (Z, T) for the meromorphic matrix function $A(z)$ at z_0 is unique up to similarity. Moreover, the sizes of the blocks in the Jordan form of T coincide with the positive partial multiplicities for $A(z)$ at z_0.*

PROOF. As in the analytic case (see the proof of Theorem 1.2.6) one can show that

$$\operatorname{Im} \operatorname{col}\big(Z(T - z_0 I)^i\big)_{i=0}^{\ell-1}$$
$$= \operatorname{Span}\{\operatorname{col}(x_k, x_{k-1}, \ldots, x_0, 0, \ldots, 0) : x_0, \ldots, x_k \text{ is a}$$
$$\text{right null chain for } A(z) \text{ at } z_0\}.$$

Once this is established, the proof follows as in the analytic case. □

The analogue of Proposition 1.2.6 it is more convenient to express in terms of null chains rather than null pairs for the meromorphic case.

THEOREM 3.2.3. *Let* $(x_{p0}, \ldots, x_{p,r_p-1})$, $p = 1, \ldots, q$ *be a set of right null chains for* $A(z)$ *at* z_0, *and assume that* $x_{10}, \ldots, x_{q0}$ *are linearly independent. Then the following statements are equivalent:*

(i) *the set* $(x_{p0}, \ldots, x_{p,r_p-1})$, $p = 1, \ldots, q$, *is canonical;*

(ii) *the sum* $r_1 + \cdots + r_q$ *is maximal possible among all sets of right null chains for* $A(z)$ *at* z_0 *with linearly independent leading vectors;*

(iii) $r_1 + \cdots + r_q$ *is equal to the sum of the positive partial multiplicities for* $A(z)$ *at* z_0. $\square$

We introduce now the notions of left null chains and left null pairs. A chain of row vectors $y_0, y_1, \ldots, y_{n-1}$ is a *left null chain* for the meromorphic matrix function $A(z)$ at z_0 if there exists a *left null function* $\psi(z)$ of order r having $\frac{1}{j!}\psi^{[j]}(z_0) = y_j$ for $0 \leq j \leq r - 1$; equivalently, the chain of column vectors $y_0^T, y_1^T, \ldots, y_{r-1}^T$ is a right null chain for $A^T(z) := A(z)^T$ at z_0. A system of left null chains

$$(3.2.7) \qquad y_{10}, y_{12}, \ldots, y_{1,r_1-1}; \cdots; y_{p0}, \ldots, y_{p,r_p-1}$$

is said to be *canonical* if there is an associated set of left null functions

$$\{\psi^{(1)}(z), \ldots, \psi^{(p)}(z)\}$$

with $\frac{1}{\alpha!}\psi^{(j)[\alpha]}(z_0) = y_{j\alpha}$ for $1 \leq j \leq p$, $0 \leq \alpha \leq r_j - 1$ which is canonical. Now let (3.2.7) be a canonical set of left null chains. For $1 \leq i \leq p$ let $R_i = \mathrm{col}(y_{i\alpha})_{\alpha=0}^{r_i-1}$. Let J_i be the $r_i \times r_i$ Jordan block with z_0 on the main diagonal and put

$$R_0 = \mathrm{col}(R_j)_{j=1}^p, \qquad J_0 = J_1^T \oplus \cdots \oplus J_p^T.$$

Then the pair (J_0, R_0) will be called a *left null pair* for $A(z)$ at z_0. Any pair of the form (S, Y), where $S = M^{-1}J_0M$, $Y = M^{-1}R_0$ for some invertible matrix M, will also be called a *left null pair* for $A(z)$ at z_0. It follows from the definition that (S, Y) is a left null pair of $A(z)$ at z_0 if and only if (Y^T, S^T) is a right null pair of $A(z)^T$ at z_0. Using this fact various properties analogous to those described above for right null pairs can be verified for left null pairs.

Observe that if (Z, T_1) and (T_2, Y) are right and left null pairs of $A(z)$ at z_0, then T_1 and T_2 are similar. Indeed, $\sigma(T_1) = \sigma(T_2) = \{z_0\}$, and the sizes of Jordan blocks in each T_1 and T_2 are just the positive partial multiplicities of $A(z)$ at z_0.

Finally, we introduce pole chains and pole pairs. By definition a chain of column vectors $(y_0, \ldots, y_{r-1})$ with $y_0 \neq 0$ is a *right pole chain* for $A(z)$ at z_0 if there exists a right pole function $\varphi(z)$ for $A(z)$ at z_0 of order r such that $\frac{1}{\alpha!}\varphi^{[\alpha]}(z_0) = y_\alpha$ for $0 \leq \alpha \leq r - 1$. Equivalently, there exist column vectors $\psi_0, \psi_1, \ldots, \psi_{q-1}$ with $\psi_0 \neq 0$

such that

(3.2.8)
$$
\begin{bmatrix}
A_{-q} & 0 & \cdots & 0 \\
A_{-q+1} & A_{-q} & \cdots & 0 \\
\vdots & \vdots & \ddots & \vdots \\
A_{-1} & A_{-2} & \cdots & A_{-q}
\end{bmatrix}
\begin{bmatrix}
\psi_0 \\
\psi_1 \\
\vdots \\
\psi_{q-1}
\end{bmatrix}
=
\begin{bmatrix}
0 \\
\vdots \\
0 \\
y_0 \\
\vdots \\
y_{r-2} \\
y_{r-1}
\end{bmatrix}
$$

where

(3.2.9)
$$
A(z) = \sum_{j=-q}^{\infty} (z - z_0)^j A_j, \qquad q \geq 0.
$$

The number r is the *length* of the chain and the leading vector y_0 in such a chain is called a *right pole vector* for $A(z)$ at z_0. Note that it is easy to see directly from this criterion that *there exists a pole-vector of $A(z)$ at z_0 if and only if z_0 is a pole of $A(z)$*, i.e. in the representation (3.2.9) $q \geq 1$ and $A_{-q} \neq 0$. Another formulation is that $(y_0, \ldots, y_{r-1})$ with $y_0 \neq 0$ is a *right pole chain* for $A(z)$ at z_0 if this chain is a right null chain of $A(z)^{-1}$ at z_0, that is, if there exists additional vectors $y_r, \ldots, y_{r+q-1}$ (q being the order of z_0 as a pole of $A(z)^{-1}$) such that $A(z)^{-1}y(z)$ is analytic at z_0 with a zero of order r at z_0, where $y(z) = \sum_{j=0}^{r+q-1}(z - z_0)^j y_j$. As for the null chains, we introduce the canonical sets of pole chains and the *right pole pair* (Z, T) and the *left pole pair* (S, Y) of $A(z)$ at z_0. So (Z, T) (resp. (S, Y)) is a right (resp. left) pole pair of $A(z)$ at z_0 if and only if (Z, T) (resp. (S, Y)) is a right (resp. left) null pair of $A(z)^{-1}$ at z_0.

3.3 NULL AND POLE TRIPLES AND SINGULAR PARTS

Let $A(z)$ be a regular meromorphic $n \times n$ matrix function, and let z_0 be a pole of $A(z)$. A triple of matrices (Z, T, Y) is called a *pole triple* of $A(z)$ at z_0 if (Z, T) is a right pole pair of $A(z)$ at z_0, (T, Y) is a left pole pair of $A(z)$ at z_0, and

$$
A(z) - Z(zI - T)^{-1}Y
$$

is analytic at z_0. In particular, (Z, T) is a null kernel pair and (T, Y) is a full range pair. Thus, $Z(zI - T)^{-1}Y$ represents the singular part of the Laurent expansion of $A(z)$ in a deleted neighborhood of z_0. The definition of a null triple is analogous: (Z, T, Y) is a *null triple* for $A(z)$ at its zero z_0 if (Z, T) is a right null pair of $A(z)$ at z_0, (T, Y) is a left null pair of $A(z)$ at z_0, and

$$
A(z)^{-1} - Z(zI - T)^{-1}Y
$$

is analytic at z_0.

THEOREM 3.3.1. *A null triple (Z, T, Y) of $A(z)$ at z_0 exists and is unique up to similarity: if (Z_j, T_j, Y_j), $j = 1, 2$ are two null triples of $A(z)$ at z_0, then there* *exists a nonsingular matrix S such that $Z_2 = Z_1 S$, $T_2 = S^{-1}T_1 S$, $Y_2 = S^{-1}Y_1$. An analogous statement holds for pole triples.*

PROOF. For the proof of existence consider first the case when $A(z)$ is in the local Smith-McMillan form:

$$R(z) = \text{diag}\big((z - z_0)^{k_1}, \ldots, (z - z_0)^{k_n}\big),$$

where $k_1 \geq \cdots \geq k_p > 0 \geq k_{p+1} \geq \cdots \geq k_n$. In this case the existence of a null triple is easily seen. For instance, assuming for notational simplicity that $p = 1$, one such null triple is given by

$$
Z = [e_1 0 \cdots 0]; \quad T =
\begin{bmatrix}
z_0 & 1 & 0 & \cdots & 0 \\
0 & z_0 & 1 & \cdots & 0 \\
\vdots & \vdots & & \ddots & \vdots \\
& & & & 1 \\
0 & 0 & & \cdots & z_0
\end{bmatrix}; \quad
Y =
\begin{bmatrix}
0 \\
\vdots \\
0 \\
e_1^T
\end{bmatrix},
$$

where the size of T is $k_1 \times k_1$.

The general case is reduced to the local Smith-McMillan form, as follows. Assume (Z, T, Y) is a null triple for $A(z)$ at z_0, and let $F(z)$ be an $n \times n$ matrix function analytic and invertible at z_0. We shall construct a null triple for $\widetilde{A}(z) := A(z)F^{-1}(z)$ at z_0 as follows. As (Z, T) is a right null pair for $A(z)$ at z_0, without loss of generality we can assume that

$$Z = [Z_1, \ldots, Z_r], \qquad Z_j = [x_0^{(j)} \cdots x_{k_j-1}^{(j)}],$$

$$T = \text{diag}(T_1, T_2, \ldots, T_r),$$

where the columns of Z_j form a right null chain of $A(z)$ at z_0 and T_j is the corresponding Jordan block with eigenvalue z_0. By Example 3.1.6 the pair $(\widetilde{Z} = [\widetilde{Z}_1 \cdots \widetilde{Z}_r], T)$ is a right null pair for $\widetilde{A}(z) = A(z)F^{-1}(z)$ corresponding to z_0 where

$$\widetilde{Z}_j = [y_0^{(j)} \cdots y_{k_j-1}^{(j)}]$$

and

(3.3.1)
$$y_q^{(j)} = \sum_{p=0}^{q} \frac{1}{p!} F^{[p]}(z_0) x_{q-p}^{(j)}.$$

We verify now that the function

$$G(z) := \big(F(z)Z - \widetilde{Z}\big)(zI - T)^{-1}$$

is analytic at z_0. We have

$$G(z) = \sum_{j=1}^{r} (F(z)Z_j - \tilde{Z}_j)(zI - T_j)^{-1}$$

$$= \sum_{j=1}^{r} \{F(z)[x_0^{(j)} \cdots x_{k_j-1}^{(j)}] - [y_0^{(j)} \cdots y_{k_j-1}^{(j)}]\}$$

$$\times \begin{bmatrix} (z-z_0)^{-1} & (z-z_0)^{-2} & \cdots & (z-z_0)^{-k_j} \\ 0 & (z-z_0)^{-1} & & \\ \vdots & \vdots & & \vdots \\ 0 & 0 & \cdots & (z-z_0)^{-1} \end{bmatrix}.$$

For $p = 1, 2, \ldots$, the coefficient of $(z - z_0)^{-p}$ in the right-hand side is

$$\sum_{j=1}^{r} \left\{ F(z_0)[0 \cdots 0 x_0^{(j)} \cdots x_{k_j-p}^{(j)}] + F'(z_0)[0 \cdots 0 x_0^{(j)} \cdots x_{k_j-p-1}^{(j)}] + \cdots \right.$$

$$\left. + \frac{1}{(k_j - p)!} F^{[k_j-p]}(z_0)[0 \cdots 0 x_0^{(j)}] - [0 \cdots 0 y_0^{(j)} \cdots y_{k_j-p}^{(j)}] \right\}$$

which is zero in view of (3.3.1). As

$$F(z)A(z)^{-1} - \tilde{Z}(zI - T)^{-1}Y$$
$$= F(z)\big(A(z)^{-1} - Z(zI - T)^{-1}Y\big) + (F(z)Z - \tilde{Z})(zI - T)^{-1}Y,$$

it follows that $(\tilde{Z}, T, Y)$ is a null triple of $\tilde{A}(z)$ at z_0.

Analogously one verifies that the existence of a null triple for $A(z)$ at z_0 implies existence of a null triple for $E(z)A(z)$ at z_0, where $E(z)$ is a matrix function analytic and invertible at z_0. It remains to use Theorem 3.1.1 and the existence of a null triple for a local Smith-McMillan form verified above.

The uniqueness of a null triple is proved in the same way as in the proof of Proposition 1.4.1. We omit the details. $\square$

The proof of Theorem 3.3.1 reveals that null and pole triples can be characterized without reference to null and pole pairs, as follows.

COROLLARY 3.3.2. *A triple of matrices (Z, T, Y) where Z is $n \times p$, T is $p \times p$ and Y is $p \times n$, is a null triple for $A(z)$ at z_0 if and only if the following conditions are satisfied:*

(i) $\sigma(T) = \{z_0\}$;

(ii) $A(z)^{-1} - Z(zI - T)^{-1}Y$ *is analytic at z_0;*

(iii) (Z, T) *and (T, Y) are null kernel and full range pairs, respectively.*

In particular, the conditions (i)–(iii) imply that (Z, T) is a right null pair of $A(z)$ at z_0, and (T, Y) is a left null pair of $A(z)$ at z_0.

An analogous statement holds for pole triples of $A(z)$ at z_0 as well.

COROLLARY 3.3.3. (a) *If (Z, T) is a right null pair for $A(z)$ at z_0, then there exists a unique $\widetilde{Y}$ such that $(T, \widetilde{Y})$ is a full range pair and $(Z, T, \widetilde{Y})$ is a null triple for $A(z)$ at z_0.*

(b) *If (T, Y) is a left null pair for $A(z)$ at z_0, then there exists unique $\widetilde{Z}$ such that $(\widetilde{Z}, T)$ is a null kernel pair and $(\widetilde{Z}, T, Y)$ is a null triple for $A(z)$ at z_0.*

PROOF. We prove part (a) only. Let (Z', T', Y') be any null triple for $A(z)$ at z_0. In particular, (Z', T') is a right null pair for $A(z)$ at z_0. By the uniqueness property of right null pairs, there exists an invertible S such that $Z = Z'S$, $T = S^{-1}T'S$. Now take $\widetilde{Y} = S^{-1}Y'$ to satisfy the requirements of the corollary.

For the uniqueness of $\widetilde{Y}$, assume that $(Z, T, \widetilde{Y}_i)$ for $i = 1, 2$ are null triples for $A(z)$ at z_0, where $\widetilde{Y}_1$ and $\widetilde{Y}_2$ are matrices with full range pairs $(T, \widetilde{Y}_1)$ and $(T, \widetilde{Y}_2)$. By Theorem 3.3.1, there is an invertible matrix S such that

$$Z = ZS; \quad T = S^{-1}TS; \quad \widetilde{Y}_1 = S^{-1}\widetilde{Y}_2.$$

Then

$$\mathrm{col}(ZT^i)^q_{i=0} = \mathrm{col}(ZT^i)^q_{i=0} \cdot S, \qquad q = 0, 1, \ldots$$

and since (Z, T) is a null kernel pair, we obtain $S = I$.   □

An analogous corollary holds, of course, for the pole triples.

Next, we use pole triples to describe the null functions of $A(z)^{-1}$ in terms of Jordan chains of the matrix T taken from a pole pair (T, Y).

THEOREM 3.3.4. *Let z_0 be a pole of $A(z)$, and let (T, Y) be a left pole pair for $A(z)$ at z_0. Let $\psi(z)$ be a right null function of $A(z)^{-1}$ at z_0 of order k, and let φ_j be the coefficients of $\varphi(z) := A(z)^{-1}\psi(z)$:*

$$(3.3.2) \qquad \varphi(z) = \sum_{j=k}^{\infty} (z - z_0)^j \varphi_j.$$

Then

$$(3.3.3) \qquad x_j = \sum_{\nu=k}^{\infty} (T - z_0 I)^{\nu-j-1} Y \varphi_\nu, \qquad j = 0, \ldots, k-1$$

is a Jordan chain for T at z_0. Conversely, if z_0 is an eigenvalue of T and $x_0, \ldots, x_{k-1}$ is a Jordan chain of T at z_0, then there is a right null function $\psi(z)$ for $A(z)^{-1}$ at z_0 with order not less than k for which (3.3.3) holds (in particular, z_0 is a pole of $A(z)$).

Note that as $\sigma(T) = \{z_0\}$ the series in (3.3.3) is actually finite.

PROOF. Observe that in view of Theorem 3.3.1 there exists a matrix Z such that (Z, T, Y) is a pole triple of $A(z)$ at z_0. By definition, vectors (3.3.3) form a Jordan chain for T at z_0 if

$$\begin{cases} (T - z_0 I)x_0 = 0, & x_0 \neq 0 \\ (T - z_0 I)x_j = x_{j-1}, & j = 1, 2, \ldots, k-1. \end{cases}$$

The last $k-1$ statements follow immediately from (3.3.3). Also

$$(T - z_0 I)x_0 = \sum_{\nu=k}^{\infty} (T - z_0 I)^{\nu} Y \varphi_{\nu}.$$

Now the Laurent series for $A(z)$ at z_0, say $A(z) = \sum_{j=-q}^{\infty} (z - z_0)^j A_j$, has the following coefficients of negative powers of $(z - z_0)$:

(3.3.4) $\qquad\qquad A_{-j} = Z(T - z_0 I)^{j-1} Y, \qquad j = 1, 2, \ldots, q,$

and it is easily seen that q is the least positive integer for which $(T - z_0 I)^q = 0$ (one checks this by passing to the Jordan form of T). Now recall that $\psi(z) = A(z)\varphi(z)$ is analytic near z_0; so equating coefficients of negative powers of $(z - z_0)$ to zero and using the fact that $(T - z_0 I)^q = 0$, we obtain for $j = 1, 2, \ldots$:

$$\begin{aligned}
0 = \sum_{\nu=k}^{q-j} A_{-\nu-j}\varphi_{\nu} &= \sum_{\nu=k}^{q-j} Z(T - z_0 I)^{\nu+j-1} Y \varphi_{\nu} \\
&= \sum_{\nu=k}^{\infty} Z(T - z_0 I)^{\nu+j-1} Y \varphi_{\nu} \\
&= Z(T - z_0 I)^{j-1}(T - z_0 I)x_0.
\end{aligned}$$

Since (Z, T) is a null kernel pair, the matrix $\mathrm{col}[ZT^i]_{i=0}^{r-1}$ is left invertible for some integer r. As

$$\begin{bmatrix} Z \\ Z(T - z_0 I) \\ \vdots \\ Z(T - z_0 I)^{r-1} \end{bmatrix} = \begin{bmatrix} I & & & \cdots & 0 \\ -z_0 & I & & \cdots & \\ \vdots & \vdots & & & \cdots \\ (-z_0)^{r-1} & \binom{r-1}{1}(-z_0)^{r-2} & \cdots & & I \end{bmatrix} \begin{bmatrix} Z \\ ZT \\ \vdots \\ ZT^{r-1} \end{bmatrix},$$

the matrix $\mathrm{col}\big(Z(T - z_0 I)^i\big)_{i=0}^{r-1}$ is left invertible as well, and since $(T - z_0 I)^s = 0$ for $s \geq q$, we obtain the left invertibility of $\mathrm{col}\big(Z(T - z_0 I)^{j-1}\big)_{j=1}^q$. It now follows that $(T - z_0 I)x_0 = 0$ as required. Finally, since $\psi(z_0) = Zx_0$ it is also true that $x_0 \neq 0$. Thus, as asserted, equations (3.3.3) do associate a Jordan chain for T with the null function $\psi(z)$.

Conversely, let $x_0, x_1, \ldots, x_{k-1}$ be a Jordan chain of T at z_0. Since (T, Y) is a full range pair, the matrix

$$[Y, (T - z_0 I)Y, \ldots, (T - z_0 I)^{m-1} Y]$$

is right invertible for some integer m. Consequently, there exist vectors $\varphi_k, \varphi_{k+1}, \ldots$, with only finitely many nonzero, such that

(3.3.5) $\qquad\qquad x_{k-1} = \sum_{j=k}^{\infty} (T - z_0 I)^{j-k} Y \varphi_j.$

The definition of a Jordan chain includes $(T - z_0 I)x_j = x_{j-1}$ for $j = 1, 2, \ldots, k - 1$ and so equations (3.3.3) follows immediately from (3.3.5). It only remains to check that $A(z)\varphi(z)$ is now a null function of $A(z)^{-1}$ at z_0, where $\varphi(z) = \sum_{j=k}^{\infty}(z - z_0)^j \varphi_j$.

Observe that $x_0 \neq 0$ and that $ZT^j x_0 = z_0^j Z x_0$ for $j = 0, 1, 2, \ldots$. As (Z, T) is a null kernel pair, it follows that $Z x_0 \neq 0$. But using (3.3.4):

$$0 \neq Z x_0 = \sum_{\nu=k}^{\infty} Z(T - z_0 I)^{\nu-1} Y \varphi_\nu = \lim_{z \to z_0} A(z)\varphi(z). \quad \square$$

If the Jordan chain $x_0, \ldots, x_{k-1}$ of T at z_0 cannot be prolonged, then $x_{k-1} \notin \text{Im}(T - z_0 I)$, and it follows from (3.3.5) that $\varphi_k \neq 0$. Thus, a maximal Jordan chain of length k determines, by means of (3.3.3), an *associated* null function $\psi(z)$ of $A(z)^{-1}$ of order k.

A statement analogous to Theorem 3.3.4 concerning pole functions holds as well.

We indicate the following fact which is essentially a corollary from Theorem 3.2.2 and the dual theorem.

COROLLARY 3.3.5. *Let $A(z)$ be a meromorphic $n \times n$ matrix function with $\det A(z) \not\equiv 0$, and let (Z_π, T_π, Y_π) and $(Z_\zeta, T_\zeta, Y_\zeta)$ be a pole triple and a zero triple of $A(z)$, respectively. Then the absolute values of negative partial multiplicities of $A(z)$ at z_0 coincide with the sizes of Jordan blocks with eigenvalue z_0 in the Jordan form of T_π, i.e. with the partial multiplicities of z_0 as an eigenvalue of T_π. The positive partial multiplicities of $A(z)$ at z_0 coincide with the partial multiplicities of z_0 as an eigenvalue of T_ζ.*

We now consider the construction of a null (or pole) triple when right and left null (or pole) pairs are given. This construction uses the singular part of the Laurent expansion of $A(z)^{-1}$ (or $A(z)$) in a deleted neighborhood of z_0, and extends the result of Theorem 1.4.3 to the meromorphic case.

THEOREM 3.3.6. *Let $A(z)$ be a meromorphic matrix function in a neighborhood of z_0, with the singular part of the Laurent series*

$$SP(A(z_0)) = \sum_{j=1}^{\nu}(z - z_0)^{-j} K_j.$$

Let (Z, T) and (T, Y) be right and left pole pairs, respectively, of $A(z)$ at z_0. Define

$$G = \begin{bmatrix} Z \\ Z(T - z_0 I) \\ \vdots \\ Z(T - z_0 I)^{\nu-1} \end{bmatrix}^{-1} \begin{bmatrix} K_1 & K_2 & \cdots & K_{\nu-1} & K_\nu \\ K_2 & K_3 & \cdots & K_\nu & \\ \vdots & \vdots & & \vdots & \\ K_{\nu-1} & K_\nu & & & \\ K_\nu & & \cdots & & 0 \end{bmatrix}$$

$$\times [Y, (T - z_0 I)Y, \ldots, (T - z_0 I)^{\nu-1} Y]^{-1},$$

where the inverse matrices are understood as some one-sided inverses from the appropriate side. (Note that in view of Theorem 2.5.4 the matrix $\mathrm{col}(Z(T - z_0 I)^i)_{i=0}^{\nu-1}$ *is left invertible, while*

$$[Y, (T - z_0 I)Y, \ldots, (T - z_0 I)^{\nu-1}Y]$$

is right invertible.) Then G *is invertible,* $GT = TG$, *and* (ZG, T, Y), *as well as* (Z, T, GY), *is a pole triple for* $A(z)$ *at* z_0. *Moreover,* G *is the unique* $p \times p$ *matrix* (p *is the size of* T) *which commutes with* T *and for which* (ZG, T, Y) *is a pole triple for* $A(z)$ *at* z_0.

PROOF. Let

$$A(z) = E(z) \, \mathrm{diag}\big((z - z_0)^{u_1}, \ldots, (z - z_0)^{u_n}\big) F(z)$$

be the local Smith-McMillan form of $A(z)$, and consider the matrix function $\widetilde{A}(z)$ whose inverse is

$$\widetilde{A}(z)^{-1} = E(z) \, \mathrm{diag}\big((z - z_0)^{\mu_1}, \ldots, (z - z_0)^{\mu_n}\big) F(z),$$

where $\mu_i = \min(0, u_i)$. Clearly, $\widetilde{A}(z)$ is analytic at z_0, and, moreover, right and left pole pairs for $A(z)$ are just right and left null pairs for $\widetilde{A}(z)$ (see Theorem 3.1.2). Also,

$$SP\big(A(z_0)\big) = SP\big(\widetilde{A}^{-1}(z_0)\big).$$

It remains now to apply Theorem 1.4.3 for $\widetilde{A}(z)$. □

Theorem 1.4.4 extends to the meromorphic case as well:

THEOREM 3.3.7. *Let* (Z, T) *and* (T, Y) *be a null kernel and a full range pair, respectively, with* $\sigma(T) = \{z_0\}$. *Then all possible singular parts at* z_0 *of meromorphic matrix functions* $A(z)$ *having* (Z, T) *and* (T, Y) *as right and left pole pairs corresponding to* z_0 *are given by the formula*

(3.3.6) $$SP\big(A(z_0)\big) = ZG(zI - T)^{-1}Y = Z(zI - T)^{-1}GY,$$

where G *is any invertible matrix commuting with* T. *The correspondence, given by* (3.3.6), *between the sets of all singular parts* $SP\big(A(z_0)\big)$ *and the set of all invertible matrices commuting with* T, *is one-to-one.*

The proof of Theorem 3.3.7 is done again by reduction to the analytic case, as in the proof of Theorem 3.3.6.

Applying Theorems 3.3.6 and 3.3.7 to $A(z)^{-1}$ instead of $A(z)$, we can obtain analogous results concerning null pairs and null triples.

3.4 LOCAL CHARACTERISTICS RELATIVE TO SEVERAL POINTS

Just as it was done in Section 1.5, one can extend many notions and results describing local properties of a meromorphic matrix function at one point to the setting of several (but finite number of) points.

Let $A(z)$ be regular $n \times n$ matrix function meromorphic in a domain $\Omega \subset \mathbf{C}$. Fix a compact set $\sigma \subset \Omega$, and let $z_1, \ldots, z_k$ be all the poles of $A(z)$ in σ. A pair of matrices (Z, T) which is similar to

$$([Z_1, Z_2, \ldots, Z_k], \quad \operatorname{diag}(T_1, \ldots, T_k))$$

where (Z_j, T_j) is a right pole pair of $A(z)$ at z_j, is called a *right pole pair* of $A(z)$ relative to σ. Analogously, left pole pairs and left and right null pairs of $A(z)$ relative to σ are defined. As in the proof of Theorem 1.5.1 one verifies that these pairs are unique up to similarity and are null kernel pairs or full range pairs, as appropriate. A triple of matrices (Z, T, Y) is called a *pole triple* of $A(z)$ relative to σ if the following conditions are satisfied:

(i) (Z, T) is a null kernel pair;

(ii) (T, Y) is a full range pair;

(iii) $\sigma(T) \subset \sigma$;

(iv) the matrix function $A(z) - Z(zI - T)^{-1}Y$ admits analytic continuation to the whole σ.

An equivalent definition of a pole triple of $A(z)$ relative to σ can be given by replacing (i) and (ii) with (v) and (vi) below:

(v) (Z, T) is a right pole pair of $A(z)$ relative to σ;

(vi) (T, Y) is a left pole pair of $A(z)$ relative to σ.

Conversely, conditions (i)–(iv) (or equivalently, (v), (vi), (iii), (iv)) characterize a pole triple relative to σ. This is the extension of Corollary 3.3.2 to several points. We record the result formally for future reference.

THEOREM 3.4.1. *A triple of matrices* (Z, T, Y), *where* Z *is* $n \times p$, T *is* $p \times p$ *and* Y *is* $p \times n$, *is a pole triple for* $A(z)$ *relative to* $\sigma \subset \mathbf{C}$ *if and only if*

(i) $\sigma(T) \subset \sigma$;

(ii) $A(z) - Z(zI - T)^{-1}Y$ *has analytic continuation to all of* σ;

(iii) (Z, T) *and* (T, Y) *are null kernel and full range pairs respectively.*

In particular, the conditions (i)–(iii) imply that (Z, T) is a right pole pair of $A(z)$ with respect to σ and that (T, Y) is a left pole pair of $A(z)$ with respect to σ.

PROOF. The necessity of conditions (i), (ii) and (iii) is clear from the definitions and the results of Section 3.3. Conversely, suppose (i), (ii) and (iii) are satisfied. We may write T as a direct sum $\operatorname{diag}(T_1, \ldots, T_k)$ where each T_j has spectrum equal to a single point z_j $(1 \leq j \leq k)$. This induces a decomposition of Z and Y as

$$Z = [Z_1 \cdots Z_k]$$

and

$$Y = \operatorname{col}(Y_1, \ldots, Y_k).$$

As in the proof of Theorem 1.5.1, we see that each (Z_j, T_j) is a null-kernel pair since (Z, T) is a null-kernel pair, and that each (T_j, Y_j) is a full range pair since (T, Y) is a full range pair. Note also that

$$Z(zI - T)^{-1}Y = \sum_{j=1}^{k} Z_j(zI - T_j)^{-1}Y_j.$$

We next verify that $A(z) - Z_j(zI - T_j)^{-1}Y_j$ is analytic at z_j for each j. Indeed, note that

$$A(z) - Z_j(zI - T_j)^{-1}Y_j = [A(z) - Z(zI - T)^{-1}Y] + \sum_{i \neq j} Z_i(zI - T_i)^{-1}Y_i.$$

By hypothesis the first term has analytic continuation to all of σ (in particular to z_j), while the second term is analytic at z_j since $\sigma(T_i) = \{z_i\}$ where $z_i \neq z_j$ for $i \neq j$. Then by Corollary 3.3.2, we conclude that (Z_j, T_j, Y_j) is a null triple for $A(z)$ at z_j. Since $A(z) - Z(zI - T)^{-1}Y$ is analytic on σ and $\sigma(T) = \{z_1, \ldots, z_k\}$, we see that the only poles of $A(z)$ in σ are at $z_1, \ldots, z_k$. We conclude from the definitions that (Z, T, Y) is a null-pole triple for $A(z)$ with respect to σ. $\square$

There is also an analogue of Corollary 3.3.3 for several points.

THEOREM 3.4.2. (a) *If (Z, T) is a right pole pair for the regular meromorphic matrix function $A(z)$ with respect to the region $\sigma \subset \mathbf{C}$, then there exists a unique $\widetilde{Y}$ such that $(T, \widetilde{Y})$ is a full range pair and $(Z, T, \widetilde{Y})$ is a pole triple for $A(z)$ with respect to σ.*

(b) *If (T, Y) is a left pole pair for $A(z)$ with respect to σ, then there exists a unique $\widetilde{Z}$ such that $(\widetilde{Z}, T)$ is a null kernel pair and $(\widetilde{Z}, T, Y)$ is a pole triple for $A(z)$ with respect to σ.*

PROOF. We consider part (a) only. If (Z, T) is a right pole pair for $A(z)$ with respect to σ, by using a similarity transformation we may assume $Z = [Z_1, \ldots, Z_k]$, and $T = \mathrm{diag}(T_1, \ldots, T_k)$ where $\sigma(T_j) = \{z_j\}$ and $z_i \neq z_j$ for $i \neq j$. Then (Z_j, T_j) is a right pole pair of $A(z)$ at z_j for $1 \leq j \leq k$. So by Corollary 3.3.3 for each j there exists a unique matrix $\widetilde{Y}_j$ such that $(T_j, \widetilde{Y}_j)$ is a full range pair and $(Z_j, T_j, \widetilde{Y}_j)$ is a pole triple for $A(z)$ at z_j. Set $\widetilde{Y} = \mathrm{col}(Y_1, \ldots, Y_k)$. Then by definition, since $z_1, \ldots, z_k$ are the only poles of $A(z)$ relative to σ, we conclude that $(Z, T, \widetilde{Y})$ is a pole triple for $A(z)$ with respect to σ. The full range property of $(T, \widetilde{Y})$ follows from that of each $(T_j, \widetilde{Y}_j)$ by an argument as in the proof of Theorem 1.5.1. If $\widehat{Y} = \mathrm{col}(\widehat{Y}_1, \ldots, \widehat{Y}_k)$ were another such matrix, then $(Z_j, T_j, \widehat{Y}_j)$ would also be a pole triple for $A(z)$ at z_j, so $\widehat{Y}_j = \widetilde{Y}_j$ for each j by uniqueness of $\widehat{Y}_j$ and hence $\widehat{Y} = \widetilde{Y}$. $\square$

Null triples of $A(z)$ relative to σ, can be defined analogously; note that null triples of $A(z)$ are precisely pole triples of $A(z)^{-1}$. Theorems 3.4.1 and 3.4.2 therefore have obvious analogue for null triples, the statements of which we leave to the reader.

Finally, we state the analogue of the local Smith-McMillan form for several points.

THEOREM 3.4.3. *Let $A(z)$ be a (regular) meromorphic $n \times n$ matrix function in a domain Ω, and let $\sigma \subset \Omega$ be compact. Then $A(z)$ admits the representation*

$$A(z) = E(z) \operatorname{diag}(d_1(z), d_2(z), \dots, d_n(z)) F(z),$$

where $E(z)$ and $F(z)$ are matrix functions analytic and invertible in σ, and $d_1(z), \dots, d_n(z)$ are rational functions with all poles and zeros in σ and such that $d_i(z)(d_{i+1}(z))^{-1}$ is a polynomial for $i = 1, \dots, n-1$. Moreover, one of the factors $E(z)$ and $F(z)$ can be chosen to be a unimodular matrix polynomial.

PROOF. We reduce to the corresponding statement for analytic matrix functions (Theorem 1.5.2).

Let $p(z)$ be a suitable scalar analytic (in Ω) function such that the matrix function $\widetilde{A}(z) := p(z) A(z)$ is also analytic in Ω (for instance, one can take $p(z)$ to be the product of denominators of the entries of $A(z)$). An application of Theorem 1.5.2 yields

$$\widetilde{A}(z) = \widetilde{E}(z) \operatorname{diag}(\varphi_1(z), \varphi_2(z), \dots, \varphi_n(z)) \widetilde{F}(z),$$

where $\widetilde{E}(z)$ and $\widetilde{F}(z)$ are analytic matrix functions (in Ω) which are invertible in σ, and $\varphi_1(z), \dots, \varphi_n(z)$ are scalar polynomials with all zeros in σ and such that $\varphi_i(z)$ is divisible by $\varphi_{i+1}(z)$, $i = 1, \dots, n-1$. Moreover, one of $\widetilde{E}(z)$ and $\widetilde{F}(z)$, say $\widetilde{F}(z)$, is a unimodular matrix polynomial. Factor $p(z) = p_1(z) p_2(z)$, where $p_2(z)$ is a polynomial with zeros in σ and $p_1(z)$ is analytic function without zeros in σ. Now put

$$F(z) = \widetilde{F}(z); \qquad d_i(z) = \varphi_i(z)(p_2(z))^{-1}$$

and

$$E(z) = (p_1(z))^{-1} \widetilde{E}(z)$$

to satisfy the requirements of Theorem 3.4.1. $\square$

3.5 LOCAL DATA AT INFINITY AND MÖBIUS TRANSFORMATIONS

In this section we consider matrix valued functions that are meromorphic at infinity and their local data.

Let $A(z)$ be $n \times n$ matrix function which is meromorphic at infinity and regular. For such function the basic notions of right and left null and pole functions at infinity are defined analogously to the definitions for the case of a point in the (finite) complex plane. For example, a $\mathbf{C}^n$-valued function $\varphi(z)$ which is analytic at infinity (i.e. $\varphi(z)$ admits a power series expansion

$$\varphi(z) = \sum_{j=0}^{-\infty} z^j \varphi_j$$

which converges in a neighborhood of infinity) is called a *right null function of $A(z)$ at infinity* of order $k > 0$ if $\varphi(\infty) \neq 0$ and if $A(z)\varphi(z)$ is analytic at ∞ with a zero of order

k at infinity. Consequently, we can define the notions of right and left null and pole pairs and null and pole triples for $A(z)$ at infinity. Another way to define these notions is simply by observing that (Z, T) is a right null pair for $A(z)$ at infinity if and only if (Z, T) is a right null pair for the meromorphic function $\widetilde{A}(z) = A(z^{-1})$ at zero, and similar statements apply to other pairs and triples associated with $A(z)$ at infinity.

We describe the behaviour of local spectral data of rational matrix functions under Möbius transformations.

Consider the Möbius transformation

(3.5.1)
$$\varphi(z) = \frac{pz + q}{rz + s},$$

where $ps - qr \neq 0$. This transformation will be considered as a map $\mathbb{C} \cup \{\infty\} \to \mathbb{C} \cup \{\infty\}$. The inverse map φ^{-1} is given by the formula

(3.5.2)
$$\varphi^{-1}(z) = \frac{sz - q}{-rz + p}.$$

We consider the case when the local data do not involve the data at infinity.

In the following theorem $A(z)$ is a (regular) meromorphic $n \times n$ matrix function in some domain $\Omega \subset \mathbb{C}$, and σ is a compact set in Ω.

THEOREM 3.5.1. *Let* (Z, T, Y) *be a pole triple for* $A(z)$ *relative to* σ, *and let* φ *be the Möbius transformation given by* (3.5.1). *Assume that* $rz - p \neq 0$ *for all* $z \in \sigma$. *Put*

$$\widetilde{A}(z) = A(\varphi(z)), \qquad \widetilde{\sigma} = \{z \mid \varphi(z) \in \sigma\}.$$

$$E = pI - rT.$$

Then

(3.5.3)
$$((ps - qr)ZE^{-1}, \varphi^{-1}(T), E^{-1}Y)$$

is a pole triple for $\widetilde{A}$ *relative to* $\widetilde{\sigma}$.

PROOF. Note that the inverse map φ^{-1} is analytic on σ. Thus, by the Riesz calculus (see Notations and Conventions), the matrix $\varphi^{-1}(T)$ is well-defined.

By the definition of a pole triple, for some matrix function $Q(z)$ analytic in $\widetilde{\sigma}$ we have

$$\widetilde{A}(z) = Z\left(\frac{zp + q}{zr + s}I - T\right)^{-1}Y + Q(z).$$

Now

$$Z\left(\frac{zp + q}{rz + s}I - T\right)^{-1}Y = (rz + s)Z[z(pI - rT) + q - sT]^{-1}Y$$

$$= (rz + s)Z[zI - \varphi^{-1}(T)]^{-1}E^{-1}Y$$

$$= Z\{r(zI - \varphi^{-1}(T)) + r\varphi^{-1}(T) + sI\}[zI - \varphi^{-1}(T)]^{-1}E^{-1}Y$$

$$= rZE^{-1}Y + (ps - qr)ZE^{-1}[zI - \varphi^{-1}(T)]^{-1}E^{-1}Y,$$

where we have used the equality $r\varphi^{-1}(T) + sI = (ps - qr)(pI - rT)^{-1}$. This proves Theorem 3.5.1. $\square$

Using the fact that for a pole triple (Z, T, Y) for $A(z)$ relative to z_0 the pairs (Z, T) and (T, Y) are right and left pole pairs for $A(z)$ relative to z_0, Theorem 3.5.1 provides also the formulas for transformation of the right and left pole pairs under φ.

Analogously one can obtain formulas for transformation of null triples under Möbius transformations. Namely, if (Z, T, Y) is a null triple of $A(z)$ relative to σ, then

$$\left((ps - qr)ZE^{-1}, \varphi^{-1}(T), (p - rT)^{-1}Y\right)$$

is a null triple of $A(\varphi(z))$ relative to $\tilde{\sigma} = \{z \mid \varphi(z) \in \sigma\}$. Here $\varphi(z)$ is the Möbius transformation given by (3.5.1).

Finally we consider the situation where the point at infinity is involved in the Möbius transformation. So let σ be a compact set in the domain of definition of $A(z)$, and let φ be the Möbius transformation given by (3.5.1), but now we drop the assumption that $rz - p \neq 0$ for all $z \in \sigma$. So $rz_0 - p = 0$ for some $z_0 \in \sigma$, and $\varphi(\infty) = z_0$. The function $\tilde{A}(z) = A(\varphi(z))$ is meromorphic at infinity, and we have seen before that the local data of $\tilde{A}(z)$ at infinity are the same as the local data of $B(z) := \tilde{A}(z^{-1})$ at zero. For the zeros and poles of $A(z)$ in σ which are different from z_0 the formulas of Theorem 3.5.1 apply.

The formulas of this section show that the singularity at infinity of a rational matrix function can be taken care of by using a Möbius transformation and thereby reducing to a singularity in the finite complex plane.

CHAPTER 4

RATIONAL MATRIX FUNCTIONS

This chapter contains a specification in detailed form of the results of Chapter 3 for rational matrix functions, and the solution of the next interpolation problem, namely to build a rational matrix function with a given null and pole structure. Unlike the scalar case, it turns out that even when the solution exists and the value at infinity is specified, it may not be unique. A particular solution can be singled out if additional information is provided. This extra information consists of an invertible matrix, called the coupling matrix, which describes the geometry of the interaction between pole and null structures at each point. If a solution exists and the poles and zeros are disjoint, then, as in the scalar case, the solution is unique. The solution of the interpolation problem is obtained in a special form called a realization for the matrix function. The theory of realization for a rational matrix function is borrowed from systems theory; a short exposition of the necessary facts concerning realization is also presented in the chapter. In this chapter we consider only the case where the rational matrix function is analytic and invertible at infinity, and the interpolation data consist of a left null pair together with a right pole pair. This combination (as well as the reversed combination of right null and left pole pair) we call standard. (Nonstandard combinations will be considered in Chapter 8.) The solution of the interpolation problem considered in this chapter, if the value at infinity is not fixed, may be multiplied on the left by any invertible constant matrix to generate another solution of the same problem. Thus we see that solvable interpolation problems associated with standard local data are matricially homogeneous.

4.1 REALIZATIONS AND MINIMALITY

In this section we present the basic results and notions concerning realizations of $n \times n$ matrix functions with rational entries. As usual, it will be assumed throughout that the rational matrix functions involved are *regular*, i.e. with determinant not identically zero.

Consider the $n \times n$ rational matrix function $W(z)$. Assuming that $W(z)$ is finite at infinity, i.e. in each entry $\frac{p_{ij}(z)}{q_{ij}(z)}$ of $W(z)$ the degree of the polynomial $p_{ij}(z)$ is less than or equal to the degree of $q_{ij}(z)$, we define a *realization* of $W(z)$ to be a representation of the form

$$(4.1.1) \qquad W(z) = D + C(zI_m - A)^{-1}B, \qquad z \notin \sigma(A),$$

where A, B, C, D are matrices of sizes $m \times m$, $m \times n$, $n \times m$, $n \times n$, respectively. Observe that $\lim_{z \to \infty}(zI - A)^{-1} = 0$. (To verify this, assume that A is in the Jordan form, or use the Neumann series expansion.) So necessarily $D = W(\infty)$, and we may identify a realization (4.1.1) with the triple of matrices (A, B, C) (for a given $W(z)$). Rational matrix functions which are finite at infinity are called *proper*.

We start with a theorem of the existence of a realization.

THEOREM 4.1.1. *Every rational $n \times n$ matrix function $W(z)$ which is finite at infinity admits a realization.*

PROOF. There exists a monic scalar polynomial $L(z) = z^\ell + \sum_{j=0}^{\ell-1} z^j a_j$ such that $L(z)W(z)$ is a (matrix) polynomial. For instance, take $L(z)$ to be a common multiple of the denominators of the entries in $W(z)$. Let $\widetilde{L}(z) = (L(z))I$ be the associated monic $n \times n$ matrix polynomial. Then (see Theorem 2.3.2) $\widetilde{L}(z)^{-1}$ admits the resolvent form:

$$(4.1.2) \qquad \widetilde{L}(z)^{-1} = Q(zI - A)^{-1}B,$$

where $Q = [\, I \ \ 0 \ \ \cdots \ \ 0 \,]$,

$$A = \begin{bmatrix} 0 & I & 0 & \cdots & 0 \\ 0 & 0 & I & \cdots & 0 \\ \vdots & \vdots & \vdots & & \vdots \\ 0 & 0 & 0 & \cdots & I \\ -A_0 & -A_1 & -A_2 & \cdots & -A_{\ell-1} \end{bmatrix} \, ; \qquad B = \begin{bmatrix} 0 \\ \vdots \\ 0 \\ I \end{bmatrix}.$$

Define the $n \times n$ matrix functions $C_1(z), \ldots, C_\ell(z)$ for $z \notin \sigma(A)$ by

$$(4.1.3) \qquad \mathrm{col}\big(C_j(z)\big)_{j=1}^\ell = (zI - A)^{-1}B.$$

From (4.1.2) we see that $C_1(z) = \widetilde{L}(z)^{-1}$. As $(zI - A)\big[\, \mathrm{col}(C_j(z))_{j=1}^\ell \,\big] = B$, the special form of A yields

$$(4.1.4) \qquad C_i(z) = z^{i-1}C_1(z), \qquad 1 \le i \le \ell.$$

Now introduce the $n \times n$ matrix polynomial $H(z) = \big(W(z) - W(\infty)\big)\widetilde{L}(z)$. As

$$\lim_{z \to \infty} H(z)\widetilde{L}(z)^{-1} = \lim_{z \to \infty}[W(z) - W(\infty)] = 0,$$

the degree of $H(z)$ is strictly less than the degree of $\widetilde{L}(z)$. So we can write

$$H(z) = \sum_{j=0}^{\ell-1} z^j H_j,$$

and, if we denote $C = [\, H_0 \ \ H_1 \ \ \cdots \ \ H_{\ell-1} \,]$, the equalities (4.1.3) and (4.1.4) imply

$$C(zI - A)^{-1}B = \sum_{j=0}^{\ell-1} H_j C_{j+1}(z) = H(z)\widetilde{L}(z)^{-1}.$$

Consequently

$$W(z) = W(\infty) + C(zI - A)^{-1}B$$

is a realization of $W(z)$. □

A realization for $W(z)$ is far from being unique. One way to see this is from our construction; note that there are many choices for $L(z)$. More generally, if (A, B, C) is a realization of $W(z)$, then so is $(\widetilde{A}, \widetilde{B}, \widetilde{C})$, where

$$(4.1.5) \qquad \widetilde{A} = \begin{bmatrix} A_{11} & A_{12} & A_{13} \\ 0 & A & A_{23} \\ 0 & 0 & A_{33} \end{bmatrix}, \qquad \widetilde{B} = \begin{bmatrix} B_1 \\ B \\ 0 \end{bmatrix}, \qquad \widetilde{C} = [\, 0 \quad C \quad C_1 \,]$$

for any matrices A_{ij}, B_1 and C_1 with suitable sizes (in other words, the matrices $\widetilde{A}$, $\widetilde{B}$, $\widetilde{C}$ are of size $s \times s$, $s \times n$, $n \times s$, respectively, and partitioned relative to the orthogonal sum $\mathbf{C}^s = \mathbf{C}^p \oplus \mathbf{C}^m \oplus \mathbf{C}^q$, where m is the size of A; for instance, A_{13} is a $p \times q$ matrix). This statement is verified immediately by observing that

$$(zI - \widetilde{A})^{-1} = \begin{bmatrix} (zI - A_{11})^{-1} & * & * \\ 0 & (zI - A)^{-1} & * \\ 0 & 0 & (zI - A_{33})^{-1} \end{bmatrix}.$$

Among all the realizations of $W(z)$ those with the properties that (C, A) is a null kernel pair and (A, B) is a full range pair will be of special interest. The next result shows that any realization "contains" a realization with these properties. To make this precise it is convenient to introduce the notions of dilation and reduction, as follows. These notions express the basic relations between (A, B, C) and $(\widetilde{A}, \widetilde{B}, \widetilde{C})$ given by the formula (4.1.5).

Let (A, B, C) be a realization of $W(z)$, and let $m \times m$ be the size of A. Given a direct sum decomposition $\mathbf{C}^m = \mathcal{L} \dotplus \mathcal{M} \dotplus \mathcal{N}$ where the subspaces $\mathcal{L}$ and $\mathcal{L} \dotplus \mathcal{M}$ are A-invariant with the property that $C \mid_{\mathcal{L}} = 0$ and $\operatorname{Im} B \subset \mathcal{L} + \mathcal{M}$, a realization $(P_{\mathcal{M}} A \mid_{\mathcal{M}}, P_{\mathcal{M}} B, C \mid_{\mathcal{M}})$, where $P_{\mathcal{M}} : \mathbf{C}^m \to \mathcal{M}$ is a projector on $\mathcal{M}$ with $\operatorname{Ker} P_{\mathcal{M}} \supset \mathcal{L}$, is called a *reduction* of (A, B, C). Note that $(P_{\mathcal{M}} A \mid_{\mathcal{M}}, P_{\mathcal{M}} B, C \mid_{\mathcal{M}})$ is again a realization for the same $W(z)$ (this can be verified in the same way as the verification that (4.1.5) is a realization of $W(z)$ if (A, B, C) is). We also say that (A, B, C) is a *dilation* of $(P_{\mathcal{M}} A \mid_{\mathcal{M}}, P_{\mathcal{M}} B, C \mid_{\mathcal{M}})$.

THEOREM 4.1.2. *Any realization (A, B, C) of $W(z)$ is the dilation of a realization (A_0, B_0, C_0) of $W(z)$ with null kernel pair (C_0, A_0) and full range pair (A_0, B_0).*

PROOF. Introduce the subspaces

$$\mathcal{K}(C, A) = \bigcap_{j=0}^{\infty} \operatorname{Ker}(C A^j); \qquad \mathcal{I}(A, B) = \sum_{j=0}^{\infty} \operatorname{Im}(A^j B).$$

It is easy to see that both $\mathcal{K}(C, A)$ and $\mathcal{I}(A, B)$ are A-invariant. The inclusions

$$\mathcal{K}(C, A) \subset \operatorname{Ker} C; \qquad \mathcal{I}(A, B) \supset \operatorname{Im} B$$

are obvious.

Now put $\mathcal{L} = \mathcal{K}(C, A)$, let $\mathcal{M}$ be a direct complement of $\mathcal{L} \cap \mathcal{I}(A, B)$ in $\mathcal{I}(A, B)$, and let $\mathcal{N}$ be a direct complement to $\mathcal{L} + \mathcal{M}$ in $\mathbf{C}^m$, where m is the size of A. Thus,

$$\mathbf{C}^m = \mathcal{L} \dotplus \mathcal{M} \dotplus \mathcal{N}.$$

The subspaces $\mathcal{L}$ and $\mathcal{L} \dotplus \mathcal{M} = \mathcal{K}(C, A) + \mathcal{I}(A, B)$ are A-invariant. Observe also that $C\mid_{\mathcal{L}} = 0$ and $\operatorname{Im} B \subset \mathcal{I}(A, B) \subset \mathcal{L} \dotplus \mathcal{M}$. Thus, to finish the proof of Theorem 4.1.2, it remains to verify that the realization of $W(z)$ given by

$$(PA\mid_{\mathcal{M}}, PB, C\mid_{\mathcal{M}}),$$

where $P: \mathbf{C}^m \to \mathcal{M}$ is the projector on $\mathcal{M}$ along $\mathcal{L} \dotplus \mathcal{N}$, has the full range and null kernel properties. We have

$$\operatorname{Ker} C\mid_{\mathcal{M}} (PA\mid_{\mathcal{M}})^j = (\operatorname{Ker} CA^j) \cap \mathcal{M}, \qquad j = 0, 1, \dots .$$

So

$$\bigcap_{j=0}^{\infty} \operatorname{Ker} C\mid_{\mathcal{M}} (PA\mid_{\mathcal{M}})^j = \bigcap_{j=0}^{\infty} (\operatorname{Ker} CA^j) \cap \mathcal{M} = \mathcal{L} \cap \mathcal{M} = \{0\}.$$

Also

$$\operatorname{Im}(PA\mid_{\mathcal{M}})^j PB = P(\operatorname{Im} A^j B).$$

Hence

$$\sum_{j=0}^{\infty} \operatorname{Im}(PA\mid_{\mathcal{M}})^j PB = P\left(\sum_{j=0}^{\infty} \operatorname{Im} A^j B\right) = P(\mathcal{I}(A, B)) = \mathcal{M},$$

because by construction $\mathcal{M} \subset \mathcal{I}(A, B)$. □

It turns out that a realization (A, B, C) for which (C, A) is a null kernel pair and (A, B) is a full range pair is essentially unique:

THEOREM 4.1.3. *Let (A_1, B_1, C_1) and (A_2, B_2, C_2) be realizations for a rational matrix function $W(z)$ for which (C_1, A_1) and (C_2, A_2) are null kernel pairs and (A_1, B_1), (A_2, B_2) are full range pairs. Then the sizes of A_1 and A_2 coincide, and there exists an invertible matrix S such that*

(4.1.6) $A_1 = S^{-1} A_2 S, \quad B_1 = S^{-1} B_2, \quad C_1 = C_2 S.$

Moreover, the matrix S is unique and is given by

(4.1.7)
$$S = [\operatorname{col}(C_2 A_2^j)_{j=0}^{p-1}]^{-L} [\operatorname{col}(C_1 A_1^j)_{j=0}^{p-1}]$$
$$= [B_2, A_2 B_2, \dots, A_2^{p-1} B_2][B_1, A_1 B_1, \dots, A_1^{p-1} B_1]^{-R}.$$

Here p is any integer for which the matrix $\operatorname{col}(C_2 A_2^j)_{j=0}^{p-1}$ is left invertible and the matrix $[B_1, A_1 B_1, \dots, A_1^{p-1} B_1]$ is right invertible and the superscript $-L$ (resp. $-R$) indicates any left (resp. right) inverse.

PROOF. We have

$$(4.1.8) \qquad W(z) = W(\infty) + C_1(zI - A_1)^{-1}B_1 = W(\infty) + C_2(zI - A_2)^{-1}B_2$$

for $z \notin \sigma(A_1) \cup \sigma(A_2)$. By developing $(zI - A_1)^{-1}$ and $(zI - A_2)^{-1}$ into Laurent series in a neighborhood of infinity and comparing coefficients in (4.1.8), we conclude that

$$C_1 A_1^j B_1 = C_2 A_2^j B_2; \qquad j = 0, 1, \dots .$$

Now define S to be the right-hand side in (4.1.7) and argue as in the proof of Proposition 1.4.1 that the sizes of A_1 and A_2 coincide, S is invertible and the equalities (4.1.6) hold.

It remains to show that an invertible S with the properties (4.1.6) is unique. Let S_1 and S_2 be two such matrices S, and put $T = S_1 S_2^{-1}$. Then

$$A_2 = T^{-1} A_2 T; \quad C_2 = C_2 T; \quad B_2 = T^{-1} B_2.$$

We see that

$$\text{col}(C_2 A_2^j)_{j=0}^{q-1} T = \text{col}(C_2 A_2^j)_{j=0}^{q-1}, \qquad q = 1, 2, \dots,$$

and since for some q the matrix $\text{col}(C_2 A_2^j)_{j=0}^{q-1}$ is left invertible, we conclude that $T = I$. $\square$

Theorems 4.1.2 and 4.1.3 allow us to deduce the following important fact.

THEOREM 4.1.4. *In a realization (A, B, C) of $W(z)$, (C, A) and (A, B) are null kernel pairs and full range pairs, respectively, if and only if the size of A is minimal among all possible realizations of $W(z)$.*

PROOF. Assume that the size m of A is minimal. By Theorem 4.1.2 there is a reduction (A', B', C') of (A, B, C) which is a realization for $W(z)$ and for which (C, A) is a null kernel pair and (A, B) is a full range pair. But because of the minimality of m the realizations (A', B', C') and (A, B, C) must coincide, and this implies that (C, A) and (A, B) are also null kernel and full range pairs, respectively.

Conversely, assume (C, A) and (A, B) are null kernel and full range pairs, respectively. Arguing by contradiction, suppose there is a realization (A', B', C') with A' of smaller size than A. By Theorem 4.1.2 there is a reduction (A'', B'', C'') of (A', B', C') with full range pair (A'', B'') and null kernel pair (C'', A''). But then the size of A'' is smaller than that of A, which contradicts Theorem 4.1.3. $\square$

Realizations of the kind described in this theorem are, naturally, called *minimal realizations* of $W(z)$. That is, they are those realizations for which the dimension of the space on which A acts is as small as possible; this dimension is called the *McMillan degree* of $W(z)$.

For rational $n \times n$ matrix functions which are not only finite (i.e. analytic) but also invertible at infinity, the following connection between the realizations of $W(z)$ and $W(z)^{-1}$ will be very useful for us.

PROPOSITION 4.1.5. *Let $W(z)$ be a rational $n \times n$ matrix function which is analytic and invertible at infinity, and let*

$$(4.1.9) \qquad W(z) = D + C(zI - A)^{-1}B$$

be a realization of $W(z)$ (so $D = W(\infty)^{-1}$ is invertible). Then

(4.1.10) $$W(z)^{-1} = D^{-1} - D^{-1}C\big(zI - (A - BD^{-1}C)\big)^{-1}BD^{-1}$$

is a realization of $W(z)$, and (4.1.9) is minimal if and only if (4.1.10) is.

PROOF. We shall prove that the product of the right-hand sides of (4.1.9) and (4.1.10) is identically I. Performing the term by term multiplication of the right hand sides of (4.1.9) and (4.1.10), we have to show that

$$C(zI - A)^{-1}BD^{-1} - C\big(zI - (A - BD^{-1}C)\big)^{-1}BD^{-1}$$
$$- C(zI - A)^{-1}BD^{-1}C\big(zI - (A - BD^{-1}C)\big)^{-1}BD^{-1} = 0$$

or

$$C(zI-A)^{-1}\big[\big(zI-(A-BD^{-1}C)\big)-(zI-A)-BD^{-1}C\big]\big(zI-(A-BD^{-1}C)\big)^{-1}BD^{-1} = 0.$$

But this equality is evident as the sum in the square brackets is zero.

To prove the assertion on minimality in Proposition 4.1.5, we need the following fact: If (C, A) is a null kernel pair, then for every matrix F of suitable size the pair $(C, A + FC)$ is again a null kernel pair. Indeed, one proves easily by induction on α that CA^{α} $(\alpha = 0, 1, \ldots)$ can be written in the form

$$CA^{\alpha} = \sum_{i=0}^{\alpha} Y_{i\alpha}C(A + FC)^{i}$$

for some matrices $Y_{0\alpha}, \ldots, Y_{\alpha\alpha}$. Consequently,

(4.1.11) $$\bigcap_{\alpha=1}^{q-1} \mathrm{Ker}\big(C(A + FC)^{\alpha}\big) \subset \bigcap_{\alpha=1}^{q-1} \mathrm{Ker}(CA^{\alpha}),$$

and since (C, A) is a null kernel pair, for some q we have $\bigcap_{\alpha=1}^{q} \mathrm{Ker}(CA^{\alpha}) = \{0\}$, and so $(C, A + FC)$ is a null kernel pair as well in view of (4.1.11). Analogously one proves that if (A, B) is a full range pair, then so is $(A + BG, B)$ for any matrix G of suitable size.

Applying these facts with $F = -BD^{-1}$, $G = -D^{-1}C$, and using Theorem 4.1.4 it follows that if (4.1.9) is minimal, then (4.1.10) is minimal as well. The converse statement can be proved similarly (or else apply the already proved statement to $W^{-1}(z)$ instead of $W(z)$). $\square$

We will need also the following multiplication formula.

PROPOSITION 4.1.6. *Let $W_1(z)$ and $W_2(z)$ be rational matrix functions with realizations*

(4.1.12) $$W_i(z) = D_i + C_i(zI - A_i)^{-1}B_i, \qquad i = 1, 2.$$

Then the product $W_1(z)W_2(z)$ has realization

(4.1.13) $$W_1(z)W_2(z) = D_1D + [C_1, D_1C_2]\left(zI - \begin{bmatrix} A_1 & B_1C_2 \\ 0 & A_2 \end{bmatrix}\right)^{-1}\begin{bmatrix} B_1D_2 \\ B_2 \end{bmatrix}.$$

The verification is straightforward using the formula

$$\begin{bmatrix} zI - A_1 & -B_1C_2 \\ 0 & zI - A_2 \end{bmatrix}^{-1} = \begin{bmatrix} (zI - A_1)^{-1} & (zI - A_1)^{-1}B_1C_2(zI - A_2)^{-1} \\ 0 & (zI - A_2)^{-1} \end{bmatrix}.$$

The realization (4.1.13) need not be minimal even if both realizations (4.1.12) are minimal (but the minimality of (4.1.13) implies the minimality of both realizations in (4.1.12)). One important particular case when the minimality of (4.1.13) is equivalent to that of (4.1.12) is when no point is simultaneously a zero of one function and a pole of the other:

COROLLARY 4.1.7. *Let $W_i(z)$ be given by minimal realizations (4.1.12) and assume in addition that D_1 and D_2 are invertible matrices. Further assume that*

$$(4.1.14) \qquad\qquad \sigma(A_1) \cap \sigma(A_2 - B_2D_2^{-1}C_2) = \emptyset$$

$$(4.1.15) \qquad\qquad \sigma(A_1 - B_1D_1^{-1}C_1) \cap \sigma(A_2) = \emptyset.$$

Then the realization (4.1.13) is minimal as well.

PROOF. By Theorem 4.1.4 it will suffice to verify that

$$(4.1.16) \qquad\qquad \left([C_1, D_1C_2], \begin{bmatrix} A_1 & B_1C_2 \\ 0 & A_2 \end{bmatrix} \right)$$

is a null kernel pair, and that

$$(4.1.17) \qquad\qquad \left(\begin{bmatrix} A_1 & B_1C_2 \\ 0 & A_2 \end{bmatrix}, \begin{bmatrix} B_1D_2 \\ B_2 \end{bmatrix} \right)$$

is a full range pair.

Denote $\widetilde{A} = \begin{bmatrix} A_1 & B_1C_2 \\ 0 & A_2 \end{bmatrix}$. The subspace

$$\mathcal{M} := \bigcap_{j=0}^{\infty} \mathrm{Ker}([C_1, D_1C_2]\widetilde{A}^j)$$

is clearly $\widetilde{A}$-invariant. This subspace is also invariant for the matrix

$$(4.1.18) \qquad\qquad \widetilde{A} - \begin{bmatrix} B_1D_1^{-1} \\ 0 \end{bmatrix} [C_1, D_1C_2]$$

because $\mathcal{M} \subset \mathrm{Ker}[C_1, D_1C_2]$. But the matrix (4.1.18) has the form $(A_1 - B_1D_1^{-1}C_1) \oplus A_2$, and because of the assumption (4.1.15) $\mathcal{M}$ is decomposed into a direct sum

$$\mathcal{M} = \begin{bmatrix} \mathcal{M}_1 \\ 0 \end{bmatrix} + \begin{bmatrix} 0 \\ \mathcal{M}_2 \end{bmatrix},$$

where $\mathcal{M}_1 \subset \mathbf{C}^{n_1}$, $\mathcal{M}_2 \subset \mathbf{C}^{n_2}$ (here n_1 and n_2 are the sizes of A_1 and A_2, respectively). Because (C_1, A_1) is a null kernel pair, the subspace $\mathcal{M}_1$ must be zero. We will prove that $\mathcal{M}_2$ is also zero, thereby completing the proof of the null kernel property of $([C_1, D_1 C_2], \tilde{A})$.

Let $y \in \mathcal{M}_2$, so

$$(4.1.19) \qquad [C_1, D_2 C_2] \tilde{A}^j \begin{bmatrix} 0 \\ y \end{bmatrix} = 0 \quad \text{for} \quad j = 0, 1, \dots .$$

It is easily seen that

$$\begin{bmatrix} A_1 & B_1 C_2 \\ 0 & A_2 \end{bmatrix}^j = \begin{bmatrix} A_1^j & \sum_{k=0}^{j-1} A_1^k B_1 C_2 A_2^{j-k-1} \\ 0 & A_2^j \end{bmatrix}$$

for $j = 1, 2, \dots$. Now (4.1.19) gives

$$D_1 C_2 y = 0;$$

$$\sum_{k=0}^{j-1} C_1 A_1^k B_1 C_2 A_2^{j-k-1} y + D_1 C_2 A_2^j y = 0, \qquad j = 1, 2, \dots .$$

These equalities imply (by induction on j) that $C_2 A_2^j y = 0$ for $j = 0, 1, \dots$ and hence $y = 0$ by the null kernel property of (C_2, A_2).

We now turn our attention to the pair (4.1.17). Consider the $\tilde{A}$-invariant subspace

$$\mathcal{N} := \sum_{j=0}^{\infty} \operatorname{Im} \tilde{A}^j \begin{bmatrix} B_1 D_2 \\ B_2 \end{bmatrix}.$$

This subspace is clearly invariant also for the matrix

$$\tilde{A} - \begin{bmatrix} B_1 D_2 \\ B_2 \end{bmatrix} [\, 0 \quad D_2^{-1} C_2 \,] = A_1 \oplus (A_2 - B_2 D_2^{-1} C_2).$$

Because of (4.1.14), each time when $\begin{bmatrix} x \\ y \end{bmatrix} \in \mathcal{N}$ (where $x \in \mathbf{C}^{n_1}$, $y \in \mathbf{C}^{n_2}$) we have also $\begin{bmatrix} x \\ 0 \end{bmatrix} \in \mathcal{N}$, $\begin{bmatrix} 0 \\ y \end{bmatrix} \in \mathcal{N}$. So to prove the full range property of (4.1.17) we have only to show that for every $y \in \mathbf{C}^{n_2}$ there is a vector in $\mathcal{N}$ of the form $\begin{bmatrix} * \\ y \end{bmatrix}$, and for every $x \in \mathbf{C}^{n_1}$ there is a vector in $\mathcal{N}$ of the form $\begin{bmatrix} x \\ * \end{bmatrix}$. The statement concerning $y \in \mathbf{C}^{n_2}$ follows easily from the full range property of (A_2, B_2). For the statement concerning $x \in \mathbf{C}^{n_1}$, write by the full range property of (A_1, B_1)

$$x = \sum_{j=0}^{p} A_1^j B_1 u_j$$

for some vectors u_j. On the other hand,

$$\sum_{j=0}^{p} \begin{bmatrix} A_1 & B_1C_2 \\ 0 & A_2 \end{bmatrix}^j \begin{bmatrix} B_1D_2 \\ B_2 \end{bmatrix} w_j$$

$$= \sum_{j=0}^{p} \begin{bmatrix} A_1^j B_1 D_2 w_j + \sum_{k=0}^{j-1} A_1^k B_1 C_2 A_2^{j-k-1} B_2 w_j \\ * \end{bmatrix}.$$

If we interchange the order of summation, we see that

$$A_1^j B_1 D_2 w_j + \sum_{k=0}^{j-1} A_1^k B_1 C_2 A_2^{j-k-1} B_2 w_j$$

$$= A_1^p B_1 D_2 w_p + A_1^{p-1}[B_1 D_2 w_{p-1} + B_1 C_2 B_2 w_p] + \cdots$$

$$+ [B_1 D_2 w_0 + B_1 C_2 B_2 w_1 + B_1 C_2 A_2 B_2 w_2 + \cdots + B_1 C_2 A_2^{p-1} B_2 w_j].$$

To show that $\begin{bmatrix} x \\ x' \end{bmatrix} \in \mathcal{N}$ for some $x' \in \mathbf{C}^{n_2}$, it suffices therefore to solve the system of equations (with unknowns w_j)

(4.1.20)
$$A_1^p B_1 u_p = A_1^p B_1 D_2 w_p$$
$$A_1^{p-1} B_1 u_{p-1} = A_1^{p-1} B_1 D_2 w_{p-1} + A_1^{p-1} B_1 C_2 B_2 w_p$$
$$\vdots$$
$$B_1 u_0 = B_1 D_2 w_0 + B_1 C_2 B_2 w_1 + B_1 C_2 A_2 B_2 w_2 + \cdots + B_1 C_2 A_2^{p-1} B_2 w_p.$$

The system (4.1.20) indeed has a solution obtained recursively: $w_p = D_2^{-1} u_p$; $w_{p-1} = D_2^{-1}(u_{p-1} - C_2 B_2 w_p)$, and in general

$$w_j = D_2^{-1}\bigg(u_j - Q_0 u_{j+1} - (Q_1 - Q_0^2)u_{j+2} - \cdots$$

$$- \bigg(Q_{p-j-1} - \sum_{\substack{\alpha+\beta=p-j-2 \\ \alpha,\beta\geq 0}} Q_\alpha Q_\beta + \sum_{\substack{\alpha+\beta+\gamma=p-j-3 \\ \alpha,\beta,\gamma\geq 0}} Q_\alpha Q_\beta Q_\gamma - \cdots$$

$$+ (-1)^{p-j-1} Q_0^{p-j}\bigg) u_p\bigg), \qquad j = 0, 1, \ldots, p,$$

where $Q_k = C_2 A_2^k B_2 D_2^{-1}$, $k = 0, 1, \ldots$.

4.2 REALIZATIONS AND LOCAL DATA

The notions of various types of local null and pole data developed in Chapter 3 apply in particular to rational matrix functions. In this section we obtain the connections of all these notions with a minimal realization for the rational matrix function $W(z)$ which is analytic and invertible at infinity. Since a rational matrix function has only finitely many poles in the whole complex plane $\mathbb{C}$ we may apply the notions of local data for several points developed in Section 3.4 for the case of a rational matrix function with the subset σ chosen to be equal to $\mathbb{C}$. A (left or right) pole or null pair for a rational matrix function with respect to $\sigma = \mathbb{C}$ we shall refer to as a *global* (left or right) pole or null pair. Similarly a null or pole triple for $W(z)$ we shall called a *global* null or pole triple. The connection of these ideas with a minimal realization for $W(z)$ is given by the following theorem.

THEOREM 4.2.1. *Let $W(z)$ be a rational matrix function with finite and invertible value $W(\infty) = D$ at infinity. Let*

$$(4.2.1) \qquad W(z) = D + C(zI - A)^{-1}B$$

be a minimal realization for $W(z)$. Then

(i) *(C, A, B) is a global pole triple of $W(z)$.*

(ii) *$(-D^{-1}C, A - BD^{-1}C, BD^{-1})$ is a global null triple of $W(z)$.*

In particular

(iii) *(C, A) is a global right pole pair of $W(z)$ while (A, B) is a global left pole pair of $W(z)$;*

and

(iv) *$(D^{-1}C, A - BD^{-1}C)$ is a global right null pair of $W(z)$ while $(A - BD^{-1}C, BD^{-1})$ is a global left null pair of $W(z)$.*

PROOF. Suppose (4.2.1) is a minimal realization for $W(z)$. Then (C, A) is a null kernel pair, (A, B) is a full range pair by Theorem 4.1.4, and

$$W(z) - C(zI - A)^{-1}B = D$$

is a constant matrix function. Assertions (i) and (iii) now follow immediately from Theorem 3.4.1. Since a null triple for $W(z)$ is just a pole triple for $W(z)^{-1}$, assertions (ii) and (iv) follow in a similar way from the minimal realization (4.1.10) for $W(z)^{-1}$ given by Proposition 4.1.5.

Also useful for us will be the following converse of Theorem 4.2.1.

THEOREM 4.2.2. *Let $W(z)$ be a rational matrix function with finite invertible value $W(\infty) = D$ at infinity. Then:*

(i) *If (C, A, B) is a global pole triple for $W(z)$, then*

$$W(z) = D + C(zI - A)^{-1}B$$

is a minimal realization for $W(z)$.

(ii) *If $(\widetilde{C}, \widetilde{A}, \widetilde{B})$ is a global null triple for $W(z)$ then*

$$W(z) = D - D\widetilde{C}(zI - (\widetilde{A} - \widetilde{B}D\widetilde{C}))^{-1}\widetilde{B}D$$

is a minimal realization for $W(z)$.

(iii) *If* (C, A) *is a global right pole pair for* $W(z)$, *then there exists a unique matrix* $\widetilde{B}$ *so that*

$$W(z) = D + C(zI - A)^{-1}\widetilde{B}$$

is a minimal realization for $W(z)$.

Similarly, if (A, B) *is a global left pole pair for* $W(z)$ *then there is a unique matrix* $\widetilde{C}$ *so that*

$$W(z) = D + \widetilde{C}(zI - A)^{-1}B$$

is a minimal realization for $W(z)$.

(iv) *If* (C, A) *is a global right null pair for* $W(z)$ *then there exists a unique matrix* $\widetilde{B}$ *so that*

$$W(z) = D - DC(zI - (A - \widetilde{B}DC))^{-1}\widetilde{B}D$$

is a minimal realization for $W(z)$. *Similarly if* (A, B) *is a global left null pair for* $W(z)$ *then there exists a unique matrix* $\widetilde{C}$ *so that*

$$W(z) = D - D\widetilde{C}(zI - (A - BD\widetilde{C}))^{-1}BD$$

is a minimal realization for $W(z)$.

PROOF. If (C, A, B) is a global pole triple for $W(z)$, then by definition $R(z) \overset{\text{def}}{=} W(z) - C(zI - A)^{-1}B$ has analytic continuation to all of $\mathbb{C}$. As $R(z)$ also has the finite value D at infinity, Liouville's Theorem implies that $R(z) \equiv D$, and hence (A, B, C) is a realization for $W(z)$. As (C, A) is a null-kernel pair and (A, B) is a full-range pair by definition of pole triple, (A, B, C) is also a minimal realization; this verifies (i). In a similar way, if $(\widetilde{C}, \widetilde{A}, \widetilde{B})$ is a global null triple, then $(\widetilde{A}, \widetilde{B}, \widetilde{C})$ is a minimal realization for $W(z)^{-1}$. By using formula (4.1.10) to compute a minimal realization for $W(z) = \left(W(z)^{-1}\right)^{-1}$, we get (ii). If (C, A) is a global right pole pair for $W(z)$, then Theorem 3.4.2(a) guarantees the existence of a unique matrix $\widetilde{B}$ for which $(C, A, \widetilde{B})$ is a global pole triple for $W(z)$. Now the first assertion in (iii) is a consequence of (i). The remaining assertions in (iii) and (iv) follow in a similar way. □

REMARK. Note that the existence of a global pole triple for $W(z)$ was established in Theorem 3.4.2. Thus Theorem 4.2.2 yields the existence of a realization for $W(z)$ with the additional properties that the associated (C, A) is a null-kernel pair and (A, B) is a full-range pair by a different route from that used in Section 4.1. □

We also connect in the next theorem minimal realizations with local data with respect to a general subset of σ.

Given an $m \times m$ matrix A and a subset $\sigma \subset \mathbb{C}$, the *Riesz projection* Q onto the A-invariant subspace corresponding to the eigenvalues in σ is, by definition, the projection on $\sum_{\lambda \in \sigma} \text{Ker}(A - \lambda I)^m$ along $\sum_{\lambda \in \mathbb{C} \backslash \sigma} \text{Ker}(A - \lambda I)^m$. In other words, Q projects onto the sum of root subspaces corresponding to the eigenvalues of A in σ along the sum of root subspaces corresponding to the eigenvalues of A outside σ.

THEOREM 4.2.3. *Let $W(z)$ be a rational matrix function with minimal realization*

$$W(z) = D + C(zI - A)^{-1}B$$

where $D = W(\infty)$ is invertible, and let σ be a subset of $\mathbb{C}$. Let Q be the Riesz projection onto the A-invariant spectral subspace associated with eigenvalues in σ, and let $Q^{\times}$ be the corresponding object for $A^{\times} \overset{\text{def}}{=} A - BD^{-1}C$. Then:

(i) $(C \mid \operatorname{Im} Q, A \mid \operatorname{Im} Q, QB)$ *is a pole triple for $W(z)$ with respect to σ.*

(ii) $(-D^{-1}C \mid \operatorname{Im} Q^{\times}, A^{\times} \mid \operatorname{Im} Q^{\times}, Q^{\times}BD^{-1})$ *is a null triple for $W(z)$ with respect to σ.*

In particular,

(iii) $(C \mid \operatorname{Im} Q, A \mid \operatorname{Im} Q)$ *is a right pole pair of $W(z)$ with respect to σ, while $(A \mid \operatorname{Im} Q, QB)$ is a left pole pair of $W(z)$ with respect to σ.*

(iv) $(D^{-1}C \mid \operatorname{Im} Q^{\times}, A^{\times} \mid \operatorname{Im} Q^{\times})$ *is a right null pair of $W(z)$ with respect to σ, while $(A^{\times} \mid \operatorname{Im} Q^{\times}, Q^{\times}BD^{-1})$ is a left null pair of $W(z)$ with respect to σ.*

PROOF. To verify (i), use a similarity transformation to write A as $A = \operatorname{diag}(A_1, A_2)$ where $\sigma(A_1) \subset \sigma$ and $\sigma(A_2) \cap \sigma = \phi$. With respect to the decomposition,
$Q = \begin{bmatrix} I & 0 \\ 0 & 0 \end{bmatrix}$, $C = [C_1, C_2]$ and $B = \operatorname{col}(B_1, B_2)$. Then $C \mid \operatorname{Im} Q$ can be identified with C_1, QB can be identified with B_1. Now note that

$$W(z) - C_1(zI - A_1)^{-1}B_1 = D + C_2(zI - A_2)^{-1}B_2$$

has analytic continuation to all of σ since $\sigma(A_2) \cap \sigma = \phi$. Also that (C_1, A_1) is a null kernel pair and (A_1, B_1) is a full range pair follows again as in the proof of Theorem 1.5.1. This verifies (i). The remaining assertions follow by using the idea in the proof of (i) to adapt the arguments in the proof of Theorem 4.2.2. $\square$

4.3 A GLOBAL INTERPOLATION PROBLEM FOR RATIONAL MATRIX FUNCTIONS

A scalar rational function is uniquely determined up to a nonzero constant factor by its zeros and poles; this is essentially a consequence of the fundamental theorem of algebra. For the matrix case we have seen that a canonical set of zero chains at each point plays the role of the zeros and similarly a canonical set of pole chains at each pole gives more complete pole information. In the scalar case, multiplying by a nonzero constant does not affect the poles and the zeros; for the matrix case let us suppose we are interested only in left-side information concerning a rational matrix function $W(z)$, i.e. in pole-null structure that is unaffected by multiplying by a constant invertible matrix on the right. It is easy to see (see Example 3.1.6) that a canonical set of left null chains at each point (or equivalently, a left null pair (A_ζ, B_ζ)) and a canonical set of right pole chains at each point (i.e. a right pole pair (C_π, A_π)) have these properties. For this reason, as was explained in the introduction to this chapter, we refer to data of this type as a *standard set of (homogeneous) interpolation data*. As poles at infinity require special consideration, we restrict ourselves in this chapter to rational matrix function $W(z)$ which are analytic and invertible at ∞. Then a natural question to ask is: does a left null pair (A_ζ, B_ζ) and a right pole pair (C_π, A_π) for a rational matrix function $W(z)$ determine $W(z)$ uniquely up to a right invertible factor? Also, given a full-range pair (A_ζ, B_ζ) and a null-kernel pair (C_π, A_π), when is there a rational matrix function $W(z)$ (regular at ∞) having (A_ζ, B_ζ) as a (global) left null pair and (C_π, A_π) as a (global) right pole pair? For the scalar case we know that the answer to the first question is yes while the answer to the second question is yes if and only if A_ζ and A_π have the same size and disjoint spectra (i.e. the number of zeros is the same as the number of poles (including multiplicities) and the set of zeros is disjoint from the set of poles). For the matrix case the answer is more complicated. We have already seen (e.g. via the local Smith-McMillan form) that it is possible in the matrix case for a point z_0 to be a pole and a zero for a function $W(z)$, so the answer to the second question is certainly more subtle in the matrix case. As the following example illustrates, in general the answer to the first question is no in the matrix case.

EXAMPLE 4.3.1. Consider the function

$$W_\alpha(z) = \begin{bmatrix} 1 & \alpha z^{-1} \\ 0 & 1 \end{bmatrix}$$

where $\alpha \neq 0$ is a complex parameter. Then for all $\alpha \neq 0$, W_α has a pole only at 0, and a canonical set of right pole chains at 0 consists of the single vector $\left\{ \begin{bmatrix} 1 \\ 0 \end{bmatrix} \right\}$.

Thus $(C_\pi, A_\pi) = \left(\begin{bmatrix} 1 \\ 0 \end{bmatrix}, 0 \right)$ is a right pole pair for $W_\alpha(z)$ for all $\alpha \neq 0$. As $W_\alpha^{-1} = \begin{bmatrix} 1 & -\alpha z^{-1} \\ 0 & 1 \end{bmatrix}$, we see similarly that a left pole pair for W_α^{-1} (i.e. a left null pair for W_α) is $(A_\zeta, B_\zeta) = (0, [0, 1])$. Thus each of the infinitely many functions W_α ($\alpha \neq 0$) has the same left null pair (A_ζ, B_ζ) and right pole pair (C_π, A_π). As $W_\alpha(\infty) = I$ for all α, any two of these functions must differ from each other by more than a constant right factor. □

The answer for the matrix case can be derived by using the connection between a left null pair and right pole pair for a rational matrix function $W(z)$ and a minimal realization for $W(z)$. We assume that $W(z)$ is analytic and invertible at infinity; by considering $W(z)W(\infty)^{-1}$ in place of $W(z)$ we may assume that the value at infinity is I. Then if (C_π, A_π) is a (global) right pole pair for $W(z)$, we know by Theorem 4.2.2 that there exists a (in fact, unique) matrix $\tilde{B}$ such that

(4.3.1) $$W(z) = I + C_\pi(zI - A_\pi)^{-1}\tilde{B}$$

is a minimal realization for $W(z)$. Also by Theorem 4.2.2, if (A_ζ, B_ζ) is a (global) left null pair for $W(z)$ then there exists a unique matrix $\tilde{C}$ for which

(4.3.2) $$W(z) = I - \tilde{C}(zI - (A_\zeta - B_\zeta\tilde{C}))^{-1}B_\zeta$$

is also a minimal realization for $W(z)$. By Theorem 4.1.3, these realizations must be similar, i.e. there must exist an invertible matrix S for which

(4.3.3) $$A_\pi = S^{-1}(A_\zeta - B_\zeta\tilde{C})S, \quad \tilde{B} = S^{-1}B_\zeta, \quad C_\pi = -\tilde{C}S.$$

From the first equation we get

$$SA_\pi - A_\zeta S = -B_\zeta \widetilde{C} S$$

while from the last equation we get

$$-B_\zeta \widetilde{C} S = B_\zeta C_\pi.$$

Then the Sylvester equation

$$SA_\pi - A_\zeta S = B_\zeta C_\pi$$

must have an invertible solution S whenever there exists a rational matrix function $W(z)$ regular at infinity having (C_π, A_π) as a global right pole pair and (A_ζ, B_ζ) as a global left null pair. This necessary condition turns out also to be sufficient. Unless stated otherwise, by a null or pole pair for $W(z)$ will be meant a global null or pole pair in the following.

THEOREM 4.3.1. *Suppose (C_π, A_π) is a null-kernel pair and (A_ζ, B_ζ) is a full-range pair of matrices. Then a necessary and sufficient condition for the existence of a rational matrix function $W(z)$ regular at infinity such that*

(i) *(C_π, A_π) is a global right pole pair for $W(z)$ and*

(ii) *(A_ζ, B_ζ) is a global left null pair for $W(z)$*

is that there exist an invertible solution S of the Sylvester equation

(4.3.4) $$SA_\pi - A_\zeta S = B_\zeta C_\pi.$$

In this case, solutions $W(z)$ of (i) and (ii) having $W(\infty) = I$ are in one-to-one correspondence with invertible solutions S of (4.3.4) according to the formula

(4.3.5) $$W(z) = I + C_\pi (zI - A_\pi)^{-1} S^{-1} B_\zeta$$

with inverse given by

(4.3.6) $$W(z)^{-1} = I - C_\pi S^{-1} (zI - A_\zeta)^{-1} B_\zeta.$$

PROOF. We have already noted in the discussion immediately preceding the theorem that whenever there exists a $W(z)$ satisfying (i) and (ii) then there exists an invertible solution S of the Sylvester equation (4.3.4). Moreover, when we substitute $\widetilde{B} = S^{-1} B_\zeta$ (from (4.3.3)) into the realization (4.3.1) for $W(z)$, we see that $W(z)$ has the form (4.3.5) for this solution S of the Sylvester equation (4.3.4). Similarly, substituting $\widetilde{C} = -C_\pi S^{-1}$ (from (4.3.3)) into the second realization (4.3.2) for W and applying Proposition 4.1.5, we come upon the realization (4.3.6) for $W(z)^{-1}$.

Conversely, suppose that S is any invertible solution of (4.3.4). We then use (4.3.5) to define a rational matrix function $W(z)$. A consequence of (4.3.4) is

(4.3.7) $$S(A_\pi - S^{-1} B_\zeta C_\pi) S^{-1} = A_\zeta$$

and hence the pair $(A_\pi - S^{-1}B_\zeta C_\pi, S^{-1}B_\zeta)$ is similar to the full-range pair (A_ζ, B_ζ) and hence is itself a full-range pair. But it is easy to see that $(A_\pi - S^{-1}B_\zeta C_\pi, S^{-1}B_\zeta)$ is full-range if and only if $(A_\pi, S^{-1}B_\zeta)$ is full-range. We can now conclude that (4.3.5) is a minimal realization of $W(z)$. Then by Theorem 4.2.1 (C_π, A_π) is a global right pole pair for $W(z)$ and $(A_\pi - S^{-1}B_\zeta C_\pi, S^{-1}B_\zeta)$ is a global left null pair for $W(z)$. As was already observed above this latter pair is similar to (A_ζ, B_ζ), and hence (A_ζ, B_ζ) is a global left null pair for $W(z)$ as well. If we calculate a minimal realization for $W(z)^{-1}$ from (4.3.5) by using Proposition 4.1.5 and then use (4.3.7), we come upon the realization (4.3.6) for $W(z)^{-1}$.

It remains only to show that distinct invertible solutions S_1 and S_2 of (4.3.4) lead via (4.3.5) to distinct rational matrix functions $W_1(z)$ and $W_2(z)$. But by Theorem 4.2.2 the right pole pair (C_π, A_π) for a given rational matrix function $W(z)$ determines a unique matrix $S^{-1}B_\zeta$ for which (4.3.5) is a minimal realization. We have therefore (arguing by contradiction and assuming that $W_1(z) \equiv W_2(z)$) the equality $S_1^{-1}B_\zeta = S_2^{-1}B_\zeta$. Now

$$S_1^{-1}A_\zeta B_\zeta = (A_\pi S_1^{-1} - S_1^{-1}B_\zeta C_\pi S_1^{-1})B_\zeta = A_\pi(S_1^{-1}B_\zeta) - (S_1^{-1}B_\zeta)C_\pi(S_1^{-1}B_\zeta)$$
$$= (A_\pi S_2^{-1} - S_2^{-1}B_\zeta C_\pi S_2^{-1})B_\zeta = S_2^{-1}A_\zeta B_\zeta.$$

By induction, one shows that $S_1^{-1}A_\zeta^j B_\zeta = S_2^{-1}A_\zeta^j B_\zeta$ for $j = 0, 1, \ldots$. In view of the full range property of (A_ζ, B_ζ) we have $S_1^{-1} = S_2^{-1}$, a contradiction with the inequality $S_1 \neq S_2$. $\square$

We close this section by illustrating Theorem 4.3.1 for the scalar case. For simplicity we assume that all prescribed zeros $z_1, \ldots, z_n$ and prescribed poles $w_1, \ldots, w_m$ are simple. The interpolation problem is to construct a scalar rational function $r(z)$ with $r(\infty) = 1$ with simple zeros at the points $z_1, \ldots, z_n$, simple poles at the points $w_1, \ldots, w_m$ and no other zeros or poles in the complex plane. By elementary scalar function theory facts we know that such a function $r(z)$ exists if and only if $n = m$ and no zero is also a pole, in which case the unique solution is given by

$$(4.3.8) \qquad r(z) = \frac{(z - z_1) \cdots (z - z_n)}{(z - w_1) \cdots (z - w_n)}.$$

On the other hand, by Example 3.1.2, prescribing that the zeros of the scalar rational function $r(z)$ occur at $z_1, \ldots, z_n$ and all be simple is the same as to prescribe that (A_ζ, B_ζ) be a global left null pair for $r(z)$, where A_ζ is the $n \times n$ matrix

$$(4.3.9) \qquad A_\zeta = \mathrm{diag}(z_1, \ldots, z_n)$$

and B_ζ is the $n \times 1$ matrix

$$(4.3.10) \qquad B_\zeta = \mathrm{col}(1, \ldots, 1).$$

Similarly (see Example 3.1.2), prescribing that the poles occur at $w_1, \ldots, w_m$ and that all these poles are simple is the same as prescribing (C_π, A_π) is a right pole pair for $r(z)$, where C_π is the $1 \times n$ matrix

$$(4.3.11) \qquad C_\pi = [1, \ldots, 1]$$

and A_π is the $m \times m$ matrix

$$(4.3.12) \qquad\qquad A_\pi = \operatorname{diag}(w_1, \ldots, w_m).$$

Theorem 4.3.1 gives a criterion for existence and uniqueness of a solution of this problem in terms of invertible solutions of a Sylvester equation and an alternative more complicated formula for the solution $r(z)$ as in (4.3.8). The following result gives a direct verification that these two approaches yield the same solution.

THEOREM 4.3.2. *Let $z_1, \ldots, z_n$ be a distinct points and $w_1, \ldots, w_m$ another set of m distinct points in $\mathbb{C}$. Define matrices C_π, A_π, A_ζ, B_ζ by (4.3.9), (4.3.10), (4.3.11) and (4.3.12). Then the Sylvester equation (4.3.4) has a solution S if and only if the set of $m + n$ points $z_1, \ldots, z_n, w_1, \ldots, w_m$ are all distinct. In this case, the solution is unique, and is invertible if and only if $n = m$. Moreover, if $n = m$ and S is the unique solution, then*

$$(4.3.13) \qquad 1 + C_\pi(zI - A_\pi)^{-1}S^{-1}B_\pi = \frac{(z - z_1) \cdots (z - z_n)}{(z - w_1) \cdots (z - w_n)}.$$

PROOF. In view of Lemma A.1.5 of the Appendix, it remains only to verify the identity (4.3.13). By the same lemma, we know that

$$(4.3.14) \qquad\qquad S^{-1} = [t_{\alpha\beta}]_{1 \le \alpha, \beta \le n},$$

where

$$(4.3.15) \qquad t_{\alpha\beta} = \frac{\prod_{1 \le j \le n}(w_j - z_\beta) \cdot \prod_{1 \le k \le n}(w_\alpha - z_k)}{\prod_{\substack{1 \le j \le n \\ j \ne \beta}}(z_\beta - z_j) \cdot \prod_{\substack{1 \le k \le n \\ k \ne \alpha}}(w_k - w_\alpha)} \cdot \frac{1}{w_\alpha - z_\beta}.$$

Plugging in the expressions (4.3.11), (4.3.12), (4.3.10) and (4.3.14) for C_π, A_π, B_ζ and S^{-1}, we see that

$$1 + C_\pi(zI - A_\pi)^{-1}S^{-1}B_\pi = 1 + \sum_{\alpha=1}^{n}\sum_{\beta=1}^{n}(z - w_\alpha)^{-1}t_{\alpha\beta}$$

$$= 1 + \sum_{\alpha=1}^{n}\left\{\sum_{\beta=1}^{n}t_{\alpha\beta}\right\}(z - w_\alpha)^{-1}.$$

On the other hand, the right side of (4.3.13) can be expanded in partial fractions

$$\frac{(z - z_1) \cdots (z - z_n)}{(z - w_1) \cdots (z - w_n)} = 1 + \sum_{\alpha=1}^{n}\left\{\frac{\prod_{1 \le j \le n}(w_\alpha - z_j)}{\prod_{\substack{1 \le j \le n \\ j \ne \alpha}}(w_\alpha - w_j)}\right\}(z - w_\alpha)^{-1}.$$

Thus (4.3.13) is verified once we show

$$(4.3.16) \qquad\qquad \sum_{\beta=1}^{n}t_{\alpha\beta} = \frac{\prod_{1 \le j \le n}(w_\alpha - z_j)}{\prod_{\substack{1 \le j \le n \\ j \ne \alpha}}(w_\alpha - w_j)}$$

for $1 \leq \alpha \leq n$. Plugging in now (4.3.15) for $t_{\alpha\beta}$, we see that the left side of (4.3.16) is

$$\sum_{\beta=1}^{n} t_{\alpha\beta} = \frac{\prod_{1 \leq k \leq n}(w_\alpha - z_k)}{\prod_{\substack{1 \leq k \leq n \\ k \neq \alpha}}(w_\alpha - w_k)} \cdot (-1)^{n-1} \cdot \sum_{\beta=1}^{n} \left\{ \frac{\prod_{\substack{1 \leq j \leq n \\ j \neq \alpha}}(w_j - z_\beta)}{\prod_{\substack{1 \leq j \leq n \\ j \neq \beta}}(z_\beta - z_j)} \right\}.$$

Thus (4.3.16) is verified once we show

(4.3.17)
$$1 = (-1)^{n-1} \sum_{\beta=1}^{n} \left\{ \frac{\prod_{\substack{1 \leq j \leq n \\ j \neq \alpha}}(w_j - z_\beta)}{\prod_{\substack{1 \leq j \leq n \\ j \neq \beta}}(z_\beta - z_j)} \right\}.$$

To verify this consider z_n as a complex variable ζ and the other parameters $z_1, \ldots, z_{n-1}$, $w_1, \ldots, w_n$ as fixed. The right side of (4.3.17) can be expressed as

$$f(\zeta) = (-1)^{n-1} \sum_{\beta=1}^{n-1} \left\{ \frac{\prod_{\substack{1 \leq j \leq n \\ j \neq \alpha}}(w_j - z_\beta)}{\prod_{\substack{1 \leq j \leq n-1 \\ j \neq \beta}}(z_\beta - z_j)} \right\} (z_\beta - \zeta)^{-1} + (-1)^{n-1} \frac{\prod_{\substack{1 \leq j \leq n \\ j \neq \alpha}}(w_j - \zeta)}{\prod_{1 \leq j \leq n-1}(\zeta - z_j)}.$$

Note that $f(\zeta)$ is a rational function in ζ with value 1 at ∞ and with only poles of order at most 1 occurring at the points $\zeta = z_1, z_2, \ldots, z_{n-1}$. Compute the residue R_β of $f(\zeta)$ at the pole $\zeta = z_\beta$ to get

$$R_\beta = \lim_{\zeta \to z_\beta} (\zeta - z_\beta)f(\zeta) = 0.$$

Thus $f(\zeta)$ is a rational function with value 1 at infinity and having no poles in the complex plane. Thus f must be identically equal to 1 and (4.3.17) (and hence also (4.3.13)) follows as desired. □

4.4 NULL-POLE COUPLING MATRIX

As we shall see in the next section, the invertible solution S of the Sylvester equation (4.3.4) may not be unique, and hence in general one needs more information than just a left null pair and a right pole pair to determine completely a rational matrix function up to a right invertible factor. To specify a particular rational matrix function $W(z)$ (say with $W(\infty) = I$) having a given left null pair (A_ζ, B_ζ) and right pole pair (C_π, A_π), one must have some way of specifying a particular invertible solution S of (4.3.4) so that $W(z)$ is the function given by (4.3.5). In the discussion preceding Theorem 4.3.1 we saw how a particular rational matrix function $W(z)$ having a given left null pair and right pole pair determines the appropriate invertible solution of (4.3.4). This suggests the following definition.

Given a rational matrix function $W(z)$ with finite and invertible value $W(\infty) = D$ at infinity and given a (global) right pole pair (C_π, A_π) and a (global) left null pair (A_ζ, B_ζ) for $W(z)$, by Theorem 4.2.2 we know that there exist unique matrices $\widetilde{C}$ and $\widetilde{B}$ for which

(4.4.1)
$$W(z) = D + C_\pi(zI - A_\pi)^{-1}\widetilde{B}$$

and

(4.4.2) $$W(z) = D - D\widetilde{C}(zI - (A_\zeta - B_\zeta D\widetilde{C}))^{-1}B_\zeta D$$

are both minimal realizations for $W(z)$. By Theorem 4.1.3, these two minimal realizations must be *similar*, i.e. there exists a unique invertible matrix S for which

(4.4.3) $$S^{-1}(A_\zeta - B_\zeta D\widetilde{C})S = A_\pi, \quad S^{-1}B_\zeta D = \widetilde{B}, \quad -D\widetilde{C}S = C_\pi.$$

From the first equation in (4.4.3), we get

$$SA_\pi - A_\zeta S = -B_\zeta D\widetilde{C}S$$

and then from the last equation in (4.4.3) we get that S is a solution of the Sylvester equation (4.3.4)

(4.4.4a) $$SA_\pi - A_\zeta S = B_\zeta C_\pi$$

which satisfies the additional conditions

(4.4.4b) $$\widetilde{B} = S^{-1}B_\zeta D$$

and

(4.4.4c) $$\widetilde{C} = -D^{-1}C_\pi S^{-1}.$$

The invertible matrix S defined by (4.4.4a)–(4.4.4c) is uniquely determined by the rational matrix function $W(z)$ and a choice of right pole pair (C_π, A_π) and left null pair (A_ζ, B_ζ) for $W(z)$; this matrix S we shall call the *null-pole coupling matrix* for $W(z)$ (associated with right pole pair (C_π, A_π) and left null pair (A_ζ, B_ζ)). The solution S of the Sylvester equation (4.4.4a) may not be unique (see Section A.1 in the Appendix) but the solution of the system (4.4.4a)–(4.4.4c) is unique. We pick out the particular solution of (4.4.4a) associated with the function $W(z)$ and refer to $(C_\pi, A_\pi; A_\zeta, B_\zeta; S)$ as a (*global*) *null-pole triple* for $W(z)$.

Note that a null-pole triple $\tau = (C_\pi, A_\pi; A_\zeta, B_\zeta; S)$ for $W(z)$ contains left-side information about $W(z)$ in the sense that if $\widetilde{W}(z) = W(z)\Gamma$ is an adjustment by an invertible constant matrix factor on the right, then τ is also a null-pole triple for $\widetilde{W}(z)$. For this reason, when not clear from the context, we will say that τ is a *left* (global) null-pole triple for $W(z)$. Intuitively the null-pole coupling matrix gives information on how the poles interact with the zeros for a rational matrix function; this extra information is crucial particularly for the case where a zero and a pole occur at the same point. We shall see an illustration of this phenomenon from a different point of view in Part III. We could have done all the above analysis starting with a left pole pair (A_π, B_π) and right null pair (C_ζ, A_ζ) for $W(z)$; we would then have come upon a *right* (global) null-pole triple $(C_\zeta, A_\zeta; A_\pi, B_\pi; S_r)$ for $W(z)$. Equivalently, $(C_\zeta, A_\zeta; A_\pi, B_\pi; S_r)$ is a right null-pole triple for $W(z)^T$ if and only if $(B_\pi^T, A_\pi^T; A_\zeta^T, C_\zeta^T; S_r^T)$ is a left null-pole triple for $W(z)$.

The next result gives an easy way to determine a null-pole triple for $W(z)$ from a minimal realization.

PROPOSITION 4.4.1. *Let*

$$W(z) = D + C(zI - A)^{-1}B$$

be a minimal realization for the rational matrix function $W(z)$ having finite and invertible value D at infinity. Then $(C, A; A - BD^{-1}C, BD^{-1}; I)$ is a left null-pole triple for $W(z)$.

PROOF. Since (A, B, C) is a minimal realization we know by Theorem 4.2.1 that $(C_\pi, A_\pi) = (C, A)$ is a right pole pair and $(A_\zeta, B_\zeta) = (A - BD^{-1}C, BD^{-1})$ is a left null pair for $W(z)$. From the minimal realization (A, B, C) it is clear that we should choose $\widetilde{B} = B$ in (4.4.1). With the above choice of (A_ζ, B_ζ) the realization (4.4.2) has the form

$$W(z) = D - D\widetilde{C}(zI - (A - BD^{-1}C - B\widetilde{C}))^{-1}B.$$

From the minimal realization (A, B, C) it is clear that $\widetilde{C} = -D^{-1}C$ is the appropriate choice of $\widetilde{C}$ and therefore the two realizations (4.4.1) and (4.4.2) are the same. By the uniqueness of the similarity between two minimal realizations, we conclude that the associated null-pole coupling matrix S is equal to I. □

We next classify the amount of nonuniqueness in a left null-pole triple for a given rational matrix function $W(z)$. Let us say that the data triple $(C_\pi, A_\pi; A_\zeta, B_\zeta; S)$, where C_π, A_π, A_ζ, B_ζ, S are matrices of sizes $n \times p$, $p \times p$, $q \times q$, $q \times n$ and $q \times p$ respectively, is similar to the data triple $(\widetilde{C}_\pi, \widetilde{A}_\pi; \widetilde{A}_\zeta, \widetilde{B}_\zeta; \widetilde{S})$ if there are invertible matrices T_1 and T_2 of compatible square sizes for which

(4.4.5a) $$\widetilde{C}_\pi = C_\pi T_1, \qquad \widetilde{A}_\pi = T_1^{-1} A_\pi T_1,$$

(4.4.5b) $$\widetilde{A}_\zeta = T_2^{-1} A_\zeta T_2, \qquad \widetilde{B}_\zeta = T_2^{-1} B_\zeta,$$

(4.4.5c) $$\widetilde{S} = T_2^{-1} S T_1.$$

THEOREM 4.4.2. *Any two global left null-pole triples $(C_\pi, A_\pi; A_\zeta, B_\zeta; S)$ and $(\widetilde{C}_\pi, \widetilde{A}_\pi; \widetilde{A}_\zeta, \widetilde{B}_\zeta; \widetilde{S})$ for the same rational matrix function $W(z)$ (with finite invertible value D at infinity) are similar. In particular, if $W(z) = D + C(zI - A)^{-1}B$ is a minimal realization for $W(z)$, then $\tau = (C_\pi, A_\pi; A_\zeta, B_\zeta; S)$ is a left null-pole triple for $W(z)$ if and only if τ has the form*

$$\tau = (CT_1, T_1^{-1}AT_1, T_2^{-1}(A - BD^{-1}C)T_2, T_2^{-1}BD^{-1}, T_2^{-1}T_1)$$

for some invertible matrices T_1 and T_2.

PROOF. Suppose $\tau = (C_\pi, A_\pi; A_\zeta, B_\zeta; S)$ and $\widetilde{\tau} = (\widetilde{C}_\pi, \widetilde{A}_\pi; \widetilde{A}_\zeta, \widetilde{B}_\zeta; \widetilde{S})$ are two null-pole triples for $W(z)$. Then in particular, both (C_π, A_π) and $(\widetilde{C}_\pi, \widetilde{A}_\pi)$ are right pole pairs for $W(z)$ and (A_ζ, B_ζ) and $(\widetilde{A}_\zeta, \widetilde{B}_\zeta)$ are both left null pairs for $W(z)$. By the uniqueness theory for these objects (see Section 3.4), there exist invertible matrices T_1 and T_2 which satisfy (4.4.5a) and (4.4.5b) respectively. To determine how $\widetilde{S}$ is connected with S, analyze the system of equations (4.4.4a)–(4.4.4c) for the two systems τ and $\widetilde{\tau}$; this gives (4.4.5c). Finally, the last assertion of the theorem follows upon combining the first part with Proposition 4.4.1. $\square$

We are now ready to characterize which data sets $\tau = (C_\pi, A_\pi; A_\zeta, B_\zeta; S)$ arise as the left null pole triple for some rational matrix function $W(z)$ with finite and invertible value at infinity; this can be considered as the solution of an interpolation problem which is more restrictive than what was considered in the previous section. Let us say that the collection of matrices $\tau = (C_\pi, A_\pi; A_\zeta, B_\zeta; S)$ is an *admissible Sylvester data set* if

   (i) (C_π, A_π) is a null-kernel pair

   (ii) (A_ζ, B_ζ) is a full-range pair

and

   (iii) S satisfies the Sylvester equation

$$SA_\pi - A_\zeta S = B_\zeta C_\pi.$$

We have already seen that a necessary condition for a collection of matrices τ to be the left null-pole triple for a rational matrix function $W(z)$ (regular at infinity) is that τ be an admissible Sylvester data set having an invertible coupling matrix S. This conditions turns out also to be sufficient.

THEOREM 4.4.3. *The collection of matrices* $\tau = (C_\pi, A_\pi; A_\zeta, B_\zeta; S)$ *is a left null-pole triple for some rational matrix function* $W(z)$ *with finite and invertible value at infinity if and only if* τ *is an admissible Sylvester data set with invertible coupling matrix* S. *In this case, the unique such function* $W(z)$ *with* $W(\infty) = I$ *is given by*

(4.4.6) $$W(z) = I + C_\pi(zI - A_\pi)^{-1}S^{-1}B_\zeta$$

with inverse given by

(4.4.7) $$W(z)^{-1} = I - C_\pi S^{-1}(zI - A_\zeta)^{-1}B_\zeta.$$

PROOF. Necessity are already been noted. To verify sufficiency, define a rational matrix function $W(z)$ by (4.4.6). We have already verified in Theorem 4.3.1 that then (C_π, A_π) is a right pole pair and (A_ζ, B_ζ) is a left null pair for $W(z)$, and that (4.4.6) and (4.4.7) are minimal realizations for $W(z)$ and $W(z)^{-1}$. It remains only to verify that S is a solution of the Sylvester equation (4.4.4a) which satisfies the additional constraints (4.4.4b) and (4.4.4c). From the minimal realization (4.4.6) it is clear that $\widetilde{B}$ as in (4.4.1) should be taken to be $S^{-1}B_\zeta$ and hence (4.4.4b) follows. From the realization (4.4.7) for $W(z)^{-1}$ we compute, using Proposition 4.1.5, a second minimal realization for $W(z)$:

$$W(z) = I + C_\pi S^{-1}\left(zI - (A_\zeta - B_\zeta C_\pi S^{-1})\right)^{-1}B_\zeta$$

from which it is clear that $\widetilde{C}$ in (4.4.2) should be taken to be $-C_\pi S^{-1}$; this verifies (4.4.4c) and completes the proof of sufficiency.

To see that (4.4.6) gives the only such matrix function $W(z)$, note that S by definition of null-pole coupling matrix must satisfy (4.4.4b), so $\widetilde{B}$ is determined by $\widetilde{B} = S^{-1}B$. Then from the definition of $\widetilde{B}$ in (4.4.1) we conclude that $W(z)$ must be given by (4.4.6). $\square$

4.5 INTERPOLATION WITH INCOMPLETE DATA: SPECIAL CASE

In this section we localize the results of the previous section. Specifically, given a subset $\sigma \subset \mathbf{C}$ and a rational matrix function $W(z)$ analytic and invertible at infinity, we associate with a right pole pair (C_π, A_π) and left null pair for $W(z)$ on σ a local null-pole coupling operator S on σ; the associated data set $(C_\pi, A_\pi; A_\zeta, B_\zeta; S)$ we will call a null-pole triple for $W(z)$ on σ. We will then characterize which triples $(C_\pi, A_\pi; A_\zeta, B_\zeta; S)$ arise as the null-pole triple for some rational matrix function on σ.

To define the local version of the null-pole coupling matrix, we proceed as follows. Let a right pole pair (C_π, A_π) and left null pair (A_ζ, B_ζ) for the rational matrix function $W(z)$ on the subset $\sigma \subset \mathbf{C}$ be given. We shall assume that $W(\infty) = I$. Choose any minimal realization

$$W(z) = I + C(zI - A)^{-1}B$$

for $W(z)$; we use the symbol $\theta = (A, B, C)$ to specify this choice of minimal realization. Let Q_θ (resp. $Q_\theta^\times$) be the Riesz projection corresponding to the eigenvalues of A (resp. $A^\times := A - BC$) in the set σ. By Theorem 4.2.3 we know that $(C \mid \operatorname{Im} Q_\theta, A \mid \operatorname{Im} Q_\theta)$ is a right pole pair for $W(z)$ on σ, and that $(A^\times \mid \operatorname{Im} Q_\theta^\times, Q_\theta^\times B)$ is a left null pair for W on σ. By the uniqueness of right pole pairs and left null pairs on σ up to similarity (see Section 3.4), there exist invertible linear transformations

$$(4.5.1) \qquad T_\pi(\theta) \colon \mathbf{C}^{n_\pi} \to \operatorname{Im} Q_\theta, \qquad T_\zeta(\theta) \colon \mathbf{C}^{n_\zeta} \to \operatorname{Im} Q_\theta^\times$$

such that

$$(4.5.2a) \qquad \left(C_\pi T_\pi^{-1}(\theta), T_\pi(\theta) A_\pi T_\pi^{-1}(\theta)\right) = \left(C \mid \operatorname{Im} Q_\theta, A \mid \operatorname{Im} Q_\theta\right)$$

$$(4.5.2b) \qquad \left(T_\zeta(\theta) A_\zeta T_\zeta^{-1}(\theta), T_\zeta(\theta) B_\zeta\right) = \left(A^\times \mid \operatorname{Im} Q_\theta^\times, Q_\theta^\times B\right).$$

Here $n_\pi \times n_\pi$ and $n_\zeta \times n_\zeta$ are the sizes of A_π and A_ζ respectively. Define $S \colon \mathbf{C}^{n_\pi} \to \mathbf{C}^{n_\zeta}$ by

$$(4.5.3) \qquad S = T_\zeta^{-1}(\theta) Q_\theta^\times T_\pi(\theta).$$

We say that S is the *null-pole coupling matrix on σ* associated with the right pole pair (C_π, A_π) and left null pair (A_ζ, B_ζ) for $W(z)$ with respect to σ.

The definition does not depend on the choice of minimal realization. Indeed, if $\widetilde{\theta} = (\widetilde{A}, \widetilde{B}, \widetilde{C})$ is another realization, then by the state space isomorphism theorem (Theorem 4.1.3), there exists an invertible matrix H such that

$$(4.5.4) \qquad \widetilde{C} = CH^{-1}, \quad \widetilde{A} = HAH^{-1}, \quad \widetilde{B} = HB.$$

By the uniqueness of the similarity mapping between the right pole pairs or between two left null pairs for the same function W on σ (see Section 3.4), it is straightforward to check that

$$(4.5.5) \qquad T_\pi(\widetilde{\theta}) = HT_\pi(\theta), \qquad T_\zeta(\widetilde{\theta}) = HT_\zeta(\theta).$$

From these identities together with $Q_{\widetilde{\theta}}^\times = HQ_\theta^\times H^{-1}$, we conclude that the definition of S does not depend on the choice of minimal realization.

Note that from the identities (4.5.2a), (4.5.2b) and (4.5.3) we have

$$
\begin{aligned}
SA_\pi - A_\zeta S &= T_\zeta(\theta)^{-1}Q_\theta^\times T_\pi(\theta)A_\pi - A_\zeta T_\zeta(\theta)^{-1}Q_\theta^\times T_\pi(\theta) \\
&= T_\zeta(\theta)^{-1}[Q_\theta^\times A - (A - BC)Q_\theta^\times]T_\pi(\theta) \\
&= T_\zeta(\theta)^{-1}[Q_\theta^\times A - Q_\theta^\times(A - BC)]T_\pi(\theta) \\
&= T_\zeta(\theta)^{-1}[Q_\theta^\times BC]T_\pi(\theta) \\
&= B_\zeta C_\pi,
\end{aligned}
$$

that is S satisfies the Sylvester equation

$$(4.5.6) \qquad SA_\pi - A_\zeta S = B_\zeta C_\pi.$$

In what follows, for σ a subset of $\mathbf{C}$, we shall call the collection of matrices $\tau = (C_\pi, A_\pi; A_\zeta, B_\zeta; S)$ a (left) *null-pole triple for* $W(z)$ *on* σ, or a (left) σ-*null-pole triple for* $W(z)$, if (C_π, A_π) is a right pole pair for $W(z)$ with respect to σ, (A_ζ, B_ζ) is a left null pair for $W(z)$ with respect to σ, and S is the associated null-pole coupling matrix on σ. From the definition it is easily seen that for any invertible matrices E and F of suitable sizes $\widetilde{\tau} = (C_\pi E, E^{-1}A_\pi E; F^{-1}A_\zeta F, F^{-1}B_\zeta; F^{-1}SE)$ is again a null-pole triple for $W(z)$ on σ whenever $\tau = (C_\pi, A_\pi; A_\zeta, B_\zeta; S)$ is. In fact, from the uniqueness of right pole pairs and left null pairs for $W(z)$ with respect to σ up to similarity, it is easy to see that any σ-null-pole triple $\widetilde{\tau}$ for $W(z)$ is obtained in this way from a given one τ. Also it is easy to check from the definitions that a left σ-null-pole triple for $W(z)$ (where W has finite and invertible value at infinity) is a left (global) null-pole triple for $W(z)$ (as defined in Section 4.4) if all poles and zeros of W are contained in the set σ.

Recall (from Section 4.4) that we say that a collection of matrices $(C_\pi, A_\pi; A_\zeta, B_\zeta; S)$ is an admissible Sylvester data set if

(i) (C_π, A_π) is a null-kernel pair

(ii) (A_ζ, B_ζ) is a full-range pair

and

(iii) S satisfies the Sylvester equation

$$SA_\pi - A_\zeta S = B_\zeta C_\pi.$$

We have already seen above that a σ-null-pole triple for a rational matrix function with value I at infinity is necessarily an admissible Sylvester data set. The next result shows that this is essentially the only restriction. The proof of this result for the special case, that the complement of σ in $\mathbb{C}$ is an infinite set, is the main goal of this section. The case of a general proper subset σ of the finite complex plane will be presented in the next section.

THEOREM 4.5.1. *Let σ be a subset of $\mathbb{C}$ such that $\mathbb{C}\backslash\sigma$ is infinite. Then the collection of matrices $\tau = (C_\pi, A_\pi; A_\zeta, B_\zeta; S)$ is a left σ-null-pole triple for some rational matrix function $W(z)$ with value I at infinity if and only if*

(i) *the eigenvalues of A_π and A_ζ are in σ and*

(ii) *τ is an admissible Sylvester data set.*

Furthermore, in this case the minimal possible McMillan degree of such a W is equal to

(4.5.7) $n_\pi + n_\zeta - \operatorname{rank} S$

where $n_\pi \times n_\pi$ is the size of A_π and $n_\zeta \times n_\zeta$ is the size of A_ζ.

Recall that the McMillan degree of $W(z)$ is the size of the matrix A taken from a minimal realization (A, B, C) of $W(z)$.

PROOF. The necessity of (i) follows from the fact that (C_π, A_π) and (A_ζ, B_ζ) must be a right pole pair and left null pair for $W(z)$ with respect to σ by definition of τ being a left σ-null-pole triple for $W(z)$. The necessity of (ii) was already noted above.

Furthermore, because of the invertibility of $T_\pi(\theta)$ and $T_\zeta(\theta)$ in (4.5.1), we have

$$
\begin{aligned}
n_\pi + n_\zeta - \operatorname{rank} S &= \operatorname{rank} Q_\theta + \operatorname{rank} Q_\theta^\times - \operatorname{rank}(Q_\theta^\times \mid \operatorname{Im} Q_\theta) \\
&= \operatorname{rank} Q_\theta^\times + \dim\{\operatorname{Im} Q_\theta \cap \operatorname{Ker} Q_\theta^\times\} \\
&\leq N
\end{aligned}
$$

where $N \times N$ is the size of the matrix A in the realization $\theta = (A, B, C)$ for $W(z)$. Since θ is minimal, we conclude that (4.5.7) is a lower bound for the McMillan degree of $W(z)$. The proof of the converse statement in Theorem 4.5.1 will follow in this section after some preliminary work.

To complete the proof of Theorem 4.5.1 we shall need some operations on admissible Sylvester data sets. If $\tau = (C_\pi, A_\pi; A_\zeta, B_\zeta; S)$ is an admissible Sylvester data set for which the eigenvalues of A_π and A_ζ are all in σ, we shall say that τ is a σ-*admissible Sylvester data set*.

Our first operation on admissible Sylvester data sets is that of direct sum. For $\nu = 1, 2$ let

(4.5.8) $\tau_\nu = (C_\pi^{(\nu)}, A_\pi^{(\nu)}; A_\zeta^{(\nu)}, B_\zeta^{(\nu)}; S_{\nu\nu})$

be a σ_ν-admissible Sylvester data set. Assume that $\sigma_1 \cap \sigma_2 = \emptyset$. We define the *direct sum* of τ_1 and τ_2 to be the collection of matrices

(4.5.9)
$$\tau_1 \oplus \tau_2 = \left([C_\pi^{(1)}, C_\pi^{(2)}], \begin{bmatrix} A_\pi^{(1)} & 0 \\ 0 & A_\pi^{(2)} \end{bmatrix} ; \right.$$
$$\left. \begin{bmatrix} A_\zeta^{(1)} & 0 \\ 0 & A_\zeta^{(2)} \end{bmatrix}, \begin{bmatrix} B_\zeta^{(1)} \\ B_\zeta^{(2)} \end{bmatrix} ; \begin{bmatrix} S_{11} & S_{12} \\ S_{21} & S_{22} \end{bmatrix} \right)$$

where S_{12} and S_{21} are the unique solutions of the Sylvester equations

(4.5.10)
$$S_{12} A_\pi^{(2)} - A_\zeta^{(1)} S_{12} = B_\zeta^{(1)} C_\pi^{(2)}$$

(4.5.11)
$$S_{21} A_\pi^{(1)} - A_\zeta^{(2)} S_{21} = B_\zeta^{(2)} C_\pi^{(1)}.$$

Note that the assumption $\sigma_1 \cap \sigma_2 = \emptyset$ guarantees that $A_\pi^{(2)}$ and $A_\zeta^{(1)}$ have no common eigenvalues, and hence (4.5.10) is uniquely solvable for S_{12} (see Section A.1 in the Appendix). A similar remark applies to (4.5.11).

PROPOSITION 4.5.2. *For $\nu = 1, 2$ let τ_ν be a σ_ν-admissible Sylvester data set. Assume $\sigma_1 \cap \sigma_2 = \emptyset$. Then the direct sum $\tau_1 \oplus \tau_2$ is a $(\sigma_1 \cup \sigma_2)$-admissible Sylvester data set.*

PROOF. Let τ_ν be given by (4.5.8) and denote $\tau_1 \oplus \tau_2$ given by (4.5.9) by $\tau_1 \oplus \tau_2 = (C_\pi, A_\pi; A_\zeta, B_\zeta; S)$. Then (C_π, A_π) is a null-kernel pair since $(C_\pi^{(\nu)}, A_\pi^{(\nu)})$ is for $\nu = 1, 2$, and (A_ζ, B_ζ) is a full range pair since each $(A_\zeta^{(\nu)}, B_\zeta^{(\nu)})$ is, by an argument as in the proof of Theorem 1.5.1. The fact that S_{12} and S_{21} are defined by (4.5.10) and (4.5.11) and that $S_{\nu\nu}$ satisfies the appropriate Sylvester equation as part of the admissible Sylvester data set τ_ν $(\nu = 1, 2)$ imply that S satisfies the Sylvester equation $SA_\pi - A_\zeta S = B_\zeta C_\pi$. Finally the eigenvalues of A_π and A_ζ are all in $\sigma_1 \cup \sigma_2$ since the eigenvalues of $A_\pi^{(\nu)}$ and $A_\zeta^{(\nu)}$ are in σ_ν for $\nu = 1, 2$. We conclude that $\tau_1 \oplus \tau_2$ is a $(\sigma_1 \cup \sigma_2)$-admissible Sylvester data set. $\square$

The second operation on data sets which we shall require is that of corestriction; in a simplified form it has already appeared in Section 4.2 (specifically Theorem 4.2.3). In general let us say that the pair of matrices (C_π, A_π) is *right admissible* if C_π has size $n \times n_\pi$ and A_π has size $n_\pi \times n_\pi$; here n is fixed throughout the discussion but n_π can depend on the pair. Similarly we say that (A_ζ, B_ζ) is a *left admissible* pair if A_ζ has size $n_\zeta \times n_\zeta$ and B has size $n_\zeta \times n$. If $(C_\pi^{(1)}, A_\pi^{(1)})$ and $(C_\pi^{(2)}, A_\pi^{(2)})$ are two right admissible pairs with $n_\pi^{(\nu)} \times n_\pi^{(\nu)}$ the size of $A_\pi^{(\nu)}$ for $\nu = 1, 2$ and $\mathcal{M}$ is a subspace of $\mathbf{C}^{n_\pi^{(1)}}$ and $\Phi : \mathcal{M} \to \mathbf{C}^{n_\pi^{(2)}}$ is a bijective linear transformation, we say that the pair $(C_\pi^{(2)}, A_\pi^{(2)})$ is the $\{\Phi, \mathcal{M}\}$-*restriction* of $(C_\pi^{(1)}, A_\pi^{(1)})$ if $\mathcal{M}$ is invariant under $A_\pi^{(1)}$ and

(4.5.12)
$$\Phi(A_\pi^{(1)} \mid \mathcal{M}) = A_\pi^{(2)} \Phi, \qquad C_\pi^{(1)} \mid \mathcal{M} = C_\pi^{(2)} \Phi.$$

Compression is defined as the dual concept. More precisely, if $(A_\zeta^{(\nu)}, B_\zeta^{(\nu)})$ is a left admissible pair where $A_\zeta^{(\nu)}$ has size $n_\zeta^{(\nu)} \times n_\zeta^{(\nu)}$ for $\nu = 1, 2$, π is a projection of $\mathbb{C}^{n_\zeta^{(1)}}$ onto a subspace $\mathcal{L}$ of $\mathbb{C}^{n_\zeta^{(1)}}$ and $\Psi \colon \mathbb{C}^{n_\zeta^{(2)}} \to \mathcal{L}$ is a bijective linear transformation, then the pair $(A_\zeta^{(2)}, B_\zeta^{(2)})$ is said to be the $\{\Psi, \pi\}$-compression of $(A_\zeta^{(1)}, B_\zeta^{(1)})$ if the kernel of π is invariant under $A_\zeta^{(1)}$ and if

$$(4.5.13) \qquad \pi(A_\zeta^{(1)} \mid \mathcal{L})\Psi = \Psi A_\zeta^{(2)}, \qquad \pi B_\zeta^{(1)} = \Psi B_\zeta^{(2)}.$$

Now suppose $\tau_1 = (C_\pi^{(1)}, A_\pi^{(1)}; A_\zeta^{(1)}, B_\zeta^{(1)}; S_1)$ is an admissible Sylvester data set. Assume that $(C_\pi^{(2)}, A_\pi^{(2)})$ is a $\{\Phi, \mathcal{M}\}$-restriction of $(C_\pi^{(1)}, A_\pi^{(1)})$ and $(A_\zeta^{(2)}, B_\zeta^{(2)})$ is a $\{\Psi, \pi\}$-compression of $(A_\zeta^{(1)}, B_\zeta^{(1)})$. Then the collection of matrices $\tau_2 = (C_\pi^{(2)}, A_\pi^{(2)}; A_\zeta^{(2)}, B_\zeta^{(2)}; S_2)$ is called the $\{(\Phi, \mathcal{M}), (\Psi, \pi)\}$-corestriction of τ_1 if in addition

$$(4.5.14) \qquad \Psi S_2 \Phi = \pi(S_1 \mid \mathcal{M}).$$

PROPOSITION 4.5.3. *The corestriction of a σ-admissible Sylvester data set is again a σ-admissible Sylvester data set.*

PROOF. Let τ_1 and its corestriction τ_2 be as in the preceding paragraph. Assume that the eigenvalues of $A_\pi^{(1)}$ and $A_\zeta^{(1)}$ are in the set σ, so τ_1 is a σ-admissible Sylvester data set. We have to prove that τ_2 is also a σ-admissible Sylvester data set. From the definitions of restriction and compression it is clear that $(C_\pi^{(2)}, A_\pi^{(2)})$ is a null-kernel pair, $(A_\zeta^{(2)}, B_\zeta^{(2)})$ is a full-range pair and that the eigenvalues of $A_\pi^{(2)}$ and $A_\zeta^{(2)}$ are in σ. It remains only to show that S_2 satisfies the associated Sylvester equation.

To verify this we use the identities (4.5.12), (4.5.13) and (4.5.14). Take $y \in \mathbb{C}^{n_\pi^{(2)}}$. Then there exists s unique $x \in \mathcal{M}$ for which $\Phi x = y$. Note that then $A_\pi^{(1)} x \in \mathcal{M}$. Thus

$$
\begin{aligned}
(S_2 A_\pi^{(2)} - A_\zeta^{(2)} S_2)y &= S_2 A_\pi^{(2)} \Phi x - A_\zeta^{(2)} S_2 \Phi x \\
&= S_2 \Phi A_\pi^{(1)} x - A_\zeta^{(2)} S_2 \Phi x \\
&= \Psi^{-1} \pi S_1 A_\pi^{(1)} x - A_\zeta^{(2)} \Psi^{-1} \pi S_1 x \\
&= \Psi^{-1} (\pi S_1 A_\pi^{(1)} x - \pi A_\zeta^{(1)} \pi S_1 x) \\
&= \Psi^{-1} (\pi S_1 A_\pi^{(1)} x - \pi A_\zeta^{(1)} S_1 x) \\
&= \Psi^{-1} \pi B_\zeta^{(1)} C_\pi^{(1)} x = B_\zeta^{(2)} C_\pi^{(2)} y
\end{aligned}
$$

as needed. Note that $\pi A_\zeta^{(1)} \pi = \pi A_\zeta^{(1)}$ because of the $A_\zeta^{(1)}$-invariance of Ker π. $\quad\square$

Let τ_1 and τ_2 be admissible Sylvester data sets as above, and let τ_2 be the $\{(\Phi, \mathcal{M}), (\Psi, \pi)\}$-corestriction of τ_1. Assume that $\mathcal{M} = \mathbb{C}^{n_\pi^{(1)}}$ and π is the identity linear transformation on $\mathbf{C}^{n_\zeta^{(1)}}$. Thus

$$\Phi\colon \mathbf{C}^{n_\pi^{(1)}} \to \mathbf{C}^{n_\pi^{(2)}}, \qquad \Psi\colon \mathbf{C}^{n_\zeta^{(2)}} \to \mathbf{C}^{n_\zeta^{(1)}}$$

are invertible linear transformations and the following identities hold true:

(4.5.15)
$$\Phi A_\pi^{(1)} \Phi^{-1} = A_\pi^{(2)}, \qquad C_\pi^{(1)} \Phi^{-1} = C_\pi^{(2)}$$

(4.5.16)
$$\Psi^{-1} A_\zeta^{(1)} \Psi = A_\zeta^{(2)}, \qquad \Psi^{-1} B_\zeta^{(1)} = B_\zeta^{(2)}$$

(4.5.17)
$$\Psi^{-1} S_1 \Phi^{-1} = S_2.$$

In this case we shall say that τ_1 and τ_2 are *similar*, or, more precisely, that τ_2 is (Φ, Ψ)-*similar* to τ_1.

This extends the notion of similarity already introduced in Theorem 4.4.2 for global left null pole triples to general admissible triples. Also, we observed above that the left σ-null-pole triples for a given W are all similar. Also any admissible Sylvester data set which is similar to a left σ-null-pole triple for W is again a left σ-null-pole triple for W.

We next derive an elementary lemma which will be useful later.

LEMMA 4.5.4. *Let τ_1, τ_2 and τ_3 be admissible Sylvester data sets. Then*

(i) *τ_3 is a corestriction of τ_2 and τ_2 is a corestriction of τ_1 implies that τ_3 is a corestriction of τ_1.*

Next, assume that $\tau_1 \oplus \tau_3$ is well-defined. Then

(ii) *$\tau_1 \oplus \tau_3$ and $\tau_3 \oplus \tau_1$ are similar,*

(iii) *τ_1 and τ_3 are corestrictions of $\tau_1 \oplus \tau_3$.*

If, in addition, τ_2 is a corestriction of τ_1, then

(iv) *$\tau_2 \oplus \tau_3$ is a corestriction of $\tau_1 \oplus \tau_3$.*

PROOF. For $\nu = 1, 2, 3$, let $\tau_\nu = (C_\pi^{(\nu)}, A_\pi^{(\nu)}; A_\zeta^{(\nu)}, B_\zeta^{(\nu)}; S_\nu)$ where $A_\pi^{(\nu)}$ has size $n_\pi^{(\nu)} \times n_\pi^{(\nu)}$ and $A_\zeta^{(\nu)}$ has size $n_\zeta^{(\nu)} \times n_\zeta^{(\nu)}$. To prove (i), assume that τ_2 is the $\{(\Phi, \mathcal{M}), (\Psi, \pi)\}$-corestriction of τ_1 and τ_3 is the $\{(E, \mathcal{N}), (F, \rho)\}$-corestriction of τ_2. Put $\mathcal{M}_0 = \Phi^{-1}\mathcal{N}$, and let π_0 be the projection of $\mathbf{C}^{n_\zeta^{(1)}}$ onto $\Psi \operatorname{Im}\rho$ along $\Psi \operatorname{Ker}\rho \oplus \operatorname{Ker}\pi$. Then it is straightforward to check that τ_3 is the $\{(E \circ \Phi, \mathcal{M}_0), (\Psi \circ F, \pi_0)\}$-corestriction of τ_1.

Next assume that $\tau_1 \oplus \tau_3$ is well-defined. Thus τ_1 is a σ_1-admissible Sylvester data set and τ_3 is a σ_3-admissible Sylvester data set with $\sigma_1 \cap \sigma_3 = \emptyset$. The statement

(ii) is trivial. To prove that τ_1 is a corestriction of $\tau_1 \oplus \tau_3$, define

$$\mathcal{M} = \mathbf{C}^{n_\pi^{(1)}} \oplus \{0\}$$

$$\Phi = [\, I \quad 0\,] : \mathcal{M} \to \mathbf{C}^{n_\pi^{(1)}}$$

$$\pi = \begin{bmatrix} I & 0 \\ 0 & 0 \end{bmatrix} : \mathbf{C}^{n_\zeta^{(1)}} \oplus \mathbf{C}^{n_\zeta^{(3)}} \to \mathbf{C}^{n_\zeta^{(1)}} \oplus \mathbf{C}^{n_\zeta^{(3)}}$$

$$\Psi = \begin{bmatrix} I \\ 0 \end{bmatrix} : \mathbf{C}^{n_\pi^{(1)}} \to \operatorname{Im} \pi$$

and then check that the $\{(\Phi, \mathcal{M}), (\Psi, \pi)\}$-corestriction of $\tau_1 \oplus \tau_3$ is precisely τ_1. $\square$

Let now W denote a rational $n \times n$ matrix function with value I at infinity. The following theorem is a refinement of Theorem 4.4.3.

THEOREM 4.5.5. *For $\nu = 1, \ldots, r$, let*

(4.5.18) $$\tau_\nu = (C_\pi^{(\nu)}, A_\pi^{(\nu)}; A_\zeta^{(\nu)}, B_\zeta^{(\nu)}; S_{\nu\nu})$$

be a σ_ν-admissible Sylvester data set. Assume that $\sigma_1, \ldots, \sigma_r$ are mutually disjoint and for $i \neq j$ let S_{ij} be the unique solution of the Sylvester equation

(4.5.19) $$S_{ij} A_\pi^{(j)} - A_\zeta^{(i)} S_{ij} = B_\zeta^{(i)} C_\pi^{(j)}.$$

Then there exists a rational matrix function W with value I at infinity such that for each ν the data set τ_ν is a σ_ν-null-pole triple for W and all poles and zeros of W belong to $\sigma_1 \cup \cdots \cup \sigma_r$ if and only if the matrix

(4.5.20) $$S = \begin{bmatrix} S_{11} & \cdots & S_{1r} \\ \vdots & & \vdots \\ S_{r1} & \cdots & S_{rr} \end{bmatrix}$$

is invertible. In this case W is uniquely determined and has the following minimal realization:

$$W(z) = I + C(zI - A)^{-1} B$$

where

(4.5.21)
$$A = \operatorname{diag}(A_\pi^{(1)}, \cdots, A_\pi^{(r)}),$$
$$B = S^{-1} \cdot \operatorname{col}(B_\zeta^{(1)}, \cdots, B_\zeta^{(r)}),$$
$$C = [C_\pi^{(1)}, \cdots, C_\pi^{(r)}].$$

For the proof we need the following lemma.

LEMMA 4.5.6. *For $\nu = 1, 2$, let τ_ν be a σ_ν-admissible Sylvester data set and let W be a rational matrix function with value I at infinity. Assume $\sigma_1 \cap \sigma_2 = \emptyset$. Then*

$\tau_1 \oplus \tau_2$ *is a left* $(\sigma_1 \cup \sigma_2)$*-null-pole triple for* W *if and only if for* $\nu = 1, 2$ *the triple* τ_ν *is a left* σ_ν*-null-pole triple for* W.

PROOF. In what follows $\nu = 1$ or $\nu = 2$. Let $\theta = (A, B, C)$ be a minimal realization for W. By Q_ν (resp. $Q_\nu^\times$) we denote the Riesz projection corresponding to the eigenvalues of A (resp. $A^\times = A - BC$) in σ_ν. Put $Q = Q_1 + Q_2$ and $Q^\times = Q_1^\times + Q_2^\times$ and let $\sigma = \sigma_1 \cup \sigma_2$. Since $\sigma_1 \cap \sigma_2 = \emptyset$, the linear transformation Q is the Riesz projection corresponding to the eigenvalues of A in σ. A similar remark holds true for $Q^\times$ with A replaced by $A^\times$.

Let τ_ν be given by (4.5.18); denote the coupling matrix of $\tau_1 \oplus \tau_2$ by S, so

$$S = \begin{bmatrix} S_{11} & S_{12} \\ S_{21} & S_{22} \end{bmatrix}$$

where S_{11} and S_{22} are the coupling matrices of τ_1 and τ_2, respectively, and S_{12} and S_{21} are defined by (4.5.10) and (4.5.11). Let $n_\pi^{(\nu)} \times n_\pi^{(\nu)}$ be the size of $A_\pi^{(\nu)}$ and $n_\zeta^{(\nu)} \times n_\zeta^{(\nu)}$ be the size of $A_\zeta^{(\nu)}$.

Assume that $\tau_1 \oplus \tau_2$ is a σ-null-pole triple for W. Then there exist invertible linear transformations

$$T_\pi : \mathbb{C}^{n_\pi^{(1)}} \oplus \mathbb{C}^{n_\pi^{(2)}} \to \operatorname{Im} Q$$

$$T_\zeta : \mathbb{C}^{n_\zeta^{(1)}} \oplus \mathbb{C}^{n_\pi^{(2)}} \to \operatorname{Im} Q^\times$$

such that

(4.5.22) $$[\, C_\pi^{(1)} \quad C_\pi^{(2)} \,] T_\pi^{-1} = C \mid \operatorname{Im} Q$$

(4.5.23) $$T_\pi \operatorname{diag}(A_\pi^{(1)}, A_\pi^{(2)}) T_\pi^{-1} = A \mid \operatorname{Im} Q$$

(4.5.24) $$T_\zeta \operatorname{col}(B_\zeta^{(1)}, B_\zeta^{(2)}) = Q^\times B$$

(4.5.25) $$T_\zeta \operatorname{diag}(A_\zeta^{(1)}, A_\zeta^{(2)}) T_\zeta^{-1} = A^\times \mid \operatorname{Im} Q^\times$$

and

(4.5.26) $$S = T_\zeta^{-1} Q^\times T_\pi.$$

Since σ_1 and σ_2 are disjoint sets the similarities in (4.5.23) and (4.5.25) imply that $T_\pi(\mathbb{C}^{n_\pi^{(1)}} \oplus 0) = \operatorname{Im} Q_1$, $T_\pi(0 \oplus \mathbb{C}^{n_\pi^{(2)}}) = \operatorname{Im} Q_2$, $T_\zeta(\mathbb{C}^{n_\zeta^{(1)}} \oplus 0) = \operatorname{Im} Q_1^\times$, $T_\zeta(0 \oplus \mathbb{C}^{n_\zeta^{(2)}}) = \operatorname{Im} Q_2^\times$ and thus T_π and T_ζ have the block diagonal form

$$T_\pi = \begin{bmatrix} T_{1,\pi} & 0 \\ 0 & T_{2,\pi} \end{bmatrix} : \mathbb{C}^{n_\pi^{(1)}} \oplus \mathbb{C}^{n_\pi^{(2)}} \to \operatorname{Im} Q_1 \dotplus \operatorname{Im} Q_2$$

and

$$T_\zeta = \begin{bmatrix} T_{1,\zeta} & 0 \\ 0 & T_{2,\zeta} \end{bmatrix} : \mathbb{C}^{n_\zeta^{(1)}} \oplus \mathbb{C}^{n_\zeta^{(2)}} \to \operatorname{Im} Q_1^\times \dotplus \operatorname{Im} Q_2^\times .$$

Note that the linear transformations $T_{\nu,\pi}$ and $T_{\nu,\zeta}$ are invertible. From (4.5.22)–(4.5.26) it follows that

$$(4.5.27) \qquad (C_\pi^{(\nu)} T_{\nu,\pi}^{-1}, T_{\nu,\pi} A_\pi^{(\nu)} T_{\nu,\pi}^{-1}) = (C \mid \operatorname{Im} Q_\nu, A \mid \operatorname{Im} Q_\nu)$$

$$(4.5.28) \qquad (T_{\nu,\zeta} A_\zeta^{(\nu)} T_{\nu,\zeta}^{-1}, T_{\nu,\zeta} B_\zeta^{(\nu)}) = (A^\times \mid \operatorname{Im} Q_\nu^\times, Q_\nu^\times B),$$

$$(4.5.29) \qquad S_{\nu\nu} = T_{\nu,\zeta}^{-1} Q_\nu^\times T_{\nu,\pi}.$$

To prove the latter identity one uses that (4.5.26) may be rewritten in the following form:

$$(4.5.30) \qquad \begin{bmatrix} T_{1,\zeta} & 0 \\ 0 & T_{2,\zeta} \end{bmatrix} \begin{bmatrix} S_{11} & S_{12} \\ S_{21} & S_{22} \end{bmatrix} \begin{bmatrix} T_{1,\pi}^{-1} & 0 \\ 0 & T_{2,\pi}^{-1} \end{bmatrix} = Q^\times \mid \operatorname{Im} Q.$$

To prove the converse, assume that for $\nu = 1, 2$ the triple τ_ν is a left σ_ν-null-pole triple for W. So there exist invertible linear transformations

$$T_{\nu,\pi} : \mathbb{C}^{n_\pi^{(\nu)}} \to \operatorname{Im} Q_\nu, \qquad T_{\nu,\zeta} : \mathbb{C}^{n_\zeta^{(\nu)}} \to \operatorname{Im} Q_\nu^\times$$

such that (4.5.27), (4.5.28) and (4.5.29) hold true. We shall prove that $T_{\nu,\pi}$ and $T_{\nu,\zeta}$ satisfy (4.5.30). In order to do this we have to show that

$$(4.5.31) \qquad T_{1,\zeta} S_{11} T_{1,\pi}^{-1} = Q_1^\times \mid \operatorname{Im} Q_1, \qquad T_{2,\zeta} S_{22} T_{2,\pi}^{-1} = Q_2^\times \mid \operatorname{Im} Q_2$$

$$(4.5.32) \qquad T_{1,\zeta} S_{12} T_{2,\pi}^{-1} = Q_1^\times \mid \operatorname{Im} Q_2, \qquad T_{2,\zeta} S_{21} T_{1,\pi}^{-1} = Q_2^\times \mid \operatorname{Im} Q_1.$$

The identities in (4.5.31) follow directly from (4.5.29). From (4.5.10) and the identities (4.5.27) and (4.5.28) we conclude that $Z = T_{1,\zeta} S_{12} T_{2,\pi}^{-1}$ is the unique solution of the Sylvester equation

$$(4.5.33) \qquad Z(A \mid \operatorname{Im} Q_2) - (A^\times \mid \operatorname{Im} Q_1^\times) Z = Q_1^\times B(C \mid \operatorname{Im} Q_2).$$

Since $A - A^\times = BC$, the solution Z of (4.5.33) is equal to $Q_1^\times \mid \operatorname{Im} Q_2$ which proves the first identity in (4.5.32). The second is proved in a similar way.

Now define $T_\pi = \operatorname{diag}(T_{1,\pi}, T_{2,\pi}) : \mathbb{C}^{n_\pi^{(1)}} \oplus \mathbb{C}^{n_\pi^{(2)}} \to \operatorname{Im} Q_1 \dotplus \operatorname{Im} Q_2$ and $T_\zeta = \operatorname{diag}(T_{1,\zeta}, T_{2,\zeta}) : \mathbb{C}^{n_\zeta^{(1)}} \oplus \mathbb{C}^{n_\zeta^{(2)}} \to \operatorname{Im} Q_1^\times \dotplus \operatorname{Im} Q_2^\times$. Then T_π and T_ζ are invertible, the identities (4.5.27) and (4.5.28) yield (4.5.22)–(4.5.25), and the formula (4.5.30) yields (4.5.26). Hence $\tau_1 \oplus \tau_2$ is a left σ-null-pole triple for W. $\quad\square$

PROOF OF THEOREM 4.5.5. Note that the matrix S in (4.5.20) is precisely the coupling matrix of the data set $\tau_1 \oplus \cdots \oplus \tau_r$. Let W be a proper rational matrix function with $W(\infty) = I$ such that all the poles and zeros of W belong to $\sigma_1 \cup \cdots \cup \sigma_r$. Assume that for each ν the data set τ_ν is a σ_ν-null-pole triple for W. Then Lemma 4.5.6 implies that $\tau_1 \oplus \cdots \oplus \tau_r$ is a $(\sigma_1 \cup \cdots \cup \sigma_r)$-null-pole triple for W. Since the poles and zeros of W are in $\sigma_1 \cup \cdots \cup \sigma_r$ in fact $\tau_1 \oplus \cdots \oplus \tau_r$ is a global null-pole triple for W, so the coupling matrix S by Theorem 4.4.3 must be invertible. We also see that W is uniquely determined and $W(z) = I + C(zI - A)^{-1}B$ with (A, B, C) given by (4.5.21). $\square$

We conclude this section with the second part of the proof of Theorem 4.5.1. We have already shown that if τ is a σ-null-pole triple for W, then τ is a σ-admissible Sylvester data set and (4.5.7) is a lower bound for the degree of W. We shall now solve the inverse problem. This will be done on the basis of the next lemma.

LEMMA 4.5.7. *Let $\tau = (C_\pi, A_\pi; A_\zeta, B_\zeta; S)$ be a σ-admissible Sylvester set, where A_π and A_ζ have sizes $n_\pi \times n_\pi$ and $n_\zeta \times n_\zeta$ respectively and n be the number of rows in C_π (= number of columns in B_ζ). Put $n_1 = \dim \mathbb{C}^{n_\zeta} / \operatorname{Im} S$, $n_2 = \dim \operatorname{Ker} S$, and let Ω be an arbitrary non-empty infinite set disjoint with σ. Then there exist matrices A_1, A_2, C_1, B_2, and S of sizes $n_1 \times n_1$, $n_2 \times n_2$, $n \times n_1$, $n_2 \times n$, and $n_2 \times n_1$ respectively such that $\tau_0 = (C_1, A_1; A_2, B_2; S_0)$ is an Ω-admissible Sylvester data set and the coupling matrix of the direct sum $\tau \oplus \tau_0$ is invertible.*

SECOND PART OF THE PROOF OF THEOREM 4.5.1. Let τ be a σ-admissible Sylvester data set. We want to show that τ is a null-pole triple for some W such that the degree of W is given by (4.5.7). To do this take τ_0 as in Lemma 4.5.7. Since the coupling matrix of $\tau \oplus \tau_0$ is invertible, we can apply Theorem 4.4.3 to show that there exists a rational matrix function W with $W(\infty) = I$ such that $\tau \oplus \tau_0$ is a global spectral triple for W. This implies (Lemma 4.5.6) that τ is a σ-spectral triple for W. Note that W has the desired degree. Indeed,

$$\delta(W) = n_\pi + n_1 = n_\pi + n_\zeta - \operatorname{rank} S.$$

PROOF OF LEMMA 4.5.7. Assume $\operatorname{Im} S \neq \mathbb{C}^{n_\zeta}$. First we will select a vector $x_0 \in \mathbb{C}^{n_\zeta} \setminus \operatorname{Im} S$ and a point $\alpha \in \Omega$ such that

(4.5.34) $$(A_\zeta - \alpha)x_0 \in \operatorname{Im} B_\zeta.$$

Since (A_ζ, B_ζ) is a null kernel pair, there exist (see e.g., Wonham [1985], Section 5.7) bases $\mathcal{B}$ and $\mathcal{E}$ in $\mathbb{C}^{n_\zeta}$ and $\mathbb{C}^n$, respectively, and a linear transformation $G: \mathbb{C}^{n_\zeta} \to \mathbb{C}^n$ such that relative to these bases the matrix of $A_\zeta - B_\zeta G$ is equal to

(4.5.35) $$J = J_{\nu_1} \oplus \cdots \oplus J_{\nu_k},$$

where

$$J_\nu = \begin{bmatrix} 0 & 1 & & & \\ & 0 & 1 & & \\ & & 0 & \ddots & \\ & & & \ddots & 1 \\ & & & & 0 \end{bmatrix}$$

is a $\nu \times \nu$ single Jordan block with 0 on the main diagonal, and all the entries of the matrix of B_ζ are zero except those in the positions $(\nu_1, 1), (\nu_1 + \nu_2, 2), \cdots, (\nu_1 + \cdots + \nu_k, k)$, which are equal to one. (This form of $A_\zeta - B_\zeta G$ and B_ζ is known as the Brunovsky canonical form.) We assume $\nu_1 \geq \nu_2 \geq \cdots \geq \nu_k$. Put $n = \nu_1 + \cdots + \nu_k$. For $\alpha \neq 0$ let $u(\alpha, i)$ be the vector in $\mathbf{C}^{n_\zeta}$ of which the j-th coordinate relative to the basis $\mathcal{B}$ in $\mathbf{C}^{n_\zeta}$ is equal to

$$\alpha^{j - (\nu_1 + \cdots + \nu_{i-1} + 1) - \nu_i}$$

if $j = \nu_1 + \cdots + \nu_{i-1} + 1, \ldots, \nu_1 + \cdots + \nu_{i-1} + \nu_i$ and is equal to zero otherwise. Then

(4.5.36) $\alpha u(\alpha, i) - (A_\zeta - B_\zeta G) u(\alpha, i) = B_\zeta e_i$

for $i = 1, \ldots, k$. Here e_i is the i-th vector in the basis $\mathcal{E}$ of $\mathbf{C}^n$. Let Ω' be the non-empty infinite set which one obtains if the point 0 (i.e., the only eigenvalue of $A_\zeta - B_\zeta G$) is deleted from Ω. Note that (4.5.34) holds if and only if $\alpha x_0 - (A_\zeta - B_\zeta G) x_0 \in \operatorname{Im} B_\zeta$. Since $B_\zeta e_1, \ldots, B_\zeta e_k$ is a basis for $\operatorname{Im} B_\zeta$, we conclude from (4.5.36) that for $\alpha \in \Omega'$

$$(\alpha - A_\zeta) x_0 \in \operatorname{Im} B_\zeta \iff x_0 \in \operatorname{Span}\{u(\alpha, i) \mid 1 \leq i \leq k\}.$$

Take ν_1 different points $\alpha_1, \ldots, \alpha_{\nu_1}$ in Ω'. Then

$$\mathbf{C}^{n_\zeta} = \operatorname{Span}\{u(\alpha_j, i) \mid i = 1, \ldots, k, j = 1, \ldots, \nu_i\}.$$

So there exist j_0 and i_0 such that $u(\alpha_{j_0}, i_0) \notin \operatorname{Im} S$. Then $x_0 = u(\alpha_{j_0}, i_0)$ and $\alpha = \alpha_{j_0}$ have the desired properties (i.e. satisfy (4.5.34)).

Let x_0 and α be as in the first paragraph of this proof. Introduce the following linear transformations:

$$\gamma: \mathbf{C} \to X_\zeta, \qquad \gamma 1 = x_0$$
$$C: \mathbf{C} \to \mathbf{C}^n, \qquad (\alpha - A_\zeta) x_0 = B_\zeta C 1.$$

Since $x_0 \neq 0$ and α is not an eigenvalue of A_ζ, the vector $C1 \neq 0$, and hence $C \neq 0$. Consider the triple

$$\hat{\tau} = \left(\begin{bmatrix} C_\pi & C \end{bmatrix}, \begin{bmatrix} A_\pi & 0 \\ 0 & \alpha \end{bmatrix}; A_\zeta, B_\zeta; \begin{bmatrix} S & \gamma \end{bmatrix} \right).$$

Put $\hat{\sigma} = \sigma \cup \{\alpha\}$. We claim that $\hat{\tau}$ is a $\hat{\sigma}$-admissible triple. For $\hat{\tau}$ condition (ii) in the definition of an admissible Sylvester data set holds true trivially. Since $C \neq 0$ and α is not an eigenvalue of A_π, condition (i) is also satisfied (cf. the proof of Theorem 1.5.1). To check the third condition note that

$$\begin{bmatrix} S & \gamma \end{bmatrix} \begin{bmatrix} A_\pi & 0 \\ 0 & \alpha \end{bmatrix} - A_\zeta \begin{bmatrix} S & \gamma \end{bmatrix} = \begin{bmatrix} SA_\pi & \alpha \gamma \end{bmatrix} - \begin{bmatrix} A_\zeta S & A_\zeta \gamma \end{bmatrix}$$
$$= \begin{bmatrix} SA_\pi - A_\zeta S & (\alpha - A_\zeta) \gamma \end{bmatrix}$$
$$= \begin{bmatrix} B_\zeta C_\pi & B_\zeta C \end{bmatrix}$$
$$= B_\zeta \begin{bmatrix} C_\pi & C \end{bmatrix}.$$

Thus $\hat{\tau}$ is a $\hat{\sigma}$-admissible Sylvester data set. Obviously,

$$(4.5.37) \qquad \operatorname{Im} S \subsetneq \operatorname{Im}[\ S \quad \gamma\ ] \subset \mathbf{C}^{n_\zeta}.$$

If the second inclusion in (4.5.37) is proper, then we can repeat for $\hat{\tau}$ the same construction as for τ, replacing Ω by $\hat{\Omega} = \Omega \backslash \{\alpha\}$. Proceeding in this way we obtain in a finite number of steps linear transformations

$$A_1 : \mathbf{C}^{n_1} \to \mathbf{C}^{n_1}, \quad C_1 : \mathbf{C}^{n_1} \to \mathbf{C}^{n}, \quad S_1 : \mathbf{C}^{n_1} \to \mathbf{C}^{n_\zeta}$$

such that the eigenvalues of A_1 are in Ω and

$$\tau_1 = \left\{ [\ C_\pi \quad C_1\ ], \begin{bmatrix} A_\pi & 0 \\ 0 & A_1 \end{bmatrix}; A_\zeta, B_\zeta; [\ S \quad S_1\ ] \right\}$$

is an admissible Sylvester data set with the extra property that $\operatorname{Im}[\ S \quad S_1\ ] = \mathbf{C}^{n_\zeta}$. Here $n_1 = n_\zeta - \operatorname{rank} S$.

Let Ω_1 be the infinite set which one obtains by deleting the eigenvalues of A_1 from Ω. Apply the result obtained in the previous paragraph to the admissible Sylvester data set

$$\tau_1^* = \left\{ B_\zeta^*, A_\zeta^*; \begin{bmatrix} A_\pi^* & 0 \\ 0 & A_1^* \end{bmatrix}, \begin{bmatrix} C_\pi^* \\ C_1^* \end{bmatrix}; \begin{bmatrix} S^* \\ S_1^* \end{bmatrix} \right\},$$

with Ω replaced by Ω_1. Here T^* denotes the adjoint of the linear transformation T. In this way we see that there exist linear transformations (where $n_2 = n_1 + n_\pi - n_\zeta$)

$$A_2 : \mathbf{C}^{n_2} \to \mathbf{C}^{n_2}, \quad B_2 : \mathbf{C}^{n} \to \mathbf{C}^{n_2},$$
$$S_2 : \mathbf{C}^{n_\pi} \to \mathbf{C}^{n_2}, \quad S_0 : \mathbf{C}^{n_1} \to \mathbf{C}^{n_2},$$

such that the eigenvalues of A_2 are in $\Omega_1 \subset \Omega$ and

$$\tau_2 = \left\{ [\ C_\pi \quad C_1\ ], \begin{bmatrix} A_\pi & 0 \\ 0 & A_1 \end{bmatrix}; \begin{bmatrix} A_\zeta & 0 \\ 0 & A_2 \end{bmatrix}, \begin{bmatrix} B_\zeta \\ B_2 \end{bmatrix}; \begin{bmatrix} S & S_1 \\ S_2 & S_0 \end{bmatrix} \right\}$$

is an admissible Sylvester data set with an invertible coupling matrix. Now put $\tau_0 = (C_1, A_1; A_2, B_2; S_0)$. Then τ_0 is an Ω-admissible Sylvester data set and $\tau \oplus \tau_0 = \tau_2$ has the desired properties. $\square$

In the next chapter we shall define σ-null-pole triples even for improper (i.e. not analytic at infinity) regular rational matrix functions. We shall need the following result concerning the extent of uniqueness in the solution of the interpolation problems in Theorem 4.5.1. The proof will be given in Chapter 12 (see Corollary 12.3.3) using techniques which are different from those of this chapter; there we shall see that the result continues to hold even for improper rational matrix functions. The result in this general form will be needed in the next chapter.

THEOREM 4.5.8. *Let W_1 and W_2 be regular rational matrix functions and let σ be a subset of $\mathbf{C}$. Then W_1 and W_2 have a common left σ-null-pole triple if and only if $W_2(z)^{-1}W_1(z)$ has no poles and no zeros in σ.*

Finally, for future reference we indicate an important fact giving the left σ-null-pole triple in terms of a minimal realization.

COROLLARY 4.5.9. *Let*

$$W(z) = D + C(zI - A)^{-1}B$$

be a minimal realization for the rational matrix function $W(z)$ having finite and invertible value D at infinity, and let $\sigma \subset \mathbf{C}$. Further, denote by P and Q the Riesz projections corresponding to the eigenvalues of A and of $A - BD^{-1}C$, respectively, in the set σ. Then

$$(C \mid \operatorname{Im} P, A \mid \operatorname{Im} P; (A - BD^{-1}C) \mid \operatorname{Im} Q, QB; Q \mid \operatorname{Im} P)$$

is a left σ-pole triple of $W(z)$.

Here Q is understood as a linear transformation onto its range.

The proof follows immediately by combining Proposition 4.4.1 and Lemma 4.5.6.

4.6 INTERPOLATION WITH INCOMPLETE DATA: GENERAL CASE

In this section we solve the same local null/pole interpolation problem as in the previous section. The difference here is that we now allow σ to be any proper subset of the complex plane $\mathbf{C}$. The proof of the previous section in fact is adequate whenever $\mathbf{C} \backslash \sigma$ has at least $[n_\pi + n_\zeta - \operatorname{rank} S] - \max\{n_\pi, n_\zeta\}$ many points. In the other cases more delicate arguments must be used. Here we prove the theorem with only the assumption that σ is a proper subset of $\mathbf{C}$.

It is convenient at this stage to introduce the formal notion of a minimal complement to a σ-admissible Sylvester data set. Therefore assume that $\tau = (C_\pi, A_\pi; A_\zeta, B_\zeta; \Gamma)$ is a σ-admissible Sylvester set with Γ not invertible, and let

$$\tau_0 = (C_{\pi 0}, A_{\pi 0}; A_{\zeta 0}, B_{\zeta 0}; \Gamma_0)$$

be an ε-admissible Sylvester data set, where $\sigma \cap \varepsilon = \emptyset$. We call τ_0 a *complement* to τ if the matrix

$$\begin{bmatrix} \Gamma & \Gamma_{12} \\ \Gamma_{21} & \Gamma_0 \end{bmatrix}$$

is square and invertible, where Γ_{12} and Γ_{21} are the unique solutions of

$$\Gamma_{12}A_{\pi 0} - A_\zeta \Gamma_{12} = B_\zeta C_{\pi 0}, \qquad \Gamma_{21}A_\pi - A_{\zeta 0}\Gamma_{21} = B_{\zeta 0}C_\pi.$$

The complement will be called *minimal* if among all complements τ_0 of τ the order of the matrix

$$\begin{bmatrix} \Gamma & \Gamma_{12} \\ \Gamma_{21} & \Gamma_0 \end{bmatrix}$$

is as small as possible. If τ_0 is a minimal complement to τ, then the function

$$W(z) = I_m + [\, C_\pi \quad C_{\pi 0}\,] \begin{bmatrix} (zI - A_\pi)^{-1} & 0 \\ 0 & (zI - A_{\pi 0})^{-1} \end{bmatrix} \begin{bmatrix} \Gamma & \Gamma_{12} \\ \Gamma_{21} & \Gamma_0 \end{bmatrix}^{-1} \begin{bmatrix} B_\zeta \\ B_{\zeta 0} \end{bmatrix}$$

is a rational matrix function with minimal possible McMillan degree with τ as its left σ-null-pole triple. The description of one such a minimal complement is given by Lemma 4.5.7. In this section we study, in particular, the case when ε consists of one point only. Let us abuse notation and denote also by ε this singleton point in $\mathbb{C}$ and not in σ.

The construction is somewhat intricate, so we first give some of the motivation behind it. Suppose that $\tau_0' = (C_1, T; S, B_2; \Gamma_0)$ is a minimal complement to $\tau = (C_\pi, A_\pi; A_\zeta, B_\zeta; \Gamma)$. Then the matrix $\widetilde{\Gamma} = \begin{bmatrix} \Gamma & \Gamma_{12} \\ \Gamma_{21} & \Gamma_0 \end{bmatrix}$ represents an invertible linear transformation with (1,1)-block entry equal to Γ and with domain and range space of smallest possible dimension with this property (the fact that $\widetilde{\Gamma}$ also satisfies a Sylvester equation will be brought into the discussion later). The most obvious choice of a linear transformation $\widetilde{\Gamma}$ enjoying these properties is

$$\widetilde{\Gamma} = \begin{bmatrix} \Gamma & \eta_\zeta \\ \rho_\pi & 0 \end{bmatrix} : \mathbb{C}^{n_\pi} \dotplus K \to \mathbb{C}^{n_\zeta} \dotplus \operatorname{Ker} \Gamma$$

where

$$K = \text{a direct complement of } \operatorname{Im} \Gamma \text{ in } \mathbb{C}^{n_\zeta}$$
$$\eta_\zeta = \text{the injection of } K \text{ into } \mathbb{C}^{n_\zeta}$$
$$\rho_\pi = \text{a projection of } \mathbb{C}^{n_\pi} \text{ onto } \operatorname{Ker} \Gamma.$$

We maintain invertibility of $\widetilde{\Gamma}$ by multiplying on the left and/or on the right by invertible linear transformations $\widetilde{Y}$ and $\widetilde{X}$:

$$\widetilde{\Gamma} \to \widetilde{Y}\widetilde{\Gamma}\widetilde{X}.$$

To maintain that the new $\widetilde{\Gamma}$ have (1,1)-block entry equal to Γ, we insist that $\widetilde{Y}$ and $\widetilde{X}$ have a block triangular structure

$$\widetilde{Y} = \begin{bmatrix} I & 0 \\ -Y & I \end{bmatrix} : \mathbb{C}^{n_\zeta} \dotplus \operatorname{Ker} \Gamma \to \mathbb{C}^{n_\zeta} \dotplus \operatorname{Ker} \Gamma$$

and

$$\widetilde{X} = \begin{bmatrix} I & -X \\ 0 & I \end{bmatrix} : \mathbb{C}^{n_\pi} \dotplus K \to \mathbb{C}^{n_\pi} \dotplus K.$$

Thus we assume that $\widetilde{\Gamma}$ has the form

(4.6.1)
$$\widetilde{\Gamma} = \begin{bmatrix} I & 0 \\ -Y & I \end{bmatrix} \begin{bmatrix} \Gamma & \eta_\zeta \\ \rho_\pi & 0 \end{bmatrix} \begin{bmatrix} I & -X \\ 0 & I \end{bmatrix}$$
$$= \begin{bmatrix} \Gamma & -\Gamma X + \eta_\zeta \\ -Y\Gamma + \rho_\pi & Y\Gamma X - Y\eta_\zeta - \rho_\pi X \end{bmatrix}$$

for some choice of $X: K \to \mathbf{C}^{n_\pi}$ and $Y: \mathbf{C}^{n_\zeta} \to \operatorname{Ker} \Gamma$. If we now define

$$\rho_\zeta = \text{the projection of } \mathbf{C}^{n_\zeta} \text{ onto } K \text{ along } \operatorname{Im} \Gamma$$

and

$$\eta_\pi = \text{the injection of } \operatorname{Ker} \Gamma \text{ into } \mathbf{C}^{n_\pi},$$

and let Γ^+ be a generalized inverse of Γ such that

$$\Gamma\Gamma^+ = I - \rho_\zeta = \text{projection of } \mathbf{C}^{n_\zeta} \text{ onto } \operatorname{Im} \Gamma \text{ along } K$$

and

$$\Gamma^+\Gamma = I - \rho_\pi = \text{projection of } \mathbf{C}^{n_\pi} \text{ onto } \operatorname{Ker} \rho_\pi \text{ along } \operatorname{Ker} \Gamma$$

then

$$\begin{bmatrix} \Gamma & \eta_\zeta \\ \rho_\pi & 0 \end{bmatrix}^{-1} = \begin{bmatrix} \Gamma^+ & \eta_\pi \\ \rho_\zeta & 0 \end{bmatrix} : \mathbf{C}^{n_\zeta} \dotplus \operatorname{Ker} \Gamma \to \mathbf{C}^{n_\pi} \dotplus K.$$

Hence

$$
\begin{aligned}
\tilde{\Gamma}^{-1} &= \begin{bmatrix} I & X \\ 0 & I \end{bmatrix} \begin{bmatrix} \Gamma^+ & \eta_\pi \\ \rho_\zeta & 0 \end{bmatrix} \begin{bmatrix} I & 0 \\ Y & I \end{bmatrix} \\
(4.6.2) \qquad &= \begin{bmatrix} \Gamma^+ + X\rho_\zeta + \eta_\pi Y & \eta_\pi \\ \rho_\zeta & 0 \end{bmatrix} \\
&\stackrel{\text{def}}{=} \begin{bmatrix} \Gamma_{11}^\times & \Gamma_{12}^\times \\ \Gamma_{21}^\times & \Gamma_{22}^\times \end{bmatrix}.
\end{aligned}
$$

If we assume that $X = (I - \rho_\pi)X$ and $Y = Y(I - \rho_\zeta)$, then we recover X and Y from $\tilde{\Gamma}^{-1}$ as

$$(4.6.3) \qquad\qquad X = (I - \rho_\pi)\Gamma_{11}^\times \eta_\zeta$$

and

$$(4.6.4) \qquad\qquad Y = \rho_\pi \Gamma_{11}^\times (I - \rho_\zeta).$$

If we next define linear transformations

$$G: \mathbf{C}^m \to \operatorname{Ker} \Gamma \quad \text{and} \quad F: K \to \mathbf{C}^m$$

by

$$(4.6.5) \qquad\qquad G = \rho_\pi \Gamma_{11}^\times (I - \rho_\zeta) B_\zeta + B_2 \qquad (= Y B_\zeta + B_2)$$

and

$$(4.6.6) \qquad\qquad F = -C_\pi (I - \rho_\pi)\Gamma_{11}^\times \eta_\zeta - C_1 \qquad (= -C_\pi X - C_1)$$

then B_2 and C_1 (taken from the minimal complement $(C_1, T; S, B_2; \Gamma_0)$ to τ) have the form

$$(4.6.7) \qquad\qquad B_2 = -YB_\zeta + G$$

$$(4.6.8) \qquad\qquad C_1 = -C_\pi X - F$$

by definition. But moreover, it turns out that we recover S and T as

$$(4.6.9) \qquad\qquad S = \rho_\pi(A_\pi - GC_\pi) \mid \operatorname{Ker}\Gamma$$

and

$$(4.6.10) \qquad\qquad T = \rho_\zeta(A_\zeta - B_\zeta F) \mid K.$$

To see (4.6.9) and (4.6.10) we now use that $\widetilde{\Gamma} = \begin{bmatrix} \Gamma & \eta_\zeta - \Gamma X \\ \rho_\pi - Y\Gamma & \Gamma_{22} \end{bmatrix}$ satisfies the Sylvester equation

$$(4.6.11) \qquad \begin{aligned} &\begin{bmatrix} \Gamma & \eta_\zeta - \Gamma X \\ \rho_\pi - Y\Gamma & \Gamma_{22} \end{bmatrix} \begin{bmatrix} A_\pi & 0 \\ 0 & T \end{bmatrix} \\ &\quad - \begin{bmatrix} A_\zeta & 0 \\ 0 & S \end{bmatrix} \begin{bmatrix} \Gamma & \eta_\zeta - \Gamma X \\ \rho_\pi - Y\Gamma & \Gamma_{22} \end{bmatrix} = \begin{bmatrix} B_\zeta \\ B_2 \end{bmatrix} \begin{bmatrix} C_\pi & C_1 \end{bmatrix}. \end{aligned}$$

Hence, for $x \in \operatorname{Ker}\Gamma$,

$$\begin{aligned} A_\pi x - Sx - GC_\pi x &= A_\pi x - Sx - \rho_\pi \Gamma_{11}^\times (I - \rho_\zeta) B_\zeta C_\pi x - \rho_\pi \Gamma_{12}^\times B_2 C_\pi x \\ &= A_\pi x - Sx - \rho_\pi \Gamma_{11}^\times (I - \rho_\zeta)(\Gamma A_\pi - A_\zeta \Gamma)x \\ &\quad + S(\rho_\pi - \rho_\pi \Gamma_{11}^\times \Gamma)x - \rho_\pi A_\pi x + \rho_\pi \Gamma_{11}^\times \Gamma A_\pi x \end{aligned}$$

where we used two consequences of (4.6.11), namely

$$B_\zeta C_\pi = \Gamma A_\pi - A_\zeta \Gamma$$

and

$$B_2 C_\pi = (\rho_\pi - Y\Gamma)A_\pi - S(\rho_\pi - Y\Gamma)$$

where Y satisfies (4.6.4). Since x is in $\operatorname{Ker}\Gamma$, this collapses to

$$A_\pi x - Sx - GC_\pi x = A_\pi x - Sx - \rho_\pi \Gamma_{11}^\times (I - \rho_\zeta)\Gamma A_\pi x + S\rho_\pi x - \rho_\pi A_\pi x + \rho_\pi \Gamma_{11}^\times \Gamma A_\pi x.$$

Since $\Gamma_{11}^\times (I - \rho_\zeta)\Gamma = \Gamma_{11}^\times \Gamma$, the third and last term cancel, and since $x = \rho_\pi x$, the second and fourth term cancel. We are left with

$$A_\pi x - Sx - GC_\pi x = (I - \rho_\pi)A_\pi x \in \operatorname{Ker}\rho_\pi.$$

Hence (4.6.9) holds as asserted. Via an analogous computation, we get for $x \in K$

$$\begin{aligned}
A_\zeta x - Tx - B_\zeta Fx &= A_\zeta x - Tx + B_\zeta C_\pi (I - \rho_\pi)\Gamma_{11}^\times x + B_\zeta C_1 x \\
&= A_\zeta x - Tx + (\Gamma A_\pi - A_\zeta \Gamma)(I - \rho_\pi)\Gamma_{11}^\times x \\
&\quad + (\eta_\zeta - \Gamma\Gamma_{11}^\times \eta_\zeta)Tx - A_\zeta(\eta_\zeta - \Gamma\Gamma_{11}^\times \eta_\zeta)x
\end{aligned}$$

where we used two consequences of (4.6.11), namely, again

$$B_\zeta C_\pi = \Gamma A_\pi - A_\zeta \Gamma$$

and also

$$B_\zeta C_1 = (\eta_\zeta - \Gamma X)A_\pi - A_\zeta(\eta_\zeta - \Gamma X)$$

where X satisfies (4.6.3). Collecting all terms together in $\operatorname{Im}\Gamma$ gives

$$\begin{aligned}
A_\zeta x - Tx - B_\zeta Fx &= \Gamma\big(A_\pi(I - \rho_\pi)\Gamma_{11}^\times x - \Gamma_{11}^\times \eta_\zeta Tx\big) + A_\zeta x - Tx \\
&\quad - A_\zeta \Gamma(I - \rho_\pi)\Gamma_{11}^\times x + \eta_\zeta Tx - A_\zeta \eta_\zeta x + A_\zeta \Gamma\Gamma_{11}^\times \eta_\zeta x.
\end{aligned}$$

Use $\eta_\zeta x = x$ and $(I - \rho_\pi)\Gamma_{11}^\times = \Gamma_{11}^\times$ to see that the last term cancels the fourth from last and that the second cancels the second from last. The remaining last two terms cancel since $Tx \in K$ by definition. Hence

$$A_\zeta x - Tx - B_\zeta Fx \in \operatorname{Im}\Gamma$$

for all $x \in K$ and (4.6.10) follows as asserted.

Our ultimate goal is to arrange that S and T have spectrum at the single point ε. From (4.6.9) and (4.6.10) we see that we cannot choose S and T arbitrarily with this property; for example, one could not choose $T = \varepsilon I$ unless $\operatorname{Im}(A_\zeta \mid K) + \operatorname{Im} B + \operatorname{Im}\Gamma = \mathbf{C}^{n_\zeta}$. It turns out that the lengths of and number of Jordan chains for S and T of the form (4.6.9) and (4.6.10) respectively are completely determined by the data of the problem. Bases in $\operatorname{Ker}\Gamma$ and in K with respect to which S and T respectively are in their respective Jordan forms must be constructed carefully. This leads to the notions of *incoming bases* and *outgoing bases* which we now describe.

Let the σ-admissible Sylvester data set $\tau = (C_\pi, A_\pi; A_\zeta, B_\zeta; \Gamma)$ be given. We define a sequence of *incoming subspaces* H_j containing $\operatorname{Im}\Gamma$ by

$$(4.6.12) \qquad H_j = \operatorname{Im}\Gamma + \operatorname{Im} B_\zeta + \operatorname{Im} A_\zeta B_\zeta + \cdots + \operatorname{Im} A_\zeta^{j-1} B_\zeta$$

for $j = 0, 1, 2, \ldots$. Of course by H_0 we mean $\operatorname{Im}\Gamma$. Obviously $H_0 \subset H_1 \subset H_2 \subset \cdots$, and the full range assumption on the pair (A_ζ, B_ζ) implies that $H_j = \mathbf{C}^{n_\zeta}$ once j is sufficiently large (here $n_\zeta \times n_\zeta$ is the size of A_ζ). Since by assumption τ is an admissible Sylvester data set, the Sylvester equation

$$(4.6.13) \qquad \Gamma A_\pi - A_\zeta \Gamma = B_\zeta C_\pi$$

is satisfied. This has as a consequence that

$$A_\zeta \operatorname{Im} \Gamma \subset \operatorname{Im} \Gamma + \operatorname{Im} B_\zeta.$$

From this it is easy to see that

$$H_1 + A_\zeta H_k = H_{k+1}, \qquad k = 0, 1, 2, \dots .$$

Observe next that changing A_ζ to $A_\zeta - \varepsilon I$ has no effect on the definition of the spaces H_k. Hence we have

(4.6.14) $$\qquad H_1 + (A_\zeta - \varepsilon I)H_k = H_{k+1}, \qquad k = 0, 1, 2, \dots .$$

We shall use these identities to construct a system of vectors

(4.6.15) $$\qquad f_{jk}, \qquad k = 1, \dots, \omega_j, \quad j = 1, \dots, s$$

with the following properties:

(1) $1 \leq \omega_1 \leq \omega_2 \leq \cdots \leq \omega_s$;

(2) $(A_\zeta - \varepsilon I)f_{jk} - f_{j,k+1} \in \operatorname{Im} \Gamma + \operatorname{Im} B_\zeta$, $k = 1, \dots, \omega_j$, where by definition $f_{j,\omega_j+1} = 0$;

(3) the vectors f_{jk}, $\min\{p : \omega_p \geq k\} \leq j \leq s$, form a basis for H_k modulo H_{k-1}.

Such a system of vectors we shall call an *incoming basis* for τ (with respect to ε).

For all such incoming bases, the integers s and $\omega_1, \dots, \omega_s$ are the same and independent of the choice of ε. In fact

(4.6.16) $$\qquad s = \dim(H_1/H_0)$$
(4.6.17) $$\qquad \omega_j = \#\{k : s - \dim(H_k/H_{k-1}) \leq j - 1\}.$$

Both formulae can be seen from condition (3). We call $\omega_1, \dots, \omega_s$ the *incoming indices* of A_ζ with respect to the pair $(\operatorname{Im} \Gamma, \operatorname{Im} B_\zeta)$. Observe that $\omega_s = \omega$ where ω is the first integer such that $H_\omega = H_{\omega+1}$ (and then also $H_\omega = \mathbf{C}^{n_\zeta}$). The incoming indices of A_ζ with respect to $(\operatorname{Im} \Gamma, \operatorname{Im} B_\zeta)$ are also given by the identity:

(4.6.18) $$\qquad \#\{j : 1 \leq j \leq s, \ \omega_j = k\} = \dim(H_k/H_{k-1}) - \dim(H_{k+1}/H_k).$$

Note that for an incoming basis $\{f_{jk}\}_{k=1,j=1}^{\omega_j, s}$ the following holds:

(3a) the vectors f_{jk}, $k = 1, \dots, \omega_j$, $j = 1, \dots, s$ form a basis for a complement of $\operatorname{Im} \Gamma$;

(3b) the vectors $f_{11}, \dots, f_{s1}$ form a basis for $\operatorname{Im} \Gamma + \operatorname{Im} B_\zeta$ modulo $\operatorname{Im} \Gamma$.

Conversely, if (4.6.15) is a system of vectors such that (1), (2), (3a) and (3b) are satisfied, then the system (4.6.15) is an incoming basis for A_ζ with respect to $(\operatorname{Im} \Gamma, \operatorname{Im} B_\zeta)$; one can readily see this by using (4.6.14).

We are now ready to explain how to construct an incoming basis. Put

$$s_k = \dim(H_1/H_0) - \dim(H_{k+1}/H_k), \qquad k = 0, 1, \ldots, \omega.$$

Obviously $s_\omega = \dim(H_1/H_0) = s$. Fix $k \in \{1, \ldots, \omega\}$. From (4.6.14) it is clear that $A_\zeta - \varepsilon I$ induces a surjective linear transformation

$$[A_\zeta - \varepsilon I] : H_k/H_{k-1} \to H_{k+1}/H_k.$$

This implies that $s_k \geq s_{k-1}$ and $\dim \operatorname{Ker}[A_\zeta - \varepsilon I] = s_k - s_{k-1}$.

Assume $[A_\zeta - \varepsilon I][f] = 0$, where $[f]$ denotes the class in H_k/H_{k-1} containing f. Then there exists $g \in [f]$ such that $(A_\zeta - \varepsilon I)g \in H_1$. Indeed from $[A_\zeta - \varepsilon I][f] = 0$ it follows that $(A_\zeta - \varepsilon I)f \in H_k$ and so by (4.6.14) there exists $h \in H_{k-1}$ such that $(A_\zeta - \varepsilon I)(f - h) \in H_1$. The vector $g = f - h$ then has the desired properties. By this result we can choose vectors f_{jk}, $s_{k-1} + 1 \leq j \leq s_k$, in H_k, linearly independent modulo H_{k-1}, such that $[f_{s_{k-1}+1}], \ldots, [f_{s_k}]$ is a basis of $\operatorname{Ker}[A_\zeta - \varepsilon I]$ and

$$(A_\zeta - \varepsilon I)f_{jk} \in H_1, \qquad s_{k-1} + 1 \leq j \leq s_k.$$

Now let $f_{s_k+1,k+1}, \ldots, f_{s,k+1}$ be a basis for H_{k+1} modulo H_k. According to (4.6.14) there exist vectors f_{jk}, $s_k + 1 \leq j \leq s$ in H_k such that

$$(A_\zeta - \varepsilon I)f_{jk} - f_{j,k+1} \in H_1, \qquad s_k + 1 \leq j \leq s.$$

Clearly, the vectors f_{jk}, $s_{k-1}+1 \leq j \leq s$, form a basis of H_k modulo H_{k-1}. By repeating the construction for $k-1$ instead of k, one can get in a finite number of steps an incoming basis. Note that $\omega_j = k$ whenever $s_{k-1} + 1 \leq j \leq s_k$.

The notion of a sequence of *outgoing subspaces* and an *outgoing basis* for A_π relative to the pair $(\operatorname{Ker}\Gamma, \operatorname{Ker}C_\pi)$ is dual to that of sequence of incoming subspaces and incoming basis just explained above. Given the σ-admissible Sylvester data set $\tau = (C_\pi, A_\pi; A_\zeta, B_\zeta; \Gamma)$ we define the sequence of *outgoing subspaces* K_j by

$$(4.6.19) \qquad K_j = \operatorname{Ker}\Gamma \cap \operatorname{Ker}C_\pi \cap \operatorname{Ker}C_\pi A_\pi \cap \cdots \cap \operatorname{Ker}C_\pi A_\pi^{j-1},$$

where $j = 0, 1, 2, \ldots$. For $j = 0$, we interpret the right side of (4.6.19) as $\operatorname{Ker}\Gamma$. Clearly $K_0 \supset K_1 \supset \cdots$ with not all inclusions proper since all spaces are finite dimensional. Since by assumption (C_π, A_π) is a null kernel pair, clearly $K_j = \{0\}$ for j sufficiently large, and if α is the smallest integer α for which $K_\alpha = K_{\alpha+1}$ then in fact $K_\alpha = \{0\}$. From the definition (4.6.19), we see that if the vector x is such that both x and $A_\pi x$ are in K_j, then x is in K_{j+1}. Conversely, if x is in K_{j+1}, then $x \in K_j$ and $A_\pi x \in \operatorname{Ker}C_\pi \cap \cdots \cap \operatorname{Ker}C_\pi A_\pi^{j-1}$; that $A_\pi x$ also is in $\operatorname{Ker}\Gamma$ follows from the Sylvester equation

$$\Gamma A_\pi - A_\zeta \Gamma = B_\zeta C_\pi$$

and hence we also have $A_\pi x \in K_j$. We have thus verified

$$K_j \cap A^{-1}K_j = K_{j+1}, \qquad j = 0, 1, 2, \ldots .$$

Let now ε be a complex number not in σ; since the definition of the spaces K_j does not change when A_π is replaced by $A_\pi - \varepsilon I$, we have

$$(4.6.20) \qquad K_j \cap (A_\pi - \varepsilon I)^{-1} K_j = K_{j+1}, \qquad j = 0, 1, 2, \ldots .$$

We shall use these identities to construct a system of vectors

$$(4.6.21) \qquad d_{jk}, \qquad k = 1, \ldots, \alpha_j; \quad j = 1, \ldots, t$$

with the following properties:

(4) $\alpha_1 \geq \alpha_2 \geq \cdots \geq \alpha_t \geq 1$;

(5) $(A_\pi - \varepsilon I) d_{jk} = d_{j,k+1}$, $k = 1, \ldots, \alpha_j - 1$, $j = 1, \ldots, t$;

(6) the vectors d_{jk}, $k = 1, \ldots, \alpha_j - m$, $\alpha_j \geq m+1$, form a basis of K_m.

Such a system of vectors we shall call an *outgoing basis* for τ (with respect to ε).

We first observe some basic properties of outgoing bases and then will show how one can be constructed. For all outgoing bases for τ the integers t and $\alpha_1, \ldots, \alpha_t$ are the same and independent of the choice of ε. In fact from (6) one sees that

$$(4.6.22) \qquad \begin{aligned} t &= \dim(K_0/K_1) \\ \alpha_j &= \#\{m \colon \dim(K_{m-1}/K_m) \geq j\}, \qquad j = 1, \ldots, t. \end{aligned}$$

We call $\alpha_1, \ldots, \alpha_t$ the *outgoing indices* of the triple τ. Observe that $\alpha_1 = \alpha$. The outgoing indices are also determined by the identity

$$(4.6.23) \qquad \#\{j \colon 1 \leq j \leq t, \ \alpha_j = k\} = \dim(K_{p-1}/K_p) - \dim(K_p/K_{p+1}).$$

For an outgoing basis $\{d_{jk}\}_{k=1,j=1}^{\alpha_j, t}$ the following holds:

(6a) the vectors d_{jk}, $k = 1, \ldots, \alpha_j$, $j = 1, \ldots, t$, form a basis of $\operatorname{Ker} \Gamma$;

(6b) the vectors d_{jk}, $k = 1, \ldots, \alpha_j - 1$, $\alpha_j \geq 2$, form a basis of $\operatorname{Ker} \Gamma \cap \operatorname{Ker} C_\pi$.

Conversely, if (4.6.21) is a system of vectors such that (4), (5), (6a) and (6b) are satisfied, then the system (4.6.21) is an outgoing basis for τ. This one can easily see by using (4.6.20).

Now we indicate how to construct an outgoing basis. Define

$$t_m = \dim(K_0/K_1) - \dim(K_m/K_{m+1}), \qquad m = 0, 1, \ldots, \alpha.$$

Clearly $t_\alpha = \dim(K_0/K_1) = t$. Putting $n_j = t_\alpha - t_{\alpha-j} = t - t_{\alpha-j}$, we have $n_j = \dim(K_{\alpha-j}/K_{\alpha-j+1})$ and

$$n_1 + \cdots + n_j = \dim K_{\alpha-j}, \qquad j = 1, \ldots, \alpha.$$

In particular, $n_1 = \dim K_{\alpha-1}$. Let $d_{11}, \ldots, d_{n_1,1}$ be a basis of $K_{\alpha-1}$. For $i = 1, \ldots, n_1$, write $d_{i2} = (A_\pi - \varepsilon I) d_{i1}$. From (4.6.20) it is clear that $d_{12}, \ldots, d_{n_1,2}$ are vectors in $K_{\alpha-2}$, linearly independent modulo $K_{\alpha-1}$. But then $n_2 = \dim(K_{\alpha-2}/K_{\alpha-1}) \geq n_1$, and we can choose vectors $d_{n_1+1,1}, \ldots, d_{n_2,1}$ such that $d_{11}, \ldots, d_{n_2,1}, d_{12}, \ldots, d_{n_1,2}$ form a basis of $K_{\alpha-2}$. Put

$$d_{i3} = (A_\pi - \varepsilon I)d_{i2}, \qquad i = 1, \ldots, n_1$$
$$d_{i2} = (A_\pi - \varepsilon I)d_{i1}, \qquad i = n_1 + 1, \ldots, n_2.$$

Again using (4.6.20) we see that $d_{13}, \ldots, d_{n_1,3}, d_{n_1+1,2}, \ldots, d_{n_2,2}$ are vectors in $K_{\alpha-3}$, linearly independent modulo $K_{\alpha-2}$. It follows that $n_3 \geq n_2$ and by choosing additional vectors $d_{n_2+1,1}, \ldots, d_{n_3,1}$ we can produce a basis of $K_{\alpha-3}$. Proceeding in this way we obtain an outgoing basis for τ in a finite number of steps.

We are now ready to define the construction of a minimal complement $\tau_0 = (C_{\pi 0}, A_{\pi 0}; A_{\zeta 0}, B_{\zeta 0}; \Gamma_0)$ of a given σ-admissible Sylvester data set $\tau = (C_\pi, A_\pi; A_\zeta, B_\zeta; \Gamma)$ such that the spectrum of $A_{\pi 0}$ and of $A_{\zeta 0}$ consists of the single point ε not in σ. To do this we let

$$f_{jk}, \qquad k = 1, \ldots, \omega_j; \quad j = 1, \ldots, s$$

be an incoming basis for τ with respect to ε and

$$d_{jk}, \qquad k = 1, \ldots, \alpha_j; \quad j = 1, \ldots, t$$

be an outgoing basis for τ with respect to ε. We collect the key properties:

(3a) the vectors f_{jk}, $k = 1, \ldots, \omega_j$, $j = 1, \ldots, s$ form a basis for a complement of $\operatorname{Im}\Gamma$;

(3b) the factors $f_{11}, \ldots, f_{s1}$ form a basis for $\operatorname{Im}\Gamma + \operatorname{Im}B_\zeta$ modulo $\operatorname{Im}\Gamma$;

(2) $(A_\zeta - \varepsilon I)f_{jk} - f_{j,k+1} \in \operatorname{Im}\Gamma + \operatorname{Im}B_\zeta$, $k = 1, \ldots, \omega_j$, where $f_{j,\omega_j+1} = 0$;

(6a) the vectors d_{jk}, $k = 1, \ldots, \alpha_j$, $j = 1, \ldots, t$ form a basis of $\operatorname{Ker}\Gamma$;

(6b) the vectors d_{jk}, $k = 1, \ldots, \alpha_j - 1$, $\alpha_j \geq 2$, form a basis of $\operatorname{Ker}\Gamma \cap \operatorname{Ker}C_\pi$;

(5) $(A_\pi - \varepsilon I)d_{jk} = d_{j,k+1}$, $k = 1, \ldots, \alpha_j - 1$, $j = 1, \ldots, t$.

We denote by K the span of the outgoing basis vectors f_{jk} ($k = 1, \ldots, \omega_j$; $j = 1, \ldots, s$); thus by (3a) K is a direct complement of $\operatorname{Im}\Gamma$ in $\mathbf{C}^{n_\zeta}$. As before, it is assumed $\alpha_1 \geq \alpha_2 \geq \cdots \geq \alpha_t$ and $\omega_1 \geq \omega_2 \geq \cdots \geq \omega_s$. Define projections $Q_{j\nu}$ in $\operatorname{Ker}\Gamma$ and $P_{j\nu}$ in K as follows. For $x = \sum_{j=1}^{s}\sum_{k=1}^{\omega_j} a_{jk}f_{jk} \in K$ set

(4.6.24)
$$P_{j\nu}x = \sum_{k=1}^{\omega_j - \nu + 1} a_{jk}f_{jk},$$

and for $x = \sum_{j=1}^{t}\sum_{k=1}^{\alpha_j} b_{jk}d_{jk} \in \operatorname{Ker}\Gamma$ put

(4.6.25)
$$Q_{j\nu}x = \sum_{k=1}^{\alpha_j - \nu + 1} b_{jk}d_{jk}.$$

Choose a complement N to $\operatorname{Ker}\Gamma$, i.e., $\mathbb{C}^{n_\pi} = N \dotplus \operatorname{Ker}\Gamma$. Let ρ_π be the projection onto $\operatorname{Ker}\Gamma$ along N, and let ρ_ζ be the projection onto K along $\operatorname{Im}\Gamma$. Further, let η_π be the injection of $\operatorname{Ker}\Gamma$ into $\mathbb{C}^{n_\pi}$, and let η_ζ be the injection of K into $\mathbb{C}^{n_\zeta}$.

THEOREM 4.6.1. *Let* $\tau = (C_\pi, A_\pi; A_\zeta, B_\zeta; \Gamma)$ *be a σ-admissible Sylvester data set. Then a minimal complement to τ is given by*

(4.6.26)
$$(-C_\pi X - F, T; S, -YB_\zeta + G; \Gamma_0).$$

Here $S\colon \operatorname{Ker}\Gamma \to \operatorname{Ker}\Gamma$ and $T\colon K \to K$ are given by

(4.6.27)
$$(S - \varepsilon)d_{jk} = d_{j,k+1} \qquad (d_{j,\alpha_j+1} := 0),$$

(4.6.28)
$$(T - \varepsilon)f_{jk} = f_{j,k+1} \qquad (f_{j,\omega_j+1} := 0).$$

Furthermore, G and F are defined as follows. Let $z_j = C_\pi d_{j,\alpha_j}$, and choose vectors y_j such that $f_{j1} - B_\zeta y_j \in \operatorname{Im}\Gamma$. Choose a subspace Y_0 such that

(4.6.29)
$$\mathbb{C}^m = \operatorname{Span}\{z_j\}_{j=1}^t \dotplus Y_0 \dotplus \operatorname{Span}\{y_j\}_{j=1}^s$$

where m is the number of rows of C_π ($=$ the number of columns of B_ζ). Then $G\colon \mathbb{C}^m \to \operatorname{Ker}\Gamma$ is defined by

$$Gz_j = \rho_\pi(A_\pi - \varepsilon)d_{j\alpha_j}, \qquad j = 1,\ldots,t,$$
$$Gy = 0, \qquad y \in Y_0 \dotplus \operatorname{Span}\{y_j\}.$$

The linear transformation $F\colon K \to \mathbb{C}^m$ is given by $Ff_{jk} = u_{jk}$, where $u_{jk} \in \operatorname{Span}\{y_i\}_{i=1}^s$ is such that

(4.6.30)
$$(A_\zeta - T)f_{jk} + B_\zeta u_{jk} \in \operatorname{Im}\Gamma.$$

Finally,

(4.6.31)
$$X = \sum_{j=1}^s \sum_{\nu=1}^{\omega_j} (A_\pi - \varepsilon I)^{-\nu}\Gamma^+(A_\zeta - B_\zeta F)(T - \varepsilon I)^{\nu-1} P_{j\nu}$$

(4.6.32)
$$Y = \sum_{j=1}^t \sum_{\nu=1}^{\alpha_j} (S - \varepsilon I)^{\nu-1} Q_{j\nu}\rho_\pi(A_\pi - GC_\pi)\Gamma^+(A_\zeta - \varepsilon I)^{-\nu},$$

and

(4.6.33)
$$\Gamma_0 = Y\Gamma X - Y\eta_\zeta - \rho_\pi X$$

where Γ^+ is a generalized inverse of Γ such that $\Gamma\Gamma^+ = I - \rho_\zeta$, $\Gamma^+\Gamma = I - \rho_\pi$, $\operatorname{Ker}\Gamma^+ = K$ and $\operatorname{Im}\Gamma^+ = N$.

PROOF. First note that $z_1, \ldots, z_t$ are linearly independent, so G is well-defined. Moreover

(4.6.34) $$\rho_\pi (A_\pi - G C_\pi)|_{Ker\,\Gamma} = S.$$

Furthermore, because of (3a) and (4.6.28), we have

$$(A_\zeta - T)f_{jk} = (A_\zeta - \varepsilon)f_{jk} - f_{j,k+1} \in \operatorname{Im}\Gamma + \operatorname{Im} B_\zeta.$$

So there exists a vector $u_{jk} \in \mathbf{C}^m$ such that $(A_\zeta - T)f_{jk} + B_\zeta u_{jk} \in \operatorname{Im}\Gamma$. Note that $\{y_j\}_{j=1}^s$ is a basis for $\mathbf{C}^m$ modulo $B_\zeta^{-1}(\operatorname{Im}\Gamma)$. So, adding to u_{jk} any vector v_{jk} with $B_\zeta v_{jk} \in \operatorname{Im}\Gamma$ we can replace u_{jk} by $u_{jk} + v_{jk}$. In particular, in this way it is clear that we can take $u_{jk} \in \operatorname{Span}\{y_j\}_{j=1}^s$ without loss of generality. Since also $y_1, \ldots, y_s$ are linearly independent, it follows that F is well-defined.

To show that τ_0 is a minimal complement by definition we must check the following. If Γ_{12} and Γ_{21} are defined to be the unique solutions of the Sylvester equations

(4.6.35) $$\Gamma_{12}T - A_\zeta \Gamma_{12} = B_\zeta(-C_\pi X - F)$$

(4.6.36) $$\Gamma_{21} A_\pi - S\Gamma_{21} = (-YB_\zeta + G)C_\pi$$

then the matrix $\widetilde{\Gamma} := \begin{bmatrix} \Gamma & \Gamma_{12} \\ \Gamma_{21} & \Gamma_0 \end{bmatrix}$ must be invertible. Moreover, $(-C_\pi X - F, T)$ should be a null kernel pair and $(S, -YB_\zeta + G)$ should be a full range pair. Thirdly, Γ_0 should satisfy the Sylvester equation

(4.6.37) $$\Gamma_0 T - S\Gamma_0 = (-YB_\zeta + G)(-C_\pi X - F).$$

All this guarantees that τ_0 is a complement to τ. For τ_0 to be a minimal complement we must also show that $\operatorname{rank}\widetilde{\Gamma}$ is as small as possible among all possible complements. But this last condition is easy: any $\widehat{\Gamma} = \begin{bmatrix} \Gamma & \widehat{\Gamma}_{12} \\ \widehat{\Gamma}_{21} & \widehat{\Gamma}_0 \end{bmatrix}$ which is invertible must satisfy $\operatorname{rank}\widehat{\Gamma} \geq n_\pi + \dim K = n_\zeta + \dim\operatorname{Ker}\Gamma = \operatorname{rank}\widetilde{\Gamma}$.

To check the invertibility of $\widetilde{\Gamma}$ we must first compute the solutions Γ_{12} and Γ_{21} of (4.6.35) and (4.6.36). In fact, by direct substitution one can verify that

(4.6.38) $$\Gamma_{12} = -\Gamma X + \eta_\zeta, \qquad \Gamma_{21} = -Y\Gamma + \rho_\pi.$$

Thus $\widetilde{\Gamma}$ has the form

(4.6.39)
$$\widetilde{\Gamma} = \begin{bmatrix} \Gamma & -\Gamma X + \eta_\zeta \\ -Y\Gamma + \rho_\pi & Y\Gamma X - Y\eta_\zeta - \rho_\pi X \end{bmatrix}$$
$$= \begin{bmatrix} I & 0 \\ -Y & I \end{bmatrix} \begin{bmatrix} \Gamma & \eta_\zeta \\ \rho_\pi & 0 \end{bmatrix} \begin{bmatrix} I & -X \\ 0 & I \end{bmatrix}.$$

Since $\begin{bmatrix} \Gamma & \eta_\zeta \\ \rho_\pi & 0 \end{bmatrix}$ is clearly invertible, it follows that $\widetilde{\Gamma}$ is invertible as required.

Before proceding to the next step we record the following identity which can be verified by direct computation:

$$(4.6.40) \qquad A_\pi X - XT = A_{12}$$

where we define A_{12} by

$$(4.6.41) \qquad A_{12} = \Gamma^+(A_\zeta - \eta_\zeta T - B_\zeta F)\eta_\zeta.$$

Since $\mathrm{Im}(A_\zeta - T - B_\zeta F)\eta_\zeta \subset \mathrm{Im}\,\Gamma$ by (4.6.30), a consequence of (4.6.40) is the identities

$$(4.6.42) \qquad \Gamma A_{12} = (A_\zeta - \eta_\zeta T - B_\zeta F)\eta_\zeta$$

$$(4.6.43) \qquad \rho_\pi A_{12} = 0.$$

A related identity is

$$(4.6.44) \qquad Y A_\zeta - SY = A_{21}$$

where A_{21} is defined by

$$(4.6.45) \qquad A_{21} = \rho_\pi(A_\pi - S\rho_\pi - GC_\pi)\Gamma^+.$$

A consequence of these definitions then is

$$(4.6.46) \qquad A_{21}\Gamma = \rho_\pi(A_\pi - S\rho_\pi - GC_\pi)$$

and

$$(4.6.47) \qquad A_{21}\eta_\zeta = 0.$$

We now show that $(-C_\pi X - F, T)$ is a null kernel pair. Suppose not. Then there is a nonzero vector x for which $x \in \mathrm{Ker}(C_\pi X + F)T^j$ for $j = 0, 1, 2, \ldots$. By considering $(T - \varepsilon I)^k x$ in place of x for an appropriate nonnegative integer k, we may assume without loss of generality that x is an eigenvector for T (necessarily with eigenvalue ε). Then in particular we have $-C_\pi X x = Fx$ and $Tx = \varepsilon x$. By (4.6.41) we have

$$(4.6.48) \qquad \begin{aligned} B_\zeta F x &= -\Gamma A_{12}x - Tx + A_\zeta x \\ &= -\Gamma A_{12}x - (\varepsilon I - A_\zeta)x. \end{aligned}$$

On the other hand, by the Sylvester equation (4.6.13) we have

$$(4.6.49) \qquad \begin{aligned} -B_\zeta C_\pi X x &= A_\zeta \Gamma X x - \Gamma A_\pi X x \\ &= A_\zeta \Gamma X x - \Gamma A_{12}x - \varepsilon \Gamma X x. \end{aligned}$$

Equating (4.6.48) and (4.6.49) gives

$$(\varepsilon I - A_\zeta)(x - \Gamma X x) = 0.$$

As ε is not in $\sigma(A_\zeta)$, this gives $x = \Gamma X x$. But $x \in K$ so this forces $x = 0$, a contradiction. Hence $(-C_\pi X - F, T)$ is a null kernel pair.

The proof that $(S, -Y B_\zeta + G)$ is a full range pair proceeds by a dual argument. Suppose not. Then there is a nonzero vector y with $y^T = y^T \rho_\pi$ for which

$$y^T S^j(-Y B_\zeta + G) = 0$$

for $j = 0, 1, 2, \ldots$. By considering $y^T(S - \varepsilon I)^k$ in place of y^T for an appropriate non-negative integer k, we may assume without loss of generality that $y^T S = \varepsilon y^T$. Then in particular we have

$$y^T S = \varepsilon y^T, \qquad y^T(-Y B_\zeta + G) = 0,$$

so

$$y^T Y B_\zeta = y^T G.$$

On the one hand, from (4.6.46),

(4.6.50)
$$\begin{aligned}
y^T G C_\pi &= -y^T A_{21} \Gamma - y^T S \rho_\pi + y^T A_\pi \\
&= -y^T A_{21} \Gamma - y^T(\varepsilon I - A_\pi).
\end{aligned}$$

On the other hand, from (4.6.13) and (4.6.44),

(4.6.51)
$$\begin{aligned}
y^T Y B_\zeta C_\pi &= y^T Y \Gamma A_\pi - y^T Y A_\zeta \Gamma \\
&= y^T Y \Gamma A_\pi - y^T(A_{21} + SY)\Gamma \\
&= y^T Y \Gamma A_\pi - y^T A_{21} \Gamma - \varepsilon y^T Y \Gamma.
\end{aligned}$$

Equating (4.6.50) and (4.6.51) gives

$$y^T(\varepsilon I - A_\pi) = y^T Y \Gamma(\varepsilon I - A_\pi).$$

Since ε is not in $\sigma(A_\pi)$ this gives

$$y^T = y^T Y \Gamma$$

and since $y^T = y^T \rho_\pi$ we get

$$y^T = y^T Y \Gamma \rho_\pi = 0,$$

a contradiction.

It remains only to verify that Γ_0 satisfies the Sylvester equation (4.6.37)

$$\Gamma_0 T - S \Gamma_0 = (-Y B_\zeta + G)(-C_\pi X - F).$$

Write out the right hand side to get

$$YB_\zeta C_\pi X - GC_\pi X + YB_\zeta F - GF.$$

From the definitions we see that $GF = 0$. Now use (4.6.13), (4.6.42) and (4.6.46) to get that

$$(-YB_\zeta + G)(-C_\pi X - F) = Y(\Gamma A_\pi - A_\zeta \Gamma)X - [-A_{21}\Gamma + \rho_\pi A_\pi - S\rho_\pi]X$$
$$+ Y[-\Gamma A_{12} + A_\zeta \eta_\zeta - \eta_\zeta T].$$

Next use (4.6.43) and (4.6.41) to get from this

$$Y\Gamma A_\pi X - YA_\zeta \Gamma X + [YA_\zeta - SY]\Gamma X - \rho_\pi A_\pi X + S\rho_\pi X$$
$$- Y\Gamma[A_\pi X - XT] + YA_\zeta \eta_\zeta - Y\eta_\zeta T$$
$$= -SY\Gamma X - \rho_\pi A_\pi X + S\rho_\pi X + Y\Gamma XT + YA_\zeta \eta_\zeta - Y\eta_\zeta T$$
$$= (Y\Gamma X - Y\eta_\zeta)T - S(Y\Gamma X - \rho_\pi X) - \rho_\pi A_\pi X + YA_\zeta \eta_\zeta.$$

From the definition (4.6.33) of Γ_0 this in turn is equal to

$$\Gamma_0 T - S\Gamma_0 + \rho_\pi XT - SY\eta_\zeta - \rho_\pi A_\pi X + YA_\zeta \eta_\zeta$$
$$= \Gamma_0 T - S\Gamma_0 + \rho_\pi(XT - A_\pi X) + (YA_\zeta - SY)\eta_\zeta.$$

But from (4.6.40) and (4.6.43),

$$\rho_\pi(XT - A_\pi X) = \rho_\pi A_{12} = 0$$

and similarly from (4.6.44) and (4.6.47)

$$(YA_\zeta - SY)\eta_\zeta = A_{21}\eta_\zeta = 0.$$

This verifies the Sylvester equation (4.6.37) as required and completes the proof of Theorem 4.6.1. $\square$

Next we compute the rational matrix function W corresponding to the Sylvester data set $\tau \oplus \tau_0$, where τ_0 is the minimal complement to τ constructed in the preceding theorem.

THEOREM 4.6.2. *Let the minimal complement τ_0 to the Sylvester data set τ be given as in Theorem 4.6.1. Then the rational matrix function $W(z)$ with global null-pole triple $\tau \oplus \tau_0$ and value I at infinity is given by*

(4.6.52)
$$W(z) = I + C_\pi(zI - A_\pi)^{-1}\{(\Gamma^+ + X\rho_\zeta)B_\zeta + \eta_\pi G\}$$
$$+ (-C_\pi X + F)(zI - T)^{-1}\rho_\zeta B_\zeta$$
$$= I + C_\pi(zI - A_\pi)^{-1}\{(\Gamma^+ + X\rho_\zeta)B_\zeta + \eta_\pi G\}$$

(4.6.53)
$$+ \sum_{\nu=1}^{\omega_1}(z - \varepsilon)^{-\nu}(-C_\pi X + F)(T - \varepsilon I)^{\nu-1}\rho_\zeta B_\zeta,$$

and

(4.6.54)
$$
\begin{aligned}
W(z)^{-1} &= I - \{C_\pi(\Gamma^+ + \eta_\pi Y) - F\rho_\zeta\}(zI - A_\zeta)^{-1}B_\zeta \\
&\quad - C_\pi(zI - S)^{-1}(-YB_\zeta + G) \\
&= I - \{C_\pi(\Gamma^+ + \eta_\pi Y) - F\rho_\zeta\}(zI - A_\zeta)^{-1}B_\zeta
\end{aligned}
$$

(4.6.55)
$$
+ \sum_{\nu=1}^{\alpha_1}(\lambda - \varepsilon)^{-\nu}C_\pi(S - \varepsilon I)^{\nu-1}(-YB_\zeta + G).
$$

PROOF. The coupling matrix for $\tau \oplus \tau_0$ by (4.6.39) is given by

$$
\tilde{\Gamma} = \begin{bmatrix} I & 0 \\ -Y & I \end{bmatrix}\begin{bmatrix} \Gamma & \eta_\zeta \\ \rho_\pi & 0 \end{bmatrix}\begin{bmatrix} I & -X \\ 0 & I \end{bmatrix}.
$$

It is easily verified that

$$
\begin{bmatrix} \Gamma & \eta_\zeta \\ \rho_\pi & 0 \end{bmatrix}^{-1} = \begin{bmatrix} \Gamma^+ & \eta_\pi \\ \rho_\zeta & 0 \end{bmatrix}
$$

and hence

$$
\begin{aligned}
\tilde{\Gamma}^{-1} &= \begin{bmatrix} I & X \\ 0 & I \end{bmatrix}\begin{bmatrix} \Gamma^+ & \eta_\pi \\ \rho_\zeta & 0 \end{bmatrix}\begin{bmatrix} I & 0 \\ Y & I \end{bmatrix} \\
&= \begin{bmatrix} \Gamma^+ + X\rho_\zeta + \eta_\pi Y & \eta_\pi \\ \rho_\zeta & 0 \end{bmatrix}.
\end{aligned}
$$

Then by Theorem 4.4.3 the function $W(z)$ is given by (4.6.52) with inverse equal to (4.6.54). To get formulas (4.6.53) and (4.6.55) we use the formulas

$$
\begin{aligned}
(zI - T)^{-1} &= \sum_{j=1}^{s}\sum_{\nu=1}^{\omega_j}\frac{(T - \varepsilon I)^{\nu-1}}{(z - \varepsilon)^\nu}P_{j\nu} \\
&= \sum_{\nu=1}^{\omega_1}\frac{(T - \varepsilon I)^{\nu-1}}{(z - \varepsilon)^\nu}
\end{aligned}
$$

and

$$
\begin{aligned}
(zI - S)^{-1} &= \sum_{j=1}^{t}\sum_{\nu=1}^{\alpha_j}\frac{(S - \varepsilon I)^{\nu-1}}{(z - \varepsilon)^\nu}Q_{j\nu} \\
&= \sum_{\nu=1}^{\alpha_1}\frac{(S - \varepsilon I)^{\nu-1}}{(z - \varepsilon)^\nu}. \quad \square
\end{aligned}
$$

Finally, we shall give an example of the simplest possible form.

EXAMPLE 4.6.1. Given two complex numbers λ and p, $p \neq \lambda$ and two nonzero vectors $x, y \in \mathbf{C}^m$, we are looking for a rational matrix function $W(z)$ such that $W(\infty) = I$, the McMillan degree of W is as small as possible, and

$$
W(p)^{-1}y = 0, \qquad x^*W(\lambda) = 0.
$$

Without loss of generality we may assume that both x and y are unit vectors ($x^*x = 1$, $y^*y = 1$). For sake of simplicity we shall assume that both p and λ are non-zero. This falls into the class of problems studied here if we put

$$A_\pi = [p], \quad A_\zeta = [\lambda], \quad B_\zeta = x^*, \quad C_\pi = y.$$

The solution $\Gamma A_\pi - A_\zeta \Gamma = B_\zeta C_\pi$ is given by $\Gamma = \left(\frac{x^*y}{p-\lambda}\right)$. In case x and y are not orthogonal, we have $\Gamma \neq 0$, and hence Γ is invertible. In this case W can be taken as follows

$$W(z) = I + \frac{p - \lambda yx^*}{z - px^*y}, \qquad W(z)^{-1} = I + \frac{\lambda - pyx^*}{z - \lambda x^*y}.$$

In case x and y are orthogonal, we have $\Gamma = 0$, so both K and $\operatorname{Ker}\Gamma$ are $\mathbf{C}$. Then (taking $\varepsilon = 0$) S and T are 0 and we may define F and G by

$$F = -\lambda x : \mathbf{C} \to \mathbf{C}^m$$
$$G = py^* : \mathbf{C}^m \to \mathbf{C}.$$

Since $\Gamma = 0$, also $\Gamma^+ = 0$ and $X = Y = 0$. Note also that $\rho_\pi = 1 = \eta_\pi$, $\rho_\zeta = 1 = \eta_\zeta$. Thus from (4.6.52) and (4.6.54) we get

(4.6.56) $$W(z) = I_m + (z - p)^{-1}pyy^* - z^{-1}\lambda xx^*$$

(4.6.57) $$W(z)^{-1} = I_m - (z - \lambda)^{-1}\lambda xx^* - z^{-1}pyy^*.$$

Of course for this simple example it is possible to check directly that $W(z)^{-1}$ is given by (4.6.57) if $W(z)$ is given by (4.6.56) and that this $W(z)$ has the requisite properties.   □

CHAPTER 5
RATIONAL MATRIX FUNCTIONS WITH NULL AND POLE STRUCTURE AT INFINITY

In this chapter we extend the theory developed in Chapter 4 for the case of rational matrix functions which may have zeros and poles at infinity. This requires a more general realization theory and allows one to solve less restrictive interpolation problems of constructing a rational matrix function with a given null and pole structure. The special case of matrix polynomials, which may be considered as rational matrix functions with only pole at infinity, is also discussed from the point of view of the general theory. As in Chapter 4 we will restrict ourselves to standard problems only which are here also matricially homogeneous.

5.1 REALIZATIONS AND MÖBIUS TRANSFORMATIONS

In Section 3.5 we discussed the behavior of a null or pole triple (and hence also of a null or pole pair) under a Möbius change of variable for a meromorphic matrix function. In this section we study the effect of a Möbius change of variable on the invariants for rational matrix functions introduced in the previous chapter, namely: realization and null-pole triple.

We first deal with realizations. Consider the Möbius transformation

$$(5.1.1) \qquad \varphi(z) = \frac{kz + \ell}{rz + s}$$

where $ks - \ell r \neq 0$. The inverse map φ^{-1} is given by

$$(5.1.2) \qquad \varphi^{-1}(z) = \frac{sz - \ell}{-rz + k}.$$

If a system of matrices $\theta = (A, B, C, D)$ gives a realization $W(z) = D + C(zI - A)^{-1}B$ for a proper (i.e., finite at infinity) rational matrix function, we refer to A as the *main operator*, B as the *input operator* and C as the *output operator* for θ. From time to time it will be convenient to think of A, B, C, D as linear transformations acting on finite dimensional linear spaces. The space on which A acts will be called the *main space* for θ (or for the pair (C, A) or for the pair (A, B)).

THEOREM 5.1.1. *Let the system of matrices* $\theta = (A, B, C, D)$ *induce a realization*

$$W(z) = D + C(zI - A)^{-1}B$$

for the proper rational $n \times n$ matrix function $W(z)$ and let φ be the Möbius transformation (5.1.1). *Assume that $T = -rA + kI$ is invertible. Then a realization for the function $W_\varphi(z) := W(\varphi(z))$ is given by the system of matrices*

$$(5.1.3) \qquad \theta_\varphi = \left(-(\ell I - sA)T^{-1}, T^{-1}B, (ks - \ell r)CT^{-1}, D + rCT^{-1}B\right).$$

The proof follows the same line of argument as the proof of Theorem 3.5.1.

Recall from Proposition 4.1.5 that if the system of matrices $\theta = (A, B, C, D)$ generates a realization $W(z) = D + C(zI - A)^{-1}B$ for the rational matrix function $W(z)$ and if D is invertible, then the associate system $\theta^\times = (A^\times, BD^{-1}, -D^{-1}C, D^{-1})$ (where $A^\times = A - BD^{-1}C$) gives a realization of $W(z)^{-1}$. The next theorem shows that the operation $\theta \to \theta^\times$ is well-defined with respect to a Möbius change of variable.

THEOREM 5.1.2. *Let φ be the Möbius transformation as in (5.1.1) and let $\theta = (A, B, C, D)$ give a realization for the rational $n \times n$ matrix function $W(z)$. Assume that $T = -rA + kI$, $T^\times = -rA^\times + kI$ (where $A^\times = A - BD^{-1}C$) and $W_\varphi(\infty) = D + rCT^{-1}B$ are invertible. Let θ_φ be given in general by (5.1.3). Then $(\theta_\varphi)^\times = (\theta^\times)_\varphi$, that is, a realization for $W_\varphi(z)^{-1} = W(\varphi(z))^{-1}$ is given by*

(5.1.4)
$$(\theta^\times)_\varphi = ((sA^\times - \ell I)T^{\times -1}, T^{\times -1}BD^{-1}, -(ks - \ell r)D^{-1}CT^{\times -1}, [D + rCT^{-1}B]^{-1}).$$

PROOF. By Theorem 5.1.1, a realization for W_φ is given by
$$\theta_\varphi = ((sA - \ell I)T^{-1}, T^{-1}B, (ks - \ell r)CT^{-1}, D + rCT^{-1}B).$$

Let us denote by H the main operator $H = (sA - \ell I)T^{-1} = \varphi^{-1}(A)$ of θ_φ and by $H^\times$ the main operator

(5.1.5) $\quad H^\times = H - (ks - \ell r)(-rA + kI)^{-1}B[D + rC(-rA + kI)^{-1}B]^{-1}C(-rA + kI)^{-1}$

of $(\theta_\varphi)^\times$. By Proposition 4.1.5, a realization for $W_\varphi(z)^{-1}$ is given by

(5.1.6)
$$(\theta_\varphi)^\times = (H^\times, T^{-1}B[D + rCT^{-1}B]^{-1},$$
$$- (ks - \ell r)[D + rCT^{-1}B]^{-1}CT^{-1}, [D + rCT^{-1}B]^{-1})$$

Use the identity

(5.1.7) $\qquad [D + rC(-rA + kI)^{-1}B]^{-1} = D^{-1} - rD^{-1}C(-rA^\times + kI)^{-1}BD^{-1}$

to rewrite (5.1.5) as

(5.1.8)
$$H^\times = H - (ks - \ell r)(-rA + kI)^{-1}B[D^{-1} - rD^{-1}C(-rA^\times + kI)^{-1}BD^{-1}]$$
$$\times C(-rA + kI)^{-1}$$
$$= H - (ks - \ell r)(-rA + kI)^{-1}BD^{-1}C(-rA + kI)^{-1}$$
$$+ (ks - \ell r)(-rA + kI)^{-1}BD^{-1}C(-rA^\times + kI)^{-1}(rBD^{-1}C)(-rA + kI)^{-1}.$$

Use the identity
$$rBD^{-1}C = r(A - A^\times)$$
$$= -(-rA + kI) + (-rA^\times + kI)$$

in the last term of (5.1.8) to rewrite (5.1.8) as

(5.1.9)
$$H^\times = H - (ks - \ell r)(-rA + kI)^{-1}BD^{-1}C(-rA + kI)^{-1}$$
$$+ [-(ks - \ell r)(-rA + kI)^{-1}BD^{-1}C(-rA^\times + kI)^{-1}$$
$$+ (ks - \ell r)(-rA + kI)^{-1}BD^{-1}C(-rA + kI)^{-1}]$$
$$= H - (ks - \ell r)(-rA + kI)^{-1}BD^{-1}C(-rA^\times + kI)^{-1}.$$

On the other hand we compute

$$\varphi^{-1}(A^\times) - \varphi^{-1}(A) = (sA^\times - \ell I)(-rA^\times + kI)^{-1} - (sA - \ell I)(-rA + kI)^{-1}.$$

Substitute $A^\times = A - BD^{-1}C$ in the first factor of the first term to get from this

$$\varphi^{-1}(A^\times) - \varphi^{-1}(A) = [(sA - \ell I)(-rA^\times + kI)^{-1} - sBD^{-1}C(-rA^\times + kI)^{-1}]$$
$$- (sA - \ell I)(-rA + kI)^{-1}.$$

Collect the terms with the common factor $(sA - \ell I)$ to get that this in turn equals

$$(sA - \ell I)[(-rA^\times + kI)^{-1} - (-rA + kI)^{-1}] - sBD^{-1}C(-rA^\times + kI)^{-1}$$
$$= (sA - \ell I)(-rA + kI)^{-1}[(-rA + kI) - (-rA^\times + kI)](-rA^\times + kI)^{-1}$$
$$- sBD^{-1}C(-rA^\times + kI)^{-1}.$$

The term in brackets collapses to $-r(A - A^\times) = -rBD^{-1}C$. Thus

$$\varphi^{-1}(A^\times) - \varphi^{-1}(A) = (sA - \ell I)(-rA + kI)^{-1}(-rBD^{-1}C)(-rA^\times + kI)^{-1}$$
$$- sBD^{-1}C(-rA^\times + kI)^{-1}$$
$$= (-rA + kI)^{-1}[(sA - \ell I)(-rBD^{-1}C)$$
$$- s(-rA + kI)BD^{-1}C](-rA^\times + kI)^{-1}.$$

The term in brackets collapses to $-(ks - \ell r)BD^{-1}C$. We conclude that

$$(5.1.10) \qquad \varphi^{-1}(A^\times) - \varphi^{-1}(A) = -(ks - \ell r)(-rA + kI)^{-1}BD^{-1}C(-rA^\times + kI)^{-1}.$$

Upon comparing (5.1.9) with (5.1.10) and remembering that $H = \varphi^{-1}(A)$, we get that $H^\times = \varphi^{-1}(A^\times)$, that is, the main operator of $(\theta_\varphi)^\times$ (5.1.6) is the same as the main operator of $(\theta^\times)_\varphi$ (5.1.4).

We next compute input operators. Again using the identity (5.1.7), we get that the input operator for $(\theta_\varphi)^\times$ (5.1.6) is given by

$$T^{-1}B[D + rCT^{-1}B]^{-1}$$
$$= (-rA + kI)^{-1}B[D^{-1} - rD^{-1}C(-rA^\times + kI)^{-1}BD^{-1}]$$
$$= (-rA + kI)^{-1}BD^{-1} - (-rA + kI)^{-1}(rBD^{-1}C)(-rA^\times + kI)^{-1}BD^{-1}.$$

Use again that $-rBD^{-1}C = (-rA^\times + kI) - (-rA + kI)$ to get from this

$$T^{-1}B[D + rCT^{-1}B]^{-1}$$
$$= (-rA + kI)^{-1}BD^{-1} + (-rA + kI)^{-1}[(-rA + kI)$$
$$- (-rA^\times + kI)](-rA^\times + kI)^{-1}BD^{-1}$$
$$= (-rA + kI)^{-1}BD^{-1} + (-rA^\times + kI)^{-1}BD^{-1} - (-rA + kI)^{-1}BD^{-1}$$
$$= T^{\times -1}BD^{-1}$$

which is the input operator for $(\theta^\times)_\varphi$ (5.1.4) as desired.

The computation for the output operators proceeds similarly. Indeed, again by (5.1.7)

(5.1.11)
$$
\begin{aligned}
[D + rCT^{-1}&B]^{-1}CT^{-1} \\
&= [D^{-1} - rD^{-1}C(-rA^\times + kI)^{-1}BD^{-1}]C(-rA + kI)^{-1} \\
&= D^{-1}C(-rA + kI)^{-1} + D^{-1}C(-rA^\times + kI)^{-1}[(-rA + kI) \\
&\quad - (-rA^\times + kI)](-rA + kI)^{-1} \\
&= D^{-1}C(-rA + kI)^{-1} + D^{-1}C(-rA^\times + kI)^{-1} - D^{-1}C(-rA + kI)^{-1} \\
&= D^{-1}C(-rA^\times + kI)^{-1} = D^{-1}CT^{\times -1}.
\end{aligned}
$$

Comparison of (5.1.4) and (5.1.6) together with (5.1.11) shows that $(\theta^\times)_\varphi$ and $(\theta_\varphi)^\times$ have the same output operator as asserted. The D-matrices are the same by inspection; hence $(\theta_\varphi)^\times = (\theta^\times)_\varphi$ so by Proposition 4.1.5 combined with Theorem 5.1.1, $(\theta^\times)_\varphi$ gives a realization of $W_\varphi(z)^{-1}$.   □

Our next goal is to describe how a Möbius transformation affects a σ-null-pole triple for a rational matrix function (see Section 4.5). The next theorem is a complement to the results of Section 3.5 concerning null and pole triples for a meromorphic matrix function. In the next theorem σ is a finite subset of $\mathbb{C}$.

THEOREM 5.1.3. *Let* $\tau = \{C_\pi, A_\pi; A_\zeta, B_\zeta; S\}$ *be a (left) σ-null-pole triple for the rational $n \times n$ matrix function W with value I at infinity, and let φ be the Möbius transformation given by (5.1.1). Assume that $rz - k \neq 0$ for all poles and zeros z of W. Put*

$$\widetilde{W}(z) = W(\varphi(z)), \qquad \widetilde{\sigma} = \{z \in \mathbb{C}: \varphi(z) \in \sigma\}.$$

Then

(5.1.12) $\widetilde{\tau} = \left((ks - \ell r)C_\pi, \varphi^{-1}(A_\pi); \varphi^{-1}(A_\zeta), B_\zeta; (rA_\zeta - kI)S(rA_\pi - kI)\right)$

is a left $\widetilde{\sigma}$-null-pole triple for $\widetilde{W}$.

PROOF. Note that the inverse map φ^{-1} is analytic on σ. Thus, by the Riesz functional calculus, the matrices $\varphi^{-1}(A_\pi)$ and $\varphi^{-1}(A_\zeta)$ are well-defined. Choose a minimal realization of W as follows:

$$W(z) = I + C(zI - A)^{-1}B.$$

Put $A^\times = A - BC$. Let Q (resp. $Q^\times$) be the Riesz projection corresponding to A (resp. $A^\times$) and σ. Then there exist invertible linear transformations $E: \mathbb{C}^{n_\pi} \to \operatorname{Im} Q$ and $F: \mathbb{C}^{n_\zeta} \to \operatorname{Im} Q^\times$ such that

$$
\begin{aligned}
C_\pi &= (C \mid \operatorname{Im} Q)E, & A_\pi &= E^{-1}(A \mid \operatorname{Im} Q)E, \\
B_\zeta &= F^{-1}(Q^\times B), & A_\zeta &= F^{-1}(A^\times \mid \operatorname{Im} Q^\times)F, \\
S &= F^{-1}(Q \mid \operatorname{Im} Q^\times)E.
\end{aligned}
$$

Here $n_\pi \times n_\pi$ and $n_\zeta \times n_\zeta$ are the sizes of A_π and A_ζ, respectively. By the assumption that $rz - k \neq 0$ for all poles and zeros z of W we may apply Theorem 5.1.1 to show that

$$\widetilde{W}(z) = \left(I + rC(-rA + kI)^{-1}B\right) + (ks - \ell r)$$
$$\times C(-rA + kI)^{-1}(zI - H)^{-1}(-rA + kI)^{-1}B,$$
$$=: \widetilde{D} + \widetilde{C}(zI - H)^{-1}\widetilde{B}$$

where $H = \varphi^{-1}(A)$. Furthermore, by Theorem 5.1.2,

$$H^\times = H - (ks - \ell r)(rA - kI)^{-1}B[I - rC(rA - kI)^{-1}B]^{-1}C(rA - kI)^{-1}$$
$$= \varphi^{-1}(A^\times),$$

$$(rA - kI)^{-1}B[I - rC(rA - kI)^{-1}B]^{-1} = (rA^\times - kI)^{-1}B.$$

From the functional calculus it follows that Q (resp. $Q^\times$) is precisely the Riesz projection corresponding to H (resp. $H^\times$) and $\tilde{\sigma}$. Put

$$\widehat{C}_\pi = (ks - \ell r)C(rA - kI)^{-1} \mid \operatorname{Im} Q,$$
$$\widehat{A}_\pi = H \mid \operatorname{Im} Q,$$
$$\widehat{B}_\zeta = Q^\times(rA^\times - kI)^{-1}B,$$
$$\widehat{A}_\zeta = H^\times \mid \operatorname{Im} Q^\times,$$
$$\widehat{S} = Q^\times \mid \operatorname{Im} Q.$$

Thus $\widehat{\tau} = \{\widehat{C}_\pi, \widehat{A}_\pi; \widehat{A}_\zeta, \widehat{B}_\zeta; \widehat{S}\}$ is a $\tilde{\sigma}$-null-pole triple for $\widetilde{W}$. Next, consider the invertible linear transformations

$$\widehat{E} = (rA - kI)E: \mathbf{C}^{n_\pi} \to \operatorname{Im} Q,$$
$$\widehat{F} = (rA^\times - kI)^{-1}F: \mathbf{C}^{n_\zeta} \to \operatorname{Im} Q^\times,$$

and put

$$\widetilde{C}_\pi = \widehat{C}_\pi \widehat{E}, \qquad \widetilde{A}_\pi = (\widehat{E})^{-1}\widehat{A}_\pi \widehat{E},$$
$$\widetilde{B}_\zeta = (\widehat{F})^{-1}\widehat{B}_\zeta, \qquad \widetilde{A}_\zeta = (\widehat{F})^{-1}\widehat{A}_\zeta \widehat{F},$$
$$\widetilde{S} = (\widehat{F})^{-1}\widehat{S}\widehat{E}.$$

By similarity, $\widetilde{\tau} := \{\widetilde{C}_\pi, \widetilde{A}_\pi; \widetilde{A}_\zeta, \widetilde{B}_\zeta; \widetilde{S}\}$ is again a $\tilde{\sigma}$-null-pole triple for $\widetilde{W}$. It remains to show that $\widetilde{\tau}$ is also given by (5.1.12). To do this we use that $H = \varphi^{-1}(A)$ commutes with $rA - kI$ and that $H^\times = \varphi^{-1}(A^\times)$ commutes with $rA^\times - kI$. Also, we shall employ the following two identities:

$$E(rA_\pi - kI) = (rA - kI)E, \qquad F(rA_\zeta - kI) = (rA^\times - kI)F.$$

Indeed

$$
\begin{aligned}
\widetilde{C}_\pi &= (ks - \ell r)C(rA - kI)^{-1}(rA - kI)E \\
&= (ks - \ell r)(C \mid \operatorname{Im} Q)E = (ks - \ell r)C_\pi, \\
\widetilde{A}_\pi &= E^{-1}(rA - kI)^{-1}(\varphi^{-1}(A) \mid \operatorname{Im} Q)(rA - kI)E \\
&= E^{-1}(\varphi^{-1}(A) \mid \operatorname{Im} Q)E = \varphi^{-1}(A_\pi), \\
\widetilde{B}_\zeta &= F^{-1}(rA^\times - kI)(rA^\times - kI)^{-1}B = F^{-1}Q^\times B = B_\zeta, \\
\widetilde{A}_\zeta &= F^{-1}(rA^\times - kI)(\varphi^{-1}(A^\times) \mid \operatorname{Im} Q^\times)(rA^\times - kI)^{-1}F \\
&= F^{-1}(\varphi^{-1}(A^\times) \mid \operatorname{Im} Q^\times)F = \varphi^{-1}(A_\zeta), \\
\widetilde{S} &= F^{-1}(rA^\times - kI)\widehat{S}(rA - kI)E \\
&= (rA_\zeta - kI)F^{-1}(Q^\times \mid \operatorname{Im} Q)E(rA_\pi - kI) \\
&= (rA_\zeta - kI)S(rA_\pi - kI).
\end{aligned}
$$

The theorem is proved. $\square$

5.2   REALIZATIONS OF IMPROPER RATIONAL MATRIX FUNCTIONS

In this section we extend the results concerning realization from Section 4.1 to rational matrix functions not necessarily finite at infinity. This is done with the aid of the notion of a local realization at infinity.

By definition, a triple of matrices $(A_\infty, B_\infty, C_\infty)$ is a *local realization of the rational matrix function* $W(z)$ *at infinity* if $(A_\infty, B_\infty, C_\infty)$ is a local realization of the function $\widetilde{W}(z) = W(z^{-1})$ at the point zero. This means that $\sigma(A_\infty) = \{0\}$ and the function

$$
\widetilde{W}(z) - C_\infty(zI - A_\infty)^{-1}B_\infty
$$

is analytic in a neighborhood of zero. In general we say that the triple (A, B, C) is a *finite realization* of the rational matrix function $W(z)$ (not necessarily finite at infinity) if $W(z) - C(zI - A)^{-1}B$ is analytic in the whole complex plane (but not necessarily at infinity).

THEOREM 5.2.1.   *There is a finite realization* (A, B, C) *for every rational matrix function* $W(z)$ *(as usual, we implicitly assume that* $\det W(z) \not\equiv 0$*). Also there is a local realization* $(A_\infty, B_\infty, C_\infty)$ *for* $W(z)$ *at infinity. If* (A, B, C) *is a finite realization and* $(A_\infty, B_\infty, C_\infty)$ *is a local realization at infinity, then there is a constant matrix* D *for which*

(5.2.1) $\qquad W(z) = C(zI - A)^{-1}B + D + zC_\infty(I - zA_\infty)^{-1}B_\infty.$

The representation $(A, B, C, D, A_\infty, B_\infty, C_\infty)$ of $W(z)$ as in (5.2.1) we refer to as a *realization (in standard form)* for the rational matrix function $W(z)$.

PROOF.   We have

$$
W(z) = U(z) + W_1(z)
$$

where $U(z)$ is a matrix polynomial and $W_1(z)$ is a rational matrix function with finite value at infinity. Applying Theorem 4.1.1 we obtain the existence of the realization (A, B, C) for $W_1(z)$. The existence of a local realization at infinity $(A_\infty, B_\infty, C_\infty)$ follows from the existence of a pole triple for $\widetilde{W}(z) = W(z^{-1})$ at zero (by Theorem 3.3.1 applied to $\widetilde{W}^{-1}$).

By definition, $W(z) - C(zI - A)^{-1}B = U(z)$ is analytic in the complex plane, while $W(z) - C_\infty(z^{-1}I - A_\infty)^{-1}B_\infty = W(z) - zC_\infty(I - zA_\infty)^{-1}B_\infty$ is analytic in a neighborhood of infinity. Taking into account that $\sigma(A_\infty) = \{0\}$ and hence $(I - zA_\infty)^{-1}$ is a matrix polynomial, we see that the function

$$W(z) - C(zI - A)^{-1}B - zC_\infty(I - zA_\infty)^{-1}B_\infty$$

is analytic in $\mathbf{C} \cup \{\infty\}$ and hence it must be constant. $\square$

For σ a subset of $\mathbf{C}$ let σ^c denote the complement of σ in $\mathbf{C} \cup \{\infty\}$. In what follows we also assume that $0 \in \sigma$. Let us say that a realization $(A, B, C, D, A_\infty, B_\infty, C_\infty)$ for a rational matrix function $W(z)$ is σ-canonical if $\sigma(A) \subset \sigma$ and $\sigma(A_\infty) = [\sigma^c]^{-1} \stackrel{\text{def}}{=} \{0\} \cup \{z \in \mathbf{C}; z^{-1} \in \sigma^c\}$. We say that $(A, B, C, D, A_\infty, B_\infty, C_\infty)$ is $minimal$ if (C, A) and (C_∞, A_∞) are null-kernel pairs and (A, B) and (A_∞, B_∞) are full-range pairs. Theorem 5.2.1 shows that $\mathbf{C}$-canonical realizations always exist. Also from the results of Chapters 3 and 4 it is clear from the construction in the proof of Theorem 5.2.1 that one can arrange that this $\mathbf{C}$-canonical realization can be taken to be minimal. For a general subset σ of $\mathbf{C}$ containing zero, it is easy to get a minimal σ-canonical realization from a minimal $\mathbf{C}$-canonical realization for a given rational matrix function $W(z)$. Indeed if $(A, B, C, D, A_\infty, B_\infty, C_\infty)$ is a $\mathbf{C}$-canonical minimal realization, write A in the form $A_1 \oplus A_2$ where $\sigma(A_1) \subset \sigma$ and $\sigma(A_2) \subset \sigma^c$. This induces a decomposition $B = \begin{bmatrix} B_1 \\ B_2 \end{bmatrix}$, $C = [C_1, C_2]$ for B and C. Then $(A_1, B_1, C_1, D, A_2^{-1} \oplus A_\infty, \text{col}(B_2, B_\infty), [C_2, C_\infty])$ is a minimal σ-canonical realization for $W(z)$. As we shall see in later chapters using σ-canonical realizations for various choices of σ gives more flexibility which is convenient in applications.

A minimal realization of the form (5.2.1) in general is not unique; however once a choice of σ is made, a minimal σ-canonical realization for a given $W(z)$ is unique up to similarity:

THEOREM 5.2.2.   *Let $W(z)$ be a rational matrix function and suppose σ is a subset of $\mathbf{C}$ containing zero. Then any two minimal σ-canonical realizations $(A, B, C, D, A_\infty, B_\infty, C_\infty)$ and $(\widetilde{A}, \widetilde{B}, \widetilde{C}, \widetilde{D}, \widetilde{A}_\infty, \widetilde{B}_\infty, \widetilde{C}_\infty)$ for $W(z)$ as in (5.2.1) are similar, i.e. there exist invertible matrices S and T for which*

$$\widetilde{C} = CS, \quad \widetilde{A} = S^{-1}AS, \quad \widetilde{B} = S^{-1}B$$

(5.2.2)
$$\widetilde{C}_\infty = C_\infty T, \quad \widetilde{A}_\infty = T^{-1}A_\infty T, \quad \widetilde{B}_\infty = T^{-1}B_\infty$$

$$\widetilde{D} = D.$$

PROOF. Suppose $(A, B, C, D, A_\infty, B_\infty, C_\infty)$ and $(\widetilde{A}, \widetilde{B}, \widetilde{C}, \widetilde{D}, \widetilde{A}_\infty, \widetilde{B}_\infty, \widetilde{C}_\infty)$

are two such realizations. Then

$$(5.2.3) \quad \begin{aligned} & C(zI - A)^{-1}B - \widetilde{C}(zI - \widetilde{A})^{-1}\widetilde{B} + D - \widetilde{D} \\ & = z\widetilde{C}_\infty(I - z\widetilde{A}_\infty)^{-1}\widetilde{B}_\infty - zC_\infty(I - zA_\infty)B_\infty. \end{aligned}$$

Note that the left side of (5.2.3) is analytic off σ (including at infinity) while the right hand side is analytic off σ^c; therefore by Liouville's Theorem they each equal a constant matrix E. Evaluating the left side at infinity gives $E = D - \widetilde{D}$ while evaluating the right side at zero gives $E = 0$; thus $D = \widetilde{D}$ and

$$C(zI - A)^{-1}B = \widetilde{C}(zI - \widetilde{A})^{-1}\widetilde{B}$$

as well as

$$C_\infty(z^{-1}I - A_\infty)^{-1}B_\infty = \widetilde{C}_\infty(z^{-1}I - \widetilde{A}_\infty)^{-1}\widetilde{B}_\infty.$$

The result now follows from the uniqueness of minimal realizations up to similarity for rational matrix functions with finite value at infinity (Theorem 4.1.3). □

A realization $W(z) = D + C(zI - A)^{-1}B$ for a proper rational matrix function $W(z)$ has the property that it displays the value D of W at infinity. On the other hand a realization of the form (5.2.1) has the disadvantage that the value of W at infinity or even at any point $z_0 \in \mathbb{C}$ is not displayed in an obvious way. The situation can be alleviated by changing the form of the realization (5.2.1). Let us say that the system of matrices $(A_F, B_F, C_F, D, A_\infty, B_\infty, C_\infty)$ is a realization for $W(z)$ *centered at the point* z_0 if

$$(5.2.4) \quad W(z) = D + (z - z_0)C_F(zI - A_F)^{-1}B_F + (z - z_0)C_\infty(I - zA_\infty)^{-1}B_\infty.$$

If we choose z_0 so that z_0 is not in the spectrum of A_F and z_0^{-1} is not in the spectrum of A_∞, then z_0 is not a pole of $W(z)$ and (5.2.4) displays D as the value of W at z_0. We may apply the terms σ-*canonical* and *minimal* to realizations centered at z_0 in the same way in which these terms were applied to realizations of the standard form (5.2.1). The next result describes how these various types of realizations are connected with each other.

THEOREM 5.2.3. (i) *Let* $W(z)$ *be a rational matrix function and suppose* $z_0 \in \mathbb{C}$ *is not a pole of* $W(z)$. *Let* σ *be a subset of* $\mathbb{C}$ *containing zero. Then there exists a minimal* σ-*canonical realization* $(A_F, B_F, C_F, A_\infty, B_\infty, C_\infty)$ *for* $W(z)$ *centered at* z_0 (*i.e.* (5.2.4) *holds with* $D = W(z_0)$).

(ii) *If* $(\widetilde{A}_F, \widetilde{B}_F, \widetilde{C}_F, \widetilde{A}_\infty, \widetilde{B}_\infty, \widetilde{C}_\infty)$ *is another such realization then there exist invertible matrices* S *and* T *for which equations* (5.2.2) *hold.*

(iii) $(A, B, C, D, A_\infty, B_\infty, C_\infty)$ *is a minimal* σ-*canonical realization for* $W(z)$ *in the standard form* (5.2.1) *if and only if* $\left(A, -(z_0I - A)^{-1}B, C, W(z_0), A_\infty, (I - z_0A_\infty)^{-1}B_\infty, C_\infty\right)$ *gives a minimal* σ-*canonical realization for* $W(z)$ *of the form* (5.2.4) *centered at* z_0.

PROOF. Statements (i) and (ii) follow from (iii) by using the results of Theorems 5.2.1 and 5.2.2 for realizations in a standard form. To prove (iii), suppose

$(A, B, C, D, A_\infty, B_\infty, C_\infty)$ is a minimal σ-canonical realization for $W(z)$ of the standard form (5.2.1). Then we compute

$$W(z) - W(z_0) = C(zI - A)^{-1}B + zC_\infty(I - zA_\infty)^{-1}B_\infty - C(z_0I - A)^{-1}B$$
$$- z_0 C_\infty(I - z_0 A_\infty)^{-1}B_\infty.$$

Note that

$$(zI - A)^{-1} - (z_0 I - A)^{-1} = -(z - z_0)(zI - A)^{-1}(z_0 I - A)^{-1}$$

while

$$z(I - zA_\infty)^{-1} - z_0(I - z_0 A_\infty)^{-1} = (z - z_0)(I - zA_\infty)^{-1}(I - z_0 A_\infty)^{-1}$$

and hence

$$W(z) - W(z_0) = -(z - z_0)C(zI - A)^{-1} \cdot (z_0 I - A)^{-1}B$$
$$+ (z - z_0)C_\infty(I - zA_\infty)^{-1}(I - z_0 A_\infty)^{-1}B_\infty$$

and thus $(A, -(z_0 I - A)^{-1}B, C, A_\infty, (I - z_0 A_\infty)^{-1}B_\infty, C_\infty)$ is a realization for $W(z)$ centered at z_0. It is easily checked that this realization centered at z_0 is minimal and σ-canonical if and only if the original realization in standard form $(A, B, C, D, A_\infty, B_\infty, C_\infty)$ is. By reversing the steps in the above computation one can get a realization in standard form from one centered at z_0. This completes the proof of the theorem. □

5.3   NULL-POLE TRIPLES AT INFINITY AND MÖBIUS TRANSFORMATIONS

In Section 3.5 we defined null and pole triple, as well as left and right null and pole pair for a meromorphic matrix function at the point infinity. In this section we introduce the notion of a null-pole coupling matrix associated with a left null pair and right pole pair at infinity for a rational matrix function and from this the notion of null-pole triple at infinity. We also extend the results of Section 5.1 to improper rational matrix functions.

Let us assume that the extended complex plane $\mathbf{C}^* = C \cup \{\infty\}$ is partitioned into two disjoint subsets σ_F and σ_∞

(5.3.1) $$\mathbf{C}^* = \sigma_F \cup \sigma_\infty$$

where $0 \in \sigma_F$ and $\infty \in \sigma_\infty$. Recall from Section 4.5 that a data set of matrices

(5.3.2) $$\tau_F = (C_{\pi,F}, A_{\pi,F}; A_{\zeta,F}, B_{\zeta,F}; S_F)$$

is said to be a σ_F-admissible Sylvester data set

(i) the spectrum of $A_{\pi,F}$ and of $A_{\zeta,F}$ is in σ

(ii) $(C_{\pi,F}, A_{\pi,F})$ is a null-kernel pair and $(A_{\zeta,F}, B_{\zeta,F})$ is a full-range pair

and

(iii) S_F satisfies the Sylvester equation $SA_{\pi,F} - A_{\zeta,F}S = B_{\zeta}C_{\pi}$.

For the case where we use the subset σ_∞ which includes the point infinity, we say that the data set

$$(5.3.3) \qquad \tau_\infty = (C_{\pi,\infty}, A_{\pi,\infty}; A_{\zeta,\infty}, B_{\zeta,\infty}; S_\infty)$$

is a σ_∞-*admissible Sylvester data set* if τ_∞ is a $(\sigma_\infty)^{-1}$-admissible Sylvester data set, where

$$(\sigma_\infty)^{-1} = \{z \in \mathbb{C} : z^{-1} \in \sigma_\infty\}.$$

Here z^{-1} for the case $z = 0$ is interpreted to be the point infinity. We emphasize that in general the definition of σ-admissible data set is dependent on whether the point infinity is included in or excluded from σ.

Now suppose that $W(z)$ is a regular rational matrix function. Then a data set τ_F as in (5.3.2) is said to be a *left σ_F-null-pole triple for W* if there is a rational matrix function $V(z)$ with $V(\infty) = I$ such that

(i) $V^{-1}(z)W(z)$ has no poles and no zeros in σ_F

and

(ii) τ_F is a left σ_F-null-pole triple for $V(z)$.

Similarly, a data set τ_∞ as in (5.3.3) is said to be a *left σ_∞-null-pole triple for $W(z)$* if there exists a rational matrix function $V(z)$ with $V(\infty) = I$ such that

(i) $V^{-1}(z)W(z^{-1})$ has no poles and no zeros in $(\sigma_\infty)^{-1}$

and

(ii) τ_∞ is a left $(\sigma_\infty)^{-1}$-null-pole triple for $V(z)$.

Note as in the case of σ-admissible Sylvester data set that the definition of σ-null-pole triple depends crucially on whether or not the point infinity is assigned to σ. By Theorem 4.5.8 we see that these definitions are independent of the choice of $V(z)$ in both cases, and that the definition of a left σ_F-null-pole triple for W for the case where $W(\infty) = I$ is consistent with that given in Section 4.5. Also from the results of Section 4.5, we see that τ_F must be a σ_F-admissible Sylvester data set in order for τ_F to be a σ_F-null-pole triple for some W, and similarly a σ_∞-null-pole triple for some W necessarily is a σ_∞-admissible Sylvester data set. The converse of this statement is the main interpolation theorem of Section 5.4.

The following theorem explains how a τ_F-null-pole triple and a τ_∞-null-pole triple for a regular rational matrix function can be related to null-pole triples for a rational matrix function with value I at infinity via a Möbius change of variable.

THEOREM 5.3.1. *Let W be a regular rational $n \times n$ matrix function such that τ_F given by (5.3.2) is a σ_F-null-pole triple and τ_∞ given by (5.3.3) is a σ_∞-null-pole triple for W. Let $z_0 \in \mathbb{C}$ be such that z_0 is neither a pole nor a zero of W. Put*

$D = W(z_0)$ and let $\varphi(z)$ and $\psi(z)$ be the Möbius transformations

(5.3.4) $$\varphi(z) = \frac{z_0 z + 1}{z}, \qquad \varphi^{-1}(z) = \frac{1}{z - z_0},$$

and

(5.3.5) $$\psi(z) = \frac{z}{z_0 z + 1}, \qquad \psi^{-1}(z) = \frac{z}{1 - z_0 z}.$$

Define the rational matrix function $\widetilde{W}$ by

$$\widetilde{W}(z) = W(\varphi(z)) D^{-1} = V(\psi(z)) D^{-1}$$

where $V(z) = W(z^{-1})$. Then $\widetilde{W}(\infty) = I$ and

(5.3.6) $$\begin{aligned}\widehat{\tau}_F = (&-C_{\pi,F}(A_{\pi,F} - z_0 I)^{-1}, (A_{\pi,F} - z_0 I)^{-1}; \\ &(A_{\zeta,F} - z_0 I)^{-1}, (A_{\zeta,F} - z_0 I)^{-1} B_{\zeta,F}; S_F)\end{aligned}$$

is a $[\varphi^{-1}(\sigma_F) \cap \mathbb{C}]$-null-pole triple for $\widetilde{W}$, and

(5.3.7) $$\begin{aligned}\widehat{\tau}_\infty = (&C_{\pi,\infty}(I - z_0 A_{\pi,\infty})^{-1}, A_{\pi,\infty}(I - z_0 A_{\pi,\infty})^{-1}; \\ &A_{\zeta,\infty}(I - z_0 A_{\zeta,\infty})^{-1}, (I - z_0 A_{\zeta,\infty})^{-1} B_{\zeta,\infty}; S_\infty)\end{aligned}$$

is a $[\psi^{-1}((\sigma_\infty)^{-1}) \cap \mathbb{C}]$-null-pole triple for $\widetilde{W}(z)$ in the (finite) sense of Section 4.5.

PROOF. Note first that since $\varphi(z) = 1/\psi(z)$ we have

$$W(\varphi(z)) D^{-1} = V(\psi(z)) D^{-1}$$

so $\widetilde{W}(z)$ is well defined. Since $\varphi(\infty) = z_0$ it is clear that $\widetilde{W}(\infty) = I$. To show that $\widehat{\tau}_F$ is a $[\varphi^{-1}(\sigma_F) \cap \mathbb{C}]$-null-pole triple for $\widetilde{W}$, we need only show that Theorem 5.1.3 applies even if W does not have the value I at infinity.

To do this, let $\tau = \tau_F$ and let φ be given by (5.1.1) as in Theorem 5.1.3. Then choose a complement τ_0 of τ_F as in Lemma 4.5.7, and let G be the rational matrix function with value I at infinity corresponding to $\tau \oplus \tau_0$ as in Theorem 4.3.3. Let ρ be the set of poles and zeros of G. We may choose the complement τ_0 such that $rz - k \neq 0$ for each $z \in \rho$. Since both W and G have τ as a σ_F-null-pole triple, the function $E(z) = W(z)^{-1} G(z)$ has no poles and no zeros on σ_F by Theorem 4.5.8. Put

$$\widetilde{G}(z) = G(\varphi(z)), \qquad \widetilde{E}(z) = E(\varphi(z)).$$

Then $\widetilde{E}$ has no poles and no zeros on $\widetilde{\sigma}_F = \varphi^{-1}(\sigma_F)$ and $\widetilde{W}(z) = \widetilde{G}(z) \widetilde{E}(z)$. Thus $\widetilde{W}(z)$ and $\widetilde{G}(z)$ have the same $\widetilde{\sigma}_F$-null-pole triples, again by Theorem 4.5.8. Note that G is analytic with value I at infinity; we conclude that Theorem 5.1.3 continues to hold for improper regular rational matrix functions. We thus conclude that $\widehat{\tau}_F$ given by (5.3.6) is a $[\varphi^{-1}(\sigma_F) \cap \mathbb{C}]$-null-pole triple for $\widetilde{W}$ as asserted.

Next note that by definition the set τ_∞ is a $(\sigma_\infty)^{-1}$-null-pole triple for V and $1 - z_0 z \neq 0$ for $z \in \sigma(A_{\zeta,\infty})$. So, again we may apply Theorem 5.1.3 using V and ψ. This yields that $\widehat{\tau}_\infty$ as in (5.3.7) is a $[\varphi^{-1}((\sigma_\infty)^{-1}) \cap \mathbb{C}]$-null-pole triple for $\widetilde{W}$ as asserted. $\square$

5.4 INTERPOLATION WITH DATA AT INFINITY

In this section we consider the following interpolation problem for rational matrix functions: given a partitioning $\mathbb{C}^* = \sigma_F \cup \sigma_\infty$ of the extended complex plane, with $0 \in \sigma$ and $\infty \in \sigma_\infty$, given a σ_F-admissible Sylvester data set τ_F as in (5.3.2) and a σ_∞-admissible Sylvester data set τ_∞ in (5.3.3), when is there a (possibly improper) regular rational matrix function $W(z)$ having τ_F as a left σ_F-null-pole triple and τ_∞ as a right σ_∞-null-pole triple? We shall obtain an explicit criterion as to when a solution exists, and when this criterion is satisfied we obtain the solution in terms of a realization centered at z_0 (see Section 5.2), where z_0 is any point in $\mathbb{C}$ not in $\sigma(A_{\pi,F}) \cup \sigma(A_{\zeta,F}) \cup \left(\sigma(A_{\pi,\infty})\right)^{-1} \cup \left(\sigma(A_{\zeta,\infty})\right)^{-1}$.

To state the solution we need the matrix

$$(5.4.1) \qquad S = \begin{bmatrix} S_F & S_{12} \\ S_{21} & S_\infty \end{bmatrix}$$

where S_{12} and S_{21} are the unique solutions of the following Stein equations:

$$(5.4.2) \qquad A_{\zeta,F} S_{12} A_{\pi,\infty} - S_{12} = B_{\zeta,F} C_{\pi,\infty}$$

$$(5.4.3) \qquad S_{21} - A_{\zeta,\infty} S_{21} A_{\pi,F} = -B_{\zeta,\infty} C_{\pi,F}$$

(see Section A.2 in the Appendix for a discussion of Stein equations). Since the set $\sigma(A_{\zeta,F}) \subset \sigma_F$ is disjoint from the set $\sigma(A_{\pi,\infty})^{-1} \subset \sigma_\infty$, (5.4.2) has a unique solution; similarly (5.4.3) has a unique solution. The main goal of this section is to prove the following two theorems.

THEOREM 5.4.1. *In order that τ_F be a left σ_F-null-pole triple and that τ_∞ be a left σ_∞-null-pole triple for a regular $n \times n$ rational matrix function it is necessary and sufficient that*

$$(5.4.4) \qquad \det \begin{bmatrix} S_F & S_{12} \\ S_{21} & S_\infty \end{bmatrix} \neq 0.$$

Here S_{12} and S_{21} are defined by (5.4.2) and (5.4.3).

THEOREM 5.4.2. *Let $z_0 \in \mathbb{C}$ and D be an invertible $n \times n$ matrix and assume that τ_F is a σ_F-admissible Sylvester data set and that τ_∞ is a σ_∞-admissible Sylvester data set. Then there exists a regular rational $n \times n$ matrix function W such that*

(i) τ_F *is a left σ_F-null-pole triple for W,*

(ii) τ_∞ *is a left σ_∞-null-pole triple for W,*

(iii) $W(z_0) = D$,

if and only if z_0 is not an eigenvalue of $A_{\pi,F}$ or $A_{\zeta,F}$, z_0^{-1} (if finite) is not an eigenvalue of $A_{\pi,\infty}$ or $A_{\zeta,\infty}$, and (5.4.4) holds true. In this case there is precisely one W with the *properties* (i), (ii) *and* (iii) *which is given by*

$$W(z) = D + (z - z_0)[\, C_{\pi,F} \quad C_{\pi,\infty} \,]$$
$$\cdot \begin{bmatrix} (zI - A_{\pi,F})^{-1} & 0 \\ 0 & (I - zA_{\pi,\infty})^{-1} \end{bmatrix} \begin{bmatrix} S_F & S_{12} \\ S_{21} & S_\infty \end{bmatrix}^{-1}$$
$$\cdot \begin{bmatrix} (A_{\zeta,F} - z_0 I)^{-1} B_{\zeta,F} D \\ (I - z_0 A_{\zeta,\infty})^{-1} B_{\zeta,\infty} D \end{bmatrix}.$$

Furthermore, this function has the following inverse:

$$W(z)^{-1} = D^{-1} - (z - z_0)[\, D^{-1}C_{\pi,F}(z_0 I - A_{\pi,F})^{-1} \quad C_{\pi,\infty}(I - z_0 A_{\pi,\infty})^{-1} \,]$$
$$\cdot \begin{bmatrix} S_F & S_{12} \\ S_{21} & S_\infty \end{bmatrix}^{-1} \begin{bmatrix} (A_{\zeta,F} - zI)^{-1} & 0 \\ 0 & (I - zA_{\zeta,\infty})^{-1} \end{bmatrix} \begin{bmatrix} B_{\zeta,F} \\ B_{\zeta,\infty} \end{bmatrix}.$$

PROOF. We shall prove Theorems 5.4.1 and 5.4.2 together. To prove necessity, suppose that W is a regular rational $n \times n$ matrix function such that τ_F is a σ_F-null-pole triple and τ_∞ is a σ_∞-null-pole triple for W, and choose $z_0 \in \mathbb{C}$ which is neither a pole nor a zero for W. Set $D = W(z_0)$. Let φ and ψ be the Möbius transformations given by (5.3.4) and (5.3.5) respectively as in Theorem 5.3.1. From Theorem 5.3.1 we see that $\widehat{\tau}_F$ as in (5.3.6) is a $[\varphi^{-1}(\sigma_F) \cap \mathbb{C}]$-null-pole triple for $\widetilde{W}$ and that $\widehat{\tau}_\infty$ as in (5.3.7) is a $[\psi^{-1}((\sigma_\infty)^{-1}) \cap \mathbb{C}]$-null-pole triple for $\widetilde{W}$, where $\widetilde{W}(z) = W(\varphi(z))D^{-1}$ is analytic with value I at infinity. Finally note that $\psi^{-1}((\sigma_\infty)^{-1}) = \varphi^{-1}(\sigma_\infty)$ so

$$[\mathbb{C} \cap \varphi^{-1}(\sigma_F)] \cup [\mathbb{C} \cap \psi^{-1}((\sigma_\infty)^{-1})] = \mathbb{C}$$

gives a partitioning of the finite complex plane $\mathbb{C}$. It follows by Proposition 4.5.2 that $\widehat{\tau}_F \oplus \widehat{\tau}_\infty$ is a global null-pole triple for $\widetilde{W}(z)$ as defined in Section 4.4. Therefore the coupling $\widetilde{S}$ matrix for $\widehat{\tau}_F \oplus \widehat{\tau}_\infty$ must be invertible. By definition this coupling matrix is given by

$$\widetilde{S} = \begin{bmatrix} S_F & S_{12} \\ S_{21} & S_\infty \end{bmatrix}$$

where S_{12} and S_{21} are determined by the Sylvester equations

(5.4.5)
$$S_{12}A_{\pi,\infty}(I - z_0 A_{\pi,\infty})^{-1} - (A_{\zeta,F} - z_0 I)^{-1}S_{12}$$
$$= (A_{\zeta,F} - z_0 I)^{-1}B_{\zeta,F}C_{\pi,\infty}(I - z_0 A_{\pi,\infty})^{-1}$$

and

(5.4.6)
$$S_{21}(A_{\pi,F} - z_0 I)^{-1} - A_{\zeta,\infty}(I - z_0 A_{\zeta,\infty})^{-1}S_{21}$$
$$= -(I - z_0 A_{\zeta,\infty})^{-1}B_{\zeta,\infty}C_{\pi,F}(A_{\pi,F} - z_0 I)^{-1}.$$

These equations can be rewritten as

$$(A_{\zeta,F} - z_0 I)S_{12}A_{\pi,\infty} - S_{12}(I - z_0 A_{\pi,\infty}) = B_{\zeta,F}C_{\pi,\infty}$$

$$(I - z_0 A_{\zeta,\infty})S_{21} - A_{\zeta,\infty}S_{21}(A_{\pi,F} - z_0 I) = -B_{\zeta,\infty}C_{\pi,F}$$

which clearly are equivalent to (5.4.2) and (5.4.3); this proves the necessity of condition (5.4.4) in Theorem 5.4.1. Furthermore the function $\widetilde{W}$ is uniquely determined by $\widetilde{\tau}_F \oplus \widetilde{\tau}_\infty$ and the condition that $\widetilde{W}(\infty) = I$. In fact, by Theorem 4.4.3,

$$\widetilde{W}(z) = I + [-C_{\pi,F}(A_{\pi,F} - z_0 I)^{-1}, C_{\pi,\infty}(I - z_0 A_{\pi,\infty})^{-1}]$$
$$\cdot \begin{bmatrix} zI - (A_{\pi,F} - z_0 I)^{-1} & 0 \\ 0 & zI - A_{\pi,\infty}(I - z_0 A_{\pi,\infty})^{-1} \end{bmatrix}^{-1}$$
$$\cdot \begin{bmatrix} S_F & S_{12} \\ S_{21} & S_\infty \end{bmatrix}^{-1} \begin{bmatrix} (A_{\zeta,F} - z_0 I)^{-1}B_{\zeta,F} \\ (I - z_0 A_{\zeta,\infty})^{-1}B_{\zeta,\infty} \end{bmatrix},$$

and

$$\widetilde{W}(z)^{-1} = I - [-C_{\pi,F}(A_{\pi,F} - z_0 I)^{-1}, C_{\pi,\infty}(I - z_0 A_{\pi,\infty})^{-1}]$$
$$\cdot \begin{bmatrix} S_F & S_{12} \\ S_{21} & S_\infty \end{bmatrix}^{-1} \begin{bmatrix} zI - (A_{\zeta,F} - z_0 I)^{-1} & 0 \\ 0 & zI - A_{\zeta,\infty}(I - z_0 A_{\zeta,\infty})^{-1} \end{bmatrix}^{-1}$$
$$\cdot \begin{bmatrix} (A_{\zeta,F} - z_0 I)^{-1}B_{\zeta,F} \\ (I - z_0 A_{\zeta,\infty})^{-1}B_{\zeta,\infty} \end{bmatrix}.$$

Since

$$W(z) = \widetilde{W}\left(\frac{1}{z - z_0}\right)D, \qquad W(z)^{-1} = D^{-1}\widetilde{W}\left(\frac{1}{z - z_0}\right)^{-1},$$

the above formulas for $\widetilde{W}(z)$ and $\widetilde{W}(z)^{-1}$ yield the formulas for $W(z)$ and $W(z)^{-1}$ in Theorem 5.4.2. To see this one uses the identities

$$[(z - z_0)^{-1}I - (A_{\pi,F} - z_0 I)^{-1}]^{-1} = (z - z_0)(A_{\pi,F} - z_0 I)(A_{\pi,F} - zI)^{-1}$$

and

$$[(z - z_0)^{-1}I - A_{\pi,\infty}(I - z_0 A_{\pi,\infty})^{-1}]^{-1} = (z - z_0)(I - z_0 A_{\pi,\infty})(I - zA_{\pi,\infty})^{-1},$$

together with the analogous identities with the index π replaced by the index ζ.

It remains to prove the sufficiency of the condition (5.4.4) in Theorem 5.4.1. Define S_{12} and S_{21} by (5.4.2) and (5.4.3) and assume that

$$\det \begin{bmatrix} S_F & S_{12} \\ S_{21} & S_\infty \end{bmatrix} \neq 0.$$

We must show that there exists a regular $n \times n$ matrix function such that τ_F is a σ_F-null-pole triple for W and τ_∞ is a σ_∞-null-pole triple for W and such that $W(z_0) = D$. Define $\widehat{\tau}_F$ and $\widehat{\tau}_\infty$ by (5.3.6) and (5.3.7) and define φ and ψ by (5.3.4) and (5.3.5). From $S_F A_{\pi,F} - A_{\zeta,F}S_F = B_\zeta C_\pi$ it follows that

$$(A_{\zeta,F} - z_0 I)S_F - S_F(A_{\pi,F} - z_0 I) = -B_\zeta C_\pi,$$

and thus

$$(5.4.7) \ \ S_F(A_{\pi,F} - z_0 I)^{-1} - (A_{\zeta,F} - z_0 I)^{-1} S_F = -(A_{\zeta,F} - z_0 I)^{-1} B_\zeta C_\pi (A_{\pi,F} - z_0 I)^{-1}.$$

By the Riesz functional calculus, the eigenvalues of $(A_{\pi,F} - z_0 I)^{-1}$ and of $(A_{\zeta,F} - z_0 I)^{-1}$ are located in $\varphi^{-1}(\sigma_F)$; this along with (5.4.7) enables us to conclude that $\hat{\tau}_F$ is a $(\varphi^{-1}(\sigma_F) \cap \mathbb{C})$-admissible Sylvester data set. Next we use $S_\infty A_{\pi,\infty} - A_{\zeta,\infty} S_\infty = B_{\zeta,\infty} C_{\pi,\infty}$ to show that $\hat{\tau}_\infty$ is a $(\psi^{-1}((\sigma_\infty)^{-1}) \cap \mathbb{C})$-admissible Sylvester data set. Note also that $\psi^{-1}((\sigma_\infty)^{-1}) = \varphi^{-1}(\sigma_\infty)$ is disjoint from $\varphi^{-1}(\sigma_F)$, since by assumption $\sigma_F \cap \sigma_\infty = \emptyset$. Thus we can form the direct sum $\hat{\tau}_F \oplus \hat{\tau}_\infty$ as in Proposition 4.5.2. The coupling matrix Γ of $\hat{\tau}_F \oplus \hat{\tau}_\infty$ is the matrix

$$\tilde{S} = \begin{bmatrix} S_F & S_{12} \\ S_{21} & S_\infty \end{bmatrix}$$

where S_{12} and S_{21} are defined by (5.4.5) and (5.4.6). We already observed in the proof of necessity that these are equivalent to (5.4.2) and (5.4.3). But then, by our hypothesis $\tilde{S}$ is invertible, so by Theorem 4.4.3 $\hat{\tau}_F \oplus \hat{\tau}_\infty$ is a global null-pole triple for a unique rational matrix function $\widetilde{W}(z)$ having finite value I at infinity. Define $W(z)$ by

$$(5.4.8) \qquad\qquad W(z) = \widetilde{W}((z - z_0)^{-1}) D$$

where D is the invertible matrix in the hypothesis of Theorem 5.4.2. Since $\widetilde{W}(\infty) = I$, we see from (5.4.8) that $W(z_0) = D$. When we use (5.4.8) to express $\widetilde{W}$ in terms of W we get

$$\widetilde{W}(z) = W(\varphi(z)) D^{-1} = V(\psi(z)) D^{-1}.$$

Theorem 5.3.1 now gives expressions for a $[\varphi^{-1}(\sigma_F) \cap \mathbb{C}]$-null-pole triple for $\widetilde{W}$ in terms of a σ_F-null-pole triple for W and for a $[\psi^{-1}((\sigma_\infty)^{-1}) \cap \mathbb{C}]$-null-pole triple for $\widetilde{W}$ in terms of a σ_∞-null-pole triple for W. Setting these equal to $\hat{\tau}_F$ and $\hat{\tau}_\infty$ respectively and backsolving, we see that τ_F is a σ_F-null-pole triple for W and that τ_∞ is a σ_∞-null-pole triple for W. $\square$

5.5 SPECIALIZATIONS FOR MATRIX POLYNOMIALS

In this section we indicate how the techniques for solving homogeneous interpolation problems for improper rational matrix functions given in this chapter can be used to recover some of the interpolation results for matrix polynomials given in Chapter 2. Specifically we shall discuss Theorem 2.3.1(b) and the left-sided version of Theorem 2.5.4 from this point of view.

THEOREM 5.5.1. (= *Theorem 2.3.1(b)*). *Let* (T, Y) *be a full-range pair of matrices of sizes* $n\ell \times n\ell$ *and* $n\ell \times n$ *respectively. Then there exists a monic* $n \times n$ *matrix polynomial* $L(z)$ *of degree* ℓ *having* (T, Y) *as a left null pair over* $\mathbb{C}$ *if and only if the* $n\ell \times n\ell$ *matrix*

$$(5.5.1) \qquad\qquad S = [T^{\ell-1}Y, \dots, TY, Y]$$

is invertible. In this case the unique solution $L(z)$ is given by

$$L(z) = z^\ell I_n + \sum_{j=0}^{\ell-1} z^j A_j$$

where

(5.5.2) $$\operatorname{col}(A_{\ell-1}, \ldots, A_1, A_0) = -S^{-1} T^\ell Y.$$

To prove this result using the techniques of this chapter, we would like to use a different form of Theorem 5.4.2, where the solution $W(z)$ of the interpolation problem is given in terms of a realization in standard form rather than in terms of a realization centered at some point z_0 (see Section 5.2 for a discussion of these various types of realization). Also in the following theorem, as one does not specify a regular point z_0 and a value D for W at z_0, the solution is not as explicit. Although the following theorem could be proved directly from Theorem 5.4.2, we postpone its proof until Chapter 15 where it will serve to illustrate a different approach to interpolation.

THEOREM 5.5.2. *Suppose that $\mathbb{C}^* = \sigma_F \cup \sigma_\infty$ is a partitioning of the extended complex plane such that $0 \in \sigma_F$ and $\infty \in \sigma_\infty$, and suppose that τ_F and τ_∞ are respectively σ_F- and σ_∞-admissible Sylvester data sets. Then there exists a regular rational $n \times n$ matrix function $W(z)$ for which*

(i) *τ_F is a left σ_F-null-pole triple for W*

and

(ii) *τ_∞ is a left σ_∞-null-pole triple for W if and only if condition (5.4.4) in Theorem 5.4.1 holds. In this case the matrix Ψ given by*

$$\Psi = \begin{bmatrix} S_F & A_{\zeta,F} S_{12} & -B_{\zeta,F} \\ A_{\zeta,\infty} S_{21} & S_\infty & -B_{\zeta,\infty} \end{bmatrix}$$

has dim Ker $\Psi = n$, and a rational matrix function $W(z)$ is regular and satisfies (i) and (ii) if and only if

(5.5.3) $$W(z) = C_{\pi,F}(zI - A_{\pi,F})^{-1} B_{\pi,F} + z C_{\pi,\infty}(I - z A_{\pi,\infty})^{-1} B_{\pi,\infty} + D_\pi,$$

where the columns of $\operatorname{col}(B_{\pi,F}, B_{\pi,\infty}, D_\pi)$ form a basis for Ker Ψ. Moreover the matrix Φ given by

$$\Phi = \begin{bmatrix} -S_F & S_{12} A_{\pi,\infty} \\ S_{21} A_{\pi,F} & -S_\infty \\ -C_{\pi,F} & -C_{\pi,\infty} \end{bmatrix}$$

has left kernel of dimension equal to n, and the inverse $W^{-1}(z)$ of $W(z)$ in (5.5.3) is given by

(5.5.4) $$W^{-1}(z) = L^{-1}[C_{\zeta,F}(zI - A_{\zeta,F})^{-1} B_{\zeta,F} + z C_{\zeta,\infty}(I - z A_{\zeta,\infty})^{-1} B_{\zeta,\infty} + D_\zeta]$$

where the rows of $[C_{\zeta,F}, C_{\zeta,\infty}, D_\zeta]$ *form a basis for the left kernel of* Φ *and where the invertible matrix* L *is given by*

$$L = -C_{\zeta,F} S_{12} B_{\pi,\infty} - C_{\zeta,\infty} S_{21} B_{\pi,F} - D_\zeta D_\pi.$$

PROOF OF THEOREM 5.5.1. A matrix polynomial $L(z)$ is simply a rational matrix function with no poles in the finite plane $\mathbb{C}$. Thus if (T, Y) is a left null pair for such a function $L(z)$, then $\{0, 0; T, Y; 0\}$ is a left null-pole triple for $L(z)$ over $\mathbb{C}$. If $L(z)$ in addition is a monic matrix polynomial, then $L(z) = (z^\ell I_n) \cdot W(z)$ where $W(z)$ is analytic and with value I_n at infinity. Therefore by Theorem 4.5.8, $L(z)$ and $z^\ell I$ have the same null-pole triple at infinity. Note that $z^\ell I$ can be realized as $z^\ell I = zC(I - zA)^{-1}B$ where $C = [0, \ldots, 0, I]$ $(n \times n\ell)$ and where A is the $n\ell \times n\ell$ matrix

$$A = \begin{bmatrix} 0 & & & 0 \\ I & 0 & & \\ & \ddots & \ddots & \\ 0 & & I & 0 \end{bmatrix}$$

and $B = \text{col}(I, 0, \ldots, 0)$. It is easily seen that (C, A) is a null-kernel pair. Since also $(z^\ell I)^{-1} = z^{-\ell} I$ is analytic at infinity, we conclude that

$$\left([0, \ldots, 0, I], \begin{bmatrix} 0 & & & 0 \\ I & 0 & & \\ & \ddots & \ddots & \\ 0 & & I & 0 \end{bmatrix}; 0, 0; 0 \right)$$

is a null-pole triple for $z^\ell I$, and hence also for the monic matrix polynomial $L(z)$, over $\{\infty\}$. By Theorem 5.4.1 there exists a rational matrix function having $(0, 0; T, Y; 0)$ as a null-pole triple over $\mathbb{C}$ and $(C, A; 0, 0; 0)$ as a null-pole triple over $\{\infty\}$ if and only if the matrix $\widehat{S} = \begin{bmatrix} S_F & S_{12} \\ S_{21} & S_\infty \end{bmatrix}$ is invertible. For our case here, S_F, S_{21} and S_∞ are all vacuous, so $\widehat{S}$ collapses to S_{12} which is the unique solution of the Stein equation

$$T\widehat{S}A - \widehat{S} = YC.$$

Since A is nilpotent of index ℓ, it is easy to see that the unique solution S of this Stein equation is given by

$$\widehat{S} = -\sum_{j=0}^{\ell-1} T^j Y C A^j$$
$$= -[0, \ldots, 0, Y] - T[0, \ldots, Y, 0] - \cdots - T^{\ell-1}[Y, 0, \ldots, 0]$$
$$= -[T^{\ell-1}Y, \ldots, TY, Y].$$

Thus $S = -\widehat{S}$ is given as in (5.5.1) as desired. Any rational matrix function having null-pole triple $(0, 0; T, Y; 0)$ over $\mathbb{C}$ and $(C, A; 0, 0; 0)$ over infinity in particular has no finite poles so is a matrix polynomial. Rather than seeking a solution $L(z)$ assuming a prescribed invertible value D at a regular point z_0 as is done in Theorem 5.4.2, we seek a monic matrix polynomial, i.e., one satisfying the normalization

$$(5.5.5) \qquad \lim_{z \to \infty} z^{-\ell} L(z) = I_n.$$

To find a $L(z)$ satisfying this condition, it is more convenient to use the less rigid formula (5.5.3) given in Theorem 5.5.2. Thus the required solution $L(z)$ we know has the form $L(z) = D + zC(I - zA)^{-1}B$ where D and B are to be determined so that the columns of $\begin{bmatrix} B \\ D \end{bmatrix}$ form a basis for Ker Ψ, where for our case Ψ collapses to

$$(5.5.6) \qquad \Psi = [T\widehat{S}, -Y].$$

Decompose $B = \operatorname{col}(A_\ell, A_{\ell-1}, \ldots, A_1)$, and $D = A_0$, where each A_j has size $n \times n$. Then the equation $\Psi \begin{bmatrix} B \\ D \end{bmatrix} = 0$ takes the form (from (5.5.6) and (5.5.1))

$$(5.5.7) \qquad [T^\ell Y, T^{\ell-1} Y, \ldots, TY, Y] \begin{bmatrix} A_\ell \\ A_{\ell-1} \\ \vdots \\ A_0 \end{bmatrix} = 0.$$

On the other hand, from the special form of C and A we can compute explicitly

$$L(z) = D + zC(I - zA)^{-1}B$$
$$= D + z \sum_{k=0}^{\ell-1} CA^k B z^k$$
$$= A_0 + zA_1 + \cdots + z^{\ell-1}A_{\ell-1} + z^\ell A_\ell.$$

Thus the monicity requirement (5.5.5) amounts to $A_\ell = I$. With this extra condition we can then use that S as in (5.5.1) is invertible to solve equation (5.5.7) uniquely for $\operatorname{col}(A_{\ell-1}, \ldots, A_0)$; the result is (5.5.2). $\square$

We next discuss the problem of constructing a regular matrix polynomial $L(z)$ having a given full range pair (T, Y) as a left null pair over $\mathbf{C}$ in the framework of this chapter. One solution is already given by Theorem 2.5.4 in Chapter 2. With no essential loss of generality we may assume that T is invertible. One approach is to reduce to the monic case (i.e. Theorem 5.5.1 or, what is the same, Theorem 2.3.1) by considering $\widetilde{L}(z) \stackrel{\text{def}}{=} z^\ell L(z^{-1})$ (ℓ = the degree of $L(z)$) in place of $L(z)$ and by using Theorem 2.4.2; indeed, this is what is done essentially in the proof of Theorem 2.5.4 in Chapter 2. An alternative approach which illustrates further the ideas of this chapter is to consider $\widehat{L}(z) = z^{-\ell} L(z)$ in place of $L(z)$, where again ℓ is the degree of $L(z)$. Then the unknown matrix function $\widehat{L}(z)$ is a rational matrix function with only pole in the extended plane at 0; by our assumption that T is invertible (so $L(0)$ is invertible), the pole structure at 0 can be computed precisely. Since ℓ is the degree of $L(z)$, it is clear that $\widehat{L}(z)$ has no pole at infinity, and hence a null-pole triple for $\widehat{L}(z)$ over $\{\infty\}$ must have the form

$$\tau_\infty = (0, 0; A_{\zeta,\infty}, B_{\zeta,\infty}; 0)$$

where $A_{\zeta,\infty}$ is nilpotent. Note that $(A_{\zeta,\infty}, B_{\zeta,\infty})$ is not specified by the data of the interpolation problem to be solved. A left pole pair for $\widehat{L}(z)$ over $\mathbb{C}$ by Theorem 4.5.8 must be the same as one for $z^{-\ell} I_n$, namely (C, A) with

(5.5.8) $$C = [0, \dots, 0, I]$$

and

(5.5.9) $$A = \begin{bmatrix} 0 & & & \\ I & 0 & & \\ & \ddots & \ddots & \\ & & I & 0 \end{bmatrix}.$$

Since the function $z^{-\ell} I$ is analytic and invertible at all nonzero points in $\mathbb{C}$, we deduce that (T, Y) is a left null pair over $\mathbb{C}$ for $\widehat{L}(z)$ as well as for $L(z)$ (again by Theorem 4.5.8). The coupling matrix S for which $(C, A; T, Y; S)$ is a null-pole triple over $\mathbb{C}$ for $\widehat{L}(z)$ is then uniquely determined by the Sylvester equation

$$SA - TS = YC$$

or equivalently by the Stein equation

$$T^{-1}SA - S = T^{-1}YC.$$

As in the proof of Theorem 5.5.1, we see that for A and C of the special form (5.5.9) and (5.5.8) we get

(5.5.10)
$$S = -\sum_{j=0}^{\ell-1} T^{-j}(T^{-1}YC)A^j$$
$$= [-T^{-\ell}Y, \dots, -T^{-2}Y, -T^{-1}Y].$$

Thus a null-pole triple for $\widehat{L}(z)$ over $\mathbb{C}$ is

$$\left([0, \dots, 0, I], \begin{bmatrix} 0 & & & 0 \\ I & \ddots & & \\ & \ddots & \ddots & \\ & & I & 0 \end{bmatrix}; T, Y; [-T^{-\ell}Y, \dots, -T^{-1}Y] \right)$$

and is completely specified by the data of the problem. The criterion for existence of a solution $\widehat{L}(z)$ from Theorem 5.4.1 is that the matrix $\widehat{S} = \begin{bmatrix} S_F & S_{12} \\ S_{21} & S_\infty \end{bmatrix}$ be invertible. For the case at hand $\widehat{S}$ collapses to

$$\widehat{S} = \begin{bmatrix} S \\ S_{21} \end{bmatrix}$$

where S is as in (5.5.10) and S_{21} is the unique solution of the Stein equation

(5.5.11) $$S_{21} - A_{\zeta,\infty} S_{21} A = -B_{\zeta,\infty} C$$

where A and C are given by (5.5.9) and (5.5.10). Again from the special form of A and C, the unique solution S_{21} of (5.5.11) can be seen to be

$$S_{21} = [-A_{\zeta,\infty}^{\ell-1} B_{\zeta,\infty}, \ldots, -A_{\zeta,\infty} B_{\zeta,\infty}, -B_{\zeta,\infty}]$$

and thus $\widehat{S}$ has the form

$$\widehat{S} = \begin{bmatrix} S \\ S_{21} \end{bmatrix} = -\begin{bmatrix} T^{-\ell}Y & \cdots & T^{-2}Y & T^{-1}Y \\ A_{\zeta,\infty}^{\ell-1} B_{\zeta,\infty} & \cdots & A_{\zeta,\infty} B_{\zeta,\infty} & B_{\zeta,\infty} \end{bmatrix}.$$

Now again by Theorem 2.4.2 there exists a choice of ℓ and of full range pair $(A_{\zeta,\infty}, B_{\zeta,\infty})$ for which $\widehat{S}$ is invertible. Once some such $(A_{\zeta,\infty}, B_{\zeta,\infty})$ is chosen, one can apply the formula in Theorem 5.4.2 to compute the unique such $\widehat{L}(z)$ having value $z_0^{-\ell}D$ at a nonzero point z_0 not in the spectrum of T, where D is a prescribed invertible matrix; upon some simplification the result is:

(5.5.12) $$\widetilde{L}(z) = z_0^{-\ell}D + (z - z_0)[z^{-\ell}I, \ldots, z^{-1}I]\widehat{S}^{-1}$$
$$\cdot \begin{bmatrix} (T - z_0 I)^{-1}Y(z_0^{-\ell}D) \\ (I - z_0 A_{\zeta,\infty})^{-1}B_{\zeta,\infty}(z_0^{-\ell}D) \end{bmatrix}.$$

Therefore, a polynomial $L(z)$ having (T, Y) as a left null pair over $\mathbb{C}$ and having value D at z_0 is

$$L(z) = z^\ell z_0^{-\ell}D + (z - z_0)z_0^{-\ell}[I, zI, \ldots, z^{\ell-1}I]\widehat{S}^{-1} \begin{bmatrix} (T - z_0 I)^{-1}YD \\ (I - z_0 A_{\zeta,\infty})^{-1}B_{\zeta,\infty}D \end{bmatrix}.$$

5.6 INTERPOLATION WITH PRESCRIBED NULL AND POLE DATA EXCEPT AT INFINITY

In this section we consider the following interpolation problem for a rational matrix function: we are given an admissible Sylvester data set τ and seek a regular rational matrix function $W(z)$ (possibly improper or value at infinity not invertible) having τ as a null-pole triple over the finite plane $\mathbb{C}$. To tackle this notion we introduce the notion of a *minimal supplement*.

Let $\tau = (C_\pi, A_\pi; A_\zeta, B_\zeta; \Gamma)$ be an admissible Sylvester data set. The admissible Sylvester data set $\tau_0 = (C_{\pi\infty}, A_{\pi\infty}; A_{\zeta\infty}, B_{\zeta\infty}; \Gamma_\infty)$ is called a *supplement* for τ if $\sigma(A_{\pi\infty}) = \sigma(A_{\zeta\infty}) = \{0\}$ and if the matrix

$$\begin{bmatrix} \Gamma & \Gamma_{12} \\ \Gamma_{21} & \Gamma_\infty \end{bmatrix}$$

is square and invertible, where Γ_{12} and Γ_{21} are the unique solutions of

(5.6.1) $$A_\zeta \Gamma_{12} A_{\pi\infty} - \Gamma_{12} = B_\zeta C_{\pi\infty}$$

(5.6.2) $$A_{\zeta\infty}\Gamma_{21}A_\pi - \Gamma_{21} = B_{\zeta\infty}C_\pi.$$

The supplement τ_0 will be called a *minimal supplement* in case the size of the matrix

$$\begin{bmatrix} \Gamma & \Gamma_{12} \\ \Gamma_{21} & \Gamma_\infty \end{bmatrix}$$

is as small as possible among all supplements. In this case the function

(5.6.3)
$$\begin{aligned}
W(z) = D + (z - \alpha)[\,C_\pi \quad C_{\pi\infty}\,] \\
\times \begin{bmatrix} (zI - A_\pi)^{-1} & 0 \\ 0 & (I - zA_{\pi\infty})^{-1} \end{bmatrix} \begin{bmatrix} \Gamma & \Gamma_{12} \\ \Gamma_{21} & \Gamma_\infty \end{bmatrix}^{-1} \\
\times \begin{bmatrix} (A_\zeta - \alpha I)^{-1}B_\zeta D \\ (I - \alpha A_{\zeta\infty})^{-1}B_{\zeta\infty}D \end{bmatrix}
\end{aligned}$$

where $\alpha \notin \sigma(A_\zeta) \cup \sigma(A_\pi)$, is a rational matrix function of minimal possible McMillan degree having τ as a null-pole triple over $\mathbb{C}$ and having value D at α (see Theorems 5.4.1 and 5.4.2).

The construction of a minimal supplement for a given Sylvester data set τ is analogous to the construction of a minimal complement given in Section 4.6. The construction hinges on finding outgoing and incoming bases at infinity for τ. A basis $\{d_{jk}\}_{k=1,j=1}^{\alpha_j,t}$ of $\operatorname{Ker}\Gamma$ is called an *outgoing basis at infinity for τ* if

(5.6.4) $$\{d_{jk}\}_{k=1,j=1}^{\alpha_j,t} \quad \text{is a basis for} \quad \operatorname{Ker}\Gamma \cap \operatorname{Ker}C_\pi,$$

(5.6.5) $$A_\pi d_{j,k+1} = d_{jk} \quad (k = 1, \ldots, \alpha_j - 1).$$

A basis $\{g_{jk}\}_{k=1,j=1}^{\omega_j,s}$ of a direct complement of $\operatorname{Im}\Gamma$ in $\mathbb{C}^{n_\pi}$ is called an *incoming basis at infinity for τ* if

(5.6.6) $$\{g_{j\omega_j}\}_{j=1}^{s} \quad \text{is a basis for a direct complement of} \quad \operatorname{Im}\Gamma \quad \text{in} \quad \operatorname{Im}\Gamma + \operatorname{Im}B_\zeta,$$

(5.6.7) $$A_\zeta g_{j,k+1} - g_{jk} \in \operatorname{Im}\Gamma + \operatorname{Im}B_\zeta, \qquad g_{j0} := 0.$$

We assume $\alpha_1 \geq \cdots \geq \alpha_t$, and $\omega_1 \geq \cdots \geq \omega_s$.

LEMMA 5.6.1. *There exist incoming and outgoing bases at infinity for τ.*

PROOF. We construct an outgoing basis at infinity for τ. Let α_1 be the largest number such that

$$H_{\alpha_1-1} := \operatorname{Ker}\Gamma \cap \operatorname{Ker}C_\pi \cap \cdots \cap \operatorname{Ker}C_\pi A_\pi^{\alpha_1-1} \neq \{0\}.$$

Pick a basis $\{d_{j\alpha_j}\}_{j=1}^{\nu}$ ($\alpha_1 = \alpha_2 = \cdots = \alpha_\nu$) in H_{α_1-1}. Note that for $j = 1, \ldots, \nu$:

$$A_\pi d_{j\alpha_j} \in \operatorname{Ker}\Gamma \cap \operatorname{Ker}C_\pi \cap \cdots \cap \operatorname{Ker}C_\pi A_\pi^{\alpha_1-2} =: H_{\alpha_1-2}.$$

Clearly, also $d_{j\alpha_j} \in H_{\alpha_1-2}$. Furthermore, the set of vectors $\{d_{j\alpha_j}, A_\pi d_{j\alpha_j}\}_{j=1}^\nu$ linearly independent. Indeed, suppose that there are scalars $\{c_j\}_{j=1}^\nu$ and $\{c_j'\}_{j=1}^\nu$, not all zero, such that $x = A_\pi y$, where

$$x = \sum_{j=1}^\nu c_j d_{j\alpha_j} \in H_{\alpha_1-1}$$

and

$$y = -\sum_{j=1}^\nu c_j' d_{j\alpha_j} \in H_{\alpha_1-1}.$$

Since both y and $A_\pi y \in H_{\alpha_1-1}$, from the definitions we see that

$$y \in \operatorname{Ker}\Gamma \cap \operatorname{Ker} C_\pi \cap \cdots \cap \operatorname{Ker} C_\pi A_\pi^{\alpha_1} = \{0\}$$

and hence $c_j' = 0$ for $j = 1, \ldots, \nu$. This then forces $x = 0$ and hence $c_j = 0$ for $j = 1, \ldots, \nu$ as well. This then verifies the linear independence of $\{d_{j\alpha_j}, A_\pi d_{j\alpha_j}\}_{j=1}^\nu$ as claimed.

Now take vectors $\{d_{j\alpha_j}\}_{j=\nu+1}^x$ ($\alpha_{\nu+1} = \cdots = \alpha_x = \alpha_1 - 1$) such that $\{d_{j\alpha_j}\}_{j=\nu+1}^x \cup \{A_\pi d_{j\alpha_j}\}_{j=1}^\nu$ is a basis for H_{α_1-2}. Put $d_{j,\alpha_j-1} = A_\pi d_{j\alpha_j}$, $j = 1, \ldots, \nu$. Note that $\{d_{j,\alpha_j-1}\}_{j=1}^x$ is a basis for a direct complement of H_{α_1-1} in H_{α_1-2}.

Suppose we have chosen in this fashion a basis $\{d_{jk}\}_{k=\nu,j=1}^{\alpha_j,\beta}$ for $H_{\nu-1}$ such that $\{d_{j\nu}\}_{j=1}^\beta$ is a basis for a direct complement of H_ν in $H_{\nu-1}$. Consider $A_\pi d_{j\nu} \in H_{\nu-2}$; these vectors are independent of $\{d_{jk}\}_{k=\nu,j=1}^{\alpha_j,\beta}$. Indeed, suppose there is a linear combination x of the $d_{j\nu}$'s such that

$$A_\pi x \in H_{\nu-1} = \operatorname{Ker}\Gamma \cap \operatorname{Ker} C_\pi \cap \cdots \cap \operatorname{Ker} C_\pi A_\pi^{\nu-1}.$$

For $x \in H_{\nu-1}$ and $A_\pi x \in H_{\nu-1}$ we get that

$$x \in H_\nu = \operatorname{Ker}\Gamma \cap \operatorname{Ker} C_\pi \cap \cdots \cap \operatorname{Ker} C_\pi A_\pi^\nu.$$

Since x is a linear combination of the $d_{j\nu}$'s, we conclude that $x = 0$ as desired.

Hence we can continue the procedure by setting $d_{j,\nu-1} = A_\pi d_{j\nu}$ and completing the set $\{d_{jk}\}_{k=\nu-1,j=1}^{\alpha_j,\beta}$ to a basis in $H_{\nu-2}$ by picking vectors $d_{j,\nu-1}$ ($j > \beta$) such that the set $\{d_{j,\nu-1}\}$ is a basis for a direct complement of $H_{\nu-1}$ in $H_{\nu-2}$.

The construction of an incoming basis at infinity for τ is done analogously.

$\square$

We shall give a minimal supplement next. Before stating the theorem let us introduce some notation. Let $\mathbb{C}^{n_\pi} = \operatorname{Ker}\Gamma \dotplus N$, $\mathbb{C}^{n_\varsigma} = \operatorname{Im}\Gamma \dotplus K$. Also we select $\Gamma^+: \mathbb{C}^{n_\varsigma} \to \mathbb{C}^{n_\pi}$ a generalized inverse of Γ such that $\operatorname{Im}\Gamma^+ = N$ and $\operatorname{Ker}\Gamma^+ = K$. Let ρ_π be the projection onto $\operatorname{Ker}\Gamma$ along N and ρ_ς the projection onto K along $\operatorname{Im}\Gamma$, so $\Gamma\Gamma^+ = (I - \rho_\varsigma)$, $\Gamma^+\Gamma = (I - \rho_\pi)$. Furthermore, η_π stands for the injection of $\operatorname{Ker}\Gamma$ in $\mathbb{C}^{n_\pi}$, and η_ς for the injection of K in $\mathbb{C}^{n_\varsigma}$.

THEOREM 5.6.2. *Let* $\tau = (C_\pi, A_\pi; A_\zeta, B_\zeta; \Gamma)$ *be a given admissible Sylvester data set, and let* $\{d_{jk}\}_{k=1,j=1}^{\alpha_j,t}$ *and* $\{g_{jk}\}_{k=1,j=1}^{\omega_j,s}$ *be outgoing and incoming bases for* τ *at infinity, respectively. Define*

(5.6.8) $$S_\infty: \operatorname{Ker}\Gamma \to \operatorname{Ker}\Gamma, \quad S_\infty d_{jk} = d_{j,k+1} \qquad (d_{j,\alpha_j+1} = 0)$$

and

(5.6.9) $$T_\infty: K \to K, \quad T_\infty g_{jk} = g_{j,k+1} \qquad (g_{j,\omega_j+1} = 0).$$

Put $z_j = C_\pi d_{j1}$, *and pick* y_j *such that* $g_{j\omega_j} - B_\zeta y_j \in \operatorname{Im}\Gamma$. *Furthermore, let* $Y_0 \subset \mathbf{C}^m$ *(here m is the number of rows in C_π which is equal to the number of columns in B_ζ) be a subspace such that*

(5.6.10) $$\mathbf{C}^m = \operatorname{Span}\{z_j\}_{j=1}^t \dotplus Y_0 \dotplus \operatorname{Span}\{y_j\}_{j=1}^s.$$

Then define $G: \mathbf{C}^m \to \operatorname{Ker}\Gamma$ *by*

(5.6.11) $$Gz_j = (S_\infty \rho_\pi A_\pi - I)d_{j1}, \qquad j = 1, \ldots, t$$

(5.6.12) $$Gy = 0, \qquad y \in Y_0 \dotplus \operatorname{Span}\{y_j\}_{j=1}^s.$$

Also define $F: K \to \mathbf{C}^m$ *as follows. Note that* $(A_\zeta T_\infty - I)K \subset \operatorname{Im}\Gamma + \operatorname{Im}B_\zeta$ *by (5.6.7), (5.6.9). Choose* $u_{jk} \in \operatorname{Span}\{y_j\}_{j=1}^s$, $u_{j\omega_j} = y_j$ *such that*

$$\rho_\zeta(A_\zeta T_\infty - I)g_{jk} = -\rho_\zeta B_\zeta u_{jk},$$

and define $Fg_{jk} = u_{jk}$.

Let X *and* Y *be given by*

(5.6.13) $$X = \sum_{j=1}^{\omega_1} A_\pi^j \Gamma^+ (A_\zeta T_\infty + B_\zeta F)\eta_\zeta T_\infty^j,$$

(5.6.14) $$Y = \sum_{j=1}^{\alpha_1} S_\infty^j (S_\infty \rho_\pi A_\pi - GC_\pi)\Gamma^+ A_\zeta^j,$$

and put

(5.6.15) $$\Gamma_0 = Y\Gamma X + Y\eta_\zeta + \rho_\pi X.$$

Then a particular minimal supplement of τ *is given by*

(5.6.16) $$\tau_0 = (-C_\pi X T_\infty - F, T_\infty; S_\infty, S_\infty Y B_\zeta + G; \Gamma_0).$$

PROOF. We first remark that G is well-defined. Indeed, since $B_\zeta z_j = B_\zeta C_\pi d_{j1} = (\Gamma A_\pi - A_\zeta \Gamma) d_{j1} = \Gamma A_\pi d_{j1} \in \operatorname{Im}\Gamma$ we easily obtain that $z_1, \ldots, z_t, y_1, \ldots, y_s$ are linearly independent. Furthermore, it is easy to see that F is well-defined as well.

By definition we have to show the following. Let Γ_{12} and Γ_{21} be the solutions of

$$(5.6.17) \qquad A_\zeta \Gamma_{12} T_\infty - \Gamma_{12} = -B_\zeta (C_\pi X T_\infty + F),$$

$$(5.6.18) \qquad S_\infty \Gamma_{21} A_\pi - \Gamma_{21} = (S_\infty Y B_\zeta + G) C_\pi.$$

Then $\widetilde{\Gamma} = \begin{bmatrix} \Gamma & \Gamma_{12} \\ \Gamma_{21} & \Gamma_0 \end{bmatrix}$ must be invertible. Moreover, $(C_\pi X T_\infty + F, T_\infty)$ should be a null kernel pair and $(S_\infty, S_\infty Y B_\zeta + G)$ a full range pair. Thirdly, Γ_0 should satisfy

$$(5.6.19) \qquad \Gamma_0 T_\infty - S_\infty \Gamma_0 = -(S_\infty Y B_\zeta + G)(C_\pi X T_\infty + F).$$

All this will in fact guarantee that τ_0 is a supplement to τ. To see it is a minimal supplement we must show that $\operatorname{rank}\widetilde{\Gamma}$ is as small as possible among all supplements to τ. But the latter is obvious: any $\widehat{\Gamma} = \begin{bmatrix} \Gamma & \widehat{\Gamma}_{12} \\ \widehat{\Gamma}_{21} & \widehat{\Gamma}_0 \end{bmatrix}$ which is invertible must satisfy $\operatorname{rank}\widehat{\Gamma} \geq n_\pi + \dim K = n_\zeta + \dim \ker \Gamma = \operatorname{rank}\widetilde{\Gamma}$.

We shall first show that $(C_\pi X T_\infty + F, T_\infty)$ is a null kernel pair. Suppose not, and let

$$0 \neq x \in \bigcap_{j\geq 0}(C_\pi X T_\infty + F)T_\infty^j.$$

Since T_∞ is nilpotent, we may replace x with $T_\infty^k x$ for an appropriate nonnegative integer k if necessary, and assume that $T_\infty x = 0$. Then we have

$$T_\infty x = 0, \qquad (C_\pi X T_\infty + F)x = 0.$$

Thus

$$T_\infty x = 0, \qquad Fx = 0.$$

Since $T_\infty x = 0$ we have $x \in \operatorname{Span}\{g_{j\omega_j}\}$. But $Fg_{j\omega_j} = y_j$ and the independence of the y_j's then forces $x = 0$, a contradiction. Similarly, $(S_\infty, S_\infty Y B_\zeta + G)$ is a full range pair. Indeed, suppose $y^T S_\infty = 0$ and $y^T(S_\infty Y B_\zeta + G) = 0$. Then $y^T G = 0$ and $y^T S_\infty = 0$, so $y^T d_{j1} = 0$. Consequently, $y^T x = 0$ for every $x \in \operatorname{Im} S_\infty$, which implies $y^T x = 0$ for every $x \in \operatorname{Ker}\Gamma$, so $y = 0$.

Next we show that

$$(5.6.20) \qquad \Gamma_{12} = \eta_\zeta + \Gamma X, \qquad \Gamma_{21} = \rho_\pi + Y\Gamma.$$

Once this is shown it is also clear that $\widetilde{\Gamma}$ is invertible, since in this case

$$(5.6.21) \qquad \widetilde{\Gamma} = \begin{bmatrix} I & 0 \\ Y & I \end{bmatrix}\begin{bmatrix} \Gamma & \eta_\zeta \\ \rho_\pi & 0 \end{bmatrix}\begin{bmatrix} I & X \\ 0 & I \end{bmatrix}$$

which is evidently invertible. To show the first identity in (5.6.20) we just insert $\eta_\zeta + \Gamma X$ in the left hand side of (5.6.17):

$$A_\zeta(\eta_\zeta + \Gamma X)T_\infty - \eta_\zeta - \Gamma X = A_\zeta T_\infty + (\Gamma A_\pi - B_\zeta C_\pi)X T_\infty - \eta_\zeta - \Gamma X$$
$$= A_\zeta T_\infty + \Gamma A_\pi X T_\infty - B_\zeta C_\pi X T_\infty - \eta_\zeta - \Gamma X.$$

Now from (5.6.13) one sees that X actually solves

$$(5.6.22) \qquad\qquad X - A_\pi X T_\infty = \Gamma^+(A_\zeta T_\infty + B_\zeta F)\eta_\zeta.$$

So the above is equal to

$$A_\zeta T_\infty + \Gamma X - \Gamma\Gamma^+(A_\zeta T_\infty + B_\zeta F)\eta_\zeta - B_\zeta C_\pi X T_\infty - \eta_\zeta - \Gamma X$$
$$= \rho_\zeta A_\zeta T_\infty - (I - \rho_\zeta)B_\zeta F\eta_\zeta - B_\zeta C_\pi X T_\infty - \eta_\zeta$$
$$= \rho_\zeta(A_\zeta T_\infty - I + B_\zeta F)\eta_\zeta - B_\zeta(C_\pi X T_\infty + F)\eta_\zeta.$$

Because of our choice of F we have

$$(5.6.23) \qquad\qquad \rho_\zeta(A_\zeta T_\infty - I + B_\zeta F)|_K = 0.$$

Hence, we have that $\eta_\zeta + \Gamma X$ satisfies (5.6.17). Inserting $\rho_\pi + Y\Gamma$ in the left hand side of (5.6.18) yields:

$$S_\infty(\rho_\pi + Y\Gamma)A_\pi - \rho_\pi - Y\Gamma = S_\infty \rho_\pi A_\pi + S_\infty Y A_\zeta \Gamma + S_\infty Y B_\zeta C_\pi - \rho_\pi - Y\Gamma.$$

From (5.6.14) we see that

$$(5.6.24) \qquad\qquad Y - S_\infty Y A_\zeta = (S_\infty \rho_\pi A_\pi - GC_\pi)\Gamma^+.$$

So the above equals

$$S_\infty \rho_\pi A_\pi + Y\Gamma - S_\infty \rho_\pi A_\pi \Gamma^+\Gamma + GC_\pi \Gamma^+\Gamma + S_\infty Y B_\zeta C_\pi - \rho_\pi - Y\Gamma$$
$$= S_\infty \rho_\pi A_\pi - S_\infty \rho_\pi A_\pi(I - \rho_\pi) + GC_\pi(I - \rho_\pi) + S_\infty Y B_\zeta C_\pi - \rho_\pi$$
$$= (S_\infty \rho_\pi A_\pi - \rho_\pi - GC_\pi)\rho_\pi + (S_\infty Y B_\zeta + G)C_\pi.$$

By our choice of G we have

$$(5.6.25) \qquad\qquad (S_\infty \rho_\pi A_\pi - GC_\pi - \rho_\pi)\rho_\pi = 0$$

so $\rho_\pi + Y\Gamma$ satisfies (5.6.18).

It remains to show that (5.6.19) holds. For this we compute

$$-(S_\infty Y B_\zeta + G)(C_\pi X T_\infty + F) = -S_\infty Y B_\zeta C_\pi X T_\infty - GC_\pi X T_\infty - S_\infty Y B_\zeta F - GF.$$

Noting that $GF = 0$, using $B_\zeta C_\pi = \Gamma A_\pi - A_\zeta \Gamma$, we have that this equals

$$(S_\infty Y A_\zeta)\Gamma X T_\infty - S_\infty Y\Gamma(A_\pi X T_\infty) - S_\infty Y B_\zeta F - GC_\pi X T_\infty.$$

We now use (5.6.22), (5.6.23) to transform this expression to

$$Y\Gamma XT_\infty - S_\infty \rho_\pi A_\pi (I - \rho_\pi) XT_\infty + GC_\pi (I - \rho_\pi) XT_\infty - S_\infty Y\Gamma X$$
$$+ S_\infty Y(I - \rho_\zeta)(A_\zeta T_\infty + B_\zeta F)\eta_\zeta - S_\infty Y B_\zeta F - GC_\pi XT_\infty$$
$$= Y\Gamma XT_\infty - S_\infty \rho_\pi A_\pi (I - \rho_\pi) XT_\infty - GC_\pi \rho_\pi XT_\infty - S_\infty Y\Gamma X$$
$$+ S_\infty Y(I - \rho_\zeta)A_\zeta T_\infty - S_\infty Y \rho_\zeta B_\zeta F\eta_\zeta$$
$$= Y\Gamma XT_\infty - S_\infty \rho_\pi A_\pi XT_\infty + (S_\infty \rho_\pi A_\pi - GC_\pi)\rho_\pi XT_\infty - S_\infty Y\Gamma X$$
$$+ S_\infty Y A_\zeta T_\infty - S_\infty Y \rho_\zeta (A_\zeta T_\infty + B_\zeta F).$$

Now use (5.6.23) and (5.6.25) to rewrite $(S_\infty \rho_\pi A_\pi - GC_\pi)\rho_\pi$ as ρ_π and $(A_\zeta T_\infty + B_\zeta F)\eta_\zeta$ as η_ζ, respectively. Further, note that by (5.6.22) we have $\rho_\pi (X - A_\pi XT_\infty) = 0$ so $\rho_\pi A_\pi XT_\infty = \rho_\pi X$. Finally, by (5.6.24) we have $(Y - S_\infty Y A_\zeta)\eta_\zeta = 0$, so $S_\infty Y A_\zeta \eta_\zeta = Y\eta_\zeta$. Thus, inserting all this in the equation above:

$$- (S_\infty Y B_\zeta + G)(C_\pi XT_\infty + F)$$
$$= Y\Gamma XT_\infty - S_\infty \rho_\pi X + \rho_\pi XT_\infty - S_\infty Y\Gamma X + Y\eta_\zeta T_\infty - S_\infty Y\eta_\zeta$$
$$= \Gamma_0 T_\infty - S_\infty \Gamma_0,$$

as desired. □

Next, we shall give a formula for a rational matrix function $W(z)$ solving the problem posed in the first paragraph of this section.

THEOREM 5.6.3. *Let the minimal supplement τ_0 of τ at infinity be given as in Theorem 5.6.2 and let α be any complex number not in $\sigma(A_\zeta) \cup \sigma(A_\pi)$. Then the rational matrix function having τ as its null-pole triple on $\mathbb{C}$, τ_0 as its null-pole triple at infinity and with value D at α is given by*

(5.6.26)
$$W(z) = D + (z - \alpha)C_\pi (zI - A_\pi)^{-1}\{[(\Gamma^+ - X\rho_\zeta - Y\eta_\pi)(A_\zeta - \alpha)^{-1}$$
$$+ (I - \alpha S_\infty)^{-1}S_\infty Y]B_\zeta + \eta_\pi (I - \alpha S_\infty)^{-1}G\}D$$
$$- (z - \alpha)\sum_{j=0}^{\omega_1 - 1} z^j (C_\pi XT_\infty + F)T_\infty^j \rho_\zeta (A_\zeta - \alpha)^{-1}B_\zeta D,$$

and

(5.6.27)
$$W(z)^{-1} = D^{-1} - (z - \alpha)D^{-1}\{C_\pi[(\alpha - A_\pi)^{-1}(\Gamma^+ - X\rho_\zeta - Y\eta_\pi)$$
$$- XT_\infty (I - \alpha T_\infty)^{-1}\rho_\zeta] - F(I - \alpha T_\infty)^{-1}\rho_\zeta\}(A_\zeta - zI)^{-1}B_\zeta$$
$$- (z - \alpha)\sum_{j=0}^{\alpha_1 - 1} z^j D^{-1}C_\pi (\alpha I - A_\pi)^{-1}\eta_\pi S_\infty^j (S_\infty Y B_\zeta + G).$$

PROOF. According to Theorem 5.4.2, $W(z)$ is given by

(5.6.28)
$$W(z) = D + (z - \alpha)[C_\pi, -C_\pi XT_\infty - F]\begin{bmatrix} (zI - A_\pi)^{-1} & 0 \\ 0 & (I - zT_\infty)^{-1} \end{bmatrix}$$
$$\times \tilde{\Gamma}^{-1}\begin{bmatrix} (A_\zeta - \alpha)^{-1}B_\zeta D \\ (I - \alpha S_\infty)^{-1}(S_\infty Y B_\zeta + G)D \end{bmatrix},$$

and
(5.6.29)
$$W(z)^{-1} = D^{-1} - (z - \alpha)D^{-1}[C_\pi(\alpha I - A_\pi)^{-1}, (-C_\pi X T_\infty - F)(I - \alpha T_\infty)^{-1}]$$
$$\times \widetilde{\Gamma}^{-1} \begin{bmatrix} (A_\zeta - zI)^{-1} & 0 \\ 0 & (I - zS_\infty)^{-1} \end{bmatrix} \begin{bmatrix} B_\zeta \\ S_\infty Y B_\zeta + G \end{bmatrix}.$$

One easily checks that

$$\widetilde{\Gamma}^{-1} = \begin{bmatrix} \Gamma^+ - X\rho_\zeta - Y\eta_\pi & \eta_\pi \\ \rho_\zeta & 0 \end{bmatrix}.$$

Inserting this expression for $\widetilde{\Gamma}^{-1}$ in (5.6.28) and (5.6.29), and using

$$(I - zT_\infty)^{-1} = \sum_{j=0}^{\omega_1 - 1} z^j T_\infty^j, \qquad (I - zS_\infty)^{-1} = \sum_{j=0}^{\alpha_1 - 1} z^j S_\infty^j,$$

yields (5.6.26), (5.6.27). □

We conclude this section with a simple example.

EXAMPLE 5.6.1. Suppose we are given two complex numbers p, λ, $p \neq \lambda$, and non-zero vectors $x, y \in \mathbb{C}^m$, and that we are looking for a rational $m \times m$ matrix function $W(z)$ such that

$$W(p)^{-1}y = 0, \qquad x^*W(\lambda) = 0.$$

Without loss of generality we may assume that x and y are unit vectors as in Example 4.6.1. Furthermore, we require W to be analytic and invertible on $\mathbb{C}$, except at the points p and λ, and we want the McMillan degree of W to be as small as possible. In Example 4.6.1 we have already discussed the case when x and y are not orthogonal, so we shall restrict ourselves to the orthogonal case. We take the notation as it was in Example 4.6.1. In that case both K and $\operatorname{Ker}\Gamma$ are equal to $\mathbb{C}$, which gives $S_\infty = T_\infty = 0$. Moreover, we have $\rho_\pi = \rho_\zeta = \eta_\pi = \eta_\zeta = 1$. Take Y_0 to be the orthogonal complement in $\mathbb{C}^m$ to $\operatorname{Span}\{x, y\}$, and define $G: \mathbb{C}^m \to \mathbb{C}$ and $F: \mathbb{C} \to \mathbb{C}^m$ by $G = y^*$, $F = x$. Then a function W as required, with the additional property that $W(\alpha) = I$, is given by

(5.6.30)
$$W(z) = I - \frac{z - \alpha}{z - p}yy^* - \frac{z - \alpha}{\lambda - \alpha}xx^*$$

(5.6.31)
$$W(z)^{-1} = I - \frac{z - \alpha}{z - \lambda}xx^* - \frac{z - \alpha}{p - \alpha}yy^*. \quad □$$

5.7 MINIMAL INTERPOLATION FOR MATRIX POLYNOMIALS

In this short section we consider the following problem. Given a full range pair (A_ζ, B_ζ), a number $\alpha \neq \sigma(A_\zeta)$ and an invertible matrix D, we seek a regular matrix polynomial $L(z)$ of minimal McMillan degree such that $L(\alpha) = D$ and L has (A_ζ, B_ζ) as its left null pair over $\mathbb{C}$. Recall that the McMillan degree of a matrix polynomial $L(z)$ is, by definition, the minimal possible size of the matrix A_∞ with $\sigma(A_\infty) = \{0\}$ in the realization

$$L(z) = D + zC_\infty(I - zA_\infty)^{-1}B_\infty$$

(see Theorem 5.2.1). Theorems 5.5.1 and 2.3.1 discussed conditions for the existence of a monic matrix polynomial with given left null pair. Also discussed in Section 5.5 and in Theorem 2.4.2 were methods for constructing a regular matrix polynomial with given left null pair for the case where a monic one may not exist; however, neither of these procedures need give a solution of minimal McMillan degree. Here also we do not insist that L be monic but we do demand minimal McMillan degree. This is just the special case of the problem treated in the previous section where τ has the special form $(0, 0; A_\zeta, B_\zeta; 0)$. By applying Theorems 5.6.2 and 5.6.3 we see that such an L can be constructed in the following way. Take a basis $\{g_{jk}\}_{k=1,j=1}^{\omega_j, s}$ in $\mathbb{C}^{n_\zeta}$ (here $n_\zeta \times n_\zeta$ is the size of A_ζ), $\omega_1 \geq \cdots \geq \omega_s$, such that

$$A_\zeta g_{j,k+1} - g_{jk} \in \mathrm{Im}\, B_\zeta \quad \text{and} \quad \{g_{j,\omega_j}\}_{j=1}^{s} \text{ is a basis of } \mathrm{Im}\, B_\zeta.$$

Define $T_\infty: \mathbb{C}^{n_\zeta} \to \mathbb{C}^{n_\zeta}$ by

$$T_\infty g_{jk} = g_{j,k+1}.$$

Choose vectors $y_j \in \mathbb{C}^m$ such that $B_\zeta y_j = g_{j,\omega_j}$ and choose $F: \mathbb{C}^{n_\zeta} \to \mathrm{Span}\{y_j\}_{j=1}^{s}$ such that

$$B_\zeta F g_{jk} = g_{jk} - A_\zeta T_\infty g_{jk}.$$

Then

(5.7.1)
$$L(z) = D - (z - \alpha) \sum_{j=0}^{\omega_1 - 1} z^j F T_\infty^j (A_\zeta - \alpha I)^{-1} B_\zeta$$

(5.7.2)
$$L(z)^{-1} = D^{-1} + (z - \alpha)D^{-1}F(I - \alpha T_\infty)^{-1}(A_\zeta - zI)^{-1}B_\zeta.$$

EXAMPLE 5.7.1. Here we present the simplest special case of the problem presented in this section. Given two complex numbers λ, α and a vector $x \neq 0$ in $\mathbb{C}^m$, we are looking for a regular $m \times m$ matrix polynomial L of minimal McMillan degree such that $L(\alpha) = I$, $x^*L(\lambda) = 0$. This is the special case when $A_\zeta = [\lambda]$ and $B_\zeta = x^*$. Without loss of generality we assume $x^*x = 1$. Then one particular choice of L is

(5.7.3)
$$L(z) = I - \frac{z - \alpha}{\lambda - \alpha}xx^*. \quad \square$$

CHAPTER 6

RATIONAL MATRIX FUNCTIONS WITH J-UNITARY VALUES ON THE IMAGINARY LINE

In this chapter we consider rational matrix functions $U(z)$ which have J-unitary values on the imaginary axis, i.e. $(U(z))^* J U(z) = J$ for any regular point z of U in $i\mathbb{R}$. Here J is a *signature matrix*, namely $J = J^*$ and $J^2 = I$. For this class of matrix functions we analyze the null-pole structure, the special properties of realizations for this class, and the appropriate interpolation problems. We also consider here the interpolation problem with standard local data and its derivatives. These problems are also matricially homogeneous, with the understanding that one is permitted to multiply by J-unitary matrices. The coupling matrix which appeared in the general interpolation problems from Chapter 4 now is Hermitian and plays an extra role. The case $J = I$ has special features and is discussed separately.

Throughout this chapter J is a fixed $m \times m$ signature matrix.

6.1 REALIZATION AND INTERPOLATION PROBLEMS WITH COMPLETE DATA

In this section we characterize the form of a minimal realization for a rational matrix function which is J-unitary on the imaginary line, give a canonical form for a null-pole triple for such a matrix function and solve the related inverse interpolation problem.

We begin with the form of a minimal realization.

THEOREM 6.1.1. *Let U be a rational $m \times m$ matrix function analytic at infinity. Then U is J-unitary on the imaginary axis if and only if a minimal realization $U(z) = D + C(zI - A)^{-1}B$ for U has the properties:*

(a) *D is J-unitary; in other words $D^* J D = J$*

(b) *There exists a unique invertible Hermitian solution H to the Lyapunov equation*

$$(6.1.1) \qquad HA + A^* H = -C^* J C$$

for which

$$(6.1.2) \qquad B = -H^{-1} C^* J D.$$

The matrix H is the null-pole coupling matrix associated with the (global) right pole pair (C, A) and (global) left null pair $(-A^*, -C^* J)$ for $U(z)$.

We note that property b) can equivalently be expressed as

b') *There exists a unique invertible Hermitian solution G of the Lyapunov equation*

$$(6.1.3) \qquad GA^* + AG = -BJB^*$$

for which

(6.1.4) $$C = -DJB^*G^{-1}.$$

The matrix G also can be identified as the null-pole coupling matrix associated with the right null pair $(-JB^*, -A^*)$ and left pole pair (A, B) of $U(z)$. Observe that the matrix H of the property (b) and the matrix G of the property (b') are related by reciprocity: $G = H^{-1}$.

      PROOF. If $U(z) = D + C(zI - A)^{-1}B$ is a minimal realization for $U(z)$ and U is J-unitary on the imaginary axis, then certainly $D = U(\infty)$ is J-unitary. Since U is J-unitary on the imaginary axis, we have

(6.1.5) $$U(z)^{-1} = JU(-\bar{z})^*J,$$

and, since a minimal realization of $U^{-1}(z)$ is

$$U^{-1}(z) = D^{-1} - D^{-1}C(zI - A^{\times})^{-1}BD^{-1}$$

with

$$A^{\times} = A - BD^{-1}C$$

(see Proposition 4.1.5), equation (6.1.5) may be written as

(6.1.6) $$D^{-1} - D^{-1}C(zI - A^{\times})^{-1}BD^{-1} = J(D^* + B^*(zI + A^*)^{-1}(-C^*))J.$$

Equation (6.1.6) is an equality between two minimal realizations of the same rational matrix function; hence by Theorem 4.1.3, there is a uniquely defined invertible matrix H such that

(6.1.7) $$JB^*H = -D^{-1}C, \quad -H^{-1}C^*J = BD^{-1}, \quad -H^{-1}A^*H = A^{\times}, D^{-1} = JD^*J.$$

By taking adjoints in each equation and rearranging, we see that H^* also satisfies (6.1.7); therefore by the uniqueness of H, $H = H^*$. It is straightforward to check that equations (6.1.7) are equivalent to (6.1.1) and (6.1.2), and thus the solution H to (6.1.1) and (6.1.2) is unique. The second realization in (6.1.6) for $U^{-1}(z)$ establishes that $(-A^*, -C^*J)$ is a left null pair for $U(z)$. By definition the null-pole coupling matrix H associated with the right pole pair (C, A) and left null pair $(-A^*, -C^*J)$ is the unique invertible matrix implementing the similarity between the realizations in (6.1.6); i.e. the solution H to equations (6.1.7).

      Conversely, suppose $U(z)$ has a minimal realization $D + C(zI - A)^{-1}B$ for which D is J-unitary and $B = -H^{-1}C^*JD$ for some invertible Hermitian solution of (6.1.1). Then we compute

$$U(z)JU(w)^* = (D + C(zI - A)^{-1}B)J(D^* + B^*(\bar{w}I - A^*)^{-1}C^*)$$
$$= DJD^* + C(zI - A)^{-1}BJD^* + DJB^*(\bar{w}I - A^*)^{-1}C^*$$
$$+ C(zI - A)^{-1}BJB^*(\bar{w}I - A^*)^{-1}C^*.$$

Replacing DJD^* by J, BJB^* by $-AH^{-1} - H^{-1}A^*$ and BJD^* by $-H^{-1}C^*$ we obtain

$$(6.1.8) \qquad U(z)JU(w)^* = J - (z + \overline{w})C(zI - A)^{-1}H^{-1}(\overline{w}I - A^*)^{-1}C^*.$$

From (6.1.8) it follows that U is J-unitary on the imaginary axis. □

There are explicit formulas for the similarity matrix H (see Theorem 4.1.3). H is given by each of the following formulas

$$H = -\left[\mathrm{col}\left(JB^*(A^*)^j\right)_{j=0}^{n-1}\right]^+ \left[\mathrm{col}\left(D^{-1}C(A^\times)^j\right)_{j=0}^{n-1}\right]$$

and

$$H = -\left[\mathrm{row}\left((A^*)^j C^* J\right)_{j=0}^{n-1}\right] \left[\mathrm{row}\left((A^\times)^j BD^{-1}\right)_0^{n-1}\right]^\dagger$$

where $n \times n$ is the size of A, the symbol $+$ indicates a left inverse and the symbol $\dagger$ indicates a right inverse.

We now give some examples of rational functions J-unitary on the imaginary line.

EXAMPLE 6.1.1. Let P be an $m \times m$ matrix such that

$$PJP^* = P = P^*$$

and let w be a non pure imaginary point. Then, the function

$$(6.1.9) \qquad U(z) = I_m - PJ + \left(\frac{z - w}{z + \overline{w}}\right)PJ$$

is J-unitary on the imaginary axis.

A typical example of P is given by

$$P = \frac{uu^*}{u^*Ju}$$

where u is a non J-neutral vector ($u^*Ju \neq 0$). □

EXAMPLE 6.1.2. Let u_1 and u_2 be two vectors such that

$$u_1^*Ju_1 = u_2^*Ju_2 = 0$$

$$u_1^*Ju_2 \neq 0$$

and define, for $i \neq j$,

$$W_{ij} = u_i(u_j^*Ju_i)^{-1}u_j^*J.$$

Let w_1 and w_2 be two non pure imaginary points. Then, the function

$$(6.1.10) \qquad U(z) = I_m + \left(\frac{z - w_2}{z + \overline{w}_1} - 1\right)W_{12} + \left(\frac{z - w_1}{z + \overline{w}_2} - 1\right)W_{21}$$

is J-unitary on the imaginary axis. □

EXAMPLE 6.1.3. Let α be a pure imaginary (or zero) number, let n be a positive integer and u be a J-neutral vector. Then, the function

$$(6.1.11) \qquad\qquad U(z) = I + \frac{iuu^*J}{(z-\alpha)^{2n}}$$

is J-unitary on the imaginary axis. □

We next seek a characterization of global left null-pole triples for rational matrix functions which are J-unitary on the imaginary axis.

LEMMA 6.1.2. *Suppose $U(z)$ is a rational matrix function regular at infinity and $(C, A_\pi; A_\zeta, B; S)$ is a global left null-pole triple for $U(z)$. Then $(-JB^*, -A_\zeta^*; -A_\pi^*, -C^*J; S^*)$ is a global left null-pole triple for $J(U(-\bar z)^*)^{-1}J$.*

PROOF. In general, the admissible Sylvester data set $(C, A_\pi; A_\zeta, B; S)$ is a global left null-pole triple for the rational matrix function $U(z)$ regular at infinity if and only if $U(z)$ has a minimal realization

$$U(z) = D + C(zI - A_\pi)^{-1}S^{-1}BD$$

with inverse given by

$$U^{-1}(z) = D^{-1} - D^{-1}CS^{-1}(zI - A_\zeta)^{-1}B$$

(see Theorem 4.4.3). For such a $U(z)$ we compute

$$J(U(-\bar z)^*)^{-1}J = J(D^*)^{-1}J + JB^*(zI + A_\zeta^*)^{-1}(S^{-1})^*C^*(D^{-1})^*J$$

and

$$\begin{aligned}
\left[J(U(-\bar z)^*)^{-1}J\right]^{-1} &= JU(-\bar z)^*J \\
&= JD^*J - JD^*B^*(S^*)^{-1}(zI + A_\pi^*)^{-1}C^*J.
\end{aligned}$$

We thus conclude that $(-JB^*, -A_\zeta^*; -A_\pi^*, -C^*J; S^*)$ is a global left null-pole triple for $J(U(-\bar z)^*)^{-1}J$. □

We next use Theorem 6.1.1 to obtain the following characterization of matrix functions with J-unitary values on the imaginary axis in terms of a null-pole triple.

THEOREM 6.1.3. *Suppose U is a rational matrix function with finite value at infinity. Then U is J-unitary on the imaginary axis if and only if*

a) $U(\infty)$ *is J-unitary*

and

b) U *has a global left null-pole triple of the form $(C, A; -A^*, -C^*J; H)$ with a Hermitian coupling matrix H*

or equivalently

b') $(-JB^*, -A_\zeta^*; -A_\pi^*, -C^*J; S^*)$ is a left null-pole triple for $U(z)$ whenever $(C, A_\pi; A_\zeta, B; S)$ is.

PROOF. We have already observed in Theorem 6.1.1 that if U is J-unitary on the imaginary axis, then $(-A^*, -C^*J)$ is a left null pair whenever (C, A) is a right pole pair, and then the associated null-pole coupling matrix H is Hermitian, and of course $U(\infty)$ is J-unitary since the imaginary axis extends to infinity. Conversely suppose (a) and (b) hold. By Lemma 6.1.2, $(C, A; -A^*, -C^*J; H)$ is also global left null-pair for $J(U(-\bar{z})^*)^{-1}J$. By Theorem 4.4.3, there is an invertible matrix X for which

$$U(z) = J(U(-\bar{z})^*)^{-1}JX.$$

Thus $U(-\bar{z})^*JU(z)$ is independent of z. If we evaluate at infinity and use (a), we get J. Thus $U(-\bar{z})^*JU(z) = J$ identically, and U is J-unitary on the imaginary axis.

We next show the equivalence of (a) and (b') with U being J-unitary on the imaginary axis. If U is J-unitary on the imaginary axis, then in particular (a) holds. Moreover, by analytic continuation $J(U(-\bar{z})^*)^{-1}J = U(z)$. Hence (b') is an immediate consequence of Lemma 6.1.2. Conversely, if (a) and (b') hold, by Lemma 6.1.2 $J(U(-\bar{z})^*)^{-1}J$ and $U(z)$ differ only by a constant invertible right multiplicative factor. Evaluating at infinity and using (a) it follows that the factor must be I. □

We now come to our first interpolation theorem for rational matrix functions with J-unitary values on the imaginary line.

THEOREM 6.1.4. Suppose (C, A) is a null-kernel pair and J is a signature matrix. Then there exists a rational matrix function $U(z)$ such that

a) $U(z)$ is J-unitary for z on the imaginary axis

and

b) (C, A) is a global right pole pair for $U(z)$

if and only if there exists an invertible Hermitian solution to the Lyapunov equation

(6.1.12) $HA + A^*H = -C^*JC.$

In this case any such $U(z)$ has the form

(6.1.13) $U(z) = D - C(zI - A)^{-1}H^{-1}C^*JD$

where H is an invertible Hermitian solution of (6.1.12) and D is any J-unitary matrix.

PROOF. The necessity direction has already been observed. Conversely, if H is an invertible Hermitian solution of (6.1.12), then by Theorem 4.4.3 the matrix function $U(z)$ defined by (6.1.13) has global left null-pole triple $(C, A; -A^*, -C^*J; H)$. In particular (C, A) is a right pole pair for $U(z)$. Moreover, by Theorem 6.1.3, if D is chosen to be J-unitary we see that $U(z)$ is J-unitary for z on the imaginary axis. □

The analogue of Theorem 6.1.4 for the left pole pairs (A, B) is as follows.

THEOREM 6.1.5. Suppose (A, B) is a full range pair and J is a signature matrix. Then there exists a rational matrix function $U(z)$ such that

a) $U(z)$ is J-unitary for z on the imaginary axis

and

b) (A, B) is a left pole pair for $U(z)$ if and only if there exists an invertible Hermitian solution G to the Lyapunov equation

$$(6.1.14) \qquad GA^* + AG = -BJB^*.$$

In this case any such $U(z)$ has the form

$$(6.1.15) \qquad U(z) = D - DJB^*G^{-1}(zI - A)^{-1}B$$

where G is any invertible Hermitian solution of (6.1.14) and D is any J-unitary matrix.

Finally the next theorem characterizes which admissible Sylvester data sets $(C, A_\pi; A_\zeta, B; S)$ can arise as the global left null-pole triple for a rational matrix function regular at infinity and J-unitary valued on the imaginary axis.

THEOREM 6.1.6. *Suppose $(C, A_\pi; A_\zeta, B; S)$ is an admissible Sylvester data set. Then $(C, A_\pi; A_\zeta, B; S)$ is the global left null-pole triple for a rational matrix function $U(z)$ which is J-unitary valued on the imaginary axis and regular at ∞ if and only if S is invertible and $(C, A_\pi; A_\zeta, B; S)$ is similar to $(-JB^*, -A_\zeta^*; -A_\pi^*, -C^*J; S^*)$. In this case any such $U(z)$ is of the form*

$$(6.1.16) \qquad U(z) = D + C(zI - A_\pi)^{-1}S^{-1}BD$$

where D is any J-unitary matrix.

PROOF. The necessity direction follows from Lemma 6.1.2 by using the uniqueness of a global left null-pole triple of $U(z)$ up to similarity. Conversely, suppose that the admissible Sylvester data set $\tau = (C, A_\pi; A_\zeta, B; S)$ is similar to $\tilde{\tau} = (-JB^*, -A_\zeta^*; -A_\pi^*, -C^*J; S^*)$. Then by Theorem 4.4.3, the rational matrix function (6.1.16) has τ as a global left null-pole triple. Since τ is similar to $\tilde{\tau}$, $\tilde{\tau}$ is also a global left null-pole triple for $U(z)$. Then by Lemma 6.1.2, if also D is J-unitary then $U(z)$ has J-unitary values on the imaginary axis. $\square$

6.2 THE ASSOCIATED HERMITIAN COUPLING MATRIX

To different minimal realizations of a given rational matrix function J-unitary on the line and analytic at infinity correspond different Hermitian coupling matrices. Here we study the invariants of the associated Hermitian matrix.

LEMMA 6.2.1. *Let U be a rational matrix function, analytic at infinity and J-unitary on the imaginary axis. Let $U(z) = D + C_i(zI - A_i)^{-1}B_i$, $i = 1, 2$, be the minimal realizations of U, with associated Hermitian coupling matrix H_i for right pole pair (C_i, A_i) and left null pair $(-A_i^*, -C_i^*J)$. Then the two minimal realizations are similar*

$$C_1 = C_2 S, \quad A_1 = S^{-1}A_2 S, \quad B_1 = S^{-1}B_2$$

for a unique invertible matrix S. Moreover

$$H_1 = S^* H_2 S.$$

In particular, the matrices H_1 and H_2 have the same number of positive and negative eigenvalues.

PROOF. The existence and uniqueness of the similarity matrix S follows from Theorem 4.1.3. Also, the two Sylvester data sets $(C_i, A_i; -A_i^*, -C_i^* J; H_i)$ are global null-pole triples for the same function and hence by Theorem 4.4.2 are similar via unique similarity matrices S_1 and S_2:

$$
\begin{aligned}
C_1 &= C_2 S_1, & A_1 &= S_1^{-1} A_2 S_1 \\
-A_1^* &= -S_2^{-1} A_2^* S_2, & -C_1^* J &= -S_2^{-1} C_2^* J \\
H_1 &= S_2^{-1} H_2 S_1.
\end{aligned}
$$

In fact S_1 is the unique invertible matrix implementing the similarity of the null-kernel pair (C_1, A_1) with (C_2, A_2) and S_2 is the unique invertible matrix implementing the similarity of the full range pair $(-A_1^*, -C_1^* J)$ with $(-A_2^*, -C_2^* J)$. We conclude that necessarily $S_1 = S$ and $S_2 = (S^*)^{-1}$. Thus it falls out that $H_1 = S^* H_2 S$ as claimed. $\square$

From Lemma 6.2.1 we see that the number of positive and negative eigenvalues of the Hermitian matrix H associated with a rational matrix function $U(z)$ analytic and invertible at infinity and J-unitary valued on the imaginary axis is independent of the choice of the right pole pair (C, A) for $U(z)$, and thus is a characteristic intrinsic to the function $U(z)$ itself. An alternative characterization of the number of negative eigenvalues of H more intrinsic to the function $U(z)$ is in terms of the number of negative squares had by a *kernel function* associated with $U(z)$.

We recall (see Krein-Langer [1970]) that an $n \times n$ matrix valued function $K(z, w)$ defined for z and w is some set E and such that $K(z, w)^* = K(w, z)$ has ν negative squares if for any positive integer r, points $w_1, \ldots, w_r$ in E and any vectors $c_1, \ldots, c_r$ in $\mathbb{C}^n$, the $r \times r$ Hermitian matrix with ij entry

(6.2.1) $$c_j^* K(w_j, w_i) c_i$$

has at most ν strictly negative eigenvalues and exactly ν strictly negative eigenvalues for some choice of $r, w_1, \ldots, w_r, c_1, \ldots, c_r$. With this definition at hand, we can now state the following theorem, which gives a characterization of the number of negative eigenvalues of the associated matrix H.

THEOREM 6.2.2. *Let U be a rational matrix function J-unitary on the imaginary line and analytic at infinity, and let $U(z) = D + C(zI_n - A)^{-1}B$ be a minimal realization of U, with associated Hermitian matrix H. Then, the number of negative eigenvalues of the matrix H is equal to the number of negative squares of each of the functions*

(6.2.2) $$K_U(z, w) = \frac{J - U(z)JU(w)^*}{z + \overline{w}} \quad \text{and} \quad \frac{J - U(w)^* JU(z)}{z + \overline{w}}.$$

Finally, let $\mathcal{K}(U)$ be the span of the functions $z \to K_U(z, w)c$ where w is in the resolvent set of A and where c is in $\mathbb{C}^m$. Then, $\mathcal{K}(U)$ is a finite dimensional space of rational functions analytic on the resolvent set of A and the dimension of $\mathcal{K}(U)$ is equal to the McMillan degree of U.

Note that the second function in (6.2.2) is alternately defined as

$$\frac{J - (U(w))^* J U(z)}{z + \overline{w}} = K_{\widetilde{U}}(\overline{w}, \overline{z}),$$

where $\widetilde{U}(z) = (U(\overline{z}))^*$.

PROOF. From the formula (6.1.8) we have for z and w in the resolvent set of A

(6.2.3) $$K_U(z, w) = C(zI_n - A)^{-1} H^{-1} (\overline{w} I_n - A^*)^{-1} C^*.$$

Let r be a positive integer and $w_1, \ldots, w_r$ be the resolvent set of A, $c_1, \ldots, c_r$ be in C^n. Then, the matrix equality

(6.2.4) $$\left(c_i^* K_U(w_i, w_j) c_j \right)_{i,j=1,r} = X^* H^{-1} X$$

with

$$X = \left[(\overline{w}_1 I - A^*)^{-1} C^* c_1, \ldots, (\overline{w}_r I - A^*)^{-1} C^* c_r \right]$$

makes it clear that the function K_U has at most ν_H negative squares, where ν_H denotes the number of negative eigenvalues of the Hermitian matrix H. The pair (C, A) is a null kernel pair, and thus we can choose a basis in C^n of the form $x_i = (\overline{w}_i - A^*)^{-1} C^* c_i$, $i = 1, \ldots, n$. In particular, $\det X \neq 0$ for $X = (x_1, \ldots, x_n)$ and the matrix $X^* H^{-1} X$ has exactly ν_H negative eigenvalues (counted with multiplicities), and thus K_U has ν_H negative squares.

The case of the function $(z + \overline{w})^{-1} (J - U(w)^* J U(z))$ is treated similarly. Equation (6.2.3) implies that any finite linear combination of functions $K_U(z, w)c$ is of the form

$$C(zI_n - A)^{-1} f$$

where f is in C^n. Thus, $\mathcal{K}(U)$ is a finite dimensional vector space of dimension at most n. From the null kernel property of the pair (C, A), we see that $C(zI_n - A)^{-1} f \equiv 0$ implies that $f = 0$ and then $\dim \mathcal{K}(U) = n$ follows. $\square$

We will denote by $\nu(U)$ the number of negative squares of either of the functions defined in (6.2.2).

The last theorem of this section deals with the product of two J-unitary rational functions. In general, if U_i $(i = 1, 2)$ are two rational matrix functions with realizations

(6.2.5) $$U_i(z) = D_i + C_i(zI - A_i)^{-1} B_i, \qquad i = 1, 2$$

then the product has the realization

$$U(z) = U_1(z) U_2(z) = D + C(zI - A)^{-1} B$$

where

(6.2.6) $$D = D_1 D_2, \quad C = [C_1, D_1 C_2], \quad B = \mathrm{col}(B_1 D_2, B_2)$$

and

(6.2.7)
$$A = \left[\begin{array}{cc} A_1 & B_1 C_2 \\ 0 & A_2 \end{array} \right].$$

The factorization $U = U_1 \cdot U_2$ is said to be *minimal* if the realization for U given by (6.2.6) and (6.2.7) is minimal whenever the realization (6.2.5) is minimal for U_i for $i = 1, 2$. The last theorem of this section deals with the minimal product of two rational matrix functions with J-unitary values on the imaginary axis.

THEOREM 6.2.3. *Let U_i be a rational matrix function with finite value at infinity and with minimal realization (6.2.5) and associated Hermitian coupling matrices H_i for $i = 1, 2$, and suppose that the factorization $U = U_1 \cdot U_2$ is minimal. Then the Hermitian matrix*

(6.2.8)
$$H = \left[\begin{array}{cc} H_1 & 0 \\ 0 & H_2 \end{array} \right]$$

is the Hermitian coupling matrix associated with the minimal realization (A, B, C, D) given by (6.2.6) and (6.2.7) for the product $U_1 U_2$. In particular any associated Hermitian coupling matrix for the product $U_1 U_2$ will be congruent to the matrix H defined by (6.2.8).

PROOF. It suffices to check that equations (6.1.1) and (6.1.2) are satisfied for A, B, C, D and H as in the statement. This is an easy computation which is omitted. The second claim of the theorem is a consequence of Lemma 6.2.1. $\square$

COROLLARY 6.2.4. *Let U_1 and U_2 be rational matrix functions analytic at infinity and with J-unitary values on the imaginary line, and suppose that the factorization $U = U_1 \cdot U_2$ is minimal. Then*

$$\nu(U_1 U_2) = \nu(U_1) + \nu(U_2).$$

6.3 *J*-UNITARY ON THE IMAGINARY AXIS WITH GIVEN NULL-POLE STRUCTURE IN THE RIGHT HALF PLANE

Given any collection of points $z_1, \ldots, z_p$ (including possibly repetitions) in the right half plane and a second collection of points $w_1, \ldots, w_q$ in the right half plane but disjoint from the first set, then there exists a scalar rational function

$$r(z) = c \frac{z - z_1}{z + \overline{z}_1} \cdots \frac{z - z_p}{z + \overline{z}_p} \frac{z + \overline{w}_1}{z - w_1} \cdots \frac{z + \overline{w}_q}{z - w_q}$$

(where $|c| = 1$) having precisely the set $z_1, \ldots, z_p$ as its zeros in the right half plane (including multiplicities) and $w_1, \ldots, w_q$ as its poles in the right half plane, and assuming values of modulus 1 on the imaginary axis. In this section we derive the analogue of this fact for matrix valued functions.

THEOREM 6.3.1. *Let $(C, A_\pi, A_\zeta, B, S)$ be an admissible Sylvester set of data such that $\sigma(A_\pi)$ and $\sigma(A_\zeta)$ are both contained in the open right half plane Π^+. Then there exists a rational matrix function $U(z)$ such that*

a) U is regular on the imaginary axis (including ∞)

b) U assumes J-unitary values on the imaginary axis

and

c) $(C, A_\pi; A_\zeta, B; S)$ is a left null-pole triple for $U(z)$ over the right half plane if and only if the Hermitian matrix

(6.3.1)
$$\widehat{S} = \begin{bmatrix} S_1 & S^* \\ S & S_2 \end{bmatrix}$$

is invertible. Here S_1 and S_2 are the unique solutions of the respective Lyapunov equations

(6.3.2)
$$S_1 A_\pi + A_\pi^* S_1 = -C^* J C$$

(6.3.3)
$$S_2 A_\zeta^* + A_\zeta S_2 = B J B^*.$$

In this case, any such matrix function $U(z)$ has a minimal realization of the form

(6.3.4)
$$U(z) = D + [C, -JB^*] \begin{bmatrix} (z - A_\pi)^{-1} & 0 \\ 0 & (z + A_\zeta^*)^{-1} \end{bmatrix} \widehat{S}^{-1} \begin{bmatrix} -C^* J \\ B \end{bmatrix} D$$

where D is any J-unitary matrix.

PROOF. By Lemma 6.1.2 $(-JB^*, -A_\zeta^*; -A_\pi^*, -C^* J; S^*)$ is a left null-pole triple for $U(z)$ over the left half plane if $(C, A_\pi; A_\zeta, B; S)$ is a left null-pole triple for $U(z)$ over the right half plane and if $U(z)$ is J-unitary on the imaginary axis. If $U(z)$ is regular on the imaginary axis then by Lemma 4.5.6, a global left null-pole triple for $U(z)$ is given by

(6.3.5)
$$\left([C, -JB^*], \begin{bmatrix} A_\pi & 0 \\ 0 & -A_\zeta^* \end{bmatrix}, \begin{bmatrix} -A_\pi^* & 0 \\ 0 & A_\zeta \end{bmatrix}, \begin{bmatrix} -C^* J \\ B \end{bmatrix}, \widehat{S} \right)$$

where $\widehat{S}$ is as in the statement of the theorem. In particular, $\widehat{S}$ necessarily is invertible since it is a global null-pole coupling matrix, and $U(z)$ is given by (6.3.4) with $D = U(\infty)$, by Theorem 4.4.3.

Conversely, suppose $\widehat{S}$ is invertible. We may then define the rational matrix function $U(z)$ by (6.3.4). We note that (6.3.2) and (6.3.3) combined with the Lyapunov equation satisfied by the coupling matrix S

$$S A_\pi - A_\zeta S = B C$$

implies that $\widehat{S}$ satisfies the Lyapunov equation

$$\widehat{S} \begin{bmatrix} A_\pi & 0 \\ 0 & -A_\zeta^* \end{bmatrix} - \begin{bmatrix} -A_\pi^* & 0 \\ 0 & A_\zeta \end{bmatrix} \widehat{S} = \begin{bmatrix} -C^* J \\ B \end{bmatrix} [C, -JB^*].$$

By inspection we see that $\widehat{S}$ is Hermitian. Therefore the realization (6.3.4) satisfies all the requirements of Theorem 6.1.1. We conclude that $U(z)$ is J-unitary on the imaginary axis. By inspection we also see that the Riesz spectral projections for $\begin{bmatrix} A_\pi & 0 \\ 0 & -A_\zeta^* \end{bmatrix}$ and $\begin{bmatrix} -A_\pi^* & 0 \\ 0 & A_\zeta \end{bmatrix}$ associated with the right half plane are $\begin{bmatrix} I & 0 \\ 0 & 0 \end{bmatrix}$ and $\begin{bmatrix} 0 & 0 \\ 0 & I \end{bmatrix}$. Therefore the restriction of the null-pole triple (6.3.5) to the right half plane is $(C, A_\pi, A_\zeta, B, \widehat{S}_{21} = S)$, and thus $(C, A_\pi, A_\zeta, B, S)$ is a left null-pole triple for $U(z)$ over the right half plane. $\square$

6.4 J-UNITARY WITH POLES ON THE IMAGINARY AXIS

In this section we show how to construct a rational matrix function $U(z)$ with J-unitary values at all regular points on the imaginary axis with prescribed null and pole data on the closed right half plane. For convenience we assume that $U(z)$ is regular at infinity; this restriction can also be removed to produce somewhat more complicated formulas by using the results of Section 5.4.

The first order of business is to characterize the form of null-pole triples over the imaginary line for such functions.

THEOREM 6.4.1. (i) *If $(C, A_\pi; A_\zeta, B; S)$ is a null-pole triple over $i\mathbb{R}$ for a rational matrix function $U(z)$ which is J-unitary valued at all regular points of $i\mathbb{R}$, then $(-JB^*, -A_\zeta^*; -A_\pi^*, -C^*J; S^*)$ is also a null-pole triple over $i\mathbb{R}$ for $U(z)$.*

(ii) *A rational matrix function $U(z)$ with J-unitary values at regular points of $i\mathbb{R}$ has a null-pole triple over $i\mathbb{R}$ of the form*

$$\tau = (C_0, A_0; -A_0^*, -C_0^*J; H)$$

with Hermitian coupling matrix H or equivalently of the form $\tau = (-JB_0^*, -A_0^*; A_0, B_0; H)$ with H Hermitian.

PROOF. If $U(z)$ is J-unitary valued at regular points of $i\mathbb{R}$, then by analytic continuation

$$U(z) = JU(z)^{*-1}J$$

identically. By Lemma 6.1.2, $(-J\widetilde{B}^*, -\widetilde{A}_\zeta; -\widetilde{A}_\pi^*, -\widetilde{C}^*J; \widetilde{S}^*)$ is a global null-pole triple for $U(z)$ whenever $(\widetilde{C}, \widetilde{A}_\pi; \widetilde{A}_\zeta, \widetilde{B}; \widetilde{S})$ is. Also note that in general $\sigma(A) \subset i\mathbb{R}$ if and only if $\sigma(-A^*) \subset i\mathbb{R}$. By restriction of a global null-pole triple for $U(z)$ we see that $(-JB^*, -A_\zeta^*; -A_\pi^*, -C^*J; S^*)$ is a null-pole triple over $i\mathbb{R}$ whenever $(C, A_\pi; A_\zeta, B; S)$ is. This establishes (i).

In particular from (i) we see that $(-A_0^*, -C_0^*J)$ is a left null pair over $i\mathbb{R}$ for $U(z)$ whenever (C_0, A_0) is a right pole pair over $i\mathbb{R}$ for $U(z)$. If H is the coupling matrix associated with right pole pair (C_0, A_0) and left null pair $(-A_0^*, -C_0^*J)$, then by (i) both $(C_0, A_0; -A_0^*, -C_0^*J; H)$ and $(C_0, A_0; -A_0^*, -C_0^*J; H^*)$ are left null-pole triples for $U(z)$. Since the coupling matrix is uniquely determined by the function and a choice of right pole pair and left null pair, we conclude that $H = H^*$. $\square$

As a consequence of Theorem 6.4.1, in prescribing a null-pole triple over $i\mathbf{R}$ for a matrix function having J-unitary values on $i\mathbf{R}$, we may assume that the prescribed admissible Sylvester data set over $i\mathbf{R}$ has the form $(C_0, A_0; -A_0^*, -C_0^*J; H)$. This leads to the following extensions of Theorem 6.3.1.

THEOREM 6.4.2. *Let* $\tau_1 = (C, A_\pi; A_\zeta, B; S)$ *and* $\tau_2 = (C_0, A_0, -A_0^*, -C_0^*J, H)$ *be admissible Sylvester data sets over the open right half plane* Π^+ *and the imaginary line* $i\mathbf{R}$ *respectively. Let* $J^* = J^{-1}$ *be a signature matrix. Introduce matrices* S_{11}, S_{21}, S_{32} *and* S_{33} *as the unique solutions of the respective Lyapunov or Sylvester equations*

$$(6.4.1) \qquad\qquad S_{11}A_\pi + A_\pi^* S_{11} = -C^* J C$$

$$(6.4.2) \qquad\qquad S_{21}A_\pi + A_0^* S_{21} = -C_0^* J C$$

$$(6.4.3) \qquad\qquad S_{32}A_0 - A_\zeta S_{32} = BC_0$$

$$(6.4.4) \qquad\qquad S_{33}A_\zeta^* + A_\zeta S_{33} = -BJB^*$$

and let S *be the Hermitian matrix given by*

$$(6.4.5) \qquad\qquad S = \begin{bmatrix} S_{11} & S_{21}^* & S^* \\ S_{21} & H & S_{32}^* \\ S & S_{32} & S_{33} \end{bmatrix}.$$

Then there exists a rational $n \times n$ *matrix function* $U(z)$ *analytic and invertible at infinity such that*

      (i) τ_1 *is a null-pole triple for* $U(z)$ *over* Π^+

      (ii) τ_2 *is a null-pole triple for* $U(z)$ *over* $i\mathbf{R}$

and

      (iii) $U(-\bar{z})^* J U(z) = J$

if and only if the matrix H *is Hermitian and the matrix* S *is invertible. In this case any such* $U(z)$ *is given by*

$$(6.4.6) \qquad\qquad U(z) = \tilde{D} + \tilde{C}(zI - \tilde{A}_\pi)^{-1}S^{-1}\tilde{B}\tilde{D}$$

with inverse

$$(6.4.7) \qquad\qquad U(z)^{-1} = \tilde{D}^{-1} - \tilde{D}^{-1}\tilde{C}S^{-1}(zI - \tilde{A}_\zeta)^{-1}\tilde{B}$$

where

$$(6.4.8) \qquad\qquad \tilde{D} = \text{ any } J\text{-unitary matrix}$$

(6.4.9) $$\widetilde{C} = [C, C_0, -JB^*]$$

(6.4.10) $$\widetilde{A}_\pi = \mathrm{diag}(A_\pi, A_0, -A_\zeta^*)$$

(6.4.11) $$\widetilde{A}_\zeta = \mathrm{diag}(-A_\pi^*, -A_0^*, A_\zeta)$$

and

(6.4.12) $$\widetilde{B} = \mathrm{col}(-C^*J, -C_0^*J, B).$$

PROOF. By Lemma 6.1.2 we see that $\tau_1^* = (-JB^*, -A_\zeta^*; -A_\pi^*, -C^*J, S^*)$ must be a left null-pole triple over the open left half plane Π^- whenever $U(z)$ satisfies (i) and (iii). Thus we have specified null-pole triples for $U(z)$ over Π^+, $i\mathbb{R}$ and Π^- when we include also condition (ii). Then we build a global Sylvester set of data $\tau = (\widetilde{C}, \widetilde{A}_\pi; \widetilde{A}_\zeta, \widetilde{B}, \mathcal{S})$ such that τ restricted to Π^+ is τ_1, τ restricted to $i\mathbb{R}$ is τ_2 and τ restricted to Π^- is τ_1^* explained in Lemma 4.5.6. In doing this we may take $\widetilde{C}$, $\widetilde{A}_\pi$, $\widetilde{A}_\zeta$, $\widetilde{B}$ as in (6.4.9)–(6.4.12). The associated coupling matrix $\mathcal{S}$ is then completely specified by the condition

(6.4.13)   the restrictions of τ to Π^+, $i\mathbb{R}$ and Π^-
               are τ_1, τ_2 and τ_1^* respectively

together with the Sylvester equation

(6.4.14) $$\mathcal{S}\widetilde{A}_\pi - \widetilde{A}_\zeta \mathcal{S} = \widetilde{B}\widetilde{C}.$$

If we let $\mathcal{S} = [S_{ij}]_{1 \leq i,j \leq 3}$, then condition (6.4.13) forces $S_{31} = S$, $S_{13} = S^*$, $S_{22} = H$. Since τ_1 and τ_2 by hypothesis are admissible Sylvester data sets, S and H satisfy Sylvester equations

(6.4.15) $$SA_\pi - A_\zeta S = BC$$

(6.4.16) $$HA_0 + A_0^*H = -C_0^*JC_0.$$

Given that (6.4.15) and (6.4.16) hold we then see that $\mathcal{S}$ satisfies (6.4.14) if and only if S_{11}, S_{21}, S_{32}, S_{33} satisfy (6.4.1)–(6.4.4). Since $\sigma(A_\pi)$ is disjoint from $\sigma(-A_\pi^*) \cup \sigma(-A_0^*)$ and $\sigma(A_\zeta)$ is disjoint from $\sigma(A_0) \cup \sigma(-A_\zeta^*)$, we see that the solutions of (6.4.1)–(6.4.4) are unique. Thus any U satisfying (i)–(iii) must have τ as a global null pole triple. If $U(z)$ is regular at infinity then necessarily by Theorem 4.4.3 $\mathcal{S}$ must be invertible.

Conversely, suppose $\mathcal{S}$ is invertible. Then by Theorem 4.4.3 there exists a rational matrix function $U(z)$ having τ as a global left null-pole triple and having invertible value at infinity. From the form of τ and by Lemma 6.1.2 we see that τ is also a global left null-pole triple for $JU(-\bar{z})^{*-1}J$. By uniqueness in Theorem 4.5.8 we see that $U(z) = JU(-\bar{z}^*)^{-1}J\Gamma$ for a constant invertible matrix Γ, i.e.

$$U(-\bar{z})^* JU(z) = \widetilde{\Gamma}$$

for all z. Factor $\widetilde{\Gamma}$ as X^*JX and replace $U(z)$ with $U(z)X^{-1}$ to conclude that we can arrange that $\widetilde{\Gamma} = J$ as desired. the formulas (6.4.6) and (6.4.7) for U and U^{-1} are also immediate consequences of the formulas in Theorem 4.4.3. □

6.5 THE CASE $J = I$

For the case where the signature matrix J is taken to be simply the identity matrix certain special features arise in the theory of the preceding sections. The one seen most immediately is that a matrix function $U(z)$ which is unitary at all regular points on the imaginary line has norm equal to 1 there, and therefore can have no poles on the imaginary line. Thus for the case $J = I$, the results of Section 6.4 contain nothing beyond those of 6.3. It is instructive to see how this fact can also be seen in the framework of null-pole triples. By Theorem 6.4.1 we know that there always exists a null-pole triple over $i\mathbf{R}$ for a $U(z)$ which is unitary valued on $i\mathbf{R}$ of the form $\tau = (C_0, A_0; -A_0^*, -C_0^*; H)$ with H Hermitian. The following result therefore illustrates in a different way why a $U(z)$ with unitary values at all regular points of the imaginary line can have no poles or zeros on the imaginary line.

THEOREM 6.5.1. *The only admissible Sylvester data set over $i\mathbf{R}$ of the form* $\tau = (C_0, A_0; -A_0^*, -C_0^*; H)$ *is* $\tau = (0, 0; 0, 0; 0)$.

PROOF. Suppose $\tau = (C_0, A_0; -A_0^*, -C_0^*, H)$ is such an admissible Sylvester data set. Then $\sigma(A_0) \subset i\mathbf{R}$ and H satisfies the Lyapunov equation

(6.5.1) $$H A_0 + A_0^* H = -C_0^* C_0.$$

Let $w_0 \in i\mathbf{R}$ be an eigenvalue of A_0 with eigenvector $x_0 \neq 0$. Then

$$x_0^* H A_0 x_0 + x_0^* A_0^* H x_0 = x_0^* H(w_0 x_0) + (\overline{w}_0 x_0^*) H x_0 = 0.$$

So by (6.5.1) we obtain $x_0^*(-C_0^* C_0) x_0 = 0$ and hence $C_0 x_0 = 0$. It follows that

$$x_0 \in \bigcap_{i=0}^{\infty} \operatorname{Ker} C_0 A_0^i,$$

a contradiction with the full kernel property of (C_0, A_0). $\square$

The next result relates the number of negative squares of the kernel function $K_U(z, w)$ associated with a matrix function $U(z)$ with unitary values on $i\mathbf{R}$ with the number of poles of $U(z)$ in the right half plane Π^+.

THEOREM 6.5.2. *Let H be the Hermitian coupling matrix associated with the matrix function $U(z)$ having unitary values on $i\mathbf{R}$. Then the number ν_H of negative eigenvalues of H is equal to the number n_π of poles of $U(z)$ in Π^+ (counting multiplicities), i.e. n_π such that the size of A_π is $n_\pi \times n_\pi$ whenever (C, A_π) is a right pole pair for $U(z)$ over Π^+.*

PROOF. Let $(C, A_\pi; A_\zeta, B; S)$ be a null pole triple for $U(z)$ over the right half plane Π^+. Then by Theorem 6.3.1 a global null-pole triple for $U(z)$ is given by

$$\left([C, -B^*], \begin{bmatrix} A_\pi & 0 \\ 0 & -A_\zeta^* \end{bmatrix}; \begin{bmatrix} -A_\pi^* & 0 \\ 0 & A_\zeta \end{bmatrix}, \begin{bmatrix} -C^* \\ B \end{bmatrix}, H \right)$$

where

$$H = \begin{bmatrix} S_1 & S^* \\ S & S_2 \end{bmatrix}$$

and S_1 and S_2 are determined by

(6.5.2)
$$S_1 A_\pi + A_\pi^* S_1 = -C^* C$$

(6.5.3)
$$S_2 A_\zeta^* + A_\zeta S_2 = BB^*.$$

From the contour integral formula (A.1.11) of the Appendix for the solution S_1 of (6.5.2) we see that S_1 is given by

(6.5.4)
$$S_1 = -\frac{1}{2\pi i} \int_\gamma (zI + A_\pi^*)^{-1} C^* C (zI - A_\pi)^{-1} dz$$

where γ is a simple closed contour in the complex plane such that $\sigma(A_\pi)$ is inside γ and $\sigma(-A_\pi^*)$ is outside γ. Since the integrand in (6.5.4) decays like $|z|^{-2}$ as $|z|$ tends to infinity, and since $\sigma(A_\pi)$ is contained in Π^+ while $\sigma(-A_\pi^*)$ is contained in Π^-, as is standard in the theory of residues, we may take $\gamma = \gamma_R$ where

$$\gamma_R = \{ix : R \geq x \geq -R\} \cup \{Re^{i\theta} : -\pi \leq \theta \leq \pi\}$$

and then let R tend to infinity to get

(6.5.5)
$$S_1 = \frac{1}{2\pi i} \int_{-\infty}^{\infty} (ixI + A_\pi^*)^{-1} C^* C (ixI - A_\pi)^{-1} (idx)$$
$$= -\frac{1}{2\pi} \int_{-\infty}^{\infty} (xI + iA_\pi)^{*-1} C^* C (xI + iA_\pi)^{-1} dx$$

from which we see that S_1 is negative semidefinite. Since (C, A_π) is also a null-kernel pair, in fact S_1 is negative definite. Indeed, for $y \in \text{Ker } S_1$ we have in view of (6.5.5) that

(6.5.6)
$$C(xI + iA_\pi)^{-1} y = 0, \qquad x \in \mathbf{R}.$$

By analytic continuation this equality holds also for $x \in \mathbf{C}$ with $|x| > \|A_\pi\|$. Develop (6.5.6) into power series in x^{-1} in a neighborhood of infinity and equate the coefficients of this series with zero to obtain $C(-iA_\pi)^j y = 0$ for $j = 0, 1, \ldots$. In view of the null kernel property of (C, A_π) we obtain $y = 0$, as claimed. By an analogous argument one can deduce that S_2 is positive definite. Now compute the Schur complement S_c of H with respect to S_1:

(6.5.7)
$$S_c = S_2 - S S_1^{-1} S^*.$$

As S_2 is positive definite and S_1 is negative definite we conclude that S_c is positive definite. Now

$$S = \begin{bmatrix} I & 0 \\ S S_1^{-1} & I \end{bmatrix} \begin{bmatrix} S_1 & 0 \\ 0 & S_c \end{bmatrix} \begin{bmatrix} I & (S S_1^{-1})^* \\ 0 & I \end{bmatrix}.$$

So S is congruent to the block diagonal matrix $\begin{bmatrix} S_1 & 0 \\ 0 & S_c \end{bmatrix}$, and we conclude that S has precisely n_π negative eigenvalues, where $n_\pi \times n_\pi$ is the size of A_π, as claimed. $\square$

The Schur complement argument used in the preceding proof also leads to the following result.

THEOREM 6.5.3. *If $\tau = (C, A_\pi; A_\zeta, B, S)$ is a Π^+-admissible Sylvester data set, then the matrix $H = \begin{bmatrix} S_1 & S^* \\ S & S_2 \end{bmatrix}$, where S_1 and S_2 are given by (6.5.2) and (6.5.3), is invertible. Hence there exists a rational matrix function $U(z)$ such that*

(i) $U(-\bar{z})^*U(z) = I$

and

(ii) τ *is a null-pole triple for $U(z)$ over Π^+.*

PROOF. In the proof of Theorem 6.5.2 it was established that H is congruent to $\begin{bmatrix} S_1 & 0 \\ 0 & S_c \end{bmatrix}$ where S_1 is negative definite and S_c is positive definite. In particular S_1 and S_c are invertible, so H is invertible. By Theorem 6.3.1, it follows that there exists a $U(z)$ with unitary values on $i\mathbf{R}$ and having τ as a null-pole triple over Π^+. Moreover there can be found an explicit formula for such a function. $\square$

CHAPTER 7

RATIONAL MATRIX FUNCTIONS WITH J-UNITARY VALUES ON THE UNIT CIRCLE

In this chapter we give the analogue of the theory from Chapter 6 concerning rational matrix functions with J-unitary values on the imaginary axis for the case J-unitary values on the unit circle. The developments here parallel those in Chapter 6 for the case of the imaginary axis. The treatment of the null and pole structure at infinity is given additional attention. As in Chapter 6, also here J is a fixed $m \times m$ signature matrix.

7.1 REALIZATION THEOREMS AND EXAMPLES

In this section we study the minimal realizations of rational matrix functions which are J-unitary on the unit circle. Let U be such a function: for any z on the unit circle where it is analytic, we have $U(z)JU(\bar{z}^{-1})^* = J$, and, by analytic continuation, the identity extends to all points z where $U(z)$ is analytic and invertible by

$$(7.1.1) \qquad U(z)JU\left(\frac{1}{\bar{z}}\right)^* = J$$

Equality (7.1.1) implies the following lemma, the proof of which is easy and will be omitted.

LEMMA 7.1.1. *Let U be a rational function J-unitary on the unit circle and invertible at the origin (resp. at the point $z \neq 0$) (resp. at infinity). Then, U is analytic and invertible at infinity (resp. at the point $\frac{1}{\bar{z}}$) (resp. at the origin).*

After this preliminary lemma we turn to the main topic of this subsection, and begin with the following theorem.

THEOREM 7.1.2. *Let U be a rational function, analytic and invertible at infinity, and let $U(z) = D + C(zI_n - A)^{-1}B$ be a minimal realization of U. Then, U is J-unitary on the unit circle if and only if*

a) U *is analytic and invertible at the origin*

b) *There exists an invertible Hermitian matrix H such that*

$$(7.1.2) \qquad \begin{bmatrix} A & B \\ C & D \end{bmatrix}^* \begin{bmatrix} H & 0 \\ 0 & -J \end{bmatrix} \begin{bmatrix} A & B \\ C & D \end{bmatrix} = \begin{bmatrix} H & 0 \\ 0 & -J \end{bmatrix}.$$

The matrix H is unique and also is the null-pole coupling matrix associated with right pole pair (C, A) and left null pair $((A^*)^{-1}, (A^*)^{-1}C^*J)$ for $U(z)$.

PROOF. We first prove the necessity of conditions a) and b). By Lemma 7.1.1, the function U is also analytic and invertible at the origin; thus, A is invertible since the given realization of U is minimal. Equation (7.1.1) may be rewritten as

$$U^{-1}(z) = JU\left(\frac{1}{\bar{z}}\right)^* J$$

and thus (with $A^\times = A - BD^{-1}C$) we have

(7.1.3) $\qquad D^{-1} - D^{-1}C(zI_n - A^\times)^{-1}BD^{-1} = J(D^* + zB^*(I_n - zA^*)^{-1}C^*)J.$

Setting $D_1 = JD^*J - JB^*(A^*)^{-1}C^*J$, equation (7.1.3) becomes
(7.1.4)
$$D^{-1} - D^{-1}C(zI_n - A^\times)^{-1}BD^{-1} = D_1 - JB^*(A^*)^{-1}(zI_n - (A^*)^{-1})^{-1}(A^*)^{-1}C^*J.$$

Letting z go to infinity, we get to $D_1 = D^{-1}$, i.e.:

(7.1.5) $\qquad\qquad\qquad D^{-1} = JD^*J - JB^*(A^*)^{-1}C^*J.$

From the second minimal realization for $U(z)^{-1}$ in (7.1.4), we see that $((A^*)^{-1}, (A^*)^{-1}$ $C^*J)$ is a left null pair for $U(z)$. Moreover, (7.1.4) is an equality between two minimal realizations of a given rational function, and thus there exists a unique invertible matrix H, which by definition is the null pole coupling matrix associated with the right pole pair (C, A) and left null pair $((A^*)^{-1}, (A^*)^{-1}C^*J)$, such that

(7.1.6) $\qquad\qquad\qquad D^{-1}C = JB^*(A^*)^{-1}H$

(7.1.7) $\qquad\qquad\qquad A^\times = H^{-1}(A^*)^{-1}H$

(7.1.8) $\qquad\qquad\qquad BD^{-1} = H^{-1}(A^*)^{-1}C^*J.$

These three equations are also satisfied by H^*, as is easily checked; by the uniqueness of the similarity matrix we get $H = H^*$. Moreover, equations (7.1.6), (7.1.8) in conjunction with equation (7.1.5) lead easily to (7.1.2).

Conversely, suppose $U(z) = D + C(zI_n - A)^{-1}B$ where A, B, C, D satisfy equation (7.1.2) for some Hermitian invertible matrix H. We compute $U(w)^*JU(z)$ for z and w in the resolvent set of A

$$\begin{aligned}U(w)^*JU(z) &= (D^* + B^*(\bar{w}I_n - A^*)^{-1}C^*)J(D + C(zI_n - A)^{-1}B) \\ &= D^*JD + B^*(\bar{w}I_n - A^*)^{-1}C^*JD + D^*JC(zI_n - A)^{-1}B \\ &\quad + B^*(\bar{w}I_n - A^*)^{-1}C^*JC(zI_n - A)^{-1}B.\end{aligned}$$

Using (7.1.2), we replace D^*JD by $J + B^*HB$, C^*JD by A^*HB and C^*JC by $-H + A^*HA$, and, after some algebra, we get to

(7.1.9) $\qquad U(w)^*JU(z) = J - (1 - z\bar{w})B^*(\bar{w}I_n - A^*)^{-1}H(zI_n - A)^{-1}B$

and thus U is J-unitary on the unit circle. $\square$

As in Chapter 6, it follows from the proof of the theorem that the matrix H associated to the minimal realizations $U(z) = D + C(zI_n - A)^{-1}B$ of the rational function U J-unitary on the unit circle is uniquely defined; it will be called the associated Hermitian matrix. It can be given by each of the following formulas (cf. Theorem 4.1.3)

$$H = [\operatorname{col}(JB^*(A^*)^{-j+1})_{j=0}^{n-1}]^{-L} \operatorname{col}(D^{-1}CA^{\times j})_{j=0}^{n-1},$$
$$H = [\operatorname{row}((A^*)^{-j+1}C^*J)_{j=0}^{n-1}][\operatorname{row}(A^{\times j}BD^{-1})_{j=0}^{n-1}]^{-R},$$

where the superscript "$-L$" (resp. "$-R$") denotes left (resp. right) inverse.

We note that equation (7.1.2) is equivalent to

(7.1.10)
$$\begin{bmatrix} A & B \\ C & D \end{bmatrix} \begin{bmatrix} H^{-1} & 0 \\ 0 & -J \end{bmatrix} \begin{bmatrix} A & B \\ C & D \end{bmatrix}^* = \begin{bmatrix} H^{-1} & 0 \\ 0 & -J \end{bmatrix}.$$

Indeed, we obtain from (7.1.2)

$$\begin{bmatrix} H^{-1} & 0 \\ 0 & J \end{bmatrix} \begin{bmatrix} A & B \\ C & D \end{bmatrix}^* \begin{bmatrix} H & 0 \\ 0 & -J \end{bmatrix} \begin{bmatrix} A & B \\ C & D \end{bmatrix} = I.$$

Consequently,

$$\begin{bmatrix} H & 0 \\ 0 & -J \end{bmatrix} \begin{bmatrix} A & B \\ C & D \end{bmatrix} \begin{bmatrix} H^{-1} & 0 \\ 0 & J \end{bmatrix} \begin{bmatrix} A & B \\ C & D \end{bmatrix}^* = I,$$

and (7.1.10) follows. Moreover, (7.1.2) may be rewritten as

(7.1.11)
$$H - A^*HA = -C^*JC$$

(7.1.12)
$$C^*JD = A^*HB$$

(7.1.13)
$$J - D^*JD = -B^*HB$$

while (7.1.10) may be written as

(7.1.14)
$$H^{-1} - AH^{-1}A^* = -BJB^*$$

(7.1.15)
$$CH^{-1}A^* = DJB^*$$

(7.1.16)
$$J - DJD^* = -CH^{-1}C^*.$$

Thus, the J-unitary on the unit circle rational matrix function $U(z) = D + C(zI_n - A)^{-1}B$ can be written also in the following forms:

$$U(z) = D(I_m + JB^*(A^*)^{-1}H(zI_n - A)^{-1}B)$$
$$= (I_m + C(zI_n - A)^{-1}H^{-1}(A^*)^{-1}C^*J)D.$$

Let A, B, C, D be matrices satisfying equation (7.1.2) for some (not necessarily invertible) Hermitian matrix H. Then $U(z) = D + C(zI_n - A)^{-1}B$ defines a rational matrix function J-unitary on the unit circle. This follows from equation (7.1.4), still valid in this case. The realization may not be minimal in general.

Similarly, if A, B, C, D satisfy (7.1.10) where H^{-1} is replaced by some Hermitian matrix Y, then $U(z) = D + C(zI_n - A)^{-1}B$ is a rational matrix function J-unitary on the unit circle, as follows from the formula

$$(7.1.17) \qquad U(z)JU(w)^* = J - (1 - z\overline{w})C(zI_n - A)^{-1}Y(\overline{w}I_n - A^*)^{-1}C^*$$

valid for z and w in the resolvent set of A.

The examples of Section 6.1 adapted to the case of the unit circle are the following.

EXAMPLE 7.1.1. With the notation of Example 6.1.1, and ν not on the unit circle, the function

$$U(z) = I_m - PJ + \left(\frac{z - \nu}{1 - z\overline{\nu}}\right)PJ$$

is J-unitary on the unit circle.   □

EXAMPLE 7.1.2. The function

$$U(z) = I_m + \left(\frac{z - \nu_2}{1 - z\overline{\nu}_1} - 1\right)W_{12} + \left(\frac{z - \nu_1}{1 - z\overline{\nu}_2} - 1\right)W_{21}$$

is J-unitary on the unit circle, where W_{ij} is defined in Example 6.1.2 and where ν_1 and ν_2 are of modulus different from 1.   □

EXAMPLE 7.1.3. Let α be on the unit circle and u be a J-neutral vector; then

$$U(z) = I_m + i\left(\frac{z - \alpha}{1 - z\alpha}\right)^n uu^*J$$

is J-unitary on the unit circle.   □

THEOREM 7.1.3. *Suppose U is a rational matrix function which is regular at infinity and let J be a signature matrix. Then there exists an invertible (constant) matrix X such that*

$$U(\overline{z}^{-1})^* JU(z) = X$$

if and only if $\left(JB^*(A_\zeta^*)^{-1}, (A_\zeta^*)^{-1}; (A_\pi^*)^{-1}, (A_\pi^*)^{-1}C^*J; S^*\right)$ is a (global) left null-pole triple for $U(z)$ whenever $(C, A_\pi; A_\zeta, B; S)$ is.

PROOF. We know by Theorem 4.4.3 that $(C, A_\pi; A_\zeta, B; S)$ is a global left null-pole triple for $U(z)$ if and only if

$$U(z) = D + C(zI - A_\pi)^{-1}S^{-1}BD$$

with inverse given by

$$U^{-1}(z) = D^{-1} - D^{-1}CS^{-1}(zI - A_\zeta)^{-1}B$$

for an invertible matrix $D = U(\infty)$. Using this criterion, one can show by direct computation that a global left null-pole triple for $J\left(U(\bar{z}^{-1})^*\right)^{-1}J$ is $\left(JB^*(A_\zeta^*)^{-1}, (A_\zeta^*)^{-1}; (A_\pi^*)^{-1}, (A_\pi^*)^{-1}C^*J; S^*\right)$. It then follows by Theorem 4.4.2 that this admissible Sylvester data set is also a global left null-pole triple for $U(z)$ if and only if $U(z)$ and $J\left(U(\bar{z}^{-1})^*\right)^{-1}J$ differ only by a right constant invertible factor. $\square$

The counterpart of Theorem 6.1.3 is

THEOREM 7.1.4. *Suppose U is a rational matrix function which is regular at infinity. Then U is J-unitary on the unit circle if and only if*

a) *for some point α where U is analytic and invertible,*

$$U(\bar{\alpha}^{-1})^* JU(\alpha) = J$$

[if $\alpha = 0$, $\bar{\alpha}^{-1}$ is interpreted to be infinity]

and

b) *U has a global left null-pole triple of the form $(C, A; (A^*)^{-1}, (A^*)^{-1}C^*J; H)$ with a Hermitian coupling matrix H*

or equivalently

b') *$\left(JB^*(A_\zeta^*)^{-1}, (A_\zeta^*)^{-1}; (A_\pi^*)^{-1}, (A_\pi^*)^{-1}C^*J; S^*\right)$ is a global left null-pole triple for $U(z)$ whenever $(C, A_\pi; A_\zeta, B; S)$ is.*

PROOF. By Theorem 7.1.3, condition (b') is equivalent to $W(z) \overset{\text{def}}{=} U(\bar{z}^{-1})^* JU(z)$ being independent of z. Thus if $W(\alpha)$ is computed to be equal to J for some point of regularity α for $U(z)$, then $W(z) = J$ identically, and U is J-unitary on the unit circle.

If $U(z)$ has a global left null-pole triple of the form $(C, A; (A^*)^{-1}, (A^*)^{-1}C^*J; H)$ with Hermitian H, then the reflected Sylvester data set $\left(JB^*(A_\zeta^*)^{-1}, (A_\zeta^*)^{-1}; (A_\pi^*)^{-1}, (A_\pi^*)^{-1}C^*J; S^*\right)$ as in (b') for this choice of null-pole pair is equal to itself, thus (b') holds. If (a) also holds, then it follows that U is J-unitary on the unit circle. Conversely, if U is J-unitary on the unit circle, then (a) follows by analytic continuation, and (b) follows by the last statement in Theorem 7.1.2. $\square$

The next theorem is the counterpart of Theorem 6.1.4. If A and C are given matrices of sizes $n \times n$ and $m \times n$ respectively, and H and J are given such that

$$H - A^*HA = -C^*JC$$

where H is an invertible Hermitian matrix and J is a signature matrix, for α, $|\alpha| = 1$, a number in the resolvent set of A we define

(7.1.18) $$D_\alpha = I + CH^{-1}(I - \alpha A^*)^{-1}C^*J$$

and

(7.1.19) $$B_\alpha = H^{-1}(A^*)^{-1}C^*J D_\alpha.$$

THEOREM 7.1.5. *Suppose (C, A) is a null-kernel pair and J is a signature matrix. Then there exists a rational matrix function $U(z)$ such that*

a) *U is regular at infinity*

b) *U is J-unitary valued and regular on the unit circle*

and

c) *(C, A) is a global right pole pair for $U(z)$ if and only if A is invertible and there exists an invertible Hermitian solution H to the Stein equation*

$$(7.1.20) \qquad H - A^* H A = -C^* J C.$$

In this case, all such $U(z)$ are given by

$$U(z) = [D_\alpha + C(zI - A)^{-1} B_\alpha] V$$

where D_α and B_α are defined by (7.1.18) and (7.1.19), H is an invertible Hermitian solution of (7.1.20), and V is an arbitrary J-unitary matrix.

PROOF. Suppose U satisfies conditions (a), (b) and (c). Then U has a minimal realization of the form

$$U(z) = D + C(zI - A)^{-1} B.$$

By Theorem 7.1.2 there exists an invertible Hermitian matrix H such that $\begin{bmatrix} A & B \\ C & D \end{bmatrix}$ is $\begin{bmatrix} H & 0 \\ 0 & -J \end{bmatrix}$-unitary. A particular consequence of this is that H satisfies the Stein equation (7.1.20).

Conversely, suppose H is an invertible Hermitian solution of the Stein equation (7.1.20). Then for any invertible matrix D, by Theorem 4.4.3 the rational matrix function

$$U(z) = D + C(zI - A)^{-1} H^{-1} (A^*)^{-1} C^* J D$$

with inverse

$$U(z)^{-1} = D^{-1} - D^{-1} C H^{-1} (zI - (A^*)^{-1})^{-1} (A^*)^{-1} C^* J$$

has the admissible Sylvester data set $(C, A; (A^*)^{-1}, (A^*)^{-1} C^* J; H)$ as a global left null-pole triple. We choose a point α on the unit circle and choose D so that $U(\alpha)^{-1} = I$; solving for D gives

$$D = I - C H^{-1} (\alpha I - (A^*)^{-1})^{-1} (A^*)^{-1} C^* J$$
$$= I + C H^{-1} (I - \alpha A^*)^{-1} C^* J = D_\alpha.$$

With this choice of D, $B = H^{-1} (A^*)^{-1} C^* J D = B_\alpha$. Then since $U(\alpha) = I$ and $|\alpha| = 1$, $U(\overline{\alpha}^{-1})^* J U(\alpha) = J$. We conclude from Theorem 7.1.4 that U is J-unitary on the unit circle. For a fixed choice of H, $(C, A; (A^*)^{-1}, (A^*)^{-1}C^*J; H)$ is a global left null-pole triple for $U(z)$ and so $U(z)$ is unique up to a constant J-unitary right factor (Theorem 4.5.8). $\square$

The following analogous theorem for left pole pairs can be proved in a similar way.

THEOREM 7.1.6. *Suppose (A, B) is a full-range pair and J is a signature matrix. Then there exists a rational matrix function $U(z)$ such that*

a) *U is regular at infinity*

b) *U is J-unitary valued on the unit circle*

and

c) *(A, B) is a global left pole pair for $U(z)$ if and only if A is invertible and there exists an invertible Hermitian solution G to the Stein equation*

$$G - AGA^* = -BJB^*.$$

In this case, one such $U(z)$ is given by

$$U(z) = D'_\alpha + C'_\alpha (zI - A)^{-1} B$$

where

$$D'_\alpha = I + JB^* (I - \alpha A^*)^{-1} G^{-1} B$$

and

$$C'_\alpha = D'_\alpha J B^* (A^*)^{-1} G^{-1}$$

and α, $|\alpha| = 1$, is in the resolvent set of A. Moreover, $U(z)$ is uniquely defined by (A, B) and G up to multiplication on the left by a constant J-unitary matrix.

The counterpart of Theorem 6.1.6 is:

THEOREM 7.1.7. *Suppose $(C, A_\pi; A_\zeta, B; S)$ is a Sylvester set of data. Then $(C, A_\pi; A_\zeta, B; S)$ is a global left null-pole triple for a rational matrix function regular at infinity and J-unitary valued on the unit circle if and only if*

(a) *S, A_π and A_ζ are invertible*

and

(b) *$(C, A_\pi; A_\zeta, B; S)$ and $(JB^*(A_\zeta^*)^{-1}, (A_\zeta^*)^{-1}; (A_\pi^*)^{-1}, (A_\pi^*)^{-1}C^*J; S^*)$ are similar.*

In this case, if α, $|\alpha| = 1$ is in the resolvent set of A_ζ, and if we set

$$D_\alpha = I - CS^{-1}(\alpha I - A_\zeta)^{-1} B$$
$$B_\alpha = S^{-1} B D_\alpha$$

then any such function $U(z)$ has the form

$$U(z) = [D_\alpha + C(zI - A_\pi)^{-1} B_\alpha]\Gamma$$

where Γ is an arbitrary J-unitary constant matrix.

PROOF. The necessity has already been observed in Theorem 7.1.3. Conversely, if (a) and (b) hold and D is any invertible matrix, then

$$U(z) = [I + C(zI - A_\pi)^{-1}S^{-1}B]D$$

with inverse given by

$$U^{-1}(z) = D^{-1}[I - CS^{-1}(zI - A_\zeta)^{-1}B]$$

has $(C, A_\pi; A_\zeta, B; S)$ is a global left null-pole triple. If α, $|\alpha| = 1$, is in the resolvent set of A_ζ, then α is also in the resolvent set of A_π (since A_ζ is similar to $(A_\pi^*)^{-1}$), hence $U^{-1}(\alpha) = D^{-1}[I - CS^{-1}(zI - A_\zeta)^{-1}B]$ is invertible. If we choose $D = D_\alpha$, then $U^{-1}(\alpha) = I$. We now conclude that U is J-unitary on the unit circle by Theorem 7.1.4.
$\square$

7.2 THE ASSOCIATED HERMITIAN COUPLING MATRIX

In this section we give the analogues of the results of Section 6.2.

LEMMA 7.2.1. *Let U be a rational matrix function, regular at infinity and J-unitary on the unit circle. Let $U(a) = D + C_i(zI - A_i)^{-1}B_i$, $i = 1, 2$, be two minimal realizations of U, with associated Hermitian coupling matrix H_i for right pole pair (C_i, A_i) and left null pair $((A_i^*)^{-1}, (A_i^*)^{-1}C_i^*J)$. Then the two minimal realizations are similar*

$$C_1 = C_2 S, \quad A_1 = S^{-1}A_2 S, \quad B_1 = S^{-1}B_2$$

for a unique invertible matrix S which has the additional property

$$H_1 = S^* H_2 S.$$

We omit the proof since it is the same as the imaginary axis case.

THEOREM 7.2.2. *Let U be a rational function analytic and invertible at infinity and J-unitary on the unit circle. Let $U(z) = D + C(zI - A)^{-1}B$ be a minimal realization of U, with associated Hermitian matrix H. Then, the number of negative eigenvalues of H is equal to the number of negative squares of each of the functions*

(7.2.1) $$K_U(z, w) = \frac{J - U(z)JU(w)^*}{1 - z\overline{w}} \quad \text{and} \quad \frac{J - U(w)^*JU(z)}{1 - z\overline{w}}.$$

Finally, let $\mathcal{K}(U)$ be the span of the functions $z \to K_U(z, w)c$ where w spans the points of analyticity of U and c spans $\mathbb{C}^n$, then $\mathcal{K}(U)$ is a finite dimensional vector space of rational functions and its dimension is equal to the McMillan degree of U.

PROOF. Formula (7.1.17) leads to the equality

(7.2.2) $$\frac{J - U(z)JU(w)^*}{1 - z\overline{w}} = C(zI - A)^{-1}H^{-1}(\overline{w}I - A^*)^{-1}C^*$$

valid for z and w in the resolvent set of A. The claim on the number of negative eigenvalues of H is then proved exactly as in Theorem 6.2.2.

From (7.2.2), we see that any linear combination of functions $\frac{J-U(z)JU(w)^*}{1-z\overline{w}}c$ is of the form $C(zI_n - A)^{-1}f$ for some vector f in $\mathbf{C}^n$ and thus $\dim \mathcal{K}(U) \leq n$ where $n \times n$ is the size of A. The null-kernel property of the pair (C, A) leads to

$$C(zI_n - A)^{-1}f \equiv 0 \Longrightarrow f = 0$$

and then to $\dim \mathcal{K}(U) = n$. The second function on (7.2.1) is treated analogously. $\square$

The function $K_U(z, w)$ given by (7.2.1) will be called the *kernel function* associated with $U(z)$. Note that the second function in (7.2.1) can be alternatively written as

$$K_{\widetilde{U}}(\overline{w}, \overline{z}) = \frac{J - U(w)^* JU(z)}{1 - z\overline{w}},$$

where $\widetilde{U}(z) = \big(U(\overline{z})\big)^*$.

As in the imaginary line case, $\nu(U)$ will denote the number of negative squares of the functions defined in (7.2.1).

THEOREM 7.2.3. *Let* U_i, $i = 1, 2$ *be two rational functions, analytic and invertible at infinity, with minimal realizations* $D_i + C_i(zI - A_i)^{-1}B_i$, $i = 1, 2$, *and associated Hermitian matrices* H_i, $i = 1, 2$. *Then the matrix* $\begin{bmatrix} H_1 & 0 \\ 0 & H_2 \end{bmatrix}$ *is the associated Hermitian matrix corresponding to the minimal realization* $D + C(zI - A)^{-1}B$ *of the product* $U_1 U_2$, *where* A, B, C, D *are given by*

$$D = D_1 D_2, \qquad C = [C_1, D_1 C_2]$$

$$A = \begin{bmatrix} A_1 & B_1 C_2 \\ 0 & A_2 \end{bmatrix}, \qquad B = \begin{bmatrix} B_1 D_2 \\ B_2 \end{bmatrix}.$$

The proof is the same as that of Theorem 6.2.3.

7.3 J-UNITARY ON THE UNIT CIRCLE WITH GIVEN NULL-POLE STRUCTURE IN THE UNIT DISK

Here we collect the analogues of the results in Section 6.3.

THEOREM 7.3.1. *Let* $(C, A_\pi; A_\zeta, B; S)$ *be an admissible Sylvester set of data such that* $\sigma(A_\pi)$ *and* $\sigma(A_\zeta)$ *are both contained in the open unit disk. Then there exists a rational matrix function* $U(z)$ *such that*

a) U *is regular at infinity and on the unit circle*

b) U *assumes J-unitary values on the unit circle*

and

c) $(C, A_\pi; A_\zeta, B; S)$ *is a left null-pole triple for* $U(z)$ *over the open unit disk* *if and only if A_π, A_ζ and $\widehat{S}$ are invertible, where $\widehat{S}$ is the Hermitian matrix given by*

$$(7.3.1) \qquad \widehat{S} = \begin{bmatrix} S_1 & S^* \\ S & S_2 \end{bmatrix}.$$

Here S_1 and S_2 are the unique solutions of the Stein equations

$$(7.3.2) \qquad S_1 - A_\pi^* S_1 A_\pi = -C^* J C$$

$$(7.3.3) \qquad S_2 - A_\zeta S_2 A_\zeta^* = B J B^*.$$

In this case any such matrix function $U(z)$ has the form

$$(7.3.4) \quad U(z) = D + [C, -JB^*] \begin{bmatrix} (zI - A_\pi)^{-1} & 0 \\ 0 & (I - zA_\zeta^*)^{-1} \end{bmatrix} \widehat{S}^{-1} \begin{bmatrix} A_\pi^{*-1} C^* J \\ B \end{bmatrix} D$$

where, for any fixed choice of α with $|\alpha| = 1$,

$$(7.3.5) \quad D = \Gamma - [C, JB^*(A_\zeta^*)^{-1}] \widehat{S}^{-1} \begin{bmatrix} (I - \alpha A_\pi)^{-1} & 0 \\ 0 & (\alpha I - A_\zeta)^{-1} \end{bmatrix} \begin{bmatrix} -C^* J \\ B \end{bmatrix} \Gamma$$

where Γ is any J-unitary matrix.

PROOF. If $U(z)$ is regular at infinity and J-unitary on the unit circle, a consequence of Theorem 7.1.3 is that $(JB^*(A_\zeta^*)^{-1}, (A_\zeta^*)^{-1}; (A_\pi^*)^{-1}, (A_\pi^*)^{-1} C^* J, S^*)$ is a left null-pole triple for $U(z)$ over the exterior $\mathcal{D}_e$ of the unit disk and $(C, A_\pi; A_\zeta, B; S)$ is a left null-pole triple for $U(z)$ on the interior of the unit disk $\mathcal{D}$. Therefore by Lemma 4.5.6, if U is regular on the unit circle then a global left null pole pair for $U(z)$ is

$$\tau = \left([C, JB^*(A_\zeta^*)^{-1}], \begin{bmatrix} A_\pi & 0 \\ 0 & (A_\zeta^*)^{-1} \end{bmatrix}; \begin{bmatrix} (A_\pi^*)^{-1} & 0 \\ 0 & A_\zeta \end{bmatrix}, \begin{bmatrix} (A_\pi^*)^{-1} C^* J \\ B \end{bmatrix}, \widehat{S} \right)$$

where $\widehat{S}$ is given as in (7.3.1). In particular it follows that $\widehat{S}$, A_π and A_ζ must be invertible.

Conversely, suppose $\widehat{S}$, A_π and A_ζ are invertible. Since S satisfies the Sylvester equation

$$S A_\pi - A_\zeta S = BC$$

and S_1 and S_2 satisfy (7.3.2) and (7.3.3), we deduce that $\widehat{S}$ satisfies

$$\widehat{S} \widehat{A}_\pi - \widehat{A}_\zeta \widehat{S} = \widehat{B}\widehat{C}$$

where

$$\widehat{A}_\pi = \begin{bmatrix} A_\pi & 0 \\ 0 & (A_\zeta^*)^{-1} \end{bmatrix}, \qquad \widehat{A}_\zeta = \begin{bmatrix} (A_\pi^*)^{-1} & 0 \\ 0 & A_\zeta \end{bmatrix},$$

$$\widehat{B} = \begin{bmatrix} (A_\pi^*)^{-1} C^* J \\ B \end{bmatrix}, \qquad \widehat{C} = [C, J B^* (A_\zeta^*)^{-1}].$$

Then by Theorem 4.4.3, if $\widehat{D}$ is any invertible matrix, the matrix function

$$U(z) = \widehat{D} + \widehat{C}(zI - \widehat{A}_\pi)^{-1} \widehat{S}^{-1} \widehat{B} \widehat{D}$$

with inverse given by

$$U^{-1}(z) = \widehat{D}^{-1} - \widehat{D}^{-1} \widehat{C} \widehat{S}^{-1} (zI - \widehat{A}_\zeta)^{-1} \widehat{B}$$

has $(\widehat{C}, \widehat{A}_\pi; \widehat{A}_\zeta, \widehat{B}; \widehat{S})$ as global left null-pole triple. If we choose $\widehat{D}$ to be D as in (7.3.5), then $U(\alpha)$ is also J-unitary. Then by Theorem 7.1.4 $U(z)$ is J-unitary on the unit circle. Moreover, the restriction of the admissible Sylvester data set $(\widehat{C}, \widehat{A}_\pi; \widehat{A}_\zeta, \widehat{B}; \widehat{S})$ to the unit disk is clearly $(C, A_\pi; A_\zeta, B; S)$, so $(C, A_\pi; A_\zeta, B; S)$ is a left null-pole triple for $U(z)$ over the unit disk. $\square$

7.4 *J*-UNITARY WITH SINGULARITY AT INFINITY

If $U(z)$ is a rational matrix function having a zero at 0 and which is J-unitary valued on the unit circle, necessarily $U(z)$ is improper (i.e., not analytic at infinity). To recover such a $U(z)$ from null-pole data we must use the more general formulas for improper rational matrix functions from Section 5.4. We shall here be content to indicate how to recover such a $U(z)$ from null-pole data on the unit disk $\mathcal{D}$. The following lemma is the starting point.

LEMMA 7.4.1. *Suppose $U(z)$ is a rational matrix function satisfying*

$$U(\bar{z}^{-1})^* J U(z) = J$$

at all regular points z for some signature matrix J. Then $\tau_\mathcal{D} = (C, A_\pi; A_\zeta, B; S)$ is a null-pole triple for $U(z)$ over the unit disk $\mathcal{D}$ if and only if $\tau_{\mathcal{D}_e} = (-JB^*, A_\zeta^*; A_\pi^*, C^*J; S^*)$ is a null pole triple for $U(z)$ over the region $\mathcal{D}_e = \{z \in \mathbb{C} : |z| > 1\} \cup \{\infty\}$ in the sense of Section 5.3.

PROOF. By Theorem 4.5.1 (see also Theorem 4.6.2) we can construct a proper rational matrix function $V(z)$ with value I at infinity which has $\tau_\mathcal{D}$ as a left null-pole triple over $\mathcal{D}$. Moreover we may assume that $V(z)$ has a minimal realization of the form

(7.4.1) $$V(z) = I + C_V(zI - A_{V,\pi})^{-1} S_V^{-1} B_V$$

with

(7.4.2) $$V(z)^{-1} = I - C_V S_V^{-1}(zI - A_{V,\zeta})^{-1} B_V$$

where C_V, $A_{V,\pi}$, $A_{V,\zeta}$, B_V, S_V have the forms

$$C_V = [C, C_e]; \quad A_{V,\pi} = A_\pi \oplus A_{\pi,e}; \quad A_{V,\zeta} = A_\zeta \oplus A_{\zeta,e}; \quad B_V = \begin{bmatrix} B \\ B_e \end{bmatrix}$$

and

$$S_V = \begin{bmatrix} S & S_{12} \\ S_{21} & S_e \end{bmatrix}$$

for appropriate matrices C_e, $A_{\pi,e}$, $A_{\zeta,e}$, B_e, S_{12}, S_{21}, S_e, where $\sigma(A_{\pi,e}) \cup \sigma(A_{\zeta,e}) \subset \mathcal{D}_e$. Then by Theorem 4.5.8 we know that $U(z) = V(z)W(z)$ for a matrix function $W(z)$ which is analytic and invertible on $\mathcal{D}$. But then also

$$U(z) = JU(\bar{z}^{-1})^{*-1}J$$
$$= JV(\bar{z}^{-1})^{*-1}J \cdot JW(\bar{z}^{-1})^{*-1}J$$

where $JW(\bar{z}^{-1})^{*-1}J$ is analytic and invertible on $\mathcal{D}_e$ (including infinity). Then by the converse direction of Theorem 4.5.8, we know that a left null-pole triple for $U(z)$ over $\mathcal{D}_e$ coincides with one for $JV(\bar{z}^{-1})^{*-1}J$. Thus is suffices to show that $\tau_{\mathcal{D}_e}$ is a left null-pole triple (in the sense of Section 5.3 with the region including infinity) for

$$\widetilde{V}(z) := JV(\bar{z}^{-1})^{*-1}J \quad \text{on} \quad \mathcal{D}_e.$$

From the realization (7.4.2) for $V(z)^{-1}$ we compute

(7.4.3) $$\widetilde{V}(z) = JV(\bar{z}^{-1})^{*-1}J = I - JB_V^*(z^{-1}I - A_{V,\zeta}^*)^{-1}S_V^{*-1}C_V^*J$$

while from the realization (7.4.1) for $V(z)$ we get

(7.4.4)
$$\widetilde{V}(z)^{-1} = JV(\bar{z}^{-1})^*J$$
$$= I + JB_V^*S_V^{*-1}(z^{-1}I - A_{V,\pi}^*)^{-1}C_V^*J.$$

Since both (7.4.3) and (7.4.4) are minimal realizations we read off from these that

$$\tau_\infty = (-JB_V^*, A_{V,\zeta}^*; A_{V,\pi}^*, C_V^*J; S_V^*)$$

is a null-pole triple for $\widetilde{V}(z)$ on the region $\mathbb{C}^* \setminus \{0\}$. Restricting this triple to the region $\{z \in \mathbb{C} : z^{-1} \in \mathcal{D}_e\} = \mathcal{D}$ then gives the null-pole triple for $\widetilde{V}(z)$ on $\mathcal{D}_e$. From the explicit form of C_V, $A_{V,\pi}$, $A_{V,\zeta}$, B_V and S_V, we see that this restriction process yields $\tau_{\mathcal{D}_e}$ as required. $\square$

We now state the main result of this section.

THEOREM 7.4.2. *Suppose $\tau_\mathcal{D} = (C, A_\pi; A_\zeta, B; S)$ is an admissible Sylvester data set over the unit disk $\mathcal{D}$, and let $J = J^* = J^{-1}$ be a given signature matrix. Define matrices S_{12} and S_{21} as the unique Hermitian solutions of the Stein equations*

(7.4.5) $$A_\zeta S_2 A_\zeta^* - S_2 = -BJB^*$$

(7.4.6) $$S_1 - A_\pi^* S_1 A_\pi = -C^* JC.$$

Then there exists a rational matrix function $U(z)$ for which

(i) $U(z)$ *has no poles on the unit circle* $\{z : |z| = 1\}$

(ii) *for all points z where $U(z)$ is analytic and invertible*

$$JU(\bar{z}^{-1})^{*-1}J = U(z)$$

and

(iii) $U(z)$ *has $\tau_\mathcal{D}$ as a (left) null-pole triple on the unit disk $\mathcal{D}$ if and only if the Hermitian matrix*

$$H = \begin{bmatrix} S_1 & S^* \\ S & S_2 \end{bmatrix}$$

is invertible. In this case the unique such function $U(z)$ satisfying the additional constraint

(iv) $U(z_0) = I$ *for a given point z_0 on the unit circle*

is given by

(7.4.7)
$$U(z) = I + (z - z_0)[C, -JB^*]$$
$$\cdot \begin{bmatrix} (zI - A_\pi)^{-1} & 0 \\ 0 & (I - zA_\zeta^*)^{-1} \end{bmatrix} H^{-1} \begin{bmatrix} (I - z_0 A_\pi^*)^{-1} C^* J \\ (A_\zeta - z_0 I)^{-1} B \end{bmatrix}$$

with inverse given by

(7.4.8)
$$U(z)^{-1} = I - (z - z_0)[C(z_0 I - A_\pi)^{-1}, -JB^*(I - z_0 A_\zeta^*)^{-1}]$$
$$\cdot H^{-1} \cdot \begin{bmatrix} (I - zA_\pi^*)^{-1} & 0 \\ 0 & (A_\zeta - zI)^{-1} \end{bmatrix} \begin{bmatrix} C^* J \\ B \end{bmatrix}.$$

PROOF. By the previous Lemma 7.4.1, if $\tau_\mathcal{D}$ is a null-pole triple on $\mathcal{D}$ for a matrix function $U(z)$ satisfying (ii), then $\tau_{\mathcal{D}_e}$ (as defined in Lemma 7.4.1) is a null-pole triple for $U(z)$ on $\mathcal{D}_e$. If no poles are allowed on the unit circle (condition (i)), then (ii) implies that $U(z)$ has no zeros on the unit circle as well. Then a null-pole triple on $\mathcal{D}$ for $U(z)$ is also a null-pole triple on the closure $\mathcal{D}^-$ of the unit disk for $U(z)$ as well. By Theorem 5.4.1 we see that a matrix function $U(z)$ having $\tau_\mathcal{D}$ as null-pole triple on $\mathcal{D}^-$ and $\tau_{\mathcal{D}_e}$ as null-pole triple on $\mathcal{D}_e$ exists if and only if the matrix

$$\widehat{S} = \begin{bmatrix} S & S_2 \\ S_1 & S^* \end{bmatrix}$$

is invertible, where S_2 and S_1 are given by (7.4.5) and (7.4.6). Since clearly $\widehat{S} = \begin{bmatrix} 0 & I \\ I & 0 \end{bmatrix} H$, we see that this is equivalent to the invertibility of H. Since S_2^* satisfies (7.4.5) whenever S_2 does and the solution of (7.4.5) is known to be unique since $\sigma(A_\zeta) \subset \mathcal{D}$ (see Section A.2 in the Appendix), we get that S_2 is Hermitian; similarly S_1

is Hermitian, so indeed H is Hermitian. The formula for $U(z)$ and $U(z)^{-1}$ follows by plugging the data $\tau_{\mathcal{D}}$ and $\tau_{\mathcal{D}_e}$ into the general realization formulas in Theorem 5.4.2; this is the unique $U(z)$ which satisfies (i) and (iii). To see that this $U(z)$ satisfies (ii) as well, use the computation in the proof of Lemma 7.4.1 to see that $\tau_{\mathcal{D}}$ and $\tau_{\mathcal{D}_e}$ are null-pole triples also for $\widetilde{U}(z) := JU(\bar{z}^{-1})^{*-1}J$ on $\mathcal{D}^-$ and $\mathcal{D}_e$ respectively. Thus by Theorem 4.5.8, $\widetilde{U}(z) = U(z)\Gamma$ for a constant invertible matrix Γ. However at z_0 both $\widetilde{U}(z)$ and $U(z)$ have the value I, so $\Gamma = I$ and (ii) follows. $\square$

The Hermitian matrix H in Theorem 7.4.2 of course depends on the choice of null-pole triple $\tau_{\mathcal{D}}$ for $U(z)$ on $\mathcal{D}$. However, if $\widetilde{\tau}_{\mathcal{D}}$ is another choice of null-pole triple for $U(z)$ on $\mathcal{D}$, then as in Theorem 4.4.2 $\widetilde{\tau}_{\mathcal{D}}$ is similar to $\tau_{\mathcal{D}}$; using this similarity of null-pole triple it is easy to show that the Hermitian matrix $\widetilde{H}$ associated with null-pole triple $\widetilde{\tau}_{\mathcal{D}}$ is congruent to the Hermitian matrix associated with null-pole triple $\tau_{\mathcal{D}}$. Thus the number of negative eigenvalues of H is independent of the choice of null-pole triple $\tau_{\mathcal{D}}$ and is intrinsic to the matrix function $U(z)$. The following gives a more intrinsic characterization of the number of negative eigenvalues of the associated Hermitian matrix H, and extends the main result of Section 7.3 to the improper case.

THEOREM 7.4.3. *Suppose H is the invertible Hermitian matrix associated with the rational matrix function satisfying conditions (i)–(iii) as in Theorem 7.4.2. Then the number ν_H of negative eigenvalues of H is the same as the number of negative squares of each of the kernel functions*

$$K_U(z,w) = \frac{J - U(z)JU(w)^*}{1 - z\overline{w}}$$

and

$$K_{\widetilde{U}}(\overline{w},\overline{z}) = \frac{J - U(w)^*JU(z)}{1 - z\overline{w}},$$

where $\widetilde{U}(z) = \left(U(\overline{z})\right)^*$. Also, if $\mathcal{K}(U)$ is the linear span of the functions $z \to K_U(z,w)c$ where w spans the points of analyticity of U and c spans $\mathbb{C}^n$, then $\mathcal{K}(U)$ is a finite dimensional vector space of rational functions with dimension equal to the McMillan degree of U.

PROOF. We verify that ν_H is the number of negative squares of K_U only; the proof for $K_{\widetilde{U}}(\overline{w},\overline{z})$ is similar. From the formula (7.4.7) for $U(z)$ in Theorem 7.4.2, we deduce the formula

$$U(w)^* = I + (\overline{w} - \overline{z}_0)[JC, -B^*] \begin{bmatrix} (I - \overline{z}_0 A_\pi)^{-1} & 0 \\ 0 & (\overline{z}_0 I - A_\zeta^*)^{-1} \end{bmatrix} H^{-1}$$
$$\cdot \begin{bmatrix} (\overline{w}I - A_\pi^*)^{-1} & 0 \\ 0 & (I - \overline{w}A_\zeta)^{-1} \end{bmatrix} \begin{bmatrix} C^* \\ -BJ \end{bmatrix}.$$

Therefore we compute

$$U(z)JU(w)^* =$$

$$J + (z - z_0)[C, -JB^*] \begin{bmatrix} (zI - A_\pi)^{-1} & 0 \\ 0 & (I - zA_\zeta^*)^{-1} \end{bmatrix} \cdot H^{-1}$$

$$\cdot \begin{bmatrix} (I - z_0 A_\pi^*)^{-1} & 0 \\ 0 & (z_0 I - A_\zeta)^{-1} \end{bmatrix} \begin{bmatrix} C^* \\ -BJ \end{bmatrix}$$

$$+ (\overline{w} - \overline{z}_0)[C, -JB^*] \begin{bmatrix} (I - \overline{z}_0 A_\pi)^{-1} & 0 \\ 0 & (\overline{z}_0 I - A_\zeta^*)^{-1} \end{bmatrix} H^{-1}$$

$$\cdot \begin{bmatrix} (\overline{w}I - A_\pi^*)^{-1} & 0 \\ 0 & (I - \overline{w}A_\zeta)^{-1} \end{bmatrix} \begin{bmatrix} C^* \\ -BJ \end{bmatrix}$$

$$+ (z - z_0)(\overline{w} - \overline{z}_0)[C, -JB^*] \begin{bmatrix} (zI - A_\pi)^{-1} & 0 \\ 0 & (I - zA_\zeta^*)^{-1} \end{bmatrix} H^{-1}$$

$$\cdot \begin{bmatrix} (I - z_0 A_\pi^*)^{-1} & 0 \\ 0 & (z_0 I - A_\zeta)^{-1} \end{bmatrix} \left\{ \begin{bmatrix} C^* J \\ -B \end{bmatrix} J[JC, -B^*] \right\}$$

$$\cdot \begin{bmatrix} (I - \overline{z}_0 A_\pi)^{-1} & 0 \\ 0 & (\overline{z}_0 I - A_\zeta^*)^{-1} \end{bmatrix} H^{-1}$$

$$\cdot \begin{bmatrix} (\overline{w}I - A_\pi^*)^{-1} & 0 \\ 0 & (I - \overline{w}A_\zeta)^{-1} \end{bmatrix} \begin{bmatrix} C^* \\ -BJ \end{bmatrix}$$

$$= J + X(z)Y(z,w)X(w)^*$$

where

$$X(z) = [C, -JB^*] \begin{bmatrix} (zI - A_\pi)^{-1} & 0 \\ 0 & (I - zA_\zeta^*)^{-1} \end{bmatrix} H^{-1}$$

$$\cdot \begin{bmatrix} I - z_0 A_\pi^*)^{-1} & 0 \\ 0 & (z_0 I - A_\zeta)^{-1} \end{bmatrix}$$

and

$$Y(z,w) = (z - z_0) \begin{bmatrix} \overline{w}I - A_\pi^* & 0 \\ 0 & I - \overline{w}A_\zeta \end{bmatrix} H \begin{bmatrix} I - \overline{z}_0 A_\pi & 0 \\ 0 & \overline{z}_0 I - A_\zeta^* \end{bmatrix}$$

$$+ (\overline{w} - \overline{z}_0) \begin{bmatrix} I - z_0 A_\pi^* & 0 \\ 0 & z_0 I - A_\zeta \end{bmatrix} H \begin{bmatrix} zI - A_\pi & 0 \\ 0 & I - zA_\zeta^* \end{bmatrix}$$

$$+ (z - z_0)(\overline{w} - \overline{z}_0) \begin{bmatrix} C^* JC & -C^* B^* \\ -BC & BJB^* \end{bmatrix}.$$

From the Stein equations (7.4.5) and (7.4.6) and since $\tau_\mathcal{D}$ is an admissible Sylvester data set, we have

$$\begin{bmatrix} C^* JC & -C^* B^* \\ -BC & BJB^* \end{bmatrix} = \begin{bmatrix} A_\pi^* S_1 A_\pi - S_1 & S^* A_\zeta^* - A_\pi^* S^* \\ A_\zeta S - S A_\pi & S_2 - A_\zeta S_2 A_\zeta^* \end{bmatrix}.$$

Recalling that $H = \begin{bmatrix} S_1 & S^* \\ S & S_2 \end{bmatrix}$ and writing out $Y(z,w)$ as

$$Y(z,w) = \begin{bmatrix} Y_{11}(z,w) & Y_{12}(z,w) \\ Y_{21}(z,w) & Y_{22}(z,w) \end{bmatrix}$$

we next compute out separately $Y_{ij}(z, w)$. For Y_{11} we have

$$
\begin{aligned}
Y_{11}(z, w) &= (z - z_0)(\overline{w}I - A_\pi^*)S_1(I - \overline{z}_0 A_\pi) \\
&\quad + (\overline{w} - \overline{z}_0)(I - z_0 A_\pi^*)S_1(zI - A_\pi) \\
&\quad + (z - z_0)(\overline{w} - \overline{z}_0)\{A_\pi^* S_1 A_\pi - S_1\} \\
&= -(1 - z\overline{w})S_1 + (1 - z\overline{w})z_0 A_\pi^* S_1 \\
&\quad + (1 - z\overline{w})\overline{z}_0 S_1 A_\pi - (1 - z\overline{w})A_\pi^* S_1 A_\pi \\
&= -(1 - z\overline{w})\{S_1 - z_0 A_\pi^* S_1 - S_1(\overline{z}_0 A_\pi) + A_\pi^* S_1 A_\pi\}
\end{aligned}
$$

while

$$
\begin{aligned}
Y_{12}(z, w) &= (z - z_0)(\overline{w}I - A_\pi^*)S^*(\overline{z}_0 I - A_\zeta^*) \\
&\quad + (\overline{w} - \overline{z}_0)(I - z_0 A_\pi^*)S^*(I - z A_\zeta^*) \\
&\quad + (z - z_0)(\overline{w} - \overline{z}_0)\{-A_\pi^* S^* + S^* A_\zeta^*\} \\
&= -(1 - z\overline{w})\overline{z}_0 S^* + (1 - z\overline{w})A_\pi^* S^* \\
&\quad + (1 - z\overline{w})S^* A_\zeta^* - (1 - z\overline{w})z_0 A_\pi^* S^* A_\zeta^* \\
&= -(1 - z\overline{w})\{\overline{z}_0 S^* - A_\pi^* S^* - S^* A_\zeta^* + z_0 A_\pi^* S^* A_\zeta^*\}.
\end{aligned}
$$

Then

$$
\begin{aligned}
Y_{21}(z, w) &= Y_{12}(w, z)^* \\
&= -(1 - z\overline{w})\{z_0 S - S A_\pi - A_\zeta S + \overline{z}_0 A_\zeta S A_\pi\}
\end{aligned}
$$

and

$$
\begin{aligned}
Y_{22}(z, w) &= (z - z_0)(I - \overline{w}A_\zeta)S_2(\overline{z}_0 I - A_\zeta^*) \\
&\quad (\overline{w} - \overline{z}_0)(z_0 I - A_\zeta)S_2(I - z A_\zeta^*) \\
&\quad + (z - z_0)(\overline{w} - \overline{z}_0)\{S_2 - A_\zeta S_2 A_\zeta^*\} \\
&= -(1 - z\overline{w})S_2 + (1 - z\overline{w})\overline{z}_0 A_\zeta S_2 \\
&\quad + (1 - z\overline{w})z_0 S_2 A_\zeta^* - (1 - z\overline{w})A_\zeta S_2 A_\zeta^* \\
&= -(1 - z\overline{w})\{S_2 - \overline{z}_0 A_\zeta S_2 - z_0 S_2 A_\zeta^* + A_\zeta S_{12} A_\zeta^*\}
\end{aligned}
$$

Putting the pieces together we conclude

$$
Y(z, w) = -(1 - z\overline{w})\begin{bmatrix} I - z_0 A_\pi^* & 0 \\ 0 & z_0 I - A_\zeta \end{bmatrix} H \begin{bmatrix} I - \overline{z}_0 A_\pi & 0 \\ 0 & \overline{z}_0 I - A_\zeta^* \end{bmatrix}.
$$

Therefore

(7.4.10)
$$
\frac{J - U(z)JU(w)^*}{1 - z\overline{w}} = \widetilde{X}(z)H^{-1}\widetilde{X}(w)^*
$$

where

$$
\widetilde{X}(z) = [C, -JB^*]\begin{bmatrix} (zI - A_\pi)^{-1} & 0 \\ 0 & (I - zA_\zeta^*)^{-1} \end{bmatrix}.
$$

Since (C, A_π) is a null kernel pair and (A_ζ, B) is a full range pair, (7.4.10) shows that the number of negative squares of $K_U(z, w)$ is the same as the number ν_H of negative eigenvalues of H, just as in the proof of Theorem 6.2.2. The claim concerning the dimension of K_U follows now also exactly as in the proof of Theorem 6.2.2. $\square$

7.5 *J*-UNITARY WITH POLES ON THE UNIT CIRCLE

In this section we extend the results of the previous section to the case where the rational matrix function $U(z)$ with J-unitary values at all regular points on the unit circle $\mathbf{T} = \{z \colon |z| = 1\}$ may have singularities on $\mathbf{T}$. We first give a characterization of the form of a null-pole triple for such a $U(z)$ over $\mathbf{T}$.

THEOREM 7.5.1. *Let $U(z)$ be a rational matrix function such that*

$$U(\bar{z}^{-1})^* J U(z) = J$$

where $J = J^* = J^{-1}$ is a signature matrix. Then

(i) $(JB^* A_\zeta^{*-1}, A_\zeta^{*-1}; A_\pi^{*-1}, A_\pi^{*-1} C^* J; S^*)$ *is a null-pole triple for $U(z)$ over* $\mathbf{T}$ *whenever $(C, A_\pi; A_\zeta, B; S)$ is.*

(ii) $U(z)$ *has a null-pole triple over* $\mathbf{T}$ *of the form $(C, A_\pi; A_\pi^{*-1}, A_\pi^{*-1} C^* J; H)$ with Hermitian coupling matrix H.*

(iii) $U(z)$ *has a null-pole triple over* $\mathbf{T}$ *of the form $(JB^* A_\zeta^{*-1}, A_\zeta^{*-1}; A_\zeta, B; G)$ with Hermitian coupling matrix G.*

PROOF. For the case where $U(z)$ has a finite, invertible value at infinity, Theorem 7.1.3 gives the result (i) for global null-pole triples; the result for null-pole triples over $\mathbf{T}$ follows by restricting a global null-pole triple. The restriction that $U(z)$ have finite, invertible value at infinity can be removed as in the proof of Lemma 7.4.1. In particular $(A_\pi^{*-1}, A_\pi^{*-1} C^* J)$ is a left null pair for $U(z)$ over $\mathbf{T}$ whenever (C, A_π) is a right pole pair over $\mathbf{T}$. If H is the coupling matrix associated with right pole pair (C, A_π) and left null pair $(A_\pi^{*-1}, A_\pi^{*-1} C^* J)$, then again by (i) both $(C, A_\pi; A_\pi^{*-1}, A_\pi^{*-1} C^* J; H)$ and $(C, A_\pi; A_\pi^{*-1}, A_\pi^{*-1} C^* J; H^*)$ are null-pole triples for $U(z)$ over $\mathbf{T}$. By the uniqueness of the coupling matrix we obtain $H = H^*$. This proves (ii) while (iii) follows similarly. □

We now give the construction of a matrix function $U(z)$ with J-unitary values on the unit circle T and prescribed null-pole data on the closed unit disk.

THEOREM 7.5.2. *Let $\tau_1 = (C, A_\pi; A_\zeta, B; S)$ and $\tau_2 = (C_0, A_0; A_0^{*-1}, A_0^{*-1} C_0^* J; H)$ be Sylvester data sets over $\mathcal{D}$ and $\mathbf{T}$ respectively, where $J = J^* = J^{-1}$ is a given signature matrix. Introduce matrices S_{11}, S_{21}, S_{32} and S_{33} as the unique solutions of the respective Sylvester-Stein equations*

(7.5.1) $$S_{11} - A_\pi^* S_{11} A_\pi = -C^* J C$$

(7.5.2) $$A_0^* S_{21} A_\pi - S_{21} = C_0^* J C$$

(7.5.3) $$S_{32} A_0 - A_\zeta S_{32} = B C_0$$

(7.5.4) $$A_\zeta S_{33} A_\zeta^* - S_{33} = -B J B^*$$

and let S be the matrix

(7.5.5) $$S = \begin{bmatrix} S_{11} & S_{21}^* & S^* \\ S_{21} & H & S_{32}^* \\ S & S_{32} & S_{33} \end{bmatrix}.$$

Then there exists a rational matrix function $U(z)$ such that

(i) τ_1 *is a null-pole triple for* $U(z)$ *over* $\mathcal{D}$

(ii) τ_2 *is a null-pole triple for* $U(z)$ *over* T

and (iii) $U(\bar{z}^{-1})^* J U(z) = J$ *if and only if H is Hermitian and the Hermitian matrix S is invertible. In this case the unique such matrix function $U(z)$ satisfying the additional requirement*

(iv) $U(z_0) = D$ *where* $z_0 \in \mathsf{T} \backslash \sigma(A_0)$ *and D is a J-unitary matrix, is given by*

$$
U(z) = D + (z - z_0)[C, C_0, -JB^*]
$$

(7.5.6)
$$
\cdot \begin{bmatrix} (zI - A_\pi)^{-1} & 0 & 0 \\ 0 & (zI - A_0)^{-1} & 0 \\ 0 & 0 & (I - zA_\zeta^*)^{-1} \end{bmatrix} S^{-1}
$$

$$
\cdot \begin{bmatrix} (I - z_0 A_\pi^*)^{-1} C^* JD \\ (I - z_0 A_0^*)^{-1} C_0^* JD \\ (A_\zeta - z_0 I)^{-1} BD \end{bmatrix}
$$

with inverse given by

$$
U(z)^{-1} = D^{-1} - (z - z_0)[D^{-1} C(z_0 I - A_\pi)^{-1}, D^{-1} C_0(z_0 I - A_0)^{-1},
$$
$$
- JB^*(I - z_0 A_\zeta^*)^{-1}]
$$

(7.5.7)
$$
\cdot S^{-1} \begin{bmatrix} (I - zA_\pi^*)^{-1} & 0 & 0 \\ 0 & (I - zA_0^*)^{-1} & 0 \\ 0 & 0 & (A_\zeta - zI)^{-1} \end{bmatrix} \begin{bmatrix} C^* J \\ C_0^* J \\ B \end{bmatrix}.
$$

PROOF. If τ_1 is a null-pole triple for $U(z)$ over $\mathcal{D}$, by Lemma 7.4.1 $\tau_1^* := (-JB^*, A_\zeta^*; A_\pi^*, C^*J; S^*)$ is a null pole triple for $U(z)$ over $\mathcal{D}_e$ (in the sense of Section 5.3 as we include infinity in $\mathcal{D}_e$). Let τ_F be the (essentially unique) admissible Sylvester data set over $\mathcal{D} \cup \mathsf{T}$ whose restriction to $\mathcal{D}$ is τ_1 and whose restriction to T is τ_2. Thus (see Lemma 4.5.6) τ_F has the form

(7.5.8) $\tau_F = \left([C, C_0], \begin{bmatrix} A_\pi & 0 \\ 0 & A_0 \end{bmatrix}; \begin{bmatrix} A_\zeta & 0 \\ 0 & A_0^{*-1} \end{bmatrix}, \begin{bmatrix} B \\ A_0^{*-1} C_0^* J \end{bmatrix}; \begin{bmatrix} S & S_{32} \\ S_{21} & H \end{bmatrix} \right)$

where S_{32} and S_{21} are determined by the condition that $\widehat{S} = \begin{bmatrix} S & S_{32} \\ S_{21} & H \end{bmatrix}$ satisfy the Sylvester equation

(7.5.9) $\widehat{S} \begin{bmatrix} A_\pi & 0 \\ 0 & A_0 \end{bmatrix} - \begin{bmatrix} A_\zeta & 0 \\ 0 & A_0^{*-1} \end{bmatrix} \widehat{S} = \begin{bmatrix} B \\ A_0^{*-1} C_0^* J \end{bmatrix} [C \quad C_0].$

By inspection we see that S_{32} and S_{21} are determined by Sylvester-Stein equations (7.5.3) and (7.5.2). Next we compute the matrix $\widetilde{S}$ given by (5.4.1) for the case where τ_F is given by (7.5.8) and $\tau_\infty = \tau_1^*$. Write $\widetilde{S}$ as the block 3×3 matrix

(7.5.10)
$$
\widetilde{S} = \begin{bmatrix} S & S_{32} & S_{33} \\ S_{21} & H & S_{23} \\ S_{11} & S_{12} & S^* \end{bmatrix}
$$

where S_{11}, S_{12}, S_{33} and S_{23} are to be determined. The determining equations for S_{11}, S_{12}, S_{33} and S_{23} are obtained by specializing equations (5.4.5) and (5.4.6) to our situation. The result is

$$(7.5.11) \qquad \begin{bmatrix} A_\zeta & 0 \\ 0 & A_0^{*-1} \end{bmatrix} \begin{bmatrix} S_{33} \\ S_{23} \end{bmatrix} A_\zeta^* - \begin{bmatrix} S_{33} \\ S_{23} \end{bmatrix} = - \begin{bmatrix} B \\ A_0^{*-1}C^*J \end{bmatrix} JB^*$$

and

$$(7.5.12) \qquad [\, S_{11} \quad S_{12} \,] - A_\pi^*[\, S_{11} \quad S_{12} \,] \begin{bmatrix} A_\pi & 0 \\ 0 & A_0 \end{bmatrix} = -C^*J[C,C_0].$$

From these equations it is easily deduced that S_{11} and S_{33} are given by (7.5.1) and (7.5.4), and that $S_{12} = S_{21}^*$, $S_{23} = S_{32}^*$. The result of Theorem 5.4.1 then is that there exists a rational matrix function $U(z)$ with τ_F as left null-pole pair over $\mathcal{D} \cup \mathsf{T}$ and τ_1^* as left null-pole pair over $\mathcal{D}_e$ if and only if the matrix $\widetilde{S}$, given by (7.5.10) together with (7.5.1)–(7.5.4) and with $S_{12} = S_{21}^*$, $S_{23} = S_{32}^*$, is invertible. Note that the matrix S given by (7.5.5) is obtained from $\widetilde{S}$ simply by interchanging the first and third block row, i.e.

$$S = \begin{bmatrix} 0 & 0 & I \\ 0 & I & 0 \\ I & 0 & 0 \end{bmatrix} \widetilde{S}.$$

Therefore invertibility of S is equivalent to that of $\widetilde{S}$. This establishes that invertibility of S is necessary for the existence of a rational matrix function $U(z)$ satisfying (i)–(iii).

Conversely, if S is invertible, then $\widetilde{S}$ is invertible so there exists a rational matrix function $U(z)$ having τ_F and τ_1^* as null-pole triples over $\mathcal{D} \cup \mathsf{T}$ and $\mathcal{D}_e$ respectively. By the form of τ_F and τ_1^*, if H is Hermitian then by computations as in the proofs of Theorem 7.1.3 and Lemma 7.4.1 we see that τ_F and τ_1^* are null-pole triples over $\mathcal{D} \cup \mathsf{T}$ and $\mathcal{D}_e$ respectively for $V(z) := JU(\bar{z}^{-1})^{*-1}J$ as well. But then by Theorem 4.5.8 there is a constant invertible matrix Γ for which

$$U(z) = V(z)\Gamma.$$

From this we get

$$U(\bar{z}^{-1})^* JU(z) = \widetilde{\Gamma}$$

with $\widetilde{\Gamma} = J\Gamma$. If it is arranged that $U(z_0)$ is J-unitary at a point $z_0 \in \mathsf{T} \backslash \sigma(A_0)$, then necessarily $\widetilde{\Gamma} = J$, i.e. $\Gamma = I$, so U satisfies (iii). Finally the formulas (7.5.6) and (7.5.7) for $U(z)$ come directly from the formulas in Theorem 5.4.2 specialized to τ_F and $\tau_\infty = \tau_1^*$ of the form as given in the statement of Theorem 7.5.2.   □

As the final result of this section we present the following extension of Theorem 7.4.3.

THEOREM 7.5.3. *Suppose S is the Hermitian matrix (7.5.5) associated with a rational matrix function $U(z)$ having J-unitary values at all regular points of the unit* *circle as in Theorem 7.5.2. Then the number ν_S of negative eigenvalues of S is the same as the number of negative squares of each of the kernel functions*

$$K_U(z, w) = \frac{J - U(z)JU(w)^*}{1 - z\overline{w}}$$

and

$$K_{\widetilde{U}}(\overline{w}, \overline{z}) = \frac{J - U(w)^* JU(z)}{1 - z\overline{w}}.$$

where $\widetilde{U}(z) = \left(U(\overline{z})\right)^*$. Also, if $\mathcal{K}(U)$ is the linear span of the functions $z \to K_U(z, w)c$ where w spans the points of analyticity of U and c spans $\mathbf{C}^n$, then $\mathcal{K}(U)$ is a finite dimensional vector space of rational vector functions with dimension equal to the McMillan degree of U.

PROOF. By a brute force calculation as in the proof of Theorem 7.4.3, one can verify that

$$\frac{J - U(z)JU(w)^*}{1 - z\overline{w}} = X(z)\mathcal{S}^{-1}X(w)^*$$

where

$$X(z) = [C, C_0, -JB^*] \begin{bmatrix} (zI - A_\pi)^{-1} & 0 & 0 \\ 0 & (zI - A_0)^{-1} & 0 \\ 0 & 0 & (I - zA_\zeta^*)^{-1} \end{bmatrix}.$$

From this identity all the assertions follow just as in the proof of Theorem 7.4.3. □

7.6 THE CASE $J = I$

As in the case in Chapter 6 in connection with matrix functions assuming J-unitary values on the imaginary line, special features arise in the results of this chapter for the case $J = I$. All these results are of a qualitative nature and can be seen for the circle case from the line case by using a conformal change of variable. For this reason we simply state here without proof the analogues of Theorems 6.5.1–6.5.3 for the circle case.

As in the line case, it is easily seen that a rational matrix function $U(z)$ with unitary values at all regular points on the unit circle T in fact can have no poles on T. A consequence of this, together with Theorem 7.5.1 is that the only Sylvester data set of T of the form $(C_0, A_0; A_0^{*-1}, A_0^{*-1}C_0^*; H)$ is vacuous. We state this result formally as follows; it can be proved directly via a Schur complement argument as was done for the imaginary line case (Theorem 6.5.1).

THEOREM 7.6.1. *The only Sylvester data set over T of the form $\tau = (C_0, A_0; A_0^{*-1}, A_0^{*-1}C_0^*; H)$ is $\tau = (0, 0; 0, 0; 0)$.*

THEOREM 7.6.2. *Let S be the Hermitian coupling matrix associated with the matrix function $U(z)$ having unitary values on T, as in Theorem 7.5.2. Then the number ν_S of negative eigenvalues of S is the same as the number n_π (counting multiplicities)* *of poles of $U(z)$ in $\mathcal{D}$, i.e. A_π has size $n_\pi \times n_\pi$ whenever (C, A_π) is a right pole pair for $U(z)$ over $\mathcal{D}$.*

THEOREM 7.6.3. *If $\tau = (C, A_\pi; A_\zeta, B; S)$ is an admissible Sylvester data set over the unit disk $\mathcal{D}$, then the matrix*

$$H = \begin{bmatrix} S_1 & S^* \\ S & S_2 \end{bmatrix},$$

where S_2 and S_1 are uniquely determined by the Stein equations

$$A_\zeta S_2 A_\zeta^* - S_2 = -BB$$
$$S_1 - A_\pi^* S_1 A_\pi = -C^* C$$

is invertible. Hence (by Theorem 7.4.2) there exists a rational matrix function $U(z)$ with unitary values on $\mathbf{T}$ having τ as a left null-pole triple over $\mathcal{D}$.

NOTES FOR PART I

CHAPTER 1   The main source for Chapter 1 is the paper of Gohberg-Rodman [1981].

The notion of the null chain in the case when the function is $zI - A$ where A is a square matrix is well known from linear algebra and is often called a Jordan chain. One of the earliest systematic uses of these chains for the nonlinear cases was by Keldysh [1951] (see also Gohberg-Marcus [1955]). Hence the name "Keldysh chains" can be found in the literature. Keldysh's results refer to infinite dimensional Hilbert spaces. A detailed exposition of these results can be found in Chapter 5 Section 9 of the book of Gohberg-Krein [1969]. Further development of the theory of chains in the infinite dimensional case for analytic and meromorphic functions appears in Marcus [1958], Gohberg [1971, 1972], Gohberg-Sigal [1971]. Null functions were first introduced by S.G. Krein and V.P. Trofimov [1969] and V.I. Macaev and Ju.A. Palant [1970].

In contrast with all the aforementioned papers, we use these results only for matrix rather than operator valued functions. In this case, the main results of this chapter are obtained from the Smith form (Smith [1861], Frobenius [1878]). See also Gantmacher [1959], Gohberg-Lancaster-Rodman [1982].

CHAPTER 2   The second chapter contains the results of Gohberg-Lancaster-Rodman [1978], Gohberg-Rodman [1978] which are summarized in Gohberg-Lancaster-Rodman [1982]. We follow mostly their exposition.

CHAPTER 3   Chapter 3 follows mostly the papers of Gohberg-Sigal [1971] and Gohberg-Rodman [1986]. The Möbius transform in Section 3.5 in connection with local spectral data was first used in Bart-Gohberg-Kaashoek [1979].

CHAPTER 4   Section 4.1 contains the now classical realization theory which was introduced in systems theory and is the basis for the state space method (see Kailath [1980], Wonham [1985], Kalman-Falb-Arbib [1965]). Section 4.2 contains results of Bart-Gohberg-Kaashoek [1979]. Section 4.3 follows the paper of Gohberg-Kaashoek-Lerer-Rodman [1984]. The results of Section 4.4 were obtained by Ball-Ran [1987a, 1987b]. The exposition in Section 4.5 follows the paper Gohberg-Kaashoek [1987]; a special case of the main result appeared in Ball-Ran [1987a, 1987b]. The results and exposition of Section 4.6 follow Gohberg-Kaashoek-Ran [1989a]; the incoming and outgoing basis for a σ-admissible Sylvester data step in a special form arising in the content of Wiener-Hopf factorization appears in Bart-Gohberg-Kaashoek [1986].

CHAPTER 5   The results of Chapter 5 appear in two different forms in Gohberg-Kaashoek [1988] and Ball-Cohen-Ran [1988]. Here we follow mostly the exposition of Gohberg-Kaashoek [1988]. Theorem 5.5.2 comes from Ball-Cohen-Ran [1988]; the latter paper had an important influence on the approach presented in Part III. Section 5.6 follows Gohberg-Kaashoek-Ran [1989a].

CHAPTERS 6 and 7   Chapters 6 and 7 follow mostly the paper Alpay-Gohberg [1988]. The results of Sections 6.4, 7.4 and 7.5 appear here for the first time. Another version of results from Sections 6.3 and 7.3 appeared in Ball-Ran [1987a, 1987b].

The function $U(z)$ which appears in Theorem 6.1.1 for the case when $H = I$ is the well known Livsic-Brodskii characteristic operator function (with $D = I$) (see Brodskii [1971]). For $H = I$, the function $U(z)$ which appears in Theorem 7.1.2 is also a characteristic function (see Brodskii-Gohberg-Krein [1978] and Sz.-Nagy-Foias [1970] for $J = I$).

The additive property of the number of negative squares stated in Corollary 6.2.4 was noticed by Sakhnovitch [1986].

PART II

HOMOGENEOUS INTERPOLATION PROBLEMS WITH OTHER FORMS OF LOCAL DATA

The part contains interpolation problems with different forms of local data. For instance we study here the interpolation problem where one is given the right null pair and right pole pair, or the problem where both right and left null and pole pairs are given. Much attention is paid to interpolation problems connected with divisibility. These problems look like problems of building a common multiple with certain properties for a set of matrix functions. An interpolation problem connected with coprime representation is also included. All these problems are homogeneous. A difference between problems considered in this part (in particular, in Chapter 8) and those considered in the first part is that the set of solutions of a problem considered in this part need not be invariant under multiplication on the right or on the left by an invertible constant matrix, as opposed to the case for the problems studied in Part I. In this part we do not go as deeply into the study of the interpolation problem as we did in Part I for the standard forms of the local data. The results of this part are not used later; they are presented here mostly for the sake of completeness.

CHAPTER 8

INTERPOLATION PROBLEMS WITH NULL AND POLE PAIRS

In Part I we were considering interpolation problems with standard combinations of the data. There we considered interpolation problems with given right null pair and left pole pair; the analogous interpolation problem where a left null pair and right pole pair are given can be reduced to the previous one by taking transpose. In this chapter we consider the interpolation problems for the remaining nonstandard combinations, namely the case where a null pair and pole pair from the same side are given. This problem differs considerably from the case of standard local data, both the procedure to find a solution and also in the properties of the solution. We also consider in this chapter the problem of interpolation when all four pairs (right and left null pair, right and left pole pair) are given. This problem can be regarded as an overdetermined version of the problem with standard data.

8.1 RATIONAL MATRIX FUNCTIONS WITH GIVEN RIGHT NULL PAIRS AND RIGHT POLE PAIRS: NECESSARY CONDITIONS

For $\nu = 1, 2$ let C_ν be an $m_\nu \times n_\nu$ matrix and A_ν an $n_\nu \times n_\nu$ matrix. We consider the following problems.

PROBLEM 1. Does there exist a rational matrix function $W(z)$ such that $W(z)$ is analytic and invertible at infinity, (C_1, A_1) is a right null pair and (C_2, A_2) is a right pole pair of $W(z)$ (with respect to $\mathbb{C}$)?

PROBLEM 2. If the answer is yes, describe all possible solutions of Problem 1.

We state first the necessary conditions for Problem 1 to admit a solution.

THEOREM 8.1.1. *Let (C_1, A_1) be a right null pair and (C_2, A_2) a right pole pair (with respect to $\mathbb{C}$) of a rational $m \times m$ matrix function $W(z)$ which is analytic and invertible at infinity. Then*

(i) *A_1 and A_2 are square matrices of equal size, $n \times n$ say, and C_1 and C_2 have size $m \times n$;*

(ii) *For some invertible matrix S of the form*

$$(8.1.1) \qquad S = \begin{bmatrix} N & F \\ 0 & D \end{bmatrix} : \mathbb{C}^n \oplus \mathbb{C}^m \to \mathbb{C}^n \oplus \mathbb{C}^m$$

we have

$$(8.1.2) \qquad NA_1 + FC_1 = A_2N; \qquad DC_1 = C_2N.$$

Moreover, S can be chosen so that in addition

(8.1.3)
$$W(z) = D + C_2(zI - A_2)^{-1}F,$$
$$W(z)^{-1} = D^{-1} + C_1(zI - A_1)^{-1}(-N^{-1}FD^{-1}).$$

PROOF. Let

(8.1.4)
$$W(z) = D + X^P(zI - J^P)^{-1}Y^P$$

and

(8.1.5)
$$W(z)^{-1} = D^{-1} + X(zI - J)^{-1}Y$$

be minimal realizations of $W(z)$ and $W(z)^{-1}$, respectively. By Theorem 4.2.1 we can assume that $(X, J) = (C_1, A_1)$ and $(X^P, J^P) = (C_2, A_2)$. Proposition 4.1.5 ensures in view of (8.1.5) that

(8.1.6)
$$W(z) = D - DC_1(zI - (A_1 - YDC_1))^{-1}YD$$

is a minimal realization for $W(z)$. As any two minimal realizations for $W(z)$ are similar (Theorems 4.1.3 and 4.1.4), by comparing (8.1.4) and (8.1.6) we deduce that there exists an invertible matrix N such that

(8.1.7)
$$DC_1 = C_2N, \qquad NA_1 - NYDC_1 = A_2N, \qquad -NYD = Y^P.$$

Put $F = Y^P$. Then $NA_1 + FC_1 = A_2N$, and so the matrix $S = \begin{bmatrix} N & F \\ 0 & D \end{bmatrix}$ satisfies (8.1.2). □

The condition (ii) of Theorem 8.1.1 is not as mysterious as it appears. It has to do with strict equivalence of matrix pencils. Two matrix pencils $zP_1 + Q_1$ and $zP_2 + Q_2$, where P_1, P_2, Q_1, Q_2 are (not necessarily square) matrices of the same size, are called *strictly equivalent* if there exist invertible matrices S and T such that

$$S(zP_1 + Q_1)T = zP_2 + Q_2, \qquad z \in \mathbb{C}.$$

The canonical form of matrix pencils under strict equivalence (the Kronecker canonical form) is a well-known classical result and can be found, for example, in Gantmacher [1959], or in Gohberg-Lancaster-Rodman [1986].

PROPOSITION 8.1.2. *Let (C_1, A_1) and (C_2, A_2) be two pairs of matrices with A_i $n \times n$ and C_i $m \times n$ $(i = 1, 2)$. The matrix pencils*

(8.1.8)
$$\begin{bmatrix} zI + A_1 \\ C_1 \end{bmatrix} \quad \text{and} \quad \begin{bmatrix} zI + A_2 \\ C_2 \end{bmatrix}$$

are strictly equivalent if and only if there exists an invertible matrix S of the form

(8.1.9)
$$S = \begin{bmatrix} N & F \\ 0 & D \end{bmatrix} : \mathbb{C}^n \oplus \mathbb{C}^m \to \mathbb{C}^n \oplus \mathbb{C}^m$$

such that (8.1.2) holds.

PROOF. If (8.1.2) holds, then we have

$$\begin{bmatrix} N & F \\ 0 & D \end{bmatrix} \begin{bmatrix} zI + A_1 \\ C_1 \end{bmatrix} = \begin{bmatrix} zI + A_2 \\ C_2 \end{bmatrix} N,$$

so the pencils are strictly equivalent. Conversely assume that

$$\begin{bmatrix} S_{11} & S_{12} \\ S_{21} & S_{22} \end{bmatrix} \begin{bmatrix} zI + A_1 \\ C_1 \end{bmatrix} = \begin{bmatrix} zI + A_2 \\ C_2 \end{bmatrix} T, \qquad z \in \mathbb{C}$$

for some invertible matrices $\begin{bmatrix} S_{11} & S_{12} \\ S_{21} & S_{22} \end{bmatrix}$ and T. Comparison of the coefficients of z shows that $S_{11} = T$ and $S_{21} = 0$. Consequently, S_{22} is invertible. $\square$

Observe that conditions (ii) in Theorem 8.1.1 and and the condition in Proposition 8.1.2 are identical.

By definition, if (C_1, A_1) is a right null pair and (C_2, A_2) is a right pole pair for a rational matrix function, then both (C_1, A_1) and (C_2, A_2) are null kernel pairs.

In general the conditions of Theorem 8.1.1 are not sufficient, i.e., if the conditions (i), (ii) of Theorem 8.1.1 on two null kernel pairs (C_1, A_1) and (C_2, A_2) are satisfied, then it does not necessarily follow that (C_1, A_1) is a right null pair and (C_2, A_2) is a right pole pair (with respect to $\mathbb{C}$) of a rational matrix function which is analytic and invertible at infinity. This follows from the next example.

EXAMPLE 8.1.1. Let

$$A_1 = A_2 = \begin{bmatrix} 0 & 1 & 0 \\ 1 & 0 & 0 \\ 0 & 0 & 0 \end{bmatrix}, \qquad C_1 = C_2 = \begin{bmatrix} 0 & 1 & 0 \\ 0 & 0 & 1 \end{bmatrix}.$$

Obviously (C_1, A_1) and (C_2, A_2) are null kernel pairs and the conditions (i), (ii) of Theorem 8.1.1 are fulfilled. Now suppose that (C_1, A_1) is a right null pair and (C_2, A_2) is a right pole pair (relative to $\mathbb{C}$) of rational matrix function $W(z)$ which is analytic and invertible at infinity. According to Theorem 8.1.1 this implies that $W(z)$ is of the form $W(z) = D + C_2(zI - A_2)^{-1}F$, where D and F are such that

(8.1.10)
$$\begin{bmatrix} N & F \\ 0 & D \end{bmatrix} \begin{bmatrix} A_1 \\ C_1 \end{bmatrix} = \begin{bmatrix} A_2 \\ C_2 \end{bmatrix} N$$

with D and N invertible. Now all $\begin{bmatrix} N & F \\ 0 & D \end{bmatrix}$ satisfying (8.1.10) are of the form

$$\begin{bmatrix} N & F \\ 0 & D \end{bmatrix} = \begin{bmatrix} a & 0 & c & 0 & b \\ 0 & a & b & 0 & c \\ 0 & 0 & d & 0 & 0 \\ 0 & 0 & 0 & a & b \\ 0 & 0 & 0 & 0 & d \end{bmatrix}, \qquad ad \neq 0.$$

So all possible $W(z)$'s can be computed. We obtain

$$W(z) = \begin{bmatrix} a & b + (cz + b)(z^2 - 1)^{-1} \\ 0 & d \end{bmatrix}, \qquad ad \neq 0.$$

But none of these $W(z)$ has (C_1, A_1) as a right null pair, because $W(0)$ is invertible and 0 is an eigenvalue of A_1. $\square$

According to Theorem 8.1.1 any $W(z)$ with (C_1, A_1) as a right null pair and (C_2, A_2) as a right pole pair can be expressed in terms of a matrix S satisfying (8.1.2). But, in general, even if there exists such a $W(z)$, it is not true that, conversely, each such S produces a solution $W(z)$. This follows from the next example.

EXAMPLE 8.1.2. Take

$$A_1 = A_2 = \begin{bmatrix} 0 & 1 \\ 0 & 0 \end{bmatrix}, \qquad C_1 = C_2 = \begin{bmatrix} 1 & 0 \\ 0 & 0 \end{bmatrix}.$$

All possible matrices S of the form (8.1.1) and satisfying (8.1.2) are given by

$$\begin{bmatrix} a & 0 & 0 & c \\ 0 & a & 0 & d \\ 0 & 0 & a & b \\ 0 & 0 & 0 & e \end{bmatrix},$$

where a, b, c, d and e are arbitrary complex parameters such that $ae \neq 0$. Applying Theorem 8.1.1 we conclude that any rational matrix function $W(z)$, which is analytic and invertible at infinity and has (C_1, A_1) as a right null pair and (C_2, A_2) as a right pole pair, must be of the form

(8.1.11) $$W(z) = \begin{bmatrix} a & b + cz^{-1} + dz^{-2} \\ 0 & e \end{bmatrix}, \qquad ae \neq 0.$$

For $d = 0$ the realization of $W(z)$, which one gets from the formulas (8.1.3) is not minimal, and so similarities with $d = 0$ do not lead to rational matrix functions with (C_1, A_1) as a right null pair. For $d \neq 0$ the realization is minimal and hence in that case (C_1, A_1) is a right null pair and (C_2, A_2) is a right pole pair of the function (8.1.11). $\square$

The next theorem shows that the conditions (i), (ii) of Theorem 8.1.1 are necessary and sufficient if A_1 and A_2 have no common eigenvalues.

THEOREM 8.1.3. *Suppose that A_1 and A_2 have no common eigenvalues. In order that (C_1, A_1) is a right null pair and (C_2, A_2) is a right pole pair (both with respect to $\mathbb{C}$) of a rational $m \times m$ matrix function $W(z)$, which is analytic and invertible at infinity, it is necessary and sufficient that the conditions (i) and (ii) of Theorem 8.1.1 are fulfilled. In that case the general form of $W(z)$ is*

(8.1.12) $$W(z) = D + C_2(zI - A_2)^{-1}F,$$

where $\begin{bmatrix} N & F \\ 0 & D \end{bmatrix}$ *is an arbitrary invertible matrix satisfying (8.1.2).*

PROOF. From Theorem 8.1.1 we know that the conditions (i) and (ii) are necessary and that $W(z)$ must have the form (8.1.12). It remains to show that (8.1.12) is the general form. Let $\begin{bmatrix} N & F \\ 0 & D \end{bmatrix}$ be an arbitrary invertible matrix satisfying (8.1.2), and let $W(z)$ be defined by (8.1.12). We have to show that (C_1, A_1) is a right null pair and (C_2, A_2) is a right pole pair of $W(z)$. We have

$$(8.1.13) \qquad NA_1 + FC_1 = A_2N, \qquad DC_1 = C_2N,$$

and hence $W(z)^{-1} = D^{-1} + C_1(zI - A_1)^{-1}(-N^{-1}FD^{-1})$. By Theorem 4.2.1 it suffices to show that this realization of $W(z)^{-1}$ and the realization (8.1.12) of $W(z)$ are minimal. To do this we use that, by our hypotheses, the matrices A_1 and A_2 have no common eigenvalues.

Let $\mathcal{V} = \cap_{i=0}^{\infty} \mathrm{Ker}(C_2 A_2^i)$. Then $\mathcal{V}$ is an A_2-invariant subspace. Further, $A_2 \mid \mathcal{V}$, the restriction of A_2 to $\mathcal{V}$, is equal to the restriction $A_2 - FD^{-1}C_2 \mid \mathcal{V}$, because $\mathcal{V} \subset \mathrm{Ker}\, C_2$. It follows from formula (8.1.13) that $A_2 - FD^{-1}C_2$ is similar to A_1. Therefore each eigenvalue of $A_2 \mid \mathcal{V}$ is an eigenvalue of A_1 and, of course, also of A_2. As A_2 and A_1 do not have common eigenvalues, we conclude that $\mathcal{V} = \{0\}$. With a similar reasoning one can prove that $\cap_{i=0}^{\infty} \mathrm{Ker}(F^* A_2^{*i}) = \{0\}$, i.e. that (A_2, F) is a full range pair. By Theorem 4.1.4 the realization (8.1.12) is minimal. The minimality of (8.1.13) follows now from Proposition 4.1.5. $\square$

Note that in Theorem 8.1.3 the conditions that (C_j, A_j) are null kernel pairs does not appear explicitly. The reason is that for the case when A_1 and A_2 have no common eigenvalues conditions (i) and (ii) of Theorem 8.1.1 imply that (C_j, A_j) are null kernel pairs.

The complete solution of problems 1 and 2 is given in Gohberg-Kaashoek-van Schagen [1982].

8.2 RATIONAL MATRIX FUNCTIONS WITH GIVEN RIGHT AND LEFT POLE AND NULL PAIRS

In this section we solve the problem of existence, and description of rational matrix functions (with value I at infinity) with given left and right null and poles pairs:

THEOREM 8.2.1. Let $(C_\zeta, A_\zeta), (C_\pi, A_\pi)$ (resp. $(A_\pi, B_\pi), (A_\zeta, B_\zeta)$) be null kernel (resp. full range) pairs, where the number of rows in C_ζ and in C_π, as well as the number of columns in B_π and in B_ζ is equal to n. Then there exists a rational $n \times n$ matrix function $W(z)$ with $W(\infty) = I$ for which $(C_\zeta, A_\zeta), (C_\pi, A_\pi), (A_\pi, B_\pi)$ and (A_ζ, B_ζ) are a right null pair, a right pole pair, a left pole pair, and a left null pair, respectively, with respect to $\mathbb{C}$ if and only if there exists an invertible matrix S_1 such that

$$(8.2.1) \qquad S_1 A_\zeta - A_\pi S_1 = -B_\pi C_\zeta,$$

and for some invertible matrices F and R the following equalities hold:

$$(8.2.2) \qquad FA_\zeta = A_\zeta F; \qquad RA_\pi = A_\pi R; \qquad B_\zeta = FS_1^{-1}B_\pi; \qquad C_\pi = C_\zeta S_1^{-1}R.$$

In such a case one may take

(8.2.3) $$W(z) = I + C_\zeta S_1^{-1}(zI - A_\pi)^{-1}B_\pi.$$

PROOF. Assume (8.2.1) and (8.2.2) hold. Define $W(z)$ by (8.2.3). As $C_\zeta S_1^{-1} = C_\pi R^{-1}$ and $RA_\pi = A_\pi R$ it follows from the null kernel property of (C_π, A_π) that $(C_\zeta S_1^{-1}, A_\pi)$ is a null kernel pair and by Theorem 4.1.4 the realization (8.2.3) is minimal. Now by Theorem 4.2.1 (A_π, B_π) is a global right pole pair of $W(z)$ and (C_π, A_π) (which is similar to $(C_\pi R^{-1}, RA_\pi R^{-1}) = (C_\zeta S_1^{-1}, A_\pi)$ is a global left pole pair of $W(z)$. Further,

$$W(z)^{-1} = I - C_\zeta S_1^{-1}(zI - (A_\pi - B_\pi C_\zeta S_1^{-1}))^{-1}B_\pi$$
$$= I - C_\zeta(zI - A_\zeta)^{-1}S_1^{-1}B_\pi,$$

and we show analogously that (C_ζ, A_ζ) is a global right null pair for $W(z)$, while (A_ζ, B_ζ) is its global left null pair.

Conversely, let $W(z)$ be a rational matrix function with $W(\infty) = I$ having the prescribed null and pole pairs. By Theorem 4.2.2 there is unique matrix $\widetilde{C}$ such that

(8.2.4) $$W(z) = I + \widetilde{C}(zI - A_\pi)^{-1}B_\pi$$

is a minimal realization for $W(z)$. By Theorem 4.2.1 $(\widetilde{C}, A_\pi)$ is a global right pole pair of $W(z)$ and hence is similar to (C_π, A_π): $\widetilde{C}R = C_\pi$; $RA_\pi = A_\pi R$ for some invertible R. Analogously, there is a unique $\widetilde{B}$ such that

(8.2.5) $$W(z)^{-1} = I + C_\zeta(zI - A_\zeta)^{-1}\widetilde{B}$$

is a minimal realization for $W(z)^{-1}$, and consequently $(A_\zeta, \widetilde{B})$ is similar to $(A_\zeta, -B_\zeta)$: $F\widetilde{B} = -B_\zeta$; $A_\zeta F = FA_\zeta$ for some invertible F. On the other hand, (8.2.4) leads in view of Proposition 4.1.5 to the minimal realization

$$W(z)^{-1} = I - \widetilde{C}(zI - (A_\pi - B_\pi \widetilde{C}))^{-1}B_\pi$$

which must be similar to (8.2.5):

$$\widetilde{C} = C_\zeta S_1^{-1}; \qquad A_\pi - B_\pi \widetilde{C} = S_1 A_\zeta S_1^{-1}; \qquad B_\pi = -S_1 \widetilde{B}$$

for some invertible matrix S_1. The equalities (8.2.2) and (8.2.3) are now easily verified. □

As we know (see Chapter 3) if (Z, T_1) and (T_2, Y) are right and left null (or pole) pairs of $W(z)$ (corresponding to C), then T_1 and T_2 are similar. So, indeed, we can assume in Theorem 8.2.1 that the same matrix A_ζ appears in both right and left null pairs, and the same A_π appears in both right and left pole pairs.

In general, a function $W(z)$ from Theorem 8.2.1 is not unique, as the following example shows.

EXAMPLE 8.2.1. Set

$$C_\zeta = C_\pi = \begin{bmatrix} 1 \\ \alpha \end{bmatrix}; \qquad B_\zeta = B_\pi = [-\alpha \quad 1]; \qquad A_\zeta = A_\pi = 0.$$

The conditions of Theorem 8.2.1 are satisfied with an arbitrary non-zero number S_1 and $F = R = S_1$. In this case formula (8.2.3) gives

$$W_{S_1}(z) = I + \frac{1}{S_1 z} \begin{bmatrix} -\alpha & 1 \\ -\alpha^2 & \alpha \end{bmatrix}.$$

Every function $W_{S_1}(z)$ (independently of S_1) has $\left(\begin{bmatrix} 1 \\ \alpha \end{bmatrix}, 0 \right)$, $\left(\begin{bmatrix} 1 \\ \alpha \end{bmatrix}, 0 \right)$, $(0, [-\alpha \quad 1])$ and $(0, [-\alpha \quad 1])$ as its right null pair, right pole pair, left pole pair and left null pair, respectively. □

One obvious case when there is at most one rational matrix function $W(z)$ with value I at infinity with given pairs (C_ζ, A_ζ), (C_π, A_π), (A_π, B_π), (A_ζ, B_ζ) as in Theorem 8.2.1 appears when $\sigma(A_\zeta) \cap \sigma(A_\pi) = \emptyset$.

CHAPTER 9

INTERPOLATION PROBLEMS FOR RATIONAL MATRIX FUNCTIONS BASED ON DIVISIBILITY

In this chapter we solve a problem of interpolation in which the data are given in the form of a number of rational matrix functions $W_1, \ldots, W_k$ which should serve as divisors of the function to be determined. We assume that the poles and zeros of the given divisors are in disjoint regions $\sigma_1, \sigma_2, \ldots, \sigma_k$ of the complex plane, and we require that the interpolant W should have the property that for each j the quotient WW_j^{-1} should have no zeros or poles in σ_j. This is a special case of the more general problem of finding a rational matrix function W which has each of the given functions $W_1, \ldots, W_k$ as minimal right divisors.

9.1 STATEMENT OF THE PROBLEM AND THE MAIN THEOREM

Consider the $n \times n$ rational matrix functions $W_1(z), \ldots, W_k(z)$ with value I at infinity, and assume that $W_i(z)$ and $W_j(z)$ (for $i \neq j$) have no common zeros and/or poles; in other words,

$$(9.1.1) \qquad \sigma(W_i) \bigcap \sigma(W_j) = \emptyset \qquad (i \neq j),$$

where we denote by $\sigma(W)$ the union of the set of zeros of W and the set of poles of W. We address in this chapter the following problem: Find an $n \times n$ rational matrix function $W(z)$ with value I at infinity such that

$$(9.1.2) \qquad W(z) = W_j(z)V_j(z), \qquad j = 1, \ldots, k,$$

where $V_1(z), \ldots, V_k(z)$ are rational matrix functions with the property that

$$(9.1.3) \qquad \sigma(V_j) \bigcap \sigma(W_j) = \emptyset \qquad \text{for} \qquad j = 1, \ldots, k.$$

This problem can be understood as an interpolation problem. Indeed, the functions $W_1(z), \ldots, W_k(z)$ can be thought of as representing the local data (more precisely, the left local data) of the unknown function $W(z)$ in a neighborhood of $\sigma(W_1), \ldots, \sigma(W_k)$, respectively.

We state now the main theorem of this chapter.

THEOREM 9.1.1. *Given $n \times n$ rational matrix functions $W_1, \ldots, W_k$ with value I at infinity and with the property (9.1.1), there is a rational matrix function W with value I at infinity such that*

$$\sigma(W_j^{-1}W) \bigcap \sigma(W_j) = \emptyset, \qquad j = 1, \ldots, k.$$

Moreover, for any given $z_0 \notin \cup_{i=1}^{k} \sigma(W_i)$ there exists $W(z)$ with the additional property that

$$\sigma(W) \subset \{z_0\} \bigcup \bigcup_{i=1}^{k} \sigma(W_i).$$

The proof of this theorem will be given in the next section.

The hypothesis that all the functions involved have value I at infinity is not essential and can be thought of as a convenient normalization. Indeed, the methods of Chapter 5 allow one to reduce the general case of $n \times n$ rational matrix functions with determinant not identically zero to the case considered in Theorem 9.1.1.

By passing in Theorem 9.1.1 to the transposed matrix functions a dual theorem is obtained.

THEOREM 9.1.2. *Let $W_1, \ldots, W_k$ be as in Theorem 9.1.1. Then for every $z_0 \notin \cup_{i=1}^{k} \sigma(W_i)$ there exists a rational matrix function W with $W(\infty) = I$ and such that*

$$\sigma(WW_j^{-1}) \bigcap \sigma(W_j) = \emptyset, \qquad j = 1, \ldots, k$$

and

$$\sigma(W) \subset \{z_0\} \bigcup \bigcup_{i=1}^{k} \sigma(W_i).$$

Finally, observe that the interpolation problem solved in this chapter is a particular case of a more general interpolation problem involving the notion of minimal factorization. Given three regular $n \times n$ rational matrix functions W_1, W_2 and W_3 such that

(9.1.4) $W_1(z) = W_2(z)W_3(z)$

for all z which are not poles of $W_2(z)$ and $W_3(z)$, the factorization (9.1.4) of $W_1(z)$ is called *minimal* at z_0 if the sum of the zero multiplicities of W_2 and W_3 at z_0 is equal to the sum of the zero multiplicities of W_1 at z_0. An equivalent definition is obtained if "zero multiplicities" is replaced by "pole multiplicities" in the above definition. With obvious modifications the definition of a minimal factorization can be applied also when $z_0 = \infty$. As minimal factorizations of rational matrix functions will appear only sporadically in this book, we refer the reader to Bart-Gohberg-Kaashoek [1979] or Gohberg-Lancaster-Rodman [1986] for more information on the minimal factorization.

We formulate now the more general interpolation problem. Given a set $\sigma \subset$ C, and given regular $n \times n$ rational matrix functions $W_1, \ldots, W_k$ find, if possible, an $n \times n$ rational matrix function W such that all factorizations $W = W_i(W_i^{-1}W)$, $i = 1, \ldots, k$ are minimal at every point $z_0 \in \sigma$. Observe that we do not assume that the sets $\sigma(W_1), \ldots, \sigma(W_k)$ are all disjoint (as in Theorem 9.1.1). In contrast with Theorem 9.1.1, a solution W to this interpolation problem does not always exist. Full solution of this problem (in terms of the null-pole triples of $W_1, \ldots, W_k$ with respect to σ), as well as some conditions which guarantee existence of a solution, are given in Ball-Gohberg-Rodman [1989e].

9.2 PROOF OF THEOREM 9.1.1

Let $W_1, \ldots, W_k$ be as in Theorem 9.1.1. Let

$$W_j(z) = I + C_j(zI - A_j)^{-1} B_j$$

be a minimal realization for W_j; $j = 1, \ldots, k$ (the matrices $A_1, \ldots, A_k$ need not be of the same size). Define the matrices

$$C = [C_1 C_2 \cdots C_k];$$

$$A = \operatorname{diag}[A_1, A_2, \ldots, A_k];$$

$$B = \begin{bmatrix} B_1 \\ B_2 \\ \vdots \\ B_k \end{bmatrix}; \qquad A^\times = \operatorname{diag}[A_1^\times, A_2^\times, \ldots, A_k^\times],$$

where $A_j^\times = A_j - B_j C_j$. It follows from Theorem 4.2.1 that the eigenvalues of A_j coincide with the poles of W_j, while the eigenvalues of $A_j^\times$ coincide with the zeros of W_j. The hypotheses of Theorem 9.1.1 ensure that

$$\sigma(A_j) \bigcap \sigma(A_k^\times) = \emptyset \qquad \text{for} \qquad j \neq k.$$

Hence there exists unique Γ_{ij} satisfying the equation

$$\Gamma_{ij} A_j - A_i^\times \Gamma_{ij} = B_i C_j \qquad (i \neq j).$$

Form the matrix

(9.2.1)
$$\Gamma = \begin{bmatrix} I & \Gamma_{12} & \cdots & \Gamma_{1k} \\ \Gamma_{21} & I & \cdots & \Gamma_{2k} \\ \vdots & \vdots & & \vdots \\ \Gamma_{k1} & \Gamma_{k2} & \cdots & I \end{bmatrix}.$$

Then

$$\Gamma A - A^\times \Gamma = BC.$$

Furthermore, $\sigma(A_i) \cap \sigma(A_j) = \emptyset$ for $i \neq j$, and since each pair (C_i, A_i) is a null-kernel pair, it follows that (C, A) is a null-kernel pair as well (see the proof of Theorem 1.5.1). Analogously, $(B, A^\times)$ is a full range pair. We conclude that

(9.2.2)
$$(C, A; A^\times, B; \Gamma)$$

is an admissible Sylvester data set. The application of Theorem 4.5.1 (with $\sigma = \cup_{i=1}^k \sigma(W_i)$) yields existence of a rational matrix function $W(z)$ with $W(\infty) = I$ whose left σ-null-pole triple is precisely (9.2.2). Moreover, by Theorem 4.6.1 for any given $z_0 \notin \cup_{i=1}^k \sigma(W_i)$, $W(z)$ can be chosen so that

$$\sigma(W) \subset \{z_0\} \bigcup \bigcup_{i=1}^k \sigma(W_i)$$

and the left $\sigma(W_i)$-null-pole triple of W is exactly $(C_i, A_i; A_i^\times, B_i; I)$. Now by Theorem 4.5.8 all the poles and zeros of $W_j^{-1} W$ are in $\left(\cup_{i \neq j} \sigma(W_i) \right) \cup \{z_0\}$, and hence

$$\sigma(W_j^{-1} W) \bigcap \sigma(W_j) = \emptyset,$$

as required.

9.3 A GENERALIZATION

In this section we present a natural generalization of Theorem 9.1.1 to the situation when the rational matrix functions are considered over a fixed subset of the complex plane (as opposed to Theorem 9.1.1 where all of the complex plane was under the consideration). A similar generalization of Theorem 9.1.2 is valid as well; however, we will not state it explicitly.

Let σ be a non-empty subset of $\mathbf{C}$.

THEOREM 9.3.1. *Let* $W_1(z), \ldots, W_k(z)$ *be regular* $n \times n$ *rational matrix functions with the property that*

$$(\sigma(W_i) \bigcap \sigma) \bigcap (\sigma(W_j) \bigcap \sigma) = \emptyset \qquad (i \neq j).$$

Then there is a regular rational matrix function $W(z)$ *such that*

$$(\sigma(W_j^{-1} W) \bigcap \sigma) \bigcap (\sigma(W_j) \bigcap \sigma) = \emptyset, \qquad j = 1, \ldots, k$$

(*in other words,*

$$W = W_j V_j,$$

where V_j *has no common poles and zeros with* W_j *in* σ). *Moreover, assuming* $\sigma \neq \mathbf{C}$, *for every* $z_0 \in \mathbf{C} \backslash \sigma$ *there is such* $W(z)$ *with the additional property that* $W(\infty) = I$ *and*

$$\sigma(W) \subset \{z_0\} \bigcup \bigcup_{i=1}^{k} \sigma(W_i).$$

PROOF. Let $(C_i, A_j; A_j^\times, B_j; \Gamma_j)$ be the left σ-null-pole triple of W_j ($j = 1, \ldots, k$). Repeat the arguments used in the proof of Theorem 9.1.1, replacing I in the block (j, j)-position in (9.2.1) with Γ_j. $\square$

CHAPTER 10

POLYNOMIAL INTERPOLATION PROBLEMS
BASED ON DIVISIBILITY

In this chapter we solve the interpolation problem which consists in finding a matrix polynomial having each of a given set of matrix polynomials as a right polynomial divisor. This problem is a polynomial analogue of the problem solved in the previous chapter, and coincides with the problem of finding a common multiple of a given collection of polynomials. The complication arising in the matrix case is mostly due to the noncommutativity of matrix multiplication.

We note that the main result on divisibility of matrix polynomials (Theorem 10.1.1) is a particular case of more general results described in Part III. However, the exposition in this chapter is on a more elementary level so that the results are accessible after reading Chapter 2.

10.1 DIVISIBILITY OF MATRIX POLYNOMIALS

Let $L(z) = \sum_{i=0}^{\ell} z^i L_i$ be a matrix polynomial whose coefficients are $n \times n$ matrices; we assume as usual that all the matrix polynomials involved are *regular*, i.e. with determinants not identically zero. Recall from Chapter 2 the notion of a right null pair (C_ζ, A_ζ) of the matrix polynomial $L(z)$ with respect to the whole complex plane. An $n \times n$ matrix polynomial $L_1(z)$ will be called a *right divisor* of an $n \times n$ matrix polynomial $L(z)$ if $L(z) = V(z)L_1(z)$ for some (necessarily regular) matrix polynomial $V(z)$. We shall express the right divisibility of matrix polynomials in terms of their right null pairs.

To this end we need the notion of restrictions and extensions of right null pairs of matrix polynomials. We put these notions in a more general framework of right admissible pairs. A pair of matrices (C, A) will be called *right admissible* if C is $n \times p$ and A is $p \times p$; in this case n will be called the *base dimension* of (C, A) and p will be called the *order* of (C, A). The most important examples of right admissible pairs in this chapter are right null pairs for matrix polynomials (these are, in addition to being admissible pairs, also null kernel pairs (Theorem 2.1.3)). Given two right admissible pairs (C, A) and (C_1, A_1) with the same base dimensions, we say that (C_1, A_1) is a *restriction* of (C, A), or, equivalently, (C, A) is an *extension* of (C_1, A_1) if $C_1 = CS$; $AS = SA_1$ for some matrix S with zero kernel (the size of S is $p \times p_1$, where p (resp. p_1) is the order of (C, A) (resp. of (C_1, A_1)); a necessary condition for existence of such an S with zero kernel is $p \geq p_1$).

We state now the main result of this section.

THEOREM 10.1.1. *Let $B(z)$ and $L(z)$ be $n \times n$ matrix polynomials with right null pairs $(C_{\zeta B}, A_{\zeta B})$ and $(C_{\zeta L}, A_{\zeta L})$, respectively (relative to $\mathbb{C}$). Then $L(z)$ is a right divisor of $B(z)$ if and only if $(C_{\zeta B}, A_{\zeta B})$ is an extension of $(C_{\zeta L}, A_{\zeta L})$.*

Recalling the definition of a right null pair for a matrix polynomial, Theorem 10.1.1 can be restated as follows:

COROLLARY 10.1.2. *An $n \times n$ matrix polynomial $L(z)$ is a right divisor of an $n \times n$ matrix polynomial $B(z)$ if and only if every right null chain of $L(z)$ (corresponding to a zero z_0) is also a right null chain of $B(z)$ corresponding to the same zero z_0.*

Two right admissible pairs (C, A) and (C_1, A_1) with the same base dimensions are called *similar* if each is a restriction of the other, or, equivalently, if $C_1 = CS$, $AS = SA_1$ for some invertible matrix S. The importance of this notion in the divisibility theory for matrix polynomials is apparent from the following corollary (actually, a particular case) of Theorem 10.1.1; it can also be seen as the specialization of Theorem 4.5.8 to polynomials with $\sigma = \mathbb{C}$.

COROLLARY 10.1.3. *Two (regular) $n \times n$ matrix polynomials $B(z)$ and $L(z)$ are left equivalent (i.e. $L(z) = S(z)B(z)$ for some matrix polynomial $S(z)$ with constant non-zero determinant) if and only if their right null pairs relative to $\mathbb{C}$ are similar.*

Indeed, the proof of Corollary 10.1.3 is obtained at once from Theorem 10.1.1 using the easily verified fact that two (regular) matrix polynomials are left equivalent if and only if each is a right divisor of the other.

The rest of this section will be geared towards the proof of Theorem 10.1.1.

We start with another description of divisibility of matrix polynomials, given by the following proposition.

PROPOSITION 10.1.4. *Let $B(z)$ and $L(z)$ be (regular) $n \times n$ matrix polynomials. Assume*

$$B(z) = \sum_{j=0}^{m} B_j z^j; \qquad B_m \neq 0,$$

and let (C_ζ, A_ζ) be a right null pair of $L(z)$ (relative to $\mathbb{C}$). Then $L(z)$ is a right divisor of $B(z)$ if and only if

$$\sum_{j=0}^{m} B_j C_\zeta A_\zeta^j = 0.$$

We shall need some preparation for the proof of Proposition 10.1.4.

First, without loss of generality we can (and will) assume that $L(z)$ is comonic, i.e., $L(0) = I$. Indeed, a matrix polynomial $L(z)$ is a right divisor of a matrix polynomial $B(z)$ if and only if the comonic polynomial $L^{-1}(\alpha)L(z + \alpha)$ is a right divisor of the comonic polynomial $B^{-1}(\alpha)B(z + \alpha)$. Here $\alpha \in \mathbb{C}$ is such that both $L(\alpha)$ and $B(\alpha)$ are nonsingular; existence of such an α is ensured by the regularity of both $L(z)$ and $B(z)$. Observe also that (C_ζ, A_ζ) is a right null pair of $L(z)$ (relative to $\mathbb{C}$) if and only if $(C_\zeta, A_\zeta + \alpha I)$ is a right null pair of $L^{-1}(\alpha)L(z + \alpha)$, see Proposition 2.5.2.

Second, we describe the process of division of $B(z)$ by $L(z)$ (with the assumption that $L(z)$ is comonic). Let ℓ be the degree of $L(z)$, and let (C_∞, A_∞) be a supplementary right null pair of $L(z)$. Put $C = [\, C_\zeta \quad C_\infty \,]$; $A = A_\zeta^{-1} \oplus A_\infty$. Then (see the proof of Theorem 2.5.3) (C, A) is a right null pair for the monic matrix polynomial $\tilde{L}(z) = z^\ell L(z^{-1})$. In particular $\mathrm{col}(CA^{i-1})_{i=1}^\ell$ is nonsingular. For $\alpha \geq 0$ and $1 \leq \beta \leq \ell$ set $F_{\alpha\beta} = CA^\alpha Z_\beta$, where

$$(10.1.1) \qquad [Z_1 Z_2 \cdots Z_\ell] = [\mathrm{col}(CA^{i-1})_{i=1}^\ell]^{-1}.$$

Further, for each $\alpha \geq 0$ and $\beta \leq 0$ or $\beta > \ell$ put $F_{\alpha\beta} = 0$. With this choice of $F_{\alpha\beta}$ the following formulas hold:

$$(10.1.2) \qquad L(z) = I - \sum_{j=1}^\infty z^j F_{\ell, \ell+1-j},$$

and

$$(10.1.3) \qquad F_{\alpha+1, \beta} = F_{\alpha\ell} F_{\ell\beta} + F_{\alpha, \beta-1}.$$

Indeed, formula (10.1.2) is an immediate consequence from Theorem 2.3.1 applied to $\tilde{L}(z)$. To prove (10.1.3) we argue as follows. From (10.1.1) we have

$$(10.1.4) \qquad I = [Z_1 \cdots Z_\ell] \begin{bmatrix} C \\ CA \\ \vdots \\ CA^{\ell-1} \end{bmatrix}.$$

Consider the composition with CA^α on the left and $A[Z_1 \cdots Z_\ell]$ on the right to obtain

$$[F_{\alpha+1,1} \cdots F_{\alpha+1,\ell}] = [F_{\alpha 1} \cdots F_{\alpha\ell}] \begin{bmatrix} F_{11} & \cdots & F_{1\ell} \\ \vdots & & \vdots \\ F_{\ell 1} & \cdots & F_{\ell\ell} \end{bmatrix}.$$

Further, upon rewriting (10.1.4) in the form

$$\mathrm{col}(CA^i)_{i=0}^{\ell-1} \cdot [Z_1 \cdots Z_\ell] = I$$

we see that $F_{ij} = \delta_{i,j-1} I$ for $i = 0, \ldots, \ell - 1$ and $j = 1, \ldots, \ell$. Hence (10.1.3) follows.

The process of division of $B(z) = \sum_{j=0}^m B_j z^j$ by $L(z)$ can now be described as follows:

$$(10.1.5) \qquad B(z) = Q_k(z) L(z) + R_k(z), \qquad k = 1, 2, \ldots,$$

where

$$Q_k(z) = \sum_{j=0}^{k-1} z^j \left(B_j + \sum_{i=0}^{j-1} B_i F_{\ell+j-1-i, \ell} \right),$$

$$(10.1.6) \qquad R_k(z) = \sum_{j=k}^{\infty} z^j \left(B_j + \sum_{i=0}^{k-1} B_i F_{\ell+k-1-i,\ell+k-j} \right),$$

and $B_j = 0$ for $j > m$. We verify (10.1.5) using induction on k. For $k = 1$ (10.1.5) is trivial. Assume that (10.1.5) holds for some $k \geq 1$. Then using the recursion (10.1.3), one sees that

$$B(z) - Q_{k+1}(z)L(z) = R_k(z) - z^k \left(B_k + \sum_{i=0}^{k-1} B_i F_{\ell+k-1-i,\ell} \right) L(z)$$

$$= \sum_{j=k+1}^{\infty} z^j \left(B_j + \sum_{i=0}^{k-1} B_i F_{\ell+k-1-i,\ell+k-j} \right)$$

$$+ z^k \left(B_k + \sum_{i=0}^{k-1} B_i F_{\ell+k-1-i,\ell} \right)$$

$$- z^k \left(B_k + \sum_{i=0}^{k-1} B_i F_{\ell+k-1-i,\ell} \right) \left(I - \sum_{j=1}^{\infty} z^j F_{\ell,\ell+1-j} \right)$$

$$= \sum_{j=k+1}^{\infty} z^j \left(B_j + \sum_{i=0}^{k-1} B_i F_{\ell+k-1-i,\ell+k-j} \right)$$

$$+ \sum_{j=k+1}^{\infty} z^j \left(B_k + \sum_{i=0}^{k-1} B_i F_{\ell+k-1-i,\ell} \right) F_{\ell,\ell+k+1-j}$$

$$= \sum_{j=k+1}^{\infty} z^j \left(B_j + B_k F_{\ell,\ell+k+1-j} + \sum_{i=0}^{k-1} B_i F_{\ell+k-i,\ell+k+1-j} \right)$$

$$= R_{k+1}(z),$$

and the induction is complete.

It is not difficult to see that $L(z)$ is a right divisor of $B(z)$ if and only if $R_i(z) \equiv 0$ for some k.

PROOF OF PROPOSITION 10.1.4. Using the definition of $F_{\alpha\beta}$ and the property $C A^{i-1} Z_j = \delta_{ij} I$ $(1 \leq i, j \leq \ell)$, rewrite (10.1.6) in the form

$$R_k(z) = \sum_{j=k+\ell}^{\infty} B_j z^j + \sum_{i=0}^{\ell+k-1} B_i C A^{\ell+k-1-i} \left(\sum_{\beta=1}^{\ell} Z_\beta z^{\ell+k-\beta} \right).$$

It follows that $R_k(z) = 0$ if and only if

$$(10.1.7) \qquad \ell + k > m, \qquad \sum_{i=0}^{m} B_i C A^{\ell+k-1-i} = 0$$

(we assume that the leading coefficient B_m of $B(z)$ is non-zero). Now suppose that $L(z)$ is a right divisor of $B(z)$. Then there exists $k \geq 1$ such that (10.1.7) holds. Because of

$C = [\, C_\zeta \quad C_\infty \,]$ and $A = A_\zeta^{-1} \oplus A_\infty$ this implies

$$\sum_{i=0}^{m} B_i C_\zeta A_\zeta^{-\ell-k+1+i} = 0.$$

Multiplying the left- and right-hand sides of this identity by $A_\zeta^{\ell+k-1}$ yields the desired formula

(10.1.8) $$\sum_{j=0}^{m} B_j C_\zeta A_\zeta^{j} = 0.$$

Conversely, suppose that (10.1.8) holds. Let ν be a positive integer such that $A_\infty^\nu = 0$. Choose $k \geq 1$ such that $\ell + k \geq m + \nu$. Multiply the left- and right-hand sides of (10.1.8) by $A_\zeta^{-\ell-k+1}$. This gives

$$\sum_{i=0}^{m} B_i C_\zeta A_\zeta^{-(\ell+k-1-i)} = 0.$$

As $C_\infty A_\infty^{\ell+k-1-i} = 0$ for $i = 0, \ldots, m$, we see that with this choice of k, formula (10.1.7) obtains. Hence $L(z)$ is a right divisor of $B(z)$.   □

Our next step towards the proof of Theorem 10.1.1 is an alternative description of extensions of right admissible pairs.

LEMMA 10.1.5. *Let (C, A) and (C_1, A_1) be right admissible pairs. If (C, A) is an extension of (C_1, A_1), then*

(10.1.9) $\mathrm{Im}[\mathrm{col}(C_1 A_1^{i-1})_{i=1}^{m}] \subset \mathrm{Im}[\mathrm{col}(C A^{i-1})_{i=1}^{m}], \qquad m \geq 1.$

Conversely, if (C_1, A_1) is a null kernel pair and (10.1.9) holds for all $m \geq 1$, then (C, A) is an extension of (C_1, A_1).

PROOF. The first part of the lemma follows easily from the definition of extension. To prove the second part, assume that (C, A) is a null kernel pair as well, and choose k so that

$$\mathrm{Ker}[\mathrm{col}(C A^{i-1})_{i=1}^{k}] = \{0\}, \qquad \mathrm{Ker}[\mathrm{col}(C_1 A_1^{i-1})_{i=1}^{k}] = \{0\}.$$

Put $\Omega = \mathrm{col}(C A^{i-1})_{i=1}^{k}$ and $\Omega_1 = \mathrm{col}(C_1 A_1^{i-1})_{i=1}^{k}$, and let $p \times p$ and $p_1 \times p_1$ be the sizes of A and A_1, respectively. From our hypothesis it follows that $\mathrm{Im}\,\Omega_1 \subset \mathrm{Im}\,\Omega$. Hence, there exists a linear transformation $S \colon \mathbb{C}^{p_1} \to \mathbb{C}^{p}$ such that $\Omega S = \Omega_1$. As $\mathrm{Ker}\,\Omega_1 = \{0\}$, we see that S is injective (i.e. with kernel zero). Further, from $\mathrm{Ker}\,\Omega = \{0\}$ it follows that S is uniquely determined. Note that $\Omega S = \Omega_1$ implies that $CS = C_1$. To prove that $AS = SA_1$, let us consider

$$\widetilde{\Omega} = \mathrm{col}(C A^{i-1})_{i=1}^{k+1}, \qquad \widetilde{\Omega}_1 = \mathrm{col}(C_1 A_1^{i-1})_{i=1}^{k+1}.$$

As before, there exists a linear transformation $\widetilde{S} \colon \mathbb{C}^{p_1} \to \mathbb{C}^{p}$ such that $\widetilde{\Omega}\widetilde{S} = \widetilde{\Omega}_1$. The last equality implies $\Omega\widetilde{S} = \Omega_1$, and thus $S = \widetilde{S}$. But then $\Omega A S = \Omega A \widetilde{S} = \Omega_1 A_1$. It follows that $\Omega(AS - SA_1) = 0$. As $\operatorname{Ker} \Omega = \{0\}$, we have $AS = SA_1$. This completes the proof of the lemma in the case when (C, A) is a null kernel pair.

If (C, A) happens not to be a null kernel pair, then repeat the previous argument with (C, A) replaced by $(C|_{\mathcal{M}}, PA|_{\mathcal{M}})$, where the subspace $\mathcal{M} \subset \mathbb{C}^p$ is a direct complement to $\mathcal{N} := \bigcap_{i \geq 0} \operatorname{Ker}(CA^i)$, and $P \colon \mathbb{C}^p \to \mathcal{M}$ is the projection on $\mathcal{M}$ along $\mathcal{N}$. To check the validity of this procedure, observe that $(C|_{\mathcal{M}}, PA|_{\mathcal{M}})$ is a null kernel right admissible pair, that

$$\operatorname{col}(CA^{i-1})_{i=1}^m = \operatorname{col}\big(C|_{\mathcal{M}}(PA|_{\mathcal{M}})^{i-1}\big)_{i=0}^m$$

for $m = 1, 2, \ldots$ (here we need the inclusion $\operatorname{Ker} C \supset \mathcal{N}$) and that (C, A) is an extension of $(C|_{\mathcal{M}}, PA|_{\mathcal{M}})$. The lemma is proved completely. $\square$

PROOF OF THEOREM 10.1.1. Suppose that $(C_{\zeta B}, A_{\zeta B})$ is an extension of $(C_{\zeta L}, A_{\zeta L})$, i.e. $C_{\zeta B}S = C_{\zeta L}$; $A_{\zeta B}S = SA_{\zeta L}$ for some injective linear transformation S. Then $C_{\zeta B}A_{\zeta B}^jS = C_{\zeta L}A_{\zeta L}^j$ for $j = 0, 1, \ldots$. Now write $B(z) = \sum_{j=0}^m B_j z^j$. Then $\sum_{j=0}^m B_j C_{\zeta B} A_{\zeta B}^j = 0$ by Theorem 2.1.3, and hence

$$\sum_{j=0}^m B_j C_{\zeta L} A_{\zeta L}^j = \left(\sum_{j=0}^m B_j C_{\zeta B} A_{\zeta B}^j\right)S = 0.$$

By Proposition 10.1.4 $L(z)$ is a right divisor of $B(z)$.

Conversely, assume that $L(z)$ is a right divisor of $B(z)$. We have to prove that $(C_{\zeta B}, A_{\zeta B})$ is an extension of $(C_{\zeta L}, A_{\zeta L})$. Without loss of generality we assume that both $A_{\zeta B}$ and $A_{\zeta L}$ are in the Jordan normal form. As $L(z)$ is a right divisor of $B(z)$, it follows easily (cf. the proof of Lemma 1.1.1) that every right null function of $L(z)$ is also a right null function for $B(z)$ (corresponding to the same z_0). Passing to the language of null chains, we obtain that every right null chain of $L(z)$ is also a right null chain of $B(z)$ (corresponding to the same z_0). In particular, this observation applies to the canonical sets of right null chains of $L(z)$, for each zero z_0, that appears as columns in the matrix $C_{\zeta L}$ (at this point we use the fact that $A_{\zeta L}$ is in the Jordan form). Combining this with Lemma 1.2.3, the following inclusions are obtained:

$$\operatorname{Im}[\operatorname{col}(C_{\zeta L} A_{\zeta L}^{i-1})_{i=1}^m] \subset \operatorname{Im}[\operatorname{col}(C_{\zeta B} A_{\zeta B}^{i-1})_{i=1}^m],$$

$m = 1, 2, \ldots$. As both $(C_{\zeta L}, A_{\zeta L})$ and $(C_{\zeta B}, A_{\zeta B})$ are null kernel pairs, by Lemma 10.1.5 we finish the proof. $\square$

We prove now a local version of Theorem 10.1.1:

THEOREM 10.1.6. Let $B(z)$ and $L(z)$ be regular $n \times n$ matrix polynomial with right null pairs $(C_{\zeta B}, A_{\zeta B})$ and $(C_{\zeta L}, A_{\zeta L})$, respectively, relative to a set σ in the complex plane. Then the function $B(z)L(z)^{-1}$ is analytic in σ if and only if $(C_{\zeta L}, A_{\zeta L})$ is a restriction of $(C_{\zeta B}, A_{\zeta B})$.

PROOF. By Theorem 2.5.4, there is a matrix polynomial $M(z)$ for which $(C_{\zeta L}, A_{\zeta L})$ is a right null pair with respect to $\mathbb{C}$. As $(C_{\zeta L}, A_{\zeta L})$ is obviously a restriction of a right null pair of $L(z)$ with respect to $\mathbb{C}$, we have by Theorem 10.1.1 that $M(z)$ is a right divisor of $L(z)$. Moreover, by taking determinants in the equality $L(z) = N(z)M(z)$ and by using the fact that the size of $A_{\zeta L}$ coincides with the degree of $\det M(z)$ as well as with the sum of multiplicities of all zeros of $\det L(z)$ in σ, it follows that $\det N(z) \neq 0$ for all $z \in \sigma$, i.e. $N(z)$ has no zeros in σ. Now clearly $B(z)L(z)^{-1}$ is analytic in σ if and only if $B(z)M(z)^{-1}$ is, and one more application of Theorem 10.1.1 finishes the proof. $\square$

Clearly all the results of this section admit left analogues. We shall state one of them (leaving out the proof) — the left analogue of Theorem 10.1.6. To this end we need the dual notion of left admissible pairs. A pair of matrices (A, B) will be called *left admissible* with *base dimension* n and *order* p if A is $p \times p$ and B is $p \times n$. Given two left admissible pairs (A, B) and (A_1, B_1) with the same base dimensions and orders p and p_1, respectively, we say that (A_1, B_1) is a *corestriction* of (A, B), or (A, B) is a *coextension* of (A_1, B_1), if there exists a surjective (i.e. right invertible) linear transformation $T: \mathbb{C}^p \to \mathbb{C}^{p_1}$ such that $B_1 = TB$; $A_1 T = TA$.

THEOREM 10.1.7. *Let $M(z)$ and $L(z)$ be regular $n \times n$ matrix polynomials with left null pairs $(A_{\zeta M}, B_{\zeta M})$ and $(A_{\zeta L}, B_{\zeta L})$ respectively, corresponding to a set σ in the complex plane. Then $L(z)$ is a left divisor of $M(z)$ with respect to σ (in other words, the rational matrix function $L(z)^{-1}M(z)$ is analytic in σ) if and only if $(A_{\zeta L}, B_{\zeta L})$ is a corestriction of $(A_{\zeta M}, B_{\zeta M})$.*

10.2 LEAST COMMON MULTIPLES

Let $L_1, \ldots, L_r$ be $n \times n$ regular matrix polynomials. A regular $n \times n$ matrix polynomial M will be called a *left common multiple* of $L_1, \ldots, L_r$ if the function $ML_1^{-1}, \ldots, ML_r^{-1}$ are all polynomials. A common left multiple M of $L_1, \ldots, L_r$ is called a *left least common multiple (l.c.m.)* of $L_1, \ldots, L_r$ if any other left common multiple N of $L_1, \ldots, L_r$ is itself a left multiple of M: $N = QM$ for some matrix polynomial Q. Analogously right common multiple and right l.c.m. of $L_1, \ldots, L_r$ are defined. We shall focus on the left common multiples only leaving out the analogous statements and their proofs concerning right common multiples.

The exposition will be based on the properties of right null pairs for the polynomials involved. We start however with some general properties of admissible pairs and their common extensions. Let $(X_1, T_1), \ldots, (X_r, T_r)$ be right admissible pairs with the same base dimension. The right admissible pair (X, T) is said to be a *common extension* of $(X_1, T_1), \ldots, (X_r, T_r)$ if (X, T) is an extension of each (X_j, T_j), $j = 1, \ldots, r$. We call (X_0, T_0) a *least common extension* of $(X_1, T_1), \ldots, (X_r, T_r)$ if any common extension of $(X_1, T_1), \ldots, (X_r, T_r)$ is an extension of (X_0, T_0).

THEOREM 10.2.1. *Let $(X_1, T_1), \ldots, (X_r, T_r)$ be null kernel right admissible pairs with the same base dimension and of orders $p_1, \ldots, p_r$, respectively. Then up to similarity there exists a unique least common extension of $(X_1, T_1), \ldots, (X_r, T_r)$.*

Put $X = [X_1, \ldots, X_r]$, $T = T_1 \oplus \cdots \oplus T_r$ and $p = p_1 + \cdots + p_r$, and let P be a projection of $\mathbb{C}^p$ along $\mathcal{M} = \bigcap_{i \geq 0} \operatorname{Ker}(XT^i)$. Then one such least common extension *is given by*

(10.2.1) $$(X|_{Im\,P}, PT|_{Im\,P}).$$

PROOF. Let X_0 be the first term in (10.2.1) and T_0 the second. By choosing some basis in $Im\,P$ we can understand (X_0, T_0) as a right admissible pair of matrices. As $\mathcal{M}$ is invariant under T and $Xu = 0$ for $u \in \mathcal{M}$ one sees that (X_0, T_0) is a null kernel pair, and

(10.2.2) $$\mathrm{Im}[\mathrm{col}(X_0 T_0^{i-1})_{i=1}^m] = \mathrm{Im}[\mathrm{col}(X T^{i-1})_{i=1}^m], \qquad m \geq 1.$$

From the definitions of X and T it is clear that

(10.2.3)
$$\mathrm{Im}[\mathrm{col}(X T^{i-1})_{i=1}^m] = \mathrm{Im}[\mathrm{col}(X_1 T_1^{i-1})_{i=1}^m]$$
$$+ \cdots + \mathrm{Im}[\mathrm{col}(X_r T_r^{i-1})_{i=1}^m], \qquad m \geq 1.$$

But then we can apply Lemma 10.1.5 to show that (X_0, T_0) is a common extension of $(X_1, T_1), \ldots, (X_r, T_r)$.

Now assume that (Y, R) is a common extension of $(X_1, T_1), \ldots, (X_r, T_r)$. Then (cf. Lemma 10.1.5) for $1 \leq j \leq r$ we have

$$\mathrm{Im}[\mathrm{col}(Y R^{i-1})_{i=1}^m] \supset \mathrm{Im}[\mathrm{col}(X_j T_j^{i-1})_{i=1}^m], \qquad m \geq 1.$$

Together with (10.2.2) and (10.2.3) this implies that

$$\mathrm{Im}[\mathrm{col}(Y R^{i-1})_{i=1}^m] \supset \mathrm{Im}[\mathrm{col}(X_0 T_0^{i-1})_{i=1}^m], \qquad m \geq 1.$$

But then we can apply the second part of Lemma 10.1.5 to show that (Y, R) is an extension of (X_0, T_0). It follows that (X_0, T_0) is a least common extension of $(X_1, T_1), \ldots, (X_r, T_r)$.

Let $(\tilde{X}_0, \tilde{T}_0)$ be another common least extension of $(X_1, T_1), \cdots, (X_r, T_r)$. Then $(\tilde{X}_0, \tilde{T}_0)$ is an extension of (X_0, T_0) and conversely (X_0, T_0) is an extension of $(\tilde{X}_0, \tilde{T}_0)$. Hence both pairs are similar. $\square$

It follows from the proof of Theorem 10.2.1 that a least common extension (X, T) of (X_i, T_i) $(1 \leq i \leq r)$ has the property that

$$\sigma(T) = \bigcup_{i=1}^r \sigma(T_i).$$

The proof of Theorem 10.2.1 shows that one should be able to characterize the least common extension of (X_i, T_i), $i = 1, \ldots, r$ in terms of the subspaces $\mathrm{Im}[\mathrm{col}(X_i T_i^{j-1})_{j=1}^m]$, much as Lemma 10.1.5 describes extensions (or restrictions). This characterization is given as follows.

THEOREM 10.2.2. *Let* $(X_0, T_0), (X_1, T_1), \ldots, (X_r, T_r)$ *be null kernel right admissible pairs. Then* (X_0, T_0) *is a least common extension of* $(X_1, T_1), \ldots, (X_r, T_r)$ *if and only if for each* $m \geq 1$ *the equality*

(10.2.4)
$$\mathrm{Im}[\mathrm{col}(X_0 T_0^{i-1})_{i=1}^m] = \mathrm{Im}[\mathrm{col}(X_1 T_1^{i-1})_{i=1}^m]$$
$$+ \cdots + \mathrm{Im}[\mathrm{col}(X_r T_r^{i-1})_{i=1}^m]$$

holds.

PROOF. Define X, T and P as in Theorem 10.2.1, and let

$$\widetilde{X}_0 = X|_{Im\,P}, \qquad \widetilde{T}_0 = PT|_{Im\,P}$$

(as in (10.2.1)). Then $(\widetilde{X}_0, \widetilde{T}_0)$ is a least common extension of $(X_1, T_1), \ldots, (X_r, T_r)$ and

$$\mathrm{Im}[\mathrm{col}(\widetilde{X}_0, \widetilde{T}_0^{i-1})_{i=1}^m] = \mathrm{Im}[\mathrm{col}(X_1 T_1^{i-1})_{i=1}^m]$$
$$+ \cdots + \mathrm{Im}[\mathrm{col}(X_r T_r^{i-1})_{i=1}^m], \qquad m \geq 1.$$

(See Theorem 10.2.1 and the first part of its proof.) Now the pair (X_0, T_0) is a least common extension of $(X_1, T_1), \ldots, (X_r, T_r)$ if and only if (X_0, T_0) and $(\widetilde{X}_0, \widetilde{T}_0)$ are similar. By Lemma 10.1.5 the last statement is equivalent to the requirement that

$$\mathrm{Im}[\mathrm{col}(X_0 T_0^{i-1})_{i=1}^m] = \mathrm{Im}[\mathrm{col}(\widetilde{X}_0 \widetilde{T}_0^{i-1})_{i=1}^m]$$

for all $m \geq 1$, and hence the proof is complete. $\square$

It is convenient to introduce the notion of index of stabilization for finite number of right admissible pairs. Given right admissible pairs $(X_1, T_1), \ldots, (X_r, T_r)$ with the same base dimension, define the index of stabilization $\mathrm{ind}\{(X_1, T_1), \ldots, (X_r, T_r)\}$ to be

$$\mathrm{ind}\{(X_1, T_1), \ldots, (X_r, T_r)\} = \min\{s \colon \mathrm{Ker}(\mathrm{col}[X_1 T_1^i, X_2 T_2^i, \ldots, X_r T_r^i]_{i=0}^{s-1})$$
$$= \mathrm{Ker}(\mathrm{col}[X_1 T_1^i, X_2 T_2^i, \ldots, X_r T_r^i]_{i=0}^{s})\}.$$

Because all matrices are finite dimensional, the index of stabilization is well-defined. A useful property of the index of stabilization is that

(10.2.5) $$\mathrm{Ker}(\mathrm{col}[X_1 T_1^i, \ldots, X_r T_r^i]_{i=0}^{s-1}) = \mathrm{Ker}(\mathrm{col}[X_1 T_1^i, \ldots, X_r T_r^i]_{i=0}^{t-1})$$

for every pair of integers s, t such that

$$\min(s, t) \geq \mathrm{ind}\{(X_1, T_1), \ldots, (X_r, T_r)\}.$$

To verify (10.2.5) we can assume that $s = \mathrm{ind}\{(X_1, T_1), \ldots, (X_r, T_r)\}$ and prove (10.2.5) by induction on t starting with $t = s + 1$. For $t = s + 1$, the equality (10.2.5) is just the definition of the index of stabilization. Assuming (10.2.5) is already proved for $t = t_0$, let

$$\mathrm{col}(x_1, \ldots, x_r) \in \mathrm{Ker}(\mathrm{col}[X_1 T_1^i, \ldots, X_r T_r^i]_{i=1}^{t_0-1}).$$

Then

$$\mathrm{col}(T_1^{t_0-s} x_1, \ldots, T_r^{t_0-s} x_r) \in \mathrm{Ker}(\mathrm{col}[X_1 T_1^i, \ldots, X_r T_r^i]_{i=0}^{s-1}),$$

hence

$$\mathrm{col}(T_1^{t_0-s} x_1, \ldots, T_r^{t_0-s} x_r) \in \mathrm{Ker}(\mathrm{col}[X_1 T_1^i, \ldots, X_r T_r^i]_{i=0}^{s})$$

and

$$\operatorname{col}(x_1,\ldots,x_r) \in \operatorname{Ker}(\operatorname{col}[X_1 T_1^i,\ldots,X_r T_r^i]_{i=0}^{t_0}).$$

This proves the inclusion

$$\operatorname{Ker}(\operatorname{col}[X_1 T_1^i,\ldots,X_r T_r^i]_{i=0}^{t_0-1}) \subset \operatorname{Ker}(\operatorname{col}[X_1 T_1^i,\ldots,X_r T_r^i]_{i=1}^{t_0}).$$

Since the opposite inclusion is obvious, we are done.

We now return to our original problem of least common multiples of matrix polynomials.

It will be convenient to assume that all matrix polynomials under consideration are comonic, i.e. with lower coefficient I (otherwise apply the transformation $L(z) \to L(\alpha)^{-1} L(z + \alpha)$ for a suitably chosen $\alpha \in \mathbb{C}$). Observe that the matrix T in a right null pair (X,T) of a comonic matrix polynomial is invertible.

THEOREM 10.2.3. *For any finite set of (regular) $n \times n$ matrix polynomial $L_1(z),\ldots,L_r(z)$ there exists a least common multiple which is unique up to multiplication on the left by a matrix polynomial with non-zero constant determinant. The minimal possible degree of such a least common multiple is equal to $s := \operatorname{ind}\{(X_1,T_1),\ldots,(X_r,T_r)\}$, where $(X_1,T_1),\ldots,(X_r,T_r)$ are right null pairs of $L_1,\ldots,L_r$ respectively. One least common multiple $L(z)$ of the minimal degree s is given by the following recipe (assuming that $L_1(z),\ldots,L_r(z)$ are comonic). Put $X = [X_1 X_2 \cdots X_r]$ and $T = T_1 \oplus \cdots \oplus T_r$, and let*

$$X_0 = X|_{\operatorname{Im} P}, \qquad T_0 = PT|_{\operatorname{Im} P},$$

where P is a projection of $\mathbb{C}^p$ ($p := \sum_{j=1}^r$ degree $\det L_j(z)$) along $\mathcal{M} := \bigcap_{i \geq 0} \operatorname{Ker}(XT^i)$. Further, let $[V_1 V_2 \cdots V_s]$ be the special left inverse of $\operatorname{col}(X_0 T_0^{1-i})_{i=1}^s$ (see Section 2.4 for this construction). Then

$$(10.2.6) \qquad L(z) = I - X_0 T_0^{-s}(V_s z + V_{s-1} z^2 + \cdots + V_1 z^s).$$

PROOF. From our hypothesis it follows that $T_1,\ldots,T_r$ are invertible. Hence the same is true for T and T_0. From the construction of the pair (X_0,T_0) it is clear that (X_0,T_0) is a null kernel pair. Moreover, $\bigcap_{i=0}^{s-1} \operatorname{Ker}(X_0 T_0^i) = \{0\}$ but $\bigcap_{i=0}^{s-2} \operatorname{Ker}(X_0 T_0^i) \neq \{0\}$ (at this point we use the property of the index of stabilization given by equality (10.2.5)). Theorem 10.2.1 shows that (X_0,T_0) is a least common extension of the right admissible pairs $(X_1,T_1),\ldots,(X_r,T_r)$. On the other hand, Theorem 10.1.1 shows that $L(z)$ is a least common multiple of $L_1,\ldots,L_r$ if and only if a right null pair of $L(z)$ coincides with the least common extension of the right null pairs of $L_1,\ldots,L_r$, i.e., coincides with (X_0,T_0). Now the assertion of Theorem 10.2.3 concerning the minimal degree of a least common multiple follows from Theorem 2.5.4, and uniqueness of a least common multiple follows from Corollary 10.1.3.

It remains to show that the matrix polynomial $L(z)$ given by (10.2.6) has (X_0,T_0) as its right null pair (relative to $\mathbb{C}$). We obtain from Theorem 2.4.5 that the monic matrix polynomial $\tilde{L}(z) = z^s L(z^{-1})$ has (X_0,T_0^{-1}) as its right null pair relative to $\mathbb{C}\backslash\{0\}$. Letting

$$\widetilde{L}(z) = z^s I + \sum_{j=0}^{s-1} N_j z^j,$$

by Proposition 2.5.1 we have

$$\sum_{j=0}^{s-1} N_j X_0 T_0^{-1} + X_0 T_0^{-s} = 0.$$

Rewrite this equality in the form

$$X_0 + N_{s-1} X_0 T_0 + \cdots + N_0 X_0 T_0^s = 0.$$

As $L(z) = I + N^{s-1} z + \cdots + N_0 z^s$, it is easy to see that Proposition 2.5.1 is applicable and yields the desired conclusion that (X_0, T_0) is a right null pair of $L(z)$ relative to $\mathbb{C}$. $\square$

We conclude this section with an indication how the theory of least common multiples developed in this section works if one replaces $\mathbb{C}$ by any subset σ of the complex plane. All the results, when properly reformulated, are valid in this framework also, so we shall state only the definition and the main theorem.

Let $L_1, \ldots, L_r$ be (regular) $n \times n$ matrix polynomials. We say that a (regular) $n \times n$ matrix polynomial M is a *left common multiple* of $L_1, \ldots, L_r$ relative to a set σ in the complex plane if the rational matrix functions $M(z)L_1(z)^{-1}, \ldots, M(z)L_r(z)^{-1}$ are analytic in σ.

THEOREM 10.2.4. *For any $n \times n$ matrix polynomials $L_1, \ldots, L_r$ there exists a least left common multiple relative to σ which is uniquely determined up to multiplication on the left by a matrix polynomial without zeros in σ. The minimal possible degree of such a least common multiple relative to σ is equal to* $\mathrm{ind}\{(X_1, T_1), \ldots, (X_r, T_r)\}$, *where* $(X_1, T_1), \ldots, (X_r, T_r)$ *are right null pairs of $L_1, \ldots, L_r$, respectively, relative to σ. Finally, one least common multiple $L(z)$ relative to σ of minimal degree is given by the recipe of Theorem 10.2.3, where (X_i, T_i) is a right null pair of L_i relative to σ (assuming that all the matrix polynomials $L_1, \ldots, L_r$ are comonic).*

We omit the proof of Theorem 10.2.4; it goes analogously to the proof of Theorem 10.2.3.

CHAPTER 11

COPRIME REPRESENTATION AND AN INTERPOLATION PROBLEM

A coprime representation of a rational matrix function $R(z)$ is a representation of the form $R(z) = N_R(z)D_R(z)^{-1}$ where N_R and D_R are matrix polynomials. A coprime representation of this form is called *right* coprime representation, in contrast to the left which has the form $R(z) = D_L(z)^{-1}N_L(z)$ for matrix polynomials D_L and N_L. The interpolation problem which we solve here is to find $R(z)$ with given right numerator N_R and given left denominator D_L.

11.1 COPRIME MATRIX FRACTIONS

In this section we shall study representations of rational matrix functions as quotients of matrix polynomials and a description of zero and pole pairs in terms of these representations.

Let $W(z)$ be a rational matrix function (with $\det W(z) \not\equiv 0$). There are many ways to write $W(z)$ in the form $D(z)^{-1}N(z)$, where $D(z)$ and $N(z)$ are $n \times n$ matrix polynomials. For instance, one could take $D(z) = p(z)I$, where $p(z)$ is any scalar polynomial divisible by denominators of all non-zero entries in $W(z)$. Further, if

$$W(z) = D(z)^{-1}N(z)$$

for some matrix polynomials $D(z)$ and $N(z)$, then obviously $W(z) = \big(M(z) D(z)\big)^{-1}M(z)N(z)$ for any regular matrix polynomial $M(z)$. We shall be interested in ways to write $W(z)$ as a quotient of matrix polynomials which are as small as possible (in the sense of divisibility of matrix polynomials). We arrive naturally at the following concept. A representation $W(z) = D_L(z)^{-1}N_L(z)$ is called a *left coprime matrix fraction* if the matrix polynomials $D_L(z)$ and $N_L(z)$ are *left coprime*, i.e.

$$N_L(z)X_1(z) + D_L(z)X_2(z) = I, \qquad z \in \mathbb{C}$$

for some matrix polynomials $X_1(z)$ and $X_2(z)$. The following well-known fact explains this terminology and provides a useful alternative description of left coprimeness.

THEOREM 11.1.1. *The following statements are equivalent for (regular) $n \times n$ matrix polynomials $A_1(z)$ and $A_2(z)$:*

(i) $A_1(z)$ *and* $A_2(z)$ *are left coprime;*

(ii) $A_1(z)$ *and* $A_2(z)$ *have no common left eigenvector corresponding to the same zero;*

(iii) *The only left common divisor of* $A_1(z)$ *and* $A_2(z)$ *is a trivial one: if* $A_1(z) = D(z)B_1(z)$, $A_2(z) = D(z)B_2(z)$ *for some* $n \times n$ *matrix polynomials* $D(z)$, $B_1(z)$ *and* $B_2(z)$, *then the determinant of* $D(z)$ *is a non-zero constant.*

In the proof of this proposition, as well as later in this chapter, the well-known Smith form will be used. For completeness this form is stated in the following lemma (for the proof the reader is referred to Gohberg-Lancaster-Rodman [1982], for example, or simply adapt the proof of the local Smith form for analytic matrix functions, namely Theorem 1.1.2). Having in mind subsequent uses we state it here in the framework of general rational matrix functions.

LEMMA 11.1.2. *Let $A(z)$ be a rational $m \times n$ matrix function (not necessarily of square size). Then $A(z)$ admits the Smith form*

$$A(z) = E(z)D(z)F(z),$$

where $E(z)$ and $F(z)$ are square size matrix polynomials with constant non-zero determinants, and $D(z)$ is a diagonal matrix (possibly bordered by zero columns and/or rows) with rational scalar functions $d_1(z) \not\equiv 0, \ldots, d_r(z) \not\equiv 0$ on the diagonal such that all the quotients $d_{i+1}(z)\big(d_i(z)\big)^{-1}$ ($i = 1, \ldots, r-1$) are polynomials. The number r of the not identically zero diagonal entries in $D(z)$ is uniquely determined by $A(z)$, and the entries $d_i(z)$ themselves are uniquely determined up to non-zero multiplicative constants. If $A(z)$ is a polynomial to start with, then $D(z)$ is also a polynomial.

PROOF OF THEOREM 11.1.1. (i) $\Rightarrow$ (ii) $\Rightarrow$ (iii) is easy. Indeed, assuming (i) we have

$$A_1(z)X_1(z) + A_2(z)X_2(z) \equiv I$$

for some matrix polynomials $X_1(z)$ and $X_2(z)$. If there was a common left eigenvector x_0 of $A_1(z)$ and $A_2(z)$ corresponding to the same zero z_0, then

$$\begin{aligned} x_0 &= x_0[A_1(z_0)X_1(z_0) + A_2(z_0)X_2(z_0)] \\ &= (x_0 A_1(z_0))X_1(z_0) + (x_0 A_2(z_0))X_2(z_0) = 0, \end{aligned}$$

a contradiction. Thus (i) $\Rightarrow$ (ii). If

$$A_1(z) = D(z)B_1(z), \qquad A_2(z) = D(z)B_2(z)$$

for some matrix polynomials $D(z)$, $B_1(z)$ and $B_2(z)$ such that $\det D(z)$ is not a constant, then a left null vector of $D(z)$ is a common left eigenvector for $A_1(z)$ and $A_2(z)$ corresponding to the same zero. This proves (ii) $\Rightarrow$ (iii).

Finally, we prove (iii) $\Rightarrow$ (i). We first show that the range of the $n \times 2n$ matrix $[\, A_1(z) \quad A_2(z) \,]$ is the whole of $\mathbb{C}^n$, for all $z \in \mathbb{C}$. Indeed, assuming the contrary, we can find $z_0 \in \mathbb{C}$ and a non-zero row vector x_0 such that

$$x_0 A_1(z_0) = x_0 A_2(z_0) = 0.$$

Then letting $\{x_0, y_1, \ldots, y_{n-1}\}$ be a basis in the n-dimensional space of row vectors, put

$$D(z) = A^{-1} \operatorname{diag}[(z - z_0), 1, \ldots, 1],$$

where A is the $n \times n$ matrix with rows $x_0, y_1, \ldots, y_{n-1}$. One verifies that $D(z)^{-1}A_1(z)$ and $D(z)^{-1}A_2(z)$ are polynomials, a contradiction with (iii) because the determinant of $D(z)$ is not constant. Having verified that

$$\operatorname{Im}[\, A_1(z) \quad A_2(z) \,] = \mathbb{C}^n; \qquad z \in \mathbb{C}$$

we see that the Smith form (Lemma 11.1.2) of the $n \times 2n$ matrix polynomial $[\, A_1(z) \quad A_2(z) \,]$ must be

$$[\, A_1(z) \quad A_2(z) \,] = E(z)[\, I \quad 0 \,]F(z).$$

Now we can easily identify a pair of matrix polynomials $X_1(z)$ and $X_2(z)$ that satisfy

$$A_1(z)X_1(z) + A_2(z)X_2(z) = I, \qquad z \in \mathbb{C},$$

namely,

$$\left[\begin{array}{c} X_1(z) \\ X_2(z) \end{array}\right] = F(z)^{-1}\left[\begin{array}{c} I \\ 0 \end{array}\right]E(z)^{-1}. \quad \square$$

We now prove existence and uniqueness (up to natural transformations) of left coprime polynomial matrix fraction representations.

THEOREM 11.1.3.   *Any rational $n \times n$ matrix function $W(z)$ with* $\det W(z) \not\equiv 0$ *admits a left coprime polynomial matrix fraction representation $W(z) =$* $D_L(z)^{-1}N_L(z)$. *The matrix polynomials $D_L(z)$ and $N_L(z)$ are uniquely determined by* $W(z)$ *up to simultaneous multiplication on the left by some matrix polynomial with non-zero constant determinant.*

PROOF. For the existence of the left coprime matrix fraction, we easily see, using the Smith form for $W(z)$ (Lemma 11.1.2), that we can assume $W(z)$ to be diagonal:

$$W(z) = \mathrm{diag}[a_1(z), \dots, a_n(z)].$$

Writing out each $a_j(z)$ as the quotient of two polynomials without common zeros, the first part of the theorem follows.

For the second part, let

(11.1.1) $$W(z) = D_{L1}(z)^{-1}N_{L1}(z) = D_{L2}(z)^{-1}N_{L2}(z)$$

be two left coprime matrix fractions of W. Then

(11.1.2) $$N_{L1}(z)X_{11}(z) + D_{L1}(z)X_{21}(z) \equiv I,$$

(11.1.3) $$N_{L2}(z)X_{12}(z) + D_{L2}(z)X_{22}(z) \equiv I,$$

for some matrix polynomials X_{11}, X_{12}, X_{21}, X_{22}. The equality (11.1.2) implies

$$D_{L1}^{-1}N_{L1}X_{11} + X_{21} = D_{L1}^{-1}$$

and hence

$$D_{L2}D_{L1}^{-1} = D_{L2}D_{L1}^{-1}N_{L1}X_{11} + D_{L2}X_{21} = N_{L2}X_{11} + D_{L2}X_{21}$$

in view of (11.1.1). So $A(z) := D_{L2}(z)D_{L1}(z)^{-1}$ is an analytic function in $\mathbb{C}$. An analogous argument shows that $A(z)^{-1}$ is also analytic in $\mathbb{C}$. So $A(z)$ must be a matrix polynomial with constant non-zero determinant. This proves the second part of Theorem 11.1.3. □

The following theorem describes left coprime matrix fractions of an $n \times n$ rational matrix function $W(z)$ with $\det W(z) \not\equiv 0$ in terms of its null and pole pairs.

THEOREM 11.1.4. Let $W(z) = L_2(z)^{-1} L_1(z)$ where $L_1(z)$ and $L_2(z)$ are $n \times n$ matrix polynomials. Then the following statements are equivalent:

(i) for every zero z_0 of $W(z)$ the right null pair (C_1, A_1) of L_1 corresponding to z_0 coincides with a right null pair of $W(z)$ corresponding to z_0;

(ii) for every pole z_0 of $W(z)$ the right null pair (C_2, A_2) of L_2 corresponding to z_0 coincides with a right pole pair of $W(z)$ corresponding to the same z_0;

(iii) $L_1(z)$ and $L_2(z)$ are left coprime.

PROOF. We start with a preliminary observation. As $L_1(z) = L_2(z)W(z)$ and $L_2(z)$ is a polynomial, it follows that any right null chain of $W(z)$ at z_0 is also a right null chain of $L_1(z)$ at z_0. Analogously, one shows that any right null chain of $W(z)^{-1}$ at z_0 is also a right null chain of $L_2(z)$ at z_0.

Suppose now that $L_1(z)$ and $L_2(z)$ are left coprime. We have to prove that (C_1, A_1) is a right null pair of $W(z)$. To this end it is sufficient to prove that the sets of right null chains of $L_1(z)$ and $W(z)$ are equal. In view of the preliminary observation we are left to prove that any right null chain of $L_1(z)$ is a right null chain of $W(z)$. So let $(x_0, \dots, x_{k-1})$ be a right null chain of $L_1(z)$ at a point z_0. By definition of a right null chain, there is a vector polynomial $\phi(z)$ such that

$$\phi(z) - \sum_{i=0}^{k-1} (z - z_0)^i x_i$$

and $L_1(z)\phi(z)$ both have a zero or order at least k at z_0. Let $X_1(z)$ and $X_2(z)$ be matrix polynomials such that $L_1(z)X_1(z) + L_2(z)X_2(z) = I$. Then we have

$$L_2(a)^{-1} L_1(z)[\phi(z) - X_1(z)L_1(z)\phi(z)] = X_2(z)L_1(z)\phi(z).$$

Now both $\phi(z) - X_1(z)L_1(z)\phi(z) - \sum_{i=0}^{k-1}(z - z_0)^i x_i$ and $X_2(z)L_1(z)\phi(z)$ have a zero of order at least k at z_0. So $(x_0, \dots, x_{k-1})$ is a right null chain of $W(z)$ at z_0 and it follows that (C_1, A_1) is a right null pair of $W(z)$ at z_0. So (iii) ⇒ (i) is proved.

Next we prove (i) ⇒ (iii). Let

(11.1.4) $$W(z) = D_L^{-1}(z)N_L(z)$$

be a left coprime matrix fraction representation for W. By the already proved part (iii) ⇒ (i) (applied to (11.1.4)) we know that $N_L(z)$ and $W(z)$ have the same right null pairs. But by (i), L_1 and W have the same right null pairs. So by Corollary 10.1.3 $L_1(z) = F(z)N_L(z)$, where $F(z)$ is an everywhere invertible matrix polynomial. Now

$$L_2(z)^{-1}L_1(z) \equiv D_L^{-1}(z)N_L(z),$$

so $L_2(z) = F(z)D_L(z)$, and L_1 and L_2 are easily seen to be left coprime because N_L and D_L are such.

Note that $W(z)^{-1} = L_1(z)^{-1}L_2(z)$. So by applying what has been proved so far to $W(z)^{-1}$ we see that the statements (ii) and (iii) are equivalent. □

All the results in this section admit analogues for *right* coprime matrix fractions. They can be obtained by applying Theorems 11.1.3 and 11.1.4 to the transposed matrix functions. Let us state these results.

Regular matrix polynomials $D_R(z)$ and $N_R(z)$ are called *right coprime* if

$$X(z)N_R(z) + Y(z)D_R(z) = I, \qquad z \in \mathbb{C}$$

for some matrix polynomials $X(z)$ and $Y(z)$. Equivalent condition can be given as in Proposition 11.1.1 (replacing there "left" by "right"). A representation $W(z) = N_R(z)D_R(z)^{-1}$ is called *right coprime matrix fraction* if the matrix polynomials $N_R(z)$ and $D_R(z)$ are right coprime.

THEOREM 11.1.5.   *Any rational $n \times n$ matrix function $W(z)$ with* $\det W(z) \not\equiv 0$ *admits a right coprime matrix fraction representation* $W(z) = N_R(z)D_R(z)^{-1}$. *The matrix polynomials $D_R(z)$ and $N_R(z)$ are uniquely determined by $W(z)$ up to simultaneous multiplication on the right by the same matrix polynomial with constant non-zero determinant.*

THEOREM 11.1.6.   *Let $W(z) = L_2(z)L_1(z)^{-1}$, where $L_1(z)$ and $L_2(z)$ are regular $n \times n$ matrix polynomials. The following statements are equivalent:*

(i) *for every zero z_0 of $W(z)$ the left null pair of L_2 corresponding to z_0 coincides with a left null pair of $W(z)$ corresponding to the same z_0;*

(ii) *for every pole z_0 of $W(z)$ the left null pair of L_1 corresponding to z_0 coincides with a left pole pair of $W(z)$ corresponding to z_0;*

(iii) *$L_1(z)$ and $L_2(z)$ are right coprime.*

11.2 LOCAL COPRIME MATRIX FRACTIONS

Coprime decompositions may be also studied upon the local viewpoint, i.e. when points belonging to a fixed subset only of the complex plane are taken into account. In this section we develop this point of view.

Let σ be a subset in the complex plane. Introduce the set $R(\sigma)$ of all $n \times n$ rational matrix functions with no poles in σ. The set $R(\sigma)$ is a ring under the natural addition and multiplication. If $\sigma = \mathbb{C}$, then $R(\sigma)$ is just the set of all $n \times n$ matrix polynomials. Regular rational matrix functions $A_1(z) \in R(\sigma)$ and $A_2(z) \in R(\sigma)$ are called *left coprime relative to σ* if there exist $X_1(z) \in R(\sigma)$ and $X_2(z) \in R(\sigma)$ such that the rational function $A_1(z)X_1(z) + A_2(z)X_2(z)$ has no zeros in σ. It is easy to see that for the case $\sigma = \mathbb{C}$ this is equivalent to the definition of left coprime matrix polynomials given in Section 11.1. The analogue of Proposition 11.1.1 holds:

PROPOSITION 11.2.1.   *The following statements are equivalent for regular $n \times n$ rational matrix functions $A_1(z)$ and $A_2(z)$ with no poles in σ:*

(i) $A_1(z)$ and $A_2(z)$ are left coprime relative to σ;

(ii) $A_1(z)$ and $A_2(z)$ have no common left eigenvector corresponding to the same zero in σ;

(iii) the only left common divisor of $A_1(z)$ and $A_2(z)$ with respect to σ is the trivial one: if $A_1(z) = D(z)B_1(z)$, $A_2(z) = D(z)B_2(z)$, where $D(z)$, $B_1(z)$ and $B_2(z)$ belong to $R(\sigma)$, then $\det D(z)$ is a non-zero constant.

PROOF. The implications (i) $\Rightarrow$ (ii) $\Rightarrow$ (iii) are verified as in the proof of Proposition 11.1.1.

The proof of (iii) $\Rightarrow$ (i) is somewhat more subtle. As in the proof of Proposition 11.1.1, we verify (assuming (iii)) that

$$\mathrm{Im}[\ A_1(z) \quad A_2(z)\ ] = \mathbf{C}^n; \qquad z \in \sigma.$$

But now an application of the Smith from (Lemma 11.1.2) yields

$$[\ A_1(z) \quad A_2(z)\ ] = E(z)[\ D(z) \quad 0\ ]F(z),$$

where $D(z) = \mathrm{diag}[d_1(z), \ldots, d_n(z)]$ and the scalar rational functions $d_1(z), \ldots, d_n(z)$ have no zeros in σ ($E(z)$ and $F(z)$ are square matrix polynomials of sizes $n \times n$ and $2n \times 2n$, respectively, with constant non-zero determinants). So by putting

$$\left[\begin{array}{c} X_1(z) \\ X_2(z) \end{array} \right] = F(z)^{-1} \left[\begin{array}{c} I \\ 0 \end{array} \right]$$

we ensure that

$$A_1(z)X_1(z) + A_2(z)X_2(z) = E(z)D(z)$$

has no zeros in σ. $\square$

Now let $W(z)$ be a regular rational $n \times n$ matrix function, and let $\sigma \subset \mathbf{C}$. A representation $W(z) = D_L(z)^{-1}N_L(z)$, where D_L and N_L are rational matrix functions without poles in σ, is called a *left coprime matrix fraction with respect to* σ if $D_L(z)$ and $N_L(z)$ are left coprime with respect to σ. From Theorem 11.1.3 it follows that a left coprime matrix fraction with respect to σ exists always (even with polynomial D_L and N_L). Analogously to the proof of Theorem 11.1.3 one verifies also these matrix fractions are unique in the following natural sense: If $W(z) = D_{L1}(z)^{-1}N_{L1}(z) = D_{L2}(z)^{-1}N_{L2}(z)$ are two left coprime matrix fractions with respect to σ, then the rational function $D_{L2}D_{L1}^{-1} = N_{L2}N_{L1}^{-1}$ and its inverse have no poles in σ.

The local version of Theorem 11.1.4 reads as follows:

THEOREM 11.2.2. *The following statements are equivalent for a representation of a regular rational $n \times n$ matrix function* $W(z) = L_2(z)^{-1}L_1(z)$, *where* $L_1(z)$ *and* $L_2(z)$ *are in* $R(\sigma)$:

(i) *the right null pair of* L_1 *relative to* σ *coincides with the right null pair of* W *relative to* σ;

(ii) *the right null pair of L_2 relative to σ coincides with the right pole pair of W relative to σ;*

(iii) $L_1(z)$ *and* $L_2(z)$ *are left coprime with respect to* σ.

PROOF. The implication (iii) $\Rightarrow$ (i) is proved as in Theorem 11.1.4, using the left coprimeness (with respect to σ) condition in the form

$$L_1(z)X_1(z) + L_2(z)X_2(z) = I, \qquad z \in \mathbb{C},$$

where $X_1(z)$ and $X_2(z)$ are rational matrix functions without poles in σ.

Assume now (i) holds, and let

$$W(z) = D_L(z)^{-1}N_L(z)$$

be a left coprime matrix fraction with respect to σ, chosen so that D_L and N_L are matrix polynomials. Applying the already proved part of Theorem 11.2.2 we obtain that N_L and W have the same right null pairs with respect to σ. But so do L_1 and W, in view of (i). Now we have that L_1 and N_L have same right null pairs with respect to σ. At this point we would like to apply Theorem 10.1.6 to the factorization

$$L_1(z) = F(z)N_L(z)$$

and conclude that the rational matrix function $F(z)$ has no poles and zeros in σ. However, Theorem 10.1.6 is not directly applicable ($N_L(z)$ is a polynomial, but $L_1(z)$ generally is not; we know only that $L_1 \in R(\sigma)$). To take care of this difficulty, let $L_1(z) = L_{12}(z)^{-1}L_{13}(z)$ be a left coprime matrix fraction of L_1 (with respect to $\mathbb{C}$). Now $L_{13}(z)$ and $N_L(z)$ are matrix polynomials with the same right null pairs with respect to σ (Theorem 11.1.4), so by Theorem 10.1.6 the function $\widetilde{F}(z) := L_{13}(z)N_L(z)^{-1}$ has no poles and zeros in σ. As $F(z) = L_{12}(z)\widetilde{F}(z)$, and the matrix polynomial $L_{12}(z)$ has no zeros in σ (because $L_1(z) \in R(\sigma)$), the function $F(z)$ has no poles and zeros in σ as well.

Comparing the two formulas

$$W(z) = L_2(z)^{-1}L_1(z); \qquad W(z) = D_L^{-1}(z)N_L(z),$$

we conclude that $L_2(z) = F(z)D_L(z)$, and L_1 and L_2 are left coprime with respect to σ because N_L and D_L are such. This proves (i) $\Rightarrow$ (iii).

Finally observe that, as in the proof of Theorem 11.1.4, the equivalence (ii) $\Leftrightarrow$ (iii) is proved by applying the already proved part of the theorem to $W(z)^{-1}$. $\square$

An analogous line of notions and results can be obtained for matrix polynomials that are right coprime with respect to σ, and for right coprime matrix fractions with respect to σ. We omit the details.

11.3 RATIONAL MATRIX FUNCTIONS WITH GIVEN LEFT NUMERATOR AND RIGHT DENOMINATOR

Let $W(z)$ be an $n \times n$ regular rational matrix function, with left coprime matrix fraction $W(z) = D_L(z)^{-1}N_L(z)$ and right coprime matrix fraction $W(z) =$

$N_R(z)D_R(z)^{-1}$. We say that $D_L(z)$ is a *left denominator* of $W(z)$, $N_L(z)$ is its *left numerator*, while $D_R(z)$ and $N_R(z)$ are *right denominator* and *right numerator* of $W(z)$, respectively. It follows from Theorems 11.1.3 and 11.1.5 that left (resp. right) numerator and left (resp. right) denominator of $W(z)$ are uniquely determined up to multiplication on the left (resp. right) by a matrix polynomial with constant non-zero determinant.

In this section we study the following problem: Given regular $n \times n$ matrix polynomials D_R and N_L, find (if possible) regular $n \times n$ matrix polynomials N_R and D_L in such a way that

$$W(z) := N_R(z)D_R(z)^{-1} = D_L(z)^{-1}N_L(z)$$

are right and left coprime matrix fraction descriptions of some rational matrix function $W(z)$ with $W(\infty) = I$.

We consider this as an interpolation problem for $W(z)$, when D_R and N_L are interpreted as the interpolation data (the condition $W(\infty) = I$ is a convenient normalization condition and really is not essential to the problem). Indeed, as Theorems 11.1.4 and 11.1.6 show, the knowledge of a left numerator of $W(z)$ amounts to the knowledge of the right null data of $W(z)$. Analogously, the knowledge of a right denominator of $W(z)$ amounts to the knowledge of the left pole data of $W(z)$.

The solution of the above mentioned problem is given by the following theorem.

THEOREM 11.3.1. *let D_R and N_L be given $n \times n$ regular matrix polynomials. Then there exists an $n \times n$ rational matrix function $W(z)$ with $W(\infty) = I$ such that D_R is a right denominator of $W(z)$ and N_L is a left numerator of $W(z)$ if and only if the equation*

(11.3.1) $$SA_\zeta - A_\pi S = BC,$$

where (C, A_ζ) is the right null pair of N_L and (A_π, B) is the left null pair of D_R, has an invertible solution S. In this case one may take

(11.3.2) $$W(z) = I - CS^{-1}(zI - A_\pi)^{-1}B.$$

The correspondence given by (11.3.2) between the set of invertible solutions S of (11.3.1) and the set of rational matrix functions $W(z)$ with the required properties, is one-to-one.

PROOF. Assume there exists $W(z)$ with the required properties, and consider the coprime matrix fraction descriptions of $W(z)$:

$$W(z) = N_R(z)D_R(z)^{-1} = D_L(z)^{-1}N_L(z)$$

with some matrix polynomials $N_R(z)$ and $D_L(z)$. Then by Theorems 11.1.6 and 11.1.4 the pair (C, A_ζ) is a right null pair for $W(z)$ (relative to $\mathbb{C}$), and (A_π, B) is a left pole pair of $W(z)$ (also relative to $\mathbb{C}$). Now use Theorem 4.4.3 (more precisely, apply this theorem to W^{-1} rather than W) to deduce the existence of a unique invertible solution of (11.3.1) for which (11.3.2) holds.

Conversely, assume that S is an invertible solution of (11.3.1). By the same Theorem 4.4.3 the rational matrix function $W(z)$ given by (11.3.2) has right null pair (C, A_ζ) (relative to $\mathbb{C}$) and left pole pair (A_π, B) (also relative to $\mathbb{C}$). Write

$$(11.3.3) \qquad W(z) = N_R(z)D_R(z)^{-1} = D_L(z)^{-1}N_L(z)$$

for some rational matrix functions N_R and D_L. We claim that actually N_R and D_L are matrix polynomials. Indeed, take a left coprime matrix fraction $W(z) = M(z)^{-1}L(z)$. Then (C, A_ζ) is the right null pair of $L(z)$ (Theorem 11.1.4). As the matrix polynomials $L(z)$ and $N_L(z)$ have the same right null pairs, we have $L(z) = U(z)N_L(z)$ for some matrix polynomial $U(z)$ with constant non-zero determinant (Corollary 10.1.3). Now

$$D_L(z) = N_L(z)W(z)^{-1} = N_L(z)L(z)^{-1}M(z) = U(z)^{-1}M(z)$$

is a matrix polynomial. Analogously one checks that N_R is a matrix polynomial. Now the coprimeness of (11.3.3) follows from Theorems 11.1.4 and 11.1.6. $\square$

Comparing with the notions and results of Chapter 4, one easily sees that $(C, A_\zeta; A_\pi, B; S)$ (in the notation of Theorem 11.3.1) is actually a left null-pole triple for $W(z)^{-1}$.

The problem which was described and solved in this section can be stated purely in terms of matrix polynomials, as follows. Given regular $n \times n$ matrix polynomials $L_1(z)$ and $L_2(z)$, find regular matrix polynomials $M_1(z)$ and $M_2(z)$ with the following properties:

(1) $M_1(z)M_2(z) = L_1(z)L_2(z)$;

(2) M_1 and L_1 are left coprime;

(3) M_2 and L_2 are right coprime;

(4) $M_1(z)^{-1}L_1(z)$ takes value I at infinity.

The solution to this problem is given by the following theorem (which is obtained from Theorem 11.3.1 by putting $N_L = L_1$; $D_R = L_2$).

THEOREM 11.3.2. *Given regular matrix polynomials L_1 and L_2, there exist regular matrix polynomials M_1 and M_2 for which (1)–(4) hold if and only if the equation*

$$(11.3.4) \qquad SA_\zeta - A_\pi S = BC$$

has an invertible solution S, where (C, A_ζ) is a right null pair of L_1 and (A_π, B) is the left null pair of L_2. In this case the formulas

$$(11.3.5) \qquad M_2(z) = [I - CS^{-1}(zI - A_\pi)^{-1}B]L_2(z),$$

$$(11.3.6) \qquad M_1(z) = L_1(z)[I + C(zI - A_\zeta)^{-1}S^{-1}B]$$

establish one-to-one correspondence between the set of invertible solutions S of (11.3.4) and the set of pairs of regular matrix polynomials (M_1, M_2) satisfying (1)–(4).

A variation of the above problem concerns the location of spectra as well, and can be stated as follows. Given regular $n \times n$ matrix polynomials $L_1(z)$ and $L_2(z)$, and given a closed simple rectifiable contour Γ in the complex plane, find a factorization

$$(11.3.7) \qquad L_1(z)L_2(z) = M_1(z)M_2(z),$$

where the regular matrix polynomials $M_1(z)$ and $M_2(z)$ are such that the zeros of M_1 and of L_2 (resp. of M_2 and of L_1) are all inside (resp. outside Γ) and the rational matrix function $M_1(z)^{-1}L_1(z)$ takes value I at infinity. Note that in this case L_1 and M_1 (resp. L_2 and M_2) are obviously left (resp. right) coprime. Using the well-known formula for the unique solution (11.3.4) in case $\sigma(A_\zeta) \cap \sigma(A_\pi) = \emptyset$ (see Section A.1 of the Appendix), we obtain the following corollary.

COROLLARY 11.3.3. *Given L_1 and L_2 as above, there exist matrix polynomials M_1 and M_2 for which (11.3.7) holds and the above spectral properties of L_1, L_2, M_1, M_2 are satisfied if and only if the matrix*

$$(11.3.8) \qquad -\frac{1}{2\pi i} \int_\Gamma (zI - A_\pi)^{-1} BC(zI - A_\zeta)^{-1} dz$$

is square and invertible (here (C, A_ζ) is a right null pair of L_1 and (A_π, B) is a left null pair of L_2). In this case such M_1 and M_2 are unique and are given by (11.3.6) and (11.3.7), respectively, where S is replaced by the matrix (11.3.8).

We consider briefly the dual problem of constructing a rational matrix function with given left denominator D_L and right numerator N_R. One easily verifies that the right numerator (denominator) for W coincides with the right denominator (numerator) for W^{-1}, and an analogous statement holds for the left numerators and denominators. Applying Theorem 11.3.1 to W^{-1}, we obtain the following result:

THEOREM 11.3.4. *Given $n \times n$ regular matrix polynomials D_L and N_R, there exists a rational matrix function $W(z)$ with $W(\infty) = I$ such that D_L is a left denominator of W and N_R is its right numerator if and only if the equation*

$$(11.3.9) \qquad TA_\pi - A_\zeta T = BC$$

has an invertible solution T, where (C, A_π) is a right null pair of D_L and (A_ζ, B) is a left null pair of N_R. In this case there is one-to-one correspondence between invertible solutions of (11.3.9) and the rational matrix function $W(z)$ with the above properties, and this correspondence is given by the formula

$$W(z) = I + C(zI - A_\pi)^{-1}T^{-1}B.$$

Observe that (in the notation of Theorem 11.3.4) $(C, A_\pi; A_\zeta, B; T)$ is actually a left null-pole triple of $W(z)$.

This problem also can be stated in terms of matrix polynomials only, and results analogous to Theorem 11.3.2 and Corollary 11.3.3 can be obtained. We leave it to the reader to state and prove these results.

11.4   RATIONAL MATRIX FUNCTIONS WITH GIVEN NU-MERATOR AND DENOMINATOR FROM THE OPPOSITE SIDES: LOCAL FRAMEWORK

In this section we consider the problem of constructing a rational matrix function with given left denominator and right numerator (or right denominator and left numerator) over a set $\sigma \neq \mathbb{C}$ in the complex plane. In contrast with the case $\sigma = \mathbb{C}$ (treated in the previous section) the solution to this problem requires the existence of solutions to the corresponding Sylvester equation, not necessarily invertibility of some such solutions.

THEOREM 11.4.1.   *Given $n \times n$ regular matrix polynomials D_L and N_R, and given a set $\sigma \neq \mathbb{C}$ in the complex plane, let (C, A_π) be a right null pair of D_L relative to σ, and let (A_ζ, B) be a left null pair of N_R relative to σ. Then there exists a rational matrix function $W(z)$ with $W(\infty) = I$ such that*

$$W(z) = N_R(z)D_R(z)^{-1} = D_L(z)^{-1}N_L(z)$$

are right and left coprime matrix fractions with respect to σ, for some rational matrix functions D_R and N_L without poles in σ, if and only if the Sylvester equation

(11.4.1) $$S A_\pi - A_\zeta S = B_\zeta C_\pi$$

admits a solution.

PROOF.   Assume that (11.4.1) has a solution S. By Theorem 4.6.1 there exists a rational matrix function $W(z)$ with $W(\infty) = I$ for which (C, A_π) is a right pole pair (relative to σ) and (A_ζ, B) is a left null pair (relative to σ). By Theorem 11.2.2 (more precisely, by its dual analogue) $W(z)$ has the required properties if we can show that the rational matrix functions $D_L(z)W(z)$ and $W(z)^{-1}N_R(z)$ have no poles in σ (indeed, in this case Theorem 11.2.2 shows that (C, A_π) is a right pole pair for W and also a right pole pair for D_L^{-1} relative to σ, and (A_ζ, B) is a left null pair for W and for N_R relative to σ).

Let

$$W(z) = \tilde{D}_L(z)^{-1}\tilde{N}_L(z)$$

be a left coprime matrix fraction of $W(z)$. By Theorem 11.1.2 $D_L(z)$ and $\tilde{D}_L(z)$ have the same right null pair corresponding to σ. As both $D_L(z)$ and $\tilde{D}_L(z)$ have no poles in σ, by Theorem 4.5.8 we conclude that $D_L(z) = F(z)\tilde{D}_L(z)$ for some rational matrix function $F(z)$ without poles and zeros in σ (or else use the argument analogous to that of the proof of Theorem 11.2.2 to reach this conclusion). Now $D_L W = F\tilde{N}_L$ clearly has no poles in σ. Analogously the absence of poles in σ for the function $W(z)^{-1}N_R(z)$ is verified.   □

We note that one can replace polynomials D_L and N_R in Theorem 11.4.1 by any regular rational matrix functions without poles in σ, with essentially the same proof.

We conclude with remark that a result analogous to Theorem 11.4.1 holds for rational matrix functions with given left numerator and right denominator with respect to σ.

NOTES FOR PART II

CHAPTER 8   Section 8.1 contains a part of the results from the paper of Gohberg-Kaashoek-van Schagen [1982]. Section 8.2 is from the paper Gohberg-Kaashoek-Lerer-Rodman [1984].

CHAPTER 9   Chapter 9 has common points with the paper of Gohberg-Kaashoek [1987]. More general results on divisibility were obtained in Ball-Gohberg-Rodman [1987, 1989e] and Gohberg-Kaashoek [1987].

CHAPTER 10   Chapter 10 presents a different point of view toward finding a common multiple of a system of polynomials; the main results were obtained in Gohberg-Kaashoek-Lerer-Rodman [1981]. See also Gohberg-Lancaster-Rodman [1982], Chapter 9. These results were extended to rational matrix functions in Ball-Gohberg-Rodman [1989e].

The results of Chapters 9 and 10 underscore the intimate connections between interpolation and factorization. The basic result (Theorem 10.1.1) can be recast in terms of the correspondence between divisors of a matrix polynomial $B(z)$ and invariant subspaces of the matrix $A_{\zeta B}$ taken from a right null pair $(C_{\zeta B}, A_{\zeta B})$ of $B(z)$. Factorization (or divisibility) results of this type, all of which bear directly on the corresponding interpolation problems, have been started in Gohberg-Lancaster-Rodman [1978a,b] for matrix polynomials and in Bart-Gohberg-Kaashoek [1979] for minimal divisibility of rational matrix functions in terms of realizations. This approach was influenced by Brodskii-Livsic [1958] and Sz.-Nagy-Foias [1970]. Later, the factorization results of this type have been obtained for minimal divisibility of rational matrix functions in terms of spectral data (Ball-Gohberg-Rodman [1987]), linear-fractional factorization of rational matrix functions (Helton-Ball [1982], Gohberg-Rubinstein [1987]), operator polynomials (Gohberg-Lancaster-Rodman [1978c], Kaashoek-van der Mee-Rodman [1983]; see also the monograph Rodman [1989]) and analytic operator valued functions (Kaashoek-van der Mee-Rodman [1982, 1983], Gohberg-Rodman [1983]).

CHAPTER 11   Chapter 11 follows Gohberg-Kaashoek-Lerer-Rodman [1984].

The concept of coprime matrix fractions is widely used in modern systems theory, where it appears as one of the many important ways to describe transfer functions of linear time invariant systems (see Kailath [1980], Vidyasagar [1985]). For a classical purely algebraic treatment of coprime matrix functions see McDuffee [1956].

PART III

SUBSPACE INTERPOLATION PROBLEMS

In this chapter we develop a new way of presenting the data for homogeneous interpolation problems. Here the data are given in the form of a null-pole subspace. This form has the advantage that it is uniquely determined by the function while the other form is only determined up to similarity. This part contains a development which is parallel to that in Part I. We first introduce and analyze the null-pole subspaces for general rational matrix functions, and especially for the case where the function is J-unitary on the unit circle or on the imaginary line. Then we solve interpolation problems where one is given null-pole subspaces. The case of data at infinity is considered separately. Generalization of Beurling-Lax invariant subspace representation theorems for rational matrix functions play here a prominent role. The approach described here can also be developed independently of Part I and in many cases is more convenient to use, as will be seen in Part V.

CHAPTER 12

NULL-POLE SUBSPACES: ELEMENTARY PROPERTIES

In this chapter is introduced a new form of local and global null-pole data. The null-pole data are now presented as a subspace of rational vector functions, called the null-pole subspace. Here we study how to express the null pole subspace via an admissible data set, via a null-pole triple, and in terms of null and pole chains and Laurent coefficients. The latter two are done in the local version only. The development of this chapter is preparatory for the interpolation problems to be solved in the succeeding chapters.

In this chapter we restrict our attention to the finite complex plane; moreover, it will be often assumed that $W(\infty) = I$; the modifications needed for the general case will be discussed in Chapter 15.

12.1 INTRODUCTION AND SIMPLE PROPERTIES

For a scalar rational function $r(z)$, the subspace $\mathcal{S}_{\{z_0\}}(r) := \{r(z)h(z) : h$ analytic at $z_0\}$ may serve as a measure of the local behaviour of r at z_0. Indeed the subspace $\mathcal{S}_{\{z_0\}}(r)$ always has the form

$$\mathcal{S}_{\{z_0\}}(r) = \{(z - z_0)^k h(z) : h \text{ rational with no pole at } z_0\}.$$

If $k > 0$, r has a zero of order k at z_0, if $k < 0$, r has a pole of order $-k$ at z_0, and if $k = 0$, r and r^{-1} are analytic at z_0. As the subspace $\mathcal{S}_{\{z_0\}}(r)$ measures the singularity of r and z_0, we call it the *null-pole subspace* for $r(z)$ at z_0.

More generally, if σ is any proper subset of the extended plane, we define the null-pole subspace $\mathcal{S}_\sigma(r)$ for $r(z)$ over σ to be

$$\mathcal{S}_\sigma(r) = \{r(z)h(z) : h \text{ rational with no poles in } \sigma\}.$$

Then $\mathcal{S}_\sigma(r) = \cap\{\mathcal{S}_{\{z_0\}}(r) : z_0 \in \sigma\}$ determines the null-pole structure of $r(z)$ over the set σ.

In this chapter we wish to obtain the analogues of these statements for the matrix valued case. Let $W(z)$ be a regular rational $n \times n$ matrix function and σ a subset of the complex plane $\mathbb{C}$. In general we let $\mathcal{R}_{m \times n}(\sigma)$ and $\mathcal{R}_n(\sigma)$ denote rational $m \times n$ matrix valued (respectively $\mathbb{C}^n$ valued) functions having no poles in σ. We define the *(left) null-pole* subspace for $W(z)$ over σ, denoted by $\mathcal{S}_\sigma(W)$, by

$$\begin{aligned}
\mathcal{S}_\sigma(W) &= W\mathcal{R}_n(\sigma) \\
&= \{W(z)h(z) : h \in \mathcal{R}_n(\sigma)\}.
\end{aligned}$$

In this introductory section we derive a few elementary properties of these subspaces. Characterizations of $\mathcal{S}_\sigma(W)$ in terms of a realization for $W(z)$, null-pole triple for $W(z)$ over σ, or in terms of a canonical set of null and pole chains for $W(z)$, will be given in succeeding sections.

PROPOSITION 12.1.1. *Let $W(z)$ be a given regular rational $n \times n$ matrix function, and $A(z)$ a given rational $n \times m$ matrix function. In order that there exist a $B(z) \in \mathcal{R}_{n \times m}(\sigma)$ such that*

$$A(z) = W(z)B(z)$$

it is necessary and sufficient that for each vector x in $\mathbf{C}^n$ the rational vector function $A(z)x$ belongs to $\mathcal{S}_\sigma(W)$.

PROOF. If $B(z) \in \mathcal{R}_{n \times m}(\sigma)$ then each column $b_j(z)$ belongs to $\mathcal{R}_n(\sigma)$, so each column $W(z)b_j(z)$ of $W(z)B(z)$ is in $\mathcal{S}_\sigma(W)$. Thus the linear combination $W(z)B(z)x$ is in $\mathcal{S}_\sigma(W)$. Conversely, if $A(z)x \in \mathcal{S}_\sigma(W)$ then each column $A(z)e_j$ is in $\mathcal{S}_\sigma(W)$, so by definition has the form $A(z)e_j = W(z)b_j(z)$ for a $b_j(z) \in \mathcal{R}_n(\sigma)$. Then $B(z) = [b_1(z), \ldots, b_m(z)] \in \mathcal{R}_{n \times m}(\sigma)$ and $A(z) = W(z)B(z)$. $\square$

As a corollary of the previous proposition we can obtain the following characterization of when two regular rational $n \times n$ matrix functions have the same left null-pole subspace.

THEOREM 12.1.2. *Let $W_1(z)$ and $W_2(z)$ be two regular rational $n \times n$ matrix functions, and let $\sigma \subset \mathbf{C}$. Then $\mathcal{S}_\sigma(W_1) = \mathcal{S}_\sigma(W_2)$ if and only if $W_1(z) = W_2(z)F(z)$ for a rational matrix function $F(z)$ such that both $F(z)$ and $F^{-1}(z)$ have no poles in σ.*

PROOF. By Proposition 12.1.1 we conclude that $W_1(z) = W_2(z)F(z)$ and $W_2(z) = W_1(z)G(z)$ where F and G are both analytic on σ. But clearly $G(z)$ must equal $F^{-1}(z)$. $\square$

As a consequence of Theorem 12.1.2 we see that if $W(z)$ has the local Smith form

$$W(z) = L(z)D(z)R(z)$$

at z_0, where L and R together with their inverses are analytic at z_0 and $D(z) = \text{diag}((z - z_0)^{\kappa_1}, \ldots, (z - z_0)^{\kappa_n})$, then the null-pole subspace $\mathcal{S}_{\{z_0\}}(W)$ for W at z_0 does not depend on the right factor $R(z)$. This also justifies the term *left* null-pole subspace for $\mathcal{S}_\sigma(W)$, as it is carrying only null-pole structural information related to the left side of W.

PROPOSITION 12.1.3. *Let σ_1 and σ_2 be subsets of $\mathbf{C}$ and let W be a regular $n \times n$ matrix function. Then*

(i) $\mathcal{S}_{\sigma_1 \cup \sigma_2}(W) = \mathcal{S}_{\sigma_1}(W) \cap \mathcal{S}_{\sigma_2}(W);$

and

(ii) $\mathcal{S}_{\sigma_1 \cap \sigma_2}(W) = \mathcal{S}_{\sigma_1}(W) + \mathcal{S}_{\sigma_2}(W).$

PROOF. These properties following from the easily verified identities

$$\mathcal{R}_n(\sigma_1 \cup \sigma_2) = \mathcal{R}_n(\sigma_1) \cap \mathcal{R}_n(\sigma_2)$$

and

$$\mathcal{R}_n(\sigma_1 \cap \sigma_2) = \mathcal{R}_n(\sigma_1) + \mathcal{R}_n(\sigma_2). \quad \square$$

For σ a subset of $\mathbb{C}$, let $\mathcal{R}_n^0(\sigma^c)$ denote the rational $\mathbb{C}^n$-valued functions having all poles in σ and vanishing at infinity. If $\mathcal{R}_n$ denotes the space of all rational vector functions, then the direct sum decomposition

$$\mathcal{R}_n = \mathcal{R}_n^0(\sigma^c) \dotplus \mathcal{R}_n(\sigma)$$

is easily seen by looking at partial fraction expansions. For W a regular rational $n \times n$ matrix function let $\mathcal{S}_{\sigma^c}^0(W) = W\mathcal{R}_n^0(\sigma^c)$. The following result is an easy consequence of the above remark. We shall obtain a converse statement in Chapter 14.

PROPOSITION 12.1.4. *Let $W(z)$ be a regular rational $n \times n$ matrix function and let σ be a subset of $\mathbb{C}$. Then*

$$\mathcal{R}_n = \mathcal{S}_{\sigma^c}^0(W) \dotplus \mathcal{S}_\sigma(W).$$

12.2   NULL-POLE SUBSPACE FOR AN ADMISSIBLE DATA SET

In order to describe the null-pole subspace of a rational matrix function in terms of a realization for the function, we will first introduce a more general notion. Recall that a pair of matrices (C, A_π) (where C is $n \times n_\pi$ and A_π is $n_\pi \times n_\pi$) is said to be a *null kernel pair* (or *observable*) if

$$\operatorname{Ker} \operatorname{col}(CA_\pi^j)_{j=0}^{n_\pi - 1} = \{0\}.$$

Also a pair of matrices (A_ζ, B) (where A_ζ is $n_\zeta \times n_\zeta$ and B is $n_\zeta \times n$) is said to be a *full range pair* (or *controllable*) if

$$\operatorname{Im}[B, A_\zeta B, \dots, A_\zeta^{n_\zeta - 1}B] = \mathbb{C}^{n_\zeta}.$$

For $K(z)$ any rational matrix function, $\operatorname{Res}_{z=z_0} K(z)$ denotes the residue of K at the point z_0, i.e. the coefficient of $(z - z_0)^{-1}$ in the Laurent expansion of $K(z)$ around z_0. Let there be given a pair of matrices (C, A_π) of sizes $n \times n_\pi$ and $n_\pi \times n_\pi$, respectively, a pair of matrices (A_ζ, B) of sizes $n_\zeta \times n_\zeta$ and $n_\zeta \times n$, respectively, and a matrix S. The whole collection

(12.2.1) $$\tau = (C, A_\pi; A_\zeta, B; S)$$

will be called a *data set* of matrices; τ will be called *admissible* if (C, A_π) is a null kernel pair, (A_ζ, B) is a full range pair and S has size $n_\zeta \times n_\pi$ (where n_ζ is the size of A_ζ and n_π is the size of A_π). If a subset σ of $\mathbb{C}$ is also specified, we say that τ is σ-admissible if also $\sigma(A_\pi) \cup \sigma(A_\zeta) \subset \sigma$. Now suppose σ is a subset of $\mathbb{C}$ which contains both $\sigma(A_\pi)$ and $\sigma(A_\zeta)$. From the data set τ we define a subspace $\mathcal{S}_\tau(\sigma)$ of $\mathcal{R}_n$ by

(12.2.2)
$$\mathcal{S}_\tau(\sigma) = \{C(zI - A_\pi)^{-1}x + h(z) : x \in \mathbb{C}^{n_\pi}, h \in \mathcal{R}_n(\sigma)$$
$$\text{such that } \sum_{z_0 \in \sigma} \operatorname{Res}_{z=z_0}(zI - A_\zeta)^{-1}Bh(z) = Sx\}.$$

Note that if (C, A_π) is vacuous $(n_\pi = 0)$, then the definition of $\mathcal{S}_\tau(\sigma)$ simplifies to

$$(12.2.3) \qquad \mathcal{S}_\tau(\sigma) = \{h \in \mathcal{R}_n(\sigma): \sum_{z_0 \in \sigma} \mathrm{Res}_{z=z_0}(zI - A_\zeta)^{-1}Bh(z) = 0\}.$$

On the other hand, if (A_ζ, B) is vacuous $(n_\zeta = 0)$, $\mathcal{S}_\tau(\sigma)$ collapses to

$$(12.2.4) \qquad \mathcal{S}_\tau(\sigma) = \{C(zI - A_\pi)^{-1}x: x \in \mathbf{C}^{n_\pi}\} \dot{+} \mathcal{R}_n(\sigma).$$

Before analyzing the structure of $\mathcal{S}_\tau(\sigma)$ we illustrate the notion of this type of null-pole subspace with some simple examples.

EXAMPLE 12.2.1.   Take C, A_π and S to be vacuous and $A_\zeta =$ $\mathrm{diag}(z_1, \ldots, z_{n_\zeta})$, $B = \mathrm{col}(b_j)_{j=1}^{n_\zeta}$, where $z_1, \ldots, z_{n_\zeta}$ are complex numbers and where each b_j is a row vector with n components. Then one can show that (A_ζ, B) is a full range pair if and only if $\{b_{j_1}, \ldots, b_{j_m}\}$ is a linearly independent set of row vectors whenever $z_{j_1} = \cdots = z_{j_m}$; in particular each $b_j \neq 0$. If $h \in \mathcal{R}_n(\sigma)$ then any pole of $(zI - A_\zeta)^{-1}Bh(\zeta)$ must occur among the points $z_1, \ldots, z_{n_\zeta}$. Then the residue of $(zI - A_\zeta)^{-1}Bh(z)$ at the point z_j has i-th component equal to 0 if $z_i \neq z_j$ and equal to $b_i h(z_i)$ if $z_i = z_j$. The sum of these residues over all poles is therefore $\mathrm{col}(b_i h(z_i))_{i=1}^{n_\zeta}$. Thus in this case $\mathcal{S}_\tau(\sigma)$ is the subspace of all functions h in $\mathcal{R}_n(\sigma)$ satisfying the homogeneous interpolation conditions

$$b_i h(z_i) = 0, \qquad 1 \leq i \leq n_\zeta. \quad \square$$

EXAMPLE 12.2.2.   Again take C, A_π and S vacuous but take A_ζ equal to a lower triangular Jordan block

$$A_\zeta = \begin{bmatrix} z_0 & & & 0 \\ 1 & z_0 & & \\ & \ddots & \ddots & \\ 0 & & 1 & z_0 \end{bmatrix}$$

of size $n_\zeta \times n_\zeta$ and again $B = \mathrm{col}(b_i)_{i=1}^{n_\zeta}$ with each b_i a row vector with n columns. In this case the pair (A_ζ, B) is a full range pair if and only if $b_1 \neq 0$. If $h \in \mathcal{R}_n(\sigma)$ then the only possible pole of $(zI - A_\zeta)^{-1}Bh(z)$ is at the eigenvalue z_0 of A. One can compute $(zI - A_\zeta)^{-1}$ explicitly as

$$(zI - A_\zeta)^{-1} = \begin{bmatrix} (z-z_0)^{-1} & & & 0 \\ (z-z_0)^{-2} & (z-z_0)^{-1} & & \\ \vdots & & \ddots & \\ (z-z_0)^{-n_\zeta} & \cdots & (z-z_0)^{-2} & (z-z_0)^{-1} \end{bmatrix}.$$

From this we see that the residue of $(zI - A_\zeta)^{-1}Bh(z)$ at the only pole z_0 in σ is given by

$$\mathrm{Res}_{z=z_0}(zI - A_\zeta)^{-1}Bh(z) = \begin{bmatrix} b_1 & & & 0 \\ b_2 & b_1 & & \\ \vdots & & \ddots & \\ b_{n_\zeta} & \cdots & & b_1 \end{bmatrix} \begin{bmatrix} h_0 \\ h_1 \\ \vdots \\ h_{n_\zeta - 1} \end{bmatrix}$$

where $h_j = \frac{1}{j!}h^{[j]}(z_0)$ is the j-th Taylor coefficient of $h(z)$ at z_0. If we set $b(z) = b_1 + b_2(z - z_0) + \cdots + b_{n_\zeta}(z - z_0)^{n_\zeta - 1}$, we see that the space $\mathcal{S}_\tau(\sigma)$ in this case is given by all vector functions $h \in \mathcal{R}_n(\sigma)$ which satisfy the homogeneous interpolation conditions

$$\frac{d^j}{dz^j}[b(z)h(z)]\mid_{z=z_0} = 0 \quad \text{for} \quad 0 \le j \le n_\zeta - 1. \quad \square$$

EXAMPLE 12.2.3. In this case suppose that A_ζ, B and S are vacuous, A_π has the diagonal form $A_\pi = \operatorname{diag}(w_1, \ldots, w_{n_\pi})$ and $C = [u_1, \ldots, u_{n_\pi}]$ where each u_j is a column vector with n components. Then, if $x = \operatorname{col}(x_j)_{j=1}^{n_\pi}$ then

$$C(zI - A_\pi)^{-1}x = (z - z_1)^{-1}x_1 u_1 + \cdots + (z - z_{n_\pi})^{-1}x_{n_\pi}u_{n_\pi}.$$

Thus in this case

$$\mathcal{S}_\tau(\sigma) = \operatorname{Span}\{(z - z_j)^{-1}u_j : j = 1, \ldots, n_\pi\} \dotplus \mathcal{R}_n(\sigma). \quad \square$$

EXAMPLE 12.2.4. Again take A_ζ, B and S vacuous but A_π equal to an $n_\pi \times n_\pi$ Jordan block

$$A_\pi = \begin{bmatrix} z_1 & 1 & & & 0 \\ & z_0 & 1 & & \\ & & \ddots & \ddots & \\ & & & & 1 \\ 0 & & & & z_0 \end{bmatrix}$$

and $C = [u_1, \ldots, u_{n_\pi}]$. Then $(zI - A_\pi)^{-1}$ has the explicit form

$$(zI - A_\pi)^{-1} = \begin{bmatrix} (z - z_0^{-1}) & (z - z_0)^{-2} & \cdots & (z - z_0)^{-n_\pi} \\ & (z - z_0)^{-1} & \ddots & \vdots \\ & & \ddots & \\ & & & (z - z_0)^{-2} \\ 0 & & & (z - z_0)^{-1} \end{bmatrix}$$

and if x is a column vector $\operatorname{col}(x_j)_{j=1}^{n_\pi}$ then

$$C(zI - A_\pi)^{-1}x = x_1(z - z_0)^{-1}u_1 + x_2((z - z_0)^{-2}u_1 + (z - z_0)^{-1}u_2) + \cdots$$
$$+ x_{n_\pi}((z - z_0)^{-n_\pi}u_1 + (z - z_0)^{-n_\pi + 1}u_2 + \cdots + (z - z_0)^{-1}u_{n_\pi}).$$

If we let $u(z)$ be the vector function

$$u(z) = u_1 + (z - z_0)u_2 + \cdots + (z - z_0)^{n_\pi - 1}u_{n_\pi}$$

then we may write $\mathcal{S}_\tau(\sigma)$ as

$$\mathcal{S}_\tau(\sigma) = \operatorname{Span}\{(z - z_0)^{-j}u(z) : 1 \le j \le n_\pi\} \dotplus \mathcal{R}_n(\sigma). \quad \square$$

EXAMPLE 12.2.5. Suppose all matrices C, A_π, A_ζ, B, S are 1×1: $C = [1]$, $A_\pi = [w_0]$, $A_\zeta = [z_0]$, $B = [1]$, $S = [s]$ where s is a number. Then $\tau = (C, A_\pi; A_\zeta, B; S)$ is admissible, $n = 1$, and $S_\tau(\sigma)$ consists of scalar rational functions k of the form

$$k(z) = (z - w_0)^{-1} x + h(z)$$

where $x \in \mathbf{C}$, h has no poles in σ and $h(z_0) = sx$. $\square$

Our first result concerning null-pole subspaces $S_\tau(\sigma)$ concerns characterizing when a given subspace $S_\tau(\sigma)$ is invariant under multiplication by scalar rational functions with no poles in σ (the ring of such functions is denoted $\mathcal{R}(\sigma)$). Note that $S_\tau(\sigma)$ is invariant under $\mathcal{R}(\sigma)$ in Examples 12.2.1, 12.2.2, 12.2.3, 12.2.4. An elementary calculation shows that $S_\tau(\sigma)$ in Example 12.2.5 is invariant under $\mathcal{R}(\sigma)$ if and only if $z_0 \neq w_0$ and $s = (w_0 - z_0)^{-1}$. In this case the invariance of $S_\tau(\sigma)$ under $\mathcal{R}(\sigma)$ is clear from the alternative characterization

$$S_\tau(\sigma) = \{k \in \text{Span}\{(z - w_0)^{-1}\} \dotplus \mathcal{R}(\sigma) : k(z_0) = 0\}.$$

The general principle is as follows.

THEOREM 12.2.1. *Let $\tau = (C, A_\pi; A_\zeta, B; S)$ be a σ-admissible data set as in* (12.2.1) *and let $S_\tau(\sigma)$ be the associated subspace given by* (12.2.2). *Then*

$$r S_\tau(\sigma) \subset S_\tau(\sigma) \quad \text{for all} \quad r \in \mathcal{R}(\sigma)$$

if and only if the matrix S satisfies the Sylvester equation

$$(12.2.5) \qquad\qquad S A_\pi - A_\zeta S = BC.$$

Equivalently, $S_\tau(\sigma)$ is invariant under the single scalar function $r(z) = z$.

Before proving Theorem 12.2.1 we need some preliminary general facts.

LEMMA 12.2.2. *Suppose C, A_π, A_ζ, B are matrices of sizes $n \times n_\pi$, $n_\pi \times n_\pi$, $n_\zeta \times n_\zeta$ and $n_\zeta \times n$ respectively.*

(i) *(C, A_π) is a null-kernel pair if and only if the map $\Gamma_\pi : \mathbf{C}^{n_\pi} \to \mathcal{R}_n$ defined by*

$$\Gamma_\pi x = C(zI - A_\pi)^{-1} x$$

has $\text{Ker}\,\Gamma_\pi = \{0\}$.

(ii) *Let σ be a subset of $\mathbf{C}$ which contains $\sigma(A_\zeta)$. Then (A_ζ, B) is a full-range pair if and only if the map $\Gamma_\zeta : \mathcal{R}_n(\sigma) \to \mathbf{C}^{n_\zeta}$ defined by*

$$\Gamma_\zeta : h(z) \to \sum_{z_0 \in \sigma} \text{Res}_{z=z_0}(zI - A_\zeta)^{-1} B h(z)$$

has $\text{Im}\,\Gamma_\zeta = \mathbf{C}^{n_\zeta}$.

PROOF. Note that for $|z|$ sufficiently large the series expansion

$$C(zI - A_\pi)^{-1} x = \sum_{j=0}^{\infty} C A_\pi^j x z^{-j-1}$$

is valid. This function is identically zero if and only if all the Laurent coefficients $CA_\pi^j x$ are 0 ($j = 0, 1, 2, \ldots$), i.e. if and only if $x \in \bigcap_{j=0}^\infty \operatorname{Ker} \operatorname{col}(CA_\pi^j)$. By the Cayley-Hamilton theorem this is equivalent to $x \in \operatorname{Ker} \operatorname{col}(CA_\pi^j)_{j=0}^{n_\pi - 1}$. This proves (i).

To prove (ii), it suffices to prove the result with $\sigma = \mathbb{C}$, i.e. with the domain space for Γ_ζ taken to be the space $\mathcal{R}_n(\mathbb{C})$ of vector polynomials $h(z) = h_0 + h_1 z + \cdots + z^q h_q$. Then $(zI - A_\zeta)^{-1} Bh(z)$ has poles in the finite plane only in $\sigma(A_\zeta) \subset \sigma$. By the residue theorem,

$$\sum_{z_0 \in \sigma} \operatorname{Res}_{z=z_0}(zI - A_\zeta)^{-1} Bh(z) = \frac{1}{2\pi i} \int_{|z|=R} (zI - A_\zeta)^{-1} Bh(z) dz$$

for any R sufficiently large. Choose R so large that the Laurent series expansion

$$(zI - A_\zeta)^{-1} = \sum_{j=0}^\infty A_\zeta^j z^{-j-1}$$

is valid for $|z| \geq R$. Then we compute

$$(zI - A_\zeta)^{-1} Bh(z) = \sum_{j=-q+1}^\infty F_j z^{-j}$$

where

$$F_j = \sum_{k=\max\{0, 1-j\}}^q A_\zeta^{j+k-1} Bh_k.$$

Integrating term by term we get

$$\frac{1}{2\pi i} \int_{|z|=R} (zI - A_\zeta)^{-1} Bh(z) dz = \sum_{j=-q+1}^\infty \frac{1}{2\pi i} \int_{|z|=R} F_j z^{-j} dz$$
$$= F_1$$

where

$$F_1 = \sum_{k=0}^q A_\zeta^k Bh_k = [B, A_\zeta B, \ldots, A_\zeta^q B] \operatorname{col}(h_k)_{k=0}^q.$$

Thus Γ_ζ as a linear transformation on vector polynomials has full image space $\mathbb{C}^{n_\zeta}$ if and only if $\operatorname{Im}[B, AB, \ldots, A^q B]$ fills out $\mathbb{C}^{n_\zeta}$ as q becomes arbitrarily large. Again by the Cayley-Hamilton theorem this is equivalent to $\operatorname{Im}[B, A_\zeta B, \ldots, A_\zeta^{n_\zeta - 1} B] = \mathbb{C}^{n_\zeta}$. By considering approximating Riemann sums we see that the image of Γ_ζ as a map on all of $\mathcal{R}_n(\sigma)$ is never larger than the smallest invariant subspace for A_ζ containing $\operatorname{Im} B$ (it is easy to see that this subspace coincides with $\operatorname{Im}[B, A_\zeta B, \ldots, A_\zeta^q B]$). In particular, $\operatorname{Im} \Gamma_\zeta$ is independent of σ as long as $\sigma(A_\zeta) \subset \sigma$. Thus Γ_ζ has full image as a map on $\mathcal{R}_n(\sigma)$ if and only if it has full image as a map on vector polynomials $\mathcal{R}_n(\mathbb{C})$. $\square$

We noted in Section 12.1 (Proposition 12.1.4) that in general $\mathcal{R}_n(\sigma)$ has the direct sum decomposition

$$\mathcal{R}_n = \mathcal{R}_n^0(\sigma^c) \dotplus \mathcal{R}_n(\sigma).$$

Let us now denote the projection of $\mathcal{R}_n$ onto $\mathcal{R}_n^0(\sigma^c)$ along $\mathcal{R}_n(\sigma)$ by $P_{\sigma^c}^0$; the complementary projection $I - P_{\sigma^c}^0$ we denote by P_σ. One consequence of Lemma 12.2.2 is that the image of the restriction $P_\sigma \mid \mathcal{S}_\tau(\sigma)$ of the projection P_σ to $\mathcal{S}_\tau(\sigma)$ is as large as possible, as long as (A_ζ, B) is a full range pair.

COROLLARY 12.2.3. *Suppose* $\tau = (C, A_\pi; A_\zeta, B; S)$ *is a* σ-*admissible data set of matrices and let* $\mathcal{S}_\tau(\sigma)$ *be the associated subspace as in* (12.2.2). *Then:*

(i) *The subspace* $\mathcal{X}_\pi := P_{\sigma^c}^0\big(\mathcal{S}_\tau(\sigma)\big)$ *is given by*

$$\mathcal{X}_\pi = \{C(zI - A_\pi)^{-1}x \colon x \in \mathbb{C}^{n_\pi}\}.$$

Moreover

$$\dim \mathcal{X}_\pi = n_\pi$$

and for each scalar rational function $r \in \mathcal{R}(\sigma)$

$$P_{\sigma^c}^0(rk) \in \mathcal{X}_\pi \quad \text{whenever} \quad k \in \mathcal{X}_\pi.$$

(ii) *The subspace* $\mathcal{S}_\tau(\sigma) \cap \mathcal{R}_n(\sigma)$ *is given by*

$$\mathcal{S}_\tau(\sigma) \cap \mathcal{R}_n(\sigma) = \{h \in \mathcal{R}_n(\sigma) \colon \sum_{z_0 \in \sigma} \mathrm{Res}_{z=z_0}(zI - A_\zeta)^{-1}Bh(z) = 0\}.$$

Moreover

$$\dim \mathcal{R}_n(\sigma)/[\mathcal{S}_\tau(\sigma) \cap \mathcal{R}_n(\sigma)] = n_\zeta$$

and for each scalar rational function $r \in \mathcal{R}(\sigma)$

$$rh \in \mathcal{S}_\tau(\sigma) \cap \mathcal{R}_n(\sigma) \quad \text{whenever} \quad h \in \mathcal{S}_\tau(\sigma) \cap \mathcal{R}_n(\sigma).$$

PROOF. By Lemma 12.2.2 we know that for each $x \in \mathbb{C}^{n_\pi}$ there exists an $h \in \mathcal{R}_n(\sigma)$ such that $\sum_{z_0 \in \sigma} \mathrm{Res}_{z=z_0}(zI - A_\zeta)^{-1}Bh(z) = Sx$. Thus for each $x \in \mathbb{C}^{n_\pi}$ there exists an $h \in \mathcal{R}_n(\sigma)$ such that $k(z) = C(zI - A_\pi)^{-1}x + h(z)$ is in $\mathcal{S}_\tau(\sigma)$. Since $\sigma(A_\pi) \subset \sigma$, the function $C(zI - A_\pi)^{-1}x \in \mathcal{R}_n^0(\sigma^c)$. Thus $(P_{\sigma^c}^0 k)(z) = C(zI - A_\pi)^{-1}x$. Also by Lemma 12.2.2, the map $\Gamma_\pi \colon x \to C(zI - A_\pi)^{-1}x$ has trivial kernel, and hence rank $\Gamma_\pi = n_\pi$. This establishes the first part of (i).

If $k(z) = C(zI - A_\pi)^{-1}x \in \mathcal{X}_\pi$, then

$$zk(z) = C(zI - A_\pi)^{-1}A_\pi x + Cx.$$

As $C(zI - A_\pi)^{-1}A_\pi x \in \mathcal{R}_n^0(\sigma^c)$ and $Cx \in \mathcal{R}_n(\sigma)$ we conclude

$$P_{\sigma^c}^0\big(zk(z)\big) = C(zI - A_\pi)^{-1}A_\pi x \in \mathcal{X}_\pi.$$

Similarly, if $w_0 \notin \sigma$, then $r(z) = (z - w_0)^{-1} \in \mathcal{R}(\sigma)$ and

$$(z - w_0)^{-1}k(z) = -C(zI - A_\pi)^{-1}(w_0I - A_\pi)^{-1}x + (z - w_0)^{-1}C(w_0I - A_\pi)^{-1}x$$

so

$$P_{\sigma^c}^0[(z - w_0)^{-1}k(z)] = -C(zI - A_\pi)^{-1}(w_0I - A_\pi)^{-1}x \in \mathcal{X}_\pi.$$

As can be seen by considering partial fraction expansions, $\mathrm{Span}\{z^j, (z - w_0)^{-k} : j, k = 0, 2, 3, \ldots, w_0 \notin \sigma\}$ equals $\mathcal{R}(\sigma)$. Thus $P_{\sigma^c}^0[rk] \in \mathcal{X}_\pi$ for $r \in \mathcal{R}(\sigma)$ and $k \in \mathcal{X}_\pi$.

The characterization of $\mathcal{S}_\tau(\sigma) \cap \mathcal{R}_n(\sigma)$ follows from the definition of $\mathcal{S}_\tau(\sigma)$. Note that this characterizes $\mathcal{S}_\tau(\sigma) \cap \mathcal{R}_n(\sigma)$ as the kernel of the map $\Gamma_\zeta \colon \mathcal{R}_n(\sigma) \to \mathbb{C}^{n_\zeta}$ arising in Lemma 12.2.2. As by Lemma 12.2.2 rank Γ_ζ is n_ζ, the codimension of $\mathrm{Ker}\,\Gamma_\zeta$ in $\mathcal{R}_n(\sigma)$ must be n_ζ.

Next, suppose that $h \in \mathcal{S}_\tau(\sigma) \cap \mathcal{R}_n(\sigma)$, so

$$\sum_{z_0 \in \sigma} \mathrm{Res}_{z=z_0}(zI - A_\zeta)^{-1}Bh(z) = 0.$$

Using the identity

$$(zI - A_\zeta)^{-1}Bzh(z) = A_\zeta(zI - A_\zeta)^{-1}Bh(z) + Bh(z)$$

we see that then

$$\sum_{z_0 \in \sigma} \mathrm{Res}_{z=z_0}(zI - A_\zeta)^{-1}Bzh(z) = 0$$

as well since the term $Bh(z)$ has no poles in σ. Thus $zh(z)$ belongs to $\mathcal{S}_\tau(\sigma) \cap \mathcal{R}_n(\sigma)$ whenever h does. Similarly, if $w_0 \notin \sigma$, use the identity

$$(zI - A_\zeta)^{-1}B(z - w_0)^{-1}h(z)$$
$$= -(w_0I - A_\zeta)^{-1}[(zI - A_\zeta)^{-1}Bh(z)] + (z - w_0)^{-1}(w_0I - A_\zeta)^{-1}Bh(z)$$

to deduce that $(z - w_0)^{-1}h(z)$ is in $\mathcal{S}_\tau(\sigma) \cap \mathcal{R}_n(\sigma)$ whenever h is. We conclude that $\mathcal{S}_\tau(\sigma) \cap \mathcal{R}_n(\sigma)$ is invariant under multiplication by any scalar rational function r in $\mathcal{S}_\tau(\sigma) \cap \mathcal{R}_n(\sigma)$. $\square$

PROOF OF THEOREM 12.2.1. Suppose $k(z) = C(zI - A_\pi)^{-1}x + h(z)$ is in $\mathcal{S}_\tau(\sigma)$. Then

$$zk(z) = C(zI - A_\pi)^{-1}A_\pi x + [Cx + zh(z)].$$

As $Cx + zh(z) \in \mathcal{R}_n(\sigma)$, the condition that $zk(z) \in \mathcal{S}_\tau(\sigma)$ reduces to

(12.2.6) $$\sum_{z_0 \in \sigma} \mathrm{Res}_{z=z_0}(zI - A_\zeta)^{-1}B[Cx + zh(z)] = SA_\pi x.$$

Note that

$$(zI - A_\zeta)^{-1}Bzh(z) = Bh(z) + A_\zeta(zI - A_\zeta)^{-1}Bh(z)$$

where

$$\sum_{z_0 \in \sigma} \text{Res}_{z=z_0} Bh(z) = 0$$

since h has no poles in σ, and

$$\sum_{z_0 \in \sigma} \text{Res}_{z=z_0} A_\zeta (zI - A_\zeta)^{-1} Bh(z) = A_\zeta Sx$$

since $k \in \mathcal{S}_\tau(\sigma)$. Finally, by the Riesz functional calculus for A_ζ,

$$\sum_{z_0 \in \sigma} \text{Res}_{z=z_0} (zI - A_\zeta)^{-1} BCx = BCx$$

since $\sigma(A_\zeta) \subset \sigma$. Thus (12.2.6) is equivalent to

$$SA_\pi x - A_\zeta Sx = BCx.$$

As $k(z)$ sweeps $\mathcal{S}_\tau(\sigma)$, x sweeps all of $\mathbb{C}^{n_\pi}$ by Corollary 12.2.3. Thus $\mathcal{S}_\tau(\sigma)$ is invariant under multiplication by z if and only if S satisfies the Sylvester equation (12.2.5).

Next choose a point $w_0 \in \mathbb{C} \backslash \sigma$. We now show that $\mathcal{S}_\tau(\sigma)$ is invariant under multiplication by $r(z) = (z - w_0)^{-1}$ if and only if S satisfies (12.2.5). Indeed, suppose $k(z) = C(zI - A_\pi)^{-1}x + h(z) \in \mathcal{S}_\tau(\sigma)$. Then use the identity

$$(z - w_0)^{-1}(zI - A_\pi)^{-1} = -(zI - A_\pi)^{-1}(w_0 I - A_\pi)^{-1} + (z - w_0)^{-1}(w_0 I - A_\pi)^{-1}$$

to compute

$$(z - w_0)^{-1} k(z) = -C(zI - A_\pi)^{-1}(w_0 I - A_\pi)^{-1} x$$
$$+ [(z - w_0)^{-1} C(w_0 I - A_\pi)^{-1} x + (z - w_0)^{-1} h(z)].$$

As the term in brackets is in $\mathcal{R}_n(\sigma)$, we see that $(z - w_0)^{-1} k(z)$ is in $\mathcal{S}_\tau(\sigma)$ if and only if

(12.2.7)
$$\sum_{z_0 \in \sigma} \text{Res}_{z=z_0} (zI - A_\zeta)^{-1} B[(z - w_0)^{-1} C(w_0 I - A_\pi)^{-1} x$$
$$+ (z - w_0)^{-1} h(z)] = -S(w_0 I - A_\pi)^{-1} x.$$

Use the identity

$$(z - w_0)^{-1}(zI - A_\zeta)^{-1} = -(zI - A_\zeta)^{-1}(w_0 I - A_\zeta)^{-1} + (z - w_0)^{-1}(w_0 I - A_\zeta)^{-1}$$

to deduce that (12.2.7) is equivalent to

$$\sum_{z_0 \in \sigma} \text{Res}_{z=z_0} \{-(zI - A_\zeta)^{-1}(w_0 I - A_\zeta)^{-1} BC(w_0 I - A_\pi)^{-1} x$$
$$+ (z - w_0)^{-1}(w_0 I - A_\zeta)^{-1} BC(w_0 I - A_\pi)^{-1} x$$
$$- (zI - A_\zeta)^{-1}(w_0 I - A_\zeta)^{-1} Bh(z)$$
$$+ (z - w_0)^{-1}(w_0 I - A_\zeta)^{-1} Bh(z)\} = -S(w_0 I - A_\pi)^{-1} x.$$

By the functional calculus for A_ζ, the first term reduces to $-(w_0 I - A_\zeta)^{-1} BC(w_0 I - A_\pi)^{-1} x$. The second and fourth terms have no poles in σ, so the sum of the residues in each term is 0. The third term is

$$-(w_0 I - A_\zeta)^{-1} \sum_{z_0 \in \sigma} \mathrm{Res}_{z=z_0}(zI - A_\zeta)^{-1} Bh(z) = -(w_0 I - A_\zeta)^{-1} Sx$$

since by assumption $k \in \mathcal{S}_\tau(\sigma)$. Thus $(z - w_0)^{-1} k(z) \in \mathcal{S}_\tau(\sigma)$ holding for all $k \in \mathcal{S}_\tau(\sigma)$ (again by Corollary 12.2.3) is equivalent to the matrix equation

$$-(w_0 I - A_\zeta)^{-1} BC(w_0 I - A_\pi)^{-1} - (w_0 I - A_\zeta)^{-1} S = -S(w_0 I - A_\pi)^{-1}.$$

It is straightforward next to show that this equation is equivalent to the Sylvester equation (12.2.5).

Reversing the argument we see that if (12.2.5) holds, then $\mathcal{S}_\tau(\sigma)$ is invariant under multiplication by powers of z and of $(z - w_0)^{-1}$ for any $w_0 \notin \sigma$. Since any $r \in \mathcal{R}(\sigma)$ is a linear combination of such elementary functions, we see that then $\mathcal{S}_\tau(\sigma)$ is invariant under multiplication by any r in $\mathcal{S}_\tau(\sigma)$. $\square$

It is also useful to understand to what extent the singular subspace $\mathcal{S}_\tau(\sigma)$ determines the admissible set τ. If $\tau = (C, A_\pi; A_\zeta, B; S)$ and $\widetilde{\tau} = (\widetilde{C}, \widetilde{A}_\pi; \widetilde{A}_\zeta, \widetilde{B}; \widetilde{S})$ are two admissible data sets we say that τ and $\widetilde{\tau}$ are *similar* if there exist invertible matrices S_1 and S_2 for which

$$\widetilde{C} = CS_1, \qquad \widetilde{A}_\pi = S_1^{-1} A_\pi S_1$$

(12.2.8)
$$\widetilde{A}_\zeta = S_2^{-1} A_\zeta S_2, \qquad \widetilde{B} = S_2^{-1} B$$
$$\widetilde{S} = S_2^{-1} S S_1.$$

THEOREM 12.2.4. *Two σ-admissible data sets $\tau = (C, A_\pi; A_\zeta, B; S)$ and $\widetilde{\tau} = (\widetilde{C}, \widetilde{A}_\pi; \widetilde{A}_\zeta, \widetilde{B}; \widetilde{S})$ determine the same null-pole subspace $\mathcal{S}_\tau(\sigma) = \mathcal{S}_{\widetilde{\tau}}(\sigma)$ if and only if τ and $\widetilde{\tau}$ are similar.*

PROOF. Suppose $\mathcal{S}_\tau(\sigma) = \mathcal{S}_{\widetilde{\tau}}(\sigma)$. Then in particular $P_{\sigma^c}^0(\mathcal{S}_\tau(\sigma)) = P_{\sigma^c}^0(\mathcal{S}_{\widetilde{\tau}}(\sigma))$. If $\Gamma_\pi: x \to C(zI - A_\pi)^{-1} x$ and $\widetilde{\Gamma}_\pi: x \to \widetilde{C}(zI - \widetilde{A}_\pi)^{-1} x$ are the maps associated with τ and $\widetilde{\tau}$ as in Lemma 12.2.2, then by Lemma 12.2.2 they both have trivial kernel and by Corollary 12.2.3 they have the same image space. Hence there must exist an invertible linear transformation S_1 on $\mathbf{C}^{n_\pi}$ for which $\widetilde{\Gamma}_\pi = \Gamma_\pi S_1$, i.e.

$$\widetilde{C}(zI - \widetilde{A}_\pi)^{-1} x = C(zI - A_\pi)^{-1} S_1 x$$

for all x in $\mathbf{C}^{n_\pi}$. Equating Laurent coefficients in the Laurent series expansions at infinity we get

$$\widetilde{C}\widetilde{A}_\pi^j x = CA_\pi^j S_1 x$$

for all x, and $j = 0, 1, 2, \ldots$. The special case $j = 0$ yields $\widetilde{C} = CS_1$, and a general consequence is

$$\widetilde{C}\widetilde{A}_\pi^j \widetilde{A}_\pi = \widetilde{C}\widetilde{A}_\pi^j (S_1^{-1} A_\pi S_1)$$

for $j = 0, 1, 2, \ldots$. Then $\widetilde{A}_\pi = S_1^{-1} A_\pi S_1$ follows from $(\widetilde{C}, \widetilde{A}_\pi)$ being a null kernel pair. Similarly, $\mathcal{S}_\tau(\sigma) = \mathcal{S}_{\widetilde{\tau}}(\sigma)$ implies that $\mathcal{R}_n(\sigma) \cap \mathcal{S}_\tau(\sigma) = \mathcal{R}_n(\sigma) \cap \mathcal{S}_{\widetilde{\tau}}(\sigma)$. Thus the maps $\Gamma_\zeta \colon h(z) \to \sum_{z_0 \in \sigma} \operatorname{Res}_{z=z_0}(zI - A_\zeta)^{-1} B h(z)$ and $\widetilde{\Gamma}_\zeta \colon h(z) \to \sum_{z_0 \in \sigma} \operatorname{Res}_{z=z_0}(zI - \widetilde{A}_\zeta)^{-1} \widetilde{B} h(z)$ defined on $\mathcal{R}_n(\sigma)$ by Lemma 12.2.2 have full image space and by Corollary 12.2.3 have the same kernel. Thus $\widetilde{\Gamma}_\zeta$ and Γ_ζ have the same image space $\mathbf{C}^{n_\zeta}$ and there must be an invertible linear transformation S_2 on $\mathbf{C}^{n_\zeta}$ for which $S_2 \widetilde{\Gamma}_\zeta = \Gamma_\zeta$. From this we deduce using the construction in the proof of Lemma 12.2.2 that

$$S_2 [\widetilde{B}, \widetilde{A}_\zeta \widetilde{B}, \ldots, \widetilde{A}_\zeta^q \widetilde{B}] = [B, A_\zeta B, \ldots, A_\zeta^q B]$$

for all $q \geq 0$. From the first block column we get $S_2 \widetilde{B} = B$. In general we deduce

$$S_2 \widetilde{A}_\zeta [\widetilde{B}, \widetilde{A}_\zeta \widetilde{B}, \ldots, \widetilde{A}_\zeta^q \widetilde{B}] = A_\zeta S_2 [\widetilde{B}, \widetilde{A}_\zeta B, \ldots, \widetilde{A}_\zeta^q \widetilde{B}].$$

Since $(\widetilde{A}_\zeta, \widetilde{B}_\zeta)$ is a full range pair we conclude that $\widetilde{A}_\zeta = S_2^{-1} A_\zeta S_2$. Finally, from

(12.2.9)
$$\mathcal{S}_\tau(\sigma) = \{ \Gamma_\pi x + h(z) \colon x \in \mathbf{C}^{n_\pi}, h \in \mathcal{R}_n(\sigma)$$
$$\text{such that } \Gamma_\zeta h = Sx \}$$

being identical to

(12.2.10)
$$\mathcal{S}_{\widetilde{\tau}}(\sigma) = \{ \widetilde{\Gamma}_\pi \widetilde{x} + h(z) \colon \widetilde{x} \in \mathbf{C}^{n_\pi}, h \in \mathcal{R}_n(\sigma)$$
$$\text{such that } \widetilde{\Gamma}_\zeta h = \widetilde{S} \widetilde{x} \}$$

together with

$$\widetilde{\Gamma}_\pi = \Gamma_\pi S_1, \qquad S_1 \widetilde{\Gamma}_\zeta = \Gamma_\zeta$$

we deduce that $\widetilde{S} = S_2^{-1} S S_1$. Thus τ and $\widetilde{\tau}$ are similar.

Conversely, if τ and $\widetilde{\tau}$ are similar it is routine to check that expressions (12.2.9) and (12.2.10) generate the same null-pole subspaces $\mathcal{S}_\tau(\sigma)$ and $\mathcal{S}_{\widetilde{\tau}}(\sigma)$. $\square$

We close this section with the analogue of Proposition 12.1.3 at the level of σ-admissible data sets. We recall from Section 4.5 the operations of direct sum and corestriction of admissible data sets; the only difference is that here we do not require that the coupling matrix of the data set satisfy the associated Sylvester equation. Suppose σ_1 and σ_2 are disjoint subsets of $\mathbf{C}$ and

(12.2.11)
$$\tau_\nu = (C^{(\nu)}, A_\pi^{(\nu)}; A_\zeta^{(\nu)}, B^{(\nu)}; S_{\nu\nu})$$

is a σ_ν-admissible data set for $\nu = 1, 2$. We define the *direct sum* $\tau_1 \oplus \tau_2$ to be the collection

(12.2.12)
$$\tau_1 \oplus \tau_2 = \left([C^{(1)}, C^{(2)}], \begin{bmatrix} A_\pi^{(1)} & 0 \\ 0 & A_\pi^{(2)} \end{bmatrix} ; \right.$$
$$\left. \begin{bmatrix} A_\zeta^{(1)} & 0 \\ 0 & A_\zeta^{(2)} \end{bmatrix}, \begin{bmatrix} B^{(1)} \\ B^{(2)} \end{bmatrix} ; \begin{bmatrix} S_{11} & S_{12} \\ S_{21} & S_{22} \end{bmatrix} \right)$$

where S_{12} and S_{21} are the unique solutions of the Sylvester equations

(12.2.13) $$S_{12}A_\pi^{(2)} - A_\zeta^{(1)}S_{12} = B^{(1)}C^{(2)}$$

(12.2.14) $$S_{21}A_\pi^{(1)} - A_\zeta^{(2)}S_{21} = B^{(2)}C^{(1)}$$

The assumption that $\sigma_1 \cap \sigma_2 = \emptyset$ implies that the spectra of $A_\pi^{(1)}$ and $A_\zeta^{(1)}$ are disjoint from the spectra of $A_\pi^{(2)}$ and $A_\zeta^{(2)}$ and hence that these equations have unique solutions S_{12} and S_{21}. From Proposition 4.5.2 it is then also clear that $\tau_1 \oplus \tau_2$ is $(\sigma_1 \cup \sigma_2)$-admissible since each τ_ν is σ_ν-admissible. The following is the analogue of the first part of Proposition 12.1.3.

PROPOSITION 12.2.5. *Suppose τ_ν as in (12.2.11) is a σ_ν-admissible data set for $\nu = 1, 2$, where σ_1 and σ_2 are disjoint subsets of* C. *Then*

$$\mathcal{S}_{\tau_1}(\sigma_1) \cap \mathcal{S}_{\tau_2}(\sigma_2) = \mathcal{S}_{\tau_1 \oplus \tau_2}(\sigma_1 \cup \sigma_2).$$

PROOF. By definition a vector function f is in $\mathcal{S}_{\tau_\nu}(\sigma_\nu)$ if and only if f has the form

(12.2.15) $$f = C^{(\nu)}(zI - A_\pi^{(\nu)})^{-1}x_\nu + h_\nu(z)$$

where $h_\nu \in \mathcal{R}_n(\sigma_\nu)$ and

(12.2.16) $$\frac{1}{2\pi i}\sum_{z_0 \in \sigma_\nu} \text{Res}_{z=z_0}(zI - A_\zeta^{(\nu)})^{-1}B^{(\nu)}h_\nu(z) = S_{\nu\nu}x_\nu$$

for $\nu = 1, 2$. Thus if f is in the intersection $\mathcal{S}_{\tau_1}(\sigma_1) \cap \mathcal{S}_{\tau_2}(\sigma_2)$, from (12.2.15) we have

$$f = C^{(1)}(zI - A_\pi^{(1)})^{-1}x_1 + h_1(z)$$
$$= C^{(2)}(zI - A_\pi^{(2)})^{-1}x_2 + h_2(z)$$

so

(12.2.17) $$h_2(z) = C^{(1)}(zI - A_\pi^{(1)})^{-1}x_1 + [h_1(z) - C^{(2)}(zI - A_\pi^{(2)})^{-1}x_2].$$

Since both $h_2(z)$ and $C^{(1)}(zI - A_\pi^{(1)})^{-1}x_1$ are analytic on σ_2, we conclude that

(12.2.18) $$k(z) := h_1(z) - C^{(2)}(zI - A_\pi^{(2)})^{-1}x_2$$

is analytic on σ_2. Since each term $h_1(z)$ and $C^{(2)}(zI - A_\pi^{(2)})^{-1}x_2$ is also analytic on σ_1, we see that $k \in \mathcal{R}_n(\sigma_1 \cup \sigma_2)$. From (12.2.18) and (12.2.15) we see that

(12.2.19) $$f = [\, C^{(1)} \quad C^{(2)} \,]\begin{bmatrix} (zI - A_\pi^{(1)})^{-1} & 0 \\ 0 & (zI - A_\pi^{(2)})^{-1} \end{bmatrix}\begin{bmatrix} x_1 \\ x_2 \end{bmatrix} + k(z).$$

From (12.2.18) and (12.2.17) we can express h_ν in terms of k and x_1, x_2 as

$$h_1(z) = C^{(2)}(zI - A_\pi^{(2)})^{-1}x_2 + k(z)$$

$$h_2(z) = C^{(1)}(zI - A_\pi^{(1)})^{-1}x_1 + k(z).$$

Thus (12.2.16) with $\nu = 1$ takes the form

(12.2.20)
$$\frac{1}{2\pi i} \sum_{z_0 \in \sigma_1} \mathrm{Res}_{z=z_0}(zI - A_\zeta^{(1)})^{-1}B^{(1)}k(z)$$
$$= S_{11}x_1 - \frac{1}{2\pi i} \sum_{z_0 \in \sigma_1} \mathrm{Res}_{z=z_0}(zI - A_\zeta^{(1)})^{-1}B^{(1)}C^{(2)}(zI - A_\pi^{(2)})^{-1}x_2$$

and from (12.2.16) with $\nu = 2$ we get

(12.2.21)
$$\frac{1}{2\pi i} \sum_{z_0 \in \sigma_2} \mathrm{Res}_{z=z_0}(zI - A_\zeta^{(2)})^{-1}B^{(2)}k(z)$$
$$- \frac{1}{2\pi i} \sum_{z_0 \in \sigma_2} \mathrm{Res}_{z=z_0}(zI - A_\zeta^{(2)})^{-1}B^{(2)}C^{(1)}(zI - A_\pi^{(1)})^{-1}x_1 + S_{22}x_2.$$

Equations (12.2.20) and (12.2.21) can be combined into the single matrix equation

(12.2.22)
$$\sum_{z_0 \in \sigma_1 \cup \sigma_2} \mathrm{Res}_{z=z_0} \begin{bmatrix} (zI - A_\zeta^{(1)})^{-1} & 0 \\ 0 & (zI - A_\zeta^{(2)})^{-1} \end{bmatrix} \begin{bmatrix} B^{(1)} \\ B^{(2)} \end{bmatrix} k(z)$$
$$= \begin{bmatrix} S_{11} & S_{12} \\ S_{22} & S_{22} \end{bmatrix} \begin{bmatrix} x_1 \\ x_2 \end{bmatrix}$$

where

(12.2.23)
$$S_{12}x_2 = - \sum_{z_0 \in \sigma_1} \mathrm{Res}_{z=z_0}(zI - A_\zeta^{(1)})^{-1}B^{(1)}C^{(2)}(zI - A_\pi^{(2)})^{-1}x_2$$

and

(12.2.24)
$$S_{21}x_1 = - \sum_{z_0 \in \sigma_2} \mathrm{Res}_{z=z_0}(zI - A_\zeta^{(2)})^{-1}B^{(2)}C^{(1)}(zI - A_\pi^{(1)})^{-1}x_1.$$

From the contour integral formula in Section A.1 of the Appendix for the solution of a Sylvester equation (in the case of uniqueness) and the Residue Theorem, we see that S_{12} and S_{21} as given by (12.2.23) and (12.2.24) are the solutions of the Sylvester equations (12.2.13) and (12.2.14). Finally from (12.2.19) and (12.2.22) it then follows that $f \in \mathcal{S}_{\tau_1 \oplus \tau_2}(\sigma_1 \cup \sigma_2)$.

Conversely one can show that any $f \in \mathcal{S}_{\tau_1 \oplus \tau_2}(\sigma_1 \cup \sigma_2)$ is in the intersection $\mathcal{S}_{\tau_1}(\sigma_1) \cap \mathcal{S}_{\tau_2}(\sigma_2)$ by reversing the steps in the argument above. □

For σ_ν a subset of $\sigma \subset \mathbb{C}$ and τ a σ-admissible data set, we define a σ_ν-corestriction $\tau \mid \sigma_\nu$ of τ as in Section 4.5.

The following is the analogue of the second part of Proposition 12.1.3. We leave the proof as an exercise for the reader.

PROPOSITION 12.2.6. *Suppose that σ_1 and σ_2 are subsets of $\sigma \subset \mathbb{C}$ and τ is a σ-admissible data set. Suppose τ_ν is a σ_ν-corestriction of τ for $\nu = 1, 2$, and τ_0 is a $(\sigma_1 \cap \sigma_2)$-corestriction of τ. Then*

$$\mathcal{S}_{\tau_0}(\sigma_1 \cap \sigma_2) = \mathcal{S}_{\tau_1}(\sigma_1) + \mathcal{S}_{\tau_2}(\sigma_2).$$

12.3 NULL-POLE SUBSPACE FROM A NULL-POLE TRIPLE

The goal of this section is to obtain the following description of the null-pole subspace for a rational matrix function in terms of a null-pole triple for the function.

THEOREM 12.3.1. *Suppose that W is a regular rational $n \times n$ matrix function, $\sigma \subset \mathbb{C}$, and $\tau = (C, A_\pi; A_\zeta, B; S)$ is a left null-pole triple for $W(z)$ over σ. Define the subspace $\mathcal{S}_\tau(\sigma)$ of $\mathcal{R}_n$ as in (12.2.2). Then $\mathcal{S}_\tau(\sigma)$ is the null-pole subspace $\mathcal{S}_\sigma(W)$ for $W(z)$ over σ.*

If we know a minimal realization

$$W(z) = I + C(zI - A)^{-1}B$$

for a rational matrix function then we know by Corollary 4.5.9 that

$$(C \mid \operatorname{Im} P, A \mid \operatorname{Im} P; A^\times \mid \operatorname{Im} P^\times, P^\times B; P^\times \mid \operatorname{Im} P)$$

is a left null pole triple for $W(z)$ over σ, where P is the Riesz projection for A associated with σ and $P^\times$ is the Riesz projection for $A^\times := A - BC$ associated with the subset σ in $\mathbb{C}$. From this we see that Theorem 12.3.1 can be used to give an explicit characterization of the null-pole subspace $\mathcal{S}_\sigma(W)$ for W in terms of a minimal realization for W.

COROLLARY 12.3.2. *Suppose $\sigma \subset \mathbb{C}$ and $W(z) = I + C(zI - A)^{-1}B$ is a minimal realization for $W(z)$ over σ. Let P and $P^\times$ be the Riesz projectors for A and $A^\times = A - BC$ associated with the subset σ. Then the left null-pole subspace $\mathcal{S}_\sigma(W)$ for W over σ is given by*

$$\mathcal{S}_\sigma(W) = \{C(zI - A)^{-1}x + h(z) : x \in \operatorname{Im} P, h \in \mathcal{R}_n(\sigma)$$
$$\text{such that } \sum_{z_0 \in \sigma} \operatorname{Res}_{z=z_0}(zI - A^\times)^{-1}P^\times Bh(z) = P^\times x\}. \quad \square$$

It is now an easy matter to prove Theorem 4.5.8.

COROLLARY 12.3.3 (= Theorem 4.5.8). *Let W_1 and W_2 be proper rational matrix functions which have the value I at infinity and let σ be a subset of $\mathbb{C}$. Then W_1*

and W_2 have a common left σ-null-pole triple if and only if $W_2(z)^{-1}W_1(z)$ has no poles and no zeros in σ.

PROOF. The condition that $W_2(z)^{-1}W_1(z)$ has no poles and no zeros in σ by Theorem 12.1.2 is equivalent to W_1 and W_2 having a common left null-pole subspace

$$\mathcal{S}_\sigma(W_1) = \mathcal{S}_\sigma(W_2).$$

By Theorem 12.3.1, $\mathcal{S}_\sigma(W_j) = \mathcal{S}_{\tau_j}(\sigma)$ where τ_j is a left σ-null-pole triple for $W_j(z)$. Now by Theorem 12.2.4 this is equivalent to a left σ-null-pole triple for $W_1(z)$ being similar to a left σ-null-pole triple for $W_2(z)$, and hence to $W_1(z)$ and $W_2(z)$ having a common left σ-null-pole triple. $\square$

Before proving Theorem 12.3.1 we first establish that $P_{\sigma^c}^0(\mathcal{S}_\sigma(W))$ and $\mathcal{S}_\sigma(W) \cap \mathcal{R}_n(\sigma)$ have the proper form as required by Corollary 12.2.3.

LEMMA 12.3.4. If (C, A_π) is a right pole pair for the rational matrix function $W(z)$ over the subset $\sigma \subset \mathbb{C}$, then

$$P_{\sigma^c}^0(\mathcal{S}_\sigma(W)) = \{C(zI - A_\pi)^{-1}x : x \in \mathbb{C}^{n_\pi}\}$$

where n_π is the size of A_π.

PROOF. Without loss of generality we assume that $W(\infty) = I$. Since (C, A_π) is a right pole pair over σ, we know that $W(z)$ has a realization of the form

$$W(z) = I + \widetilde{C}(zI - \widetilde{A}_\pi)^{-1}\widetilde{B}$$

such that $C = \widetilde{C} \mid \operatorname{Im} P_\pi$, $A_\pi = \widetilde{A}_\pi \mid \operatorname{Im} P_\pi$, and P_π is the Riesz projection for $\widetilde{A}_\pi$ associated with σ. We now compute for $h(z) \in \mathcal{R}_n(\sigma)$ the expression

$$W(z)h(z) = [I + \widetilde{C}(zI - \widetilde{A}_\pi)^{-1}\widetilde{B}]h(z).$$

Let γ be any simple closed rectifiable contour in the complex plane such that $\sigma(A_\pi)$ is in the interior of γ and $\sigma(\widetilde{A}_\pi)\backslash\sigma(A_\pi)$ and all poles of $h(z)$ are in the exterior of γ. In this case, by using partial fraction expansion and the residue theorem it is easy to see that, for z in the interior of γ

$$P_\sigma[Wh](z) = \sum_{t_0 \in \sigma} \operatorname{Res}_{t=t_0}[(t-z)^{-1}W(t)h(t)]$$

$$= \frac{1}{2\pi i}\int_\gamma (t-z)^{-1}[I + \widetilde{C}(tI - \widetilde{A}_\pi)^{-1}\widetilde{B}]h(t)dt$$

and thus

$$P_{\sigma^c}^0[Wh](z) = (I - P_\sigma)[Wh](z)$$

$$= \frac{1}{2\pi i}\int_\gamma (t-z)^{-1}\widetilde{C}[(zI - \widetilde{A}_\pi)^{-1} - (tI - \widetilde{A}_\pi)^{-1}]\widetilde{B}h(t)dt$$

$$= \frac{1}{2\pi i}\int_\gamma \widetilde{C}(zI - \widetilde{A}_\pi)^{-1}(t - \widetilde{A}_\pi)^{-1}tBh(t)dt$$

$$= C(zI - A)^{-1}P_\pi\frac{1}{2\pi i}\int_\gamma (tI - \widetilde{A}_\pi)^{-1}\widetilde{B}h(t)dt$$

has the desired form. Conversely, if x is any vector in $\mathbf{C}^{n_\pi}$, then since $(\widetilde{A}, \widetilde{B})$ is full range pair, we can find by Lemma 12.2.2 an $h \in \mathcal{R}_n(\sigma)$ for which

$$x = \frac{1}{2\pi i} \int_\gamma (tI - \widetilde{A}_\pi)^{-1} \widetilde{B} h(t) dt.$$

Then certainly $Wh \in \mathcal{S}_\sigma(W)$, and the computation above shows that $P^0_{\sigma^c}[Wh](z) = C(zI - A)^{-1}x$ for z inside γ, and hence for all z by analytic continuation. $\square$

The next result shows that the subspace $\mathcal{S}_\sigma(W) \cap \mathcal{R}_n(\sigma)$ is determined by a left null pair (A_ζ, B) for $W(z)$ over σ.

LEMMA 12.3.5. *Suppose (A_ζ, B) is a left null pair for $W(z)$ over the subset $\sigma \subset \mathbf{C}$. Then*

$$\mathcal{S}_\sigma(W) \cap \mathcal{R}_n(\sigma) = \{k \in \mathcal{R}_n(\sigma): \sum_{z_0 \in \sigma} \mathrm{Res}_{z=z_0}[(zI - A_\zeta)^{-1} Bk(z)] = 0\}.$$

PROOF. Without loss of generality we again assume $W(\infty) = I$. Since (A_ζ, B) is a left null pair for $W(z)$ over σ we know by Theorem 4.2.3 that $W^{-1}(z)$ has a minimal realization

$$W^{-1}(z) = I - \widetilde{C}(zI - \widetilde{A}_\zeta)^{-1} \widetilde{B}$$

such that

$$\widetilde{A}_\zeta \mid \mathrm{Im}\, P_\zeta = A_\zeta, \qquad P_\zeta \widetilde{B} = B$$

where P_ζ is the Riesz projection for $\widetilde{A}_\zeta$ associated with σ. Suppose $k(z) \in \mathcal{S}_\sigma(W) \cap \mathcal{R}_n(\sigma)$. Thus we have both $k(z) \in \mathcal{R}_n(\sigma)$ and $W^{-1}(z)k(z) \in \mathcal{R}_n(\sigma)$, or $P^0_{\sigma^c}[W^{-1}k] = 0$. Let γ be a simple closed rectifiable contour such that $\sigma(A_\zeta)$ is inside γ and the poles of k and the set $\sigma(\widetilde{A}_\zeta)\backslash\sigma(A_\zeta)$ are outside γ. Then for z inside γ

$$P_\sigma[W^{-1}k](z) = \sum_{t_0 \in \sigma} \mathrm{Res}_{t=t_0}[(t-z)^{-1}W^{-1}(t)k(t)]$$

$$= \frac{1}{2\pi i} \int_\gamma (t-z)^{-1}W^{-1}(t)k(t)dt.$$

Thus

$$0 = P^0_{\sigma^c}[W^{-1}k](z)$$

$$= (I - P_\sigma)[W^{-1}k](z)$$

$$= \frac{1}{2\pi i} \int_\gamma (t-z)^{-1}\widetilde{C}[(zI - \widetilde{A}_\zeta)^{-1} - (tI - \widetilde{A}_\zeta)^{-1}]\widetilde{B}k(t)dt$$

$$= \widetilde{C}(zI - \widetilde{A}_\zeta)^{-1}\frac{1}{2\pi i}\int_\gamma (tI - \widetilde{A}_\zeta)^{-1}Bk(t)dt$$

$$= \widetilde{C}(zI - \widetilde{A}_\zeta)^{-1}P_\zeta\frac{1}{2\pi i}\int_\gamma (tI - A_\zeta)^{-1}Bk(t)dt.$$

Since $(\widetilde{C}, \widetilde{A}_\zeta)$ is a null-kernel pair, we conclude by Lemma 12.2.2

$$0 = \frac{1}{2\pi i} \int_\gamma (tI - A_\zeta)^{-1} Bk(t)dt$$

$$= \sum_{z_0 \in \sigma} \text{Res}_{z=z_0}[(zI - A_\zeta)^{-1} Bk(z)].$$

Conversely, if $k \in \mathcal{R}_n(\sigma)$ satisfies this condition, then the above computation shows that $W^{-1}k \in \mathcal{R}_n(\sigma)$ so $k \in \mathcal{S}_\sigma(W) \cap \mathcal{R}_n(\sigma)$. $\square$

We are now ready for the proof of Theorem 12.3.1. Note that the special case where $W(z)$ has no zeros in σ follows immediately from Lemma 12.3.4, and the special case where $W(z)$ has no poles in σ follows directly from Lemma 12.3.5.

PROOF OF THEOREM 12.3.1. As usual we assume that $W(\infty) = I$. By Theorem 4.4.3, as $(C, A_\pi; A_\zeta, B; S)$ is a left null-pole triple for $W(z)$ over σ, $W(z)$ has a minimal realization

$$W(z) = I + \widetilde{C}(zI - \widetilde{A}_\pi)^{-1} \widetilde{S}^{-1} \widetilde{B}$$

and $W^{-1}(z)$ has a minimal realization

(12.3.1) $$W^{-1}(z) = I - \widetilde{C}\widetilde{S}^{-1}(zI - \widetilde{A}_\zeta)^{-1} \widetilde{B}$$

such that

$$\widetilde{C}|\operatorname{Im} P_\pi = C; \qquad \widetilde{A}_\pi|\operatorname{Im} P_\pi = A_\pi$$

$$\widetilde{A}_\zeta|\operatorname{Im} P_\zeta = A_\zeta; \quad P_\zeta \widetilde{B} = B; \quad P_\zeta \widetilde{S}|\operatorname{Im} P_\pi = S$$

where P_π (resp. P_ζ) is the Riesz projection for $\widetilde{A}_\pi$ (resp. $\widetilde{A}_\zeta$) associated with σ. We also know that $\widetilde{S}$ satisfies the Sylvester equation

(12.3.2) $$\widetilde{S}\widetilde{A}_\pi - \widetilde{A}_\zeta \widetilde{S} = \widetilde{B}\widetilde{C}.$$

Suppose now that $k(z) = W(z)h(z)$ is in $\mathcal{S}_\sigma(W)$, for some $h \in \mathcal{R}_n(\sigma)$. By the computation in the proof of Lemma 12.3.4,

(12.3.3) $$[P^0_{\sigma^c}(Wh)](z) = C(zI - A_\pi)^{-1}x$$

where

(12.3.4) $$x = \frac{1}{2\pi i} \int_\gamma (tI - \widetilde{A}_\pi)^{-1} P_\pi \widetilde{S}^{-1} \widetilde{B}h(t)dt$$

where γ is a simple closed rectifiable contour having $\sigma(A_\pi) \cup \sigma(A_\zeta)$ inside and $[\sigma(\widetilde{A}_\pi)\backslash\sigma(A_\pi)] \cup [\sigma(\widetilde{A}_\zeta)\backslash\sigma(A_\zeta)]$ and the poles of $h(z)$ outside. From (12.3.3) and (12.3.4), we see that we must show the equality

(12.3.5) $$\frac{1}{2\pi i} \int_\gamma (zI - A_\zeta)^{-1} B[P_\sigma k](z)dz = Sx.$$

Compute

$$\widetilde{B}[P_\sigma k](z) = \widetilde{B}\{W(z)h(z) - [P_{\sigma^c}^0(Wh)](z)\}$$
$$= \widetilde{B}[I + \widetilde{C}(zI - \widetilde{A}_\pi)^{-1}\widetilde{S}^{-1}\widetilde{B}]h(z) - \widetilde{B}\widetilde{C}(zI - \widetilde{A}_\pi)^{-1}P_\pi x.$$

Use the Sylvester equation (12.3.2) to write this as

(12.3.6)
$$\widetilde{B}[P_\sigma k](z) = \widetilde{B}h(z) + [\widetilde{S}(\widetilde{A}_\pi - zI) - (\widetilde{A}_\zeta - zI)\widetilde{S}]$$
$$\cdot \{(zI - \widetilde{A}_\pi)^{-1}\widetilde{S}^{-1}\widetilde{B}h(z) - (zI - \widetilde{A}_\pi)^{-1}P_\pi x\}$$
$$= \widetilde{B}h(z) - \widetilde{B}h(z) + (zI - \widetilde{A}_\zeta)\widetilde{S}(zI - \widetilde{A}_\pi)^{-1}\widetilde{S}^{-1}\widetilde{B}h(z)$$
$$+ \widetilde{S}P_\pi x - (zI - \widetilde{A}_\zeta)\widetilde{S}(zI - \widetilde{A}_\pi)^{-1}P_\pi x.$$

Therefore for z inside γ, by (12.3.6) we have

$$\frac{1}{2\pi i}\int_\gamma (zI - A_\zeta)^{-1}B[P_\sigma k](z)dz = \frac{1}{2\pi i}\int_\gamma P_\zeta(zI - \widetilde{A}_\zeta)^{-1}\widetilde{B}[P_\sigma k](z)dz$$
$$= \frac{1}{2\pi i}\int_\gamma P_\zeta\widetilde{S}(zI - \widetilde{A}_\pi)^{-1}\widetilde{S}^{-1}\widetilde{B}h(z)dz$$
$$+ \frac{1}{2\pi i}\int_\gamma P_\zeta(zI - \widetilde{A}_\zeta)^{-1}\widetilde{S}P_\pi x - \frac{1}{2\pi i}\int_\gamma P_\zeta\widetilde{S}(zI - \widetilde{A}_\pi)^{-1}P_\pi x.$$

Using (12.3.4) to simplify the first term and the Riesz functional calculus for the remaining two terms gives

$$\frac{1}{2\pi i}\int_\gamma (zI - A_\zeta)^{-1}B[P_\sigma k](z)dz = P_\zeta\widetilde{S}x + P_\zeta\widetilde{S}P_\pi x - P_\zeta\widetilde{S}P_\pi x$$
$$= P_\zeta\widetilde{S}x$$
$$= Sx$$

which verifies (12.3.5) as needed. Thus each function $k(z) \in \mathcal{S}_\sigma(W)$ satisfies the characterization of Theorem 12.3.1.

Conversely, suppose that h has the form $k(z) = C(zI - A_\pi)^{-1}x + k_+(z)$ where $k_+(z) \in \mathcal{R}_n(\sigma)$ satisfies

(12.3.7)
$$\frac{1}{2\pi i}\int_\gamma (zI - A_\pi)^{-1}Bk_+(z)dz = Sx.$$

To show that $k \in \mathcal{S}_\sigma(W)$, we must show that $W^{-1}k \in \mathcal{R}_n(\sigma)$, i.e.

(12.3.8)
$$P_{\sigma^c}^0[W^{-1}(z)C(zI - A_\pi)^{-1}x] + P_{\sigma^c}^0[W^{-1}(z)k_+(z)] = 0.$$

By a computation as in Lemma 12.3.5, for z inside a suitable contour γ,

(12.3.9)
$$[P_{\sigma^c}^0(W^{-1}k_+)](z) = -\widetilde{C}\widetilde{S}^{-1}(zI - \widetilde{A}_\zeta)^{-1}P_\zeta\frac{1}{2\pi i}\int_\gamma (tI - A_\zeta)^{-1}Bk_+(t)dt$$
$$= -\widetilde{C}\widetilde{S}^{-1}P_\zeta(zI - A_\zeta)^{-1}Sx,$$

where the realization (12.3.1) was used. On the other hand, using the same realization (12.3.1) again, we compute

$$P^0_{\sigma^c}[W^{-1}(z)C(zI - A_\pi)^{-1}x]$$
$$= C(zI - A_\pi)^{-1}x - P^0_{\sigma^c}[\widetilde{C}\widetilde{S}^{-1}(zI - \widetilde{A}_\zeta)^{-1}\widetilde{B}\widetilde{C}P_\pi(zI - A_\pi)^{-1}x].$$

Use (12.3.2) again to get from this

$$P^0_{\sigma^c}[W^{-1}(z)C(zI - A_\pi)^{-1}x]$$
$$= C(zI - A_\pi)^{-1}x - P^0_{\sigma^c}\widetilde{C}\widetilde{S}^{-1}(zI - A_\zeta)^{-1}[\widetilde{S}(\widetilde{A}_\pi - zI) - (\widetilde{A}_\zeta - zI)\widetilde{S}]P_\pi(zI - A_\pi)^{-1}x$$
$$= C(zI - A_\pi)^{-1}x + P^0_{\sigma^c}\widetilde{C}\widetilde{S}^{-1}(zI - A_\zeta)^{-1}\widetilde{S}P_\pi x - P^0_{\sigma^c}\widetilde{C}P_\pi(zI - A_\pi)^{-1}x.$$

Since $C(zI - A_\pi)^{-1}x \in \mathcal{R}^0_n(\sigma^c)$ and $x \in \operatorname{Im} P_\pi$, the first and last terms cancel, and we are left with

(12.3.10)
$$P^0_{\sigma^c}[W^{-1}(z)C(zI - A_\pi)^{-1}x] = P^0_{\sigma^c}\widetilde{C}\widetilde{S}^{-1}(zI - A_\zeta)^{-1}\widetilde{S}P_\pi x$$
$$= \widetilde{C}\widetilde{S}^{-1}P_\zeta(zI - A_\zeta)^{-1}Sx.$$

Combining (12.3.9) and (12.3.10) gives (12.3.8) as needed. $\square$

12.4 LOCAL NULL-POLE SUBSPACE IN TERMS OF LAURENT COEFFICIENTS

In this section we consider only the case where the subset σ is a singleton $\sigma = \{z_0\}$ and show how to express the null-pole subspace $\mathcal{S}_{\{z_0\}}(W)$ at z_0 for a regular rational matrix function in terms of Laurent coefficients of W and W^{-1} of their respective Laurent series at z_0.

We thus consider as given a regular rational matrix function $W(z)$ with Laurent expansion

(12.4.1)
$$W(z) = \sum_{j=-p}^{\infty} W_j(z - z_0)^j$$

and with inverse having the Laurent expansion

(12.4.2)
$$W(z)^{-1} = \sum_{j=-q}^{\infty} \widetilde{W}_j(z - z_0)^j.$$

We assume that p is the order of z_0 as a pole of W and q is the order of z_0 as a pole of W^{-1}; thus $p \geq 0$, $q \geq 0$ and $W_{-p} \neq 0$ if $p > 0$ and $\widetilde{W}_{-q} \neq 0$ if $q > 0$. If $(C, A_\pi; A_\zeta, B; S)$ is a null-pole triple for $W(z)$ at z_0, then by Theorem 12.3.1 we know that the null-pole subspace $\mathcal{S}_{\{z_0\}}(W) = W \cdot \mathcal{R}_n(\{z_0\})$ for W at z_0 is given by

(12.4.3)
$$\mathcal{S}_{\{z_0\}}(W) = \{C(zI - A_\pi)^{-1}x + h(z): x \in \mathbb{C}^{n_\pi},$$
$$h \in \mathcal{R}_n(\{z_0\}), \operatorname{Res}_{z=z_0}(zI - A_\zeta)^{-1}Bh(z) = Sx\},$$

where n_π is the size of A_π. It will be convenient to describe elements of $\mathcal{S}_{\{z_0\}}(W)$ in terms of their Laurent expansions $f(z) = \sum_{j=-\infty}^\infty f_j(z - z_0)^j$ as z_0.

PROPOSITION 12.4.1. *Let f be a rational vector function in $\mathcal{R}_n$ with Laurent expansion*

$$f(z) = \sum_{j=-\infty}^\infty f_j(z - z_0)^j$$

and let $\tau = (C, A_\pi; A_\zeta, B; S)$ be a null pole triple for W at z_0. Then $f \in \mathcal{S}_{\{z_0\}}(W)$ if and only if

(i) $f_j = 0$ *for $j < -p$;*

(ii) $\operatorname{col}(f_{-p}, \cdots, f_{-1}) = [\operatorname{col}(C(A_\pi - z_0 I)^{p-j})_{j=1}^p] x$ *for a vector x in $\mathbb{C}^{n_\pi}$ such that $[B, \ldots, (A_\zeta - z_0 I)^{q-1} B] \operatorname{col}(f_0, \ldots, f_{q-1}) = S x$.*

PROOF. Since $\sigma(A_\pi) = \{z_0\}$, we have

$$(zI - A_\pi)^{-1} = ((z - z_0)I - (A_\pi - z_0 I))^{-1}$$
$$= \sum_{j=1}^p (A_\pi - z_0 I)^{j-1}(z - z_0)^{-j}.$$

If $f \in \mathcal{S}_{\{z_0\}}(W)$, from the representation (12.4.3) we conclude that $f_j = 0$ for $j < -p$ and

(12.4.4) $$\operatorname{col}(f_{-p}, \ldots, f_{-1}) = [\operatorname{col}(C(A_\pi - z_0 I)^{p-j})_{j=1}^p] x.$$

Similarly, since $\sigma(A_\zeta) = \{z_0\}$ we have

$$(zI - A_\zeta)^{-1} = \sum_{j=1}^q (A_\zeta - z_0 I)^{j-1}(z - z_0)^{-j}.$$

If $h(z) = \sum_{j=0}^\infty h_j(z - z_0)^j$ is analytic at z_0, then the residue of $(zI - A_\zeta)^{-1} h(z)$ at z_0 is easily seen to be

$$\operatorname{Res}_{z=z_0}(zI - A_\zeta)^{-1} h(z) = [B, (A_\zeta - z_0 I)B, \ldots, (A_\zeta - z_0 I)^{q-1} B] \operatorname{col}(h_0, \ldots, h_{q-1}).$$

Proposition 12.4.1 now follows immediately from the representation (12.4.3) for $\mathcal{S}_{\{z_0\}}(W)$. $\square$

For W a regular rational matrix function with a pole of order p and a zero of order q at z_0, let us denote by $\mathcal{N}_{\{z_0\}}(W)$ the space of columns vectors described in Proposition 12.4.1:

(12.4.5) $$\mathcal{N}_{\{z_0\}}(W) = \{\operatorname{col}(f_{-p}, \ldots, f_{q-1}): \sum_{j=-p}^{q-1} f_j(z - z_0)^j \in \mathcal{S}_{\{z_0\}}(W)\}.$$

We refer to $\mathcal{N}_{\{z_0\}}(W)$ as the *coefficient form of the local null-pole subspace for W at z_0*, or for short *coefficient null-pole subspace*. Thus by definition we have
(12.4.6)

$$\mathcal{S}_{\{z_0\}}(W) = \left\{ \sum_{j=-p}^{q-1} f_j(z - z_0)^j : \mathrm{col}(f_{-p}, \ldots, f_{q-1}) \in \mathcal{N}_{\{z_0\}}(W) \right\} + (z - z_0)^q \mathcal{R}_n(\{z_0\})$$

and by Proposition 12.4.1

(12.4.7)

$$\mathcal{N}_{\{z_0\}}(W) = \{\mathrm{col}(f_{-p}, \ldots, f_{q-1}):$$

$$\mathrm{col}(f_{-p}, \ldots, f_{-1}) = \big[\mathrm{col}\big(C(A_\pi - z_0 I)^{p-j}\big)_{j=1}^p\big] x$$

$$\text{for an } x \in \mathbf{C}^{n_\pi} \text{ for which}$$

$$[B, \ldots, (A_\zeta - z_0 I)^{q-1} B]\,\mathrm{col}(f_0, \ldots, f_{q-1}) = Sx\}$$

where $(C, A_\pi; A_\zeta, B; S)$ is a left null-pole triple for W at z_0.

We next obtain characterizations of $\mathcal{N}_{\{z_0\}}(W)$ in terms of the Laurent coefficients (W_j) and $(\widetilde{W}_j)$ for W and W^{-1} at z_0. For a given set of matrices $K_1, K_2, \ldots, K_\nu$ of compatible sizes, let us introduce the notation $\Delta(K_1, \ldots, K_\nu)$ for the $\nu \times \nu$ block lower triangular Toeplitz matrix

$$\Delta(K_1, \ldots, K_\nu) = \begin{bmatrix} K_1 & & \cdots & 0 \\ K_2 & K_1 & & \vdots \\ \vdots & \ddots & \ddots & \\ K_\nu & \cdots & K_2 & K_1 \end{bmatrix}.$$

PROPOSITION 12.4.2. *Let $W(z)$ be a regular rational matrix function with W and W^{-1} having Laurent expansions at z_0 given by (12.4.1) and (12.4.2), and let $\mathcal{N}_{\{z_0\}}(W)$ be the coefficient null-pole subspace (12.4.5) for W at z_0. Then*

$$\mathrm{Im}\,\Delta(W_{-p}, \ldots, W_{q-1}) = \mathcal{N}_{\{z_0\}}(W) = \mathrm{Ker}\,\Delta(\widetilde{W}_{-q}, \ldots, \widetilde{W}_{p-1}).$$

PROOF. Suppose $f(z) = \sum_{j=-p}^{\infty} f_j(z - z_0)^j$ has the form $f(z) = W(z)g(z)$ where $g(z) = \sum_{j=0}^{\infty} g_j(z - z_0)^j$ is rational and analytic at z_0. Then multiplication of series expansions gives

$$\mathrm{col}(f_{-p}, \ldots, f_{q-1}) = \Delta(W_{-p}, \ldots, W_{q-1})\,\mathrm{col}(g_0, \ldots, g_{p+q-1}).$$

This identity implies $\mathrm{Im}\,\Delta(W_{-p}, \ldots, W_{q-1}) = \mathcal{N}_{\{z_0\}}(W)$. Similarly, $f(z) = \sum_{j=-p}^{\infty} f_j(z - z_0)^j$ being in $\mathcal{S}_{\{z_0\}}(W)$ means that $g(z) = W(z)^{-1} f(z)$ has Laurent expansion $g(z) = \sum_{j=-\infty}^{\infty} g_j(z - z_0)^j$ with $g_j = 0$ for $j < 0$. Again by multiplication of series we see that

$$\mathrm{col}(g_{-p-q}, \ldots, g_{-1}) = \Delta(\widetilde{W}_{-q}, \ldots, \widetilde{W}_{p-1})\,\mathrm{col}(f_{-p}, \ldots, f_{q-1}).$$

This identity then implies that

$$\mathcal{N}_{\{z_0\}}(W) = \operatorname{Ker} \Delta(\widetilde{W}_{-q}, \ldots, \widetilde{W}_{p-1}). \quad \square$$

Note as a corollary to Proposition 12.4.2 we have that

$$\operatorname{Im} \Delta(W_{-p}, \ldots, W_{q-1}) = \operatorname{Ker} \Delta(\widetilde{W}_{-q}, \ldots, \widetilde{W}_{p-1}).$$

We have seen in Chapters 1 and 3 that the null structure of W at z_0 is related to the singular part $\sum_{j=-q}^{-1} \widetilde{W}_j (z - z_0)^j$ of the Laurent expansion of W^{-1} at z_0, and the pole structure of W at z_0 is related to the singular part $\sum_{j=-p}^{-1} W_j (z - z_0)^j$ of the Laurent expansion of W at z_0. The null-pole subspace $\mathcal{S}_{\{z_0\}}(W)$ on the other hand, as follows from Theorem 12.3.1 and its converse Theorem 14.1.3 to come, gives information equivalent to that contained in a null-pole triple for W at z_0; in addition to the (right) pole structure information (C, A_π) and (left) null structure information (A_ζ, B), also included is the so-called coupling information from the coupling matrix S. This point is brought out in Proposition 12.4.2; the singular part of the Laurent expansion of neither W nor W^{-1} alone is not enough to recover the null-pole subspace (unless we are in the much simpler situation where either $p = 0$ or $q = 0$). It turns out, not surprisingly, that the singular parts of $W(z)$ and $W(z)^{-1}$ taken together do not give enough information to recover the null-pole subspace. The next theorem describes exactly what extra information is required. To state the result we need to introduce the matrix Γ given by

$$(12.4.8) \qquad \Gamma = - \begin{bmatrix} \widetilde{W}_{-q+p} & \cdots & \widetilde{W}_{-q+1} \\ \vdots & & \vdots \\ \widetilde{W}_{p-1} & \cdots & \widetilde{W}_0 \end{bmatrix} \cdot \Delta(W_{-p}, \ldots, W_{-1})$$

where $\widetilde{W}_j$ $(j \geq -q)$ are the Laurent coefficients for $W(z)^{-1}$ as in (12.4.2) and W_j $(j \geq -p)$ are the Laurent coefficients for $W(z)$ as in (12.4.1).

THEOREM 12.4.3. *Let $W(z)$ be a rational matrix function with Laurent expansion (12.4.1) and with $W(z)^{-1}$ having Laurent expansion (12.4.2) at the point $z_0 \in \mathbb{C}$, and let Γ be the matrix defined by (12.4.8). Then the coefficient null-pole subspace $\mathcal{N}_{\{z_0\}}(W)$ for W at z_0 is given by:*

$$\mathcal{N}_{\{z_0\}}(W) = \{\operatorname{col}(f_{-p}, \ldots, f_{-1}) \colon \operatorname{col}(f_{-p}, \ldots, f_{-1}) = \Delta(W_{-p}, \ldots, W_{-1})x$$

$$\textit{for some } x \in \mathbb{C}^{n_\pi} \textit{ for which } \Delta(\widetilde{W}_{-q}, \ldots, \widetilde{W}_{-1}) \operatorname{col}(f_0, \ldots, f_{q-1}) = \Gamma x\}.$$

PROOF. Proposition 12.4.2 identified $\mathcal{N}_{\{z_0\}}(W)$ in two ways

$$(12.4.9) \qquad \mathcal{N}_{\{z_0\}}(W) = \operatorname{Im} \Delta(W_{-p}, \ldots, W_{q-1})$$

and

$$(14.2.10) \qquad \mathcal{N}_{\{z_0\}}(W) = \operatorname{Ker} \Delta(\widetilde{W}_{-q}, \ldots, \widetilde{W}_{p-1}).$$

In particular, from (12.4.9) we see that a given sequence $\mathrm{col}(f_{-p}, \ldots, f_{-1})$ has a continuation $\mathrm{col}(f_0, \ldots, f_{q-1})$ for which the combined sequence $\mathrm{col}(f_{-p}, \ldots, f_{q-1})$ is in $\mathcal{N}_{\{z_0\}}(W)$ if and only if $\mathrm{col}(f_{-p}, \ldots, f_{-1}) = \Delta(W_{-p}, \ldots, W_{-1})x$ for some $x \in \mathbf{C}^{n\pi}$. Given that $\mathrm{col}(f_{-p}, \ldots, f_{-1})$ is of this form, a continuation $\mathrm{col}(f_0, \ldots, f_{q-1})$ is such that the combined sequence $\mathrm{col}(f_{-p}, \ldots, f_{q-1})$ is in $\mathcal{N}_{\{z_0\}}(W)$ if and only if

$$(12.4.11) \qquad \Delta(\widetilde{W}_{-q}, \ldots, \widetilde{W}_{p-1}) \left[\begin{array}{c} \Delta(W_{-p}, \ldots, W_{-1})x \\ \mathrm{col}(f_0, \ldots, f_{q-1}) \end{array} \right] = 0.$$

Note that $\Delta(\widetilde{W}_{-q}, \ldots, \widetilde{W}_{p-1})$ has the block decomposition

$$\Delta(\widetilde{W}_{-q}, \ldots, \widetilde{W}_{p-1}) = \left[\begin{array}{cc} \Delta_1 & 0 \\ \Delta_3 & \Delta_2 \end{array} \right]$$

where

$$\Delta_1 = \Delta(\widetilde{W}_{-q}, \ldots, \widetilde{W}_{-q+p-1})$$
$$\Delta_2 = \Delta(\widetilde{W}_{-q}, \ldots, \widetilde{W}_{-1})$$

and

$$\Delta_3 = \left[\begin{array}{ccc} \widetilde{W}_{-q+p} & \cdots & \widetilde{W}_{-q+1} \\ \vdots & \ddots & \vdots \\ \widetilde{W}_{p-1} & \cdots & \widetilde{W}_0 \end{array} \right].$$

Thus (12.4.11) is equivalent to

$$(12.4.12) \qquad \Delta_1 \cdot \Delta(W_{-p}, \ldots, W_{-1})x = 0$$

$$(12.4.13) \qquad \Delta_3 \cdot \Delta(W_{-p}, \ldots, W_{-1})x + \Delta_2 \cdot \mathrm{col}(f_0, \ldots, f_{q-1}) = 0.$$

In fact (12.4.12) is automatic; this is a consequence of the identity

$$\mathrm{Im}\, \Delta(W_{-p}, \ldots, W_{q-1}) = \mathrm{Ker}\, \Delta(\widetilde{W}_{-q}, \ldots, \widetilde{W}_{p-1})$$

which came out of Proposition 12.4.2. The equation (12.4.13) reduces to

$$\Delta(\widetilde{W}_{-q}, \ldots, \widetilde{W}_{-1}) \cdot \mathrm{col}(f_0, \ldots, f_{q-1}) = \Gamma x.$$

This completes the proof of Theorem 12.4.3. □

12.5 LOCAL NULL-POLE SUBSPACE IN TERMS OF NULL AND POLE CHAINS

Recall from Chapters 1 and 3 that a left null pair (A_ζ, B) for a rational matrix function $W(z)$ at a point z_0 arose from a canonical set of left null chains or of null functions at z_0, while a right pole pair (C, A_π) arose from a canonical set of right pole chains or pole functions at z_0. Later in Chapter 4, we used the theory of realization to introduce the null-pole coupling matrix S associated with the (left) null pair (A_ζ, B) and (right) pole pair (C, A_π) and this turned out to give complete information about the null-pole structure of $W(z)$ at z_0 on the left side. In this section we use the null-pole subspace characterization of the coupling matrix S (Theorem 12.3.1 or (12.4.7)) to give an alternative description of S in terms of null and pole chains or null and pole functions.

Suppose now that $\{x_i(z): i = 1, \ldots, d\}$ is a canonical set of left null functions and $\{u_k(z): k = 1, \ldots, e\}$ is a canonical set of right pole functions for $W(z)$ at z_0. Thus each $x_i(z)$ is a row vector function with a Taylor expansion

$$\sum_{j=1}^{\infty} x_{ij}(z - z_0)^{j-1}$$

at z_0 and each $u_k(z)$ is a column vector function with Taylor expansion of the form

$$u_k(z) = \sum_{j=1}^{\infty} u_{kj}(z - z_0)^{j-1}$$

at z_0. We suppose that $x_i(z)$ has order μ_i as a null function, so the order q of z_0 as a pole of $W(z)^{-1}$ is $q = \max\{\mu_i: i = 1, \ldots, d\}$, and that ν_k is the order of u_k as a pole function for $W(z)$ at z_0, so the order p of z_0 as a pole of $W(z)$ is $p = \max\{\nu_k: k = 1, \ldots, e\}$.

Associated with the left null function $x_i(z)$ is the left chain $(x_{i1}, x_{i2}, \ldots, x_{i\mu_i})$. From these data define matrices X_i and $A_{\zeta i}$ by

(12.5.1) $$X_i = \text{col}(x_{i1}, \ldots, x_{i\mu_i})$$

and

(12.5.2) $$A_{\zeta i} = z_0 I_{\mu_i} + J_{\mu_i}^T$$

where J_{μ_i} is the $\mu_i \times \mu_i$ nilpotent Jordan block

$$J_{\mu_i} = \begin{bmatrix} 0 & 1 & & \\ & 0 & \ddots & \\ & & \ddots & 1 \\ & & & 0 \end{bmatrix}.$$

If we then form

(12.5.3) $$X = \text{col}(X_i)_{i=1}^{d}$$

and

(12.5.4) $$A_\zeta = \text{diag}(A_{\zeta 1}, \ldots, A_{\zeta d})$$

then (see Section 1.2) (A_ζ, X) is a left null pair for $W(z)$ at z_0.

Similarly associated with the right pole function $u_k(z)$ is the right pole chain $(u_{k1}, \ldots, u_{k\nu_k})$. From these data define matrices U_k and $A_{\pi k}$ by

$$(12.5.5) \qquad\qquad U_k = [u_{k1}, \ldots, u_{k\nu_k}]$$

and

$$(12.5.6) \qquad\qquad A_{\pi k} = z_0 I_{\nu_k} + J_{\nu_k}.$$

Then define matrices U and A_π by

$$(12.5.7) \qquad\qquad U = [U_1, \ldots, U_e]$$

and

$$(12.5.8) \qquad\qquad A_\pi = \operatorname{diag}(A_{\pi 1}, \ldots, A_{\pi e}).$$

Then (U, A_π) is a right pole pair for $W(z)$ at z_0.

Associated with this left null pair (A_ζ, X) and right pole pair (U, A_π) is a coupling matrix S. From Proposition 12.4.1 we see how the coupling matrix S combines with the pole pair (U, A_π) and null pair (A_ζ, X) to describe the coefficient null-pole subspace $\mathcal{N}_{\{z_0\}}(W)$:

$$(12.5.9)$$
$$\mathcal{N}_{\{z_0\}}(W) = \{\operatorname{col}(f_{-p}, \ldots, f_{q-1}) \colon \operatorname{col}(f_{-p}, \ldots, f_{-1}) = [\operatorname{col}(U(A_\pi - z_0 I)^{p-j})_{j=1}^p] x$$
$$\text{for an } x \text{ such that } [X, \ldots, (A_\zeta - z_0 I)^{q-1} X] \operatorname{col}(f_0, \ldots, f_{p-1}) = Sx\}.$$

The goal of this section is to obtain an alternative characterization of $\mathcal{N}_{\{z_0\}}(W)$ completely in terms of canonical sets of null functions and of pole functions for $W(z)$ at z_0; this will then lead to a characterization of S in terms of pole and null functions.

To do this we introduce the notion of *augmented left null chain* for $W(z)$ at z_0. As was emphasized in Chapter 3 if $W(z)$ has both a zero and a pole at z_0 and $x_{i1}, \ldots, x_{i\mu_i}$ is a null chain of order μ_i for $W(z)$ at z_0, then the succeeding Taylor coefficients $x_{i,\mu_i+1}, \ldots, x_{i,\mu_i+p}$ for an associated null function $x_i(z) = \sum_{j=1}^\infty x_{ij}(z-z_0)^{j-1}$ cannot be taken arbitrarily; what is true is that x_{ij} can be taken arbitrarily if $j > \mu_i + p$ (where p is the order of z_0 as a pole of $W(z)$). The collection of $\mu_i + p$ row vectors $(x_{i1}, \ldots, x_{i,\mu_i+p})$ associated with the left null function $x_i(z) = \sum_{j=1}^\infty x_{ij}(z-z_0)^{j-1}$ of order μ_i we call an *augmented left null chain of length* μ_i. We may then form a *canonical set of augmented left null chains* $(x_{11}, \ldots, x_{1,\mu_1+p}, \ldots, x_{d1}, \ldots, x_{d,\mu_d+p})$ just as is done for null chains. It may be necessary however to include additional augmented null chains of a negative or zero length $\mu_i \le 0$. By this we mean a chain $(x_{i1}, \ldots, x_{i,\mu_i+p})$ for which $x_{i1} \ne 0$ and the associated row vector function $x_i(z) = \sum_{j=1}^{\mu_i+p} x_{ij}(z-z_0)^{j-1}$ has the property that $x_i(z)W(z)$ has a pole of order at most $-\mu_i < p$. The minimum possible length for an augmented null function is then $-p+1$. The *augmented eigenspace* for $W(z)$ at z_0 is defined to be the span of all beginning vectors x_{i1} in an augmented null chain $(x_{i1}, \ldots, x_{i,\mu_i+p})$.

A collection of augmented null chains $(x_{11}, \ldots, x_{1,\mu_1+p}; \ldots; x_{\delta 1}, \ldots, x_{\delta,\mu_\delta+p})$ is said to be a *canonical set of augmented null chains* if the augmented null chain $(x_{i1}, \ldots, x_{i,\mu_i+p})$ has maximal length subject to the side condition that x_{i1} is not in $\text{Span}\{x_{11}, \ldots, x_{i-1,1}\}$, and if $\text{Span}\{x_{11}, \ldots, x_{\delta 1}\}$ is equal to the entire augmented eigenspace for $W(z)$ at z_0. The following proposition should make all these notions more transparent. The proof is straightforward from the definitions.

PROPOSITION 12.5.1. *Let $W(z)$ be a rational matrix function with a pole at z_0 of order p and let $V(z) = (z - z_0)^p W(z)$. Then*

(i) $(x_1, \ldots, x_{\mu+p})$ *is an augmented null chain of length μ for $W(z)$ at z_0 if and only if $(x_1, \ldots, x_{\mu+p})$ is a null chain for $V(z)$ of length $\mu + p$.*

(ii) *The collection of vectors $(x_{11}, \ldots, x_{1,\mu_1+p}; \ldots; x_{\delta 1}, \ldots, x_{\delta,\mu_\delta+p})$ is a complete canonical set of augmented left null chains for $W(z)$ at z_0 if and only if it is also a canonical set of left null chains for $V(z)$ at z_0.*

The auxiliary rational matrix function $V(z) = (z - z_0)^p W(z)$ appears in another way in connection with the analysis of the null-pole structure of $W(z)$. Namely, if $W(z)^{-1}$ has a pole of order q while $W(z)$ has a pole of order p, then one easily sees that $V(z)$ is analytic at z_0 while $V(z)^{-1}$ has a pole of order $p + q$. As the Taylor series for $V(z)$ is just a shift of the Laurent series for $W(z)$ at z_0, we see that the coefficient null-pole subspaces for $W(z)$ and $V(z)$ at z_0 are identical:

$$(12.5.10) \qquad \mathcal{N}_{\{z_0\}}(W) = \mathcal{N}_{\{z_0\}}(V).$$

The only difference is in the indexing; elements of $\mathcal{N}_{\{z_0\}}(W)$ are indexed from $-p$ to $q - 1$ while elements of $\mathcal{N}_{\{z_0\}}(V)$ are indexed from 0 to $p + q - 1$.

We next wish to apply (12.5.9) (with V in place of W) to get a description of $\mathcal{N}_{\{z_0\}}(V)$; from (12.5.10) this then will also give a description of $\mathcal{N}_{\{z_0\}}(W)$. From the canonical set $(x_{11}, \ldots, x_{1\mu_1}, \ldots, x_{d1}, \ldots, x_{d\mu_d})$ of left null chains for $W(z)$ at z_0, we may add additional vectors as necessary (including additional augmented chains of negative length if required) to get a complete canonical set

$$(12.5.11) \qquad (x_{11}, \ldots, x_{1,\mu_1+p}; \ldots; x_{\delta 1}, \ldots, x_{\delta,\mu_\delta+p})$$

of augmented left null chains for $W(z)$ at z_0. From these we form matrices

$$(12.5.12) \qquad X_i^{(a)} = \text{col}(x_{i1}, \ldots, x_{i,\mu_i+p})$$

and

$$(12.5.13) \qquad A_{\zeta_i}^{(a)} = z_0 I_{\mu_i+p} + J_{\mu_i+p}^T.$$

These in turn lead to matrices $X^{(a)}$ and $A_\zeta^{(a)}$ given by

$$X^{(a)} = \text{col}(X_1^{(a)}, \ldots, X_\delta^{(a)})$$

and

$$A_\zeta^{(a)} = \mathrm{diag}(A_{\zeta 1}^{(a)}, \ldots, A_{\zeta \delta}^{(a)}).$$

Then since by Proposition 12.5.1 $(x_{11}, \ldots, x_{1,\mu_1+p}; \ldots; x_{\delta 1}, \ldots, x_{\delta,\mu_\delta+p})$ is a canonical set of left null chains for $V(z)$, we see that $(A_\zeta^{(a)}, X^{(a)})$ is a left null pair for $V(z)$ at z_0. As $V(z)$ has no pole at z_0, the right pole pair and coupling matrix in a null-pole triple for $V(z)$ at z_0 is trivial. Then (12.5.9) applied to $V(z)$ and combined with the observation (12.5.10) gives us

$$(12.5.14) \qquad \mathcal{N}_{\{z_0\}}(W) = \{\mathrm{col}(f_{-p}, \ldots, f_{q-1}) : [X^{(a)}, \ldots, (A_\zeta^{(a)} - z_0 I)^{p+q-1} X^{(a)}]$$
$$\cdot \mathrm{col}(f_{-p}, \ldots, f_{q-1}) = 0\}.$$

Finally, given a canonical set (12.5.11) of augmented left null chains of $W(z)$ at z_0, and given a canonical set

$$(u_{11}, \ldots, u_{1\nu_1}; \ldots; u_{e1}, \ldots, u_{e\nu_e})$$

of right pole chains of $W(z)$ at z_0, for $1 \leq i \leq d$ and $1 \leq k \leq e$ let S_{ik} be the $\mu_i \times \nu_k$ matrix

$$(12.5.15) \qquad S_{ij} = - \begin{bmatrix} x_{i,p+1} & \cdots & x_{i2} \\ \vdots & & \vdots \\ x_{i,\mu_i+p} & \cdots & x_{i,\mu_1+1} \end{bmatrix} \begin{bmatrix} 0 & 0 & \cdots & 0 \\ \vdots & \vdots & & \vdots \\ 0 & 0 & \cdots & u_{k1} \\ \vdots & \vdots & & \vdots \\ 0 & u_{k1} & \cdots & u_{k,\nu_k-1} \\ u_{k1} & u_{k2} & \cdots & u_{k,\nu_k} \end{bmatrix}$$

and let S be the $d \times e$ block matrix

$$(12.5.16) \qquad S = [S_{ik}] \qquad (1 \leq i \leq d, 1 \leq k \leq e).$$

We then have the following result.

THEOREM 12.5.2. *Let $W(z)$ be a rational matrix function with a pole of order p and with inverse having a pole of order q at z_0. Let $(x_{11}, \ldots, x_{1,\mu_1+p}; \ldots; x_{d1}, \ldots, x_{d,\mu_\alpha+p})$ be a collection of left augmented null chains associated with the canonical set $(x_{11}, \ldots, x_{1\mu_1}; \ldots; x_{d1}, \ldots, x_{d\mu_d})$ of left null chains for $W(z)$ at z_0, and let $(u_{11}, \ldots, u_{1\nu_1}; \ldots; u_{e1}, \ldots, u_{e\nu_e})$ be a canonical set of right pole chains for $W(z)$ at z_0. Define matrices U, A_π, A_ζ, X and S by (12.5.5)–(12.5.8), (12.5.1)–(12.5.4) and (12.5.15)–(12.5.16). Then the collection $(U, A_\pi; A_\zeta, X; S)$ is a left null-pole triple for $W(z)$ at z_0.*

PROOF. We have already established that (U, A_π) is a right pole pair and (A_ζ, X) is a left null pair for $W(z)$ at z_0; it remains only to show that S is the associated null-pole coupling matrix.

To do this we need only verify that $\mathcal{N}_{\{z_0\}}(W)$ has the form (12.5.9) with S given by (12.5.15)–(12.5.16). We already established in the discussion above that $\mathcal{N}_{\{z_0\}}(W)$ has the form (12.5.14). Thus it remains only to deduce (12.5.9) from (12.5.14).

Since (U, A_ζ) is a right pole pair for $W(z)$ at z_0, we already know by Proposition 12.4.1 that

$$\mathrm{col}(f_{-p}, \ldots, f_{-1}) = \left[\mathrm{col}(U(A_\pi - z_0 I)^{p-j})_{j=1}^p\right] x$$

for some vector x whenever $\mathrm{col}(f_{-p}, \ldots, f_{q-1})$ is in $\mathcal{N}_{\{z_0\}}(W)$. From (12.5.13) we see that then

$$[X^{(a)}, \ldots, (A_\zeta^{(a)} - z_0 I)^{p+q-1} X^{(a)}] \left[\begin{array}{c} \mathrm{col}(U(A_\pi - z_0 I)^{p-j})_{j=1}^p x \\ \mathrm{col}(f_0, \ldots, f_{q-1}) \end{array}\right] = 0.$$

We rewrite this equality in the form

$$[X^{(a)}, \ldots, (A_\zeta^{(a)} - z_0 I)^{p-1} X^{(a)}][\mathrm{col}(U(A_\pi - z_0 I)^{p-1})_{j=1}^p] x$$
$$+ [(A_\zeta^{(a)} - z_0 I)^p X^{(a)}, \ldots, (A_\zeta^{(a)} - z_0 I)^{p+q-1} X^{(a)}] \mathrm{col}(f_0, \ldots, f_{q-1}) = 0.$$

Thus we have

(12.5.17)
$$[(A_\zeta^{(a)} - z_0 I)^p X^{(a)}, \ldots, (A_\zeta^{(a)} - z_0 I)^{p+q-1} X^{(a)}] \mathrm{col}(f_0, \ldots, f_{q-1})$$
$$= -[X^{(a)}, \ldots, (A_\zeta^{(a)} - z_0 I)^{p-1} X^{(a)}][\mathrm{col}(U(A_\pi - z_0 I)^{p-j})_{j=1}^p] x.$$

Write x as $\mathrm{col}(x_1, \ldots, x_e)$ where $x_k \in \mathbb{C}^{\nu_k}$. Then (12.5.17) amounts to the system of equations
(12.5.18)
$$[(A_{\zeta i}^{(a)} - z_0 I)^p X_i^{(a)}, \ldots, (A_{\zeta i}^{(a)} - z_0 I)^{p+q-1} X_i^{(a)}] \mathrm{col}(f_0, \ldots, f_{q-1})$$
$$= -\sum_{k=1}^e [X_i^{(a)}, \ldots, (A_{\zeta i}^{(a)} - z_0 I)^{p-1} X_i^{(a)}][\mathrm{col}(U_k(A_{\pi k} - z_0 I)^{p-j})_{j=1}^p] x_k.$$

Assuming first that $i \le d$ so $\mu_i \ge 1$, note that the first p rows of the matrix $[(A_{\zeta i}^{(a)} - z_0 I)^p X_i^{(a)}, \ldots, (A_{\zeta i}^{(a)} - z_0 I)^{p+q-1} X_i^{(a)}]$ are zero and the remaining μ_i rows form the matrix $[X_i, \ldots, (A_{\zeta i} - z_0 I)^{q-1} X_i]$. Therefore (12.5.18) takes the form

(12.5.19)
$$\left[\begin{array}{c} 0 \\ [X_i, \ldots, (A_{\zeta i} - z_0 I)^{q-1} X_i] \end{array}\right] \mathrm{col}(f_0, \ldots, f_{q-1}) = \sum_{k=1}^e \left[\begin{array}{c} T_{1k}^{(i)} \\ T_{2k}^{(i)} \end{array}\right] x_k$$

where

$$T_{1k}^{(i)} = -[\; I_p \quad 0\;][X_i^{(a)}, \ldots, (A_{\zeta i}^{(a)} - z_0 I)^{p-1} X_i^{(a)}][\mathrm{col}(U_k(A_{\pi k} - z_0 I)^{p-j})_{j=1}^p]$$

and

$$T_{2k}^{(i)} = -[\; 0 \quad I_{\mu_i}\;][X_i^{(a)}, \ldots, (A_{\zeta i}^{(a)} - z_0 I)^{p-1} X_i^{(a)}][\mathrm{col}(U_k(A_{\pi k} - z_0 I)^{p-j})_{j=1}^p].$$

More explicitly, we have

$$(12.5.20) \qquad T_{1k}^{(i)} = - \begin{bmatrix} x_{i1} & & 0 \\ \vdots & \ddots & \\ x_{ip} & \cdots & x_{i1} \end{bmatrix} \begin{bmatrix} 0 & 0 & \cdots & 0 \\ \vdots & \vdots & & \vdots \\ 0 & 0 & \cdots & u_{k1} \\ \vdots & \vdots & & \vdots \\ 0 & u_{k1} & \cdots & u_{k,\nu_k-1} \\ u_{k1} & u_{k2} & \cdots & u_{k,\nu_k} \end{bmatrix}$$

and

$$(12.5.21) \qquad T_{2k}^{(i)} = S_{ik}.$$

Let $x_i(z) = \sum_{j=1}^{\infty} x_{ij}(z-z_0)^{j-1}$ be a left null function associated with the left augmented null chain $(x_{i1}, \ldots, x_{i,\mu_i+p})$ and $u_k(z) = \sum_{j=0}^{\infty} u_{kj}(z - z_0)^{j-1}$ be the right pole function associated with right pole chain $(u_{k1}, \ldots, u_{k\nu_k})$. Then there is an analytic vector function $\varphi_k(z)$ for which

$$(12.5.22) \qquad W(z)\varphi_k(z) = (z - z_0)^{-\nu_k} u_k(z).$$

Since $x_i(z)W(z)$ is a row vector function with a zero of order μ_i at z_0 and φ_k is analytic at z_0, we see from (12.5.22) that the scalar function

$$(12.5.23) \qquad \rho_{ik}(z) = x_i(z)u_k(z)$$

is analytic at z_0 with a zero at z_0 of order $\mu_i + \nu_k$. If we write out what this means in terms of Taylor coefficients of $x_i(z)$ and $u_k(z)$, we see that

$$(12.5.24) \qquad T_{1k}^{(i)} = 0.$$

If $d < i \le \delta$ (i.e. if $\mu_i \le 0$), then $T_{2k}^{(i)}$ is vacuous, $T_{1k}^{(i)}$ has only $\mu_i + p$ rows but still p columns and is given by

$$(12.5.25) \qquad T_{1k}^{(i)} = - \begin{bmatrix} x_{i1} & \cdots & & 0 \\ \vdots & \ddots & & \vdots \\ x_{i,\mu_i+p} \cdots x_{i1} & 0 & \cdots & 0 \end{bmatrix} \begin{bmatrix} 0 & \cdots & 0 \\ \vdots & & \vdots \\ 0 & \cdots & u_{k1} \\ \vdots & \ddots & \vdots \\ u_{k1} & \cdots & u_{k\nu_k} \end{bmatrix}$$

and (12.5.19) has the form

$$(12.5.26) \qquad 0 = \sum_{k=1}^{e} T_{1k}^{(i)} x_k.$$

By a variation on the argument above, one can show that $T_{1k}^{(i)} = 0$ in this case as well, so (12.5.26) is automatic.

Finally, from (12.5.19), (12.5.21) and (12.5.24) we conclude that $\mathcal{N}_{\{z_0\}}(W)$ has the form (12.5.9) as required. $\square$

REMARK 12.5.1. Since $(U, A_\pi; A_\zeta, X; S)$ is a null-pole triple for $W(z)$ at z_0 in Theorem 12.5.2, we know (see Theorem 4.4.3 or Theorem 12.2.1) that S satisfies the Sylvester equation

$$SA_\pi - A_\zeta S = XU.$$

This equation in turn is equivalent to

$$S_{ik}A_{\pi k} - A_{\zeta i}S_{ik} = X_i U_k$$

holding for all i and k. Since $A_{\pi k}$ and $A_{\zeta i}$ are single Jordan blocks with the common eigenvalue z_0, this means that the matrices $X_i U_k$ and S_{ik} must have the special form described in Theorem A.1.2. That S_{ik} and $X_i U_k$ indeed do have the appropriate form can also be checked directly; to do this one again uses the fact that the scalar function $\rho_{ik}(z)$ given by (12.5.23) has a zero of order $\mu_i + \nu_k$ at z_0. $\square$

REMARK 12.5.2. Theorem 12.5.2 also explains to what extent a left null chain $(x_{i1}, \ldots, x_{i\mu_i})$ for $W(z)$ at z_0 determines an augmented null chain $(x_{i1}, \ldots, x_{i,\mu_i+p})$. Indeed we have already seen in Chapter 4 that once the left null pair (A_ζ, X) and the right pole pair (U, A_π) are specified, then the associated coupling matrix S is uniquely determined. The null pair (A_ζ, X) is built from a canonical set of (nonaugmented) null chains $(x_{11}, \ldots, x_{1\mu_1}; \ldots; x_{d1}, \ldots, x_{d\mu_d})$ and the pole pair (U, A_π) is built from a canonical set of pole chains $(u_{11}, \ldots, u_{1\nu_1}; \ldots; u_{e1}, \ldots, u_{e\nu_e})$. Any continuation $(\tilde{x}_{i,\mu_1+1}, \ldots, \tilde{x}_{i,\mu_i+p})$ of $(x_{i1}, \ldots, x_{i\mu_i})$ to an augmented null chain $(x_{i1}, \ldots, x_{i\mu_i}, x_{i,\mu_i+1}, \ldots, x_{i,\mu_i+p})$ must be such that the associated matrix S_{ik} defined by (12.5.15) remains the same for all k for each i. Conversely, if the continuations $(\tilde{x}_{i,\mu_i+1}, \ldots, \tilde{x}_{i,\mu_i+p})$ have this property for each i, then we can reverse the steps in the proof of Theorem 12.5.2 to show that $\mathcal{N}_{\{z_0\}}(W)$ has the form (12.5.14) with $(\tilde{x}_{i,\mu_i+1}, \ldots, \tilde{x}_{i,\mu_i+p})$ in place of $(x_{i,\mu_i+1}, \ldots, x_{i,\mu_i+p})$. This then implies that $(x_{i1}, \ldots, x_{i\mu_i}, \tilde{x}_{i,\mu_i+1}, \ldots, \tilde{x}_{i,\mu_i+p})$ is a left augmented null chain for W for each i. Thus a necessary and sufficient condition for an augmentation $(\tilde{x}_{i,\mu_i+1}, \ldots, \tilde{x}_{i,\mu_i+p})$ of $(x_{i,1}, \ldots, x_{i,\mu_i})$ to give rise to an augmented null chain for W is that the matrices S_{ik} in (12.5.15) be preserved. $\square$

REMARK 12.5.3. It is possible to work out a dual form of Theorem 12.5.2 where the coupling matrix S is expressed in terms of a canonical set of left null chains and a canonical set of right augmented pole chains. That these two expressions for S are the same again relies on the fact that ρ_{ik} in (12.5.23) has a zero at z_0 or order $\mu_i + \nu_k$.

CHAPTER 13

NULL-POLE SUBSPACE FOR MATRIX FUNCTIONS WITH J-UNITARY VALUES ON THE IMAGINARY AXIS OR UNIT CIRCLE

In this chapter we study rational matrix functions which have J-unitary values on the imaginary line or the unit circle, where J is a fixed signature matrix (recall that a square matrix J is called a signature matrix if $J^2 = I$, $J = J^*$). A basic tool for the study of such functions is the Redheffer transform which gives a connection between J-unitary and unitary valued functions. The main goal is to establish the connection between structural properties of the associated null-pole subspace and the equivalence of certain analyticity properties with the number of negative squares of an associated kernel function (Theorems 13.2.3 and 13.2.4); these results are needed in Part V where J-unitary valued functions are used to give linear fractional maps which parametrize all solutions of an interpolation problem.

13.1 J-UNITARY VERSUS UNITARY MATRIX FUNCTIONS

In this chapter we assume that Δ is either the open right half plane Π^+ or the open unit disk $\mathcal{D}$. We denote the boundary of Δ (i.e. the imaginary axis including infinity if $\Delta = \Pi^+$ or the unit circle if $\Delta = \mathcal{D}$) by $\partial\Delta$. We fix nonnegative integers M and N and let J be the $(M+N) \times (M+N)$ signature matrix

$$(13.1.1) \qquad J = \begin{bmatrix} I_M & 0 \\ 0 & -I_N \end{bmatrix}.$$

In this chapter we will be considering rational $(M+N) \times (M+N)$ matrix functions $\theta(z)$ for which θ is J-unitary valued for all regular points on $\partial\Delta$:

$$(13.1.2) \qquad \left(\theta(z)\right)^* J\theta(z) = J, \qquad z \in \partial\Delta, \quad z \text{ not a pole of } \theta.$$

Equivalently, by analytic continuation we see that (13.1.2) is equivalent to

$$(13.1.3a) \qquad \left(\theta(-\bar{z})\right)^* J\theta(z) = J \quad \text{if} \quad \Delta = \Pi^+$$

$$(13.1.3b) \qquad \left(\theta(\bar{z}^{-1})\right)^* J\theta(z) = J \quad \text{if} \quad \Delta = \mathcal{D}.$$

We get more detailed information by partitioning θ conformally with the partitioning (13.1.1) of J:

$$(13.1.4) \qquad \theta(z) = \begin{bmatrix} \theta_{11}(z) & \theta_{12}(z) \\ \theta_{21}(z) & \theta_{22}(z) \end{bmatrix}$$

where $\theta_{11}, \theta_{12}, \theta_{21}, \theta_{22}$ have sizes $M \times M$, $M \times N$, $N \times M$ and $N \times N$ respectively.

As a preparation for the main results to be presented in the next section, in the initial discussion we concentrate on J-unitary *constant* matrices $\theta = \begin{bmatrix} \theta_{11} & \theta_{12} \\ \theta_{21} & \theta_{22} \end{bmatrix}$. Thus suppose that $\theta = [\theta_{ij}]_{1 \leq i,j \leq 2}$ is a J-unitary matrix where J is given by (13.1.1). The J-unitary property of θ implies the relations

(13.1.5a)
$$\theta_{11}\theta_{11}^* - \theta_{12}\theta_{12}^* = I_M$$

(13.1.5b)
$$\theta_{11}\theta_{21}^* - \theta_{12}\theta_{22}^* = 0$$

(13.1.5c)
$$\theta_{21}\theta_{21}^* - \theta_{22}\theta_{22}^* = -I_N.$$

In particular, from (13.1.5c) $\theta_{22}\theta_{22}^* = I + \theta_{21}\theta_{21}^*$, so θ_{22} is invertible.

Now let x_1, x_2, y_1, y_2 be variables ($x_2, y_1 \in \mathbb{C}^M$, $y_2, x_1 \in \mathbb{C}^N$) which are related according to the relation

(13.1.6)
$$\begin{bmatrix} y_1 \\ x_1 \end{bmatrix} = \begin{bmatrix} \theta_{11} & \theta_{12} \\ \theta_{21} & \theta_{22} \end{bmatrix} \begin{bmatrix} x_2 \\ y_2 \end{bmatrix}.$$

Define the matrix $U = \begin{bmatrix} U_{11} & U_{12} \\ U_{21} & U_{22} \end{bmatrix}$ by the equality

(13.1.7)
$$\begin{bmatrix} y_1 \\ y_2 \end{bmatrix} = \begin{bmatrix} U_{11} & U_{12} \\ U_{21} & U_{22} \end{bmatrix} \begin{bmatrix} x_1 \\ x_2 \end{bmatrix}.$$

Since θ_{22} is invertible, U is well-defined and is given by

(13.1.8)
$$\begin{bmatrix} U_{11} & U_{12} \\ U_{21} & U_{22} \end{bmatrix} = \begin{bmatrix} \theta_{12}\theta_{22}^{-1} & \theta_{11} - \theta_{12}\theta_{22}^{-1}\theta_{21} \\ \theta_{22}^{-1} & -\theta_{22}^{-1}\theta_{21} \end{bmatrix}.$$

When U and θ are related in this way we say that U is the Redheffer transform of θ and write $U = \mathcal{R}[\theta]$. Since θ is J-unitary, we deduce from (13.1.6) that

$$\|y_1\|^2 - \|x_1\|^2 = \|x_2\|^2 - \|y_2\|^2.$$

Rewrite this equation as

$$\|y_1\|^2 + \|y_2\|^2 = \|x_1\|^2 + \|x_2\|^2.$$

As x_1 and x_2 are arbitrary, we deduce from (13.1.7) that U is unitary:

$$U^*U = I.$$

The converse holds by reversing the argument. We thus have established the following.

LEMMA 13.1.1. *If $U = \mathcal{R}[\theta]$ is the Redheffer transform of θ, then θ is J-unitary if and only if U is unitary.*

We next consider the case where the matrices θ and U are related by $U = \mathcal{R}[\theta]$ and θ is J-contractive, i.e.,

$$\theta^* J \theta \leq J.$$

In terms of the vectors x_2, y_2, y_1, x_1, by (13.1.6) this means

(13.1.9) $$\|y_1\|^2 - \|x_1\|^2 \leq \|x_2\|^2 - \|y_2\|^2.$$

This can be rewritten as

(13.1.10) $$\|y_1\|^2 + \|y_2\|^2 \leq \|x_1\|^2 + \|x_2\|^2.$$

As x_1 and x_2 are arbitrary, then by (13.1.7) we conclude that U is a contraction in the usual sense

$$U^* U \leq I.$$

We have established the first part of the following.

PROPOSITION 13.1.2. (a) *If* $U = \mathcal{R}[\theta]$ *is the Redheffer transform of* θ, *then* θ *is* J-contractive if and only if U *is contractive.*

(b) *If* $\theta = \begin{bmatrix} \theta_{11} & \theta_{12} \\ \theta_{21} & \theta_{22} \end{bmatrix}$ *is* J-contractive, then

(i) $\theta^* = \begin{bmatrix} \theta_{11}^* & \theta_{21}^* \\ \theta_{12}^* & \theta_{22}^* \end{bmatrix}$ *is* J-contractive, and

(ii) θ_{22} *is invertible and* $\|\theta_{22}^{-1}\theta_{21}\| < 1$.

PROOF. Suppose that $\theta = \begin{bmatrix} \theta_{11} & \theta_{12} \\ \theta_{21} & \theta_{22} \end{bmatrix}$ is a J-contraction. Then in particular

$$\theta_{12}^* \theta_{12} - \theta_{22}^* \theta_{22} = [0, I] \begin{bmatrix} \theta_{11}^* & \theta_{21}^* \\ \theta_{12}^* & \theta_{22}^* \end{bmatrix} \begin{bmatrix} I & 0 \\ 0 & -I \end{bmatrix} \begin{bmatrix} \theta_{11} & \theta_{12} \\ \theta_{21} & \theta_{22} \end{bmatrix} \begin{bmatrix} 0 \\ I \end{bmatrix} \leq -I.$$

Thus

$$\theta_{12}^* \theta_{12} + I \leq \theta_{22}^* \Theta_{22}$$

and since θ_{22} is square, necessarily θ_{22} is invertible. This means that the system of equations (13.1.6) as in the J-unitary case can be rearranged in the form (13.1.7) with $U = \begin{bmatrix} U_{11} & U_{12} \\ U_{21} & U_{22} \end{bmatrix}$ given by (13.1.8). By the equivalence of (13.1.9) and (13.1.10), we see that θ is J-contractive if and only if U is contractive, and statement (a) follows. But a matrix U is contractive (in the usual sense) if and only if its adjoint U^* is contractive. Hence the J-contractiveness of $\begin{bmatrix} \theta_{11} & \theta_{12} \\ \theta_{21} & \theta_{22} \end{bmatrix}$ leads to the contractiveness of

$$\begin{bmatrix} \theta_{12}\theta_{22}^{-1} & \theta_{11} - \theta_{12}\theta_{22}^{-1}\theta_{21} \\ \theta_{22}^{-1} & -\theta_{22}^{-1}\theta_{21} \end{bmatrix}^* = \begin{bmatrix} \theta_{22}^{*-1}\theta_{12}^* & \theta_{22}^{*-1} \\ \theta_{11}^* - \theta_{21}^*\theta_{22}^{*-1}\theta_{12}^* & -\theta_{21}^*\theta_{22}^{*-1} \end{bmatrix}.$$

Then also the matrix

$$
\begin{bmatrix} 0 & -I \\ I & 0 \end{bmatrix}
\begin{bmatrix} \theta_{22}^{*-1}\theta_{12}^{*} & \theta_{22}^{*-1} \\ \theta_{11}^{*} - \theta_{21}^{*}\theta_{22}^{*-1}\theta_{12}^{*} & -\theta_{21}^{*}\theta_{22}^{*-1} \end{bmatrix}
\begin{bmatrix} 0 & I \\ -I & 0 \end{bmatrix}
$$
$$
= \begin{bmatrix} \theta_{21}^{*}\theta_{22}^{*-1} & \theta_{11}^{*} - \theta_{21}^{*}\theta_{22}^{*-1}\theta_{12}^{*} \\ \theta_{22}^{*-1} & -\theta_{22}^{*-1}\theta_{12}^{*} \end{bmatrix}
$$
$$
= \mathcal{R}[\theta^{*}]
$$

is contractive. Thus by part (a) we conclude that θ^{*} is J-contractive, and part (i) of (b) follows. Finally, from the J-contractiveness of $\begin{bmatrix} \theta_{11}^{*} & \theta_{21}^{*} \\ \theta_{12}^{*} & \theta_{22}^{*} \end{bmatrix}$ we get

$$
\theta_{21}\theta_{21}^{*} - \theta_{22}\theta_{22}^{*} = \begin{bmatrix} 0 & I \end{bmatrix}
\begin{bmatrix} \theta_{11} & \theta_{12} \\ \theta_{21} & \theta_{22} \end{bmatrix}
\begin{bmatrix} I & 0 \\ 0 & -I \end{bmatrix}
\begin{bmatrix} \theta_{11}^{*} & \theta_{21}^{*} \\ \theta_{12}^{*} & \theta_{22}^{*} \end{bmatrix}
\begin{bmatrix} 0 \\ I \end{bmatrix} \leq -I.
$$

Therefore

$$
\theta_{22}^{-1}\theta_{21}\theta_{21}^{*}\theta_{22}^{*-1} \leq I - \theta_{22}^{-1}\theta_{22}^{*-1}
$$

so $\|\theta_{21}^{*}\theta_{22}^{*-1}x\| < \|x\|$ for all $x \neq 0$. Thus $\|\theta_{22}^{-1}\theta_{21}\| < 1$ as asserted. $\square$

We now return to considering a rational matrix valued function $\theta(z)$. If

$$
\theta(z) = \begin{bmatrix} \theta_{11}(z) & \theta_{12}(z) \\ \theta_{21}(z) & \theta_{22}(z) \end{bmatrix}
$$

and $\theta(z)$ is J-unitary on $\partial\Delta$, then from (13.1.5c) we see that $\theta_{22}(z)$ is invertible at each regular point of $\partial\Delta$, so in particular $\det \theta_{22}(z)$ does not vanish identically. We may then define the rational matrix function $U(z)$ by $U(z) = \mathcal{R}[\theta(z)]$. By Lemma 13.1.1, $\theta(z)$ being J-unitary on $\partial\Delta$ is equivalent to $U(z)$ being unitary on $\partial\Delta$. By Lemma 13.1.2, $\theta(z)$ is J-contractive on Δ if and only if $U(z)$ is contractive on Δ. However the connection is even more far reaching.

Let $\rho_{\Delta}(z, w)$ be defined by

(13.1.11a) $\qquad\qquad \rho_{\Delta}(z, w) = z + \overline{w} \quad \text{if} \quad \Delta = \Pi^{+}$

(13.1.11b) $\qquad\qquad \rho_{\Delta}(z, w) = 1 - z\overline{w} \quad \text{if} \quad \Delta = \mathcal{D}.$

THEOREM 13.1.3. *Let $\theta(z)$ and $U(z)$ be rational matrix functions such that $U(z) = \mathcal{R}[\theta(z)]$. Then $\theta(z)$ is J-unitary on $\partial\Delta$ if and only if $U(z)$ is unitary on $\partial\Delta$. Also the four kernel functions*

(13.1.12) $\qquad \widehat{K}_{\theta}(z, w) = \dfrac{J - \theta(w)^{*}J\theta(z)}{\rho_{\Delta}(z, w)}, \qquad K_{\theta}(z, w) = \dfrac{J - \theta(z)J\theta(w)^{*}}{\rho_{\Delta}(z, w)}$

and

(13.1.13) $\qquad \widehat{K}_{U}'(z, w) = \dfrac{I - U(w)^{*}U(z)}{\rho_{\Delta}(z, w)}, \qquad K_{U}'(z, w) = \dfrac{I - U(z)U(w)^{*}}{\rho_{\Delta}(z, w)}$

have the same number of negative squares on Δ.

PROOF. That $\theta(z)$ is J-unitary on $\partial\Delta$ if and only if $U(z)$ is unitary on $\partial\Delta$ is an immediate consequence of Lemma 13.1.1. To prove the second assertion, choose a positive integer r, points $w_1, \ldots, w_r$ in Δ and vectors $\begin{bmatrix} u_1 \\ v_1 \end{bmatrix}, \ldots, \begin{bmatrix} u_r \\ v_r \end{bmatrix}$ in $\mathbb{C}^{M+N}$. Define rational vector functions $\begin{bmatrix} y_j(z) \\ x_j(z) \end{bmatrix}$ by

$$\begin{bmatrix} y_j(z) \\ x_j(z) \end{bmatrix} = \theta(z) \begin{bmatrix} u_j \\ v_j \end{bmatrix}$$

for $j = 1, \ldots, r$. Then note

$$
\begin{aligned}
&[\, u_i^* \quad v_i^* \,] \widehat{K}_\theta(w_j, w_i) \begin{bmatrix} u_j \\ v_j \end{bmatrix} \\
&= [u_i^* u_j - v_i^* v_j - y_i(w_i)^* y_j(w_j) + x_i(w_i)^* x_j(w_j)]/\rho_\Delta(w_i, w_j) \\
&= [u_i^* u_j + x_i(w_i)^* x_j(w_j) - v_i^* v_j - y_i(w_i)^* y_j(w_j)]/\rho_\Delta(w_i, w_j) \\
&= [\, x_i(w_i)^* \quad u_i^* \,] \widehat{K}'_U(w_j, w_i) \begin{bmatrix} x_j(w_j) \\ u_j \end{bmatrix}
\end{aligned}
$$

where we again used the connection (13.1.6) and (13.1.7) between θ and U. From this it follows that $\widehat{K}_\theta$ and $\widehat{K}'_U$ have the same number of negative squares on Δ. The statement of the theorem concerning the other two kernels follows from Theorem 7.5.3 and a suitable generalization (allowing singularities on the imaginary axis) of Theorem 6.2.2. □

COROLLARY 13.1.4. *Let* $\theta(z)$ *and* $U(z)$ *be rational matrix functions with* $U = \mathcal{R}[\theta]$, *and suppose that* $\theta(z)$ *is* J-unitary on $\partial\Delta$. *Then the number* ν *of negative squares had by each of the kernels* $\widehat{K}_\theta(z, w)$ *and* $K_\theta(z, w)$ *given by* (13.1.12) *is equal to the number of poles had by* $U(z)$ *in* Δ, *i.e., the size of the matrix* A *where* (C, A) *is a pole pair for* $U(z)$ *over the region* Δ.

PROOF. This follows immediately from Theorems 6.2.2 (suitably generalized) and 6.5.2 if $\Delta = \Pi^+$ and their analogues Theorems 7.5.3 and 7.6.2 if $\Delta = \mathcal{D}$. □

COROLLARY 13.1.5. *Let* $\theta(z) = \begin{bmatrix} \theta_{11}(z) & \theta_{12}(z) \\ \theta_{21}(z) & \theta_{22}(z) \end{bmatrix}$ *be a rational* $(M + N) \times (M + N)$ *matrix function which is* J-unitary valued on $\partial\Delta$. *Then the following statements are equivalent.*

(i) *The kernel function*

$$\widehat{K}_\theta(z, w) = \frac{J - \theta(w)^* J \theta(z)}{\rho_\Delta(z, w)}$$

is positive definite on Δ.

(i′) *The kernel function*

$$K_\theta(z, w) = \frac{J - \theta(z) J \theta(w)^*}{\rho_\Delta(z, w)}$$

is positive definite on Δ.

(ii) *$\theta(z)$ is J-contractive at all regular points z for θ in Δ.*

(iii) *The four rational matrix functions $\theta_{12}\theta_{22}^{-1}$, $\theta_{11} - \theta_{12}\theta_{22}^{-1}\theta_{21}$, θ_{22}^{-1} and $-\theta_{22}^{-1}\theta_{21}$ have no poles in Δ.*

PROOF. If $\hat{K}_\theta$ is a positive definite kernel function on Δ, then each matrix of the form

$$\left[\frac{J - \theta(z_i)^* J \theta(z_j)}{\rho_\Delta(z_j, z_i)} \right]_{1 \le i, j \le p}$$

is positive definite for any choice of p points $z_1, \ldots, z_p$ in Δ where θ is analytic; the particular case $p = 1$ implies that θ is J-contractive at all regular points in Δ. Conversely, if θ is J-contractive at all regular points in Δ then by Proposition 13.1.2 $U = \mathcal{R}[\theta]$ is contractive on Δ and so can have no pole in Δ; then from Corollary 13.1.4 we see that $\hat{K}_\theta$ is a positive definite kernel function on Δ. We conclude (i) $\iff$ (ii) in Corollary 13.1.5.

By Corollary 13.1.4 again, $\hat{K}_\theta$ is positive definite on Δ if and only if $U = \mathcal{R}[\theta]$ has no poles in Δ, that is, if and only if the four block entries $\theta_{12}\theta_{22}^{-1}$, $\theta_{11} - \theta_{12}\theta_{22}^{-1}\theta_{21}$, θ_{22}^{-1} and $-\theta_{22}^{-1}\theta_{21}$ have no poles in Δ. This establishes (i) $\iff$ (iii). Similar arguments apply to the kernel function K_θ. $\square$

13.2 NEGATIVE SQUARES AND NULL-POLE SUBSPACES

Let $\theta(z)$ be a rational $(M + N) \times (M + N)$ rational matrix function which is J-unitary on $\partial\Delta$. In Corollary 13.1.5 we saw how the J-contractiveness of θ is measured by analyticity properties of the four functions $\theta_{12}\theta_{22}^{-1}$, $\theta_{11} - \theta_{12}\theta_{22}^{-1}\theta_{21}$, θ_{22}^{-1} and $-\theta_{22}^{-1}\theta_{21}$. In this section we show how the analyticity of the first three functions is captured by geometric properties of the null-pole subspace $\mathcal{S}_{\overline{\Delta}}(\theta)$ for θ over the closure $\overline{\Delta}$ of Δ. Thus when the null-pole subspace $\mathcal{S}_{\overline{\Delta}}(\theta)$ has these special properties, one need only check that the function $\theta_{22}^{-1}\theta_{21}$ has no poles in $\overline{\Delta}$ to show that θ is J-contractive on $\overline{\Delta}$. More generally, the number of negative squares of the kernel functions $K'_U(z, w)$ and $\hat{K}'_U(z, w)$ given by (13.1.13) on $\overline{\Delta}$ equals the number of poles of $\theta_{22}^{-1}\theta_{21}$ in $\overline{\Delta}$.

Before describing the result we recall how the null-pole subspace $\mathcal{S}_{\overline{\Delta}}(\theta)$ for θ over $\overline{\Delta}$ is determined by a null-pole triple $(C, A_\pi; A_\zeta, B; S)$ for θ over $\overline{\Delta}$. Partition C as $\begin{bmatrix} C_+ \\ C_- \end{bmatrix}$ and B as $[B_+, B_-]$ conformally with the partitioning $\theta = \begin{bmatrix} \theta_{11} & \theta_{12} \\ \theta_{21} & \theta_{22} \end{bmatrix}$ of θ. Then by Theorem 12.3.1 the null-pole subspace $\mathcal{S}_{\overline{\Delta}}(\theta)$ is characterized as

(13.2.1)
$$\mathcal{S}_\Delta(\theta) = \left\{ \begin{bmatrix} C_+ \\ C_- \end{bmatrix} (zI - A_\pi)^{-1} x + \begin{bmatrix} h_+(z) \\ h_-(z) \end{bmatrix} : \right.$$
$$x \in \mathbb{C}^{n_\pi}, h_+ \in \mathcal{R}_M(\overline{\Delta}), h_- \in \mathcal{R}_N(\overline{\Delta}) \text{ such that}$$
$$\left. \frac{1}{2\pi i} \int_{\partial\Delta} (zI - A_\zeta)^{-1} [B_+, B_-] \begin{bmatrix} h_+(z) \\ h_-(z) \end{bmatrix} dz = Sx \right\}.$$

If $\mathcal{N}$ is a subspace of $\mathcal{R}_M$, then $\mathcal{N} \oplus 0$ denotes the subspace $\mathcal{N} \oplus 0 = \left\{ \begin{bmatrix} h \\ 0 \end{bmatrix} : h \in \mathcal{N} \right\}$ of $\mathcal{R}_{M+N}$. We let $P_{0 \oplus \mathcal{R}_N(\overline{\Delta})}$ denote the projection of $\mathcal{R}_{M+N}(\overline{\Delta})$ onto $0 \oplus \mathcal{R}_N(\overline{\Delta})$ along $\mathcal{R}_M(\overline{\Delta}) \oplus 0$. For W a rational $m \times n$ matrix function and σ any subset of $\mathbf{C}$. We refer to $P_{\mathcal{R}_M^0(\sigma^c)}(W \mathcal{R}_N(\sigma))$ as the *pole subspace* of W over σ. We now describe how analyticity properties of the various elements of $U(z)$ are connected with the geometry of the null-pole subspace $S_{\overline{\Delta}}(\theta) = \theta \mathcal{R}_{M+N}(\overline{\Delta})$ for a rational matrix function $\theta(z)$ with J-unitary values on $\partial \Delta$.

LEMMA 13.2.1. *Suppose* $\theta = \begin{bmatrix} \theta_{11} & \theta_{12} \\ \theta_{21} & \theta_{22} \end{bmatrix}$ *is a rational* $(M+N) \times (M+N)$

matrix function which is J-unitary on $\partial \Delta$, $\left(\begin{bmatrix} C_+ \\ C_- \end{bmatrix}, A_\pi \right)$ *is a right pole pair for* $\theta(z)$

over $\overline{\Delta}$, *and* $S_{\overline{\Delta}}(\theta) = \theta \mathcal{R}_{M+N}(\overline{\Delta})$ *is the null pole subspace for* $\theta(z)$ *over* $\overline{\Delta}$. *Then the following are equivalent:*

(i) $S_{\overline{\Delta}}(\theta) \cap [\mathcal{R}_M \oplus 0] = S_{\overline{\Delta}}(\theta) \cap [\mathcal{R}_M(\overline{\Delta}) \oplus 0]$.

(ii) *If* $x \in \mathbf{C}^{n_\pi}$ *is such that* $C_-(zI - A_\pi)^{-1}x = 0$ *and* $Sx = \sum_{z_0 \in \overline{\Delta}} \mathrm{Res}_{z=z_0}(zI - A_\zeta)^{-1} B_+ h(z)$ *for some* $h \in \mathcal{R}_M(\overline{\Delta})$, *then* $x = 0$.

(iii) *The pole subspaces for* $\begin{bmatrix} \theta_{11} - \theta_{12}\theta_{22}^{-1}\theta_{21} \\ -\theta_{22}^{-1}\theta_{21} \end{bmatrix}$ *and for* $-\theta_{22}^{-1}\theta_{21}$ *over* $\overline{\Delta}$ *have the same dimension:*

$$\dim P_{\mathcal{R}_{M+N}^0(\overline{\Delta}^c)} \left\{ \begin{bmatrix} \theta_{11} - \theta_{12}\theta_{22}^{-1}\theta_{21} \\ -\theta_{22}^{-1}\theta_{21} \end{bmatrix} \mathcal{R}_M(\overline{\Delta}) \right\}$$

$$= \dim P_{\mathcal{R}_N^0(\overline{\Delta}^c)} \{ -\theta_{22}^{-1}\theta_{21} \mathcal{R}_M(\overline{\Delta}) \}.$$

PROOF. We first show the equivalence of (i) and (ii). From (13.2.1) we deduce that

$$S_{\overline{\Delta}}(\theta) \cap [\mathcal{R}_M \oplus 0] =$$

$$\left\{ \begin{bmatrix} C_+ \\ 0 \end{bmatrix} (zI - A_\pi)^{-1}x + \begin{bmatrix} h_+ \\ 0 \end{bmatrix} : C_-(zI - A_\pi)^{-1}x = 0, \right.$$

$$\left. h_+ \in \mathcal{R}_M(\overline{\Delta}) \text{ such that } \sum_{z_0 \in \overline{\Delta}} \mathrm{Res}_{z=z_0}(zI - A_\zeta)^{-1} B_+ h_+(z) = Sx \right\}$$

while

$$S_\Delta(\theta) \cap [\mathcal{R}_M(\overline{\Delta}) \oplus 0] = \left\{ \begin{bmatrix} h_+ \\ 0 \end{bmatrix} : h_+ \in \mathcal{R}_M(\overline{\Delta}) \right.$$

$$\left. \text{such that } \sum_{z_0 \in \overline{\Delta}} (zI - A_\zeta)^{-1} B_+ h_+(z) = 0 \right\}.$$

Using that $\left(\begin{bmatrix} C_+ \\ C_- \end{bmatrix}, A_\pi \right)$ is a null kernel pair it is now easy to see that these subspaces are the same if and only if condition (ii) holds.

We next argue that (i) and (iii) are equivalent. For this analysis we use the definition of the null-pole subspace in terms of θ

$$(13.2.2) \qquad S_{\overline{\Delta}}(\theta) = \left\{ \begin{bmatrix} \theta_{11}g_1 + \theta_{12}g_2 \\ \theta_{21}g_1 + \theta_{22}g_2 \end{bmatrix} : \begin{bmatrix} g_1 \\ g_2 \end{bmatrix} \in \mathcal{R}_{M+N}(\overline{\Delta}) \right\}.$$

We thus compute

$$S_{\overline{\Delta}}(\theta) \cap [\mathcal{R}_M \oplus 0] = \left\{ \begin{bmatrix} \theta_{11}g_1 + \theta_{12}g_2 \\ 0 \end{bmatrix} : \begin{bmatrix} g_1 \\ g_2 \end{bmatrix} \in \mathcal{R}_{M+N}(\overline{\Delta}) \right.$$
$$\left. \text{such that } \theta_{21}g_1 + \theta_{22}g_2 = 0 \right\}.$$

From $\theta_{21}g_1 + \theta_{22}g_2 = 0$ we can solve for g_2:

$$g_2 = -\theta_{22}^{-1}\theta_{21}g_1.$$

Then we may write

$$S_{\overline{\Delta}}(\theta) \cap [\mathcal{R}_M \oplus 0] = \left\{ \begin{bmatrix} \theta_{11} - \theta_{12}\theta_{22}^{-1}\theta_{21} \\ 0 \end{bmatrix} g_1 : g_1 \in \mathcal{R}_M(\overline{\Delta}) \right.$$
$$\left. \text{such that } -\theta_{22}^{-1}\theta_{21}g_1 \in \mathcal{R}_N(\overline{\Delta}) \right\}.$$

The condition $S_{\overline{\Delta}}(\theta) \cap [\mathcal{R}_M \oplus 0] = S_{\overline{\Delta}}(\theta) \cap [\mathcal{R}_M(\overline{\Delta}) \oplus 0]$ is thus equivalent to:

$$g_1 \in \mathcal{R}_M(\overline{\Delta}), -\theta_{22}^{-1}\theta_{21}g_1 \in \mathcal{R}_N(\overline{\Delta})$$
$$\implies [\theta_{11} - \theta_{12}\theta_{22}^{-1}\theta_{21}]g_1 \in \mathcal{R}_M(\overline{\Delta}).$$

It is easy to see that this last condition in turn is equivalent to (iii). $\square$

LEMMA 13.2.2. *Suppose* $\theta = \begin{bmatrix} \theta_{11} & \theta_{12} \\ \theta_{21} & \theta_{22} \end{bmatrix}$ *is a rational* $(M+N) \times (M+N)$ *matrix function which is* J-*unitary on* $\partial\Delta$, $(A_\zeta, [B_+, B_-])$ *is a left null pair for* $\theta(z)$ *over* $\overline{\Delta}$, *and* $S = S_{\overline{\Delta}}(\theta) = \theta\mathcal{R}_{M+N}(\overline{\Delta})$ *is a null-pole subspace for* $\theta(z)$ *over* $\overline{\Delta}$. *Then the following are equivalent:*

(i) $P_{0 \oplus \mathcal{R}_N(\overline{\Delta})}[S \cap \mathcal{R}_{M+N}(\overline{\Delta})] = 0 \oplus \mathcal{R}_N(\overline{\Delta})$.

(ii) (A_ζ, B_+) *is a full range pair.*

(iii) *The pole subspace for* $\begin{bmatrix} \theta_{12}\theta_{22}^{-1} \\ \theta_{22}^{-1} \end{bmatrix}$ *over* $\overline{\Delta}$ *is contained in the pole sub-space for* $\begin{bmatrix} \theta_{11} - \theta_{12}\theta_{22}^{-1}\theta_{21} \\ -\theta_{22}^{-1}\theta_{21} \end{bmatrix}$ *over* $\overline{\Delta}$:

$$P_{\mathcal{R}_{M+N}^0(\overline{\Delta}^c)} \left\{ \begin{bmatrix} \theta_{12}\theta_{22}^{-1} \\ \theta_{22}^{-1} \end{bmatrix} \mathcal{R}_N(\overline{\Delta}) \right\}$$
$$\subset P_{\mathcal{R}_{M+N}^0(\overline{\Delta}^c)} \left\{ \begin{bmatrix} \theta_{11} - \theta_{12}\theta_{22}^{-1}\theta_{21} \\ -\theta_{22}^{-1}\theta_{21} \end{bmatrix} \mathcal{R}_M(\overline{\Delta}) \right\}.$$

PROOF. Using (13.2.1) we see that

$$P_{0 \oplus \mathcal{R}_N(\overline{\Delta})}[\mathcal{S} \cap \mathcal{R}_{M+N}(\overline{\Delta})] = \left\{ \begin{bmatrix} 0 \\ h_- \end{bmatrix} : h_- \in \mathcal{R}_N(\overline{\Delta}) \text{ such that} \right.$$

$$\sum_{z_0 \in \overline{\Delta}} \operatorname{Res}_{z=z_0}(zI - A_\zeta)^{-1} B_- h_-(z) = - \sum_{z_0 \in \overline{\Delta}} \operatorname{Res}_{z=z_0}(zI - A_\zeta)^{-1} B_+ h_+(z)$$

$$\left. \text{for some } h_+ \in \mathcal{R}_M(\overline{\Delta}) \right\}.$$

This space equals all of $0 \oplus \mathcal{R}_N(\overline{\Delta})$ if and only if $\operatorname{Im} \mathcal{R}[A_\zeta, B_-] \subset \operatorname{Im} \mathcal{R}[A_\zeta, B_+]$, where in general we have set $\mathcal{R}[A_\zeta, \widetilde{B}]$ equal to the map

$$h \in \mathcal{R}_K(\overline{\Delta}) \to \sum_{z_0 \in \overline{\Delta}} \operatorname{Res}_{z=z_0}(zI - A_\zeta)^{-1} \widetilde{B} h(z)$$

where K is the number of columns of $\widetilde{B}$. But since $(A_\zeta, [B_+, B_-])$, as a left null pair for $\theta(z)$ over $\overline{\Delta}$, by definition is a full range pair, by Lemma 12.2.2 we know that $\operatorname{Im} \mathcal{R}[A_\zeta, [B_+, B_-]] = \mathbb{C}^{n_\zeta}$. We have just seen that condition (i) is equivalent to $\operatorname{Im} \mathcal{R}[A_\zeta, B_-] \subset \operatorname{Im} \mathcal{R}[A_\zeta, B_+]$, i.e. to $\operatorname{Im} \mathcal{R}[A_\zeta, B_+] = \operatorname{Im} \mathcal{R}[A_\zeta, [B_+, B_-]]$. Thus (i) is equivalent to $\operatorname{Im} \mathcal{R}[A_\zeta, B_+] = \mathbb{C}^{n_\zeta}$, which again by Lemma 12.2.2 is equivalent to (A_ζ, B_+) being a full range pair. This shows that (i) and (ii) are equivalent.

We next argue that (i) and (iii) are equivalent. From (13.2.2) we deduce that

$$P_{0 \oplus \mathcal{R}_n(\overline{\Delta})}[\mathcal{S} \cap \mathcal{R}_{M+N}(\overline{\Delta})] =$$
$$0 \oplus \{\theta_{21} g_1 + \theta_{22} g_2 \in \mathcal{R}_N(\overline{\Delta}) : g_1 \in \mathcal{R}_M(\overline{\Delta}), g_2 \in \mathcal{R}_N(\overline{\Delta}), \theta_{11} g_1 + \theta_{12} g_2 \in \mathcal{R}_M(\overline{\Delta})\}.$$

In order that this to be all of $0 \oplus \mathcal{R}_N(\overline{\Delta})$, any $h \in \mathcal{R}_N(\overline{\Delta})$ must be expressible as

$$(13.2.3) \qquad\qquad h = \theta_{21} g_1 + \theta_{22} g_2$$

where $g_1 \in \mathcal{R}_M(\overline{\Delta})$, $g_2 \in \mathcal{R}_N(\overline{\Delta})$ are subject to the added restriction

$$(13.2.4) \qquad\qquad \theta_{11} g_1 + \theta_{12} g_2 \in \mathcal{R}_M(\overline{\Delta}).$$

Rewrite (13.2.3) as

$$(13.2.5) \qquad\qquad \theta_{22}^{-1} h = \theta_{22}^{-1} \theta_{21} g_1 + g_2$$

from which we may solve for g_2 and rewrite (13.2.4) as

$$(13.2.6) \qquad\qquad (\theta_{11} - \theta_{12}\theta_{22}^{-1}\theta_{21})g_1 + \theta_{12}\theta_{22}^{-1} h \in \mathcal{R}_M(\overline{\Delta}).$$

Thus (13.2.3) can be expressed as:

For each $h \in \mathcal{R}_N(\overline{\Delta})$ there is a $g_1 \in \mathcal{R}_M(\overline{\Delta})$ for which

$$(\theta_{11} - \theta_{12}\theta_{22}^{-1}\theta_{21})g_1 + \theta_{12}\theta_{22}^{-1} h \in \mathcal{R}_M(\overline{\Delta})$$

and

$$g_2 \overset{\text{def}}{=} \theta_{22}^{-1} h - \theta_{22}^{-1} \theta_{21} g_1 \in \mathcal{R}_N(\overline{\Delta}).$$

Rewrite this condition in column form as:

For each $h \in \mathcal{R}_N(\overline{\Delta})$ there is a $g_1 \in \mathcal{R}_M(\overline{\Delta})$ for which

$$\begin{bmatrix} \theta_{12}\theta_{22}^{-1} \\ \theta_{22}^{-1} \end{bmatrix} h + \begin{bmatrix} \theta_{11} - \theta_{12}\theta_{22}^{-1}\theta_{21} \\ -\theta_{22}^{-1}\theta_{21} \end{bmatrix} g_1 \in \mathcal{R}_{M+N}(\overline{\Delta}).$$

This last condition is clearly equivalent to (iii). $\square$

The next theorem puts Lemmas 13.2.1 and 13.2.2 together. For $\theta(z)$ a rational $(M + N) \times (M + N)$ matrix function with J-unitary values at all regular points of $\partial\Delta$ (including infinity if $\Delta = \Pi^+$), we let H denote the invertible Hermitian coupling matrix associated with a null-pole triple τ_1 for θ over Δ and a symmetrized null-pole triple τ_2 for θ over $\partial\Delta$ as in Theorem 6.4.2 (if $\Delta = \Pi^+$) or in Theorem 7.5.2 (if $\Delta = \mathcal{D}$). Also, we may assume that a null-pole triple τ for $\theta(z)$ over $\overline{\Delta}$ has the form $\tau_1 \oplus \tau_2$ where τ_1 is a null-pole triple for θ over Δ and τ_2 is a symmetrized null-pole triple for θ over $\partial\Delta$.

THEOREM 13.2.3. *Suppose $\theta = \begin{bmatrix} \theta_{11} & \theta_{12} \\ \theta_{21} & \theta_{22} \end{bmatrix}$ is a rational matrix function J-unitary on $\partial\Delta$ ($J = I_M \oplus -I_N$), $\left(\begin{bmatrix} C_+ \\ C_- \end{bmatrix}, A_\pi; A_\zeta, [B_+, B_-], S \right)$ is a left null-pole triple for $\theta(z)$ over $\overline{\Delta}$, H is the (global) Hermitian coupling matrix associated with θ, and $S = S_{\overline{\Delta}}(\theta)$ is the null-pole subspace for $\theta(z)$ over $\overline{\Delta}$. Let U be the Redheffer transform $U = \mathcal{R}[\theta]$ of θ as in (13.1.8). Then the following are equivalent:*

(i) $S \cap [\mathcal{R}_M \oplus 0] = S \cap [\mathcal{R}_M(\overline{\Delta}) \oplus 0]$ *and* $P_{0 \oplus \mathcal{R}_N(\Delta)}[S \cap \mathcal{R}_{M+N}(\Delta)] = 0 \oplus \mathcal{R}_N(\Delta)$.

(ii) (C_-, A_π) *is a null kernel pair and* (A_ζ, B_+) *is a full range pair.*

(iii) *The dimension of the pole subspace for $U(z)$ equals the dimension of the pole subspace for $-\theta_{22}^{-1}(z)\theta_{21}(z)$ over $\overline{\Delta}$:*

$$\dim P_{\mathcal{R}_{M+N}^0(\overline{\Delta}^c)}\{U\mathcal{R}_{N+M}(\overline{\Delta})\} = \dim P_{\mathcal{R}_N^0(\overline{\Delta}^c)}\{-\theta_{22}^{-1}\theta_{21}\mathcal{R}_M(\overline{\Delta})\}.$$

(iv) *The number of negative eigenvalues of the Hermitian coupling matrix H equals the dimension of the pole subspace for $-\theta_{22}^{-1}\theta_{21}$ over $\overline{\Delta}$ ($=$ the number of poles of $-\theta_{22}^{-1}\theta_{21}$ in $\overline{\Delta}$ counted with multiplicities).*

PROOF. Suppose (i) holds. Then by Lemma 13.2.2 (A_ζ, B_+) is a full range pair and by Lemma 13.2.1, the conditions $C_-(zI - A_\pi)^{-1}x = 0$ and $Sx \in \operatorname{Im} \mathcal{R}[A_\zeta, B_+]$ ($x \in \mathbb{C}^{n_\pi}$) imply $x = 0$. However, since (A_ζ, B_+) is a full range pair, by Lemma 12.2.2(ii) $\mathcal{R}[A_\zeta, B_+]$ has a full image space. Thus $C_-(zI - A_\pi)^{-1}x = 0$ alone implies that $x = 0$, so (C_-, A_π) is a null kernel pair by Lemma 12.2.2(i). Thus (i) $\Longrightarrow$ (ii). Conversely, if (ii) holds, then we see that (i) follows from the implications (ii) $\Longrightarrow$ (i) in Lemmas 13.2.1 and 13.2.2.

It is easy to check that conditions (iii) in Lemmas 13.2.1 and 13.2.2 holding simultaneously is equivalent to condition (iii) in Theorem 13.2.3. Thus (i) and (iii) in Theorem 13.2.3 can be seen to be equivalent by using the equivalence of (i) and (iii) in Lemmas 13.2.1 and 13.2.2. Finally, by Theorem 6.4.3 (if $\Delta = \Pi^+$) or Theorem 7.5.3 (if $\Delta = \mathcal{D}$), the number ν of negative eigenvalues of H^{-1} is the same as the number of negative squares for the kernel function $\widehat{K}_\theta(z,w) = \dfrac{J - \theta(w)^* J\theta(z)}{\rho_\Delta(z,w)}$. By Theorem 13.1.3 this is the same as the number of negative squares of the kernel function

$$\widehat{K}'_U(z,w) = \frac{I - U(w)^* U(z)}{\rho_\Delta(z,w)} \quad \text{over } \Delta.$$

For a kernel function of this form, by Corollary 13.1.4 the number of negative squares is the same as the number of poles of $U(z)$ in Δ, i.e., the dimension of the pole subspace for $U(z)$ over Δ. In this way we see that (iii) and (iv) in Theorem 13.2.3 are equivalent. $\square$

We should also be interested in identifying the number of poles of $-\theta_{22}^{-1}\theta_{21}$ in $\overline{\Delta}$ under more general circumstances. The following result will be used in Chapter 20.

THEOREM 13.2.4. *Suppose* $\theta = \begin{bmatrix} \theta_{11} & \theta_{12} \\ \theta_{21} & \theta_{22} \end{bmatrix}$ *is a rational matrix function with J-unitary values on* $\partial\Delta$ *($J = I_M \oplus -I_N$),* $\tau = \left(\begin{bmatrix} C_+ \\ C_- \end{bmatrix}, A_\pi; A_\zeta, [B_+, B_-], S \right)$ *is a left null-pole triple for* $\theta(z)$ *over* $\overline{\Delta}$, *and* $S = S_{\overline{\Delta}}(\theta)$ *is the null-pole subspace for* $\theta(z)$ *over* $\overline{\Delta}$. *Then the following are equivalent:*

(i) $P_{0 \oplus \mathcal{R}_N(\overline{\Delta})} S = 0 \oplus \mathcal{R}_N(\overline{\Delta})$.

(ii) $\operatorname{Im} \mathcal{R}[A_\zeta, B_-] \subset \operatorname{Im}[S, -\mathcal{R}[A_\zeta, B_+]]$.

(iii) *The pole subspace for* θ_{22}^{-1} *is contained in the pole subspace for* $\theta_{22}^{-1}\theta_{21}$ *over* Δ.

(iv) *The dimension of the pole subspace of* $\theta_{22}^{-1}\theta_{21}$ *over* $\overline{\Delta}$ *is the same as the number of negative squares had by the kernel function*

$$K(z,w) = \frac{-I_N - \theta_{21}(z)\theta_{21}(w)^* + \theta_{22}(z)\theta_{22}(w)^*}{\rho_\Delta(z,w)}$$

$$= [\, 0 \quad I_N \,] \frac{J - \theta(z)J\theta(w)^*}{\rho_\Delta(z,w)} \begin{bmatrix} 0 \\ I_N \end{bmatrix}.$$

PROOF. (i) $\iff$ (ii) follows in a routine way from the characterization of the left null-pole subspace in terms of a left null-pole triple over $\overline{\Delta}$ (see Theorem 12.3.1). Next we use the characterization of the null-pole subspace S as $S = \theta \mathcal{R}_{M+N}(\overline{\Delta})$ to see that

$$P_{0 \oplus \mathcal{R}_N(\overline{\Delta})} S = 0 \oplus \{ [\theta_{21}, \theta_{22}] \mathcal{R}_{M+N}(\overline{\Delta}) \cap \mathcal{R}_N(\overline{\Delta}) \}.$$

Thus $P_{0 \oplus \mathcal{R}_N(\overline{\Delta})} S = 0 \oplus \mathcal{R}_N(\overline{\Delta})$ if and only if

$$[\theta_{21}, \theta_{22}] \mathcal{R}_{M+N}(\overline{\Delta}) \supset \mathcal{R}_N(\overline{\Delta})$$

or

$$[\theta_{22}^{-1}\theta_{21}, I]\mathcal{R}_{M+N}(\overline{\Delta}) \supset \theta_{22}^{-1}\mathcal{R}_N(\overline{\Delta}).$$

This last condition is clearly equivalent to (iii). Finally, to see that (iii) and (iv) are equivalent, note that (iii) is equivalent to the number of poles of $[\theta_{22}^{-1}\theta_{21}, \theta_{22}^{-1}]$ being the same as the number of poles of $\theta_{22}^{-1}\theta_{21}$ in $\overline{\Delta}$. But by Corollary 13.1.5, the number of poles of $[\theta_{22}^{-1}\theta_{21}, \theta_{22}^{-1}]$ is the same as the number of negative squares over Δ for the kernel function

$$k(z,w) = \frac{1}{\rho_\Delta(z,w)}\left\{I - [\theta_{22}^{-1}(z)\theta_{21}(z), \theta_{22}^{-1}(z)]\left[\begin{array}{c}\theta_{21}(w)^*\theta_{22}^{-1}(w)^* \\ \theta_{22}^{-1}(w)^*\end{array}\right]\right\}.$$

This kernel function in turn has the same number of negative squares as its conjugate

$$K(z,w) = \theta_{22}(z)k(z,w)\theta_{22}(w)^*$$
$$= \frac{1}{\rho_\Delta(z,w)}[\theta_{22}(z)\theta_{22}(w)^* - \theta_{21}(z)\theta_{21}(w)^* - I_N].$$

In this way we see that (iii) and (iv) are equivalent. $\square$

CHAPTER 14

SUBSPACE INTERPOLATION PROBLEMS

In this chapter we solve the interpolation problem with given null-pole subspace for the case where no data is prescribed at infinity. The construction is based on the interpolation theorems from Part I. The interpolation theorems in Chapters 6 and 7 for J-unitary matrix functions lead to indefinite metric versions of the Beurling-Lax theorem for rational matrix functions.

14.1 INTERPOLATION OF A GIVEN NULL-POLE SUBSPACE

In Chapter 12, for a subset σ of $\mathbb{C}$ we introduced σ-admissible data sets τ and associated to each such τ a null-pole subspace $\mathcal{S}_\tau(\sigma)$, a subspace (in the sense of vector space over $\mathbb{C}$) of the space $\mathcal{R}_n$ of rational vector functions. We also introduced the null-pole subspace $\mathcal{S}_\sigma(W) = W\mathcal{R}_n(\sigma)$ of a regular rational $n \times n$ matrix function W and showed that it can be identified as the null-pole subspace $\mathcal{S}_\tau(\sigma)$ associated with a null-pole triple τ for W over σ. From Chapter 4 we know that a σ-admissible data set $\tau = (C_\pi, A_\pi; A_\zeta, B_\zeta, S)$ is the null-pole triple over σ for some rational matrix function W if and only if τ is a σ-admissible *Sylvester* data set, i.e. the Sylvester equation

$$SA_\pi - A_\zeta S = B_\zeta C_\pi$$

is satisfied. The goal of this section is to obtain characterizations of subspaces $\mathcal{S}$ of $\mathcal{R}_n$ arise as null-pole subspaces $\mathcal{S}_\tau(\sigma)$ for some σ-admissible data set τ or for the null-pole triple τ for some rational matrix function W. These results in turn lead to analogues of Beurling-Lax and Ball-Helton invariant subspace representation theorems for rational functions which will be presented in later sections of the chapter.

Given a scalar rational function $r(z)$ we denote by $M_{r(z)}$ the operator of multiplication by $r(z)$; so $(M_{r(z)}h)(z) = r(z)h(z)$ for $h \in \mathcal{R}_n$.

By Corollary 12.2.3 we know that necessary conditions for a subspace $\mathcal{S}$ of $\mathcal{R}_n$ to have the form $\mathcal{S} = \mathcal{S}_\tau(\sigma)$ for a σ-admissible data set τ are that

(14.1.1) $P_{\sigma^c}^0(\mathcal{S})$ is invariant under $P_{\sigma^c}^0 M_{r(z)} \mid \mathcal{R}_n^0(\sigma^c)$ for each $r \in \mathcal{R}(\sigma)$

(14.1.2) $\dim P_{\sigma^c}^0(\mathcal{S}) < \infty$

(14.1.3) $\mathcal{R}_n(\sigma) \cap \mathcal{S}$ is invariant under $M_{r(z)}$ for each $r \in \mathcal{R}(\sigma)$

and

(14.1.4) $\dim \mathcal{R}_n(\sigma)/[\mathcal{R}_n(\sigma) \cap \mathcal{S}] < \infty$

Our first result is that these conditions are also sufficient for S to have the form $S_\tau(\sigma)$.

THEOREM 14.1.1. *A subspace (over* C*)* S *of* $\mathcal{R}_n$ *is of the form* $S = S_\tau(\sigma)$ *for a σ-admissible data set τ if and only if S satisfies conditions* (14.1.1)–(14.1.4).

In proving Theorem 14.1.1 (and in the chapter in general), it will be convenient to consider a σ-admissible data set (or a null-pole triple) as a collection of linear transformations $(C, A_\pi; A_\zeta, B; S)$ acting on various finite dimensional subspaces:

$$C: \mathcal{X}_\pi \to \mathbf{C}^n, \quad A_\pi: \mathcal{X}_\pi \to \mathcal{X}_\pi, \quad A_\zeta: \mathcal{X}_\zeta \to \mathcal{X}_\zeta, \quad B: \mathbf{C}^n \to \mathcal{X}_\zeta, \quad S: \mathcal{X}_\pi \to \mathcal{X}_\zeta.$$

One recovers the matrix form of the data set by choosing bases for $\mathcal{X}_\pi$ and $\mathcal{X}_\zeta$ and moreover all matrix forms of the data similar to a given matrix form arise via an appropriate change of basis in $\mathcal{X}_\pi$ and $\mathcal{X}_\zeta$. This point of view gives a coordinate free definition of σ-admissible data which is useful for the constructions of this chapter.

In proving Theorem 14.1.1 we shall come up with specific linear transformations C, A_π, A_ζ, B, S for which $\tau = (C, A_\pi; A_\zeta, B; S)$ is a σ-admissible data set such that the subspace S is equal to $S_\tau(\sigma)$. To define these linear transformations, first define subspaces $\mathcal{X}_\pi$ and $\mathcal{X}_\zeta$ by

$$(14.1.5) \qquad \qquad \mathcal{X}_\pi = P_{\sigma^c}^0(S)$$

and $\mathcal{X}_\zeta$ is any direct complement of $\mathcal{R}_n(\sigma) \cap S$ in $\mathcal{R}_n(\sigma)$:

$$(14.1.6) \qquad \qquad \mathcal{R}_n(\sigma) = \mathcal{X}_\zeta \dotplus [\mathcal{R}_n(\sigma) \cap S].$$

By (14.1.2) and (14.1.4) both $\mathcal{X}_\pi$ and $\mathcal{X}_\zeta$ are finite dimensional. Let $P_\zeta: \mathcal{R}_n(\sigma) \to \mathcal{X}_\zeta$ be the projection of $\mathcal{R}_n(\sigma)$ onto $\mathcal{X}_\zeta$ along $\mathcal{R}_n(\sigma) \cap S$. Denote by M_z the operator $h(z) \to zh(z)$ of multiplication by z on $\mathcal{R}_n$. Next introduce linear transformations $C: \mathcal{X}_\pi \to \mathbf{C}^n$, $A_\pi: \mathcal{X}_\pi \to \mathcal{X}_\pi$, $A_\zeta: \mathcal{X}_\zeta \to \mathcal{X}_\zeta$, $B: \mathbf{C}^n \to \mathcal{X}_\zeta$ by

$$(14.1.7) \qquad \qquad Ck = \sum_{z_0 \in \sigma} \operatorname{Res}_{z=z_0} k(z)$$

$$(14.1.8) \qquad \qquad A_\pi k = P_{\sigma^c}^0 M_z k$$

$$(14.1.9) \qquad \qquad A_\zeta h = P_\zeta M_z h$$

and

$$(14.1.10) \qquad \qquad Bx = P_\zeta x.$$

By definition of $\mathcal{X}_\pi$, for each $k \in \mathcal{X}_\pi$ there exists an $h \in \mathcal{R}_n(\sigma)$ such that $k + h \in S$. If h_1 and h_2 are two elements of $\mathcal{R}_n(\sigma)$ such that $k + h_j \in S$ for $j = 1, 2$, then the difference $h_1 - h_2 = (k + h_1) - (k + h_2)$ is in $\mathcal{R}_n(\sigma) \cap S$. We conclude that, by definition of $\mathcal{X}_\zeta$ as a direct complement to $\mathcal{R}_n(\sigma) \cap S$ in $\mathcal{R}_n(\sigma)$, for each $k \in \mathcal{X}_\pi$ there exists a unique h

in $\mathcal{X}_\zeta$, which we denote as $T(k)$, such that $k + T(k) \in \mathcal{S}$. In this way $\mathcal{S}$ determines an linear transformation (or angle operator) $T \colon \mathcal{X}_\pi \to \mathcal{X}_\zeta$ for which

$$(14.1.11) \qquad\qquad k + T(k) \in \mathcal{S} \text{ for } k \in \mathcal{X}_\pi.$$

The following result gives an explicit σ-admissible data set τ for which $\mathcal{S} = \mathcal{S}_\tau(\sigma)$ in Theorem 14.1.1.

THEOREM 14.1.2. *Let $\mathcal{S}$ be a subspace of $\mathcal{R}_n$ which satisfies conditions (14.1.1)–(14.1.4) and let $\tau = (C, A_\pi; A_\zeta, B; T)$ be the collection of linear transformations defined by (14.1.7)–(14.1.11), where $\mathcal{X}_\pi$ and $\mathcal{X}_\zeta$ are given by (14.1.5) and (14.1.6). Then τ is a σ-admissible data set for which $\mathcal{S} = \mathcal{S}_\tau(\sigma)$.*

The analogous characterization of null-pole subspaces for a rational matrix function is as follows.

THEOREM 14.1.3. *A subspace $\mathcal{S}$ of $\mathcal{R}_n$ is the null-pole subspace $\mathcal{S}_\tau(W) = W\mathcal{R}_n(\sigma)$ for a rational matrix function W over the subset $\sigma \subset \mathbf{C}$ if and only if $\mathcal{S}$ satisfies conditions (14.1.2) and (14.1.4) together with*

$$(14.1.12) \qquad\qquad \mathcal{S} \text{ is invariant under } M_{r(z)} \text{ for each } r \in \mathcal{R}(\sigma).$$

In this case the set of linear transformations $\tau = (C, A_\pi; A_\zeta, B; T)$ defined by (14.1.7)–(14.1.11) is a σ-admissible Sylvester data set and $\mathcal{S} = \mathcal{S}_\sigma(W)$ for any regular rational matrix function W which has τ as a null-pole triple over σ.

Necessity in Theorem 14.1.1 has already been observed. If $\mathcal{S} = \mathcal{S}_\sigma(W)$ then by Theorem 12.3.1 $\mathcal{S} = \mathcal{S}_\tau(\sigma)$ where τ is a null-pole triple for W over σ, so certainly conditions (14.1.2) and (14.1.4) are necessary in Theorem 14.1.3. Since $\mathcal{R}_n(\sigma)$ is invariant under $M_{r(z)}$ for each r in $\mathcal{R}(\sigma)$ and each such $M_{r(z)}$ commutes with multiplication by the matrix function W, it is clear that $\mathcal{S}_\sigma(W) = W \cdot \mathcal{R}_n(\sigma)$ is invariant under $M_{r(z)}$ for $r \in \mathcal{R}(\sigma)$; thus necessity of (14.1.12) in Theorem 14.1.3 is clear as well. To prove sufficiency in Theorem 14.1.1 it suffices to verify that the formula for τ in Theorem 14.1.2 does the job. Theorem 14.1.3 will then also follow after some minor elaboration. The proof proceeds via a sequence of elementary lemmas.

LEMMA 14.1.4. *Suppose $\mathcal{S} \subset \mathcal{R}_n$ satisfies (14.1.1)–(14.1.4), and define A_π and A_ζ by (14.1.8) and (14.1.9). Then both A_π and A_ζ have spectrum contained in σ.*

PROOF. If $w \notin \sigma$ then $r(z) = (z - w)^{-1} \in \mathcal{R}(\sigma)$. Thus, the identity of linear transformations

$$\mathcal{P}^0_{\sigma^c} M_{(z-w)^{-1}} = P^0_{\sigma^c} M_{(z-w)^{-1}} P^0_{\sigma^c}$$

holds on $\mathcal{R}_n$.

Also by assumption $\mathcal{X}_\pi$ is invariant under $P^0_{\sigma^c} M_{(z-w)^{-1}} \mid \mathcal{X}_\pi$, since $(z - w)^{-1} \in \mathcal{R}(\sigma)$. From these properties it is easy to check that $P^0_{\sigma^c} M_{(z-w)^{-1}} \mid \mathcal{X}_\pi$ is the inverse of $A_\pi - wI$, so $w \notin \sigma(A_\pi)$.

Similarly, if $w \notin \sigma$, then $(z - w)^{-1} \in \mathcal{R}(\sigma)$ so by hypothesis $\mathcal{R}_n(\sigma) \cap \mathcal{S}$ is invariant under $M_{(z-w)^{-1}}$. This implies the identity of linear transformations

$$P_\zeta M_{(z-w)^{-1}} = P_\zeta M_{(z-w)^{-1}} P_\zeta$$

on $\mathcal{R}_n(\sigma)$. From this it is easy to check that $P_\zeta M_{(z-w)^{-1}} \mid \mathcal{X}_\zeta$ is the inverse of $A_\zeta - wI$, so $w \notin \sigma(A_\zeta)$ either. $\square$

LEMMA 14.1.5. *Suppose* $\mathcal{S} \subset \mathcal{R}_n$ *satisfies* (14.1.1)–(14.1.4) *and let* A_ζ *and* B *be defined by* (14.1.9) *and* (14.1.10). *Then for any vector* u *in* $\mathbf{C}^n$ *and number* $w \notin \sigma$ *we have*

$$(wI - A_\zeta)^{-1}Bu = P_\zeta(wI - M_z)^{-1}u.$$

PROOF. For $w \notin \sigma$, certainly $(wI - M_z)^{-1}u$ is in $\mathcal{R}_n(\sigma)$ for each $u \in \mathbf{C}^n$, so $(wI - M_z)^{-1}u$ is in the domain of P_ζ. By analytic continuation it suffices to check that the result holds for $|w|$ sufficiently large. For $|w|$ large we use the series expansions

$$(wI - A_\zeta)^{-1}Bu = \sum_{j=0}^{\infty}(A_\zeta^j Bu)w^{-j-1}$$

$$P_\zeta(wI - M_z)^{-1}u = \sum_{j=0}^{\infty}(P_\zeta M_z^j u)w^{-j-1}.$$

Since $\operatorname{Ker} P_\zeta = \mathcal{R}_n(\sigma) \cap \mathcal{S}$ is invariant under M_z we have

$$A_\zeta^j Bu = P_\zeta M_z^j P_\zeta u = P_\zeta M_z^j u$$

from which the result follows. $\square$

LEMMA 14.1.6. *Suppose* $\mathcal{S} \subset \mathcal{R}_n$ *satisfies* (14.1.1)–(14.1.4) *and* A_ζ, *B are defined by* (14.1.9) *and* (14.1.10). *Then the projection* P_ζ *of* $\mathcal{R}_n(\sigma)$ *onto* $\mathcal{X}_\zeta$ *along* $\mathcal{R}_n(\sigma) \cap \mathcal{S}$ *is given by*

$$P_\zeta h = \sum_{w_0 \in \sigma} \operatorname{Res}_{w=w_0}(wI - A_\zeta)^{-1}Bh(w).$$

PROOF. Note that by the Cauchy integral formula

$$h(z) = \left[\frac{1}{2\pi i}\int_\gamma (wI - M_z)^{-1}h(w)dw\right](z)$$

where γ is a simple closed contour such that h is analytic on and inside γ and z is inside γ. We also assume that $\sigma(A_\zeta)$ is inside γ. Then by Lemma 14.1.5 we get

$$P_\zeta h = P_\zeta\left[\frac{1}{2\pi i}\int_\gamma (wI - M_z)^{-1}h(w)dw\right]$$

$$= \frac{1}{2\pi i}\int_\gamma P_\zeta[(wI - M_z)^{-1}h(w)]dw$$

$$= \frac{1}{2\pi i}\int_\gamma (wI - A_\zeta)^{-1}Bh(w)dw.$$

The result now follows from the residue theorem. $\square$

LEMMA 14.1.7. *Suppose* $S \subset \mathcal{R}_n$ *satisfies* (14.1.1)–(14.1.4) *and* C *and* A_π *are defined by* (14.1.7) *and* (14.1.8), *where* $\mathcal{X}_\pi$ *is given by* (14.1.5). *Then if* $w \notin \sigma(A_\pi)$ *and* $k \in \mathcal{X}_\pi$, *it follows that* k *is analytic at* w *and that* $k(w)$ *is given by*

$$k(w) = C(wI - A_\pi)^{-1}k.$$

PROOF. Let $k \in \mathcal{X}_\pi$. As $\mathcal{X}_\pi \subset \mathcal{R}_n^0(\sigma^c)$, all poles of k are in σ and $k(\infty) = 0$. Suppose $w \notin \sigma(A_\pi)$ and that w is not a pole of k. Choose a simple closed contour γ with $\sigma(A_\pi)$ and all poles of k inside γ while w is outside γ. Then the Cauchy integral formula takes the form

$$k(w) = -\frac{1}{2\pi i} \int_\gamma (z - w)^{-1}k(z)dz$$

or in operator form

$$k(w) = \frac{1}{2\pi i} \int_\gamma [M_{(w-z)^{-1}}k](z)dz.$$

Since all poles of the integrand inside γ are in σ, this in turn equals

$$k(w) = \sum_{z_0 \in \sigma} \operatorname{Res}_{z=z_0} M_{(w-z)^{-1}}k.$$

Since $P_\sigma M_{(w-z)^{-1}}k$ has all poles in σ and vanishes at infinity,

$$\sum_{z_0 \in \sigma} \operatorname{Res}_{z=z_0} P_\sigma M_{(w-z)^{-1}}k = 0.$$

Thus

$$k(w) = \sum_{z_0 \in \sigma} \operatorname{Res}_{z=z_0} P_{\sigma^c}^0 M_{(w-z)^{-1}}k$$

$$= C(wI - A_\pi)^{-1}k.$$

By analytic continuation, all poles of k are actually in $\sigma(A_\pi)$ and the formula holds as long as w is not in $\sigma(A_\pi)$. $\square$

PROOF OF THEOREM 14.1.2. Since the map $\Gamma_\pi \colon \mathcal{X}_\pi \to \mathcal{X}_\pi$ given by $\Gamma_\pi \colon k \to C(zI - A_\pi)^{-1}k$ is the identity by Lemma 14.1.7, certainly Γ_π has trivial kernel; hence (C, A_π) is a null kernel pair by Lemma 12.2.2. Similarly the map $\Gamma_\zeta \colon \mathcal{R}_n(\sigma) \to \mathcal{X}_\zeta$ defined by $\Gamma_\zeta \colon h(z) \to \sum_{w_0 \in \sigma} \operatorname{Res}_{w=w_0}(wI - A_\zeta)^{-1}Bh(w)$ is equal to the projection P_ζ of $\mathcal{R}_n(\sigma)$ onto $\mathcal{X}_\zeta$ along $\mathcal{R}_n(\sigma) \cap S$, and hence has image equal to the whole space $\mathcal{X}_\zeta$; thus (A_ζ, B) is a full range pair by Lemma 12.2.2. By Lemma 14.1.4 we know that A_π and A_ζ have spectra in σ. Thus $\tau = (C, A_\pi; A_\zeta, B; T)$ is a σ-admissible data set. It remains only to show that we recover S as $S = S_\tau(\sigma)$.

By the definition (14.1.11) of T, we have that S can be expressed as

$$S = (I + T)\mathcal{X}_\pi + [S \cap \mathcal{R}_n(\sigma)].$$

Equivalently, since $\operatorname{Im} T \subset \mathcal{X}_\zeta \subset \mathcal{R}_n(\sigma)$ while $\mathcal{X}_\pi \subset \mathcal{R}_n^0(\sigma^c)$, we have

(14.1.13) $$S = \{k + h \colon k \in \mathcal{X}_\pi, h \in \mathcal{R}_n(\sigma) \text{ such that } P_\zeta h = Tk\}.$$

By Lemma 14.1.7,

$$\mathcal{X}_\pi = \{C(zI - A_\pi)^{-1}x \colon x \in \mathcal{X}_\pi\}$$

while by Lemma 14.1.6

$$P_\zeta h = \frac{1}{2\pi i} \sum_{w_0 \in \sigma} \mathrm{Res}_{w=w_0}(wI - A_\zeta)^{-1}Bh$$

and by Lemma 14.1.7 the map $x \to C(zI - A_\pi)^{-1}x$ is the identity on $\mathcal{X}_\pi$. With these identifications made, we see that formula (14.1.13) for S is also just the definition of $\mathcal{S}_\tau(\sigma)$, so $S = \mathcal{S}_\tau(\sigma)$ as asserted. $\square$

PROOF OF THEOREM 14.1.3. Note that the invariance of S under M_r implies that $\mathcal{X}_\pi$ is invariant under $P_{\sigma^c}^0 M_r \mid \mathcal{R}_n^0(\sigma^c)$ and that $\mathcal{R}_n(\sigma) \cap S$ is invariant under $M_{r(z)}$ for each $r \in \mathcal{R}(\sigma)$. Thus all hypotheses of Theorems 14.1.1 and 14.1.2 are satisfied. Therefore, if $\tau = (C, A_\pi; A_\zeta, B; T)$ is defined by (14.1.7)–(14.1.11) then τ is σ-admissible and $S = \mathcal{S}_\tau$. The hypothesis that S itself is invariant under M_r for all r in $\mathcal{R}(\sigma)$ implies by Theorem 12.2.1 that in fact τ is a σ-admissible Sylvester data set. By Theorem 4.6.2 we know that then there exists a rational matrix function W having τ as a left null-pole triple over σ. Finally by Theorem 12.3.1 it follows that $S = \mathcal{S}_\tau(\sigma)$ is the null-pole subspace $\mathcal{S}_\sigma(W) = W\mathcal{R}_n(\sigma)$ for W. $\square$

14.2   A RATIONAL INDEFINITE METRIC BEURLING-LAX THEOREM

In this section we present some variations on Theorem 14.1.3 which can be viewed as a version of the invariant subspace representation theorem due to Beurling and Lax for shift invariant subspaces of Hardy space (over the unit disk or the right half plane) adapted to the purely algebraic setting of rational functions. The Hardy space H_n^2 (over the unit disk $\mathcal{D}$), viewed as a subspace of the Lebesgue space L_n^2 of $\mathbf{C}^n$-valued norm-squared integrable functions on the unit circle $\partial\mathcal{D}$, can be defined as the closure of $\mathcal{R}_n(\overline{\mathcal{D}})$ in L_n^2. A closed subspace $S \subset L_n^2$ is said to be *shift-invariant* if $M_z S \subset S$, *simply invariant* if $\bigcap_{k \geq 0}(M_z)^k S = \{0\}$, and *full range* if $\bigcup_{k<0}(M_z)^k S$ is dense in L_n^2. One version of the Beurling-Lax theorem is: *a full range simply invariant shift invariant (closed) subspace S of L_n^2 has the form*

$$S = \theta \cdot H_n^2$$

where θ is a measurable $n \times n$ matrix function for which $\theta(z)$ is unitary for almost every z on the boundary $\partial\mathcal{D}$ of the unit disk $\mathcal{D}$.

An adaptation of this result to rational functions can be had as follows. First note that $H_n^2 \cap \mathcal{R}_n = \mathcal{R}_n(\overline{\mathcal{D}})$ and $L_n^2 \cap \mathcal{R}_n = \mathcal{R}_n(\partial\mathcal{D})$. For S a subspace of $\mathcal{R}_n(\partial\mathcal{D})$, the appropriate analogue of *closed* and *shift invariant* we take to be invariance under $M_{r(z)}$ for each (scalar-valued) r in $\mathcal{R}(\overline{\mathcal{D}})$ together with

$$z \in \partial\mathcal{D} \Rightarrow \mathrm{Span}\{f(z) \colon f \in S\} = \mathbf{C}^n.$$

As suggested by Corollary 12.2.3 we take the analogue of *full-range* to be

$$\dim S/(S \cap \mathcal{R}_n(\overline{\mathcal{D}})) < \infty$$

and the analogue of *simple invariance* to be

$$\dim P^0_{\mathcal{R}_n(\overline{\mathcal{D}}^c)}(\mathcal{S}) < \infty.$$

The following then is a purely algebraic form of the Beurling-Lax theorem in the setting of rational functions. The proof will be given later in this section.

THEOREM 14.2.1. *A subspace* $\mathcal{S} \subset \mathcal{R}_n(\partial \mathcal{D})$ *is of the form*

$$\mathcal{S} = \theta \cdot \mathcal{R}_n(\overline{\mathcal{D}})$$

where $\theta \in \mathcal{R}_{n \times n}(\partial \mathcal{D})$ *has unitary values on* $\partial \mathcal{D}$ *if and only if*

(i) $M_{r(z)}\mathcal{S} \subset \mathcal{S}$ *for each* $r \in \mathcal{R}(\overline{\mathcal{D}})$;

(ii) $z \in \partial \mathcal{D} \Rightarrow \mathrm{Span}\{f(z) \colon f \in \mathcal{S}\} = \mathbb{C}^n$;

(iii) $\dim \mathcal{S}/(\mathcal{S} \cap \mathcal{R}_n(\overline{\mathcal{D}})) < \infty$;

and

(iv) $\dim P^0_{\mathcal{R}_n(\overline{\mathcal{D}}^c)}(\mathcal{S}) < \infty$.

An extension of the Beurling-Lax theorem to the setting of an indefinite metric was recently obtained by Ball and Helton. For this result we take $n = M + N$ and let J be the signature matrix $J = I_M \oplus -I_N$. For $\mathcal{S}$ a subset of L^2_n let $\mathcal{S}^{\perp J}$ denote the J-orthogonal complement

$$\mathcal{S}^{\perp J} = \{h \in L^2_n \colon \langle Jh, g \rangle_{L^2_n} = 0 \text{ for all } g \in \mathcal{S}\},$$

where $\langle \cdot, \cdot \rangle_{L^2_n}$ is the scalar product in L^2_n. Then the indefinite metric Beurling-Lax theorem due to Ball-Helton roughly[1] is as follows: a (closed) subspace $\mathcal{S}$ of L^2_n has the representation $\mathcal{S} = \theta \cdot H^2_n$ where θ is a measurable $n \times n$ matrix function such that $\theta(z)$ is J-unitary for almost every $z \in \partial \mathcal{D}$, if and only if $\mathcal{S}$ is full range, simply shift invariant, and

(14.2.1) $$L^2_n = \mathcal{S}^{\perp J} \dotplus \mathcal{S}.$$

For the rational version of this result we express the nondegeneracy condition (14.2.1) in a different form; we will see the equivalence of this form to one exactly analogous to (14.2.1) in the next section. For $\mathcal{S}$ a subspace of $\mathcal{R}_n(\partial \mathcal{D})$ satisfying conditions (i), (ii) and (iii), define $\tau = (C, A_\pi; A_\zeta, B; T)$ by (14.1.7)–(14.1.11), where we take $\sigma = \overline{\mathcal{D}}$. It turns out that then A_π and A_ζ have spectra in $\mathcal{D}$ so (see Section A.2) we may define

---

[1] We say roughly since in general θ may not be bounded in norm but only have matrix entries in L^2 in which case we have $\mathcal{S} = L^2_n$-closure of $\theta \cdot H^\infty_n$ (where H^∞_n consists of functions in H^2_n uniformly bounded on $\partial \mathcal{D}$). Then the precise statement has the same flavor but is more technical to state; for details see Ball-Helton [1983]. The extra technicalities do not enter into the rational version to be discussed here.

linear transformations $T_2: \mathcal{X}_\zeta \to \mathcal{X}_\zeta$ and $T_1: \mathcal{X}_\pi \to \mathcal{X}_\pi$ as the unique solutions of the equations

$$(14.2.2) \qquad\qquad A_\zeta T_2 A_\zeta^* - T_2 = -BJB^*$$

$$(14.2.3) \qquad\qquad T_1 - A_\pi^* T_1 A_\pi = -C^* J C.$$

We finally define a selfadjoint linear transformation $H: \mathcal{X}_\pi \oplus \mathcal{X}_\zeta \to \mathcal{X}_\pi \oplus \mathcal{X}_\zeta$ by

$$(14.2.4) \qquad\qquad H = \begin{bmatrix} T_1 & T^* \\ T & T_2 \end{bmatrix}.$$

(The adjoint is with respect to any convenient (but fixed) positive definite inner products on $\mathcal{X}_\pi$ and $\mathcal{X}_\zeta$). Then the rational indefinite metric Beurling-Lax theorem is as follows.

THEOREM 14.2.2. *A subspace $S \subset \mathcal{R}_n(\partial D)$ is of the form*

$$S = \theta \cdot \mathcal{R}_n(\overline{D})$$

where $\theta \in \partial D$ has J-unitary values on ∂D if and only if

(i) $M_r S \subset S$ *for each $r \in \mathcal{R}(\overline{D})$;*

(ii) $z \in \partial D \Rightarrow \mathrm{Span}\{f(z) : f \in S\} = \mathbb{C}^n$;

(iii) $\dim S/(S \cap \mathcal{R}_n(\overline{D})) < \infty$;

(iv) $\dim P^0_{\mathcal{R}_n(\overline{D}^c)}(S) < \infty$;

and

(v) *the linear transformation H defined by (14.2.2)–(14.2.4) is invertible in* $\mathcal{X}_\pi \oplus \mathcal{X}_\zeta$.

We remark that our proof of Theorems 14.2.1 and 14.2.2 are constructive in that we can write down explicit formulas for the representing function θ in terms of operators associated with the given subspace S. Explicitly we have the following.

THEOREM 14.2.3. (a) *Suppose the subspace $S \subset \mathcal{R}_n(\partial D)$ satisfies conditions (i)–(v) in Theorem 14.2.2. Define linear transformation C, A_π, A_ζ, B, T, T_2, T_1 and H by (14.1.7)–(14.1.11) (with $\sigma = \overline{D}$) and (14.2.1)–(14.2.4). Let $\theta(z)$ be the rational $n \times n$ matrix function defined by*

$$(14.2.5) \quad \begin{aligned} \theta(z) = I + (z - z_0)[C, -JB^*] \\ \cdot \begin{bmatrix} (zI - A_\pi)^{-1} & 0 \\ 0 & (I - zA_\zeta^*)^{-1} \end{bmatrix} H^{-1} \begin{bmatrix} (I - z_0 A_\pi^*)^{-1} C^* J \\ (A_\zeta - z_0 I)^{-1} B \end{bmatrix} \end{aligned}$$

where z_0 is an arbitrary point on the unit circle. Then $\theta(z)$ is J-unitary for $|z| = 1$ and

$$S = \theta \cdot \mathcal{R}_n(\overline{D}).$$

(b) *If $J = I$ in statement* (a), *then condition* (v) *in Theorem* 14.2.2 *is automatic, and* $\theta(z)$ *defined by* (14.2.5) (*with* $J = I$) *has unitary values on the unit circle.*

PROOF OF THEOREMS 14.2.1–14.2.3. Necessity of conditions (i), (iii) and (iv) in Theorems 14.2.1 and 14.2.2 is clear from Theorem 14.1.3. If $S = \theta R_n(\overline{D})$ with θ rational, then θ has no poles on ∂D since by assumption $S \subset R_n(\partial D)$. If θ is J-unitary valued on ∂D, then θ can have no zeros on ∂D either (by Theorem 7.5.1 for example), so the necessity of condition (ii) in Theorems 14.2.1 and 14.2.2 follows. Also by Theorem 14.1.3 $\tau = (C, A_\pi; A_\zeta, B; T)$ defined by (14.1.7)–(14.1.11) is a null-pole triple for θ over $\overline{D}$. Since, as we observed above, θ has no poles or zeros on ∂D, the necessity of (v) in Theorem 14.2.2 follows from Theorem 7.4.2.

Conversely, suppose $S \subset R_n(\partial D)$ satisfies conditions (i)–(iv) in Theorem 14.2.2. By Theorem 14.1.3, $S = \theta R_n(\overline{D})$ for any rational matrix function θ for which $\tau = (C, A_\pi; A_\zeta, B; T)$ as defined by (14.1.7)–(14.1.11) is a null-pole triple over $\overline{D}$. Since $S \subset R_n(\partial D)$, we have $\sigma(A_\pi) \subset D$, and by (ii) $\sigma(A_\zeta) \subset D$. By condition (v), the hypotheses of Theorem 7.4.2 hold true, so there exists such a matrix function θ with the additional property that θ is J-unitary valued on ∂D. Moreover the formula (7.4.7) for such a $\theta(z)$ reduces to (14.2.5). This proves part (a) of Theorem 14.2.3 and sufficiency in Theorem 14.2.2. Finally part (b) of Theorem 14.2.3 and sufficiency in Theorem 14.2.1 follows from Theorem 7.6.3. $\square$

Theorems 14.2.1–14.2.3 have obvious analogues where the unit disk D is replaced by the open right half plane Π^+. A technical difference is that $R_n(\overline{\Pi^+})$ is not invariant under M_z (since $r(z) = z$ has a pole at infinity). All the theory nevertheless applies if one works with $\sigma = \overline{\Pi^+}\backslash\{\infty\}$ rather than with $\sigma = \overline{\Pi^+}$. For the explicit formulas for H and $\theta(z)$ in (14.2.4) and (14.2.5) one should of course use the results of Chapter 6 rather than of Chapter 7. As we shall not need these results explicitly we leave the details to the reader.

Another variation is to remove the restriction that $S \subset R_n(\partial D)$ and seek a characterization of subspaces $S \subset R_n$ having a representation $S = \theta R_n(\overline{D})$ with $\theta(z)$ J-unitary at all regular points of ∂D. In this case it is possible for A_π and A_ζ defined by (14.1.8) and (14.1.9) (with $\sigma = \overline{D}$) to have some spectrum on ∂D. For this situation let us decompose X_π and X_ζ as

$$X_\pi = X_{\pi_i} \oplus X_{\pi_0}$$

and

$$X_\zeta = X_{\zeta_i} \oplus X_{\zeta_0}$$

so that with respect to these decompositions

(14.2.6) $$A_\pi = A_{\pi_i} \oplus A_{\pi_0}$$

and

(14.2.7) $$A_\zeta = A_{\zeta_i} \oplus A_{\zeta_i}$$

where $\sigma(A_{\pi_i}) \cap \sigma(A_{\zeta_i}) \subset D$ and $\sigma(A_{\pi_0}) \cup \sigma(A_{\zeta_0}) \subset \partial D$. With respect to these decompositions of X_π and X_ζ, the linear transformations B, C and T defined by (14.1.10),

(14.1.7) and (14.1.11) take the form

(14.2.8)
$$B = \begin{bmatrix} B_i \\ B_0 \end{bmatrix}, \qquad C = [C_i, C_0]$$

and

(14.2.9)
$$T = \begin{bmatrix} T_{ii} & T_{i0} \\ T_{0i} & T_{00} \end{bmatrix}.$$

Endow the four spaces $\mathcal{X}_{\pi_i}$, $\mathcal{X}_{\pi_0}$, $\mathcal{X}_{\zeta_i}$, $\mathcal{X}_{\zeta_0}$ with any convenient fixed positive definite inner products. We shall also require the existence of a unitary linear transformation $U: \mathcal{X}_{\pi_0} \to \mathcal{X}_{\zeta_0}$ for which

(14.2.10)
$$U^{-1} A_{\zeta_0} U = A_{\pi_0}^{*-1}$$

(14.2.11)
$$U^{-1} B_0 = A_0^{*-1} C_0^* J$$

and

(14.2.12)
$$H \overset{\text{def}}{=} U^{-1} T_{00} \text{ is self-adjoint on } \mathcal{X}_{\pi_0}.$$

Define linear transformations T_{11}, T_{21}, T_{32} and T_{33} as the unique solutions of the respective Sylvester-Stein equations

(14.2.13)
$$T_{11} - A_{\pi_i}^* T_{11} A_{\pi_i} = -C_i^* J C_i$$

(14.2.14)
$$A_{\pi_0}^* T_{21} A_{\pi_0} - T_{21} = C_0^* J C$$

(14.2.15)
$$T_{32} A_{\pi_0} - A_{\zeta_i} T_{32} = B_i C_0$$

and

$$A_{\zeta_i} T_{33} A_{\zeta_i}^* - T_{33} = B_i J B_i^*.$$

Finally define $\mathcal{S}: \mathcal{X}_{\pi_i} \oplus \mathcal{X}_{\pi_0} \oplus \mathcal{X}_{\zeta_i} \to \mathcal{X}_{\pi_i} \oplus \mathcal{X}_{\pi_0} \oplus \mathcal{X}_{\zeta_i}$ by

(14.2.16)
$$\mathcal{S} = \begin{bmatrix} T_{11} & T_{21}^* & T^* \\ T_{21} & H & T_{32}^* \\ T & T_{32} & T_{33} \end{bmatrix}.$$

The result then is as follows.

THEOREM 14.2.4. *A subspace $\mathcal{S} \subset \mathcal{R}_n$ has the form $\mathcal{S} = \theta \cdot \mathcal{R}_n(\overline{\mathcal{D}})$ for a rational $n \times n$ matrix function θ with J-unitary values at all regular points of $\partial \mathcal{D}$ if and only if*

(i) $M_r \mathcal{S} \subset \mathcal{S}$ for each $r \in \mathcal{R}(\overline{\mathcal{D}})$;

(ii) $\dim \mathcal{S}/(\mathcal{S} \cap \mathcal{R}_n(\overline{\mathcal{D}})) < \infty$;

(iii) $\dim P^0_{\mathcal{R}_n(\overline{\mathcal{D}}^c)}(\mathcal{S}) < \infty$;

(iv) if C, A_π, A_ζ, B, T as defined by (14.1.7)–(14.1.11) are decomposed as in (14.2.6)–(14.2.9), then there exists a unitary operator $U: \mathcal{X}_{\pi_0} \to \mathcal{X}_{\zeta_0}$ for which (14.2.10)–(14.2.12) hold true;

and

(v) the operator S defined by (14.2.13)–(14.2.16) is Hermitian and invertible.

If conditions (i)–(v) hold, then θ can be taken to be

$$\theta(z) = I + (z - z_0)[C_i, C_0, -JB_i^*] \begin{bmatrix} (zI - A_{\pi_i})^{-1} & 0 & 0 \\ 0 & (zI - A_{\pi_0})^{-1} & 0 \\ 0 & 0 & (I - zA_{\zeta_i}^*)^{-1} \end{bmatrix}$$
$$\cdot S^{-1} \begin{bmatrix} (I - z_0 A_{\pi_i}^*)^{-1} C_i^* J \\ (I - z_0 A_{\pi_0}^*)^{-1} C_0^* J \\ (A_{\zeta_i} - z_0 I)^{-1} B_i \end{bmatrix},$$

where z_0 is a point on the unit circle not in $\sigma(A_{\pi_0})$ but otherwise arbitrary.

PROOF. This result follows in essentially the same ways as for Theorems 14.2.2 and 14.2.3 by making use of Theorem 7.5.2 in place of Theorem 7.4.2. $\square$

14.3 A RATIONAL BALL-HELTON INVARIANT SUBSPACE REPRESENTATION THEOREM

In this section we present a rational analogue of another invariant subspace representation theorem due to Ball-Helton [1984]. This theorem differs from the Beurling-Lax theorem in that one wishes to represent a pair of subspaces $\mathcal{S}$ and $\mathcal{S}_\infty$ simultaneously and in that the representer W is allowed to assume any invertible values (not necessarily unitary or J-unitary) on the unit circle.

This theorem roughly[2] says the following: a pair of (closed) subspaces $\mathcal{S}$ and $\mathcal{S}_\infty$ contained in L_n^2 can be represented simultaneously as

$$\mathcal{S} = W \cdot H_n^2$$

and

$$\mathcal{S}_\infty = W \cdot H_n^{2\perp}$$

---

[2] Again as for the indefinite metric Beurling-Lax theorem in Section 14.2, the representer W may turn out to be unbounded on the unit circle. Nevertheless the matrix entries of W are in L^2. Then we have more precisely $\mathcal{S} = L_n^2$-closure of $W \cdot H_n^\infty$ and $\mathcal{S}_\infty = L_n^2$-closure of $W \cdot (L_n^\infty \cap H_n^{2\perp})$ and the precise if and only if statement involves some additional technicalities which do not enter into our rational version of the theorem.

for a measurable matrix function θ with invertible values almost everywhere on ∂D if and only if

(i) S *is full range and simply invariant for the multiplication operator (forward shift)* M_z *on* L_n^2;

(ii) S_∞ *is full range and simply invariant for the multiplication operator (backward shift)* $M_{z^{-1}}$ *on* L_n^2;

and

(iii) $L_n^2 = S_\infty \dotplus S$.

For the analogue of this result for the rational case, we replace the closed unit disk $\overline{D}$ with an arbitrary subset σ of C. The analogue of H_n^2 becomes $\mathcal{R}_n(\sigma)$, L_n^2 becomes $\mathcal{R}_n$ and $H_n^{2\perp}$ becomes $\mathcal{R}_n^0(\sigma^c)$, the space of rational vector functions with all poles in σ and vanishing at infinity. The analogue of the direct sum decomposition $L_n^2 = H_n^{2\perp} \dotplus H_n^2$ becomes the decomposition

$$\mathcal{R}_n = \mathcal{R}_n^0(\sigma^c) \dotplus \mathcal{R}_n(\sigma).$$

The rational analogue of the above Ball-Helton theorem then is as follows.

THEOREM 14.3.1. *Let S and S_∞ be two subspaces of $\mathcal{R}_n$ and suppose σ is a subset of C. Then there exists a regular rational $n \times n$ matrix function W for which*

$$S = W\mathcal{R}_n(\sigma) \quad \text{and} \quad S_\infty = W\mathcal{R}_n^0(\sigma^c)$$

if and only if

(i) S *is invariant under* $M_{r(z)}$ *for each* $r \in \mathcal{R}(\sigma)$, $\dim P_{\sigma^c}^0(S) < \infty$ *and* $\dim \mathcal{R}_n(\sigma)/(\mathcal{R}_n(\sigma) \cap S) < \infty$;

(ii) S_∞ *is invariant under* $M_{r(z)}$ *for each* $r \in \mathcal{R}(\sigma^c)$, $\dim P_\sigma(S_\infty) < \infty$ *and* $\dim \mathcal{R}_n^0(\sigma^c)/(\mathcal{R}_n^0(\sigma^c) \cap S_\infty) < \infty$;

and

(iii) $\mathcal{R}_n = S_\infty \dotplus S$.

We shall prove Theorem 14.3.1 in this section only for the case where $S_\infty = \mathcal{R}_n^0(\sigma^c)$. The theorem will be proved in full generality in the next chapter after we incorporate data at infinity into the description of null-pole subspaces.

PROOF OF THEOREM 14.3.1 FOR THE CASE $S_\infty = \mathcal{R}_n^0(\sigma^c)$. In this case the condition (ii) in Theorem 14.3.1 holds true trivially. The necessity of (i) follows from Theorem 14.1.3 and the necessity of (iii) was already noted in Proposition 12.1.4.

Conversely suppose (i), (ii) and (iii) are satisfied with $S_\infty = \mathcal{R}_n^0(\sigma^c)$. Define $\tau = (C, A_\pi; A_\zeta, B; T)$ by (14.1.7)–(14.1.11). Then by Theorem 14.1.3 any matrix function θ having τ as a null-pole triple over D represents S as $S = W\mathcal{R}_n(\sigma)$. The problem is to find such a θ which has the additional property that $\mathcal{R}_n^0(\sigma^c) = W\mathcal{R}_n^0(\sigma^c)$. This condition in turn is easily seen to be equivalent to W having no poles and no zeros in σ^c (including infinity). Thus W must be regular at infinity and have τ as a global null-pole triple. By

Theorem 4.4.3 such a W exists if and only if the coupling matrix T is invertible, in which case the unique such W with the value I at infinity is given by

$$(14.3.1) \qquad W(z) = I + C(zI - A_\pi)^{-1} T^{-1} B.$$

Thus to prove Theorem 14.3.1 for the special case where $\mathcal{S}_\infty = \mathcal{R}_n^0(\sigma)$, we need only show that the condition

$$(14.3.2) \qquad \mathcal{R}_n^0(\sigma^c) \dotplus \mathcal{S} = \mathcal{R}_n$$

implies that T is invertible.

In fact we shall see that (14.3.2) and the invertibility of T are equivalent. By definition, $\mathcal{X}_\pi \subset \mathcal{R}_n^0(\sigma^c)$, $\mathcal{R}_n(\sigma) = \mathcal{X}_\zeta \dotplus [\mathcal{S} \cap \mathcal{R}_n(\sigma)]$, $T \colon \mathcal{X}_\pi \to \mathcal{X}_\zeta$ and

$$(14.3.3) \qquad \mathcal{S} = (I + T)\mathcal{X}_\pi + [\mathcal{S} \cap \mathcal{R}_n(\sigma)].$$

From representation (14.3.3) we observe

$$(14.3.4) \qquad \mathcal{S} \cap \mathcal{R}_n^0(\sigma^c) = \operatorname{Ker} T.$$

Thus $\operatorname{Ker} T = \{0\}$ if and only if $\mathcal{S} \cap \mathcal{R}_n^0(\sigma^c) = \{0\}$. Also from (14.3.3)

$$(14.3.5) \qquad \mathcal{R}_n^0(\sigma^c) + \mathcal{S} = \mathcal{R}_n^0(\sigma^c) + \operatorname{Im} T + [\mathcal{S} \cap \mathcal{R}_n(\sigma)].$$

Since $\operatorname{Im} T \subset \mathcal{X}_\zeta$ and $\mathcal{R}_n^0(\sigma) = \mathcal{X}_\zeta \dotplus [\mathcal{S} \cap \mathcal{R}_n(\sigma)]$, we see from (14.3.5) that $\mathcal{R}_n^0(\sigma^c) + \mathcal{S}$ equals $\mathcal{R}_n = \mathcal{R}_n^0(\sigma^c) \dotplus \mathcal{R}_n(\sigma)$ if and only if $\operatorname{Im} T = \mathcal{X}_\zeta$ (i.e. T is onto). Thus invertibility of T is equivalent to (14.3.2) as asserted. $\square$

CHAPTER 15

SUBSPACE INTERPOLATION WITH DATA AT INFINITY

In this chapter we give the modifications needed to define null-pole subspaces and solve subspace interpolation problems when a nontrivial collection of data is present at infinity.

15.1 NULL-POLE SUBSPACE WITH DATA AT INFINITY

In Chapters 12–14 for the most part we have considered null-pole subspaces $\mathcal{S}_\sigma(W) = W\mathcal{R}_n(\sigma)$ for a rational matrix function W only over a subset σ of the finite complex plane $\mathbf{C}$. In this chapter we develop the analogous theory for null-pole subspaces $\mathcal{S}_{\sigma_\infty}(W)$ over a subset σ_∞ of the extended complex plane $\mathbf{C}^*$ which may include the point infinity. More generally, we may consider σ_∞-admissible data sets τ_∞ and associated null pole subspaces $\mathcal{S}_{\tau_\infty}(\sigma_\infty)$ as in Chapter 12. The only restriction is that σ_∞ not be equal to the entire extended complex plane $\mathbf{C}^*$; for convenience we shall usually assume that the point 0 is an omitted point. For reasons which will become clear in the next section, it is useful to consider the null-pole subspace $\mathcal{S}_{\sigma_\infty}^0(W) \stackrel{\text{def}}{=} W\mathcal{R}_n^0(\sigma_\infty)$ rather than $\mathcal{S}_{\sigma_\infty}(W) \stackrel{\text{def}}{=} W\mathcal{R}_n(\sigma_\infty)$ if σ_∞ includes the point ∞; here by $\mathcal{R}_n^0(\sigma_\infty)$ we mean all rational vector functions with no poles in σ_∞ and with value 0 at ∞.

Suppose now that $W(z)$ is a given (possibly improper) regular rational matrix function and let σ_∞ be a subset of $\mathbf{C}^*$ which omits 0. Let σ_∞^* be defined by

$$\sigma_\infty^* = \{z \in \mathbf{C} : z^{-1} \in \sigma_\infty\}$$

(where $0^{-1} = \infty$). Let $\mathcal{S}_V(\sigma_\infty^*) := V\mathcal{R}_n(\sigma_\infty^*)$ be the null-pole subspace of the function $V(z) := W(z^{-1})$ over $\sigma_\infty^* \subset \mathbf{C}$. Then clearly from the definitions we have

$$\mathcal{S}_{\sigma_\infty}^0(W) = \{z^{-1}h(z^{-1}) : h \in \mathcal{S}_{\sigma_\infty^*}(V)\}.$$

By definition a left null-pole triple $(C_{\pi,\infty}, A_{\pi,\infty}; A_{\zeta,\infty}, B_\zeta; S_\infty)$ for W over σ_∞ is just a (left) null-pole triple for V over σ_∞^* (see Section 4.5). Thus the form of a null pole subspace over σ_∞ can be deduced via a change of variable from the form of a null-pole subspace over $\sigma_\infty^* \subset \mathbf{C}$. This suggests the following definitions.

Let $\tau_\infty = (C_\infty, A_{\pi,\infty}; A_{\zeta,\infty}, B_\infty; S_\infty)$ be a collection of matrices, and let σ_∞ be a subset of $\mathbf{C}^*$ which includes infinity and omits 0. Then we say that τ_∞ is a σ_∞-admissible set if τ_∞ is σ_∞^*-admissible in the sense introduced in Section 12.2; namely, if $(C_\infty, A_{\pi,\infty})$ is a null-kernel pair, $(A_{\zeta,\infty}, B)$ is a full range pair, S has size $n_{\zeta\infty} \times n_{\pi\infty}$ compatible with the sizes $n_{\zeta\infty} \times n_{\zeta\infty}$ and $n_{\pi\infty} \times n_{\pi\infty}$ of $A_{\zeta,\infty}$ and $A_{\pi,\infty}$ respectively, and $\sigma(A_{\pi,\infty}) \cup \sigma(A_{\zeta,\infty})$ is contained in σ_∞^*. The null-pole subspace $\mathcal{S}_{\tau_\infty}^0(\sigma_\infty)$ associated with this σ_∞-admissible data set is then defined to be

(15.1.1) $$\mathcal{S}_{\tau_\infty}^0(\sigma_\infty) = \{z^{-1}h(z^{-1}) : h \in \mathcal{S}_{\sigma_\infty^*}(\tau_\infty)\}$$

where $\mathcal{S}_{T\infty}(\sigma_\infty^*)$ is defined by (12.2.2):

$$\mathcal{S}_{T\infty}(\sigma_\infty^*) = \{C_\infty(zI - A_{\pi,\infty})^{-1}x + \hat{h}(z): x \in \mathbf{C}^{n\pi,\infty}, \hat{h} \in \mathcal{R}_n(\sigma_\infty^*)$$

(15.1.2)
$$\text{such that} \sum_{z_0 \in \sigma_\infty^*} \text{Res}_{z=z_0}(zI - A_{\zeta,\infty})^{-1}B_\infty\hat{h}(z) = S_\infty x\}.$$

The following theorem gives a more convenient expression for $\mathcal{S}_{T\infty}^0(\sigma_\infty)$.

THEOREM 15.1.1. *Suppose σ_∞ is a subset of $\mathbf{C}^*$ containing infinity and omitting 0 and $\tau_\infty = (C_\infty, A_{\pi,\infty}; A_{\zeta,\infty}, B_\infty; S_\infty)$ is a σ_∞-admissible data set. Then the associated null-pole subspace $\mathcal{S}_{T\infty}^0(\sigma_\infty)$ is given by*

$$\mathcal{S}_{T\infty}^0(\sigma_\infty) = \{C_\infty(I - zA_{\pi,\infty})^{-1}x + h(z): x \in \mathbf{C}^{n\pi,\infty}, h \in \mathcal{R}_n^0(\sigma_\infty),$$

$$\sum_{z_0 \in \sigma_\infty^c} \text{Res}_{z=z_0}(I - zA_{\zeta,\infty})^{-1}B_\infty h(z) = S_\infty x\}.$$

PROOF. From (15.1.1) and (15.1.2) we have

$$\mathcal{S}_{T\infty}^0(\sigma_\infty) = \{C_\infty(I - zA_{\pi,\infty})^{-1}x + h(z): x \in \mathbf{C}^{n\pi,\infty}, h \in \mathcal{R}_n^0(\sigma_\infty)$$

$$\text{such that} \sum_{z_0 \in \sigma_\infty^*} \text{Res}_{z=z_0}(zI - A_{\zeta,\infty})^{-1}B_\infty z^{-1}h(z^{-1}) = S_\infty x\}.$$

To obtain the characterization in the theorem, we need only show

$$\sum_{z_0 \in \sigma_\infty^*} \text{Res}_{z=z_0}(zI - A_{\zeta,\infty})^{-1}B_\infty z^{-1}h(z^{-1})$$

(15.1.3)
$$= \sum_{z_0 \in \sigma_\infty^c} \text{Res}_{z=z_0}(I - zA_{\zeta,\infty})^{-1}B_\infty h(z).$$

To show this, let γ be a contour such that the function $z^{-1}h(z^{-1})$ is analytic on and inside γ and such that $\sigma(A_{\zeta,\infty})$ is inside γ; such a contour exists since $z^{-1}h(z^{-1})$ is analytic on σ_∞^* and $\sigma(A_{\zeta,\infty})$ is contained in σ_∞^*. Then by the Residue Theorem the left side of (15.1.3) is given by the contour integral

(15.1.4)
$$\frac{1}{2\pi i}\int_\gamma (zI - A_{\zeta,\infty})^{-1}B_\infty z^{-1}h(z^{-1})dz.$$

We may also assume that 0 is inside γ. Then we may make a change of variable $z \to z^{-1}$ in the contour integral (15.1.4) to get

$$\frac{1}{2\pi i}\int_{\gamma^*} (z^{-1}I - A_{\zeta,\infty})^{-1}B_\infty zh(z)dz^{-1} = -\frac{1}{2\pi i}\int_{\gamma^*} (I - zA_{\zeta,\infty})^{-1}B_\infty h(z)dz$$

where γ^* is the image of γ under $z \to z^{-1}$. Since 0 was inside γ, we see that infinity is inside γ^*, i.e., γ^* has clockwise rather than counterclockwise orientation. Letting $\tilde{\gamma}^*$ be γ^* but with reverse orientation, we get that (15.1.4) equals

(15.1.5)
$$\frac{1}{2\pi i}\int_{\tilde{\gamma}^*} (I - zA_{\zeta,\infty})^{-1}B_\infty h(z)dz.$$

By the original construction of γ, it is easy to see that all poles of the integrand $(I - zA_{\zeta,\infty})^{-1}B_\infty h(z)$ in (15.1.5) which are in σ_∞ (including ∞) are outside γ; therefore all poles inside $\tilde{\gamma}^*$ are actually in σ_∞^c. Now by the Residue Theorem (15.1.5) equals the right side of (15.1.3) as needed. $\square$

An immediate corollary is the analogue of Corollary 12.2.3 for null-pole subspaces $\mathcal{S}_{\tau_\infty}(\sigma_\infty)$ over subsets $\sigma_\infty \subset \mathbb{C}^*\backslash\{0\}$. Recall that $\mathcal{R}_n$ has a direct sum decomposition

$$\mathcal{R}_n = \mathcal{R}_n^0(\sigma_\infty) \dotplus \mathcal{R}_n(\sigma_\infty^c).$$

Let $P_{\sigma_\infty}^0$ denote the projection of $\mathcal{R}_n$ onto $\mathcal{R}_n^0(\sigma_\infty)$ along $\mathcal{R}_n(\sigma_\infty^c)$, and let $P_{\sigma_\infty^c} := I - P_{\sigma_\infty}^0$ denote the complementary projection.

COROLLARY 15.1.2. Suppose $\sigma_\infty \subset \mathbb{C}^*\backslash\{0\}$ and $\tau_\infty = (C_\infty, A_{\pi,\infty}; A_{\zeta,\infty}, B_\infty; S_\infty)$ is a σ_∞-admissible data set. Let $\mathcal{S}_{\tau_\infty}^0(\sigma_\infty)$ be the associated null-pole subspace as in (15.1.1). Then:

(i) The subspace $\mathcal{X}_{\pi,\infty} = P_{\sigma_\infty^c}\left(\mathcal{S}_{\tau_\infty}^0(\sigma_\infty)\right)$ is given by

$$\mathcal{X}_{\pi,\infty} = \{C_\infty(I - zA_{\pi,\infty})^{-1}x: x \in \mathbb{C}^{n_{\pi,\infty}}\}.$$

Moreover,

$$\dim \mathcal{X}_{\pi,\infty} = n_{\pi,\infty}, \text{ where } n_{\pi,\infty} \times n_{\pi,\infty} \text{ is the size of } A_{\pi,\infty}$$

and for each scalar rational function $r \in \mathcal{R}_n(\sigma_\infty)$ we have the inclusion

$$P_{\sigma_\infty^c}(rk) \in \mathcal{X}_{\pi,\infty} \text{ whenever } k \in \mathcal{X}_{\pi,\infty}.$$

(ii) The subspace $\mathcal{S}_{\tau_\infty}^0(\sigma_\infty) \cap \mathcal{R}_n^0(\sigma_\infty)$ is given by

$$\mathcal{S}_{\tau_\infty}^0(\sigma_\infty) \cup \mathcal{R}_n^0(\sigma_\infty) = \{h \in \mathcal{R}_n^0(\sigma_\infty): \sum_{z_0 \in \sigma_\infty^c} \text{Res}_{z=z_0}(I - zA_{\zeta,\infty})^{-1}B_\infty h(z) = 0\}.$$

Moreover,

$$\dim \mathcal{R}_n^0(\sigma_\infty)/[\mathcal{S}_{\tau_\infty}^0(\sigma_\infty) \cap \mathcal{R}_n^0(\sigma_\infty)] = n_{\zeta,\infty}, \text{ where } n_{\zeta,\infty} \times n_{\zeta,\infty} \text{ is the size of } A_{\zeta,\infty}$$

and for each scalar rational function $r \in \mathcal{R}_n(\sigma_\infty)$,

$$rh \in \mathcal{S}_{\tau_\infty}^0(\sigma_\infty) \cap \mathcal{R}_n^0(\sigma_\infty) \text{ whenever } h \in \mathcal{S}_{\tau_\infty}^0(\sigma_\infty) \cap \mathcal{R}_n^0(\sigma_\infty).$$

PROOF. The results follow from the characterization of $\mathcal{S}_{\tau_\infty}^0(\sigma_\infty)$ in Theorem 15.1.1 in a way completely analogous to the proof of Corollary 12.2.3. Hence we omit the details. $\square$

We now illustrate null-pole subspaces at infinity with some examples.

EXAMPLE 15.1.1. Suppose $\sigma_\infty = \{\infty\}$, $\tau_\infty = (0, 0; A_{\zeta,\infty}, B_\infty; 0)$ where $A_{\zeta,\infty}$ is the $m \times m$ lower triangular Jordan block with the zero eigenvalue, and $B_\infty =$

$\mathrm{col}(b_{\infty,j})_{j=1}^m$. Then if $h(z) \in \mathcal{R}_n^0(\{\infty\})$ has Laurent expansion at infinity given by $h(z) = \sum_{k=1}^\infty h_k z^{-k}$, then

$$(I - zA_{\zeta,\infty})^{-1} B_\infty h(z) = \left[\sum_{j=0}^{m-1} z^j A_{\zeta,\infty}^j \right] B_\infty \cdot \sum_{k=1}^\infty h_k z^{-k}$$

$$= \sum_{k=2-m}^\infty \left\{ \sum_{j=0}^{m-1} A_{\zeta,\infty}^j B_\infty h_{j+k} \right\} z^{-k}.$$

The sum of all the residues of this expression in the finite plane by the Residue Theorem can be seen to be the coefficient of z^{-1}, namely

$$\sum_{z_0 \in \mathbb{C}} \mathrm{Res}(I - zA_{\zeta,\infty})^{-1} B_\infty h(z) = \sum_{j=0}^{m-1} A_{\zeta,\infty}^j B_\infty h_{j+1}.$$

The k-th row (where $1 \le k \le m$) of this last expression is

$$\sum_{j=1}^k b_{\infty,k+1-j} h_j.$$

Thus the subspace $\mathcal{S}_{\sigma_\infty}^0(\{\infty\})$ is characterized as the space of all rational functions $h(z)$ with Laurent expansion at infinity of the form

$$h(z) = \sum_{j=1}^\infty h_j z^{-j}$$

for which

$$\sum_{j=1}^k b_{\infty,k+1-j} h_j = 0 \quad \text{for} \quad 1 \le k \le m.$$

If we set $b_\infty(z) = \sum_{j=0}^{m-1} b_{\infty,j+1} z^{-j}$, this last condition is equivalent to

$$b_\infty(z) h(z) \text{ has a zero at infinity of order } m+1. \quad \square$$

EXAMPLE 15.1.2. Suppose $\sigma_\infty = \{\infty\}$ and $\tau_\infty = (C_\infty, A_{\pi,\infty}; 0, 0; 0)$ where $A_{\pi,\infty}$ is the $m \times m$ upper triangular Jordan block with the zero eigenvalue, and $C_\infty = [y_{1,\infty}, \ldots, y_{m,\infty}]$. For $x = \mathrm{col}(x_j)_{j=1}^m$, one can verify that

$$C_\infty (I - zA_{\pi,\infty})^{-1} x = \sum_{j=0}^{m-1} C_\infty A_{\pi,\infty}^j x z^j$$

$$= y_{\infty,1} x_1 + [y_{\infty,2} + y_{\infty,1} z] x_2 + \cdots$$

$$+ [y_{\infty,m} + y_{\infty,m-1} z + \cdots + y_{\infty,1} z^{m-1}] x_m.$$

Thus, if we set $y(z) = y_{\infty,m} + y_{\infty,m-1}z + \cdots + y_{\infty,1}z^{m-1}$ then the null-pole subspace can be identified as

$$\mathcal{S}^0_{\tau_\infty}(\{\infty\}) = \mathrm{Span}\{z^{-j}y(z) : 0 \leq j \leq m-1\} + \mathcal{R}^0_n(\{\infty\}). \quad \square$$

EXAMPLE 15.1.3. Suppose $\sigma_\infty = \{\infty\}$ and $\tau_\infty = (C_\infty, A_{\pi,\infty}; A_{\zeta,\infty}, B_\infty; S_\infty)$ is given by

$$C_\infty = \mathrm{col}(y_j)^n_{j=1}$$
$$A_{\pi,\infty} = A_{\zeta,\infty} = \text{ the zero linear transformation on } \mathbf{C}$$
$$B_\infty = [b_1, \ldots, b_n]$$
$$S_\infty = s\colon \mathbf{C} \to \mathbf{C}.$$

Then $\mathcal{S}^0_{\tau_\infty}(\{\infty\})$ is given by

$$\mathcal{S}^0_{\tau_\infty}(\{\infty\}) = \{\alpha y + z^{-1}h_1 : \alpha \in \mathbf{C}, h_1 \in \mathbf{C}^n$$
$$\text{such that } B_\infty h_1 = \alpha s\} + z^{-1}\mathcal{R}^0_n(\{\infty\}). \quad \square$$

Since the null-pole subspace $\mathcal{S}^0_{\tau_\infty}(\sigma_\infty)$ by definition (15.1.1) is a simple transform of a null pole subspace $\mathcal{S}_{\tau_\infty}(\sigma^*_\infty)$ of finite types as discussed earlier in Chapter 12, all results from Chapter 12 concerning null-pole subspaces over sets $\sigma \subset \mathbf{C}$ have easy analogues for null-pole subspaces over sets $\sigma_\infty \subset \mathbf{C}^*\backslash\{0\}$. We now list some of these.

THEOREM 15.1.3. *If $\sigma_\infty \subset \mathbf{C}^*\backslash\{0\}$ and $\tau_\infty = (C_\infty, A_{\pi,\infty}; A_{\zeta,\infty}, B_\infty; S_\infty)$ is a σ_∞-admissible data set, then the null-pole subspace $\mathcal{S}^0_{\tau_\infty}(\sigma_\infty)$ is invariant under $\mathcal{R}(\sigma_\infty)$, i.e.*

$$r\mathcal{S}^0_{\tau_\infty}(\sigma_\infty) \subset \mathcal{S}^0_{\tau_\infty}(\sigma_\infty) \text{ for all scalar rational functions}$$
$$r \text{ with no poles in } \sigma_\infty(\text{including } \infty)$$

if and only if S_∞ satisfies the Sylvester equation

$$S_\infty A_{\pi,\infty} - A_{\zeta,\infty}S_\infty = B_\infty C_\infty.$$

PROOF. See Theorem 12.2.1. $\square$

THEOREM 15.1.4. *If $\sigma_\infty \subset \mathbf{C}^*\backslash\{0\}$ and τ_∞ and $\tilde{\tau}_\infty$ are two σ_∞-admissible data sets, then $\mathcal{S}^0_{\tau_\infty}(\sigma_\infty) = \mathcal{S}^0_{\tilde{\tau}_\infty}(\sigma_\infty)$ if and only if τ_∞ and $\tilde{\tau}_\infty$ are similar.*

PROOF. See Theorem 12.2.4. $\square$

The null-pole subspace $\mathcal{S}^0_{\sigma_\infty}(W) = W\mathcal{R}^0_n(\sigma_\infty)$ for a regular rational matrix function W can be expressed in terms of a null-pole triple for W over σ_∞ (as defined in Section 4.5) according to the following theorem.

THEOREM 15.1.5. *Suppose that σ_∞ is a subset of $\mathbf{C}^*\backslash\{0\}$, W is a regular rational matrix function, and $\tau_\infty = (C_\infty, A_{\pi,\infty}; A_{\zeta,\infty}, B_\infty; S_\infty)$ is a null-pole triple for W over σ_∞. Then τ_∞ is σ_∞-admissible and*

$$\mathcal{S}^0_{\sigma_\infty}(W) = \mathcal{S}^0_{\tau_\infty}(\sigma_\infty).$$

PROOF. Note that by definition τ_∞ is a null-pole triple for $V(z) := W(z^{-1})$ over σ_∞^*. By Theorem 12.3.1

$$S_{\sigma_\infty^*}(V) = S_{\tau_\infty}(\sigma_\infty^*).$$

Now use that by definition (15.1.1) $S_{\tau_\infty}^0(\sigma_\infty)$ is the image of $S_{\tau_\infty}(\sigma_\infty^*)$ under the transformation $h(z) \to z^{-1}h(z^{-1})$ to draw the desired conclusion. $\square$

15.2 INTERPOLATION OF A NULL-POLE SUBSPACE WITH DATA AT INFINITY

In this section we describe which subspaces S of $\mathcal{R}_n$ arise as null-pole subspaces for a σ_∞-admissible data set or for a rational matrix function over the set σ_∞, where σ_∞ includes the point infinity. For σ_∞ a subset of $\mathbf{C}^* \backslash \{0\}$ containing infinity, we let $\sigma_\infty^c \subset \mathbf{C}$ be the complement of σ_∞ in $\mathbf{C}^*$. Recall that we have the direct sum decomposition

$$\mathcal{R}_n = \mathcal{R}_n^0(\sigma_\infty) \dotplus \mathcal{R}_n(\sigma_\infty^c)$$

and that we let $P_{\sigma_\infty^c}$ denote the projection of $\mathcal{R}_n$ onto $\mathcal{R}_n(\sigma_\infty^c)$ along $\mathcal{R}_n^0(\sigma_\infty)$, and $P_{\sigma_\infty}^0 = I - P_{\sigma_\infty^c}$ the complementary projection.

Corollary 15.1.2 gave some necessary conditions for a subspace S of $\mathcal{R}_n$ to be of the form $S = S_{\tau_\infty}^0(\sigma_\infty)$ for a σ_∞-admissible data set τ_∞, where $\infty \in \sigma_\infty \subset \mathbf{C}^* \backslash \{0\}$. These conditions are also sufficient.

THEOREM 15.2.1. *Let σ_∞ be a subset of $\mathbf{C}^* \backslash \{0\}$ containing the point at infinity. Then a subspace S of $\mathcal{R}_n$ is the null-pole subspace $S_{\tau_\infty}^0(\sigma_\infty)$ for a σ_∞-admissible data set τ_∞ if and only if*

(i) the subspace $P_{\sigma_\infty^c}(S)$ is finite dimensional and is invariant under $P_{\sigma_\infty^c} M_{r(z)} \mid \mathcal{R}_n(\sigma_\infty^c)$ for each $r \in \mathcal{R}(\sigma_\infty)$;

(ii) the subspace $\mathcal{R}_n^0(\sigma_\infty) \cap S$ has finite codimension in $\mathcal{R}_n^0(\sigma_\infty)$ and is invariant under $M_{r(z)}$ for each r in $\mathcal{R}(\sigma_\infty)$.

To get explicit formulas for the σ_∞-admissible data set τ_∞ in Theorem 15.2.1, we let $\mathcal{X}_{\pi,\infty}$ be the subspace

$$(15.2.1) \qquad \mathcal{X}_{\pi,\infty} = P_{\sigma_\infty^c}(S)$$

and we let $\mathcal{X}_{\zeta,\infty}$ be any direct complement of $\mathcal{R}_n^0(\sigma_\infty) \cap S$ in $\mathcal{R}_n^0(\sigma_\infty)$:

$$(15.2.2) \qquad \mathcal{R}_n^0(\sigma_\infty) = \mathcal{X}_{\zeta,\infty} \dotplus [\mathcal{R}_n^0(\sigma_\infty) \cap S].$$

Define linear transformations $C_\infty : \mathcal{X}_{\pi,\infty} \to \mathbf{C}^n$, $A_{\pi,\infty} : \mathcal{X}_{\pi,\infty} \to \mathcal{X}_{\pi,\infty}$, $A_{\zeta,\infty} : \mathcal{X}_{\zeta,\infty} \to \mathcal{X}_{\zeta,\infty}$, and $B_\infty : \mathbf{C}^n \to \mathcal{X}_{\zeta,\infty}$ by

$$(15.2.3) \qquad C_\infty h = h(0) \quad \text{for} \quad h \in \mathcal{X}_{\pi,\infty}$$

$$(15.2.4) \qquad A_{\pi,\infty} = P_{\sigma_\infty^c} M_{z^{-1}} \mid \mathcal{X}_{\pi,\infty}$$

(15.2.5) $$A_{\zeta,\infty} = P_\zeta M_{z^{-1}} \mid \mathcal{X}_{\zeta,\infty}$$

and

(15.2.6) $$B_\infty u = P_\zeta M_{z^{-1}} u \quad \text{for} \quad u \in \mathbf{C}^n$$

where P_ζ is the projection of $\mathcal{R}_n^0(\sigma_\infty)$ onto $\mathcal{X}_{\zeta,\infty}$ along $[\mathcal{R}_n^0(\sigma_\infty) \cap \mathcal{S}]$. Finally, from the definition (15.2.1) and the decomposition (15.2.2), the subspace $\mathcal{S}$ uniquely determines a linear transformation $T_\infty \colon \mathcal{X}_{\pi,\infty} \to \mathcal{X}_{\zeta,\infty}$ for which

(15.2.7) $$\mathcal{S} = (I + T_\infty)\mathcal{X}_{\pi,\infty} \dotplus [\mathcal{R}_n^0(\sigma_\infty) \cap \mathcal{S}].$$

By using these formulas we have the following result.

THEOREM 15.2.2. *Let σ_∞ be a subset of $\mathbf{C}^* \backslash \{0\}$ containing the point at infinity and suppose the subspace $\mathcal{S}$ of $\mathcal{R}_n$ satisfies conditions (i) and (ii) in Theorem 15.2.1. Then $\mathcal{S} = \mathcal{S}_{\tau_\infty}^0(\sigma_\infty)$ where $\tau_\infty = (C_\infty, A_{\pi,\infty}; A_{\zeta,\infty}, B_\infty; T_\infty)$ is given by (15.2.3)–(15.2.7).*

We have already observed in Theorem 15.1.3 that a null-pole subspace $\mathcal{S}_{\tau_\infty}^0(\sigma_\infty)$ associated with a σ_∞-admissible data set τ_∞ is invariant under $M_{r(z)}$ for all r in $\mathcal{R}(\sigma_\infty)$ if and only if τ_∞ is a Sylvester data set. On the other hand we know from Theorem 4.6.2 combined with Theorem 5.3.1 that a σ_∞-admissible Sylvester data set is the null-pole triple over σ_∞ for some regular rational matrix function W, in which case (by Theorem 15.1.5) $\mathcal{S}_{\tau_\infty}^0(\sigma_\infty)$ is the null-pole subspace $\mathcal{S}_{\sigma_\infty}^0(W) = W \cdot \mathcal{R}_n^0(\sigma_\infty)$ for W over σ_∞. This leads to the following result.

THEOREM 15.2.3. *Let σ_∞ be a subset of $\mathbf{C}^* \backslash \{0\}$ containing infinity and let $\mathcal{S}$ be a subspace of $\mathcal{R}_n$. Then $\mathcal{S}$ is the null-pole subspace*

$$\mathcal{S}_{\sigma_\infty}^0(W) = W \cdot \mathcal{R}_n^0(\sigma_\infty)$$

over σ_∞ for some rational matrix function W if and only if

  (i) *$M_{r(z)}\mathcal{S} \subset \mathcal{S}$ for each $r \in \mathcal{R}(\sigma_\infty)$;*

  (ii) *$\dim P_{\sigma_\infty^c}(\mathcal{S}) < \infty$;*

and

  (iii) *$\dim \mathcal{R}_n^0(\sigma_\infty)/(\mathcal{R}_n^0(\sigma_\infty) \cap \mathcal{S}) < \infty$.*

In this case W may be chosen to be any regular rational matrix function for which $\sigma_\infty = (C_\infty, A_{\pi,\infty}; A_{\zeta,\infty}, B_\infty; T_\infty)$ is a null-pole triple over σ_∞.

All the results in Theorems 15.2.1–15.2.3 follow from the correspondence

$$\mathcal{S}_{\tau_\infty}^0(\sigma_\infty) = \{z^{-1}h(z^{-1}) \colon h \in \mathcal{S}_{\tau_\infty}(\sigma_\infty^*)\}$$

between a null-pole subspace $\mathcal{S}_{\tau_\infty}^0(\sigma_\infty)$ of infinite type and the corresponding null-pole subspace $\mathcal{S}_{\tau_\infty}(\sigma_\infty^*)$ of finite type, together with the corresponding results in Section 14.1 for null-pole subspaces of finite type. For this reason we omit formal proofs of Theorems 15.2.1–15.2.3.

15.3   A RATIONAL BALL-HELTON THEOREM: THE GENERAL CASE

The goal of this section is to use the theory of null pole subspaces (including the case of data at infinity as developed in this chapter) to prove Theorem 14.3.1 for the general case. This will then lead to a proof of Theorem 5.5.2 as promised earlier. We will also present a generalization of this theorem to a setting where one is given a whole collection $S_1, S_2, \ldots, S_n, S_\infty$ of subspaces of R_n rather than just a pair S and S_∞.

The necessity of conditions (i), (ii) and (iii) in Theorem 14.3.1 is an immediate consequence of Theorems 14.1.1 and 15.2.1 and Proposition 12.1.4. We therefore need only prove sufficiency in Theorem 14.3.1.

Let then (S_∞, S) be a pair of subspaces satisfying conditions (i), (ii) and (iii) in Theorem 14.3.1. We assume that σ is a subset of $\mathbf{C}$ containing 0. Define subspaces $\mathcal{X}_{\pi,F}$ and $\mathcal{X}_{\pi,\infty}$ by

$$(15.3.1) \qquad \mathcal{X}_{\pi,F} = P^0_{\sigma^c}(S)$$

$$(15.3.2) \qquad \mathcal{X}_{\pi,\infty} = P_\sigma(S_\infty)$$

and choose subspaces $\mathcal{X}_{\zeta,F}$ and $\mathcal{X}_{\zeta,\infty}$ so that

$$(15.3.3) \qquad \mathcal{R}_n(\sigma) = \mathcal{X}_{\zeta,F} \dotplus [\mathcal{R}_n(\sigma) \cap S\}$$

and

$$(15.3.4) \qquad \mathcal{R}^0_n(\sigma^c) = \mathcal{X}_{\zeta,\infty} \dotplus [\mathcal{R}^0_n(\sigma^c) \cap S_\infty].$$

Define linear transformations $C_F \colon \mathcal{X}_{\pi,F} \to \mathbf{C}^n$, $A_{\pi,F} \colon \mathcal{X}_{\pi,F} \to \mathcal{X}_{\pi,F}$, $A_{\zeta,F} \colon \mathcal{X}_{\zeta,F} \to \mathcal{X}_{\zeta,F}$, and $B_F \colon \mathbf{C}^n \to \mathcal{X}_{\zeta,F}$ by

$$(15.3.5) \qquad C_F h = \sum_{z_0 \in \sigma} \operatorname{Res}_{z=z_0} h(z) \quad \text{for} \quad h \in \mathcal{X}_{\pi,F}$$

$$(15.3.6) \qquad A_{\pi,F} = P^0_{\sigma^c} M_z \mid \mathcal{X}_{\pi,F}$$

$$(15.3.7) \qquad A_{\zeta,F} = P_{\zeta,F} M_z \mid \mathcal{X}_{\zeta,F}$$

$$(15.3.8) \qquad B_F u = P_{\zeta,F} u \quad \text{for} \quad u \in \mathbf{C}^n$$

where $P_{\zeta,F}$ is the projection of $\mathcal{R}_n(\sigma)$ onto $\mathcal{X}_{\zeta,F}$ along $\mathcal{R}_n(\sigma) \cap S$. Similarly define linear transformations $C_\infty \colon \mathcal{X}_{\pi,\infty} \to \mathbf{C}^n$, $A_{\pi,\infty} \colon \mathcal{X}_{\pi,\infty} \to \mathcal{X}_{\pi,\infty}$, $A_{\zeta,\infty} \colon \mathcal{X}_{\zeta,\infty} \to \mathcal{X}_{\zeta,\infty}$, $B_\infty \colon \mathbf{C}^n \to \mathcal{X}_{\zeta,\infty}$ by

$$(15.3.9) \qquad C_\infty k = k(0) \quad \text{for} \quad k \in \mathcal{X}_{\pi,\infty}$$

$$(15.3.10) \qquad A_{\pi,\infty} = P_\sigma M_{z-1} \mid \mathcal{X}_{\pi,\infty}$$

$$(15.3.11) \qquad A_{\zeta,\infty} = P_{\zeta,\infty} M_{z-1} \mid \mathcal{X}_{\zeta,\infty}$$

and

$$(15.3.12) \qquad B_\infty u = P_{\zeta,\infty} M_{z-1} u \quad \text{for} \quad u \in \mathbb{C}^n$$

where $P_{\zeta,\infty}$ is the projection of $\mathcal{R}_n^0(\sigma^c)$ onto $\mathcal{X}_{\zeta,\infty}$ along $\mathcal{R}_n^0(\sigma^c) \cap \mathcal{S}_\infty$. From (15.3.1)–(15.3.4) we see that there are unique linear transformations $T_F \colon \mathcal{X}_{\pi,F} \to \mathcal{X}_{\zeta,F}$ and $T_\infty \colon \mathcal{X}_{\pi,\infty} \to \mathcal{X}_{\zeta,\infty}$ for which

$$(15.3.13) \qquad \mathcal{S} = (I + T_F)\mathcal{X}_{\pi,F} + [\mathcal{R}_n(\sigma) \cap \mathcal{S}\}$$

and

$$(15.3.14) \qquad \mathcal{S}_\infty = (I + T_\infty)\mathcal{X}_{\pi,\infty} + [\mathcal{R}_n^0(\sigma^c) \cap \mathcal{S}\}.$$

By Theorem 14.1.3 we know that $\tau_F = (C_F, A_{\pi,F}; A_{\zeta,F}, B_F; T_F)$ defined by (15.3.5)–(15.3.8) and (15.3.13) is a σ-admissible Sylvester data set if $\mathcal{S}$ satisfies (i) in Theorem 14.3.1, and that any rational matrix function W for which τ_F is a null-pole triple over σ is such that $\mathcal{S} = W\mathcal{R}_n(\sigma)$. Similarly, by Theorem 15.2.3 we know that if $\mathcal{S}_\infty$ satisfies (ii) in Theorem 14.3.1, then $\tau_\infty = (C_\infty, A_{\pi,\infty}; A_{\zeta,\infty}, B_\infty; T_\infty)$ is a σ^c-admissible Sylvester data set (in the sense of Section 15.1) and that any rational matrix function W for which τ_∞ is a null-pole triple over σ^c has the property that $\mathcal{S}_\infty = W\mathcal{R}_n^0(\sigma^c)$. To prove Theorem 14.3.1, we must show that if in addition condition (iii) holds, then there exists a rational matrix function W which simultaneously has τ_F as a null-pole triple over σ and τ_∞ as a null-pole triple over σ^c. But by Theorem 5.4.1 this is the case if and only if the linear transformation $T \colon \mathcal{X}_{\pi,F} \oplus \mathcal{X}_{\pi,\infty} \to \mathcal{X}_{\zeta,F} \oplus \mathcal{X}_{\zeta,\infty}$ defined by

$$(15.3.15) \qquad T = \begin{bmatrix} T_F & T_{12} \\ T_{21} & T_\infty \end{bmatrix},$$

where T_{12} and T_{21} are the unique solutions of the Stein equations

$$(15.3.16) \qquad A_{\zeta,F} T_{12} A_{\pi,\infty} - T_{12} = B_F C_\infty$$

$$(15.3.17) \qquad T_{21} - A_{\zeta,\infty} T_{21} A_{\pi,F} = -B_\infty C_F,$$

is invertible (see Theorem A.2.1). Thus the general case of Theorem 14.3.1 is completely proved once we establish the following general lemma.

LEMMA 15.3.1. *Let σ be a subset of $\mathbb{C}$ with $0 \in \sigma$ and suppose $\tau_F = (C_F, A_{\pi,F}; A_{\zeta,F}, B_F; S_F)$ is a σ-admissible data set and $\tau_\infty = (C_\infty, A_{\pi,\infty}; A_{\zeta,\infty}, B_\infty; S_\infty)$ is a σ^c-admissible data set of linear transformations, where $A_{\pi,F}, A_{\zeta,F}, A_{\pi,\infty}, A_{\zeta,\infty}$ are linear transformations acting on the finite dimensional* spaces $\mathcal{X}_{\pi,F}$, $\mathcal{X}_{\zeta,F}$, $\mathcal{X}_{\pi,\infty}$, $\mathcal{X}_{\zeta,\infty}$ respectively. Define linear transformations S_{12}, S_{21} and S by

(15.3.18)
$$A_{\zeta,F}S_{12}A_{\pi,\infty} - S_{12} = B_F C_\infty$$

(15.3.19)
$$S_{21} - A_{\zeta,\infty}S_{21}A_{\pi,F} = -B_\infty C_F$$

and

(15.3.20)
$$S = \begin{bmatrix} S_F & S_{12} \\ S_{21} & S_\infty \end{bmatrix}.$$

Then the matching condition

$$\mathcal{R}_n = S^0_{\tau_\infty}(\sigma^c) \dotplus S_{\tau_F}(\sigma)$$

holds if and only if S is invertible.

PROOF. Introduce linear transformations $\Omega_F: \mathcal{X}_{\pi,F} \to \mathcal{R}^0_n(\sigma^c)$ by

$$\Omega_F x = C_F(zI - A_{\pi,F})^{-1}x$$

and $\Theta_F: \mathcal{R}_n(\sigma) \to \mathcal{X}_{\zeta,F}$ by

$$\Theta_F h = \sum_{z_0 \in \sigma} \operatorname{Res}_{z=z_0}(zI - A_{\zeta,F})^{-1}B_F h(z).$$

Then by (12.2.2),

$$S_{\tau_F}(\sigma) = \{\Omega_F x + h : x \in \mathcal{X}_{\pi,F}, h \in \mathcal{R}_n(\sigma)$$
$$\text{such that } \Theta_F h = S_F x\}.$$

Introduce $\Omega_\infty: \mathcal{X}_{\pi,\infty} \to \mathcal{R}_n(\sigma)$ by

$$\Omega_\infty x = C_\infty(I - zA_{\pi,\infty})^{-1}x$$

and $\Theta_\infty: \mathcal{R}^0_n(\sigma^c) \to \mathcal{X}_{\zeta,\infty}$ by

$$\Theta_\infty h = \sum_{z_0 \in \sigma} \operatorname{Res}_{z=z_0}(I - zA_{\zeta,\infty})^{-1}B_\infty h(z).$$

Then by Theorem 15.1.1,

(15.3.21)
$$S^0_{\tau_\infty}(\sigma^c) = \{\Omega_\infty x + h : x \in \mathcal{X}_{\pi,\infty}, h \in \mathcal{R}^0_n(\sigma^c)$$
$$\text{such that } \Theta_\infty h = S_\infty x\}.$$

We next check that $S_{12} = -\Theta_F\Omega_\infty$ satisfies the Stein equation (15.3.18) and hence is the unique solution. Indeed, from the identity

$$-A_{\zeta,F}(zI - A_{\zeta,F})^{-1}B_F C_\infty(I - zA_{\pi,\infty})^{-1}A_{\pi,\infty}$$
$$+(zI - A_{\zeta,F})^{-1}B_F C_\infty(I - zA_{\pi,\infty})^{-1}$$
$$= B_F C_\infty(I - zA_{\pi,\infty})^{-1}A_{\pi,\infty} + (zI - A_{\zeta,F})^{-1}B_F C_\infty$$

and using the explicit form of $\Theta_F \Omega_\infty$

$$\Theta_F \Omega_\infty = \sum_{z_0 \in \sigma} \text{Res}_{z=z_0} (zI - A_{\zeta,F})^{-1} B_F C_\infty (I - zA_{\pi,\infty})^{-1}$$

we see that

$$-A_{\zeta,F}\Theta_F \Omega_\infty A_{\pi,\infty} + \Theta_F \Omega_\infty = \sum_{z_0 \in \sigma} \text{Res}_{z=z_0} \{ B_F C_\infty (I - zA_{\pi,\infty})^{-1} A_{\pi,\infty}$$
$$+ (zI - A_{\zeta,F})^{-1} B_F C_\infty \}.$$

Since $\sigma(A_{\pi,\infty}) \subset (\sigma^c)^*$, the first term inside the braces is analytic on σ and so the sum of its residues for points in σ is zero. On the other hand $\sigma(A_{\zeta,F}) \subset \sigma$ so by the Riesz functional calculus

$$\sum_{z_0 \in \sigma} \text{Res}_{z=z_0} (zI - A_{\zeta,F})^{-1} B_F C_\infty = B_F C_\infty.$$

We conclude that $S_{12} = -\Theta_F \Omega_\infty$ is the solution of (15.3.18) as claimed.

A completely analogous calculation gives that $S_{21} = -\Theta_\infty \Omega_F$ is the unique solution of (15.3.19). This time use the identity

$$- (I - zA_{\zeta,\infty})^{-1} B_\infty C_F (zI - A_{\pi,F})^{-1}$$
$$+ A_{\zeta,\infty}(I - zA_{\zeta,\infty})^{-1} B_\infty C_F (zI - A_{\pi,F})^{-1} A_{\pi,F}$$
$$= -A_{\zeta,\infty}(I - zA_{\zeta,\infty})^{-1} B_\infty C_F - B_\infty C_F (zI - A_{\pi,F})^{-1}.$$

Thus the linear transformation S given by (15.3.20) can be identified as

$$(15.3.22) \qquad S = \begin{bmatrix} S_F & -\Theta_F \Omega_\infty \\ -\Theta_\infty \Omega_F & S_\infty \end{bmatrix}.$$

Next we show that $\mathcal{S}_{T_F}(\sigma) \cap \mathcal{S}^0_{T_\infty}(\sigma^c) = \{0\}$ if and only if S as in (15.3.20) has zero kernel. Indeed suppose S has zero kernel and the function h is in $\mathcal{S}_{T_F}(\sigma) \cap \mathcal{S}^0_{T_\infty}(\sigma^c)$. Then by (15.3.20) and (15.3.21)

$$(15.3.23) \qquad \Omega_F x + h_F = h = h_\infty + \Omega_\infty y$$

for some $x \in \mathcal{X}_{\pi,F}$, $y \in \mathcal{X}_{\pi,\infty}$, $h_F \in \mathcal{R}_n(\sigma)$ and $h_\infty \in \mathcal{R}^0_n(\sigma^c)$ where

$$\Theta_F h_F = S_F x$$

and

$$\Theta_\infty h_\infty = S_\infty y.$$

The left side and the right side of (15.3.23) give two decompositions of h with respect to the direct sum decomposition

$$\mathcal{R}_n = \mathcal{R}^0_n(\sigma^c) \dotplus \mathcal{R}_n(\sigma).$$

Equating respective components, we see that (15.3.23) is equivalent to

(15.3.24) $\Omega_F x = h_\infty$

and

(15.3.25) $h_F = \Omega_\infty y.$

From (15.3.24) we get

$$\Theta_\infty \Omega_F x = S_\infty y$$

while (15.3.25) implies

$$S_F x = \Theta_F \Omega_\infty y.$$

Equations (15.3.24) and (15.3.25) therefore yield the system of equations

$$\begin{bmatrix} S_F & -\Theta_F \Omega_\infty \\ -\Theta_\infty \Omega_F & S_\infty \end{bmatrix} \begin{bmatrix} x \\ y \end{bmatrix} = \begin{bmatrix} 0 \\ 0 \end{bmatrix},$$

i.e. $\begin{bmatrix} x \\ y \end{bmatrix}$ is in $\operatorname{Ker} S = \{0\}$. Thus x and y are both zero and then from (15.3.23) $h = 0$, so $\mathcal{S}_{T_F}(\sigma) \cap \mathcal{S}^0_{T_\infty}(\sigma^c) = \{0\}$. Conversely, suppose $\mathcal{S}_{T_F}(\sigma) \cap \mathcal{S}^0_{T_\infty}(\sigma^c) = \{0\}$ and $\begin{bmatrix} x \\ y \end{bmatrix} \in \operatorname{Ker} S$. Since $(A_{\zeta,F}, B_F)$ and $(A_{\zeta,\infty}, B_\infty)$ are full range pairs, by Lemma 12.2.2 we can find $h_F \in \mathcal{R}_n(\sigma)$ and $h_\infty \in \mathcal{R}^0_n(\sigma^c)$ so that $\Theta_F h_F = S_F x$ and $\Theta_\infty h_\infty = S_\infty y$. Then by reversing the analysis in the argument above, we get that $h \overset{\text{def}}{=} \Omega_F x + h_F = h_\infty + \Omega_\infty y \in \mathcal{S}_{T_F}(\sigma) \cap \mathcal{S}^0_{T_\infty}(\sigma^c) = \{0\}$. Then in particular the $\mathcal{R}^0_n(\sigma^c)$-component $\Omega_F x$ of h is zero and the $\mathcal{R}_n(\sigma)$-component $\Omega_\infty y$ of h is zero. Since $(C_F, A_{\pi,F})$ and $(C_\infty, A_{\pi,\infty})$ are null kernel pairs, by Lemma 12.2.2 this forces $x = 0$ and $y = 0$, so $\operatorname{Ker} S = \{0\}$.

Next we show that $\mathcal{R}_n = \mathcal{S}_{T_F}(\sigma) + \mathcal{S}^0_{T_\infty}(\sigma^c)$ if and only if S is onto. Indeed, suppose $\mathcal{R}_n = \mathcal{S}_{T_F}(\sigma) + \mathcal{S}^0_{T_\infty}(\sigma^c)$ so, for every $g_\infty \in \mathcal{R}^0_n(\sigma^c)$ and $g_F \in \mathcal{R}_n(\sigma)$ the function $g_\infty + g_F \in \mathcal{R}_n$ has a representation

$$g_\infty + g_F = (\Omega_F x + h_F) + (h_\infty + \Omega_\infty y)$$
$$= (\Omega_F x + h_\infty) + (h_F + \Omega_\infty y)$$

for an $h_F \in \mathcal{R}_n(\sigma)$ and $h_\infty \in \mathcal{R}^0_n(\sigma^c)$ with $\Theta_F h_F = S_F x$ and $\Theta_\infty h_\infty = S_\infty y$. Equating $\mathcal{R}^0_n(\sigma^c)$ and $\mathcal{R}_n(\sigma)$ components gives

(15.3.26) $g_\infty = \Omega_F x + h_\infty$

and

(15.3.27) $g_F = h_F + \Omega_\infty y.$

Multiply (15.3.26) by Θ_∞ and (15.3.27) by Θ_F to get

(15.3.28a) $\Theta_\infty g_\infty = \Theta_\infty \Omega_F x + S_\infty y$

and

$$(15.3.28b) \qquad\qquad \Theta_F g_F = S_F x + \Theta_F \Omega_\infty y,$$

i.e.

$$(15.3.29) \qquad\qquad \begin{bmatrix} \Theta_F g_F \\ -\Theta_\infty g_\infty \end{bmatrix} = S \begin{bmatrix} x \\ -y \end{bmatrix}.$$

Since $(A_{\zeta,F}, B_F)$ and $(A_{\zeta,\infty}, B_\infty)$ are full range pairs, by Lemma 12.2.2 Θ_F and Θ_∞ are onto; we conclude that S is onto.

Conversely, suppose S is onto and $g_F \in \mathcal{R}_n(\sigma)$, $g_\infty \in \mathcal{R}_n^0(\sigma^c)$. As S is onto we can solve (15.3.29) for x and y. This leads to the system of equations (15.3.28). Again since $(A_{\zeta,F}, B_F)$ and $(A_{\zeta,\infty}, B_\infty)$ are full range pairs, we can solve $\Theta_F h_F = S_F x$ and $\Theta_\infty h_\infty = S_\infty y$ for $h_F \in \mathcal{R}_n(\sigma)$ and $h_\infty \in \mathcal{R}_n^0(\sigma^c)$. Then (15.3.29) can be written as

$$(15.3.30a) \qquad\qquad \Theta_\infty(g_\infty - \Omega_F x - h_\infty) = 0$$

and

$$(15.3.30b) \qquad\qquad \Theta_F(g_F - h_F - \Omega_\infty y) = 0.$$

By (15.3.20) and (15.3.21), (15.3.30) simply says

$$g_\infty - \Omega_F x - h_\infty \in \mathcal{S}_{T_\infty}^0(\sigma^c) \cap \mathcal{R}_n^0(\sigma^c)$$

and

$$g_F - h_F - \Omega_\infty y \in \mathcal{S}_{T_F}(\sigma) \cap \mathcal{R}_n(\sigma).$$

Also by (15.3.20) and (15.3.21) and the construction of h_F and h_∞,

$$\Omega_F + h_F \in \mathcal{S}_{T_F}(\sigma)$$

and

$$h_\infty + \Omega_\infty y \in \mathcal{S}_{T_\infty}^0(\sigma^c).$$

Putting all the pieces together gives

$$\begin{aligned} g_\infty + g_F &= [(g_\infty - \Omega_F x - h_\infty) + (h_\infty + \Omega_\infty y)] \\ &\quad + [(g_F - h_F - \Omega_\infty y) + (\Omega_F x + h_F)] \\ &\in \mathcal{S}_{T_\infty}^0(\sigma^c) + \mathcal{S}_{T_F}(\sigma) \end{aligned}$$

so $\mathcal{S}_{T_\infty}^0(\sigma^c) + \mathcal{S}_{T_F}(\sigma) = \mathcal{R}_n$. This completes the proof of Lemma 15.3.1 and hence also of Theorem 14.3.1. $\square$

By Theorem 5.4.2 we have the following explicit formula for the interpolating function in Theorem 14.3.1.

THEOREM 15.3.2. *Suppose that σ is a subset of $\mathbb{C}$ containing 0 and that the pair of subspaces (S_∞, S) of $\mathcal{R}_n$ satisfies conditions (i), (ii) and (iii) of Theorem 14.3.1. Define linear transformations C_F, $A_{\pi,F}$, $A_{\zeta,F}$, B_F, T_F, C_∞, $A_{\pi,\infty}$, $A_{\zeta,\infty}$, B_∞, T_∞, T_{12}, T_{21} and T by (15.3.5)–(15.3.17). Choose a complex number z_0 not in $\sigma(A_{\pi,F}) \cup \sigma(A_{\zeta,F}) \cup \sigma(A_{\pi,\infty})^{-1} \cup \sigma(A_{\zeta,\infty})^{-1}$ and let D be an arbitrary invertible $n \times n$ matrix. Then the unique rational matrix function W for which*

(i) $S = W\mathcal{R}_n(\sigma)$;

(ii) $S_\infty = W\mathcal{R}_n^0(\sigma)$;

and

(iii) $W(z_0) = D$

is given by

$$W(z) = D + (z - z_0)[C_F, C_\infty] \cdot \begin{bmatrix} (zI - A_{\pi,F})^{-1} & 0 \\ 0 & (I - zA_{\pi,\infty})^{-1} \end{bmatrix}$$
$$\cdot T^{-1} \cdot \begin{bmatrix} (A_{\zeta,F} - z_0 I)^{-1} B_F D \\ (I - z_0 A_{\zeta,\infty})^{-1} B_\infty D \end{bmatrix}$$

with inverse given by

$$W(z) = D^{-1} - (z - z_0)[D^{-1}C_F(z_0 I - A_{\pi,F})^{-1}, C_\infty(I - z_0 A_{\pi,\infty})^{-1}]$$
$$\cdot T^{-1} \cdot \begin{bmatrix} (A_{\zeta,F} - zI)^{-1} & 0 \\ 0 & (I - zA_{\zeta,\infty})^{-1} \end{bmatrix} \begin{bmatrix} B_F \\ B_\infty \end{bmatrix} .$$

The following lemma gives a more geometric way of computing the interpolating function W in Theorem 14.3.1. Rather then starting with a general pair of subspaces S_∞ and S as in Theorem 14.3.1, we shall start with a σ-admissible Sylvester data set τ_F and a σ^c-admissible Sylvester data set τ_∞ and assume as already given that $S = S_\tau(\sigma)$ and $S_\infty = S_{\tau_\infty}^0(\sigma^c)$.

LEMMA 15.3.3. *Suppose $0 \in \sigma \subset \mathbb{C}$, τ_F is a σ-admissible Sylvester data set and τ_∞ is a σ^c-admissible Sylvester data set, and suppose that the matrix S given by (15.3.20) is invertible. Then the subspace*

$$\mathcal{L} = M_z S_{\tau_\infty}^0(\sigma^c) \cap S_{\tau_F}(\sigma)$$

has dimension n over $\mathbb{C}$. Moreover the $n \times n$ matrix function $W(z)$ is regular and has τ_F as a σ-null-pole triple and τ_∞ as a σ^c-null-pole triple if and only if the columns of W form a basis for the subspace $\mathcal{L}$ over $\mathbb{C}$.

PROOF. By Theorems 12.3.1 and 15.1.5 W is regular and has τ_F as a σ-null-pole triple and τ_∞ as a σ^c-null-pole triple if and only if

$$S_{\tau_\infty}^0(\sigma^c) = W\mathcal{R}_n^0(\sigma^c)$$

and

$$S_{\tau_F}(\sigma) = W\mathcal{R}_n(\sigma).$$

If W is regular then the multiplication operator $M_W : h \to Wh$ on $\mathcal{R}_n$ is one-to-one and onto. Note also that

$$\mathcal{L} = M_z S^0_{\tau_\infty}(\sigma^c) \cap \mathcal{S}_{\tau_F}(\sigma)$$
$$= M_z M_W \mathcal{R}^0_n(\sigma^c) \cap W \mathcal{R}_n(\sigma)$$
$$= M_W \left(M_z \mathcal{R}^0_n(\sigma^c) \cap \mathcal{R}_n(\sigma) \right)$$
$$= M_W \mathbf{C}^n.$$

Thus $\mathcal{L} = M_W \mathbf{C}^n$ and $\dim \mathcal{L} = \dim \mathbf{C}^n = n$. Since the columns of W are just the images under M_W of the standard basis vectors in $\mathbf{C}^n$, we see that the columns of W form a basis for $\mathcal{L}$.

For the converse, suppose that the columns of W form a basis for $\mathcal{L}$. Let $\widetilde{W}$ be any regular rational matrix function having τ_F as a σ-null-pole triple and τ_∞ as a σ^c-null-pole triple as guaranteed by Theorem 5.4.2. By the first part of the proof the columns of $\widetilde{W}$ also form a basis for $\mathcal{L}$ over $\mathbf{C}$. But then there is an invertible (constant) $n \times n$ matrix Γ for which

$$W(z) = \widetilde{W}(z)\Gamma.$$

Thus τ_F is also a σ-null-pole triple for W, and τ_∞ is also a σ^c-null-pole triple for W. $\quad\square$

The next lemma calculates the key subspace $\mathcal{L}$ in terms of the associated Sylvester data sets.

LEMMA 15.3.4. *Let* $0 \in \sigma \subset \mathbf{C}$ *and let* τ_F *and* τ_∞ *be* σ-*admissible (respectively* σ^c-*admissible) Sylvester data sets. Then the function* $h \in \mathcal{R}_n$ *is in* $\mathcal{L} = M_z S^0_{\tau_\infty}(\sigma^c) \cap \mathcal{S}_{\tau_F}(\sigma)$ *if and only if*

$$h(z) = C_F(zI - A_{\pi,F})^{-1}x + u + zC_\infty(I - zA_{\pi,\infty})^{-1}y$$

where $\Psi \begin{bmatrix} x \\ y \\ u \end{bmatrix} = 0$ *and the linear transformation* Ψ *is given by*

$$\Psi = \begin{bmatrix} S_F & A_{\zeta,F}S_{12} & -B_F \\ A_{\zeta,\infty}S_{21} & S_\infty & -B_\infty \end{bmatrix}.$$

PROOF. Introduce the linear transformations Θ_F, Ω_F, Θ_∞, Ω_∞ as in Lemma 15.3.1, so

$$\mathcal{S}_{\tau_F}(\sigma) = \{\Omega_F x + h_F : x \in \mathcal{X}_{\pi,F}, h_F \in \mathcal{R}_n(\sigma)$$
$$\text{such that } \Theta_F h_F = S_F x\}$$

and

$$S^0_{\tau_\infty}(\sigma^c) = \{h_\infty + \Omega_\infty y : h_\infty \in \mathcal{R}_n(\sigma^c), y \in \mathcal{X}_{\pi,\infty}$$
$$\text{such that } \Theta_\infty h_\infty = S_\infty y\}.$$

Now suppose that $h \in \mathcal{L}$. Then in particular $h \in \mathcal{S}_{\tau_F}(\sigma)$ so h has the form

(15.3.31) $$h = \Omega_F x + h_F$$

where

(15.3.32)
$$\Theta_F h_F = S_F x.$$

Also $z^{-1}h(z) \in z^{-1}\mathcal{L} \subset S^0_{\sigma^c}(\tau_\infty)$, so

(15.3.33)
$$M_{z^{-1}}h = h_\infty + \Omega_\infty y$$

where

(15.3.34)
$$\Theta_\infty h_\infty = S_\infty y.$$

Note that

$$z^{-1}h_F(z) = z^{-1}h_F(0) + z^{-1}\big(h_F(z) - h_F(0)\big)$$

where the first term is in $\mathcal{R}^0_n(\sigma^c)$ and the second term is in $\mathcal{R}_n(\sigma)$. Thus

$$P^0_{\sigma^c}[M_{z^{-1}}h_F](z) = z^{-1}u,$$

where $u = h_F(0)$. Apply $P^0_{\sigma^c}M_{z^{-1}}$ to (15.3.31) and use (15.3.33) to get

$$\begin{aligned} h_\infty &= P^0_{\sigma^c}M_{z^{-1}}h \\ &= M_{z^{-1}}\Omega_F x + M_{z^{-1}}u. \end{aligned}$$

Substitute this back into (15.3.33) and solve for h to get

(15.3.35)
$$\begin{aligned} h &= \Omega_F x + u + M_z\Omega_\infty y \\ &= C_F(zI - A_{\pi,F})^{-1}x + u + zC_\infty(I - zA_{\pi,\infty})^{-1}y. \end{aligned}$$

Thus h has the form demanded in the lemma. It remains to show that $\Psi\begin{bmatrix} x \\ y \\ u \end{bmatrix} = 0.$

We show first that

(15.3.36)
$$[\,S_F \quad A_{\zeta,F}S_{12} \quad -B_F\,]\begin{bmatrix} x \\ y \\ u \end{bmatrix} = 0.$$

Note that by (15.3.32) $S_F x = \Theta_F h_F$ where h_F is seen to be

$$h_F = u + M_z\Omega_\infty y$$

from (15.3.31) and (15.3.35). We thus have

(15.3.37)
$$S_F x = \Theta_F u + \Theta_F M_z\Omega_\infty y.$$

From the definition of Θ_F it is easy to see that

(15.3.38)
$$\Theta_F u = B_F u.$$

We compute

$$(zI - A_{\zeta,F})^{-1} B_F z C_\infty (I - zA_{\pi,\infty})^{-1}$$
$$= B_F C_\infty (I - zA_{\pi,\infty})^{-1} + A_{\zeta,F} (zI - A_{\zeta,F})^{-1} B_F C_\infty (I - zA_{\pi,\infty})^{-1}.$$

From this we get

(15.3.39)
$$\Theta_F M_z \Omega_\infty = A_{\zeta,F} \Theta_F \Omega_\infty$$
$$= -A_{\zeta,F} S_{12}.$$

Combine now (15.3.37), (15.3.38) and (15.3.39) to get (15.3.36). Conversely, if (15.3.36) holds, one can reverse the argument to show that then h defined by (15.3.35) is in $\mathcal{S}_{T_F}(\sigma)$.

By an analogous argument one can show that

(15.3.40)
$$\begin{bmatrix} A_{\zeta,\infty} S_{21} & S_\infty & -B_\infty \end{bmatrix} \begin{bmatrix} x \\ y \\ u \end{bmatrix} = 0$$

is equivalent to h defined by (15.3.35) being in $M_z \mathcal{S}_{T_\infty}^0(\sigma^c)$. Indeed, $h \in M_z \mathcal{S}_{T_\infty}^0(\sigma^c)$ is equivalent to

(15.3.41)
$$M_{z^{-1}} h = M_{z^{-1}} \Omega_F x + M_{z^{-1}} u + \Omega_\infty y$$

being in $\mathcal{S}_{T_\infty}^0(\sigma^c)$. This in turn is equivalent to

(15.3.42)
$$\Theta_\infty [M_{z^{-1}} \Omega_F x + M_{z^{-1}} u] = S_\infty y.$$

One easily sees that

(15.3.43)
$$\Theta_\infty M_{z^{-1}} = B_\infty.$$

Next use the identity

$$(I - zA_{\zeta,\infty})^{-1} B_\infty z^{-1} C_F (zI - A_{\pi,F})^{-1}$$
$$= z^{-1} B_\infty C_F (zI - A_{\pi,F})^{-1} + A_{\zeta,\infty} (I - zA_{\zeta,\infty})^{-1} B_\infty C_F (zI - A_{\pi,F})^{-1}$$

to deduce

$$\Theta_\infty M_{z^{-1}} \Omega_F = \sum_{z_0 \in \sigma} \mathrm{Res}_{z=z_0} z^{-1} B_\infty C_F (zI - A_{\pi,F})^{-1} + A_{\zeta,\infty} \Theta_\infty \Omega_F.$$

The function in the first term has all poles in σ and a double zero at infinity, so the first term is zero and we have

(15.3.44)
$$\Theta_\infty M_{z^{-1}} \Omega_F = A_{\zeta,\infty} \Theta_\infty \Omega_F.$$

Finally, recall that

(15.3.45)
$$S_{21} = -\Theta_\infty \Omega_F.$$

From (15.3.43), (15.3.44), (15.3.45) we see that (15.3.42) is exactly equivalent to (15.3.40) as claimed. □

As a corollary of the preceding two lemmas we pick up the proof of Theorem 5.5.2 promised in Chapter 5.

PROOF OF THEOREM 5.5.2. We are given a σ-admissible Sylvester data set

$$\tau_F = (C_{\pi,F}, A_{\pi,F}; A_{\zeta,F}, B_{\zeta,F}; S_F)$$

and a σ_∞-admissible Sylvester data set

$$\tau_\infty = (C_{\pi,\infty}, A_{\pi,\infty}; A_{\zeta,\infty}, B_{\zeta,\infty}; S_\infty),$$

where $\sigma \cup \sigma_\infty = \mathbf{C}^*$ is a disjoint partitioning of $\mathbf{C}^* = \mathbf{C} \cup \{\infty\}$ with $0 \in \sigma$ and $\infty \in \sigma_\infty$. Define the subspace $\mathcal{L}$ by

$$\mathcal{L} \stackrel{\text{def}}{=} M_z \mathcal{S}^0_{\tau_\infty}(\sigma^c) \cap \mathcal{S}_{\tau_F}(\sigma).$$

By Lemma 15.3.4 the matrix

(15.3.46)
$$\Psi = \begin{bmatrix} S_F & A_{\zeta,F}S_{12} & -B_{\zeta,F} \\ A_{\zeta,\infty}S_{21} & S_\infty & -B_{\infty,F} \end{bmatrix}$$

has a null space with dimension n. Here S_{12} and S_{21} are defined as the unique solutions of the Stein equations

(15.3.47)
$$A_{\zeta,F}S_{12}A_{\pi,\infty} - S_{12} = B_{\zeta,F}C_{\pi,\infty}$$

(15.3.48)
$$S_{21} - A_{\zeta,\infty}S_{21}A_{\pi,F} = -B_{\zeta,\infty}C_{\pi,F}.$$

Let us define matrices $B_{\pi,F}$, $B_{\pi,\infty}$ and D_∞ (of sizes $n_{\pi,F} \times n$, $n_{\pi,\infty} \times n$ and $n \times n$ respectively) so that the columns of $\mathrm{col}(B_{\pi,F}, B_{\pi,\infty}, D_\infty)$ form a basis for $\mathrm{Ker}\ \Psi$. Then again by Lemma 15.3.4, a basis for the subspace $\mathcal{L}$ is given by the columns of the matrix function

(15.3.49)
$$W(z) = [C_{\pi,F}(zI - A_{\pi,F})^{-1}, zC_{\pi,\infty}(I - zA_{\pi,\infty})^{-1}, I] \begin{bmatrix} B_{\pi,F} \\ B_{\pi,\infty} \\ D_\pi \end{bmatrix}.$$

By Lemma 15.3.3, it follows that $W(z)$ as in (15.3.49) is a rational matrix function having τ_F as a left null-pole subspace over σ and τ_∞ as a null-pole subspace over σ_∞.

The formula for W^{-1} follows from a dual analysis. In general, if $\tau = (C_0, A_0; A_1, B_1; T)$ is a global left null-pole triple for a rational matrix function $R(z)$, then $\tilde{\tau} = (B_1^T, A_1^T; A_0^T, C_0^T; -T^T)$ is a left null pole triple for $(W(z)^{-1})^T$. For the case where $W(z)$ is finite with value I at infinity, this can be seen from Theorem 4.3.1; the general case follows via the reduction to the case $R(\infty) = I$ as in Section 5.3. By restricting a given null-pole triple to a given subset σ of $\mathbf{C}$ the same result applies for σ-null-pole triples when $\sigma \subset \mathbf{C}$. Finally, via the change of variable $z \to z^{-1}$ the result also holds for σ_∞-null-pole triples, where $\sigma_\infty \subset \mathbf{C}^*$ contains the point at infinity. Hence, if $W(z)$ is given by (15.3.49), then

$$\widetilde{\tau}_F = (B^T_{\zeta,F}, A^T_{\zeta,F}; A^T_{\pi,F}, C^T_{\pi,F}; -S^T_F)$$

is a left σ-null-pole triple for $[W(z)^{-1}]^T$ and

$$\widetilde{\tau}_\infty = (B^T_{\zeta,\infty}, A^T_{\zeta,\infty}; A^T_{\pi,\infty}, C^T_{\pi,\infty}; -S^T_\infty)$$

is a left σ_∞-null-pole triple for $[W(z)^{-1}]^T$.

On the other hand, we can obtain another formula for a rational matrix function $V(z)$ having $\widetilde{\tau}_F$ as a σ-null-pole triple and having $\widetilde{\tau}_\infty$ as a σ_∞-null-pole triple as in the first part of the proof. Specifically, if we define matrices $C_{\zeta,F}$, $C_{\zeta,\infty}$ and D_ζ so that the columns of $\mathrm{col}(C^T_{\zeta,F}, C^T_{\zeta,\infty}, D^T_\zeta)$ form a basis for the kernel of

(15.3.50)
$$\widetilde{\Psi} \stackrel{\text{def}}{=} \begin{bmatrix} -S^T_F & A^T_{\pi,F} & -C^T_{\pi,F} \\ A^T_{\pi,\infty} & -S^T_\infty & -C^T_{\pi,\infty} \end{bmatrix},$$

or equivalently, the rows of $[C_{\zeta,F}, C_{\zeta,\infty}, D_\zeta]$ form a basis for the left kernel of $\Phi \stackrel{\text{def}}{=} \widetilde{\Psi}^T$, then

(15.3.51)
$$V(z) = [B^T_{\zeta,F}(zI - A^T_{\zeta,F})^{-1}, zB^T_{\zeta,\infty}(I - zA^T_{\zeta,\infty})^{-1}, I] \begin{bmatrix} C^T_{\zeta,F} \\ C^T_{\zeta,\infty} \\ D^T_\zeta \end{bmatrix}$$

also has $\widetilde{\tau}_F$ as σ-null-pole triple and $\widetilde{\tau}_\infty$ as σ_∞-null-pole triple. By the uniqueness result Corollary 12.3.3 and Liouville's theorem, we conclude that

(15.3.52)
$$[W(z)^{-1}]^T = V(z)\Gamma$$

where Γ is a nonsingular constant matrix. Taking transposes in (15.3.52) gives

(15.3.53)
$$W(z)^{-1} = \Gamma^T V(z)^T.$$

It remains only to compute the constant matrix Γ^T. Note that $L \stackrel{\text{def}}{=} (\Gamma^T)^{-1}$ is given by

(15.3.54)
$$L = V(z)^T W(z).$$

Introduce matrix functions $H_\pi(z)$ and $H_\zeta(z)$ by

$$H_\pi(z) = \mathrm{diag}(zI - A_{\pi,F}, z^{-1}(I - zA_{\pi,\infty}), I_n)$$

and

$$H_\zeta(z) = \mathrm{diag}(zI - A_{\zeta,F}, z^{-1}(I - zA_{\zeta,\infty}), I_n).$$

Then, from (15.3.49) and (15.3.51) we see that (15.3.54) takes the form

$$(15.3.55) \qquad L = [C_{\zeta,F}, C_{\zeta,\infty}, D_{\zeta}] H_{\zeta}(z)^{-1} Q H_{\pi}(z)^{-1} \begin{bmatrix} B_{\pi,F} \\ B_{\pi,\infty} \\ D_{\pi} \end{bmatrix}$$

where

$$(15.3.56) \qquad Q = \begin{bmatrix} B_{\zeta,F} \\ B_{\zeta,\infty} \\ I \end{bmatrix} [\, C_{\pi,F} \quad C_{\pi,\infty} \quad I \,].$$

From the Stein equations (15.3.47), (15.3.48) together with the Sylvester equations

$$S_F A_{\pi,F} - A_{\zeta,F} S_F = B_{\zeta,F} C_{\pi,F}$$

$$S_\infty A_{\pi,\infty} - A_{\zeta,\infty} S_\infty = B_{\zeta,\infty} C_{\pi,\infty}$$

associated with the admissible Sylvester data sets τ_F and τ_∞, one sees that

$$(15.3.57) \qquad Q = \begin{bmatrix} S_F A_{\pi,F} - A_{\zeta,F} S_F & A_{\zeta,F} S_{12} A_{\pi,\infty} - S_{12} & B_{\zeta,F} \\ A_{\zeta,\infty} S_{21} A_{\pi,F} - S_{21} & S_\infty A_{\pi,\infty} - A_{\zeta,\infty} S_\infty & B_{\zeta,\infty} \\ C_{\pi,F} & C_{\pi,\infty} & I \end{bmatrix}.$$

Using the definition of Ψ (15.3.46) and of $\Phi = \tilde{\Psi}^T$ where $\tilde{\Psi}$ is given by (15.3.50), an entry-by-entry computation shows that

$$-Q = \begin{bmatrix} I & 0 \\ 0 & I \\ 0 & 0 \end{bmatrix} \Psi H_{\pi}(z) + H_{\zeta}(z) \Phi \begin{bmatrix} I & 0 & 0 \\ 0 & I & 0 \end{bmatrix} + H_{\zeta}(z) \begin{bmatrix} 0 & S_{12} & 0 \\ S_{21} & 0 & 0 \\ 0 & 0 & I \end{bmatrix}.$$

Using the expression (15.3.55) for L and remembering that

$$[C_{\zeta,F}, C_{\zeta,\infty}, D_{\zeta}] \Phi = 0$$

and

$$\Psi \begin{bmatrix} B_{\pi,F} \\ B_{\pi,\infty} \\ D_{\pi} \end{bmatrix} = 0$$

by the way $C_{\zeta,F}$, $C_{\zeta,\infty}$, D_{ζ}, $B_{\pi,F}$, $B_{\pi,\infty}$, D_{π} were constructed, we get

$$L = -[C_{\zeta,F}, C_{\zeta,\infty}, D_{\zeta}] \left\{ H_{\zeta}(z)^{-1} \begin{bmatrix} I & 0 \\ 0 & I \\ 0 & 0 \end{bmatrix} \Psi \right.$$

$$\left. + \Phi \begin{bmatrix} I & 0 & 0 \\ 0 & I & 0 \end{bmatrix} H_{\pi}(z)^{-1} + \begin{bmatrix} 0 & S_{12} & 0 \\ S_{21} & 0 & 0 \\ 0 & 0 & I \end{bmatrix} \right\} \begin{bmatrix} B_{\pi,F} \\ B_{\pi,\infty} \\ D_{\pi} \end{bmatrix}$$

$$= -[C_{\zeta,F}, C_{\zeta,\infty}, D_{\zeta}] \begin{bmatrix} 0 & S_{12} & 0 \\ S_{21} & 0 & 0 \\ 0 & 0 & I \end{bmatrix} \begin{bmatrix} B_{\pi,F} \\ B_{\pi,\infty} \\ D_{\pi} \end{bmatrix}$$

$$= -C_{\zeta,F} S_{12} B_{\pi,\infty} - C_{\zeta,\infty} S_{21} B_{\pi,F} - D_{\zeta} D_{\pi}.$$

By (15.3.53) and the identification $L = (\Gamma^T)^{-1}$, we get the desired formula for $W(z)^{-1}$.
□

We close this section with the following generalization of Theorem 14.3.1 concerning simultaneous representation of a collection of null-pole subspaces.

THEOREM 15.3.5. *Let $\mathcal{S}_1, \ldots, \mathcal{S}_p$ and $\mathcal{S}_\infty$ be a collection of subspaces of $\mathcal{R}_n$ and suppose that*

$$\mathbf{C}^* = \sigma_1 \dot{\cup} \cdots \dot{\cup} \sigma_p \dot{\cup} \sigma_\infty$$

is a disjoint partitioning of the extended complex plane for which σ_∞ contains infinity but does not contain 0. Then there is a rational matrix function W such that

(a) $\mathcal{S}_j = W\mathcal{R}_n(\sigma_j)$ *for* $1 \le j \le p$

and

(b) $\mathcal{S}_\infty = W\mathcal{R}_n^0(\sigma_\infty)$

if and only if

(i) *for $j = 1, \ldots, p$, $M_{r(z)}\mathcal{S}_j \subset \mathcal{S}_j$ for each $r \in \mathcal{R}(\sigma_j)$, $P_{\sigma_j^c}^0(\mathcal{S}_r)$ is finite dimensional and $\mathcal{R}_n(\sigma_j) \cap \mathcal{S}_j$ has finite codimension in $\mathcal{R}_n(\sigma_j)$;*

(ii) *$M_{r(z)}\mathcal{S}_\infty \subset \mathcal{S}_\infty$ for each $r \in \mathcal{R}(\sigma_\infty)$, $P_{\sigma_\infty^c}(\mathcal{S}_\infty)$ is finite dimensional and $\mathcal{R}_n^0 \cap \mathcal{S}_\infty$ has finite codimension in $\mathcal{R}_n^0(\sigma_\infty)$*

and

(iii) $\mathcal{R}_n = [\bigcap_{j=1}^p \mathcal{S}_j] \dot{+} \mathcal{S}_\infty$.

PROOF. Necessity of (i) and (ii) is clear from Theorem 14.1.3 and Theorem 15.2.3. To see the necessity of (iii), note that if there exists a (regular) rational matrix function for which (a) and (b) hold, then

$$\bigcap_{j=1}^p \mathcal{S}_j = \bigcap_{j=1}^p W \cdot \mathcal{R}_n(\sigma_j)$$

$$= W \cdot \bigcap_{j=1}^p \mathcal{R}_n(\sigma_j)$$

where clearly

$$\bigcap_{j=1}^p \mathcal{R}_n(\sigma_j) = \mathcal{R}_n(\sigma_1 \cup \cdots \cup \sigma_p).$$

Now necessity of (iii) follows from Theorem 14.3.1.

Conversely suppose that (i)–(iii) hold. Let $\tau_j = (C_j, A_{\pi,j}; A_{\zeta,j}, B_j; T_j)$ be the σ_j-admissible Sylvester data set given by (14.1.7)–(14.1.11) (with $\mathcal{S}_j$ in place of $\mathcal{S}$ and σ_j in place of σ) for $j = 1, \ldots, p$, and let $\tau_\infty = (C_\infty, A_{\pi,\infty}; A_{\zeta,\infty}, B_\infty; T_\infty)$ be the σ_∞-admissible Sylvester data set given by (15.3.9)–(15.3.12) and (15.3.14). To produce a

W which satisfies (a) and (b), by Theorems 14.1.3 and 15.2.2 the problem is to produce a W for which τ_j is a σ_j-null-pole triple for $j = 1, \ldots, p$ and for which τ_∞ is a σ_∞-null-pole triple. By Lemma 4.5.6, τ_j is a σ_j-null-pole triple for W for each j if and only if the direct sum $\tau_1 \oplus \cdots \oplus \tau_p$ is a $(\sigma_1 \cup \cdots \cup \sigma_p)$-null-pole triple for W. By Proposition 12.2.5,

$$\bigcap_{j=1}^{p} \mathcal{S}_j = \bigcap_{j=1}^{p} \mathcal{S}_{\tau_j}(\sigma_j)$$
$$= \mathcal{S}_{\tau_1 \oplus \cdots \oplus \tau_p}(\sigma_1 \cup \cdots \cup \sigma_p).$$

Once these preliminaries are established we see that Theorem 15.3.5 collapses to Theorem 14.3.1. □

NOTES FOR PART III

CHAPTER 12   The explicit description of the null-pole subspace of a rational matrix function in terms of null and pole data appears in Ball-Ran [1987a] in a different form with continuing refinements in Ball-Gohberg-Rodman [1987] and Ball-Cohen-Ran [1988]; one can see special cases appearing already in earlier work of Gohberg-Lancaster-Rodman [1982] and Gohberg-Rodman [1981]. The use of the null-pole subspace as a tool for the study of functions (especially in the context of various kinds of factorization problems) originates with Beurling's theorem and its extensions and elaborations (see Rosenblum-Rovnyak [1985], Sz.-Nagy-Foias [1970], Ball-Helton [1983]); this work deals more generally with nonrational functions associated with Hardy spaces. The characterization of the null-pole subspace in terms of Laurent coefficients and augmented null and pole chains presented in Sections 12.4 and 12.5 is essentially from Ball-Ran [1987a].

CHAPTER 13   The idea of the main results of this chapter (Theorems 13.2.3 and 13.2.4) are taken from Ball-Helton [1988a]; the more elementary presentation here completely in the context of rational functions is new.

CHAPTER 14   The rational forms of the Beurling-Lax and Ball-Helton invariant subspace representation theorems appear here for the first time. In its original form (Beurling [1949]) Beurling's theorem concerned shift invariant subspaces of the Hardy space H^2 over the unit disk. Later operator theoretic extensions (Lax [1959], Halmos [1961], Rovnyak [1963], Helson [1964]) as well as the extensions to Lie groups other than the unitary group (Ball-Helton [1983, 1986a, 1986c]) and the more general form with no symmetries (Ball-Helton [1984]) were all in the setting of H^p or L^p spaces.

CHAPTER 15   The study of improper rational matrix functions via the associated null-pole subspace was initiated in Ball-Cohen-Ran [1988]. There can be found the proof of Theorem 5.5.2 as presented here. The use of the "wandering subspace" appears in Halmos' proof (Halmos [1961]) of the Beurling-Lax theorem and was adapted later by Ball-Helton [1984] to prove generalized Beurling-Lax theorems.

The underlying idea of this part is to study null and pole structure of rational matrix functions through the study of associated modules over a ring of scalar functions. In the infinite dimensional setting this indeed is the main theme of the Beurling-Lax and Ball-Helton approach to the study of factorization and interpolation. Other variants of the module approach to the study of rational matrix functions are the polynomial model approach of Fuhrmann [1976, 1981] and the work of Conte, Perdon, Sain and Wyman (see, e.g., Wyman-Sain [1981, 1987], Wyman-Sain-Conte-Perdon [1989] and references there).

PART IV

NONHOMOGENEOUS INTERPOLATION PROBLEMS

The simplest nonhomogeneous interpolation problem for scalar functions is the Lagrange interpolation problem of finding a polynomial of smallest possible degree taking on given values at some given points. More sophisticated is the Lagrange-Sylvester problem where the values of the function together with values of some derivatives are prescribed at certain points. In this part, we consider a few generalizations of this problem for the matrix valued case. The simplest problem is that of finding a matrix polynomial with prescribed matrix values at certain points. This problem can be solved elementwise and hence reduces to the scalar case. More complicated is the tangential Lagrange problem which also can be transformed to the previous problem. We consider also more sophisticated interpolation problems, namely, the problem where the interpolation conditions are expressed in a very general contour integral form, and the divisor problem. The solution in both cases requires techniques developed in the previous chapters. In this part we also consider the partial realization problem which is an interpolation problem where one is required to find a rational matrix function of minimal possible degree with its value and the values of a few of its derivatives at infinity prescribed. This part does not develop a general theory in a systematic way as is done in Part I but rather is a sampling of problems and solutions based on techniques developed in Part I. This part also provides preparation for Part V where the same type of interpolation problems are studied subject to some additional constraints. Some of the important features occurring there, e.g. a linear fractional parametrization of the set of all solutions, occur already here.

CHAPTER 16

INTERPOLATION PROBLEMS FOR MATRIX POLYNOMIALS AND RATIONAL MATRIX-FUNCTIONS

In this chapter we consider a few different interpolation problems for matrix polynomials and rational matrix valued functions. We have in mind the tangential Lagrange and Lagrange-Sylvester interpolation, contour integral interpolation and interpolation with a given set of divisions and remainders. Connections among all these problems are established and in each case a complete solution is suggested.

16.1 TANGENTIAL LAGRANGE-SYLVESTER INTERPOLATION

In this chapter we consider matrix analogues of some of the classical nonhomogeneous interpolation problems. One approach to finding a solution is to reduce the problem to a system of independent scalar problems; From our point of view of matrix interpolation theory, this is not so interesting. More difficult is to describe all solutions or to find a solution of minimal possible degree; as we shall see, the problem in this form can be solved by using the theory of homogeneous matrix interpolation developed in Part I of this book. In this first section we introduce the simplest problems and give the connections with classical scalar problems.

Perhaps the simplest classical nonhomogeneous interpolation problem is that of Lagrange. We are given n distinct points in the plane $z_1, \ldots, z_n$ and n complex numbers $w_1, \ldots, w_n$ and seek a polynomial $p(z)$ for which $p(z_j) = w_j$ for $j = 1, \ldots, n$. A solution of degree at most $n - 1$ exists and is unique, and is given by

$$p(z) = \sum_{k=1}^{n} w_k \frac{\prod_{j \neq k}(z - z_j)}{\prod_{j \neq k}(z_k - z_j)}$$

(see e.g. Davis [1975]). Two matrix analogues which come to mind are:

(MLI) MATRIX LAGRANGE INTERPOLATION PROBLEM. *Given n distinct points $z_1, \ldots, z_n$ in $\mathbb{C}$ and n, $M \times N$ matrices $M_1, \ldots, M_n$, find an $M \times N$ matrix polynomial $L(z)$ for which $L(z_j) = M_j$ for $j = 1, \ldots, n$.*

This problem is not too interesting from the point of view of the theory of matrices, if the goal is simply to find at least one solution. Namely, if the unknown $L(z)$ has matrix entries $\ell_{ij}(z)$ and if $x_{ij}^{(k)}$ are the matrix entries of X_k ($1 \leq i \leq M$, $1 \leq j \leq N$), then the matrix Lagrange interpolation problem is to find MN scalar polynomials $\ell_{ij}(z)$ for which

(16.1.1) $$\ell_{ij}(z_k) = x_{ij}^{(k)}$$

for $k = 1, \ldots, n$ ($1 \leq i \leq M$, $1 \leq j \leq N$). One finds a scalar polynomial ℓ_{ij} satisfying (16.1.1) for $1 \leq k \leq n$ for each i, j, and then $L(z) = [\ell_{ij}(z)]$ is a matrix polynomial which solves the matrix problem.

More interesting is the following matrix generalization of Lagrange interpolation.

(TLI) TANGENTIAL LAGRANGE INTERPOLATION. *Given n nonzero row vectors $x_1, \ldots, x_n$ of size $1 \times M$, n row vectors $y_1, \ldots, y_n$ of size $1 \times N$ and m distinct points $z_1, \ldots, z_n$ in $\mathbb{C}$, find an $M \times N$ matrix polynomial $L(z)$ for which*

$$x_k L(z_k) = y_k$$

for $k = 1, \ldots, n$.

Note that in the tangential Lagrange interpolation problem the matrix value $L(z_j)$ is not prescribed, but only the value in a certain direction. On the surface this appears to be a true matrix interpolation problem. However it too can be reduced to a system of scalar problems in the following way, namely, for each $k = 1, \ldots, n$, choose a matrix M_k for which $x_k M_k = y_k$; this is certainly possible since by assumption x_k is a nonzero row vector. Then we generate solutions of the tangential Lagrange interpolation problem by solving a problem of the form (LI):

$$L(z_k) = M_k, \qquad k = 1, \ldots, n.$$

In this way we solve the problem with a $M \times N$ matrix polynomial of degree at most $n - 1$. However the solution certainly depends on the choice of matrices M_k satisfying $x_k M_k = y_k$. Conceivably a different choice of solution M_k would lead to an interpolant having additional described properties (e.g. a still lower degree).

A routine extension of the TLI problem is to take the vectors x_k to be full row rank matrices (still with M columns) and to take y_k to be matrices of compatible sizes. As a particular case one can take x_k to be the $M \times M$ identity matrix for each k; then the TLI problem reduces exactly to the LI problem.

The Lagrange-Sylvester interpolation problem involves derivatives as well. Namely, given distinct complex numbers $z_1, \ldots, z_n$ and a collection of numbers indexed in the form $w_{11}, \ldots, w_{1m_1}, w_{22}, \ldots, w_{2m_2}, \ldots, w_{n1}, \ldots, w_{nm_n}$, find a polynomial $\ell(z)$ for which

(16.1.2) $$\frac{d^{\alpha-1} \ell}{dz^{\alpha-1}}(z_k) = w_{k\alpha} \quad \text{for} \quad 1 \le k \le n, 1 \le \alpha \le m_k.$$

The solution is well-known and is given by the Lagrange-Sylvester interpolation polynomial $\ell(z)$. A concise way to write down $\ell(z)$ goes as follows: Introduce the matrix

$$V = \begin{bmatrix} X_1 & X_2 & \cdots & X_k \\ X_1 T_1 & X_2 T_2 & \cdots & X_k T_k \\ \vdots & \vdots & & \vdots \\ X_1 T_1^{q-1} & X_2 T_2^{q-1} & \cdots & X_k T_k^{q-1} \end{bmatrix},$$

where

$$X_j = \begin{bmatrix} 1 & 0 & \cdots & 0 \end{bmatrix}$$

is the $1 \times m_j$ row; T_j is the $m_j \times m_j$ upper triangular Jordan block with the eigenvalue z_j; and $q = m_1 + m_2 + \cdots + m_n$. One can show that V is invertible. Then the solution $\ell(z)$ to the interpolation problem (16.2.2) is given by

$$\ell(z) = [w_{11} \cdots w_{1m_1} \cdots w_{21} \cdots w_{2m_2} \cdots w_{n1} \cdots w_{nm_n}]V^{-1} \begin{bmatrix} 1 \\ z \\ z^2 \\ \vdots \\ z^{q-1} \end{bmatrix}.$$

For a proof of this formula in the more general matrix setting see Section 16.6. The matrix Lagrange-Sylvester interpolation problem is:

(LSI)   *Given distinct complex numbers $z_1, \ldots, z_N$ and a collection of $M \times N$ matrices indexed as $M_{11}, \ldots, M_{1m_1}; \ldots; M_{n1}, \ldots, M_{1m_n}$, find an $M \times N$ matrix polynomial $L(z)$ for which*

$$\left. \frac{d^{\alpha-1}}{dz^\alpha} L(z) \right|_{z=z_k} = M_{k\alpha}, \qquad 1 \le k \le n, 1 \le \alpha \le m_k.$$

Clearly the LSI problem is equivalent to $M \cdot N$ independent scalar Lagrange-Sylvester problems, and hence is not of much interest from the point of view of matrix function theory.

The tangential Lagrange-Sylvester interpolation problem we take to be the following:

(TLSI)   *Given n distinct complex numbers $z_1, \ldots, z_n$ given n row vector polynomials $x_1(z), \ldots, x_n(z)$ of size $1 \times M$ and given n row vector polynomials $y_1(z), \ldots, y_n(z)$ of size $1 \times N$, find an $M \times N$ matrix polynomial $L(z)$ for which*

(16.1.3) $$\left. \frac{d^{\alpha-1}}{dz^{\alpha-1}} \{x_k(z)L(z)\} \right|_{z=z_k} = \left. \frac{d^{\alpha-1}}{dz^{\alpha-1}} y_k(z) \right|_{z=z_k}$$

for $1 \le k \le n$, $1 \le \alpha \le m_k$.

If we let $x_k(z)$ and $y_k(z)$ to have Taylor expansions

$$x_k(z) = \sum_{j=1}^{\infty} x_{kj}(z - z_k)^{j-1}$$

and

$$y_k(z) = \sum_{j=1}^{\infty} y_{kj}(z - z_k)^{j-1}$$

at z_k, and the unknown matrix polynomial $L(z)$ to have Taylor form

$$L(z) = \sum_{j=1}^{n} L_{kj}(z - z_k)^{j-1}$$

(where thus $L_{kj} = \frac{1}{(j-1)!} \frac{d^{j-1}}{dz^{j-1}} \{L(z)\}\Big|_{z=z_k}$) then the interpolation conditions (16.1.3) can be expressed in terms of Taylor coefficients as

$$(16.1.4) \qquad \begin{bmatrix} x_{k1} & 0 & \cdots & 0 \\ x_{k2} & x_{k1} & \cdots & 0 \\ \vdots & \vdots & & \vdots \\ x_{km_k} & x_{k,m_k-1} & \cdots & x_{k1} \end{bmatrix} \begin{bmatrix} L_{k1} \\ \vdots \\ L_{km_k} \end{bmatrix} = \begin{bmatrix} y_{k1} \\ \vdots \\ y_{km_k} \end{bmatrix}$$

for $1 \le k \le n$. Since $x_{k1} \ne 0$, we can for each k find matrices $M_{k1}, \ldots, M_{km_k}$ for which

$$(16.1.5) \qquad \begin{bmatrix} x_{k1} & & 0 \\ \vdots & \ddots & \\ x_{km_k} & \cdots & x_{k1} \end{bmatrix} \begin{bmatrix} M_{k1} \\ \vdots \\ M_{km_k} \end{bmatrix} = \begin{bmatrix} y_{k1} \\ \vdots \\ y_{km_k} \end{bmatrix}.$$

If we then solve the LSI problem of finding a matrix polynomial $L(z)$ for which

$$\frac{d^{j-1}}{dz^{j-1}} \{L(z)\}\Big|_{z=z_k} = (j-1)! M_{kj}, \qquad 1 \le k \le n, 1 \le j \le m_k$$

then any such $L(z)$ will be a solution of the original TLSI problem.

By allowing the row vector polynomials $x_k(z)$ and $y_k(z)$ to be matrix polynomials of compatible sizes in (16.1.3), we can see that the LSI problem itself as a special case of TLSI.

Of course the solution of the TLSI problem depends very much on the choice of solutions $M_{k1}, \ldots, M_{km_k}$ of (16.1.5). More insight is had by solving the problem in a more intrinsically matricial way. This will be done in the succeeding sections of this chapter.

The TLSI problem as stated above will be referred to as the left problem, because the vector functions $x_k(z)$ are multiplied on the left in formula (16.1.3). The right counterpart of the TLSI interpolation conditions (16.1.3) is

$$\frac{d^{\alpha-1}}{dz^{\alpha-1}} \{L(z)x_k(z)\}\Big|_{z=z_k} = \frac{d^{\alpha-1}}{dz^{\alpha-1}} y_k(z)\Big|_{z=z_k},$$

where $x_k(z)$ and $y_k(z)$ are now column vector polynomials. Clearly, the right TLSI interpolation problem is really nothing essentially new because $L(z)$ solves the right TLSI problem if and only if the transposed polynomial $L(z)^T$ solves the left TLSI problem (with transposed $x_k(z)$ and $y_k(z)$).

So far we have discussed nonhomogeneous interpolation problems only for polynomials. We shall also consider such problems for rational matrix functions. For example, a general tangential Lagrange-Sylvester interpolation problem for rational matrix functions is

(RTLSI)   *Given a subset $\sigma \subset \mathbf{C}$, distinct points $z_1, \ldots, z_n$ in σ; rational $1 \times M$ row vector functions $x_1(z), \ldots, x_n(z)$ with $x_k(z_k) \ne 0$ and rational $1 \times N$ row* *vector functions* $y_1(z), \ldots, y_n(z)$, *find* (*at least one or all*) *rational* $M \times N$ *matrix functions* $W(z)$ *with no poles in* σ *for which*

$$(16.1.6) \qquad \frac{d^{j-1}}{dz^{j-1}}\{x_k(z)W(z)\}\bigg|_{z=z_k} = \frac{d^{j-1}}{dz^{j-1}}y_k(z)$$

for $1 \le k \le n$, $1 \le j \le m_k$.

By allowing $x_k(z)$ and $y_k(z)$ to be more general rational matrix functions, of course we can pick up the more general rational tangential Lagrange-Sylvester interpolation problem. By taking $\sigma = \mathbf{C}$, we obtain all the polynomial interpolation problems (LI, TLI, LSI, TLSI) mentioned above. The RTSLI problem can be considered as the most general nonhomogeneous interpolation problem discussed so far.

Another useful remark is that the vector functions $x_j(z)$ and $y_j(z)$ can be taken to be polynomials (indeed, in the formulas (16.1.6) we see only the first M_k coefficients in the Taylor series of $x_k(z)$ and $y_k(z)$ at z_0). We will take advantage of this observation later on.

Of course one can consider the transposed version of RTSLI where $x_k(z)$ is a column vector function multiplying $W(z)$ on the right, and $y_k(z)$ is a column vector function (right tangential interpolation). More complicated are problems where one has both right and left tangential interpolation conditions simultaneously. A very general class of problems of this type will be discussed in Sections 16.8-16.10.

16.2 LAGRANGE-SYLVESTER INTERPOLATION IN CONTOUR INTEGRAL FORM

In this section we present a more compact form of the rational tangential Lagrange-Sylvester interpolation (RTLSI) problem introduced in the previous section. Using results on homogeneous interpolation presented in Part I of this book, we will then obtain a linear fractional description of the set of all solutions.

It will be convenient to use the notation $\mathrm{Res}_{z=z_0}\, f(z)$ to denote the residue (i.e. the coefficient of $(z-z_0)^{-1}$ in the Laurent series for $f(z)$ in a deleted neighborhood of z_0) of the rational function $f(z)$ at z_0. The function $f(z)$ can be scalar or matrix valued. We state the contour integral interpolation (CII) problem in the following form:

(CII) *Given matrices* A_ζ, B_+ *and* B_- *of sizes* $n_\zeta \times n_\zeta$, $n_\zeta \times M$ *and* $n_\zeta \times N$, *respectively, and a subset* $\sigma \subset \mathbf{C}$ *such that* $\sigma(A_\zeta) \subset \sigma$, *find a rational* $M \times N$ *matrix function* $W(z)$ *having no poles in* σ *for which*

$$(16.2.1) \qquad \sum_{z_0 \in \sigma} \mathrm{Res}_{z=z_0}(zI - A_\zeta)^{-1}B_+W(z) = -B_-.$$

Equivalently, (16.2.1) can be expressed as

$$(16.2.2) \qquad \frac{1}{2\pi i}\int_\gamma (zI - A_\zeta)^{-1}B_+W(z)dz = -B_-$$

where the simple closed rectifiable contour γ is chosen so that W is analytic on and inside γ and so that $\sigma(A_\zeta)$ is inside γ.

The minus sign in the right hand sides of (16.2.1) and (16.2.2) is inserted as a convenience useful for subsequent applications.

There is no essential loss of generality in assuming that the pair (A_ζ, B_+) as in the CII problem is a full range pair. Indeed, let

$$\mathcal{M} = \text{Im}[B_+, A_\zeta B_+, \ldots, A_\zeta^{n_\zeta - 1} B_+] \subset \mathbb{C}^{n_\zeta},$$

and let P be any projection of $\mathbb{C}^{n_\zeta}$ onto $\mathcal{M}$. Observe that $\mathcal{M}$ is the smallest A_ζ-invariant subspace containing $\text{Im}\, B_+$. So

$$\text{Im}\big((zI - A_\zeta)^{-1} B_+\big) \subset \mathcal{M}$$

for every $z \notin \sigma(A_\zeta)$ and consequently

$$\sum_{z_0 \in \sigma} \text{Res}_{z=z_0} (zI - A_\zeta)^{-1} B_+ W(z) = \sum_{z_0 \in \sigma} \text{Res}_{z=z_0} (zI - A_{\zeta 0})^{-1} B_{+0} W(z),$$

where $A_{\zeta 0} = A_\zeta | \mathcal{M}$; $B_{+0} = P B_+$. It remains to note that $(A_{\zeta 0}, B_{+0})$ is a full range pair. In particular, an obvious necessary condition for existence of a solution $W(z)$ to the CII problem is that $\text{Im}\, B_- \subset \mathcal{M}$. In the sequel it will often be assumed that the pair (A_ζ, B_+) in the CII problem is a full range pair.

Before developing the general theory we discuss several examples.

EXAMPLE 16.2.1.   Suppose $z_1, \ldots, z_n$ are n distinct points in σ and $M_1, \ldots, M_n$ are n given $M \times N$ matrices. Define matrices A_ζ, B_+ and B_- by

$$A_\zeta = \text{diag}(z_1 I_M, \ldots, z_n I_M)$$

$$B_+ = \text{col}(I_M, \ldots, I_M)$$

and

$$B_- = -\text{col}(M_1, \ldots, M_n).$$

Then it is easily seen that (16.2.1) (or equivalently (16.2.2)) collapses to the interpolation condition

$$W(z_k) = M_k$$

for $1 \leq k \leq n$. Thus the rational matrix CII problem for this case collapses to the rational matrix Lagrange interpolation problem.   □

EXAMPLE 16.2.2.   Suppose we are given a point z_0 in σ and a collection of $1 \times m$ row vectors $x_1, \ldots, x_m$ with $x_1 \neq 0$ and a collection $y_1, \ldots, y_m$ of $1 \times N$ row vectors. Define matrices A_ζ, B_+ and B_- of sizes $m \times m$, $m \times M$ and $m \times N$ by

$$A_\zeta = \begin{bmatrix} z_0 & 0 & \cdots & & & 0 \\ 1 & z_0 & & & & 0 \\ 0 & & & & & \\ \vdots & & \ddots & & \ddots & \vdots \\ 0 & 0 & \cdots & 0 & 1 & z_0 \end{bmatrix}$$

$$B_+ = \text{col}(x_1, \ldots, x_m)$$

and

$$B_- = -\text{col}(y_1, \ldots, y_m).$$

Since $x_1 \neq 0$ it can be shown that (A_ζ, B_+) is a full range pair. The resolvent $(zI - A_\zeta)^{-1} = ((z - z_0)I - (A_\zeta - z_0 I))^{-1} = \sum_{j=1}^m (A_\zeta - z_0 I)^{j-1} (z - z_0)^{-j}$ can be written down explicitly as

$$(zI - A_\zeta)^{-1} = \begin{bmatrix} (z-z_0)^{-1} & 0 & \cdots & 0 \\ (z-z_0)^{-2} & (z-z_0)^{-1} & \cdots & 0 \\ \vdots & \ddots & \ddots & \vdots \\ (z-z_0)^{-m} & \cdots & (z-z_0)^{-2} & (z-z_0)^{-1} \end{bmatrix}.$$

If $W(z) = \sum_{j=1}^\infty W_j (z - z_0)^{j-1}$ is a rational $M \times N$ matrix function which is analytic at z_0 then it is not difficult to see that

$$\text{Res}_{z=z_0}(zI - A_\zeta)^{-1} B_+ W(z) = \begin{bmatrix} x_1 & 0 & \cdots & 0 \\ x_2 & x_1 & \cdots & \\ \vdots & \ddots & \ddots & \vdots \\ x_m & \cdots & x_2 & x_1 \end{bmatrix} \begin{bmatrix} W_1 \\ W_2 \\ \vdots \\ W_m \end{bmatrix}.$$

From this we see that the CII problem for this case reduces to the rational tangential Lagrange-Sylvester interpolation problem at the single point z_0, with interpolation conditions

$$\frac{d^{j-1}}{dz^{j-1}}\{x(z)W(z)\}\Big|_{z=z_0} = \frac{d^{j-1}}{dz^{j-1}} y(z)\Big|_{z=z_0}$$

where

$$x(z) = \sum_{j=1}^m x_j (z - z_0)^{j-1}$$

and

$$y(z) = \sum_{j=1}^m y_j (z - z_0)^{j-1}. \quad \square$$

EXAMPLE 16.2.3. Suppose $z_1, \ldots, z_n$ are n distinct points in σ, $x_1, \ldots, x_n$ are nonzero $1 \times M$ row vectors and $y_1, \ldots, y_n$ are $1 \times N$ row vectors. Let $A_\zeta = \text{diag}(z_1, \ldots, z_n)$, $B_+ = \text{col}(x_1, \ldots, x_n)$ and $B_- = -\text{col}(y_1, \ldots, y_n)$. Since each $x_k \neq 0$, it can be shown that (A_ζ, B_+) is a full range pair. The CII problem in this case coincides with the TLI problem of finding a W in $\mathcal{R}_{M \times N}(\sigma)$ for which $x_k W(z_k) = y_k$ for $k = 1, \ldots, n$. $\square$

EXAMPLE 16.2.4. Let the points $z_1, \ldots, z_n$ in σ, the $1 \times M$ rational row functions $x_1(z), \ldots, x_n(z)$ and $1 \times N$ rational row functions $y_1(z), \ldots, y_n(z)$ be as in the RTLSI problem. Write the Taylor series

$$x_j(z) = \sum_{k=1}^\infty x_{jk}(z - z_j)^{k-1}; \qquad y_j(z) = \sum_{k=1}^\infty y_{jk}(z - z_j)^{k-1}.$$

Introduce the matrices

$$A_\zeta = \mathrm{diag}(A_{\zeta 1}, \ldots, A_{\zeta n}),$$
$$B_+ = \mathrm{col}(B_{+1}, \ldots, B_{+n}), \qquad B_- = -\,\mathrm{col}(B_{-1}, \ldots, B_{-n})$$

where for each $k = 1, \ldots, n$, $A_{\zeta k}$ is the $n_k \times n_k$ matrix

$$A_{\zeta k} = \begin{bmatrix} z_k & 0 & \cdots & & 0 \\ 1 & z_k & & & \\ \vdots & \ddots & \ddots & & \vdots \\ 0 & & & & 0 \\ 0 & 0 & \cdots & 1 & z_k \end{bmatrix},$$

$$B_{+k} = \mathrm{col}(x_{k1}, \ldots, x_{km_k})$$

and

$$B_{-k} = \mathrm{col}(y_{k1}, \ldots, y_{km_k}).$$

The pair (A_ζ, B_+) turns out to be full range since the points $z_1, \ldots, z_n$ are distinct and the row vectors $x_{11}, \ldots, x_{n1}$ are nonzero. The CII problem with this choice of A_ζ, B_+, B_- is identical to the RTLSI interpolation discussed in the previous section.   □

We show now that the general one-sided Lagrange-Sylvester interpolation problem is equivalent to the contour integral interpolation problem.

Let there be given distinct points $z_1, \ldots, z_r$ in the complex plane. For every z_α two collections of vector polynomials $x_{\alpha 1}(z), \ldots, x_{\alpha k_\alpha}(z)$ of size $1 \times M$ and vector polynomials $y_{\alpha 1}(z), \ldots, y_{\alpha k_\alpha}(z)$ of size $1 \times N$ are given. We consider the following interpolation problem:

Find a rational $M \times N$ matrix function $W(z)$ analytic at each point z_α ($\alpha = 1, \ldots, r$) such that the interpolation conditions

(16.2.3)
$$\left. \frac{d^{i-1}}{dz^{i-1}} \{x_{\alpha j}(z) W(z)\} \right|_{z=z_\alpha} = \left. \frac{d^{i-1}}{dz^{i-1}} y_{\alpha j}(z) \right|_{z=z_\alpha},$$
$$i = 1, \ldots, \beta_{\alpha j}; j = 1, \ldots, k_\alpha; \alpha = 1, \ldots, r$$

are satisfied (here $\beta_{\alpha j}$ are given positive integers depending on α and j).

THEOREM 16.2.1. *For a suitable choice of matrices* A_ζ, B_+ *and* B_- *(with* $\sigma(A_\zeta) = \{z_1, z_2, \ldots, z_r\}$*) the interpolation conditions* (16.2.3) *take the form*

(16.2.4)
$$\sum_{\alpha=1}^r \mathrm{Res}_{z=z_\alpha}(zI - A_\zeta)^{-1} B_+ W(z) = -B_-,$$

or, equivalently, the form

$$\frac{1}{2\pi i} \int_\gamma (zI - A_\zeta)^{-1} B_+ W(z) = -B_-,$$

where γ is a suitable contour such that the points $z_1, \ldots, z_r$ are inside γ and all poles of $W(z)$ are outside γ. Conversely, given matrices A_ζ with $\sigma(A_\zeta) \subset \{z_1, z_2, \ldots, z_r\}$, B_+ and B_- of sizes $n_\zeta \times n_\zeta$, $n_\zeta \times M$ and $n_\zeta \times N$, respectively, there exist for $\alpha = 1, \ldots, r$ collections of vector polynomials $x_{\alpha 1}(z), \ldots, x_{\alpha k_\alpha}(z)$ (of size $1 \times M$) and $y_{\alpha 1}(z), \ldots, y_{\alpha k_\alpha}(z)$ (of size $1 \times N$) such that a rational $M \times N$ matrix function $W(z)$ with no poles in the set $\{z_1, \ldots, z_\alpha\}$ satisfies (16.2.4) if and only if $W(z)$ satisfies (16.2.3).

PROOF. It is easily seen that the proof of the direct statement of Theorem 16.2.1 is reduced to the case when $r = 1$ (i.e., there is only one point z_1) and $k_1 = 1$ (i.e., only two polynomials $x_{11}(z)$ and $y_{11}(z)$ are given). Indeed, assume that for fixed j and α we have found suitable matrices $A_\zeta^{(j\alpha)}$, $B_+^{(j\alpha)}$ and $B_-^{(j\alpha)}$ such that $\sigma(A_\zeta^{(j\alpha)}) = \{z_\alpha\}$ and the condition (16.2.3) (with fixed j and α) is equivalent to

$$\mathrm{Res}_{z=z_\alpha}\left(zI - A_\zeta^{(j\alpha)}\right)^{-1} B_+^{(j\alpha)} W(z) = -B_-^{(j\alpha)},$$

where $W(z)$ is analytic at each point $z_1, \ldots, z_r$. Then put

$$A_\zeta = \mathrm{diag}\left(\mathrm{diag}(A_\zeta^{(11)}, \ldots, A_\zeta^{(k_1 1)}), \ldots, \mathrm{diag}(A_\zeta^{(1r)}, \ldots, A_\zeta^{(k_r r)})\right)$$

$$B_+ = \mathrm{col}\left(\mathrm{col}(B_+^{(j\alpha)})_{j=1}^{k_\alpha}\right)_{\alpha=1}^{r}$$

$$B_- = \mathrm{col}\left(\mathrm{col}(B_-^{(j\alpha)})_{j=1}^{k_\alpha}\right)_{\alpha=1}^{r},$$

and with this choice of A_ζ, B_+, B_- the interpolation conditions (16.2.3) are equivalent to (16.2.4).

Thus, consider the case $r = 1$ and $k_1 = 1$. Let $m = \beta_{11}$.

Define matrices A_ζ, B_+ and B_- of sizes $m \times m$, $m \times M$ and $M \times N$ respectively by

$$A_\zeta = \begin{bmatrix} z_1 & 0 & \cdots & & & 0 \\ 1 & z_1 & & & & \vdots \\ 0 & 1 & & & & \\ \vdots & \vdots & \ddots & \ddots & & 0 \\ 0 & 0 & \cdots & 1 & z_1 \end{bmatrix}$$

$$B_+ = \mathrm{col}(x_1, \ldots, x_m)$$

and

$$B_- = -\mathrm{col}(y_1, \ldots, y_m),$$

where

$$(16.2.5) \qquad x_q = \frac{1}{(q-1)!}\left(x_{11}(z)\right)^{[q-1]}\Big|_{z=z_1} \; ; \qquad y_q = \frac{1}{(q-1)!}\left(y_{11}(z)\right)^{[q-1]}\Big|_{z=z_1}$$

for $q = 1, \ldots, m$ (we denote by $(f(z))^{[q]}$ the q-th derivative of the vector valued function $f(z)$).

As in Example 16.2.2, if $W(z) = \sum_{j=1}^{\infty} W_j(z - z_1)^{j-1}$ is a rational $M \times N$ matrix function which is analytic at z_1 then we see that

$$\mathrm{Res}_{z=z_1}(zI - A_\zeta)^{-1}B_+W(z) = \begin{bmatrix} x_1 & 0 & \cdots & 0 \\ x_2 & x_1 & \cdots & 0 \\ \vdots & \ddots & \ddots & \vdots \\ x_m & \cdots & x_2 & x_1 \end{bmatrix} \begin{bmatrix} W_1 \\ W_2 \\ \vdots \\ W_m \end{bmatrix}.$$

From this we see that the interpolation problem (16.2.3) indeed takes the form (16.2.4).

Conversely, assume matrices A_ζ, B_+, B_- of sizes $n_\zeta \times n_\zeta$, $n_\zeta \times M$, $n_\zeta \times N$, respectively, are given. To construct the required collections of vector polynomials, we observe first that (upon replacing A_ζ by $S^{-1}A_\zeta S$, B_+ by $S^{-1}B_+$ and B_- by $S^{-1}B_-$ for some invertible matrix S) the matrix A_ζ can be taken to be in the lower triangular Jordan normal form. As in the first part of the proof, we can further assume that A_ζ is just one block:

$$A_\zeta = \begin{bmatrix} z_0 & 0 & \cdots & 0 \\ 1 & z_0 & \cdots & 0 \\ \vdots & \ddots & z_0 & \vdots \\ & & & 0 \\ 0 & \cdots & 1 & z_0 \end{bmatrix}$$

where z_0 is one of the points $z_1, \ldots, z_r$, say $z_0 = z_1$. Define vector polynomials $x_{11}(z)$ and $y_{11}(z)$ by formulas (16.2.5), where m is the size of A_ζ. Then the previous argument shows that (16.2.4) is equivalent to

$$\frac{d^{i-1}}{dz^{i-1}}\{x_{11}W(z)\}\Big|_{z=z_1} = \frac{d^{i-1}}{dz^{i-1}}\{y_{11}(z)\}\Big|_{z=z_1}$$

for $i = 1, \ldots, m$, and we are done. $\square$

We will be mostly interested in the case when the pair (A_ζ, B_+) is a full range pair. It is not difficult to see from the proof of Theorem 16.2.1 that this happens if and only if for each z_α the vectors $x_{\alpha 1}(z_\alpha), \ldots, x_{\alpha k_\alpha}(z_\alpha)$ are linearly independent.

In a similar vein we study the dual problem which is formulated as follows:

(CII$_d$)   *Given matrices C_+, C_-, A_π of sizes $M \times n_\pi$, $N \times n_\pi$ and $n_\pi \times n_\pi$, respectively, and a subset $\sigma \subset \mathbf{C}$ such that $\sigma(A_\pi) \subset \sigma$. Find a rational $M \times N$ matrix function $W(z)$ having no poles in σ such that*

(16.2.6) $$\sum_{z_0 \in \sigma} \mathrm{Res}_{z=z_0} W(z)C_-(zI - A_\pi)^{-1} = C_+.$$

Again, the interpolation condition (16.2.6) can be interpreted as

$$\frac{1}{2\pi i} \int_\tau W(z)C_-(zI - A_\pi)^{-1}dz = C_+$$

for a suitable contour τ (such that all the eigenvalues of A_π are inside τ and all the poles of $W(z)$ are outside τ).

The CII_d problem can be easily reduced to a similar problem with (C_-, A_π) a null kernel pair. Indeed, given C_-, A_π as in CII_d, let $\mathcal{M} = \cap_{i \geq 0} \text{Ker}(C_- A_\pi^i)$. Clearly, $\mathcal{M}$ is an A_π-invariant subspace and $C_- \mid_{\mathcal{M}} = 0$. Letting $\mathcal{N}$ be any direct complement to $\mathcal{M}$ in $\mathbf{C}^{n_\pi}$, denote $\widetilde{C}_- = C \mid_{\mathcal{N}}$; $\widetilde{A}_\pi = P A_\pi \mid_{\mathcal{N}}$, where P is the projection on $\mathcal{N}$ along $\mathcal{M}$. We have

$$C_-(zI - A_\pi)^{-1}x = \widetilde{C}_-(zI - \widetilde{A}_\pi)^{-1}Px, \qquad x \in \mathbf{C}^{n_\pi}.$$

So a necessary condition for existence of $M \times N$ rational function $W(z)$ without poles in σ and (16.2.6) is that $C_+x = 0$ for all $x \in \mathcal{M}$. When this condition is satisfied, (16.2.6) is equivalent to

$$\sum_{z_0 \in \sigma} \text{Res}_{z=z_0} W(z)\widetilde{C}_-(zI - \widetilde{A}_\pi)^{-1} = C_+ \mid_{\mathcal{N}}.$$

It remains to observe that $(\widetilde{C}_-, \widetilde{A}_\pi)$ is a null kernel pair.

The equivalence of CII_d and the one-sided Lagrange-Sylvester interpolation problem is established analogously to the CII problem. Thus, let there be given distinct points $w_1, \ldots, w_s$ in the complex plane, and for every w_γ two collections of vector polynomials $u_{\gamma 1}(z), \ldots, u_{\gamma \ell_\gamma}(z)$ (of size $N \times 1$) and $v_{\gamma 1}(z), \ldots, v_{\gamma \ell_\gamma}(z)$ (of size $M \times 1$) be given. Consider the interpolation problem of finding a rational $M \times N$ matrix function $W(z)$ without poles in the set $\{w_1, \ldots, w_s\}$ and such that the equalities

$$(16.2.7) \qquad \frac{d^{i-1}}{dz^{i-1}}\{W(z)u_{\gamma j}(z)\}\Big|_{z=w_\gamma} = \frac{d^{i-1}}{dz^{i-1}}v_{\gamma j}(z)\Big|_{z=w_\gamma}$$

hold for $i = 1, \ldots, \delta_{\gamma j}$; $j = 1, \ldots, \ell_\gamma$; $\gamma = 1, \ldots, s$ (here $\delta_{\gamma j}$ are given positive integers depending on γ and j).

THEOREM 16.2.2. *For a suitable choice of matrices A_π, C_+ and C_- (with $\sigma(A_\pi) = \{w_1, \ldots, w_s\}$) the interpolation conditions (16.2.7) take the form*

$$(16.2.8) \qquad \sum_{\gamma=1}^{s} \text{Res}_{z=w_\gamma} W(z)C_-(zI - A_\pi)^{-1} = C_+,$$

or, equivalently, the form

$$\frac{1}{2\pi i}\int_\tau W(z)C_-(zI - A_\pi)^{-1}dz = C_+$$

for a suitable contour τ (such that the points $w_1, \ldots, w_s$ are inside τ but the poles of $W(z)$ are outside τ). Conversely, given matrices A_π (with $\sigma(A_\pi) \subset \{w_1, \ldots, w_s\}$), C_- and C_+ of sizes $n_\pi \times n_\pi$, $N \times n_\pi$, $M \times n_\pi$, respectively, there exist for $\gamma = 1, \ldots, s$ collections of vector polynomials $u_{\gamma 1}(z), \ldots, u_{\gamma \ell_\gamma}(z)$ and $v_{\gamma 1}(z), \ldots, v_{\gamma \ell_\gamma}(z)$ such that a *rational $M \times N$ function $W(z)$ without poles in $\{w_1, \ldots, w_s\}$ satisfies (16.2.8) if and only if it satisfies (16.2.7).*

We omit the full details of the proof of Theorem 16.2.2 and just remark that in case $\gamma = 1$ and $\ell_1 = 1$ the matrices A_π, C_+ and C_- are given as follows (here $\delta = \delta_{11}$):

$$C_+ = [v_1, \ldots, v_\delta]; \qquad C_- = [u_1, \ldots, u_\delta],$$

where

$$(16.2.9) \qquad v_q = \frac{1}{(q-1)!} (v_{11}(z))^{[q-1]} \Big|_{z=w_1} ; \qquad u_q = \frac{1}{(q-1)!} (u_{11}(z))^{[q-1]} \Big|_{z=w_1},$$

and

$$A_\pi = \begin{bmatrix} w_1 & 1 & 0 & \cdots & 0 \\ 0 & w_1 & 1 & & \vdots \\ \vdots & \ddots & \ddots & \ddots & 1 \\ 0 & 0 & 0 & \cdots & w_1 \end{bmatrix}$$

is the $\delta \times \delta$ Jordan block with eigenvalue w_1.

Note that (C_-, A_π) is a null kernel pair if and only if for each γ the vectors $u_{\gamma 1}(w_\gamma), \ldots, u_{\gamma \ell_\gamma}(w_\gamma)$ are linearly independent.

In the sequel it will be more convenient for us to work with the CII and CII_d problems rather than with the original one-sided Lagrange-Sylvester interpolation problems.

16.3   LAGRANGE-SYLVESTER INTERPOLATION IN THE DIVISOR-REMAINDER FORM

In this short section we show that the general solution of contour interpolation problem (16.2.4) can be expressed in the divisor-remainder form provided one solution to the CII problem is known.

THEOREM 16.3.1. *Suppose A_ζ, B_+, B_- are given $n_\zeta \times n_\zeta$, $n_\zeta \times M$ and $n_\zeta \times N$ matrices, σ is a subset of $\mathbb{C}$ containing $\sigma(A)$, and (A_ζ, B_+) is a full range pair. Suppose ψ is a regular rational matrix function in $\mathcal{R}_{M \times M}(\sigma)$ for which (A_ζ, B_+) is a left null pair over σ. Let $K(z) \in \mathcal{R}_{M \times N}(\sigma)$ be any particular solution of the associated interpolation condition (16.2.4). Then a rational $M \times N$ matrix function $W \in \mathcal{R}_{M \times N}(\sigma)$ is a solution of (16.2.4) if and only if W has the form*

$$(16.3.1) \qquad W = K + \psi Q$$

for some $Q \in \mathcal{R}_{M \times N}(\sigma)$.

PROOF. Note that if W and K both satisfy (16.2.4) then the difference $W - K$ satisfies the homogeneous interpolation condition

$$(16.3.2) \qquad \sum_{z_0 \in \sigma} \text{Res}_{z=z_0} (zI - A_\zeta)^{-1} B_+ \big(W(z) - K(z) \big) = 0.$$

From Theorem 12.3.1 applied to the left null-pole subspace $S_\sigma(\psi)$ for ψ over σ we see that (16.3.2) is equivalent to

$$(W - K)x \in S_\sigma(\overline{\psi})^{\rightarrow}$$

for each $x \in \mathbb{C}^N$. By Proposition 12.1.1 this in turn is equivalent to the factorization

$$W - K = \psi Q$$

for some $Q \in R_{M \times N}(\sigma)$. □

Note that (A_ζ, B_+) was assumed to be full range; we have seen already that this assumption can be made without loss of generality.

The formula (16.3.1) can be interpreted in terms of divisors and remainders: here ψ is a left divisor of W with quotient Q and remainder K. For the case of matrix polynomials such interpretation is especially convenient, and will be used later in this chapter.

The theorem dual to Theorem 16.3.1 looks as follows.

THEOREM 16.3.2. *Let C_+, C_-, A_π and σ be given as in* (CII$_d$), *and assume that (C_-, A_π) is a null kernel pair. Suppose ψ is a regular rational matrix function in $R_{N \times N}(\sigma)$ for which (C_-, A_π) is a right null pair over σ. Let $K(z) \in R_{M \times N}(\sigma)$ be any particular solution of* (16.2.6). *Then a rational matrix function $W \in R_{M \times N}(\sigma)$ is a solution of* (16.2.6) *if and only if W has the form $W = K + Q\psi$ for some $Q \in R_{M \times N}(\sigma)$.*

We omit the proof of Theorem 16.3.2 as it is analogous to that of Theorem 16.3.1.

16.4 SOLUTIONS OF THE ONE-SIDED CONTOUR INTERPOLATION PROBLEM

We start our main result which describes all solutions of CII.

THEOREM 16.4.1. *Let A_ζ, B_+, B_- be matrices as in the CII problem for a subset σ of $\mathbb{C}$ and assume that (A_ζ, B_+) is a full range pair. Then interpolation functions $W \in R_{M \times N}(\sigma)$ satisfying* (16.2.1) *exist. If $\theta = \begin{bmatrix} \theta_{11} & \theta_{12} \\ \theta_{21} & \theta_{22} \end{bmatrix}$ is any rational $(M+N) \times (M+N)$ matrix function having no poles in σ and having $(A_\zeta, [B_+, B_-])$ as a left null pair over σ, then a matrix function W in $R_{M \times N}(\sigma)$ satisfies the interpolation condition* (16.2.1) *if and only if*

$$W = (\theta_{11} Q_1 + \theta_{12} Q_2)(\theta_{21} Q_1 + \theta_{22} Q_2)^{-1}$$

for some

$$Q_1 \in R_{M \times N}(\sigma) \quad \text{and} \quad Q_2 \in R_{N \times N}(\sigma)$$

such that $\theta_{21} Q_1 + \theta_{22} Q_2$ has no zeros in σ.

When $\theta_{21} = 0$ and $\theta_{22} = I$, the linear fractional map in Theorem 16.4.1 assumes a much simpler form because one can choose $Q_2 = I$ and Q_1 any matrix function with no poles in σ. In this case, we obtain the form $W = K + \psi Q$ as in Theorem 16.3.1 with $K = \theta_{12}\theta_{22}^{-1}$, $\psi = \theta_{11}$. Conversely, if a particular solution K to the CII problem (16.2.1) is known (for example, by plugging an appropriate pair Q_1, Q_2 into the formula for W in Theorem 16.4.1), one can construct a linear fractional (in fact, affine) representation for W with $\theta_{21} = 0$, $\theta_{22} = I$.

PROOF OF THEOREM 16.4.1. Suppose that the matrices A_ζ, B_+, B_- are as in the CII problem. Then if $W \in \mathcal{R}_{M \times N}(\sigma)$ is a solution of equation (16.2.1) then $\begin{bmatrix} W \\ I_N \end{bmatrix} \in \mathcal{R}_{M \times N}(\sigma)$ is a solution of the homogeneous interpolation condition

$$(16.4.1) \qquad \sum_{z_0 \in \sigma} \text{Res}_{z=z_0}(zI - A_\zeta)^{-1}[B_+, B_-]\begin{bmatrix} W(z) \\ I_N \end{bmatrix} = 0.$$

Since (A_ζ, B_+) by hypothesis is a full range pair, certainly $(A_\zeta, [B_+, B_-])$ is a full range pair. Suppose $\theta(z) \in \mathcal{R}_{(M+N) \times (M+N)}(\sigma)$ is any regular matrix function having $(A_\zeta, [B_+, B_-])$ as a left null pair over σ, i.e., θ is such that $(0, 0; A_\zeta, [B_+, B_-]; 0)$ a null-pole triple for θ over σ; such a θ exists by Theorem 5.6.3 (and even can be taken to be a polynomial).

Then by Theorem 12.3.1 combined with Proposition 12.1.1, the equality (16.4.1) is equivalent to

$$(16.4.2) \qquad \begin{bmatrix} W \\ I_N \end{bmatrix} = \theta \begin{bmatrix} Q_1 \\ Q_2 \end{bmatrix}$$

where

$$\begin{bmatrix} Q_1 \\ Q_2 \end{bmatrix} \in \mathcal{R}_{(M+N) \times N}(\sigma).$$

If we decompose θ as

$$(16.4.3) \qquad \theta(z) = \begin{bmatrix} \theta_{11}(z) & \theta_{12}(z) \\ \theta_{21}(z) & \theta_{22}(z) \end{bmatrix}$$

where the blocks θ_{11}, θ_{12}, θ_{21}, θ_{22} have sizes $M \times M$, $M \times N$, $N \times M$ and $N \times N$ respectively, then (16.4.2) may be written as

$$\begin{bmatrix} W \\ I \end{bmatrix} = \begin{bmatrix} \theta_{11}Q_1 + \theta_{12}Q_2 \\ \theta_{21}Q_1 + \theta_{22}Q_2 \end{bmatrix}.$$

So W is given by

$$(16.4.4) \qquad W = \theta_{11}Q_1 + \theta_{12}Q_2$$

where $Q_1 \in \mathcal{R}_{M \times N}(\sigma)$ and $Q_2 \in \mathcal{R}_{N \times N}(\sigma)$ are subject to the side constraint

$$(16.4.5) \qquad I = \theta_{21}Q_1 + \theta_{22}Q_2.$$

Conversely, suppose that Q_1 in $\mathcal{R}_{M \times N}(\sigma)$ and $Q_2 \in \mathcal{R}_{N \times N}(\sigma)$ satisfy (16.4.5) and define $W \in \mathcal{R}_{M \times N}(\sigma)$ by (16.4.4). Then $\begin{bmatrix} W \\ I \end{bmatrix} = \theta \begin{bmatrix} Q_1 \\ Q_2 \end{bmatrix}$ so $\begin{bmatrix} W \\ I \end{bmatrix} x \in S_\sigma(\theta)$ for each x in $\mathbb{C}^M$. From Theorem 12.3.1 we conclude that $\begin{bmatrix} W \\ I \end{bmatrix}$ satisfies (16.4.1), and thus W is a solution of the CII problem.

A less rigid way to write (16.4.4) and (16.4.5) is as

$$(16.4.6) \qquad W = (\theta_{11} Q_1 + \theta_{12} Q_2)(\theta_{21} Q_1 + \theta_{22} Q_2)^{-1}$$

with $Q_1 \in \mathcal{R}_{M \times N}(\sigma)$ and $Q_2 \in \mathcal{R}_{N \times N}(\sigma)$ such that the function

$$(16.4.7) \qquad [\theta_{21} Q_1 + \theta_{22} Q_2](z)$$

has no zeros in σ.

Indeed, if the pair (Q_1, Q_2) satisfies (16.4.7) then the pair $(\widetilde{Q}_1, \widetilde{Q}_2)$ given by

$$\widetilde{Q}_1 = Q_1(\theta_{21} Q_1 + \theta_{22} Q_2)^{-1}, \qquad \widetilde{Q}_2 = Q_2(\theta_{21} Q_1 + \theta_{22} Q_2)^{-1}$$

has no poles in σ and satisfies (16.4.5). Moreover the W arising from (16.4.6) with the pair (Q_1, Q_2) is the same as the W arising from (16.4.4) with the pair $(\widetilde{Q}_1, \widetilde{Q}_2)$. Once we establish that pairs (Q_1, Q_2) of rational matrix functions analytic on σ and satisfying (16.4.5) or (16.4.7) exist, Theorem 16.4.1 will be proved.

To achieve this it suffices to show that

$$[\theta_{21} \theta_{22}] \mathcal{R}_{M+N}(\sigma) = \mathcal{R}_N(\sigma).$$

Since θ is known to be analytic on σ, one containment $([\theta_{21} \theta_{22}] \mathcal{R}_{M+N}(\sigma) \subset \mathcal{R}_N(\sigma))$ is clear. Conversely, suppose $h_2 \in \mathcal{R}_N(\sigma)$. To show that $h_2 \in [\theta_{21} \theta_{22}] \mathcal{R}_{M+N}(\sigma)$ it suffices to show that there exists an $h_1 \in \mathcal{R}_M(\sigma)$ such that $\begin{bmatrix} h_1 \\ h_2 \end{bmatrix}$ is in $\theta \mathcal{R}_{M+N}(\sigma)$. By Theorem 12.3.1 this is the case if and only if

$$\sum_{z_0 \in \sigma} \operatorname{Res}_{z=z_0}(zI - A_\zeta)^{-1}[B_+, B_-] \begin{bmatrix} h_1(z) \\ h_2(z) \end{bmatrix} = 0.$$

As h_2 is known while h_1 is the unknown, the problem is to solve for $h_1 \in \mathcal{R}_M(\sigma)$:

$$(16.4.8) \qquad \sum_{z_0 \in \sigma} \operatorname{Res}_{z=z_0}(zI - A_\zeta)^{-1} B_+ h_1(z) = y$$

where

$$(16.4.9) \qquad y = - \sum_{z_0 \in \sigma} \operatorname{Res}_{z=z_0}(zI - A_\zeta)^{-1} B_- h_2(z).$$

Since (A_ζ, B_+) by assumption is a full range pair, Lemma 12.2.2 implies that (16.4.8) is always solvable for $h_1 \in \mathcal{R}_M(\sigma)$. This completes the proof of Theorem 16.4.1. $\quad \square$

We conclude this section with the analogue of Theorem 16.4.1.

THEOREM 16.4.2. *Let A_π, C_-, C_+ be matrices as in the CII_d problem and assume that (C_-, A_π) is a null kernel pair. Then there exist interpolating functions $W \in \mathcal{R}_{M \times N}(\sigma)$ satisfying (16.2.6). If $\theta = \begin{bmatrix} \theta_{11} & \theta_{12} \\ \theta_{21} & \theta_{22} \end{bmatrix}$ is any rational $(M+N) \times (M+N)$ matrix function having no poles in σ and having $\left(\begin{bmatrix} C_- \\ -C_+ \end{bmatrix}, A_\pi \right)$ as a right null pair over σ, then $W \in \mathcal{R}_{M \times N}(\sigma)$ satisfies (16.2.6) if and only if*

$$W = (Q_1 \theta_{12} + Q_2 \theta_{22})^{-1}(Q_1 \theta_{11} + Q_2 \theta_{21})$$

for some $Q_1 \in \mathcal{R}_{M \times N}(\sigma)$ and $Q_2 \in \mathcal{R}_{N \times N}(\sigma)$ such that the rational function $Q_1 \theta_{12} + Q_2 \theta_{22}$ has no zeros in σ.

We omit the proof of Theorem 16.4.2.

16.5   INTERPOLATION WITH SEVERAL CONTOUR INTE-GRAL CONDITIONS

We consider in this section the interpolation problem for rational matrix functions the data for which are given in terms of several equations involving residues with respect to points in some non-empty subset σ of the complex plane.

To set up the framework, introduce a finite set of collections of matrices

$$(A_{\zeta k}, B_{+k}, B_{-k}), \qquad k = 1, \ldots, n$$

where, for each k, $(A_{\zeta k}, B_{+k}, B_{-k})$ are as follows:

$A_{\zeta k}$, B_{+k}, B_{-k} *have sizes $n_{\zeta k} \times n_{\zeta k}$, $n_{\zeta k} \times M$, $n_{\zeta k} \times N$ respectively, $(A_{\zeta k}, B_{+k})$ is a full range pair and $\sigma(A_{\zeta k}) \subset \sigma$.*

The problem is to give conditions which guarantee the existence of a rational $M \times N$ matrix function $W(z)$ which is analytic on σ and satisfies the n interpolation conditions:

(16.5.1) $$\sum_{z_0 \in \sigma} \mathrm{Res}_{z=z_0}(zI - A_{\zeta k})^{-1}B_{+k}W(z) = -B_{-k}, \qquad k = 1, \ldots, n.$$

Equivalently, equalities (16.5.1) can be written in terms of contour integrals:

$$\frac{1}{2\pi i} \int_\gamma (zI - A_{\zeta k})^{-1}B_{+k}W(z) = -B_{-k}; \qquad k = 1, \ldots, n,$$

where γ is a suitable contour in $\mathbb{C}$ such that $\cup_{k=1}^n \sigma(A_{\zeta k})$ is inside γ and all poles of $W(z)$ are outside γ. Furthermore, if such functions $W(z)$ do exist, the next problem is to describe all of them. This interpolation problem will be termed the *simultaneous contour integral interpolation problem* (SCII).

We first illustrate this problem with some simple examples.

EXAMPLE 16.5.1. Suppose $z_1, \ldots, z_n$ are distinct points in σ, $A_{\zeta k} = [z_k]$, B_{+k} is the nonzero $1 \times M$ row vector x_k and B_{-k} is the $1 \times N$ row vector $-y_k$. Then the interpolation condition (16.5.1) reduces to a single tangential interpolation condition

$$x_k W(z_k) = y_k$$

for each k. Thus the SCII interpolation problem is to solve for a W which satisfies each of these basic interpolation conditions simultaneously. This is a particular case of the one-sided Lagrange interpolation problem TLI described earlier. $\square$

EXAMPLE 16.5.2. Suppose $A_{\zeta 1} = A_{\zeta 2}$ is the 1×1 matrix $[z_0]$, $B_{+1} = B_{+2}$ is the $1 \times M$ row vector x and B_{-1} and B_{-2} are given by $1 \times N$ row vectors y_1 and y_2. Then it is clear that the associated (SCII) problem has a solution if and only if $y_1 = y_2$. $\square$

The trivial Example 16.5.2 makes clear that some compatibility condition on the collection of matrices $\{A_{\zeta k}, B_{+k}, B_{-k} : k = 1, \ldots, n\}$ is required for the SCII problem to have solutions. To describe the compatibility condition we need the notion of the least common coextension which will be introduced now (this notion is the dual to the notion of least common extension introduced and used in Chapter 10).

An ordered pair of matrices (A, B) where A is $p \times p$ and B is $p \times n$ will be called *left admissible pair* with *base dimension* n and *order* p.

Two left pairs (A_1, B_1) and (A_2, B_2) are said to be *similar* if

(16.5.2) $$A_1 S = S A_2, \qquad B_1 = S B_2$$

for some nonsingular matrix S (in particular, the orders of similar left pairs are necessarily equal). If (16.5.2) holds with a surjective (or, what is the same, right invertible) linear transformation S we say that (A_1, B_1) is a *corestriction* of (A_2, B_2); in this case we say also that (A_2, B_2) is a *coextension* of (A_1, B_1). If (A_2, B_2) is a full range pair then so are all its corestrictions, and if (A_1, B_1) is a corestriction of (A_2, B_2), then the surjective linear transformation S satisfies (16.5.2) is uniquely determined (indeed, (16.5.2) implies

$$[B_1, A_1 B_1, \ldots, A_1^{q-1} B_1] = S[B_2, A_2 B_2, \ldots, A_2^{q-1} B_2]; \qquad q = 1, 2, \ldots$$

and since $[B_2, A_2 B_2, \ldots, A_2^{q-1} B_2]$ is surjective for large q, the uniqueness of S follows). To emphasize this fact, we say that (A_1, B_1) is an *S-corestriction* of (A_2, B_2) and (A_2, B_2) is an *S-coextension* of (A_1, B_1).

Clearly, (A, B) is a left admissible pair if and only if the transposed pair (B^T, A^T) is a right admissible pair (see Chapter 10 for this notion and its properties). Moreover, (A_2, B_2) is a coextension of (A_1, B_1) if and only if (B_2^T, A_2^T) is an extension of (B_1^T, A_1^T). This and similar observations allow us to obtain results concerning coextensions of left admissible pairs from the corresponding results in Chapter 10 concerning extensions of right admissible pairs. In particular, we need the notion of least common coextensions. Given left admissible pairs $(A_1, B_1), \ldots, (A_r, B_r)$ with the same base dimension, we say that a left admissible pair (A, B) is a *least common coextension* of $(A_1, B_1), \ldots, (A_r, B_r)$ if (A, B) is a coextension of each (A_i, B_i) $(i = 1, \ldots, r)$, and any left admissible pair which is a coextension of each (A_i, B_i) $(i = 1, \ldots, r)$ is in turn a coextension of (A, B). If all (A_i, B_i) $(i = 1, \ldots, r)$ are full range pairs, then there exists a least common coextension and it is unique up to similarity (cf. Theorem 10.2.1).

We now return to the SCII problem. Here is our main result expressing the solutions of SCII in terms of least common coextensions.

THEOREM 16.5.1. *Suppose a collection of matrices $A_{\zeta k}$, B_{+k}, B_{-k} ($k = 1, \ldots, n$) is given, where for each k, $A_{\pi k}$, B_{+k}, B_{-k} have respective sizes $n_{\zeta k} \times n_{\zeta k}$, $n_{\zeta k} \times M$, $n_{\zeta k} \times N$ and $(A_{\zeta k}, B_{+k})$ is a full range pair. Let (A_ζ, B_+) be a least common coextension of the collection $\{(A_{\zeta k}, B_{+k}): k = 1, \ldots, n\}$ of full range left pairs, with associated surjective linear maps $T_k \colon \mathbf{C}^{n_\zeta} \to \mathbf{C}^{n_{\zeta k}}$ (so (A_ζ, B_+) is a T_k-coextension of $(A_{\zeta k}, B_{+k})$, $k = 1, \ldots, n$; here n_ζ is the size of A_ζ). Then the SCII problem associated with $(A_{\zeta k}, B_{+k}, B_{-k}; k = 1, \ldots, n)$ has a solution if and only if the system of equation*

$$(16.5.3) \qquad T_k B_- = B_{-k}, \qquad k = 1, \ldots, n$$

has a solution linear map $B_- \colon \mathbf{C}^n \to \mathbf{C}^{n_\zeta}$. In this case the solution B_- of (16.5.3) is unique and an $M \times N$ rational matrix function W satisfies (16.5.1) if and only if W satisfies the single contour integral interpolation condition

$$(16.5.4) \qquad \sum_{z_0 \in \sigma} \mathrm{Res}_{z=z_0}(zI - A_\zeta)^{-1} B_+ W(z) = -B_-.$$

PROOF. We start with the proof of the second part of the theorem. Suppose that W satisfies the condition (16.5.1) and let (A_ζ, B_+) be a least common coextension of the collection of full range pairs $\{(A_{\zeta k}, B_{+k}): k = 1, \ldots, n\}$. Define the matrix B_- by (16.5.4). Then we have for each k

$$(16.5.5) \qquad T_k \left(\sum_{z_0 \in \sigma} \mathrm{Res}_{z=z_0}(zI - A_\zeta)^{-1} B_+ W(z) \right) = -T_k B_-.$$

By definition, T_k satisfies the equalities

$$(16.5.6) \qquad A_{\zeta k} T_k = T_k A_\zeta; \qquad B_{+k} = T_k B_+.$$

So the left hand side of (16.5.5) is

$$\sum_{z_0 \in \sigma} \mathrm{Res}_{z=z_0}(zI - A_{\zeta k})^{-1} B_{+k} W(z).$$

Since W by assumption satisfies (16.5.1) this expression is equal to $-B_{-k}$ and therefore B_- is a solution of (16.5.3). Note also that W is a solution of (16.5.4) by definition of

B_-. That B_- is unique follows from $\cap_{k=1}^n \operatorname{Ker} T_k = \{0\}$; this in turn is a consequence of (A_ζ, B_+) being a least common coextension of $\{(A_{\zeta k}, B_{+k}): k = 1, \ldots, n\}$. Indeed, by the left analogue of Theorem 10.2.2 we have

$$(16.5.7) \qquad \operatorname{Ker}[B_+, A_\zeta B_+, \ldots, A_\zeta^{m-1} B_+] = \bigcap_{k=1}^n \operatorname{Ker}[B_{+k}, A_{\zeta k} B_{+k}, \ldots, A_{\zeta k}^{m-1} B_{+k}]$$

for $m = 1, 2, \ldots$. Using (16.5.6) rewrite the right-hand side of (16.5.7) in the form

$$\bigcap_{k=1}^n \operatorname{Ker}(T_k[B_+, A_\zeta B_+, \ldots, A_\zeta^{m-1} B_+]).$$

Hence (16.5.7) takes the form

$$(16.5.8) \qquad \operatorname{Ker}[B_+, A_\zeta B_+, \ldots, A_\zeta^{m-1} B_+] = \operatorname{Ker} \begin{bmatrix} T_1 \\ T_2 \\ \vdots \\ T_n \end{bmatrix} [B_+, A_\zeta B_+, \ldots, A_\zeta^{m-1} B_+].$$

Taking m so large that

$$\sum_{i=0}^{m-1} \operatorname{Im}(A_\zeta^i B_+) = \mathbf{C}^{n_\zeta},$$

the equality (16.5.8) easily implies

$$Y_1 T_1 + Y_2 T_2 + \cdots + Y_n T_n = I$$

for some linear transformations $Y_1, \ldots, Y_n$. Now if there is a solution B_- to (16.5.3) it must be given by

$$B_- = Y_1 B_{-1} + \cdots + Y_n B_{-n}.$$

So B_- is unique.

Conversely, suppose that the system of equations (16.5.3) has a (unique) solution B_- and that W is any solution of the single interpolation condition (16.5.4).

Multiplying (16.5.4) on the left by T_k gives (16.5.5) for each k. By (16.5.6) this is equivalent to the interpolation condition (16.5.1).

The first part of the theorem follows now from Theorem 16.4.1, which asserts (in particular) existence of an $M \times N$ rational matrix function $W(z)$ without poles in σ which satisfies (16.5.4). $\square$

Several remarks concerning Theorem 16.5.1 and its proof are in order. Firstly, the left analogue of Theorem 10.2.1 shows that (A_ζ, B_+) is a full range pair and $\sigma(A_\zeta) \subset \cup_{k=1}^n \sigma(A_{\zeta k})$. In particular, $\sigma(A_\zeta) \subset \sigma$. Now we can apply the results of Section 16.4 to describe all $M \times N$ rational matrix functions $W(z)$ which are analytic in σ and satisfy (16.5.4).

Secondly, the proof of Theorem 16.5.1 shows that (A_ζ, B_+) need not be a least common coextension; any common coextension (A'_ζ, B'_+) of $\{(A_{\zeta k}, B_{+k}); k = 1, \ldots, n\}$ which is a full range pair with $\sigma(A'_\zeta) \subset \sigma$ will do in place of (A_ζ, B_+). It is not difficult to see from the definitions of a least common coextension that the following holds: Assume equations (16.5.3) have a solution B_- for a least common coextension (A_ζ, B_+) of $\{(A_{\zeta k}, B_{+k}): k = 1, \ldots, n\}$ with the associated surjective linear maps T_k. Then for any other full range left pair (A'_ζ, B'_+) which is a common extension of $\{(A_{\zeta k}, B_{+k}): k = 1, \ldots, n\}$ with associated maps T'_k the corresponding equations

$$T'_k B'_- = B_{-k}, \qquad k = 1, \ldots, n$$

also have a solution B'_-. The advantage of using a least common coextension (A_ζ, B_+) in Theorem 16.5.1 is that the order n_ζ of (A_ζ, B_+) is as small as possible, and hence the equation (16.5.4) is as simple as possible.

A second advantage is that the solution B_- of (16.5.3) is unique only if (A_ζ, B_+) is the least common coextension of $\{(A_{\zeta k}, B_{+k}): k = 1, \ldots, n\}$. This has the consequence that when (A_ζ, B_+) is not the least common extension, one would have to consider solutions of (16.5.4) for all choices of solutions B_- of (16.5.3) in order that the problems (16.5.1) and (16.5.4) be equivalent.

Finally, note that the full range assumption on the left pairs $(A_{\zeta k}, B_{+k})$ can be dropped. A reduction of the SCII problem with left pairs $(A_{\zeta k}, B_{\zeta k})$ to the SCII problem with full range left pairs $(A_{\zeta k}, B_{\zeta k})$ can be done as in Section 16.2 (see the remark after the statement of the CII problem).

As a corollary to Theorem 16.5.1 we see that the SCII problem can always be solved if the matrices $A_{\zeta k}$ $(k = 1, \ldots, n)$ have disjoint spectra.

COROLLARY 16.5.2. *Suppose $A_{\zeta k}$, B_{+k}, B_{-k} $(k = 1, \ldots, n)$ is a collection of matrices as in Theorem 16.5.1 and suppose that $\sigma(A_{\zeta k}) \cap \sigma(A_{\zeta j}) = \emptyset$ for $k \neq j$. Then the SCII problem has a solution.*

PROOF. It is not difficult to see that $\sigma(A_{\zeta k}) \cap \sigma(A_{\zeta j}) = \emptyset$ $(k \neq j)$ implies that the left pair

(16.5.9)
$$\begin{bmatrix} A_{\zeta 1} & 0 & \cdots & 0 \\ 0 & A_{\zeta 2} & \cdots & 0 \\ \vdots & \vdots & \ddots & \vdots \\ 0 & 0 & \cdots & A_{\zeta n} \end{bmatrix}, \qquad \begin{bmatrix} B_{+1} \\ B_{+2} \\ \vdots \\ B_{+n} \end{bmatrix}$$

is a full range pair. Now the left analogue of Theorem 10.2.2 tells us that (16.5.9) is actually a least common coextension of $\{(A_{\zeta k}, B_{+k}): k = 1, \ldots, n\}$. The associated surjective maps are easy to guess:

$$T_k = [\, 0 \quad \cdots \quad 0 \quad I \quad 0 \quad \cdots \quad 0 \,]$$

(with I on the k-th position). In this situation

$$B_- = \mathrm{col}[B_{-1}, \ldots, B_{-n}]$$

is a solution of (16.5.3) for each k, so by Theorem 16.5.1 the SCII problem has solutions.
□

We consider the dual SCII interpolation problem as well. Given a non-empty set $\sigma \subset \mathbf{C}$, and a finite collection of right pairs of matrices $\{(C_{-k}, A_{\pi k}): k = 1, \ldots, n\}$ with the properties that each $(C_{-k}, A_{\pi k})$ is a null kernel pair with sizes $N \times n_{\pi k}$ and $n_{\pi k} \times n_{\pi k}$, respectively, and $\sigma(A_{\pi k}) \subset \sigma$ for all $k = 1, \ldots, n$. Further, matrices $C_{+1}, \ldots, C_{+n}$ of sizes $M \times n_{\pi 1}, \ldots, M \times n_{\pi n}$, respectively, are given. The *dual simultaneous contour integral interpolation* problem (in short, SCII$_d$) is to find an $M \times N$ rational matrix function $W(z)$ which is analytic in σ and which satisfies the interpolation conditions

$$(16.5.10) \qquad \sum_{z_0 \in \sigma} \mathrm{Res}_{z=z_0} W(z) C_{-k}(zI - A_{\pi k})^{-1} = C_{+k}; \qquad k = 1, \ldots, n.$$

Again, (16.5.10) can be rewritten using the contour integrals.

A result dual to Theorem 16.5.1 holds for the SCII problem. We state this result but omit the proof.

THEOREM 16.5.3. *Suppose a collection of matrices $A_{\pi k}$, C_{-k}, C_{+k} ($k = 1, \ldots, n$) is given, with the above properties. Let (C_-, A_π) be a least common extension of the collection $\{(C_{-k}, A_{\pi k}): k = 1, \ldots, n\}$, with associated injective linear maps $S_k: \mathbf{C}^{n_{\pi k}} \to \mathbf{C}^{n_{\pi k}}$ (so $C_- S_k = C_{-k}$, $A_\pi S_k = S_k A_{\pi k}$; here n_π is the size of A_π). Then the SCII$_d$ problem (16.5.10) with $M \times N$ rational matrix function $W(z)$ analytic in σ has a solution if and only if the system of equation*

$$C_+ S_k = C_{+k} \qquad (k = 1, \ldots, n)$$

has a solution linear map $C_+: \mathbf{C}^{n_\pi} \to \mathbf{C}^M$. In this case, an $M \times N$ rational matrix function W satisfies (16.5.10) if and only if W satisfies the single contour integral interpolation condition

$$\sum_{z_0 \in \sigma} \mathrm{Res}_{z=z_0} W(z) C_-(zI - A_\pi)^{-1} = C_+.$$

All remarks made after Theorem 16.5.1 apply also to Theorem 16.5.3 mutatis mutandis. An analogue of Corollary 16.5.2 is valid as well:

COROLLARY 16.5.4. *Let $A_{\pi k}$, C_{+k}, C_{-k} ($k = 1, \ldots, n$) be as in Theorem 16.5.3, and suppose that $\sigma(A_{\pi k}) \cap \sigma(A_{\pi j}) = \emptyset$ for $k \neq j$. Then the SCII$_d$ problem has a solution.*

16.6   INTERPOLATION OF MATRIX POLYNOMIALS WITH MONIC DIVISORS AND THE VANDERMONDE MATRIX

In this and the next two sections we consider non-homogeneous interpolation problems for matrix polynomials. For this particular class of rational matrix functions it is more natural to consider the formulation in terms of divisors and remainders (as in Theorem 16.3.1). Although one could obtain the matrix polynomial results from the general theory developed in the previous sections, we shall solve the basic interpolation problems for matrix polynomials using a different technique that is tailored specifically for polynomials.

We start with the case of monic (i.e. with leading coefficient I) matrix polynomials. Let $L_1(z), \ldots, L_r(z)$ given monic $n \times n$ matrix polynomials of degrees $p_1, \ldots, p_r$, respectively, and for $i = 1, \ldots, r$ let $R_i(z)$ be a given $n \times n$ matrix polynomial of degree $< p_i$. The problem is to find a (not necessarily regular) matrix polynomial $A(z)$ such that

$$(16.6.1) \qquad A(z) = S_i(z)L_i(z) + R_i(z), \qquad i = 1, \ldots, r$$

for some matrix polynomials $S_1(z), \ldots, S_r(z)$. If all $R_i(z)$ are identically zero, this problem is just the problem of constructing a (not necessarily regular) common multiple of $L_1(z), \ldots, L_r(z)$.

The solution of this interpolation problem is based on the notion of the Vandermonde matrix. Given monic matrix polynomials $L_1(z), \ldots, L_r(z)$ with right null pairs $(X_1, T_1), \ldots, (X_r, T_r)$ (corresponding to the whole complex plane), let

$$(16.6.2) \qquad U_i = \big(\operatorname{col}(X_i T_i^j)_{j=0}^{k_i-1}\big)^{-1},$$

where k_i is the degree of L_i. The invertibility of the matrices appearing in (16.6.2) is ensured by Theorem 2.2.1. The *monic Vandermonde matrix* $V_m(L_1, \ldots, L_r)$ of $L_1, \ldots, L_r$ with index m is defined by

$$V_m(L_1, \ldots, L_r) = \begin{bmatrix} X_1 U_1 & X_2 U_2 & \cdots & X_r U_r \\ X_1 T_1 U_1 & X_2 T_2 U_2 & \cdots & X_r T_r U_r \\ \vdots & \vdots & & \vdots \\ X_1 T_1^{m-1} U_1 & X_2 T_2^{m-1} U_2 & \cdots & X_r T_r^{m-1} U_r \end{bmatrix}.$$

It is easy to see that the products $X_i T_i^p U_i$ do not depend on the choice of the right null pair (X_i, T_i). In fact, we have the following formula for the products $X T^m U_\beta$, where (X, T) is a right null pair of the monic matrix polynomial

$$z^\ell I + \sum_{j=0}^{\ell-1} z^j A_j$$

and

$$[V_1 \cdots V_\ell] = \big(\operatorname{col}(X T^{j-1})_{j=1}^\ell\big)^{-1}:$$

$$(16.6.3)$$
$$\begin{cases} X T^\alpha V_\beta = \delta_{\alpha,\beta-1} I & (0 \le \alpha \le \ell-1; 1 \le \beta \le \ell) \\ X T^\ell V_\beta = -A_{\beta-1} & (1 \le \beta \le \ell) \\ X T^{\ell+\rho} V_\beta = \sum_{k=1}^\rho \left[\sum_{q=1}^k \sum_{\substack{i_1+\cdots+i_q=k \\ i_j>0}} \prod_{j=1}^q (-A_{\ell-i_j}) \right] (-A_{\beta+k-\rho-1}) + (-A_{\beta-\rho-1}) \end{cases}$$

for $\rho \geq 1$ and $1 \leq \beta \leq \ell$. This formula is not difficult to prove by induction on ρ.

We now state the main result on solutions of the interpolation problem (16.6.1) described in terms of the Vandermonde matrix (we denote by $\text{row}(Z_i)_{i=1}^r$ the block row matrix $[Z_1 Z_2 \cdots Z_r]$):

THEOREM 16.6.1. *The interpolation problem (16.6.1) has a solution $A(z)$ of degree $\leq m$ if and only if*

$$(16.6.4) \qquad \text{Ker}\, V_m(L_1, \ldots, L_r) \subset \text{Ker}\, \text{row}([R_{i0}R_{i1} \cdots R_{i,p_i-1}])_{i=1}^r,$$

where R_{ij} are the coefficients of the polynomials R_i:

$$R_i(z) = \sum_{j=0}^{p_i-1} R_{ij} z^j, \qquad 0 \leq j \leq p_i - 1;\ 1 \leq i \leq r.$$

Moreover, if (16.6.4) holds, then a solution $A(z)$ is given by the formula

$$(16.6.5) \qquad A(z) = \text{row}([R_{i0}R_{i1} \cdots R_{i,p_i-1}])_{i=1}^r [V_m(L_1, \ldots, L_r)]^+ \text{col}(z^i I)_{i=0}^m,$$

where $[V_m(L_1, \ldots, L_r)]^+$ is some generalized inverse of $V_m(L_1, \ldots, L_r)$. Furthermore, this solution is unique if and only if the rows of $V_m(L_1, \ldots, L_r)$ are linearly independent.

PROOF. We need a general formula concerning division of matrix polynomials.

Let $L_2(z) = z^k I - X T^k (U_1 + U_2 z + \cdots + U_k z^{k-1})$ be a monic matrix polynomial as in the formula (2.3.1). So (X, T) is a right null pair of $L_1(z)$ corresponding to $\mathbf{C}$, and

$$[\text{col}(X T^i)_{i=0}^{k-1}]^{-1} = [U_1 U_2 \cdots U_k].$$

Let $A(z) = \sum_{j=0}^m z^j A_j$ be an arbitrary matrix polynomial. Then

$$(16.6.6) \qquad A(z) = Q(z) L_1(z) + R(z),$$

where

$$Q(z) = A_{m-k} z^{m-k} + \sum_{q=0}^{m-k-1} \left(\sum_{\alpha=0}^{m-(q+1)} A_{\alpha+q+1} X T^\alpha U_k \right) z^q,$$

$$(16.6.7)$$

$$R(z) = \sum_{j=1}^k \left(\sum_{\beta=0}^m A_\beta X T^\beta U_j \right) z^{j-1}.$$

This formula is simply a rewriting of the formulas (10.1.5)–(10.1.6) applied to a comonic matrix polynomial $L(z) = z^k L_1(z^{-1})$. In other words, if the remainder is $R(z) = \sum_{j=0}^{k-1} \widetilde{R}_j z^j$, then

$$(16.6.8) \qquad [A_0 A_1 \cdots A_m] \text{col}(X T^i U_j)_{i=0}^m = \widetilde{R}_{j-1}, \qquad j = 1, \ldots, k.$$

Applying this formula to the interpolation problem (16.6.1), and using the definition of the Vandermonde matrix, we obtain from (16.6.8) that $A(z) = \sum_{j=0}^m z^j A_j$ satisfies (16.6.1) if and only if

$$
\begin{aligned}
(16.6.9) \quad & [A_0 A_1 \cdots A_m] V_m(L_1, \ldots, L_r) \\
& = [R_{10} R_{11} \cdots R_{1,p_1-1} R_{20} R_{21} \cdots R_{2,p_2-1} \cdots R_{r0} R_{r1} \cdots R_{r,p_r-1}]
\end{aligned}
$$

where $R_i(z) = \sum_{j=0}^{p_i-1} R_{ij} z^j$. All the statements of Theorem 16.6.1 follow now easily from formula (16.6.9). □

Let us illustrate the procedure described in the previous theorem in the scalar case. Consider the following problem of interpolation: Determine a scalar polynomial $q(z)$ of minimal degree such that $q(z_j) = \alpha_j$ $(j = 1, \ldots, s)$, where $z_1, \ldots, z_s$ are different complex numbers and $\alpha_1, \ldots, \alpha_s$ are given numbers. By (16.6.5) the solution is:

$$
q(z) = [\alpha_1 \alpha_2 \cdots \alpha_s]
\begin{bmatrix}
1 & 1 & \cdots & 1 \\
z_1 & z_2 & \cdots & z_s \\
\vdots & \vdots & & \vdots \\
z_1^{s-1} & z_2^{s-1} & \cdots & z_s^{s-1}
\end{bmatrix}^{-1}
\begin{bmatrix}
1 \\
z \\
\vdots \\
z^{s-1}
\end{bmatrix}.
$$

Note that this is just another representation for the Lagrange interpolation polynomial.

Another classical interpolation problem reads as follows: Determine a scalar polynomial $q(z)$ of minimal degree such that

$$
q^{(i)}(z_j) = \alpha_{ji} (i = 0, \ldots, \ell_j - 1; j = 1, \ldots, s).
$$

Here $z_1, \ldots, z_s$ are different complex numbers and α_{ji}, $i = 0, \ldots, \ell_j - 1$; $j = 1, \ldots, s$ are given numbers. To solve this problem one has to find a scalar polynomial $q(z)$ such that the remainder of $q(z)$ after division by $p_j(z) = (z - z_j)^{p_j}$ is equal to

$$
r_j(z) = \sum_{i=0}^{\ell_j - 1} \frac{1}{i!} \alpha_{ji} (z - z_j)^i = \sum_{i=0}^{\ell_j - 1} r_{ji} z^j.
$$

Put $m = \ell_1 + \ell_2 + \cdots + \ell_s$, and let V_m denote the monic Vandermonde matrix $V_m(r_1, \ldots, r_s)$. From Theorem 16.6.2 below we know that V_m is invertible, and from formula (16.6.5) we see that

$$
q(z) = [r_{10} \cdots r_{1,\ell_1-1} r_{20} \cdots r_{2,\ell_2-1} \cdots r_{s0} \cdots r_{s,\ell_s-1}] V_m^{-1} \operatorname{col}(z^i)_{i=0}^{m-1}
$$

is the desired solution. Note that this polynomial is another representation for the Lagrange-Sylvester interpolation polynomial.

In view of Theorem 16.6.1 the invertibility properties of the Vandermonde matrix are important in the context of the interpolation problems. We present a basic result along these lines.

THEOREM 16.6.2. Let $L_1, \ldots, L_r$ be monic $n \times n$ matrix polynomials of degrees $k_1, \ldots, k_r$, respectively, with $\sigma(L_i) \cap \sigma(L_j) = \emptyset$ for $i \neq j$. Then for some m the

Vandermonde matrix $V_m(L_1, \ldots, L_r)$ *is left invertible. If, in addition,* $L_1, \ldots, L_r$ *have a common left multiple which is a monic* $n \times n$ *matrix polynomial of degree* $m = k_1 + \cdots + k_r$, *then* $V_m(L_1, \ldots, L_r)$ *is invertible.*

Let us indicate immediately one corollary: if $L_1, \ldots, L_r$ are monic matrix polynomials with commuting coefficients and disjoint spectra, then $V_m(L_1, \ldots, L_r)$ is invertible, where m is the sum of the degrees of $L_1, \ldots, L_r$.

PROOF. Denoting by (X_i, T_i) the right null pair of L_i (with respect to C), for the first part of the theorem we have to prove that

$$(16.6.10) \qquad \bigcap_{m=0}^{\infty} \mathrm{Ker}[X_1 T_1^m, \ldots, X_r T_r^m] = \{0\}.$$

By arguing as in the proof of Theorem 1.5.1 the equality (16.6.10) follows.

For the second part of Theorem 16.6.2, let (X, T) be the right null pair of a common left multiple $L(z)$ of $L_1(z), \ldots, L_r(z)$ (here $L(z)$ is a monic matrix polynomial of degree $m = k_1 + \cdots + k_r$). By the division Theorem 10.1.1, (X, T) is an extension of each (X_i, T_i) $(i = 1, \ldots, r)$. So there exist T-invariant subspaces $\mathcal{M}_1, \ldots, \mathcal{M}_r$ such that $(X \mid_{\mathcal{M}_i}, T \mid_{\mathcal{M}_i})$ is similar to (X_i, T_i), i.e.

$$(16.6.11) \qquad X_i S_i = X \mid_{\mathcal{M}_i}; \qquad S_i^{-1} T_i S_i = T \mid_{\mathcal{M}_i}$$

for some invertible linear transformation $S_i \colon \mathcal{M}_i \to \mathbf{C}^{nk_i}$). As $\sigma(T_i) \cap \sigma(T_j) = \emptyset$ for $i \neq j$, and consequently $\sigma(T \mid_{\mathcal{M}_i}) \cap \sigma(T \mid_{\mathcal{M}_j}) = \emptyset$ for $i \neq j$, it is easy to see that the sum $\mathcal{M}_1 + \cdots + \mathcal{M}_r$ is in fact a direct sum. A dimension count shows that

$$\mathcal{M}_1 + \cdots + \mathcal{M}_r = \mathbf{C}^{nm}.$$

So actually (X, T) is similar to the pair

$$(16.6.12) \qquad ([X_1 X_2 \cdots X_r], \mathrm{diag}(T_1, T_2, \ldots, T_r)).$$

As $\mathrm{col}(XT^i)_{i=0}^{\ell-1}$ is an invertible matrix, the similarity of (X, T) and (16.6.12) together with the definition of $V_m(L_1, \ldots, L_r)$ finishes the proof. $\square$

COROLLARY 16.6.3. *If* $\sigma(L_i) \cap \sigma(L_j) = \emptyset$ *for* $i \neq j$, *then the interpolation problem* (16.6.1) *has a solution* $A(z)$ *for any choice of* $R_i(z)$.

PROOF. Indeed, from Theorem 16.6.2 it follows that for m large enough $V_m(L_1, \ldots, L_r)$ is left invertible, and hence by Theorem 16.6.1 a solution exists. $\square$

This corollary can be viewed as a generalization to monic matrix polynomials of the following classical fact (Chinese remainder theorem): if $p_1(z), \ldots, p_r(z)$ are pairwise coprime (scalar) polynomials with complex coefficients, then for every set of r polynomials $x_1(z), \ldots, x_r(z)$ there is a polynomial $x(z)$ such that the difference $x(z) - x_i(z)$ is divisible by $p_i(z)$ for $i = 1, \ldots, r$.

In connection with Corollary 16.6.3 the question of finding criteria for the interpolation problem (16.6.1) to have a solution for any choice of R_i arises naturally. A full answer to this question is given by the following result.

THEOREM 16.6.4. *Let $L_1, \ldots, L_r$ be monic matrix polynomials of degrees $p_1, \ldots, p_r$, respectively. The following statements are equivalent:*

(i) *for every set of matrix polynomials $R_1, \ldots, R_r$ of degrees less than $p_1, \ldots, p_r$, respectively, there exists a matrix polynomial A such that*

$$(16.6.13) \qquad A(z) = S_i(z)L_i(z) + R_i(z), \qquad i = 1, \ldots, r$$

for some matrix polynomials $S_1, \ldots, S_r$;

(ii) *for every set of $R_1, \ldots, R_n$ in (i), there exists a matrix polynomial A of degree $\leq n(p_1 + \cdots + p_r)$ satisfying (16.6.13);*

(iii) $\operatorname{Ker} V_\nu(L_1, \ldots, L_r) = \{0\}$ *for some $\nu \geq 1$;*

(iv) $\operatorname{Ker} V_\nu(L_1, \ldots, L_r) = \{0\}$ *for some $\nu \leq n(p_1 + \cdots + p_r)$;*

(v) *for every $z_0 \in \bigcup_{i=1}^r \sigma(L_i)$, the subspaces*

$$\operatorname{Ker} L_1(z_0), \ldots, \operatorname{Ker} L_r(z_0)$$

are linearly independent (i.e., $x_1 + \cdots + x_r = 0$, $x_i \in \operatorname{Ker} L_i(z_0)$, $i = 1, \ldots, r$, implies $x_1 = \cdots = x_r = 0$.)

If the statements (i)–(v) (or one of them) are true, then for any choice of $R_i(z)$ the minimal degree m of a matrix polynomial $A(z)$ for which (16.6.13) holds is equal to

$$\min\{\nu \geq 1 \mid \operatorname{Ker} V_\nu(L_1, \ldots, L_r) = \{0\}\}$$

and one such $A(z)$ with minimal degree is given by formula

$$A(z) = \operatorname{row}([R_{i0} R_{i1} \cdots R_{i,p_i-1}])_{i=1}^r \left(V_m(L_1, \ldots, L_r)\right)^{-L} \operatorname{col}(Z^i I)_{i=0}^m,$$

where $R_i(z) = \sum_{j=0}^{p_i-1} R_{ij} z^j$ and $\left(V_m(L_1, \ldots, L_r)\right)^{-L}$ is some left inverse of $V_m(L_1, \ldots, L_r)$.

The full proof of Theorem 16.6.4 requires some technical notions and results beyond the scope of this book. Because of that, the proof is omitted here, and we refer the interested reader to Gohberg-Kaashoek-Lerer-Rodman [1982], where a full proof can be found.

We conclude this section with a remark that the interpolation problem from the other side

$$(16.6.14) \qquad A(z) = L_i(z)S_i(z) + R_i(z), \qquad i = 1, \ldots, r$$

can be studied by the same methods. Here $L_1(z), \ldots, L_r(z)$ are given monic $n \times n$ matrix polynomials of degrees $p_1, \ldots, p_r$, respectively, and $R_1(z), \ldots, R_r(z)$ are given matrix polynomials of degrees strictly less than $p_1, \ldots, p_r$, respectively. In fact, the interpolation problem (16.6.14) can be reduced to (16.6.13) by taking transposes. A similar remark applies to the interpolation problem with regular divisors (16.7.1) treated in the next section.

16.7 INTERPOLATION OF MATRIX POLYNOMIALS WITH REGULAR DIVISORS

In this section we consider interpolation problems of the type studied in the previous section for the case when the divisors L_i are regular (not necessarily monic) matrix polynomials. We shall assume, without essential loss of generality, that the given polynomials L_i are comonic, i.e. with lower coefficient I.

Here is the precise statement of the interpolation problem to be considered: Given comonic $n \times n$ matrix polynomials $L_1, \ldots, L_r$ of degrees $p_1, \ldots, p_r$, respectively, and given $n \times n$ matrices $R_{i1}, \ldots, R_{ip_i}$, $i = 1, \ldots, r$. Find a (not necessarily regular) $n \times n$ matrix polynomial $A(z)$ of degree $\leq \nu$ such that

$$(16.7.1) \qquad A(z) = S_i(z)L_i(z) + R_i(z), \qquad i = 1, \ldots, r,$$

for some matrix polynomials $S_1(z), \ldots, S_r(z)$ of degree $\leq \nu - p_1, \ldots, \leq \nu - p_r$ respectively, where

$$(16.7.2) \qquad R_i(z) = \sum_{j=\nu-p_i+1}^{\nu} z^j R_{i,j-\nu+p_i}.$$

If all R_{ij} are zeros, then this interpolation problem is just the problem of finding a common left multiple.

The solution to the interpolation problem (16.7.1)–(16.7.2) will be described again using a Vandermonde type matrix, slightly different from the Vandermonde matrix for monic matrix polynomials introduced in the previous section. So, we introduce the notion of *comonic Vandermonde matrix* $\widetilde{V}_\nu(L_1, \ldots, L_r)$ defined by the equality

$$\widetilde{V}_\nu(L_1, \ldots, L_r) = V_\nu(\widetilde{L}_1, \ldots, \widetilde{L}_r),$$

where

$$\widetilde{L}_j(z) = z^{p_j} L_j(z^{-1}),$$

and $V_\nu(\widetilde{L}_1, \ldots, \widetilde{L}_r)$ stands for the Vandermonde matrix of the monic matrix polynomials $\widetilde{L}_1, \ldots, \widetilde{L}_r$ defined in Section 16.6. We use the notation $\widetilde{V}_\nu(L_1, \ldots, L_r)$ to distinguish the notion of comonic Vandermonde matrix from the notion of the Vandermonde matrix (as introduced in Section 16.6). Formulas (16.6.3) show that $\widetilde{V}_\nu(L_1, \ldots, L_r)$ is uniquely defined by the coefficients $L_1, \ldots, L_r$ (and by ν).

It turns out that for the interpolation problem (16.7.1)–(16.7.2) the comonic Vandermonde matrix plays a role similar to the Vandermonde matrix in the preceding section. First observe that the equality (16.7.1) implies the following formula:

$$(16.7.3) \qquad [A_\nu A_{\nu-1} \cdots A_0]\widetilde{V}_\nu(L_1, \ldots, L_r) = \mathrm{row}([R_{ip_i} R_{ip_i-1} \cdots R_{i1}])_{i=1}^r,$$

where $A(z) = \sum_{i=1}^\nu z^i A_i$. One can obtain (16.7.3) from (16.7.1) by passing to the monic polynomials $z^\nu A(z^{-1})$, $z^{\nu-p_i} S_i(z^{-1})$, $z^{p_i} L_i(z^{-1})$ and $z^\nu R_i(z^{-1})$ and applying formula (16.6.9) for these monic polynomials. From (16.7.3) we deduce easily the following result.

THEOREM 16.7.1. *The interpolation problem* (16.7.1)–(16.7.2) *has a solution* $A(z)$ *of degree* $\leq \nu$ *if and only if*

$$\operatorname{Ker} \widetilde{V}_\nu(L_1, \ldots, L_r) \subset \operatorname{Ker} \operatorname{row}([R_{i,p_i}, R_{i,p_i-1}, \ldots, R_{i1}])_{i=1}^r.$$

If this condition holds, a solution $A(z)$ *is given by the formula:*

$$A(z) = \operatorname{row}([R_{ip_i} \cdots R_{i1}])_{i=1}^r [\widetilde{V}_\nu(L_1, \ldots, L_r)]^+ \operatorname{col}(z^{\nu-i})_{i=0}^\nu,$$

where $[\widetilde{V}_\nu(L_1, \ldots, L_r)]^+$ *is a generalized inverse of* $\widetilde{V}_\nu(L_1, \ldots, L_r)$. *This solution is unique if and only if the rows of* $\widetilde{V}_\nu(L_1, \ldots, L_r)$ *are linerly independent.*

Necessary and sufficient conditions for the interpolation problem for comonic matrix polynomials to have a solution for any choice of $R_i(z)$ are given in the next theorem.

THEOREM 16.7.2. *Let* $L_1, \ldots, L_r$ *be comonic matrix polynomials of degrees* $p_1, \ldots, p_r$ *respectively. The following statements are equivalent.*

(i) *for every set of matrix polynomials* $R_1, \ldots, R_r$ *of type* (16.7.2), *there exists a matrix polynomial* A *such that*

$$(16.7.4) \qquad A(z) = S_i(z)L_i(z) + R_i(z), \qquad i = 1, \ldots, r$$

for some polynomials $S_1(z), \ldots, S_r(z)$ *of degrees* $\leq \nu - p_1, \ldots, \leq \nu - p_r$ *respectively;*

(ii) $\operatorname{Ker} \widetilde{V}_\nu(L_1, \ldots, L_r) = \{0\}$ *for some* $\nu \geq 1$;

(iii) *for every* $z_0 \in \bigcup_{i=1}^r \sigma(L_i)$, *the subspaces*

$$\operatorname{Ker} L_1(z_0), \ldots, \operatorname{Ker} L_r(z_0)$$

are linearly independent, and the subspaces

$$\operatorname{Ker} B_1, \ldots, \operatorname{Ker} B_r,$$

are linearly independent as well, where B_i *is the leading coefficient of* L_i, $i = 1, \ldots, r$. *If conditions* (i)–(iii) *hold, then for any choice of* $R_1, \ldots, R_r$ *the minimal degree* m *of a matrix polynomial* A *for which* (16.7.4) *holds is equal to*

$$\min\{\nu \geq 1 \mid \operatorname{Ker} \widetilde{V}_\nu(L_1, \ldots, L_r) = \{0\}\},$$

and one such polynomial A *of minimal degree* m *is given by the formula:*

$$(16.7.5) \qquad A(z) = \operatorname{row}([R_{ip_i} \cdots R_{i1}])_{i=1}^r \left(\widetilde{V}_m(L_1, \ldots, L_r)\right)^{-L} \operatorname{col}(z^{m-i}I)_{i=0}^\nu,$$

where $\left(\widetilde{V}_m(L_1, \ldots, L_r)\right)^{-L}$ *is a left inverse of* $\widetilde{V}_m(L_1, \ldots, L_r)$.

As in Theorem 16.6.4 one can require in (i) and (ii) that $\nu \leq n(p_1 + \cdots + p_r)$.

We omit the proof of Theorem 16.7.2 and mention only that a full proof can be found in Gohberg-Lerer-Rodman [1980] and it is analogous to the proof of Theorem 16.6.4 (found in Gohberg-Kaashoek-Lerer-Rodman [1982]).

The following corollary of Theorem 16.7.2 deserves to be mentioned.

COROLLARY 16.7.3. *Let $L_1, \ldots, L_r$ be comonic matrix polynomials with leading coefficients $B_1, \ldots, B_r$, respectively. Suppose that $\sigma(L_i) \cap \sigma(L_j) = \emptyset$ for $i \neq j$, and $\operatorname{Ker} B_1, \ldots, \operatorname{Ker} B_r$ are linearly independent.*

Then for m sufficiently large, there exists a solution A of the interpolation problem (16.7.1)–(16.7.2), for any choice of R_{ij}. One of the solutions is given by the formula (16.7.5).

16.8 TWO-SIDED LAGRANGE-SYLVESTER INTERPOLATION: INTRODUCTION AND EXAMPLES

In this and in the following sections we consider tangential Lagrange-Sylvester interpolation problems for rational matrix functions, when the interpolation data are given from both the right and left sides simultaneously (in contrast with the previous sections in this chapter, when the interpolation data is given from one side only). To illustrate this distinction and a new point of view adopted here, consider the simple left tangential interpolation problem: For $z_1, \ldots, z_n$ n distinct points in a subset σ of the complex plane $\mathbf{C}$, $x_1, \ldots, x_n$ nonzero row vectors in $\mathbf{C}^{1 \times M}$ and $y_1, \ldots, y_n$ row vectors in $\mathbf{C}^{1 \times N}$, find a rational matrix function $W(z)$ with no poles in σ for which

$$(16.8.1) \qquad\qquad x_i W(z_i) = y_i$$

for $i = 1, \ldots, n$. The analogous right tangential interpolation problem is,

$$(16.8.2) \qquad\qquad W(w_j) u_j = v_j$$

$(j = 1, \ldots, m)$, where $w_1, \ldots, w_m$ are given distinct points in σ, $u_1, \ldots, u_m$ are given nonzero column vectors in $\mathbf{C}^{N \times 1}$ and $v_1, \ldots, v_m$ are given column vectors in $\mathbf{C}^{M \times 1}$. When considering problems of finding rational matrix functions which satisfy both conditions (16.8.1) and (16.8.2), in the case that some point z_i coincides with a point w_j it turns out to be fitting to impose a third type of interpolation condition

$$(16.8.3) \qquad\qquad x_i W'(\xi_{ij}) u_j = \rho_{ij}$$

for all i, j for which $\xi_{ij} = z_i = w_j$, where ρ_{ij} are given complex numbers.

These conditions can be expressed more compactly by using contour integrals as follows. Let

$$A_\zeta = \begin{bmatrix} z_1 & 0 & \cdots & 0 \\ 0 & z_2 & \cdots & 0 \\ \vdots & \vdots & \ddots & \vdots \\ 0 & 0 & \cdots & z_n \end{bmatrix}; \qquad B_+ = \begin{bmatrix} x_1 \\ x_2 \\ \vdots \\ x_n \end{bmatrix}; \qquad B_- = - \begin{bmatrix} y_1 \\ y_2 \\ \vdots \\ y_n \end{bmatrix};$$

$$A_\pi = \begin{bmatrix} w_1 & 0 & \cdots & 0 \\ 0 & w_2 & \cdots & 0 \\ \vdots & \vdots & \ddots & \vdots \\ 0 & 0 & \cdots & w_m \end{bmatrix}; \quad C_- = \begin{bmatrix} u_1 \\ \vdots \\ u_m \end{bmatrix}; \quad C_+ = \begin{bmatrix} v_1 \\ \vdots \\ v_m \end{bmatrix}.$$

Then the conditions (16.8.1) and (16.8.2) can be written in the contour integral form

(16.8.4)
$$\frac{1}{2\pi i} \int_\gamma (zI - A_\zeta)^{-1} B_+ W(z) = -B_-$$

and

(16.8.5)
$$\frac{1}{2\pi i} \int_\gamma W(z) C_- (zI - A_\pi)^{-1} = C_+$$

where γ is a suitable rectifiable contour such that all the points z_j and w_i are inside γ and all poles of $W(z)$ are outside γ. To describe the condition (16.8.3) introduce the $n \times m$ matrix $S = [S_{ij}]_{1 \leq i \leq n; 1 \leq j \leq m}$ defined by

$$S_{ij} = \frac{-x_i v_j + y_i u_j}{z_i - w_j} \quad \text{if} \quad z_i \neq w_j$$

$$S_{ij} = \rho_{ij} \quad \text{if} \quad z_i = w_j.$$

It is easy to see that (16.8.3) is equivalent to

(16.8.6)
$$\frac{1}{2\pi i} \int_\gamma (zI - A_\zeta)^{-1} B_+ W(z) C_- (zI - A_\pi)^{-1} = S.$$

We now turn to the description of the general two-sided Lagrange-Sylvester interpolation problem. As in the example above, this involves simultaneous consideration of a tangential Lagrange-Sylvester interpolation problem for rational matrix functions with the data given on the right, an analogous problem with the left data, and an interpolation condition of the new type (two-sided) each time the two interpolation problems above "intersect". (In the above example, "intersect" means $z_i = w_j$.) So we start our formulation with the two one-sided Lagrange-Sylvester interpolation problems RTLSI (as described in Section 16.1).

Let there be given distinct points $z_1, \ldots, z_r$ in the complex plane. For every z_α two collections of vector polynomials $x_{\alpha 1}(z), \ldots, x_{\alpha k_\alpha}(z)$ of size $1 \times M$ and vector polynomials $y_{\alpha 1}(z), \ldots, y_{\alpha k_\alpha}(z)$ of size $1 \times N$ are given. The first interpolation problem is: Find a rational $M \times N$ matrix function $W(z)$ analytic at each point z_α ($\alpha = 1, \ldots, r$) such that the interpolation conditions

(16.8.7)
$$\frac{d^{i-1}}{dz^{i-1}} \{x_{\alpha j}(z) W(z)\} \Big|_{z=z_\alpha} = \frac{d^{i-1}}{dz^{i-1}} y_{\alpha j}(z) \Big|_{z=z_\alpha},$$
$$i = 1, \ldots, \beta_{\alpha j}; j = 1, \ldots, k_\alpha; \alpha = 1, \ldots, r$$

are satisfied (here $\beta_{\alpha j}$ are given positive integers depending on α and j).

For the second interpolation problem, let there be given distinct points $w_1, \ldots, w_s$ in the complex plane, and for every w_γ two collections of vector polynomials $u_{\gamma 1}(z), \ldots, u_{\gamma \ell_\gamma}(z)$ (of size $N \times 1$) and $v_{\gamma 1}(z), \ldots, v_{\gamma \ell_\gamma}(z)$ (of size $M \times 1$) be given. The problem is then to find a rational $M \times N$ matrix function $W(z)$ without poles in the set $\{w_1, \ldots, w_s\}$ and such that the equalities

$$(16.8.8) \qquad \frac{d^{i-1}}{dz^{i-1}}\{W(z)u_{\gamma j}(z)\}\Big|_{z=w_\gamma} = \frac{d^{i-1}}{dz^{i-1}}v_{\gamma j}(z)\Big|_{z=w_\gamma}$$

hold for $i = 1, \ldots, \delta_{\gamma j}$; $j = 1, \ldots, \ell_\gamma$; $\gamma = 1, \ldots, s$ (here $\delta_{\gamma j}$ are given positive integers depending on γ and j).

Finally, we proceed to the simultaneous consideration of the interpolation problems (16.8.7) and (16.8.8). If some z_α is equal to some w_γ, say $\xi_{\alpha\gamma} = z_\alpha = w_\gamma$ then an additional interpolation condition will be imposed. To state these conditions, we introduce the vector polynomials

$$x_{\alpha j}^{(p)}(z) = \sum_{f=0}^{p-1} \frac{1}{f!}(x_{\alpha j}(z))^{[f]}\Big|_{z=z_\alpha}(z-z_\alpha)^f$$

(i.e., $x_{\alpha j}^{(p)}(z)$ is obtained from $x_{\alpha j}(z)$ by keeping its first p terms in the Taylor expansion around z_α) and analogously

$$u_{\gamma j}^{(p)}(z) = \sum_{f=0}^{p-1} \frac{1}{f!}(u_{\lambda j}(z))^{[f]}\Big|_{z=w_\gamma}(z-w_\gamma)^f.$$

The extra interpolation conditions are

$$(16.8.9) \qquad \frac{1}{(f+g-1)!}\{x_{\alpha i}^{(f)}(z)W(z)u_{\gamma j}^{(g)}(z)\}\Big|_{z=\xi_{\alpha\gamma}} = S_{ij\alpha\gamma}^{(fg)},$$

where $f = 1, \ldots, \beta_{\alpha i}$; $g = 1, \ldots, \delta_{\gamma j}$; $i = 1, \ldots, \ell_\gamma$, and $S_{ij\alpha\gamma}^{(fg)}$ are prescribed numbers. We emphasize that conditions (16.8.9) are imposed for every pair of indices α, γ such that $z_\alpha = w_\gamma$ (and we denote their common value by $\xi_{\alpha\gamma}$).

It turns out that in general the simultaneous interpolation problem (16.8.7), (16.8.8), (16.8.9) does not always have a solution (assuming suitable full range and null kernel conditions, as in Theorems 16.4.1 and 16.4.2), in contrast with the interpolation problem (16.8.7) or with the interpolation problem (16.8.8). The questions concerning existence of a solution $W(z)$ to the simultaneous interpolation problem will be dealt with in the next section. In the next theorem we use the term "consistent" to indicate that the interpolation problem at hand admits a solution $W(z)$ with no poles in σ.

As in the case of one-sided Lagrange-Sylvester interpolation, the conditions (16.8.7)–(16.8.9) can be expressed in the contour integral form, as follows. Recall that by $\text{Res}_{z=z_0} f(z)$ we denote the residue of a rational matrix or vector valued function $f(z)$ at z_0.

THEOREM 16.8.1. *Let a consistent simultaneous interpolation problem* (16.8.7)–(16.8.9) *be given. Let σ be the set of all points z_α and w_γ. Then there exist matrices A_ζ, B_+, B_-, C_+, C_-, A_π, S of sizes $n_\zeta \times n_\zeta$, $n_\zeta \times M$, $n_\zeta \times N$, $M \times n_\pi$, $N \times n_\pi$, $n_\pi \times n_\pi$, $n_\zeta \times n_\pi$ respectively, with $\sigma(A_\zeta) \cup \sigma(A_\pi) \subset \sigma$, such that an $M \times N$ rational matrix function $W(z)$ with no poles in σ satisfies* (16.8.7)–(16.8.9) *if and only if it satisfies conditions* (16.8.10)–(16.8.12):

$$(16.8.10) \qquad \sum_{\alpha=1}^{r} \mathrm{Res}_{z=z_\alpha}(zI - A_\zeta)^{-1}B_+W(z) = -B_-$$

$$(16.8.11) \qquad \sum_{\gamma=1}^{s} \mathrm{Res}_{z=w_\gamma} W(z)C_-(zI - A_\pi)^{-1} = C_+$$

$$(16.8.12) \qquad \sum_{z_0 \in \sigma} \mathrm{Res}_{z=z_0}(zI - A_\zeta)^{-1}B_+W(z)C_-(zI - A_\pi)^{-1} = S.$$

Conversely, given matrices A_ζ, B_+, B_-, C_+, C_-, A_π, S of the sizes as above with $\sigma(A_\zeta) \cup \sigma(A_\pi) \subset \sigma$ and for which the interpolation problem (16.8.10)–(16.8.12) *is consistent, then there exist vector polynomials $x_{\alpha i}(z)$, $y_{\alpha i}(z)$, $u_{\gamma j}(z)$ and $v_{\gamma j}(z)$ of suitable sizes and numbers $S_{ij\alpha\gamma}^{(fg)}$ (prescribed for each pair $z_\alpha = w_\gamma$, where $1 \le f \le \beta_{\alpha i}$, $1 \le g \le \delta_{\gamma j}$, $1 \le i \le k_\alpha$, $1 \le j \le \ell_\gamma$) such that a rational $M \times N$ matrix function $W(z)$ with no poles in σ satisfies* (16.8.10)–(16.8.12), *if and only if $W(z)$ satisfies* (16.8.7)–(16.8.9).

Again, the formulas (16.8.10), (16.8.11), (16.8.12) can be written in the equivalent contour integral form (16.8.13), (16.8.14), (16.8.15), respectively:

$$(16.8.13) \qquad \frac{1}{2\pi i} \int_\gamma (zI - A_\zeta)^{-1}B_+W(z) = -B_-$$

$$(16.8.14) \qquad \frac{1}{2\pi i} \int_\tau W(z)C_-(zI - A_\pi)^{-1} = C_+$$

$$(16.8.15) \qquad \frac{1}{2\pi i} \int_\xi (zI - A_\zeta)^{-1}B_+W(z)C_-(zI - A_\pi)^{-1} = S.$$

Here γ, τ and ξ are suitable simple rectifiable contours such that all the poles of $W(z)$ are outside each contour, and, moreover, the points $z_1, \ldots, z_r$ are inside γ, the points $w_1, \ldots, w_s$ are inside τ and the points $z_1, \ldots, z_r, w_1, \ldots, w_s$ are inside ξ.

PROOF OF THEOREM 16.8.1. We start the proof with the converse statement. Suppose that we are given the matrices A_ζ, B_+, B_-, C_+, C_-, A_π, S as in Theorem

16.8.1. Without loss of generality (by applying similarity transformations to A_ζ and A_π and transforming B_+, B_-, C_+, C_- and S accordingly) we may assume that A_ζ is in the lower triangular Jordan form and A_π is in the upper triangular Jordan form. The proof then easily is reduced to the case when A_ζ and A_π consist of one Jordan block each. Write

$$C_+ = [v_1, \ldots, v_q]; \qquad C_- = [u_1, \ldots, u_q];$$

$$A_\pi = \begin{bmatrix} w_\gamma & 1 & & 0 \\ & w_\gamma & \ddots & \\ & & \ddots & 1 \\ 0 & & & w_\gamma \end{bmatrix}; \qquad A_\zeta = \begin{bmatrix} z_\alpha & & & 0 \\ 1 & z_\alpha & & \\ & \ddots & \ddots & \\ 0 & & 1 & z_\alpha \end{bmatrix};$$

$$B_+ = \mathrm{col}(x_1, \ldots, x_m); \qquad B_- = -\mathrm{col}(y_1, \ldots, y_m)$$

and put

$$v(z) = \sum_{f=1}^{q} v_f (z - w_\gamma)^{f-1}; \qquad u(z) = \sum_{f=1}^{q} u_f (z - w_\gamma)^{f-1};$$

$$x(z) = \sum_{f=1}^{m} x_f (z - z_\alpha)^{f-1}; \qquad y(z) = \sum_{f=1}^{m} y_f (z - z_\alpha)^{f-1};$$

(here A_π is $q \times q$ and A_ζ is $m \times m$). By Theorems 16.2.1 and 16.2.2 we have that for a rational matrix function $W(z)$ without poles at z_α or w_γ the conditions (16.8.10) and (16.8.11) are equivalent to

$$\left. \frac{d^{i-1}}{dz^{i-1}} \{x(z)W(z)\} \right|_{z=z_\alpha} = \left. \frac{d^{i-1}}{dz^{i-1}} y(z) \right|_{z=z_\alpha}, \qquad i = 1, \ldots, m$$

and

$$\left. \frac{d^{i-1}}{dz^{i-1}} \{W(z)u(z)\} \right|_{z=w_\gamma} = \left. \frac{d^{i-1}}{dz^{i-1}} v(z) \right|_{z=w_\gamma}, \qquad i = 1, \ldots, q$$

respectively.

We consider two cases: (1) $w_\gamma \neq z_\alpha$; (2) $w_\gamma = z_\alpha$. In the first case there is no condition of type (16.8.9), so we are done in this case. Assume now $\xi = w_\gamma = z_\alpha$. The (i,j)-th entry of the matrix $T(z) \overset{\text{def}}{=} (zI - A_\zeta)^{-1} B_+ W(z) C_- (zI - A_\pi)^{-1}$ works out to be

$$T_{ij}(z) = \sum_{k=1}^{i} \sum_{\ell=1}^{j} (z - \xi)^{-(i+1-k)} x_k W(z) u_\ell (z - \xi)^{-(j+1-\ell)}$$

$$= \sum_{k=0}^{i} \sum_{\ell=0}^{j} x_k W(z) u_\ell (z - \xi)^{-(i+j+2-k-\ell)}.$$

By assumption $W(z)$ is analytic at ξ. Hence

(16.8.16) $$\mathrm{Res}_{z=\xi}\, T_{ij}(z) = \sum_{k=1}^{i} \sum_{\ell=1}^{j} x_k W_{i+j+1-k-\ell} u_\ell$$

where W_j denotes the j-th Taylor coefficient

$$W_j = \frac{1}{j!} W^{[j]}(\xi)$$

of W at ξ. It it easily seen that

$$\operatorname{Res}_{z=\xi} T_{ij}(z) = \frac{1}{(i+j-1)!} \frac{d^{i+j-1}}{dz^{i+j-1}} \{x^{(i)}(z)W(z)u^{(j)}(z)\}\Big|_{z=\xi}$$

where, as in the discussion preceding Theorem 16.8.1, we denote by $x^{(i)}(z)$ the polynomial obtained from $x(z)$ by keeping the first i terms in the Taylor expansion of $x(z)$ around ξ. So the condition (16.8.9) is verified.

Assume now that the consistent simultaneous interpolation conditions (16.8.7)–(16.8.9) are given, and we have to construct the matrices A_ζ, B_+, B_-, C_+, C_-, A_π, S with the properties required in Theorem 16.8.1. For the simplicity of notation assume $r = 1$, $k_1 = 1$, $s = 1$, $\ell_1 = 1$, and let $m = \beta_{11}$, $\delta = \delta_{11}$. As in the proofs of Theorems 16.2.1 and 16.2.2 define

$$A_\zeta = \begin{bmatrix} z_1 & 0 & \cdots & 0 \\ 1 & z_1 & \cdots & 0 \\ 0 & & & \vdots \\ \vdots & \ddots & \ddots & \\ 0 & \cdots & 0 & \cdots & 1 & z_1 \end{bmatrix}; \quad B_+ = \operatorname{col}(x_1, \ldots, x_m); \quad -B_- = \operatorname{col}(y_1, \ldots, y_m),$$

where

$$x_q = \frac{1}{(q-1)!} (x_{11}(z))^{[q-1]}\Big|_{z=z_1};$$

$$y_q = \frac{1}{(q-1)!} (y_{11}(z))^{[q-1]}\Big|_{z=z_1}$$

for $q = 1, \ldots, m$ (we denote by $(f(z))^{[q]}$ the q-th derivative of the vector valued function $f(z)$). Furthermore, define

$$A_\pi = \begin{bmatrix} w_1 & 1 & 0 & \cdots & 0 \\ 0 & w_1 & 1 & \cdots & 0 \\ & & & & \vdots \\ \vdots & \vdots & & \ddots & 1 \\ 0 & 0 & \cdots & & w_1 \end{bmatrix}; \quad C_+ = [v_1, \ldots, v_\delta]; \quad C_- = [u_1, \ldots, u_\delta],$$

where

$$r_q = \frac{1}{(q-1)!} (r_{11}(z))^{[q-1]}\Big|_{z=w_1}; \quad u_q = \frac{1}{(q-1)!} (u_{11}(z))^{[q-1]}\Big|_{z=w_1}.$$

We have to define S. If $\xi = z_1 = w_1$, define $S = [S_{ij}]_{1 \le i \le m, 1 \le j \le \delta}$ where

$$S_{ij} = \frac{1}{(i+j-1)!} \frac{d^{i+j-1}}{dz^{i+j-1}} \{x_{11}^{(i)}(z)W(z)u_{11}^{(j)}(z)\}\Big|_{z=\xi}.$$

Using the calculation done in the first part of the proof we see that this choice of S satisfies the requirements of Theorem 16.8.1. Assume now $z_1 \neq w_1$, and let

$$S = \text{Res}_{z=z_1}(zI - A_\zeta)^{-1}B_+W(z)C_-(zI - A_\pi)^{-1}$$
$$+ \text{Res}_{z=w_1}(zI - A_\zeta)^{-1}B_+W(z)C_-(zI - A_\pi)^{-1}$$

where $W(z)$ is any rational function analytic at z_1 and w_1 and satisfying the interpolation conditions (16.8.7)–(16.8.9). We compute (where γ is a suitable contour such that z_1 and w_1 are inside γ and all poles of $W(z)$ are outside γ):

$$SA_\pi - A_\zeta S = \frac{1}{2\pi i}\int_\gamma (zI - A_\zeta)^{-1}B_+W(z)C_-(zI - A_\pi)^{-1}A_\pi dz$$

$$- \frac{1}{2\pi i}\int_\gamma A_\zeta(zI - A_\zeta)^{-1}B_+W(z)C_-(zI - A_\pi)^{-1} dz$$

$$= -\frac{1}{2\pi i}\int_\gamma (zI - A_\zeta)^{-1}B_+W(z)C_- dz$$

$$+ \frac{1}{2\pi i}\int_\gamma (zI - A_\zeta)^{-1}B_+W(z)C_- z dz$$

$$+ \frac{1}{2\pi i}\int_\gamma B_+W(z)C_-(zI - A_\pi)^{-1} dz$$

$$- \frac{1}{2\pi i}\int_\gamma z(zI - A_\zeta)^{-1}B_+W(z)C_-(zI - A_\pi)^{-1} dz$$

$$= \left(\frac{-1}{2\pi i}\int_\gamma (zI - A_\zeta)^{-1}B_+W(z)dz\right)C_-$$

$$+ B_+\left(\frac{1}{2\pi i}\int_\gamma W(z)C_-(zI - A_\pi)^{-1} dz\right)$$

$$= B_-C_- + B_+C_+.$$

As $\sigma(A_\pi) \cap \sigma(A_\zeta) = \emptyset$, there is unique S such that

$$SA_\pi - A_\zeta S = B_-C_- + B_+C_+.$$

This choice of S obviously satisfies the requirements of Theorem 16.8.1. $\square$

As we have seen in this section, the general two-sided Lagrange-Sylvester interpolation problem (16.8.7)–(16.8.9) is equivalent to the interpolation problem (16.8.10)–(16.8.12) given in the residue (or contour integral form). It will be more convenient for us to work with the interpolation problem in the latter formulation, and this formulation will be adopted throughout the rest of this chapter.

16.9 THE TWO-SIDED CONTOUR INTEGRAL INTERPOLATION PROBLEM IN THE DIVISOR-REMAINDER FORM

As for the one-sided contour integral interpolation problem, also solutions of the TSCII (two-sided contour integral interpolation) problem admit a characterization in the divisor-remainder form. In this section we obtain such a characterization.

Let us state first formally the TSCII problem (it is essentially the same as the interpolation problem (16.8.10)–(16.8.12) described in the previous section):

Given matrices A_ζ, B_+, B_-, C_+, C_-, A_π, S of sizes $n_\zeta \times n_\zeta$, $n_\zeta \times M$, $n_\zeta \times N$, $M \times n_\pi$, $N \times n_\pi$, $n_\pi \times n_\pi$ and $n_\zeta \times n_\pi$, respectively, and given a set $\sigma \subset \mathbb{C}$ such that $\sigma(A_\zeta) \cup \sigma(A_\pi) \subset \sigma$. Find a rational $M \times N$ matrix function $W(z)$ having no poles in σ and which satisfies the interpolation conditions

$$(16.9.1) \qquad \sum_{z_0 \in \sigma} \mathrm{Res}_{z=z_0}(zI - A_\zeta)^{-1}B_+W(z) = -B_-;$$

$$(16.9.2) \qquad \sum_{z_0 \in \sigma} \mathrm{Res}_{z=z_0} W(z)C_-(zI - A_\pi)^{-1} = C_+;$$

$$(16.9.3) \qquad \sum_{z_0 \in \sigma} \mathrm{Res}_{z=z_0}(zI - A_\zeta)^{-1}B_+W(z)C_-(zI - A_\pi)^{-1} = S.$$

The rest of this chapter will be devoted to the study of the TSCII problem.

The data set for this problem consists of a collection of matrices

$$(16.9.4) \qquad w = (C_+, C_-, A_\pi; A_\zeta, B_+, B_-; S).$$

We shall say that the data set w given by (16.9.4) is a *σ-admissible Sylvester* TSCII *data set if*:

(i) C_+, C_-, A_π *have sizes* $M \times n_\pi$, $N \times n_\pi$ *and* $n_\pi \times n_\pi$ *respectively*, (C_-, A_π) *is a null kernel pair, and* $\sigma(A_\pi) \subset \sigma$;

(ii) A_ζ, B_+, B_- *have sizes* $n_\zeta \times n_\zeta$, $n_\zeta \times M$ *and* $n_\zeta \times N$ *respectively*, (A_ζ, B_+) *is a full range pair, and* $\sigma(A_\zeta) \subset \sigma$

and

(iii) *the matrix S satisfies the Sylvester equation*

$$(16.9.5) \qquad SA_\pi - A_\zeta S = B_+C_+ + B_-C_-.$$

The following lemma shows that there is no loss of generality in the discussion of (TSCII) problems in assuming that the data set is σ-admissible.

LEMMA 16.9.1. *Suppose* $w_e = (C_{+e}, C_{-e}, A_{\pi e}; A_{\zeta e}, B_{+e}, B_{-e}; S_e)$ *is a collection of matrices of sizes* $M \times n_{\pi e}$, $N \times n_{\pi e}$, $n_{\pi e} \times n_{\pi e}$, $n_{\zeta e} \times n_{\zeta e}$, $n_{\zeta e} \times M$, $n_{\zeta e} \times N$

and $n_{\zeta e} \times n_{\pi e}$ respectively, and there exists a rational matrix function $W(z)$ analytic on $\sigma(A_{\pi e}) \cup \sigma(A_{\zeta e})$ satisfying the interpolation conditions

$$(16.9.6) \qquad \sum_{z_0 \in \sigma} \operatorname{Res}_{z=z_0} (zI - A_{\zeta e})^{-1} B_{+e} W(z) = -B_{-e};$$

$$(16.9.7) \qquad \sum_{z_0 \in \sigma} \operatorname{Res}_{z=z_0} W(z) C_{-e} (zI - A_{\pi e})^{-1} = C_{+e};$$

$$(16.9.8) \qquad \sum_{z_0 \in \sigma} \operatorname{Res}_{z=z_0} (zI - A_{\zeta e})^{-1} B_{+e} W(z) C_{-e} (zI - A_{\pi e})^{-1} = S_e;$$

(here $\sigma \in$ C). There exists σ-admissible Sylvester TSCII data set $w = (C_+, C_-, A_\pi; A_\zeta, B_+, B_-; S)$ for which the set of solutions $W(z) \in \mathcal{R}_{M \times N}(\sigma)$ of the TSCII problem (16.9.1)–(16.9.3) corresponding to w is identical to the set of solutions $W(z) \in \mathcal{R}_{M \times N}(\sigma)$ of the TSCII problem (16.9.6)–(16.9.8) corresponding to w_e.

PROOF. Let $\mathcal{X}_\pi$ be a subspace of $\mathbf{C}^{n_{\pi e}}$ which is complementary to $\mathcal{K} := \operatorname{Ker} \operatorname{col}(C_{-e}, \dots, C_{-e} A_{\pi e}^{n_{\pi e}-1})$, let P_π be a projection onto $\mathcal{X}_\pi$ along $\mathcal{K}$, and define

$$C_- = C_{-e} \mid \mathcal{X}_\pi, \qquad A_\pi = P_\pi A_{\pi e} \mid \mathcal{X}_\pi.$$

Then

$$C_{-e}(zI - A_{\pi e})x = 0$$

if $x \in \mathcal{K}$, so a necessary condition for a solution to (16.9.7) to exist in that $C_{+e} \mid \mathcal{K} = 0$. Similarly if a solution to (16.9.8) exists, then necessarily $S_e \mid \mathcal{X}_\pi = 0$. Thus if solutions exist, the class of solutions is not affected by considering $C_-, A_\pi, C_+ = C_{+e} \mid \mathcal{X}_\pi$ and $\tilde{S}_e = S_e \mid \mathcal{X}_\pi$ in place of C_{-e}, $A_{\pi e}$ and S_e; clearly by construction (C_-, A_π) is a null kernel pair.

By a dual analysis we see that there is no loss of generality to consider $A_\zeta = A_{\zeta e} \mid \mathcal{X}_\zeta$, $B_+ = P_\zeta B_{+e}$, $B_- = P_\zeta B_{-e}$ and $S = P_\zeta \tilde{S}_e$ where

$$\mathcal{X}_\zeta = \operatorname{Im} \left[B_{+e}, A_{\zeta e} B_{+e}, \dots, A_{\zeta e}^{n_{\zeta e}-1} B_{+e} \right] \subset \mathbf{C}^{n_{\zeta e}}$$

and P_ζ is a projection of $\mathbf{C}^{n_{\zeta e}}$ onto $\mathcal{X}_\zeta$. Here the compatibility condition (which follows from (16.9.6)) required for solutions to exist at all is that

$$P_\zeta B_{-e} = B_{-e}.$$

Again by construction, (A_ζ, B_+) is a full range pair. Since (16.9.6)–(16.9.8) involve only residues in σ, by further restriction to appropriate spectral invariant subspaces of A_π and A_ζ, it is no loss of generality also to assume that $\sigma(A_\pi) \cup \sigma(A_\zeta) \subset \sigma$.

We thus consider the data set

$$w = (C_+, C_-, A_\pi; A_\zeta, B_+, B_-; S)$$

where (C_-, A_π) is a null kernel pair, (A_ζ, B_+) is a full range pair, and $\sigma(A_\pi) \cup \sigma(A_\zeta) \subset \sigma$. It remains only to show that if solutions of (16.9.1)–(16.9.3) exist, then necessarily S satisfies the Sylvester equation (16.9.5). Thus suppose that $W \in \mathcal{R}_{M \times N}(\sigma)$ satisfies (16.9.1)–(16.9.3). Choose a simple closed contour γ such that W is analytic on and inside γ and such that $\sigma(A_\pi) \cup \sigma(A_\zeta)$ is inside γ. Then (16.9.1)–(16.9.3) can be expressed in contour integral form

$$(16.9.6') \qquad \frac{1}{2\pi i} \int_\gamma (zI - A_\zeta)^{-1} B_+ W(z) dz = -B_-$$

$$(16.9.7') \qquad \frac{1}{2\pi i} \int_\gamma W(z) C_- (zI - A_\pi)^{-1} dz = C_+$$

and

$$(16.9.8') \qquad \frac{1}{2\pi i} \int_\gamma (zI - A_\zeta)^{-1} B_+ W(z) C_- (zI - A_\pi)^{-1} dz = S.$$

Using the expression (16.9.8') for S we have

$$SA_\pi - A_\zeta S = \left[\frac{1}{2\pi i} \int_\gamma (zI - A_\zeta)^{-1} B_+ W(z) C_- (zI - A_\pi)^{-1} dz \right] A_\pi$$

$$- A_\zeta \left[\frac{1}{2\pi i} \int_\gamma (zI - A_\zeta)^{-1} B_+ W(z) C_- (zI - A_\pi)^{-1} dz \right].$$

Bringing A_π and A_ζ inside the integrals and adding and subtracting $z(zI - A_\zeta)^{-1} B_+ W(z) C_- (zI - A_\pi)^{-1}$ inside the integral gives

$$SA_\pi - A_\zeta S = \frac{1}{2\pi i} \int_\gamma (zI - A_\zeta)^{-1} B_+ W(z) C_- (zI - A_\pi)^{-1} (A_\pi - zI) dz$$

$$+ \frac{1}{2\pi i} \int_\gamma (zI - A_\zeta)(zI - A_\zeta)^{-1} B_+ W(z) C_- (zI - A_\pi)^{-1} dz$$

$$= -\frac{1}{2\pi i} \int_\gamma (zI - A_\zeta)^{-1} B_+ W(z) C_- dz$$

$$+ \frac{1}{2\pi i} \int_\gamma B_+ W(z) C_- (zI - A_\pi)^{-1} dz.$$

Now using the interpolation conditions (16.9.6') and (16.9.7') we get that

$$SA_\pi - A_\zeta S = B_+ C_+ + B_- C_-$$

as asserted. □

As a corollary of Lemma 16.9.1 it follows that the interpolation condition (16.9.3) is a consequence of (16.9.1) and (16.9.2) if $\sigma(A_\pi) \cap \sigma(A_\zeta) = \emptyset$. Specifically we have:

THEOREM 16.9.2. *Suppose* $w = (C_+, C_-, A_\pi; A_\zeta, B_+, B_-; S)$ *is a σ-admissible Sylvester TSCII data set such that* $\sigma(A_\pi) \cap \sigma(A_\zeta) = \emptyset$. *Then* $W(z) \in \mathcal{R}_{M+N}(\sigma)$ *satisfies the interpolation condition (16.9.3) automatically if it satisfies (16.9.1) and (16.9.2).*

PROOF. For the case where $\sigma(A_\pi) \cap \sigma(A_\zeta) = \emptyset$, the matrix S is uniquely determined by the condition that is satisfies the Sylvester equation (16.9.5). On the other hand, if $W(z)$ is any solution (16.9.1) and (16.9.2) a computation in the proof of Lemma 16.9.1 shows that then

$$\widetilde{S} := -\frac{1}{2\pi i} \int_\gamma (zI - A_\zeta)^{-1} B_+ W(z) C_- (zI - A_\pi)^{-1} dz$$

also satisfies the Sylvester equation (16.9.5). By uniqueness, we conclude that $S = \widetilde{S}$, i.e., that the interpolation condition (16.9.3) is also satisfied. □

We now state and prove the main result of this section, a description of the solutions to the TSCII problem in the divisor-remainder form.

THEOREM 16.9.3. *Suppose* $w = (C_+, C_-, A_\pi; A_\zeta, B_+, B_-; S)$ *is a σ-admissible Sylvester TSCII data set. Suppose that ψ is a regular rational $M \times M$ matrix function analytic on σ having (A_ζ, B_+) as a left null pair over σ, that φ is a regular rational $N \times N$ matrix function analytic on σ having (C_-, A_π) as a right null pair over σ, and that $K \in \mathcal{R}_{M \times N}(\sigma)$ is any particular solution of the TSCII problem associated with w. Then a matrix function W in $\mathcal{R}_{M \times N}(\sigma)$ is also a solution of the TSCII problem (16.9.1)–(16.9.3) if and only if*

$$W = K + \psi Q \varphi$$

for some $Q \in \mathcal{R}_{M \times N}(\sigma)$.

For the proof we need the following general lemma which will also be useful later.

LEMMA 16.9.4. *Suppose* $w = (C_+, C_-, A_\pi; A_\zeta, B_+, B_-; S)$ *is a σ-admissible Sylvester TSCII data set and $W \in \mathcal{R}_{M \times N}(\sigma)$. Then*

(i) *W satisfies the interpolation condition (16.9.1) if and only if the function*

$$T_1(z) := (zI - A_\zeta)^{-1} [B_+ W(z) + B_-]$$

is analytic on σ;

(ii) *W satisfies the interpolation condition (16.9.2) if and only if the function*

$$T_2(z) := [W(z) C_- - C_+] (zI - A_\pi)^{-1}$$

is analytic on σ;

  (iii) *If* B_-, C_+ *and* S *are all zero then* W *satisfies* (16.9.1)–(16.9.3) *if and only if* $T_1(z)$ *and* $T_2(z)$ *together with*

$$T_3(z) := (zI - A_\zeta)^{-1}B_+W(z)C_-(zI - A_\pi)^{-1}$$

are analytic on σ.

  PROOF. One direction is trivial; if $T_j(z)$ is analytic on σ, then the sum of its residues over σ is zero; this in turn gives (16.9.1), (16.9.2), (16.9.3) respectively for $j = 1, 2, 3$. Conversely for (i), note that W satisfies (16.9.1) if and only if $T_1(z)$ has the sum of its residues over σ equal to zero. From the identity

$$zT_1(z) = [B_+W(z) + B_-] + A_\zeta T_1(z)$$

we see that $zT_1(z)$ (and hence $(z - z_0)T_1(z)$ for any z_0 in σ) has the sum of its residues over σ equal to zero whenever T_1 does, since by assumption W is analytic on σ. By induction we see that for any scalar polynomial $p(z)$ the sum of the residues of $p(z)T_1(z)$ over S is equal to zero. This implies that in fact T_1 is analytic on σ. To prove the converses of (ii) and (iii), apply similar arguments using the identities

$$zT_2(z) = [W(z)C_- - C_+] + T_2(z)A_\pi$$

and (if C_+, B_- and S are all zero)

$$zT_3(z) = B_+T_2(z) + A_\zeta T_3(z)$$

or

$$zT_3(z) = T_1(z)C_- + T_3(z)A_\pi. \quad \square$$

  PROOF OF THEOREM 16.9.3. If W and K are both in $\mathcal{R}_{M\times N}(\sigma)$ and satisfy (16.9.1)–(16.9.3) then $F := W - K \in \mathcal{R}_{M\times N}(\sigma)$ satisfies the homogeneous interpolation conditions

(16.9.9)
$$\sum_{z_0 \in \sigma} \text{Res}_{z=z_0}(zI - A_\zeta)^{-1}B_+F(z) = 0$$

(16.9.10)
$$\sum_{z_0 \in \sigma} \text{Res}_{z=z_0} F(z)C_-(zI - A_\pi)^{-1} = 0$$

and

(16.9.11)
$$\sum_{z_0 \in \sigma} \text{Res}_{z=z_0}(zI - A_\zeta)^{-1}B_+F(z)C_-(zI - A_\pi)^{-1} = 0.$$

Theorem 16.9.3 will follow once we show that $F \in \mathcal{R}_{M\times N}(\sigma)$ satisfies (16.9.9)–(16.9.11) if and only if F has the form $F = \psi Q\varphi$ for a $Q \in \mathcal{R}_{M\times N}(\sigma)$.

By Lemma 16.9.4 we know that an F in $\mathcal{R}_{M \times N}(\sigma)$ satisfies (16.9.9)–(16.9.11) if and only if the three functions

$$(16.9.12) \qquad\qquad T_1(z) = (zI - A_\zeta)^{-1}B_+F(z)$$

$$(16.9.13) \qquad\qquad T_2(z) = F(z)C_-(zI - A_\pi)^{-1}$$

and

$$(16.9.14) \qquad\qquad T_3(z) = (zI - A_\zeta)^{-1}B_+F(z)C_-(zI - A_\pi)^{-1}$$

are analytic on σ.

Recall from Theorem 3.4.2 that since (A_ζ, B_+) is a left null pair for $\psi(z)$ over σ there exists a matrix $\widetilde{C}$ for which the rational matrix function

$$w_\psi(z) = \psi(z)^{-1} - \widetilde{C}(zI - A_\zeta)^{-1}B_+$$

is analytic on σ. Similarly, since (C_-, A_π) is a right null pair for $\varphi(z)$ over σ there is a matrix $\widetilde{B}$ for which

$$w_\varphi(z) = \varphi(z)^{-1} - C_-(zI - A_\pi)^{-1}\widetilde{B}$$

is analytic on σ. Multiplying out we have

$$
\begin{aligned}
Q(z) \stackrel{\text{def}}{=}\ & \psi(z)^{-1}F(z)\varphi(z)^{-1} \\
=\ & [w_\psi(z) + \widetilde{C}(zI - A_\zeta)^{-1}B_+]F(z)[C_-(zI - A_\pi)^{-1}\widetilde{B} + w_\varphi(z)] \\
=\ & w_\psi(z)F(z)C_-(zI - A_\pi)^{-1}\widetilde{B} + \widetilde{C}(zI - A_\zeta)^{-1}B_+F(z)C_-(zI - A_\pi)^{-1}\widetilde{B} \\
& + w_\psi(z)F(z)w_\varphi(z) + \widetilde{C}(zI - A_\zeta)^{-1}B_+F(z)w_\varphi(z) \\
=\ & w_\psi(z)T_2(z)\widetilde{B} + \widetilde{C}T_3(z)\widetilde{B} + w_\psi(z)F(z)w_\varphi(z) + \widetilde{C}T_1(z)w_\varphi(z).
\end{aligned}
$$

Thus if $F \in \mathcal{R}_{M \times N}(\sigma)$ satisfies (16.9.9)–(16.9.11), then as noted above T_1, T_2, T_3 are analytic on σ so $Q = \psi^{-1}F\varphi^{-1}$ is analytic on σ and $F = \psi Q \varphi$ has the desired form. Conversely, suppose that $F = \psi Q \varphi$ with $Q \in \mathcal{R}_{M \times N}(\sigma)$. Since (A_ζ, B_+) is a left null pair for $\psi(z)$ over σ, $(zI - A_\zeta)^{-1}B_+\psi(z)$ is analytic on σ by Theorem 1.5.4. By the dual theorem, since (C_-, A_π) is a right null pair for $\varphi(z)$ over σ, the function $\varphi(z)C_-(zI - A_\pi)^{-1}$ is analytic on σ. Plugging in $F = \psi Q \varphi$, we deduce that T_1, T_2 and T_3 given by (16.9.12)–(16.9.14) are all analytic on σ and hence by Lemma 16.9.4 the equalities (16.9.9)–(16.9.11) are satisfied. This completes the proof. $\square$

16.10 EXISTENCE AND LINEAR FRACTIONAL DESCRIPTION OF SOLUTIONS TO THE TWO-SIDED CONTOUR INTEGRAL INTERPOLATION PROBLEM

We continue to consider the TSCII problem introduced in the previous section.

The description of the solutions to the TSCII problem in the divisor-remainder form (Theorem 16.9.3) was based on the existence of one solution. In particular, it does not follow from this theorem that solutions $W \in \mathcal{R}_{M \times N}(\sigma)$ to the problem (16.9.1)–(16.9.3) exist for any σ-admissible Sylvester TSCII data set w.

The next theorem, which is our main result on the TSCII problem, shows that this is indeed the case. Moreover, it turns out that all the solutions can be described in terms of the linear fractional map defined by a rational matrix function with a given null-pole triple. (For the definition and properties of a null-pole triple see Chapters 4, 12.)

THEOREM 16.10.1. *Let σ be a subset of $\mathbf{C}$ and let $w = (C_+, C_-, A_\pi; A_\zeta, B_+, B_-; S)$ be a σ-admissible Sylvester TSCII data set. There exist rational matrix functions W in $\mathcal{R}_{M \times N}(\sigma)$ which are solutions of the associated TSCII problem (16.9.1)–*

(16.9.3). Moreover if $\theta = \begin{bmatrix} \theta_{11} & \theta_{12} \\ \theta_{21} & \theta_{22} \end{bmatrix}$ is any rational $(M + N) \times (M + N)$ matrix function having the set

$$(16.10.1) \qquad \widetilde{w} = \left(\begin{bmatrix} C_+ \\ C_- \end{bmatrix}, A_\pi; A_\zeta, [B_+, B_-]; S \right)$$

as a (left) null-pole triple over σ and if φ^{-1} is a rational $M \times M$ matrix function having the set

$$(16.10.2) \qquad \widetilde{w}_- = (C_-, A_\pi; 0, 0; 0)$$

as a (left) null-pole triple over σ, then $W \in \mathcal{R}_{M \times N}(\sigma)$ satisfies the TSCII problem (16.9.1)–(16.9.3) if and only if $W(z)$ has the following form: There exist rational matrix function $Q_1 \in \mathcal{R}_{M \times N}(\sigma)$ and $Q_2 \in \mathcal{R}_{N \times N}(\sigma)$ for which the function

$$(16.10.3) \qquad \varphi(\theta_{21} Q_1 + \theta_{22} Q_2)$$

has no zeros and poles in σ, such that

$$(16.10.4) \qquad W = (\theta_{11} Q_1 + \theta_{12} Q_2)(\theta_{21} Q_1 + \theta_{22} Q_2)^{-1}.$$

Observe that the existence of θ and φ is ensured by Theorem 4.6.1.

When $\theta_{21} = 0$, a remark analogous to the observation made after Theorem 16.4.1 applies to the TSCII problem as well. In this case again the linear fractional map in Theorem 16.4.1 assumes a much simpler form because one can choose $Q_2 = I$ and Q_1 any matrix function with no poles in σ. In this case the linear fractional representation for W in Theorem 16.4.1 reduces to the affine form $W = K + \psi Q \varphi$ as in Theorem 16.9.3 with $K = \theta_{12} \theta_{22}^{-1}$, $\psi = \theta_{11}$, $\varphi = \theta_{22}^{-1}$. Conversely, if a particular solution K to the TSCII problem is known (for example by plugging in an appropriate Q_1, Q_2 into the formula (16.10.2)) then one can construct a linear fractional (in fact, affine) representation for W with $\theta_{21} = 0$ by Theorem 16.9.3.

The rest of this section will be devoted to the proof of Theorem 16.10.1.

We first verify that pairs of functions $Q_1(z)$, $Q_2(z)$ as in the statement of the theorem always exist.

LEMMA 16.10.2. *Let* $\theta(z) = \begin{bmatrix} \theta_{11}(z) & \theta_{12}(z) \\ \theta_{21}(z) & \theta_{22}(z) \end{bmatrix}$ *be a rational* $(M+N) \times$ $(M+N)$-*matrix function having the set* $\widetilde{w}$ *given by (16.10.1) as a null-pole triple over* σ. *Let* φ *be a rational* $N \times N$ *matrix function such that* $\widetilde{w}_-$ *given by (16.10.2) is a null-pole triple over* σ *for* φ^{-1}. *Then pairs of rational matrix functions* $(Q_1(z), Q_2(z))$ *analytic on* σ *exist for which the function*

$$\varphi(\theta_{21}Q_1 + \theta_{22}Q_2)$$

has no poles and zeros in σ.

PROOF. We show that there exist $Q_1(z) \in \mathcal{R}_{M \times N}(\sigma)$ and $Q_2(z) \in \mathcal{R}_{N \times N}(\sigma)$ for which

$$\varphi(\theta_{21}Q_1 + \theta_{22}Q_2) = I,$$

that is

$$\varphi^{-1} = \theta_{21}Q_1 + \theta_{22}Q_2.$$

To show that it is sufficient to prove

$$(16.10.5) \qquad \varphi^{-1}\mathcal{R}_N(\sigma) = [\, \theta_{21} \quad \theta_{22} \,]\mathcal{R}_{M+N}(\sigma).$$

Since φ^{-1} has $\widetilde{w}_-$ as a null-pole triple over σ, we know from Theorem 12.3.1 that the associated null-pole subspace over σ is given by

$$(16.10.6) \qquad \varphi^{-1}\mathcal{R}_N(\sigma) = \{C_-(zI - A_\pi)^{-1}x + f : x \in \mathbf{C}^{n_\pi}, f \in \mathcal{R}_N(\sigma)\}.$$

On the other hand we may write

$$[\, \theta_{21} \quad \theta_{22} \,]\mathcal{R}_{M+N}(\sigma) = [0 \; I]\theta\mathcal{R}_{M+N}(\sigma)$$

where the null-pole subspace $\theta\mathcal{R}_{M+N}(\sigma)$ over σ can be computed from its null-pole triple $\widetilde{w}$ over σ as

$$(16.10.7) \qquad \theta\mathcal{R}_{M+N}(\sigma) = \left\{ \begin{bmatrix} C_+ \\ C_- \end{bmatrix}(zI - A_\pi)^{-1}y + \begin{bmatrix} h_1(z) \\ h_2(z) \end{bmatrix} : \right.$$
$$y \in \mathbf{C}^{n_\pi}, h_1 \in \mathcal{R}_M(\sigma), h_2 \in \mathcal{R}_N(\sigma) \text{ are such that}$$
$$\left. \sum_{z_0 \in \sigma} \mathrm{Res}_{z=z_0}(zI - A_\zeta)^{-1}[\, B_+ \quad B_- \,]\begin{bmatrix} h_1(z) \\ h_2(z) \end{bmatrix} = Sy \right\}.$$

From the characterizations (16.10.6) and (16.10.7) it is trivial that $[\, \theta_{21} \quad \theta_{22} \,]\mathcal{R}_{M+N}(\sigma)$ $\subset \varphi^{-1}\mathcal{R}_N(\sigma)$. For the reverse containment, the problem then is to show that for each $k_- \in \varphi^{-1}\mathcal{R}_N(\sigma)$ there exists a $k_+ \in \mathcal{R}_N$ for which $\begin{bmatrix} k_+ \\ k_- \end{bmatrix} \in \theta\mathcal{R}_{M+N}(\sigma)$, i.e., for each $x \in \mathbf{C}^{n_\pi}$ and $f \in \mathcal{R}_N(\sigma)$, we need to find an $h_1 \in \mathcal{R}_M(\sigma)$ for which the associated vector function $\begin{bmatrix} C_+ \\ C_- \end{bmatrix} (zI - A_\pi)^{-1} x + \begin{bmatrix} h_1(z) \\ h_2(z) \end{bmatrix}$ is in $\theta \mathcal{R}_{M+N}(\sigma)$. By the characterization (16.10.7) of $\theta \mathcal{R}_{M+N}(\sigma)$ this is equivalent to $h_1 \in \mathcal{R}_M(\sigma)$ satisfying

$$\sum_{z_0 \in \sigma} \mathrm{Res}_{z=z_0}(zI - A_\zeta)^{-1}B_+h_1(z) = Sx - \sum_{z_0 \in \sigma} \mathrm{Res}_{z=z_0}(zI - A_\zeta)^{-1}B_-f(z).$$

Since (A_ζ, B_+) by assumption is a full range pair, by Lemma 12.2.2 we can solve for h_1. This completes the verification of (16.10.5) as needed. □

　　　We next present a reformulation of the interpolation conditions (16.9.1)–(16.9.3) which will be useful for the proof of Theorem 16.10.1.

　　　THEOREM 16.10.3.　*Let $S_{\widetilde{w}}(\sigma)$ be the subspace associated (as in (12.2.2) with the σ-admissible Sylvester data set (16.10.1). Further, let $S_{\widetilde{w}_-}(\sigma)$ be the subspace associated with the σ-admissible Sylvester data set (16.10.2). Then the rational matrix function $W \in \mathcal{R}_{M \times N}(\sigma)$ satisfies the interpolation conditions (16.9.1)–(16.9.3) if and only if*

$$\begin{bmatrix} W(z) \\ I \end{bmatrix} h_-(z) \in S_{\widetilde{w}}(\sigma)$$

whenever $h_-(z) \in S_{\widetilde{w}_-}(\sigma)$.

　　　The proof relies on the following preliminary lemma.

　　　LEMMA 16.10.4.　*Suppose that A_ζ and A_π are $n_\zeta \times n_\zeta$ and $n_\pi \times n_\pi$ matrices respectively with $\sigma(A_\zeta) \cup \sigma(A_\pi)$ contained in the interior σ of a simple closed contour γ, and let X be a $n_\zeta \times n_\pi$ matrix. Then*

$$(16.10.8) \qquad \frac{1}{2\pi i} \int_\gamma (zI - A_\zeta)^{-1} X (zI - A_\pi)^{-1} dz = 0.$$

　　　PROOF.　Without loss of generality we may suppose that A_ζ and A_π are in Jordan canonical form. In addition we need only show that

$$\frac{1}{2\pi i} \int_\gamma P(zI - A_\zeta)^{-1} X (zI - A_\pi)^{-1} Q dz = 0$$

where P is an arbitrary Riesz projection for A_ζ and Q is an arbitrary Riesz projection for A_π. With this reduction we may assume that $\sigma(A_\zeta)$ consists of a single point $z_0 \in \sigma$ and $\sigma(A_\pi)$ consists of a single point $w_0 \in \sigma$. In general, note that if Y is the left side of (16.10.8) then

$$Y A_\pi - A_\zeta Y = \frac{1}{2\pi i} \int_\gamma (zI - A_\zeta)^{-1} X (zI - A_\pi)^{-1}(A_\pi - zI) dz$$

$$- \frac{1}{2\pi i} \int_\gamma (A_\zeta - zI)(zI - A_\zeta)^{-1} X (zI - A_\pi)^{-1} dz$$

$$= -\frac{1}{2\pi i} \int_\gamma (zI - A_\zeta)^{-1} X dz + \frac{1}{2\pi i} \int_\gamma X (zI - A_\pi)^{-1} dz$$

$$= -X + X = 0$$

by the Riesz functional calculus. Thus, $Y = 0$ if $z_0 \neq w_0$. If $z_0 = w_0$ proceed as follows. Note that

$$(zI - A_\zeta)^{-1} = \sum_{j=1}^{n_\zeta} (z - z_0)^{-j}(A_\zeta - z_0 I)^{j-1}$$

and

$$(zI - A_\pi)^{-1} = \sum_{j=1}^{n_\pi} (z - z_0)^{-j}(A_\pi - z_0 I)^{j-1}$$

and hence the product $(zI - A_\zeta)^{-1}X(zI - A_\pi)^{-1}$ has zero residue at $z_0 = w_0$, so $Y = 0$ in this case as well. $\square$

PROOF OF THEOREM 16.10.3.   By definition the null-pole subspace $\mathcal{S}_{\widetilde{w}_-}(\sigma)$ consists of all rational $\mathbf{C}^{N \times 1}$ vector functions h_- of the form

(16.10.9) $h_-(z) = C_-(zI - A_\pi)^{-1}x + k_-(z)$

where $x \in \mathbf{C}^{n_\pi}$ and $k_- \in \mathcal{R}_N(\sigma)$ are arbitrary. Similarly the null-pole subspace $\mathcal{S}_{\widetilde{w}}(\sigma)$ consists of all rational $\mathbf{C}^{(M+N) \times 1}$-vector functions $f(z)$ of the form

(16.10.10) $f(z) = \begin{bmatrix} C_+ \\ C_- \end{bmatrix}(zI - A_\pi)^{-1}y + \begin{bmatrix} h_1(z) \\ h_2(z) \end{bmatrix}$

where $y \in \mathbf{C}^{n_\pi}$ and $\begin{bmatrix} h_1 \\ h_2 \end{bmatrix} \in \mathcal{R}_{M+N}(\sigma)$ are subject to the restriction

(16.10.11) $\sum_{z_0 \in \sigma} \mathrm{Res}_{z=z_0}(zI - A_\zeta)^{-1}[\, B_+ \quad B_- \,]\begin{bmatrix} h_1(z) \\ h_2(z) \end{bmatrix} = Sy.$

Now suppose that $W \in \mathcal{R}_{M \times N}(\sigma)$ satisfies the interpolation conditions (16.9.1)–(16.9.3), and let h_- as in (16.10.9) be a generic element of $\mathcal{S}_{\widetilde{w}_-}(\sigma)$. By Lemma 16.9.4, the function

$$W(z)C_-(zI - A_\pi)^{-1}x - C_+(zI - A_\pi)^{-1}x$$

has no poles in σ. By assumption $W(z)$ and $k_-(z)$ have no poles in σ, so we conclude that

(16.10.12) $h_1(z) := W(z)\big(C_-(zI - A_\pi)^{-1}x + k_-(z)\big) - C_+(zI - A_\pi)^{-1}x$

has no poles in σ. Note also from (16.10.9) and (16.10.12) that $\begin{bmatrix} W(z) \\ I_N \end{bmatrix} h_-(z)$ has the form

$$\begin{bmatrix} f_1(z) \\ f_2(z) \end{bmatrix} = \begin{bmatrix} W(z) \\ I_N \end{bmatrix} h_-(z) = \begin{bmatrix} C_+ \\ C_- \end{bmatrix}(zI - A_\pi)^{-1}x + \begin{bmatrix} h_1(z) \\ k_-(z) \end{bmatrix}.$$

To check that $\begin{bmatrix} W(z) \\ I_N \end{bmatrix} h_-(z)$ is in $\mathcal{S}_{\widetilde{w}}(\sigma)$ it remains only to check (16.10.11) with

$$\begin{bmatrix} h_1 \\ h_2 \end{bmatrix} = \begin{bmatrix} h_1 \\ k_- \end{bmatrix}$$

and $y = x$. Plugging in the definition (16.10.12) of h_1, we see that (16.10.11) follows once we show that two equalities

$$(16.10.13) \qquad \frac{1}{2\pi i} \int_\gamma (zI - A_\zeta)^{-1} B_+ (W(z)C_- - C_+)(zI - A_\pi)^{-1} x \, dz = Sx$$

and

$$(16.10.14) \qquad \frac{1}{2\pi i} \int_\gamma (zI - A_\zeta)^{-1} (B_+ W(z) + B_-) k_-(z) dz = 0$$

are satisfied for a simple closed contour γ containing the spectrum of A_ζ and A_π in its interior and for which all poles of $W(z)$ are in its exterior. By Lemma 16.10.4 we have

$$\frac{1}{2\pi i} \int_\gamma (zI - A_\zeta)^{-1} B_+ C_+ (zI - A_\pi)^{-1} x \, dz = 0$$

and hence (16.10.13) collapses to the interpolation condition (16.9.3) satisfied by $W(z)$. Also (16.10.14) is an immediate consequence of the interpolation condition (16.9.1). We have thus verified that $\begin{bmatrix} W(z) \\ I \end{bmatrix} h_-(z) \in \mathcal{S}_{\widetilde{w}}(\sigma)$ whenever $h_- \in \mathcal{S}_{\widetilde{w}_-}(\sigma)$ if W satisfies the interpolation conditions (16.9.1)–(16.9.3).

Conversely, suppose the rational $M \times N$-matrix function $W(z)$ has the property that $\begin{bmatrix} W(z) \\ I \end{bmatrix} h_-(z) \in \mathcal{S}_{\widetilde{w}}(\sigma)$ whenever $h_- \in \mathcal{S}_{\widetilde{w}_-}(\sigma)$. We first check that $W \in \mathcal{R}_{M \times N}(\sigma)$ as follows. To see this we must argue that $W\mathcal{R}_N(\sigma) \subset \mathcal{R}_M(\sigma)$. This follows immediately from our hypothesis $\begin{bmatrix} W \\ I \end{bmatrix} \mathcal{S}_{\widetilde{w}_-}(\sigma) \subset \mathcal{S}_{\widetilde{w}}(\sigma)$ and the fact that $\mathcal{R}_N(\sigma) \subset \mathcal{S}_{\widetilde{w}_-}(\sigma)$ if we show that

$$\mathcal{S}_{\widetilde{w}}(\sigma) \cap \begin{bmatrix} \mathcal{R}_M \\ \mathcal{R}_N(\sigma) \end{bmatrix} \subset \mathcal{R}_{M+N}(\sigma).$$

But an element of $\mathcal{S}_{\widetilde{w}}(\sigma)$ by definition has the form

$$g(z) = \begin{bmatrix} C_+ \\ C_- \end{bmatrix} (zI - A_\pi)^{-1} x + \begin{bmatrix} h_1(z) \\ h_2(z) \end{bmatrix}$$

with $x \in \mathbb{C}^{n_\pi}$ and $\begin{bmatrix} h_1 \\ h_2 \end{bmatrix} \in \mathcal{R}_{M+N}(\sigma)$. Such a function g is in $\begin{bmatrix} \mathcal{R}_M \\ \mathcal{R}_N(\sigma) \end{bmatrix}$ if and only if $C_-(zI - A_\pi)^{-1} x = 0$, which by Lemma 12.2.2 happens precisely when $x = 0$. Hence $g = \begin{bmatrix} h_1 \\ h_2 \end{bmatrix} \in \mathcal{R}_{M+N}(\sigma)$ as required.

We next verify the interpolation conditions (16.9.1)–(16.9.3) for $W(z)$. For any $x \in \mathbf{C}^{n_\pi}$ the function $h_-(z) = C_-(zI - A_\pi)^{-1}x$ belongs to $\mathcal{S}_{\widetilde{w}_-}(\sigma)$ and hence the function

$$\begin{bmatrix} W(z) \\ I \end{bmatrix} C_-(zI - A_\pi)^{-1}x$$

is in $\mathcal{S}_{\widetilde{w}}(\sigma)$, and consequently it must have the form

$$\begin{bmatrix} C_+ \\ C_- \end{bmatrix}(zI - A_\pi)^{-1}y + \begin{bmatrix} h_1(z) \\ h_2(z) \end{bmatrix}$$

with $\begin{bmatrix} h_1 \\ h_2 \end{bmatrix} \in \mathcal{R}_{M+N}(\sigma)$. From the equality

$$C_-(zI - A_\pi)^{-1}x = C_-(zI - A_\pi)^{-1}y + h_2(z)$$

and the null-kernel property of (C_-, A_π), we conclude that $y = x$ and $h_2(z) = 0$. Then the equality

$$W(z)C_-(zI - A_\pi)^{-1}x = C_+(zI - A_\pi)^{-1}y + h_1(z)$$

implies that $\big(W(z)C_- - C_+\big)(zI - A_\pi)^{-1}x = h_1(z)$ has no poles in σ. Since x is arbitrary, by Lemma 16.9.4 we conclude that W satisfies (16.9.2). Moreover, in order that

$$\begin{bmatrix} W(z) \\ I \end{bmatrix} C_-(zI - A_\pi)^{-1}x$$

be in $\mathcal{S}_{\widetilde{w}}(\sigma)$ the equality

$$\frac{1}{2\pi i}\int_\gamma (zI - A_\zeta)^{-1}[\; B_+ \quad B_- \;]\begin{bmatrix} W(z) \\ I \end{bmatrix} C_-(zI - A_\pi)^{-1}x\,dz = Sx$$

must be satisfied, where γ is an appropriate contour as in the first part of the proof. Using Lemma 16.10.4 we see that this collapses to the interpolation condition (16.9.3).

Finally choose $h_- \in \mathcal{R}_N(\sigma) \subset \mathcal{S}_{\widetilde{w}_-}(\sigma)$. Since $W(z)$ has no poles in σ, we see that $\begin{bmatrix} W(z) \\ I \end{bmatrix} h_-(z)$ has no poles in σ. The condition that $\begin{bmatrix} W(z) \\ I \end{bmatrix} h_-(z) \in \mathcal{S}_{\widetilde{w}}(\sigma)$ then translates to

$$\frac{1}{2\pi i}\int_\gamma (zI - A_\zeta)^{-1}[\; B_+ \quad B_- \;]\begin{bmatrix} W(z) \\ I \end{bmatrix} h_-(z)\,dz = 0.$$

Since h_- is arbitrary in $\mathcal{R}_N(\sigma)$, we get that W satisfies the interpolation condition (16.9.1). This completes the proof of Theorem 16.10.3. $\square$

We are now ready for the proof of Theorem 16.10.1. Suppose $W \in \mathcal{R}_{M \times N}(\sigma)$ satisfies the interpolation conditions (16.9.1)–(16.9.3). We shall show that there exist $Q_1 \in \mathcal{R}_{M \times N}(\sigma)$ and $Q_2 \in \mathcal{R}_{N \times N}(\sigma)$ for which

(16.10.15)
$$\begin{bmatrix} W \\ I_N \end{bmatrix} \varphi^{-1} = \theta\begin{bmatrix} Q_1 \\ Q_2 \end{bmatrix}.$$

From (16.10.15) it will then follow that

$$W\varphi^{-1} = \theta_{11}Q_1 + \theta_{12}Q_2 \quad \text{and} \quad \varphi^{-1} = \theta_{21}Q_1 + \theta_{22}Q_2,$$

so

(16.10.16) $$W = (\theta_{11}Q_1 + \theta_{12}Q_2)(\theta_{21}Q_1 + \theta_{22}Q_2)^{-1}$$

has the form as described in Theorem 16.10.1.

To verify that (16.10.15) holds for some $Q_1 \in \mathcal{R}_{M \times N}(\sigma)$ and $Q_2 \in \mathcal{R}_{N \times N}(\sigma)$ it is enough to verify that

$$\begin{bmatrix} W \\ I_N \end{bmatrix} \varphi^{-1} \mathcal{R}_N(\sigma) \subset \theta \mathcal{R}_{M+N}(\sigma).$$

But by construction $\varphi^{-1}\mathcal{R}_N(\sigma) = \mathcal{S}_{\widetilde{w}_-}(\sigma)$ and $\theta\mathcal{R}_{M+N}(\sigma) = \mathcal{S}_{\widetilde{w}}(\sigma)$. Hence (16.10.15) is an immediate consequence of Theorem 16.10.3.

Next suppose that $W \in \mathcal{R}_{M \times N}$ has the form (16.10.16), where $\varphi(\theta_{21}Q_1 + \theta_{22}Q_2)$ has no poles and no zeros in σ. Then we may write

$$\begin{bmatrix} W \\ I_N \end{bmatrix} (\theta_{21}Q_1 + \theta_{22}Q_2) = \theta \begin{bmatrix} Q_1 \\ Q_2 \end{bmatrix}.$$

Post multiply by $(\theta_{21}Q_1 + \theta_{22}Q_2)^{-1}\varphi^{-1}$ to rewrite this as

(16.10.17) $$\begin{bmatrix} W \\ I_N \end{bmatrix} \varphi^{-1} = \theta \begin{bmatrix} Q_1 \\ Q_2 \end{bmatrix} (\theta_{21}Q_1 + \theta_{22}Q_2)^{-1}\varphi^{-1}.$$

As $\varphi(\theta_{21}Q_1 + \theta_{22}Q_2)$ has no zeros in σ and $\begin{bmatrix} Q_1 \\ Q_2 \end{bmatrix}$ has no poles in σ, we have

(16.10.18) $$\begin{bmatrix} Q_1 \\ Q_2 \end{bmatrix} (\theta_{21}Q_1 + \theta_{22}Q_2)^{-1}\varphi^{-1} \mathcal{R}_N(\sigma) \subset \mathcal{R}_{M+N}(\sigma).$$

Then (16.10.17) and (16.10.18) combine to give

$$\begin{bmatrix} W \\ I_N \end{bmatrix} \varphi^{-1} \mathcal{R}_N(\sigma) \subset \theta \mathcal{R}_{M+N}(\sigma).$$

Thus, by the other direction in Theorem 16.10.3 we obtain that W belongs to $\mathcal{R}_{M \times N}(\sigma)$ and satisfies the interpolation conditions (16.9.1)–(16.9.3). This completes the proof of the linear fractional characterization of solutions in Theorem 16.10.1. Finally the existence assertion now follows from Lemma 16.10.2.

The proof of Theorem 16.10.1 is complete. □

The following lemma (the proof of which uses the same approach as the proof of Theorem 16.10.1) will be needed later.

LEMMA 16.10.5. *Suppose* $\tau = (C_+, C_-, A_\pi; A_\zeta, B_+, B_-; S)$ *is a* σ-*admissible Sylvester TSCII data set. Suppose* K, ψ, φ, *are rational matrix functions for which*

(i) K *is a solution of the TSCII problem over* σ *associated with the data set* τ.

(ii) $(0, 0; A_\zeta, B_+; 0)$ *is a* σ-*null-pole triple for* ψ.

(iii) $(C_-, A_\pi; 0, 0; 0)$ *is a* σ-*null-pole triple for* φ^{-1}.

Then the function $L(z) = \begin{bmatrix} \psi(z) & K(z)\varphi^{-1}(z) \\ 0 & \varphi^{-1}(z) \end{bmatrix}$ *has* $\left(\begin{bmatrix} C_+ \\ C_- \end{bmatrix}, A_\pi; A_\zeta, [\, B_+ \quad B_- \,]; S \right)$ *as a* σ-*null-pole triple.*

PROOF. By Theorem 12.3.1 we need only show that the null-pole subspace $L \cdot \mathcal{R}_{M+N}(\sigma)$ for L over σ is characterized as the null-pole subspace associated with the data set τ, namely

$$
(16.10.19) \quad L \cdot \mathcal{R}_{M+N}(\sigma) = \left\{ \begin{bmatrix} C_+ \\ C_- \end{bmatrix} (zI - A_\pi)^{-1} x + \begin{bmatrix} h_1(z) \\ h_2(z) \end{bmatrix} : \right.
$$

$$
x \in \mathbf{C}^{n_\pi}, \; h_1 \in \mathcal{R}_M(\sigma), \; h_2 \in \mathcal{R}_N(\sigma) \text{ such that}
$$

$$
\left. \sum_{z_0 \in \sigma} \operatorname{Res}_{z=z_0} (zI - A_\zeta)^{-1} [\, B_+ \quad B_- \,] \begin{bmatrix} h_1(z) \\ h_2(z) \end{bmatrix} = Sx \right\}
$$

We denote by $\mathcal{S}_\tau(\sigma)$ the right-hand side of (16.10.19).

Suppose first that

$$
f = \begin{bmatrix} \psi & K\varphi^{-1} \\ 0 & \varphi^{-1} \end{bmatrix} \begin{bmatrix} k_1 \\ k_2 \end{bmatrix}
$$

is a generic element of $L\mathcal{R}_{M+N}(\sigma)$, where $k_1 \in \mathcal{R}_M(\sigma)$, $k_2 \in \mathcal{R}_N(\sigma)$. Since $(C_-, A_\pi; 0, 0; 0)$ is a σ-null-pole triple for φ^{-1}, by Theorem 12.3.1 we know that

$$
\varphi^{-1}(z)k_2(z) = C_-(zI - A_\pi)^{-1} x + g(z)
$$

for some $x \in \mathbf{C}^{n_\pi}$, $g \in \mathcal{R}_n(\sigma)$. Thus f is the sum of three terms

$$
f = \alpha_1 + \alpha_2 + \alpha_3
$$

where

$$
(16.10.20) \quad \alpha_1(z) = \begin{bmatrix} \psi(z)k_1(z) \\ 0 \end{bmatrix}
$$

$$
(16.10.21) \quad \alpha_2(z) = \begin{bmatrix} K(z) \\ I \end{bmatrix} C_-(zI - A_\pi)^{-1} x
$$

and

(16.10.22)
$$\alpha_3(z) = \left[\begin{array}{c} K(z) \\ I \end{array} \right] g(z).$$

We shall show $f \in \mathcal{S}_\tau(\sigma)$ by showing $\alpha_j \in \mathcal{S}_\tau(\sigma)$ for each $j = 1, 2, 3$.

Since $(0, 0; A_\zeta, B_+; 0)$ is a σ-null-pole triple for $\psi(z)$ by Theorem 1.5.3 the function $(zI - A_\zeta)^{-1}B_+\psi(z)$ is analytic on σ, and hence so also is $(zI - A_\zeta)^{-1}B_+\psi(z)k_1(z)$. Thus

$$\sum_{z_0 \in \sigma} \mathrm{Res}_{z=z_0}(zI - A_\zeta)^{-1}[\, B_+ \quad B_- \,]\alpha_1(z) = 0.$$

This combined with $\alpha_1 \in \mathcal{R}_{M+N}(\sigma)$ implies that $\alpha_1 \in \mathcal{S}_\tau(\sigma)$.

We next check that α_2 given by (16.10.21) is in $\mathcal{S}_\tau(\sigma)$. To do this define $h(z)$ by

(16.10.23)
$$K(z)C_-(zI - A_\pi)^{-1}x = C_+(zI - A_\pi)^{-1}x + h(z).$$

Verification that α_2 is in $\mathcal{S}_\tau(\sigma)$ amounts to checking

(16.10.24)
$$h \in \mathcal{R}_M(\sigma)$$

together with

(16.10.25)
$$\sum_{z_0 \in \sigma} \mathrm{Res}_{z=z_0}(zI - A_\zeta)[\, B_+ \quad B_- \,]\left[\begin{array}{c} h(z) \\ 0 \end{array} \right] = Sx.$$

By Lemma 16.9.4(ii) since K satisfies the interpolation condition (16.9.2), the function

$$T_2(z) = K(z)C_-(zI - A_\pi)^{-1} - C_+(zI - A_\pi)^{-1}$$

is analytic on σ; hence so also is $h(z) = T_2(z)x$, and (16.10.24) follows. Condition (16.10.25) follows from (16.10.23) combined with the interpolation condition (16.9.3) together with the equality

$$\sum_{z_0 \in \sigma} \mathrm{Res}_{z=z_0}(zI - A_\zeta)B_+C_+(zI - A_\pi)^{-1}x = 0.$$

(This equality is ensured by Lemma 16.10.4.) We conclude that $\alpha_2 \in \mathcal{S}_\tau(\sigma)$.

Finally, to show that α_3 given by (16.10.22) is in $\mathcal{S}_\tau(\sigma)$, note from Lemma 16.9.4 that

$$T_1(z) = (zI - A_\zeta)^{-1}B_+K(z) + (zI - A_\zeta)^{-1}B_-$$

is analytic on σ since K satisfies the interpolation condition (16.9.1). Thus so also is

$$(zI - A_\zeta)^{-1}[\, B_+ \quad B_- \,]\alpha_3(z) = T_1(z)g(z)$$

so necessarily

$$\sum_{z_0 \in \sigma} \text{Res}_{z=z_0}(zI - A_\zeta)^{-1}[\ B_+ \quad B_-\ ]\alpha_3(z) = 0.$$

This establishes that $\alpha_3 \in \mathcal{S}_\tau(\sigma)$. We conclude finally that $L\mathcal{R}_{M+N}(\sigma) \subset \mathcal{S}_\tau(\sigma)$.

Conversely, suppose

(16.10.26) $$f = \begin{bmatrix} C_+ \\ C_- \end{bmatrix}(zI - A_\pi)^{-1}x + \begin{bmatrix} h_1(z) \\ h_2(z) \end{bmatrix}$$

is generic element of $\mathcal{S}_\tau(\sigma)$. By Theorem 12.3.1 again (applied to φ^{-1}) we know that there is $x \in \mathbb{C}^{n_\pi}$ and a $k_2 \in \mathcal{R}_N(\sigma)$ for which

(16.10.27) $$C_-(zI - A_\pi)^{-1}x + h_2(z) = \varphi^{-1}(z)k_2(z).$$

If we show that there is a $k_1 \in \mathcal{R}_M(\sigma)$ for which

(16.10.28) $$C_+(zI - A_\pi)^{-1}x + h_1(z) - K(z)\varphi^{-1}(z)k_2(z) = \psi(z)k_1(z),$$

then it follows that

$$f = L\begin{bmatrix} k_1 \\ k_2 \end{bmatrix} \in L\mathcal{R}_{M+N}(\sigma)$$

as required. From (16.10.27) we have

$$C_+(zI - A_\pi)^{-1}x + h_1(z) - K(z)\varphi^{-1}(z)k_2(z)$$
(16.10.29) $$= [C_+(zI - A_\pi)^{-1}x - K(z)C_-(zI - A_\pi)^{-1}x] + [h_1(z) - K(z)h_2(z)]$$
$$= \beta_1(z) + \beta_2(z)$$

where β_1 and β_2 correspond to the terms in brackets. In the proof that $\alpha_2(z)$ given by (16.10.21) is in $\mathcal{S}_\tau(\sigma)$ we established that

(16.10.30) $$\sum_{z_0 \in \sigma} \text{Res}_{z=z_0}(zI - A_\zeta)^{-1}B_+\beta_1(z) = -Sx.$$

Note next that since $K(z)$ satisfies the interpolation condition (16.9.1) we know that $(zI - A_\zeta)^{-1}B_+K(z) + (zI - A_\zeta)^{-1}B_-$ is analytic on σ by Lemma 16.9.4. Hence

$$\sum_{z_0 \in \sigma} \text{Res}_{z=z_0}(zI - A_\zeta)^{-1}B_+\beta_2(z)$$

(16.10.31) $$= \sum_{z_0 \in \sigma} \text{Res}_{z=z_0}[(zI - A_\zeta)^{-1}B_+\beta_2(z) + (zI - A_\zeta)^{-1}B_+K(z)h_2(z)$$
$$\qquad + (zI - A_\zeta)^{-1}B_-h_2(z)]$$
$$= \sum_{z_0 \in \sigma} \text{Res}_{z=z_0}(zI - A_\zeta)^{-1}[\ B_+ \quad B_-\ ]\begin{bmatrix} h_1(z) \\ h_2(z) \end{bmatrix} = Sx$$

where we used that f given by (16.10.26) is in $\mathcal{S}_\tau(\sigma)$ in the last step. Combining (16.10.30) and (16.10.31) gives

$$\sum_{z_0 \in \sigma} \text{Res}_{z=z_0}(zI - A_\zeta)^{-1}\big[B_+\big(\beta_1(z) + \beta_2(z)\big)\big] = 0.$$

By Theorem 12.3.1, since $(0, 0; A_\zeta, B_+; 0)$ is a σ-null-pole triple for $\psi(z)$, this establishes that $\beta_1 + \beta_2 \in \psi\mathcal{R}_M(\sigma)$, and (16.10.28) follows as required. $\square$

16.11 TWO-SIDED CONTOUR INTEGRAL INTERPOLATION PROBLEM WITH PRESCRIBED NUMBER OF POLES

In this section we consider an extension of the TSCII problem where one relaxes the analyticity requirements imposed on solutions; this extension, while of some interest in its own right, also serves as preparation for the type of interpolation problems to be considered in Chapters 19–21.

Let us suppose that we are given a subset σ of C and a σ-admissible Sylvester TSCII data set $\omega = (C_+, C_-, A_\pi; A_\zeta, B_+, B_-; S)$. In Theorem 16.10.1 we have set down a linear fractional description of the set of all rational $M \times N$ matrix functions $W(z)$ which are analytic on σ and which satisfy the associated interpolation conditions (16.9.1)–(16.9.3). In this section we would like to include in the description rational matrix functions F analytic on $\sigma(A_\pi) \cup \sigma(A_\zeta)$ and satisfying the interpolation conditions (16.9.1)–(16.9.3) which may have up to some prescribed number κ poles (counting multiplicities) in σ. The multiplicity of a pole z_0 for a rational matrix function W we take to be the negative of the sum of the negative partial multiplicities in the local Smith-McMillan form for $W(z)$ at z_0. The result, a natural extension of Theorem 16.10.1, is as follows.

THEOREM 16.11.1. *Let σ be a subset of C and let $\omega = (C_+, C_-, A_\pi; A_\zeta, B_+, B_-; S)$ be a σ-admissible Sylvester TSCII data set, and let $\theta = \begin{bmatrix} \theta_{11} & \theta_{12} \\ \theta_{21} & \theta_{22} \end{bmatrix}$ and φ be as in Theorem 16.10.1. Then the set of rational matrix functions W analytic on $\sigma(A_\pi) \cup \sigma(A_\zeta)$ and satisfying the TSCII conditions (16.9.1)–(16.9.3) which have at most κ poles (counting multiplicities) in σ consists exactly of those functions W having a representation as*

$$W = (\theta_{11}Q_1 + \theta_{12}Q_2)(\theta_{21}Q_1 + \theta_{22}Q_2)^{-1}$$

for some rational matrix functions $Q_1 \in \mathcal{R}_{M \times N}(\sigma)$ and $Q_2 \in \mathcal{R}_{N \times N}(\sigma)$ for which the rational matrix function $\varphi(\theta_{21}Q_1 + \theta_{22}Q_2)$ is analytic in σ and $\det(\varphi(\theta_{21}Q_1 + \theta_{22}Q_2))$ has at most κ zeros (counting multiplicities) in σ, none of which occur in $\sigma(A_\pi) \cup \sigma(A_\zeta)$.

A special case is when θ is chosen to have the block upper triangular form $\theta = \begin{bmatrix} \psi & K\varphi^{-1} \\ 0 & \varphi^{-1} \end{bmatrix}$. Then the resulting parametrization is

$$W = K + \psi Q \varphi$$

where Q is a rational $M \times N$ matrix function analytic on $\sigma(A_\pi) \cup \sigma(A_\zeta)$ (equal to the zeros of ψ and φ in σ) and having at most κ poles in σ.

The proof of Theorem 16.11.1 is based on the following generalization of Theorem 16.10.3.

THEOREM 16.11.2. *Let $\sigma \subset$ C, let $\omega = (C_-, C_+, A_\pi; A_\zeta, B_+, B_-; S)$ be a σ-admissible Sylvester TSCII data set, let φ be a regular rational $N \times N$ matrix function having (C_-, A_π) as a right null pair over σ and let $S_{\widetilde{w}}(\sigma)$ be the subspace corresponding to the Sylvester data set*

$$\widetilde{w} = \left(\begin{bmatrix} C_+ \\ C_- \end{bmatrix}, A_\pi; A_\zeta, [\, B_+ \quad B_- \,]; S \right)$$

as in (12.2.2). *Then a rational $M \times N$ matrix function W is analytic on σ, satisfies the interpolation conditions* (16.9.1)–(16.9.3) *and has at most κ poles (counting multiplicities) in σ if and only if there is a rational $N \times N$ matrix function $\Psi(z)$ analytic on σ with $\det \Psi(z)$ having at most κ zeros in σ (none of which are in $\sigma(A_\pi) \cup \sigma(A_\zeta)$) for which*

$$\begin{bmatrix} W(z) \\ I \end{bmatrix} \varphi(z)^{-1} \Psi(z) h_-(z) \in \mathcal{S}_{\widetilde{w}}(\sigma)$$

as long as $h_- \in \mathcal{R}_N(\sigma)$.

PROOF. For brevity let us set $\sigma_0 = \sigma(A_\pi) \cup \sigma(A_\zeta)$. Suppose W is a solution of (16.9.1)–(16.9.3) analytic on σ_0 with at most κ poles in σ. Since W is analytic on σ_0, Theorem 16.10.3 gives that

(16.11.1) $$\begin{bmatrix} W \\ I \end{bmatrix} \varphi^{-1} \mathcal{R}_N(\sigma_0) \subset \mathcal{S}_{\widetilde{w}}(\sigma_0).$$

Since W has at most κ poles in σ we can find a regular rational $N \times N$ matrix function $\Phi(z)$ analytic on σ with $\det \Phi(z)$ having at most κ zeros in σ (but none in σ_0) for which $W\Phi$ is analytic on σ. Thus

(16.11.2) $$\begin{bmatrix} W \\ I \end{bmatrix} \Phi \mathcal{R}_N(\sigma \backslash \sigma_0) \subset \mathcal{R}_N(\sigma \backslash \sigma_0).$$

By the elementary Proposition 12.2.5

$$\mathcal{S}_{\widetilde{w}}(\sigma) = \mathcal{S}_{\widetilde{w}}(\sigma_0) \cap \mathcal{R}_N(\sigma \backslash \sigma_0).$$

Similarly, by Proposition 12.1.3

$$\varphi^{-1} \mathcal{R}_N(\sigma) = \varphi^{-1} \mathcal{R}_N(\sigma_0) \cap \mathcal{R}_N(\sigma \backslash \sigma_0)$$

and since $\det \Phi$ has at most κ zeros in $\sigma \backslash \sigma_0$, it is easy to see that $\Phi \mathcal{R}_N(\sigma \backslash \sigma_0)$ has codimension at most κ in $\mathcal{R}_N(\sigma \backslash \sigma_0)$. We conclude that

$$\widetilde{S} := \varphi^{-1} \mathcal{R}_N(\sigma_0) \cap \Phi \mathcal{R}_N(\sigma \backslash \sigma_0)$$

has codimension at most κ in $\varphi^{-1} \mathcal{R}_N(\sigma)$. Moreover from Theorem 14.1.3 it is easy to see that $\widetilde{S}$ has the form

$$\widetilde{S} = \widetilde{\Psi} \mathcal{R}_N(\sigma)$$

for a regular rational $N \times N$ matrix function $\widetilde{\Psi}$. From

$$\kappa \geq \dim \varphi^{-1} \mathcal{R}_N(\sigma) / \widetilde{\Psi} \mathcal{R}_N(\sigma)$$
$$= \dim \mathcal{R}_N(\sigma) / \varphi \widetilde{\Psi} \mathcal{R}_N(\sigma)$$

we conclude that $\Psi := \varphi \widetilde{\Psi}$ is analytic on σ and $\det \widetilde{\Psi}$ has at most κ zeros in σ. By the construction we also see that none of these zeros can be in σ_0. If we intersect (16.11.1) with (16.11.2) we get

$$\begin{bmatrix} W \\ I \end{bmatrix} \varphi^{-1} \Psi \mathcal{R}_N(\sigma) \subset \mathcal{S}_{\widetilde{w}}(\sigma)$$

as desired.

Conversely, suppose there is a $\Psi(z)$ analytic on σ such that $\det \Psi(z)$ has up to κ zeros in σ and none in σ_0 for which

$$\begin{bmatrix} W \\ I \end{bmatrix} \varphi^{-1}\Psi\mathcal{R}_N(\sigma) \subset S_{\widetilde{w}}(\sigma).$$

Localization to σ_0 gives

$$\begin{bmatrix} W \\ I \end{bmatrix} \varphi^{-1}\Psi\mathcal{R}_N(\sigma_0) \subset S_{\widetilde{w}}(\sigma_0).$$

As $\Psi\mathcal{R}_N(\sigma_0) = \mathcal{R}_N(\sigma_0)$ since Ψ is analytic and invertible on σ_0, we see from Theorem 16.10.3 that W is analytic on σ_0 and satisfies the interpolation conditions (16.9.1)–(16.9.3). Localization to $\sigma \backslash \sigma_0$ gives

$$\begin{bmatrix} W \\ I \end{bmatrix} \varphi^{-1}\Psi\mathcal{R}_N(\sigma \backslash \sigma_0) \subset \mathcal{R}_N(\sigma \backslash \sigma_0).$$

Now $\widetilde{\Psi} := \varphi^{-1}\Psi$ is analytic with at most κ zeros in $\sigma \backslash \sigma_0$ and the above implies that $W\widetilde{\Psi}$ is analytic on $\sigma \backslash \sigma_0$. We conclude that W has at most κ poles in $\sigma \backslash \sigma_0$, as needed. □

We are now ready for the proof of Theorem 16.11.1.

PROOF OF THEOREM 16.11.1. By Theorem 16.11.2 a rational matrix function W is as specified in Theorem 16.11.1 if and only if

(16.11.3) $$\begin{bmatrix} W \\ I \end{bmatrix} \varphi^{-1}\Psi\mathcal{R}_N(\sigma) \subset S_{\widetilde{w}}(\sigma)$$

for some matrix function $\Psi \in \mathcal{R}_{N \times N}(\sigma)$ with $\det \Psi$ having at most κ zeros in σ (none in σ_0). By construction the $2N \times 2N$ matrix function θ is chosen so that

$$S_{\widetilde{w}}(\sigma) = \theta\mathcal{R}_{M+N}(\sigma).$$

Thus, W satisfies (16.11.3) if and only if there exist rational matrix functions $Q_1 \in \mathcal{R}_{M \times N}(\sigma)$ and $Q_2 \in \mathcal{R}_{N \times N}(\sigma)$ for which

$$\begin{bmatrix} W \\ I \end{bmatrix} \varphi^{-1}\Psi = \begin{bmatrix} \theta_{11}Q_1 + \theta_{12}Q_2 \\ \theta_{21}Q_1 + \theta_{22}Q_2 \end{bmatrix}.$$

Solving for W and Ψ gives

$$W = (\theta_{11}Q_1 + \theta_{12}Q_2)(\theta_{21}Q_1 + \theta_{22}Q_2)^{-1}$$

and

$$\Psi = \varphi(\theta_{21}Q_1 + \theta_{22}Q_2).$$

Theorem 16.11.1 now follows. □

CHAPTER 17

PARTIAL REALIZATION AS AN INTERPOLATION PROBLEM

The problem considered in this chapter consists in finding a rational matrix function analytic at infinity of the smallest possible McMillan degree with prescribed values of itself and a few of its derivatives at infinity. A full solution of this problem is obtained and the minimal degree is computed. There is also described here a procedure for finding the solution.

17.1 STATEMENT OF THE PROBLEM AND INTRODUCTION

Let $R(z)$ be $m \times p$ rational matrix function which is analytic at infinity and has value 0 at infinity. Then we have the Laurent series at infinity for $R(z)$:

$$(17.1.1) \qquad\qquad R(z) = \sum_{i=1}^{\infty} M_i z^{-i}.$$

We consider the following interpolation problem (called the *partial realization* problem in the engineering literature): Given matrices $M_1, \ldots, M_r$, find a rational matrix function $R(z)$ of minimal possible McMillan degree for which $M_1, \ldots, M_r$ are the first r coefficients in the Laurent series (17.1.1). Recall (Section 4.1) that the *McMillan degree* of $R(z)$ is defined as the minimal possible size of the matrix A in representations of the form

$$(17.1.2) \qquad\qquad R(z) = C(zI - A)^{-1} B$$

(so C is $m \times q$, A is $q \times q$ and B is $q \times p$).

We say that a system of matrices (A, B, C) is a *realization* of the sequence $M_1, \ldots, M_r$, if for the rational matrix function $R(z)$ given by (17.1.2) the function $\left(R(z) - \sum_{i=1}^{r} M_i z^{-i}\right) z^{r+1}$ is analytic at infinity, or, equivalently, if

$$(17.1.3) \qquad\qquad C A^{i-1} B = M_i, \qquad i = 1, \ldots, r.$$

A realization (A, B, C) of $(M_1, \ldots, M_r)$ is called *minimal* if the size of A minimal among all realizations of $(M_1, \ldots, M_r)$.

It is easy to see that realizations of $M_1, \ldots, M_r$ exist always. For example, let

$$C = [\, I \quad 0 \quad \cdots \quad 0 \,]; \qquad A = \begin{bmatrix} 0 & I & 0 & \cdots & 0 \\ 0 & 0 & I & \cdots & 0 \\ \vdots & \vdots & \vdots & & \vdots \\ 0 & 0 & 0 & \cdots & I \\ 0 & 0 & 0 & \cdots & 0 \end{bmatrix}; \qquad B = \begin{bmatrix} M_1 \\ M_2 \\ \vdots \\ M_r \end{bmatrix}.$$

One checks easily that equalities (17.1.3) are satisfied for this choice of (A, B, C). consequently, minimal realizations of $M_1, \ldots, M_r$ exist as well.

It follows from Theorem 4.1.4 that in the minimal realization (A, B, C) of $(M_1, \ldots, M_r)$ the pair (C, A) is a null kernel pair and the pair (A, B) is a full range pair. The converse is false in general (in other words, not every realization (A, B, C) of $(M_1, \ldots, M_r)$ with null kernel (C, A) and full range (A, B) is minimal). The following simple example illustrates the point.

EXAMPLE 17.1.1. Let $M_1 = 1$, $M_2 = 0$ (complex numbers). Then

$$C = [\, 1 \quad 0 \,]; \qquad A = \begin{bmatrix} 0 & 1 \\ 1 & 0 \end{bmatrix}; \qquad B = \begin{bmatrix} 1 \\ 0 \end{bmatrix}$$

provides a realization for (M_1, M_2) with null kernel pair (C, A) and full range pair (A, B). However this realization is not minimal. One minimal realization for (M_1, M_2) would be $C = 1$, $A = 0$, $B = 1$. $\square$

In the next section we state and prove the main results on characterization of minimal realizations for a given sequence $(M_1, \ldots, M_r)$.

17.2 DESCRIPTION OF MINIMAL PARTIAL REALIZATIONS

We prove here the following two main theorems.

THEOREM 17.2.1. *A realization* $\Sigma = (A, B, C)$ *of a sequence of* $m \times p$ *matrices* $M_1, \ldots, M_r$ *is minimal if and only if the following conditions are satisfied:*

(i) $\mathrm{Ker} \begin{bmatrix} C \\ CA \\ \vdots \\ CA^{r-1} \end{bmatrix} = \{0\};$

(ii) $\mathrm{Im}[\, B \quad AB \quad \cdots \quad A^{r-1}B \,] = \mathbb{C}^q$ *(here* q *is the size of* A*);*

(iii) $\mathrm{Ker} \begin{bmatrix} C \\ CA \\ \vdots \\ CA^{j-1} \end{bmatrix} \subset \mathrm{Im}[\, B \quad AB \quad \cdots \quad A^{r-j-1}B \,]$ *for* $j = 1, \ldots, r-1.$

The second theorem concerns the *degree* of $M_1, \ldots, M_r$ (denoted $\delta(M_1, \ldots, M_r)$), i.e., the smallest possible size of a matrix A in a realization (A, B, C) of $(M_1, \ldots, M_r)$.

THEOREM 17.2.2. *The degree* $\delta(M_1, \ldots, M_r)$ *is given by the formula*

(17.2.1) $$\delta(M_1, \ldots, M_r) = \sum_{i+j=r+1} \mathrm{rank}\, H_{ij} - \sum_{i+j=r} \mathrm{rank}\, H_{ij},$$

where

(17.2.2) $$H_{ij} = \begin{bmatrix} M_1 & \cdots & M_j \\ \vdots & & \vdots \\ M_i & \cdots & M_{i+j-1} \end{bmatrix}, \qquad i+j \le r+1.$$

Proofs of Theorems 17.2.1 and 17.2.2 will be obtained as a consequence of the compression algorithm which we will now describe.

Let $\Sigma = (A, B, C)$ be a realization of a given sequence of $m \times p$ matrices $M_1, \ldots, M_r$. We present three basic operations that "compress" Σ to another realization $\Sigma_0 = (A_0, B_0, C_0)$ of the same $(M_1, \ldots, M_r)$ in such a way that generally the size of A_0 is smaller than the size of A. We need the following subspaces to work with:

$$N_j(\Sigma) = \bigcap_{\nu=0}^{j-1} \mathrm{Ker}(C A^\nu);$$

$$R_j(\Sigma) = \sum_{\nu=0}^{j-1} \mathrm{Im}(A^\nu B);$$

for $j = 1, 2, \ldots$.

PROPOSITION 17.2.3. *Let* $\Sigma = (A, B, C)$ *be a realization of* $(M_1, \ldots, M_r)$, *where* A *is* $q \times q$. *Take* $1 \leq j \leq r - 1$. *Choose a subspace* $\mathcal{X}_1$ *in* $N_j(\Sigma)$ *such that* $\mathcal{X}_1$ *is a direct complement to* $R_{r-j}(\Sigma)$ *in* $R_{r-j}(\Sigma) + N_j(\Sigma)$, *and let* $\mathcal{X}_0$ *be a direct complement to* $\mathcal{X}_1$ *in* $\mathbf{C}^q$ *such that* $\mathcal{X}_0 \supset R_{r-j}(\Sigma)$. *Write the partitionings of* A, B *and* C *relative to the direct sum decomposition* $\mathcal{X}_1 \dot{+} \mathcal{X}_0 = \mathbf{C}^q$:

$$(17.2.3) \qquad A = \begin{bmatrix} A_{11} & A_{10} \\ A_{01} & A_{00} \end{bmatrix}; \qquad B = \begin{bmatrix} B_1 \\ B_0 \end{bmatrix}; \qquad C = [\, C_1 \quad C_0 \,]$$

(e.g., $A_{10} \colon \mathcal{X}_0 \to \mathcal{X}_1$; $B_0 \colon \mathbf{C}^p \to \mathcal{X}_0$*). Then the system* $\Sigma_0 = (A_{00}, B_0, C_0)$ *is a realization of* $(M_1, \ldots, M_r)$ *and*

$$(17.2.4) \qquad N_j(\Sigma_0) \subset R_{r-j}(\Sigma_0).$$

PROOF. We have $R_{r-j}(\Sigma) \subset \mathcal{X}_0$, and consequently $\mathrm{Im}\, B \subset \mathcal{X}_0$ and $B_1 = 0$. By induction on ν one proves

$$(17.2.5) \qquad A^\nu B = \begin{bmatrix} 0 \\ A_{00}^\nu B_0 \end{bmatrix}, \qquad \nu = 0, 1, \ldots, r - j - 1.$$

Next, since $\mathcal{X}_1 \subset \mathrm{Ker}\, C$, we have $C_1 = 0$ and therefore $CA = [C_0 A_{01}, C_0 A_{00}]$. From the inclusion $\mathcal{X}_1 \subset \mathrm{Ker}\, CA$ it follows that $C_0 A_{01} = 0$. In general, again by induction on ν one proves

$$(17.2.6) \qquad CA^\nu = [\, 0 \quad C_0 A_{00}^\nu \,], \qquad \nu = 0, \ldots, j - 1.$$

The equalities (17.2.5) and (17.2.6) imply that

$$C_0 A_{00}^{\nu-1} B_0 = C A^{\nu-1} B = M_\nu, \qquad \nu = 1, \ldots, r - 1,$$

$$C_0 A_{00}^{r-1} B_0 = [\, 0 \quad C_0 A_0^{j-1} \,] \begin{bmatrix} A_{11} & A_{10} \\ A_{01} & A_{00} \end{bmatrix} \begin{bmatrix} 0 \\ A_{00}^{r-j-1} B_0 \end{bmatrix}$$

$$= C A^{j-1} A A^{r-j-1} B = C A^{r-1} B = M_r.$$

Hence $\Sigma_0 = (A_0, B_0, C_0)$ is a realization of $(M_1, \ldots, M_r)$.

It remains to prove the inclusion (17.2.4). Take $x_0 \in N_j(\Sigma_0)$ (in particular, $x_0 \in \mathcal{X}_0$). Then (17.2.6) implies $\begin{bmatrix} 0 \\ x_0 \end{bmatrix} \in N_j(\Sigma)$. But we have

$$N_j(\Sigma) \subset R_{r-j}(\Sigma) \dot{+} \mathcal{X}_1; \qquad R_{r-j}(\Sigma) \subset \mathcal{X}_0.$$

It follows that $\begin{bmatrix} 0 \\ x_0 \end{bmatrix} \in R_{r-j}(\Sigma)$, and (17.2.5) shows that $x_0 \in R_{r-j}(\Sigma_0)$. $\quad\square$

The method of Proposition 17.2.3 allows us to reduce the size of a realization Σ (unless the condition (17.2.4) is satisfied for Σ). We need an analogous reduction for the two extreme cases not covered in Proposition 17.2.3, namely $j = 0$ and $j = r$. It is convenient to treat these cases separately in the two following propositions.

PROPOSITION 17.2.4. *Let $\Sigma = (A, B, C)$ be a realization of $(M_1, \ldots, M_r)$, where $q \times q$ is the size of A. Let $\mathcal{X}_0$ be a direct complement to $N_1(\Sigma)$ in $\mathbf{C}^q$, and let*

$$A = \begin{bmatrix} A_{11} & A_{10} \\ A_{01} & A_{00} \end{bmatrix}; \qquad B = \begin{bmatrix} B_1 \\ B_0 \end{bmatrix}; \qquad C = [\, C_1 \quad C_0 \,]$$

be the partitions of A, B and C with respect to the direct sum decomposition $N_r(\Sigma) \dot{+} \mathcal{X}_0 = \mathbf{C}^q$. Then the system $\Sigma_0 = (A_{00}, B_0, C_0)$ is a realization of $(M_1, \ldots, M_r)$ and

$$(17.2.7) \qquad\qquad N_r(\Sigma_0) = \{0\}.$$

PROOF. Since $N_r(\Sigma) \subset \operatorname{Ker} CA^j$ ($j = 0, 1, \ldots, r - 1$), one sees as in the proof of Proposition 17.2.5 that

$$(17.2.8) \qquad\qquad CA^j = [\, 0 \quad CA_{00}^j \,], \qquad j = 0, 1, \ldots, r - 1.$$

Thus $M_j = CA^{j-1}B = C_0 A_{00}^{j-1} B_0$ for $j = 1, \ldots, r$, and Σ_0 is a realization of $(M_1, \ldots, M_r)$. Now, let $x_0 \in N_r(\Sigma_0)$. Then (17.2.8) yields $CA^j x_0 = C_0 A_{00} x_0 = 0$ for $j = 0, 1, \ldots, r - 1$. Thus $x_0 \in N_r(\Sigma)$. But $N_r(\Sigma) \cap \mathcal{X}_0 = \{0\}$, and hence $x_0 = 0$. $\quad\square$

PROPOSITION 17.2.5. *Let $\Sigma = (A, B, C)$ be a realization of $(M_1, \ldots, M_r)$, where the size of A is $q \times q$. Let $\mathcal{X}_1$ be a direct complement to $R_r(\Sigma)$ in $\mathbf{C}^q$, and partition*

$$A = \begin{bmatrix} A_{11} & A_{10} \\ A_{01} & A_{00} \end{bmatrix}; \qquad B = \begin{bmatrix} B_1 \\ B_0 \end{bmatrix}; \qquad C = [\, C_1 \quad C_0 \,]$$

according to $\mathcal{X}_1 \dot{+} R_r(\Sigma) = \mathbf{C}^q$. Then the system $\Sigma_0 = (A_{00}, B_0, C_0)$ is a realization of $(M_1, \ldots, M_r)$ and

$$(17.2.9) \qquad\qquad R_r(\Sigma_0) = R_r(\Sigma).$$

PROOF. As in the proof of Proposition 17.2.3 we see that because of the inclusions $\operatorname{Im} A^j B \subset R_r(\Sigma)$ ($j = 0, \ldots, r - 1$), the equalities

$$(17.2.10) \qquad\qquad A^j B = \begin{bmatrix} 0 \\ A_{00}^j B_0 \end{bmatrix}, \qquad j = 0, 1, \ldots, r - 1$$

hold true. It follows that $C_0 A_{00}^{j-1} B_0 = C A^{j-1} B = M_j$ for $j = 1,\ldots,r$. So Σ_0 is a realization of $(M_1,\ldots,M_r)$. From $R_r(\Sigma_0) \subset R_r(\Sigma)$ and (17.2.10) we conclude that

$$R_r(\Sigma) = \operatorname{Im} \begin{bmatrix} 0 & 0 & \cdots & 0 \\ B_0 & A_{00} B_0 & \cdots & A_{00}^{r-1} B_0 \end{bmatrix},$$

which proves (17.2.9). □

We are ready now for the proofs of Theorems 17.2.1 and 17.2.2. Let $M_1,\ldots,M_r$ be a given sequence of $m \times p$ matrices. A repeated application of the operations on realizations described in Propositions 17.2.3–17.2.5 shows that any realization $\Sigma = (A,B,C)$ of $(M_1,\ldots,M_r)$ can be compressed in a finite number of steps to a realization $\Sigma_0 = (A_0,B_0,C_0)$ of $(M_1,\ldots,M_r)$ such that the size $q \times q$ of A_0 is smaller than or equal to the size of A, and Σ_0 has the following additional properties:

(a) $N_r(\Sigma_0) = \{0\}$;

(b) $R_r(\Sigma_0) = \mathbf{C}^q$;

(c) $N_j(\Sigma_0) \subset R_{r-j}(\Sigma_0)$; $j = 1,\ldots,r-1$.

In particular, the conditions (a)–(c) hold true for any minimal realization Σ_0 of $(M_1,\ldots,M_r)$.

Now let H_{ij} be the block Hankel matrix defined by (17.2.2). Take a realization $\Sigma = (A,B,C)$ of $(M_1,\ldots,M_r)$. Then

$$H_{ij} = \begin{bmatrix} C \\ CA \\ \vdots \\ CA^{i-1} \end{bmatrix} [\,B \quad AB \quad \cdots \quad A^{j-1}B\,], \qquad i+j \le r+1.$$

It follows that

(17.2.11) $\operatorname{rank} H_{ij} = \dim\{R_j(\Sigma) + N_i(\Sigma)\} - \dim N_i(\Sigma).$

Let ρ be the number defined by the right hand side of (17.2.1):

$$\rho = \sum_{i+j=r+1} \operatorname{rank} H_{ij} - \sum_{i+j=r} \operatorname{rank} H_{ij}.$$

Using (17.2.11) we check that

(17.2.12) $\rho = \dim\{R_r(\Sigma) + N_1(\Sigma)\} - \dim N_r(\Sigma) - \sum_{i=1}^{r-1} \dim \dfrac{R_{r-i}(\Sigma) + N_i(\Sigma)}{R_{r-i}(\Sigma) + N_{i+1}(\Sigma)}.$

So

$$\rho \le \dim\{R_r(\Sigma) + N_1(\Sigma)\} \le q,$$

where $q \times q$ is the size of A. Since Σ is an arbitrary realization of $(M_1,\ldots,M_r)$ we conclude that $\rho \le \delta(M_1,\ldots,M_r)$.

According to the first paragraph after Proposition 17.2.5 we may assume that Σ satisfies the conditions (a)–(c). But then (17.2.12) reduces to $\rho = q$. Hence ρ is equal to the smallest possible size of the matrix A in a realization $\Sigma = (A, B, C)$ of $(M_1, \ldots, M_r)$, which proves Theorem 17.2.2. Furthermore, we see that Σ is minimal whenever conditions (a)–(c) are satisfied. We already know that conditions (a)–(c) are necessary for minimality. Thus Theorem 17.2.1 is also proved. $\square$

We remark also that by a repeated application of the operations described in Propositions 17.2.3–17.2.5 any realization of $M_1, \ldots, M_r$ may be compressed to a minimal realization.

We have seen in this chapter that minimal realizations of a given sequence $(M_1, \ldots, M_r)$ behave in many aspects as minimal realizations of rational matrix functions studied in Chapter 4. For example, minimality is characterized by conditions (i)–(iii) which are of the same type as full-range and null-kernel conditions. Also, from any given realization one can obtain a minimal one by a suitable compression procedure. There is, however, an important difference: unlike minimal realizations for a rational matrix function, the minimal realizations for a sequence $(M_1, \ldots, M_r)$ need not be unique up to similarity. Similarity is understood here in a natural sense: realizations $\Sigma = (A, B, C)$ and $\Sigma' = (A', B', C')$ are called *similar* if

$$C' = CS, \qquad A' = S^{-1}AS, \qquad B' = S^{-1}B$$

for some invertible matrix S. Clearly, if $\Sigma = (A, B, C)$ is a realization of $(M_1, \ldots, M_r)$, then every Σ' similar to S is also a realization of the same $(M_1, \ldots, M_r)$. Also, if Σ is minimal, then so is any realization similar to Σ. However, there may be in general several minimal realizations of $(M_1, \ldots, M_r)$ not similar to each other. Without going into details, we just note the following necessary and sufficient condition of the uniqueness of a minimal realization of $(M_1, \ldots, M_r)$ up to similarity:

(17.2.13)
$$\min\{i \mid \operatorname{rank} H_{i,r-i} = \operatorname{rank} H_{i+1,r-1}; 1 \le i \le r-1\}$$
$$+ \min\{j \mid \operatorname{rank} H_{r-j,j} = \operatorname{rank} H_{r-j,j+1}; 1 \le j \le r-1\} \le r,$$

where H_{ij} are the block Hankel matrices defined by (17.2.2). The general problem of describing all the minimal realizations of a given sequence $(M_1, \ldots, M_r)$ remains open.

17.3 EXAMPLE

We conclude this chapter with an example illustrating the ideas and results of the previous section.

Consider the following 1×2 matrices:

$$M_1 = [\, 0 \quad 0 \,], \qquad M_2 = [\, 0 \quad 1 \,],$$
$$M_3 = [\, 0 \quad 2 \,] \quad \text{and} \quad M_4 = [\, 1 \quad 4 \,].$$

A realization of the sequence M_1, M_2, M_3, M_4 is provided by the matrices

$$(17.3.1) \qquad A_1 = \begin{bmatrix} 0 & 1 & 0 & 0 \\ 0 & 0 & 1 & 0 \\ 0 & 0 & 0 & 1 \\ 1 & 0 & 0 & 2 \end{bmatrix}, \qquad B_1 = \begin{bmatrix} 0 & 1 \\ 0 & 0 \\ 1 & 0 \\ 0 & 0 \end{bmatrix}, \qquad C_1 = [\, 0 \quad 0 \quad 0 \quad 1 \,].$$

In fact, since $\delta(M_1, M_2, M_3, M_4)$ (defined by (17.2.1)) is equal to 4, the realization (17.3.1) is minimal in view of Theorem 17.2.2. The inequality (17.2.13) is not satisfied (there is no i between 1 and $r - 1$ for which rank $H_{i,r-i} = $ rank $H_{i+1,r-i}$), so according to the general result, the minimal realization of (M_1, M_2, M_3, M_4) is not unique up to similarity. An example of a minimal realization of (M_1, M_2, M_3, M_4) which is not similar to (17.3.1) is given by

$$C_2 = [\, 1 \quad 0 \quad 0 \quad 0 \,]; \qquad A_2 = \begin{bmatrix} 0 & 1 & 0 & 0 \\ 0 & 0 & 1 & 0 \\ 0 & 0 & 0 & 1 \\ 0 & 0 & 0 & 0 \end{bmatrix}, \qquad B_2 = \begin{bmatrix} 0 & 0 \\ 0 & 1 \\ 0 & 2 \\ 1 & 4 \end{bmatrix}.$$

On the other hand,

$$\delta(M_1, M_2, M_3) = 2,$$

and hence (A_1, B_1, C_1) is a realization of (M_1, M_2, M_3) but not a minimal one. We shall use the compression algorithm described in the previous section to produce a minimal realization starting with (A_1, B_1, C_1).

From Theorem 17.2.1 we know that a realization (A, B, C) of (M_1, M_2, M_3) is a minimal realization if and only if the following four conditions hold true:

(i) Ker $\begin{bmatrix} C \\ CA \\ CA^2 \end{bmatrix} = \{0\}$,

(ii) Ker $\begin{bmatrix} C \\ CA \end{bmatrix} \subset \mathrm{Im}\, B$,

(iii) Ker $C \subset \mathrm{Im}[\, B \quad AB \,]$,

(iv) Im$[\, B \quad AB \quad A^2 B \,] = \mathbf{C}^q$,

where q is the size of A. It is easy to see that for (A_1, B_1, C_1) given by (17.3.1) the conditions (iii) and (iv) are fulfilled but (i) and (ii) are not. Actually, we have (where e_1, e_2 and e_3 are the first three vectors in the standard basis in $\mathbf{C}^4$):

(17.3.2) $\qquad$ Ker $\begin{bmatrix} C_1 \\ C_1 A_1 \\ C_1 A_1^2 \end{bmatrix} = \mathrm{Span}\{e_3\} \neq \{0\}$,

(17.3.3) $\qquad$ Ker $\begin{bmatrix} C_1 \\ C_1 A_1 \end{bmatrix} = \mathrm{Span}\{e_2, e_3\} \not\subset \mathrm{Im}\, B_1 = \mathrm{Span}\{e_1, e_3\}$.

It follows from (17.3.2) that we can reduce the size of A_1 using the procedure described in Proposition 17.2.4. This yields the following new realization of M_1, M_2, M_3:

(17.3.4) $\qquad A_2 = \begin{bmatrix} 0 & 1 & 0 \\ 0 & 0 & 0 \\ 1 & 0 & 2 \end{bmatrix}, \qquad B_2 = \begin{bmatrix} 0 & 1 \\ 0 & 0 \\ 0 & 0 \end{bmatrix}, \qquad C_2 = [\, 0 \quad 0 \quad 1 \,].$

The realization (A_2, B_2, C_2) is still not a minimal realization of M_1, M_2, M_3, because the size of A_2 is three and $\delta(M_1, M_2, M_3) = 2$. Note that for the system (A_2, B_2, C_2) condition (i) holds true, but the other three conditions which we need for minimality are violated. So after the first compression the conditions (iii) and (iv), which were fulfilled by the first realization (A_1, B_1, C_1), do not hold any more.

We continue now the compression procedure and apply Proposition 17.2.5 to (A_2, B_2, C_2). Note that

$$\text{Im}[\ B_2 \quad A_2 B_2 \quad A_2^2 B_2\ ] = \text{Span} \left\{ \begin{bmatrix} 1 \\ 0 \\ 0 \end{bmatrix}, \begin{bmatrix} 0 \\ 0 \\ 1 \end{bmatrix} \right\},$$

and hence we can reduce the size of A_2. This yields the following system:

$$A_3 = \begin{bmatrix} 0 & 1 \\ 1 & 2 \end{bmatrix}, \qquad B_3 = \begin{bmatrix} 0 & 1 \\ 0 & 0 \end{bmatrix}, \qquad C_3 = [\ 0 \quad 1\ ],$$

which is a minimal realization of M_1, M_2, M_3.

NOTES FOR PART IV

CHAPTER 16   The whole material of this chapter with the exception of Sections 16.6, 16.7 and 16.11 is taken from Ball-Gohberg-Rodman [1989d] (see also Ball-Gohberg-Rodman [1989c]). The material in Sections 16.6 and 16.7 is from Gohberg-Kaashoek-Rodman [1978a] (see also Gohberg-Kaashoek-Lerer-Rodman [1982] and the monograph Gohberg-Lancaster-Rodman [1982]).

Interpolation problems for operator polynomials (i.e., polynomials whose co-efficients are linear bounded operators in a Banach rather than matrices) analogous to those considered in Sections 16.6 and 16.7 were studied in Gohberg-Kaashoek-Rodman [1978b]. A systematic exposition of this and other interpolation problems for operator polynomials is found in Rodman [1989]. The use of Vandermonde matrices (with operator entries) in the study of operator polynomials was initiated in Markus-Mereutsa [1973].

Simultaneous two-sided contour integral interpolation (the two-sided ana-logue of the problem considered in Section 16.5) was studied in Ball-Gohberg-Rodman [1989f]. The solution of this problem is based on the notion of common minimal multiples of rational matrix functions developed in Ball-Gohberg-Rodman [1989e].

CHAPTER 17   The whole problem of partial realization which is analyzed in Chapter 17 was introduced by Kalman (Kalman [1971], Kalman-Falb-Arbib [1969]) for the case when the matrices B and C are a column and a row, respectively (i.e., $R(z)$ is a scalar rational function). Since then, the partial realization was studied in many papers. Here we present the results of Gohberg-Kaashoek-Lerer [1987]; we essentially follow this exposition.

We mention briefly the main research avenues in studying partial realizations. Recursive algorithms for computing partial realizations are given in Anderson-Brasch-Lopresti [1975] and in Dickinson-Morf-Kailath [1974]. An important feature of these algorithms is the step-by-step approach, so that a computation of new partial realiza-tions with one additional coefficient M_i given does not necessitate repeating the whole computation anew. At each step a system of linear equations with block Hankel matrix has to be solved. A transparent exposition of these algorithms, together with a canonical choice of a partial realization at each step is given in van Barel-Bultheel [1989]. Deep con-nections between the partial realization problem on the one hand, and linear fractional representations, continued fractions and the Euclidean algorithm are investigated, start-ing with the pioneering paper by Kalman [1979], in Fuhrmann [1983], Antoulas [1986], and (for the scalar case only) in Gragg-Lindquist [1983]. Several papers are devoted to various parametrizations of solutions of the partial realization problem. In Anderson-Brasch-Lopresti [1975] a parametrization is obtained as a by-product of the algorithm developed there. Another parametrization of minimal partial realizations in terms of the input-output canonical form is given in Bosgra [1983]. Canonical forms and invariants of minimal partial realizations were studied also in Bistritz [1983].

We note also that the partial realization problem also appears (after change in the variable $z \to z^{-1}$) as the Padé, or rational, approximation problem. The latter problem is: Given a formal power series $\sum_{i=0}^{\infty} M_i z^i$ with matrix coefficients M_i, find a rational matrix function in the coprime matrix fraction form whose Taylor series coincides with $\sum_{i=0}^{\infty} M_i z^i$ (up to a fixed number of terms). There is rich literature on the Padé approximation, especially in the scalar case; we mention the monographs Baker-Graves-Morris [1981], Gilewicz [1978] that contain classical material on Padé approximants, and the monograph Bultheel [1987] that contains some of the recent developments. See also a review of some applications of Padé approximation in systems theory (Bultheel [1983]).

More generally one can consider the problem of classifying the possible McMillan degrees (including a formula for the minimal possible McMillan degree) for rational (or polynomial) matrix functions solving some set of tangential Lagrange-Sylvester interpolation conditions. One of the earliest papers on this problem is Belevitch [1970]. In this context the Hankel matrix is replaced by the Loewner matrix of divided differences which turns out often to be identical with our coupling matrix S in the two-sided contour integral formulation of the problem. For a generic case of full-matrix-value interpolation, Anderson-Antoulas [1989] obtained a realization formula for minimal degree interpolants (scalar version of this problem was studied in a previous paper Antoulas-Anderson [1986]). Antoulas-Ball-Kang-Willems [1989] established connections between the admissible McMillan degrees and reachability and controllability indices of matrix pairs constructed from the data for the bitangential interpolation problem. Ball-Kang [1989] obtained a realization formula for a matrix polynomial of McMillan degree at most $n-1$ which satisfies n simple left tangential interpolation conditions, as well as extensions of this result to more general two-sided contour integral interpolation problems.

In this context we mention the following important and still unsolved problem introduced by Kalman [1979, 1982] for the scalar case. Given $M_1, \ldots, M_k$ square size matrices, find when there is a rational matrix function $W(z)$ such that its Taylor series at infinity starts with $M_1 + M_2 z^{-1} + \cdots + M_k z^{-k+1}$ and $W(z)$ is positive definite on the imaginary line, and describe all such functions (when they exist) of minimal McMillan degree.

PART V

NONHOMOGENEOUS INTERPOLATION PROBLEMS
WITH METRIC CONSTRAINTS

The main new common feature of the nonhomogeneous interpolation problem in this part is the imposition of an additional metric constraint. Most of the attention is paid to the various types of Nevanlinna-Pick interpolation and its generalizations. We consider the simplest Nevanlinna-Pick problem as well as its most sophisticated form, the Nudelman form, where the interpolation conditions are given in terms of contour integrals. Separately is considered the problem where a wider class of solutions is considered; in this version, solutions are allowed to have a prescribed number of poles in the underlying region (this we call the Takagi-Nudelman problem). Special attention is paid to the case where interpolation conditions are given on the boundary of the underlying domain (sometimes called Loewner interpolation problem). The remaining problems considered in this part are those of Nehari, Nehari-Takagi and Caratheodory-Toeplitz. All problems are treated systematically using the tools and methods developed in preceding parts of this book for homogeneous interpolation problems. The full description of solution is given each time in a linear fractional form with coefficients arising as solutions of a homogeneous interpolation problem.

CHAPTER 18

MATRIX NEVANLINNA-PICK
INTERPOLATION AND GENERALIZATIONS

This chapter contains the tangential Nevanlinna-Pick interpolation problem for rational matrix valued functions and its generalizations.

The exposition is arranged by increasing level of complexity. We start with the classical scalar Nevanlinna-Pick interpolation problem in Section 18.1 and continue in Sections 18.2 and 18.3 with the simplest tangential matrix Nevanlinna-Pick problem. The general one-sided problem is stated and illustrated by several examples in Section 18.4. Sections 18.5 and 18.6 contain the solution of the two-sided Nudelman problem, the statements of the main results and proofs. This problem represents a far-reaching generalization of the classical Nevanlinna-Pick interpolation problem. The solution of the Nudelman problem is based on the results concerning Lagrange-Sylvester contour integral interpolation problem given in Chapter 16.

18.1 CLASSICAL NEVANLINNA-PICK INTERPOLATION

The classical Nevanlinna-Pick interpolation problem is as follows. We are given a collection of points $z_1, \ldots, z_n$ in a domain Δ contained in the complex plane $\mathbb{C}$ together with a collection of complex numbers $w_1, \ldots, w_n$. Let us say that a scalar function f analytic on Δ interpolates if

$$(18.1.1) \qquad f(z_j) = w_j, \qquad 1 \leq j \leq n.$$

The problem is to describe all interpolating functions f for which

$$(18.1.2) \qquad \sup_{z \in \Delta} |f(z)| < 1.$$

In the spirit of this book we will be interested only in rational solutions $f(z)$. The data set $w = \{z_j, w_j : 1 \leq j \leq n\}$ we call the data set for the interpolation problem. The set of all rational interpolating functions (i.e. functions satisfying (18.1.2)) we denote by $I(w)$. Denote by $\mathcal{BR}(\Delta)$ the set of all rational functions f with no poles on Δ for which $\sup_{z \in \Delta} |f(z)| < 1$. With this notation, the problem is to describe the set $I(w) \cap \mathcal{BR}(\Delta)$. We will consider the two cases where Δ is the open unit disk $\mathcal{D}$ and where Δ is the open right half plane Π^+. The criterion for the existence of a solution to exist involves a matrix $\Lambda(w)$ (usually called the Pick matrix) which is built from the data set w. Define $\Lambda(w)$ by

$$(18.1.3)$$
$$\Lambda(w) = \left[\frac{1 - \overline{w}_i w_j}{1 - \overline{z}_i z_j} \right]_{1 \leq i,j \leq n} \qquad \text{if} \quad \Delta = \mathcal{D}$$

$$\Lambda(w) = \left[\frac{1 - \overline{w}_i w_j}{\overline{z}_i + z_j} \right]_{1 \leq i,j \leq n} \qquad \text{if} \quad \Delta = \Pi^+.$$

The result is as follows.

THEOREM 18.1. *There exists a solution* $f \in I(w) \cap BR(\Delta)$ *of the interpolation problem if and only if the associated Pick matrix* $\Lambda(w)$ *given by (18.1.3) is positive definite. In this case there is a rational* 2×2 *matrix function* $[\Theta_{ij}(z)]_{1 \le i,j \le 2}$ *such that the infinitely many solutions* $f(z)$ *of the classical Nevanlinna-Pick interpolation problem are described by*

$$f(z) = [\Theta_{11}(z)g(z) + \Theta_{12}(z)][\Theta_{21}(z)g(z) + \Theta_{22}(z)]^{-1}$$

where g *is any function in* $BR(\Delta)$, *and*

$$\Theta(z) = \begin{bmatrix} \Theta_{11}(z) & \Theta_{12}(z) \\ \Theta_{21}(z) & \Theta_{22}(z) \end{bmatrix}$$

is a particular rational 2×2 *matrix function. For* $\Delta = \Pi^+$ *the function* $\Theta(z)$ *is given by*
(18.1.4)

$$\Theta(z) = I + \begin{bmatrix} w_1 \cdots w_n \\ 1 \cdots 1 \end{bmatrix} \begin{bmatrix} (z-z_1)^{-1} & & 0 \\ & \ddots & \\ 0 & & (z-z_n)^{-1} \end{bmatrix} \Lambda(w)^{-1} \begin{bmatrix} -\overline{w}_1 & 1 \\ -\overline{w}_2 & 1 \\ \vdots & \vdots \\ -\overline{w}_n & 1 \end{bmatrix}.$$

For $\Delta = \mathcal{D}$ *the function* $\Theta(z)$ *is given by*

$$\Theta(z) = I + (z-1) \begin{bmatrix} w_1 \cdots w_n \\ 1 \cdots 1 \end{bmatrix} \begin{bmatrix} (z-z_1)^{-1} & & 0 \\ & \ddots & \\ 0 & & (z-z_n)^{-1} \end{bmatrix}$$

(18.1.5)
$$\times \Lambda(w)^{-1} \begin{bmatrix} (1-\overline{z}_1)^{-1} & & 0 \\ & \ddots & \\ 0 & & (1-\overline{z}_n)^{-1} \end{bmatrix} \begin{bmatrix} \overline{w}_1 & -1 \\ \overline{w}_2 & -1 \\ \vdots & \vdots \\ \overline{w}_n & -1 \end{bmatrix}.$$

The result of Theorem 18.1.1 will serve as a model and also as a very simple special case of the matrix generalizations of the classical Nevanlinna-Pick interpolation problem that will be discussed in this chapter.

18.2   TANGENTIAL NEVANLINNA-PICK INTERPOLATION WITH SIMPLE MULTIPLICITIES: INTRODUCTION AND MAIN RESULT

We consider here the simplest case of a generalization of classical Nevanlinna-Pick problem to the matrix case. In the sequel Δ will denote either the open right half plane $\Pi^+ = \{z \in \mathbf{C} : \operatorname{Re} z > 0\}$ or the unit disk $\mathcal{D} = \{z \in \mathbf{C} : |z| < 1\}$. Then $\partial\Delta$ denotes the boundary of Δ

$$\partial\Delta = \{ix : x \in \mathbf{R}\} \cup \{\infty\} \quad \text{if} \quad \Delta = \Pi^+$$
$$\partial\Delta = \{z \in \mathbf{C} : |z| = 1\} \quad \text{if} \quad \Delta = \mathcal{D}.$$

We denote by $\overline{\Delta}$ the closure of Δ in the extended complex plane

$$\overline{\Delta} = \{z \in \mathbb{C} : z + \overline{z} \geq 0\} \cup \{\infty\} \quad \text{if} \quad \Delta = \Pi^+$$
$$\overline{\Delta} = \{z \in \mathbb{C} : |z| \leq 1\} \quad \text{if} \quad \Delta = \mathcal{D}.$$

We assume that we are given n distinct points $z_1, z_2, \ldots, z_n$ in Δ, n nonzero row vectors $x_1, \ldots, x_n \in \mathbb{C}^{1 \times M}$ and n row vectors $y_1, \ldots, y_n \in \mathbb{C}^{1 \times N}$. The problem is to find a rational $M \times N$ matrix function $F(z)$ such that

(i) $F(z)$ is analytic on Δ

(ii) $\sup_{z \in \Delta} \|F(z)\| < 1$

and

(iii) $x_i F(z_i) = y_i$ for $1 \leq i \leq n$.

In the case $\Delta = \mathcal{D}$ we assume that none of the points $z_1, z_2, \ldots, z_n$ is zero to avoid some extraneous technical complications. We have already seen in Chapter 16 (Theorem 16.4.1 applied to Example 16.2.3) that the set of all interpolants analytic on Δ (i.e. functions F satisfying (i) and (iii)) have a linear fractional parametrization

$$(18.2.1a) \qquad F = (\Theta_{11} G_1 + \Theta_{12} G_2)(\Theta_{21} G_1 + \Theta_{22} G_2)^{-1}$$

where G_1, G_2 is an arbitrary pair of matrix functions such that $\det(\Theta_{21} G_1 + \Theta_{22} G_2)$ has no zeros in Δ and where $\Theta = \begin{bmatrix} \Theta_{11} & \Theta_{12} \\ \Theta_{21} & \Theta_{22} \end{bmatrix}$ is an $(M+N) \times (M+N)$ matrix function built from the interpolation data. This will be our starting point when we tackle the general matrix Nevanlinna-Pick interpolation problem in Section 18.5.

From experience with the classical case (Theorem 18.1.1), we expect that the matrix function Θ needed for the parametrization (18.2.1) can be specially taylored so that matrix functions $F(z)$ satisfying (ii) in addition to (i) and (iii) are given by (18.2.1a), where the pair (G_1, G_2) has the special form (G, I) with G satisfying (i) and (ii). In this introductory section, we *assume* that solutions F of (i), (ii), (iii) have a parametrization of this form

$$(18.2.1b) \qquad F = (\Theta_{11} G + \Theta_{12})(\Theta_{21} G + \Theta_{22})^{-1}$$

(where G satisfies (i) and (ii)), and deduce in an elementary way (without reference to the machinery developed in earlier chapters) what properties the $(M+N) \times (M+N)$ matrix function Θ must have. In order that F satisfy (i) for *any* choice of G satisfying (i) and (ii), a natural property to impose on Θ is

(I) $\qquad\qquad\qquad\qquad \Theta(z)$ is analytic in $\overline{\Delta}$.

Indeed, if Θ satisfies (I) then the analyticity of F on Δ (F as in (18.2.1b)) holds for any G satisfying (i) and (ii) for which we can show $\det(\Theta_{21} G + \Theta_{22})$ does not vanish on Δ; we will return to this side issue later.

The linear fractional formula (18.2.1b) itself can be rewritten in the more linear form as

$$(18.2.2) \qquad \begin{bmatrix} F \\ I \end{bmatrix} = \begin{bmatrix} \Theta_{11} & \Theta_{12} \\ \Theta_{21} & \Theta_{22} \end{bmatrix} \begin{bmatrix} G \\ I \end{bmatrix} (\Theta_{21} G + \Theta_{22})^{-1}.$$

From this we get the identity

$$F^*F - I = [\, F^* \quad I \,] \begin{bmatrix} I & 0 \\ 0 & -I \end{bmatrix} \begin{bmatrix} F \\ I \end{bmatrix}$$

(18.2.3)

$$= X^*[\, G^* \quad I \,]\Theta^*J\Theta \begin{bmatrix} G \\ I \end{bmatrix} X$$

where $X = (\Theta_{21}G + \Theta_{22})^{-1}$ and $J = I_M \oplus -I_N$. Next, if we assume that

(18.2.4)
$$\Theta(z)^* J\Theta(z) = J$$

for certain values of z, then we may simplify (18.2.3) further to

$$F(z)^* F(z) - I = X(z)^* \big(G(z)^* G(z) - I\big)X(z).$$

If we assume that $X(z)$ is invertible for all z under consideration, this leads to

$$F(z)^* F(z) - I < 0 \Leftrightarrow G(z)^* G(z) - I < 0$$

or equivalently

$$\|F(z)\| < 1 \Leftrightarrow \|G(z)\| < 1.$$

If instead of (18.2.4) we have the weaker condition

(18.2.5)
$$\Theta(z)^* J\Theta(z) \le J$$

then from (18.2.3) we are still able to conclude

$$F(z)^* F(z) - I \le X(z)^* \big(G(z)^* G(z) - I\big)X(z)$$

so we have the implication in one direction

(18.2.6)
$$\|G(z)\| < 1 \Rightarrow \|F(z)\| < 1.$$

One cannot expect a nonconstant matrix function to satisfy (18.2.4) on all of Δ and maintain analyticity. Indeed, if (18.2.4) holds for all $z \in \Delta$, then in particular $\Theta(z)$ is invertible and

$$\overline{(\Theta(z))} = (\Theta(z))^{*T} = J\Theta(z)^{-1T}J$$

is analytic in Δ as well. However, an analytic function whose complex conjugate is analytic as well, must be a constant. As a reasonable compromise we shall impose

(II)
$$\Theta(z)^* J\Theta(z) = J \quad \text{for} \quad z \in \partial\Delta$$
$$\Theta(z)^* J\Theta(z) \le J \quad \text{for} \quad z \in \Delta.$$

In any case we shall attempt to reverse the implication (18.2.6) only when F also satisfies the interpolation conditions (iii).

We next seek additional conditions to impose on Θ so that F as in (18.2.1) in addition satisfies (iii) when G satisfies (i) and (ii). We rewrite (iii) in the form of homogeneous interpolation conditions on the block column function $\begin{bmatrix} F(z) \\ I \end{bmatrix}$:

(18.2.7) $$[x_i, -y_i] \begin{bmatrix} F(z_i) \\ I \end{bmatrix} = 0.$$

By (18.2.2) this becomes

(18.2.8) $$[x_i, -y_i]\Theta(z_i) \begin{bmatrix} G(z_i) \\ I \end{bmatrix} X(z_i) = 0 \quad \text{for} \quad 1 \leq i \leq n.$$

As $G(z)$ may be chosen to have any value at z_i and we eventually hope to show that the associated function $X(z) = (\Theta_{21}(z)G(z) + \Theta_{22}(z))^{-1}$ is analytic at z_i, to guarantee (18.2.5) it is natural to impose

(III) $$[x_i, -y_i]\Theta(z_i) = 0 \quad \text{for} \quad 1 \leq i \leq n$$

(recall that $x_i \neq 0$).

If Θ satisfies additional constraints of the form in (III) at additional points z_i in Δ then any F arising as in (18.2.1) from a G satisfying (i) and (ii) we would expect to solve additional interpolation constraints of the form (iii). Also if $u(z)$ is a row vector function analytic at z_i with $u(z_i) = [x_i, -y_i]$ and if the zero of $u(z)\Theta(z)$ at z_i is not simple, then the associated $F(z)$ will satisfy higher order interpolation conditions at z_i in addition to the simple one in (iii). Similarly, if $[u_i, -v_i]$ is linearly independent of $[x_i, -y_i]$ and satisfies a condition of the form in (III), then the $F(z)$ generated by (18.2.1b) necessarily satisfies independent interpolation conditions in addition to those listed in (iii). As we want to construct Θ so that (18.2.1b) parameterizes *all* solutions F of (i), (ii) and (iii) we impose:

(IV)   det $\Theta(z)$ has only simple zeros in Θ occurring at $z_1, \ldots, z_n$

together with the strengthened form of (III):
(III)
The left kernel $\text{Ker}_\ell \, \Theta(z_i)$ defined by $\text{Ker}_\ell \, \Theta(z_i) = \{u \in \mathbf{C}^{1 \times (M+N)} : u\Theta(z_i) = 0\}$

is the one dimensional subspace spanned by $[x_i, -y_i]$ for $1 \leq i \leq n$.

This completes a heuristic argument to show that if Θ satisfies (I)–(IV) and F and G are related as in (18.2.1b), then F satisfies (i), (ii), and (iii) if and only if G satisfies (i) and (ii). Gaps in the argument concern the analyticity of $X(z) = (\Theta_{21}(z)G(z) + \Theta_{22}(z))^{-1}$ in Θ for any G satisfying (i) and (ii) and an analysis as to when the implication (18.2.6) can be reversed. In fact these gaps can be plugged as the following theorem shows. We postpone the formal proof to the next sections.

THEOREM 18.2.1. *Suppose there exists a rational $(M + N) \times (M + N)$ matrix function $\Theta(z)$ which satisfies (I)–(IV):*

(I) $\Theta(z)$ *is analytic on* $\overline{\Delta}$
(II) $\Theta(z)^* J\Theta(z) = J$ *for* $z \in \partial\Delta$; $\Theta(z)^* J\Theta(z) \leq J$ *for* $z \in \Delta$
(III) $\text{Ker}_\ell \, \Theta(z_i) = \text{Span}\{[x_i, -y_i]\}$ *for* $1 \leq i \leq n$
(IV)   det $\Theta(z)$ *has only simple zeros in* Δ *at* $z_1, \ldots, z_n$.

Then there exists $M \times N$ matrix functions $F(z)$ which satisfy (i)–(iii):

(i) $F(z)$ *is analytic on* Δ

(ii) $\sup_{z \in \Delta} \|F(z)\| < 1$

(iii) $x_i F(z_i) = y_i$ *for* $1 \le i \le n$.

Moreover, given $\Theta(z)$ *satisfying* (I)–(IV), $F(z)$ *is a solution of* (i)–(iii) *if and only if* $F(z)$ *is given by* (18.2.1b) *for a* $G(z)$ *which is analytic in* Δ *and satisfies* $\sup_{z \in \Delta} \|G(z)\| < 1$.

The question of existence of a rational matrix function satisfying conditions (I)–(IV) is settled by the following theorem. It is actually a special case of Theorem 6.1.5 combined with 6.2.2 for $\Delta = \Pi_+$, and of Theorem 7.1.6 combined with 7.2.2 if $\Delta = \mathcal{D}$. We will give a complete elementary direct proof for the simple case needed here in the next section.

THEOREM 18.2.2. *There exists a rational matrix function satisfying conditions* (I)–(IV) *in Theorem 18.2.1 if and only if the Hermitian matrix*

$$(18.2.9) \qquad \Lambda = \left[\frac{x_i x_j^* - y_i y_j^*}{\rho_\Delta(z_i, z_j)} \right]_{1 \le i, j \le n},$$

where $\rho_\Delta(z, w) = 1 - z\overline{w}$ *if* $\Delta = \mathcal{D}$ *and* $\rho_\Delta(z, w) = z + \overline{w}$ *if* $\Delta = \Pi^+$, *is positive definite. In this case one such matrix function* $\Theta(z)$ *is given as follows: If* $\Delta = \Pi^+$, *then*

$$(18.2.10) \qquad \begin{aligned} \Theta(z) &= \begin{bmatrix} I_M & 0 \\ 0 & I_N \end{bmatrix} + \begin{bmatrix} x_1^* & \cdots & x_n^* \\ y_1^* & \cdots & y_n^* \end{bmatrix} \\ &\times \begin{bmatrix} (z + \overline{z}_1)^{-1} & & 0 \\ & \ddots & \\ 0 & & (z + \overline{z}_n)^{-1} \end{bmatrix} \Lambda^{-1} \begin{bmatrix} x_1 & -y_1 \\ \vdots & \vdots \\ x_n & -y_n \end{bmatrix}. \end{aligned}$$

If $\Delta = \mathcal{D}$, *then choose any number* α *with* $|\alpha| = 1$, *and put*

$$(18.2.11) \qquad \begin{aligned} \Theta(z) &= D + \begin{bmatrix} x_1^* & \cdots & x_n^* \\ y_1^* & \cdots & y_n^* \end{bmatrix} \begin{bmatrix} (1 - z\overline{z}_1)^{-1} & & 0 \\ & \ddots & \\ 0 & & (1 - z\overline{z}_n)^{-1} \end{bmatrix} \\ &\times \Lambda^{-1} \begin{bmatrix} x_1 & -y_1 \\ \vdots & \vdots \\ x_n & -y_n \end{bmatrix} D, \end{aligned}$$

where

$$(18.2.12) \qquad \begin{aligned} D &= I + \begin{bmatrix} x_1^* & \cdots & x_n^* \\ y_1^* & \cdots & y_n^* \end{bmatrix} \begin{bmatrix} \overline{z}_1^{-1} & & 0 \\ & \ddots & \\ 0 & & \overline{z}_n^{-1} \end{bmatrix} \\ &\times \Lambda^{-1} \begin{bmatrix} (\alpha - z_1)^{-1} & & 0 \\ & \ddots & \\ 0 & & (\alpha - z_n)^{-1} \end{bmatrix} \begin{bmatrix} x_1 & -y_1 \\ \vdots & \vdots \\ x_n & -y_n \end{bmatrix}. \end{aligned}$$

Observe that our simplifying assumption that all z_j are different from 0 (in the case $\Delta = \mathcal{D}$) was used essentially in the formula (18.2.12); this restriction is only for technical convenience to illustrate the main ideas for a simple case.

By combining Theorems 18.2.1 and 18.2.2 we see that positive definiteness of Λ in (18.2.9) is sufficient for the existence of many solutions of the tangential interpolation problem (i)–(iii). It turns out that it is also necessary.

THEOREM 18.2.3. *A necessary and sufficient condition for the tangential Nevanlinna-Pick interpolation problem* (i)–(iii) *to have a solution is that the matrix Λ in* (18.2.9) *be positive definite. In this case one such solution $F(z)$ is given by the following formulas: If $\Delta = \Pi^+$ put*

$$F(z) = [X^*(zI + A_\zeta^*)^{-1}\Lambda^{-1}Y][I + Y^*(zI + A_\zeta^* - \Lambda^{-1}YY^*)^{-1}\Lambda^{-1}Y]$$

where

$$X = \begin{bmatrix} x_1 \\ \vdots \\ x_n \end{bmatrix}, \qquad Y = \begin{bmatrix} y_1 \\ \vdots \\ y_n \end{bmatrix}$$

and

$$A_\zeta = \begin{bmatrix} z_1 & & 0 \\ & \ddots & \\ 0 & & z_n \end{bmatrix}.$$

If $\Delta = \mathcal{D}$ define $F(z)$ by

$$F(z) = \left(d_{12} + [\, x_1^* \;\cdots\; x_n^* \,] \begin{bmatrix} (1 - z\bar{z}_1)^{-1} & & 0 \\ & \ddots & \\ 0 & & (1 - z\bar{z}_n)^{-1} \end{bmatrix} \right.$$
$$\times \Lambda^{-1} \begin{bmatrix} x_1 & -y_1 \\ \vdots & \vdots \\ x_n & -y_n \end{bmatrix} \begin{bmatrix} d_{12} \\ d_{22} \end{bmatrix} \right)$$
$$\times \left(d_{22} + [\, y_1^* \;\cdots\; y_n^* \,] \begin{bmatrix} (1 - z\bar{z}_1)^{-1} & & 0 \\ & \ddots & \\ 0 & & (1 - z\bar{z}_n)^{-1} \end{bmatrix} \right.$$
$$\times \Lambda^{-1} \begin{bmatrix} x_1 & -y_1 \\ \vdots & \vdots \\ x_n & -y_n \end{bmatrix} \begin{bmatrix} d_{12} \\ d_{22} \end{bmatrix} \right)^{-1},$$

where $\begin{bmatrix} d_{12} \\ d_{22} \end{bmatrix}$ *is the second column of D (given by* (18.2.12)).

We remark that the formula for $F(z)$ in Theorem 18.2.3 is obtained by simply plugging in $G = 0$ in (18.2.1b) (so $F = \Theta_{12}\Theta_{22}^{-1}$) and then using the formula for $\Theta(z)$

((18.2.10) if $\Delta = \Pi^+$ or (18.2.11) if $\Delta = \mathcal{D}$) in Theorem 18.2.2. The only remaining advance of Theorem 18.2.3 over the combined statements of Theorems 18.2.1 and 18.2.2 is that the existence of solutions F of (i), (ii) and (iii) leads to the positive definiteness of the matrix (18.2.9). We will not give a separate elementary proof of this result; in Section 18.4 it will be proved in the context of a much more general interpolation problem.

We conclude this section with a specialization of Theorem 18.2.3 to the scalar case. In this case we may as well take the vectors x_j all to be the number 1, and the vectors y_j are simply complex numbers w_j ($1 \le j \le n$). The matrix $\Lambda = \Lambda(w)$ specializes to

$$\Lambda = \left[\frac{1 - w_i \overline{w}_j}{1 - z_i \overline{z}_j}\right]_{1 \le i,j \le n} \qquad \text{if} \quad \Delta = \mathcal{D}$$

$$\Lambda = \left[\frac{1 - w_i \overline{w}_j}{z_i + \overline{z}_j}\right]_{1 \le i,j \le n} \qquad \text{if} \quad \Delta = \Pi^+$$

which is the classical Pick matrix. The matrix function $\Theta(z)$ is then given by

$$\Theta(z) = \begin{bmatrix} 1 & 0 \\ 0 & 1 \end{bmatrix} + \begin{bmatrix} -1 & \cdots & -1 \\ \overline{w}_1 & \cdots & \overline{w}_n \end{bmatrix} \begin{bmatrix} (z + \overline{z}_1)^{-1} & & 0 \\ & \ddots & \\ 0 & & (z + \overline{z}_n)^{-1} \end{bmatrix}$$

$$\times \Lambda^{-1} \begin{bmatrix} 1 & -w_1 \\ \vdots & \vdots \\ 1 & -w_n \end{bmatrix}$$

if $\Delta = \Pi^+$, and

$$\Theta(z) = D + \begin{bmatrix} -1 & \cdots & -1 \\ \overline{w}_1 & \cdots & \overline{w}_n \end{bmatrix} \begin{bmatrix} (1 - z\overline{z}_1)^{-1} & & 0 \\ & \ddots & \\ 0 & & (1 - z\overline{z}_n)^{-1} \end{bmatrix} \Lambda^{-1} \begin{bmatrix} 1 & -w_1 \\ \vdots & \vdots \\ 1 & -w_n \end{bmatrix} D$$

if $\Delta = \mathcal{D}$, where

$$D = \begin{bmatrix} 1 & 0 \\ 0 & 1 \end{bmatrix} + \begin{bmatrix} -1 & \cdots & -1 \\ \overline{w}_1 & \cdots & \overline{w}_n \end{bmatrix} \begin{bmatrix} \overline{z}_1^{-1} & & 0 \\ & \ddots & \\ 0 & & \overline{z}_n^{-1} \end{bmatrix}$$

$$\times \Lambda^{-1} \begin{bmatrix} (\alpha - z_1)^{-1} & & 0 \\ & \ddots & \\ 0 & & (\alpha - z_n)^{-1} \end{bmatrix} \begin{bmatrix} 1 & -w_1 \\ \vdots & \vdots \\ 1 & -w_n \end{bmatrix}$$

and $|\alpha| = 1$ (we continue to assume that z_j are different from 0). With this explicit choice of $\Theta(z)$, any rational solution f satisfying $f(z_j) = w_j$ and $\sup\{|f(z)|: z \in \Delta\} < 1$ is given by

$$f(z) = \left(\Theta_{11}(z)g(z) + \Theta_{12}(z)\right)\left(\Theta_{21}(z)g(z) + \Theta_{22}(z)\right)^{-1}$$

where g is any rational function with $\sup\{|g(z)|: z \in \Delta\} < 1$.

18.3   TANGENTIAL NEVANLINNA-PICK INTERPOLATION WITH SIMPLE MULTIPLICITIES: PROOFS

In this and the next sections we provide proofs for the main results announced in the preceding section. In line with the elementary description of these results the proofs given here will be kept also at an elementary level. This section will be devoted to the proof of Theorem 18.2.2.

We start with the following basic lemma which relates null and pole structure at a given point to the Laurent expansion of the function or of its inverse at that point for the simplest case; a thorough analysis of the general case is the main topic of Chapter 3. Recall that given a $p \times q$ matrix A we denote by $\mathrm{Ker}_\ell(A)$ the set of all $u \in \mathbb{C}^{1 \times p}$ such that $uA = 0$.

LEMMA 18.3.1. *Suppose $W(z)$ is a rational $p \times p$ matrix function analytic at the point z_0 in $\mathbb{C}$, φ_0 is a nonzero $1 \times p$ row vector and ψ_0 is a nonzero $p \times 1$ column vector.*

(a) *Then $\det W(z)$ has a simple zero at z_0 and*

$$\mathrm{Ker}_\ell W(z_0) = \mathrm{Span}\{\varphi_0\}$$

if and only if the Laurent expansion for $W(z)^{-1}$ at z_0 has the form

$$W(z)^{-1} = \frac{\chi_0 \varphi_0}{z - z_0} + [\text{analytic at } z_0]$$

for some nonzero $p \times 1$ column vector χ_0.

(b) *Then $\det W(z)$ has a simple zero at z_0 and*

$$\mathrm{Ker}\, W(z_0) = \mathrm{Span}\{\psi_0\}$$

if and only if the Laurent expansion for $W(z)^{-1}$ at z_0 has the form

$$W(z)^{-1} = \frac{\psi_0 \gamma_0}{z - z_0} + [\text{analytic at } z_0]$$

for some nonzero $1 \times p$ row vector ψ_0.

PROOF. Note that (b) is simply the transposed version of (a) so we need proved only (a). Suppose first that $\det W(z)$ has a simple zero at z_0 and $\mathrm{Ker}_\ell W(z_0) = \mathrm{Span}\{\varphi_0\}$. Let $W(z)^{-1}$ have Laurent expression

$$W(z)^{-1} = \sum_{j=-q}^{\infty} (z - z_0)^j \widetilde{W}_j$$

where $\widetilde{W}_{-q} \neq 0$; since $W(z_0)$ is not invertible, certainly $q \geq 1$. As $\widetilde{W}_{-q} \neq 0$ we can choose a row vector α_0 so that $\alpha_0 \widetilde{W}_{-q} \neq 0$. Then we have

(18.3.1) $$\alpha_0 W(z)^{-1} = (z - z_0)^{-q} \gamma_0(z)$$

where γ_0 is analytic at z_0 and $\gamma_0(z_0) = \alpha_0\widetilde{W}_{-q} \neq 0$. Rewrite (18.3.1) as

(18.3.2) $(z - z_0)^q\alpha_0 = \gamma_0(z)W(z).$

In particular $\gamma_0(z_0)W(z_0) = 0$. Since by assumption $\text{Ker}_\ell\, W(z_0) = \text{Span}\{\varphi_0\}$ we conclude that

(18.3.3) $\alpha_0\widetilde{W}_{-q} = \gamma_0(z_0) = c\varphi_0$

for some nonzero complex number c. Since $\gamma_0(z_0) \neq 0$ we may choose $p - 1$ row vectors $\gamma_2, \gamma_3, \ldots, \gamma_p$ so that

$$\det\big(\text{col}(\gamma_0(z_0), \gamma_2, \ldots, \gamma_p)\big) \neq 0.$$

If the zero of $\gamma_0(z)W(z)$ at z_0 were not simple, then

$$\det\{[\text{col}(\gamma_0(z), \gamma_2, \ldots, \gamma_p)] \cdot W(z)\} = \det[\text{col}(\gamma_0(z), \gamma_2, \ldots, \gamma_p)] \cdot \det W(z)$$

would have a higher order zero at z_0. Since by assumption $\det W(z)$ has a simple zero at z_0, we conclude that the zero of $\gamma_0(z)W(z)$ at z_0 must be simple. Thus $q = 1$ from (18.3.2). Finally, since α_0 was chosen to be any row vector for which $\alpha_0\widetilde{W}_{-1} \neq 0$, we conclude from (18.3.3) that

$$\text{Im}_\ell\, \widetilde{W}_{-1} = \text{Span}\{\varphi_0\}.$$

Here and elsewhere we denote by $\text{Im}_\ell\, X$ the left image of an $m \times n$ matrix X

$$\text{Im}_\ell\, X = \{xX \colon x \text{ is a } 1 \times m \text{ row}\}.$$

From this it is elementary that there is a nonzero $p \times 1$ column vector χ_0 such that $\widetilde{W}_{-1} = \chi_0\varphi_0$ as needed.

Conversely, suppose $W(z)^{-1}$ has a Laurent expansion at z_0 of the form

$$W(z)^{-1} = \frac{\chi_0\varphi_0}{(z - z_0)} + [\text{analytic at } z_0]$$

for a nonzero column vector χ_0. Multiply both sides on the right by $(z - z_0)W(z)$ to deduce

(18.3.4) $(z - z_0)I = \chi_0\varphi_0 W(z) + (z - z_0)[\text{analytic at } z_0]W(z).$

By assumption $W(z)$ itself is analytic at z_0. Thus we may evaluate (18.3.4) at z_0 to deduce that $\chi_0\varphi_0 W(z_0) = 0$. Since χ_0 is a nonzero column vector, this in turn gives $\varphi_0 W(z_0) = 0$. If $\widetilde{\varphi}_0$ is any other row vector for which $\widetilde{\varphi}_0 W(z_0) = 0$, the first part of the proof gives that $\widetilde{\varphi}_0 \in \text{Im}_\ell\, \text{Res}_{z=z_0}\, W(z)^{-1} = \text{Span}\{\varphi_0\}$ as asserted. Next rewrite (18.3.4) as

$$(z - z_0)I = \big(\chi_0\varphi_0 + (z - z_0)[\text{analytic at } z_0]\big) \cdot W(z)$$
$$=: V(z)W(z).$$

Taking determinants of both sides gives

$$(18.3.5) \qquad (z - z_0)^p = \det V(z) \cdot \det W(z).$$

Since $\chi_0 \varphi_0$ is a rank 1 matrix, it is easy to see that $\det V(z)$ has a zero at z_0 of order at least $p - 1$. Thus from (18.3.5) we deduce that $\det W(z)$ has a zero at z_0 of order at most 1. Since $\det W(z_0) = 0$, we get that $\det W(z)$ has a zero at z_0 of order exactly equal to 1. □

    We state the interpolation problem which we now wish to analyze. We are given distinct points $z_1, z_2, \ldots, z_n \in \mathbf{C}$ and nonzero $1 \times p$ row vectors $\varphi_1, \varphi_2, \ldots, \varphi_n$, as well as additional distinct points $w_1, w_2, \ldots, w_n \in \mathbf{C}$ and nonzero $p \times 1$ column vectors $\psi_1, \psi_2, \ldots, \psi_n$. To avoid unwanted (in this section) complications we assume that the set of points $\{z_1, \ldots, z_n\}$ is disjoint from the set $\{w_1, \ldots, w_n\}$. The problem is to construct a rational $p \times p$ matrix function $W(z)$ with value I at infinity such that

(18.3.6i)   For $i = 1, \ldots, n$, $W(z)$ is analytic at z_i, $\det W(z)$ has a simple zero at z_i, and $\mathrm{Ker}_\ell W(z_i) = \mathrm{Span}\{\varphi_i\}$.

(18.3.6ii)   For $j = 1, \ldots, n$, $W(z)^{-1}$ is analytic at w_j, $\det W(z)^{-1}$ has a simple zero at w_j, and $\mathrm{Ker}\, W(w_j)^{-1} = \mathrm{Span}\{\psi_j\}$.

(18.3.6iii)   The poles of $W(z)$ occur only at $w_1, w_2, \ldots, w_n$. The poles of $W(z)^{-1}$ occur only at $z_1, z_2, \ldots, z_n$.

This is the simplest case of the global interpolation problem for rational matrix functions discussed in Section 4.3; we will give here a direct proof using elementary complex analysis without recourse to the language of realizations or null and pole pairs.

    If $W(z)$ is to be a solution of (18.3.6), from (18.3.6i) and Lemma 18.3.1(a) we know that $W(z)^{-1}$ has a simple pole at z_i with residue at z_i of the form $\chi_i \varphi_i$ for some nonzero column vector χ_i. By (18.3.6iii) we are requiring that $W(z)^{-1}$ has no other poles, and we also require that $W(\infty) = I$. Thus by Liouville's Theorem we know that a solution W of (18.3.6) must be such that its inverse $W(z)^{-1}$ has the partial fraction expansion

$$(18.3.7) \qquad W(z)^{-1} = I - \sum_{i=1}^{n} (z - z_i)^{-1} \chi_i \varphi_i$$

where now the nonzero column vectors $\chi_1, \ldots, \chi_n$ are the unknowns. Similarly, from (18.3.6ii), (18.3.6iii) and Lemma 18.3.1(b) (applied to W^{-1} rather that to W), we deduce that a solution W must itself have a partial fraction expansion

$$(18.3.8) \qquad W(z) = I + \sum_{j=1}^{n} (z - w_j)^{-1} \psi_j \gamma_j$$

for some nonzero row vectors $\gamma_1, \gamma_2, \ldots, \gamma_n$ to be determined. To examine the consequence of (18.3.7) and (18.3.8) holding for the same rational matrix function $W(z)$, we need to be able to compute the inverse of a rational matrix function given in the form of a partial fraction expansion with simple poles. This is given by the following computational lemma, the verification of which is a straightforward (note also that this lemma is actually a special case of Proposition 4.1.5).

LEMMA 18.3.2. *Suppose $z_1, \ldots, z_n$ are distinct points in $\mathbb{C}$, $\chi_1, \ldots, \chi_n$ are nonzero $p \times 1$ column vectors, $\varphi_1, \ldots, \varphi_n$ are nonzero $1 \times p$ row vectors, and $V(z)$ is the rational $p \times p$ matrix function given by*

$$V(z) = I - \sum_{i-1}^{n} (z - z_i)^{-1} \chi_i \varphi_i.$$

Then

(18.3.9) $$V(z)^{-1} = I + C_\chi \big(zI - (A_\zeta + B_\varphi C_\chi) \big)^{-1} B_\varphi$$

where

$$C_\chi = [\chi_1, \chi_2, \ldots, \chi_n]$$

(18.3.10) $$A_\zeta = \begin{bmatrix} z_1 & & 0 \\ & \ddots & \\ 0 & & z_n \end{bmatrix}, \qquad B_\varphi = \begin{bmatrix} \varphi_1 \\ \varphi_2 \\ \vdots \\ \varphi_n \end{bmatrix}.$$

From Lemma 18.3.2 we conclude that $W(z) = V(z)^{-1}$ is given by (18.3.9) and (18.3.10) if $W(z)^{-1}$ is given by (18.3.7). Thus for the interpolation problem (18.3.6i)–(18.3.6iii) to have a solution, there must be a choice of nonzero column vectors $\chi_1, \ldots, \chi_n$ and nonzero row vectors $\gamma_1, \ldots, \gamma_n$ for which $V(z)^{-1}$ given by (18.3.9) and (18.3.10) is the same as $W(z)$ given by (18.3.8). In particular the n poles of $W(z)$ in (18.3.8), namely $w_1, w_2, \ldots, w_n$, must be the same as the poles of $V(z)^{-1}$ in (18.3.9). But the poles of $V(z)^{-1}$ clearly occur at the n eigenvalues of the $n \times n$ matrix $A_\zeta + B_\varphi C_\chi$. Thus the eigenvalues of $A_\zeta + B_\varphi C_\chi$ must consist precisely of the n complex numbers $w_1, \ldots, w_n$, i.e., $A_\zeta + B_\varphi C_\chi$ must be similar to the diagonal matrix

$$A_\pi = \begin{bmatrix} w_1 & & & 0 \\ & w_2 & & \\ & & \ddots & \\ 0 & & & w_n \end{bmatrix}.$$

Hence there must exist an invertible $n \times n$ matrix S such that

(18.3.11) $$A_\pi + B_\varphi C_\chi = S A_\pi S^{-1}.$$

Moreover, by using the similarity it is easy to convert the expression (18.3.9) for $V(z)^{-1}$ to a partial fraction form to make comparison with (18.3.8) more convenient.

Specifically, plug (18.3.11) into (18.3.9) to get

$$
\begin{aligned}
V(z)^{-1} &= I + C_\chi (zI - SA_\pi S^{-1})^{-1} B_\varphi \\
&= I + C_\chi S(zI - A_\pi)^{-1} S^{-1} B_\varphi \\
&= I + C_\chi S(zI - A_\pi)^{-1} S^{-1} B_\varphi \\
&= I + \sum_{j=1}^{n} (z - w_j)^{-1} \widetilde{\psi}_j \widetilde{\gamma}_j
\end{aligned}
$$

(18.3.12)

where $\widetilde{\psi}_j$ is the j-th column of $C_\chi S$ and $\widetilde{\gamma}_j$ is the j-th row of $S^{-1} B_\varphi$.

It is easy to see that equality of $V(z)^{-1}$ given by (18.3.9) with $W(z)$ given by (18.3.8) forces $\widetilde{\psi}_j = c_j \psi_j$ and $\widetilde{\gamma}_j = c_j^{-1} \gamma_j$ for nonzero complex numbers $c_1, \ldots, c_n$. As the γ_j's are unknown, we may rescale γ_j if necessary so that we have the equality $\widetilde{\psi}_j = \psi_j$, that is

(18.3.13) $$C_\chi S = C_\psi.$$

When we plug (18.3.13) into (18.3.11) and rearrange, we get

(18.3.14) $$SA_\pi - A_\zeta S = B_\varphi C_\psi.$$

By computing the (i, j)-th entry of each side of (18.3.14), we see that the matrix $S = [s_{ij}]_{1 \le i,j \le n}$ satisfies (18.3.14) if and only if s_{ij} $(1 \le i, j \le n)$ satisfies the system of scalar equations

$$s_{ij} w_j - z_i s_{ij} = \varphi_i \psi_j.$$

Trivially the unique solution s_{ij} of this system is given by

$$s_{ij} = \varphi_i \psi_j / (w_j - z_i).$$

We have thus arrived at a good piece of the following result. This is a special case of the main result (Theorem 4.3.1) of Section 4.3.

THEOREM 18.3.3. *There exists a rational matrix function $W(z)$ with value I at infinity satisfying (18.3.6i)–(18.3.6iii) if and only if the matrix S is given by*

(18.3.15) $$S = \left[\frac{\varphi_i \psi_j}{w_j - z_i} \right]_{1 \le i,j \le n}$$

is invertible. In this case the unique such $W(z)$ is given by

(18.3.16) $$W(z) = I + \sum_{j=1}^{n} (z - w_j)^{-1} \psi_j \gamma_j$$

with inverse given by

(18.3.17) $$W(z)^{-1} = I - \sum_{i=1}^{n} (z - z_i)^{-1} \chi_i \varphi_i$$

$$where \quad \begin{bmatrix} \gamma_1 \\ \vdots \\ \gamma_n \end{bmatrix} = S^{-1} \begin{bmatrix} \varphi_1 \\ \vdots \\ \varphi_n \end{bmatrix} \quad and$$

$$[\chi_1, \ldots, \chi_n] = [\psi_1, \ldots, \psi_n] \cdot S^{-1}.$$

PROOF. The argument preceding the statement of the theorem gives that necessarily the unique solution S (18.3.15) of (18.3.14) is invertible if a solution $W(z)$ of (18.3.6i)–(18.3.6iii) with value I at infinity exists. Moreover, from (18.3.12) we have $W(z) = V(z)^{-1} = I + C_\chi S(zI - A_\pi)^{-1}S^{-1}B_\varphi$. Combining this with (18.3.13) gives the desired formula (18.3.16) for $W(z)$. Also from (18.3.7) we see that $W(z)^{-1} = I - C_\chi(zI - A_\zeta)^{-1}B_\varphi$. When this is combined with (18.3.13) we get the indicated formula (18.3.17) for $W(z)^{-1}$.

Conversely, suppose that the matrix S given by (18.3.15) is invertible. Define a matrix function $W(z)$ by (18.3.16). To see that $W(z)^{-1}$ is given by (18.3.17) we simply observe that (18.3.16) and (18.3.17) can be rewritten in the form

$$W(z) = I + C_\psi(zI - A_\pi)^{-1}S^{-1}B_\varphi$$

and

$$W(z)^{-1} = I - C_\psi S^{-1}(zI - A_\pi)^{-1}B_\varphi,$$

where

$$A_\pi = \begin{bmatrix} w_1 & & 0 \\ & \ddots & \\ 0 & & w_n \end{bmatrix}, \quad A_\zeta = \begin{bmatrix} z_1 & & 0 \\ & \ddots & \\ 0 & & z_n \end{bmatrix},$$

$$B_\varphi = \begin{bmatrix} \varphi_1 \\ \vdots \\ \varphi_n \end{bmatrix}, \quad C_\psi = [\psi_1, \ldots, \psi_n],$$

and apply Lemma 18.3.2. To show that $W(z)$ satisfies (18.3.6i)–(18.3.6iii), by Lemma 18.3.1 and the partial fraction representations (18.3.16) and (18.3.17) for $W(z)$ and $W(z)^{-1}$ it remains only to show that the rows of $S^{-1}B_\varphi$ and the columns of $C_\psi S^{-1}$ are nonzero. Thus suppose that the i-th row of $S^{-1}B_\varphi$ were zero. Then also the i-th row of $S^{-1}B_\varphi C_\psi$ would be zero. But from (18.3.14) we have

$$S^{-1}B_\varphi C_\psi = S^{-1}(SA_\pi - A_\zeta S)$$
$$= A_\pi - S^{-1}A_\zeta S.$$

Since the i-th row of A_π is $w_i e_i^T$ (where e_i is the i-th standard basis vector $[0, \ldots, 0, 1, 0, \ldots, 0]^T$ in $\mathbf{C}^n$), our hypothesis forces the i-th row of $S^{-1}A_\zeta S$ to be $w_i e_i^T$. But this property is enough to imply

$$e_i^T \cdot S^{-1}A_\zeta S = w_i e_i^T,$$

and hence w_i is an eigenvalue of $S^{-1}A_\zeta S$. As the eigenvalues of $S^{-1}A_\zeta S$ are the same as the eigenvalues of A_ζ, namely $z_1, z_2, \ldots, z_n$, this leads to a contradiction. Thus necessarily the rows of $S^{-1}B_\varphi$ are nonzero as needed. By an analogous argument one can show that the columns of $C_\psi S^{-1}$ are all nonzero. This completes the proof of Theorem 18.3.3. □

We are now ready to tackle the problem of constructing a rational matrix function $\Theta(z)$ satisfying the conditions (I)–(IV) of the previous section needed to complete the solution of the tangential Nevanlinna-Pick interpolation problem discussed there. It will be convenient to consider first the case $\Delta = \Pi^+$.

We begin with the following variation of problem (18.3.6i)–(18.3.6iii). We are given n nonzero $1 \times p$ row vectors $\varphi_1, \ldots, \varphi_n$ and n distinct points $z_1, \ldots, z_n$ in the open right half plane Π^+. We denote by $\overline{\Pi^+}$ the closure of Π^+ in the extended complex plane $(\overline{\Pi^+} = \{z \in \mathbb{C} : \operatorname{Re} z \geq 0\} \cup \{\infty\})$. The problem is to find a rational $p \times p$ matrix function $\Theta(z)$ with value I at infinity such that

(18.3.18i)   For $i = 1, \ldots, n$, $\Theta(z)$ is analytic at z_i, $\det \Theta(z)$ has
              a simple zero at z_i and $\operatorname{Ker}_\ell \Theta(z_i) = \operatorname{Span}\{\varphi_i\}$

(18.3.18ii) $\Theta(z)$ is analytic in $\overline{\Pi^+}$ and the
              only zeros of $\det \Theta(z)$ in $\overline{\Pi^+}$ are the points $z_1, z_2, \ldots, z_n$

(18.3.18iii) At all points z and $-\bar{z}$ of analyticity of $\Theta(z)$
              the identity $\Theta(-\bar{z})^* J \Theta(z) = J$ holds.

Here $J = J^* = J^{-1}$ is a given $p \times p$ signature matrix. The following result is just a simple special case of Theorem 6.3.1; we give a simple elementary proof not relying on the machinery used in Chapter 6.

THEOREM 18.3.4. *Let the points $z_i, \ldots, z_n$ in Π^+ and nonzero row vectors $\varphi_1, \ldots, \varphi_n$ be given. Then there exists a rational matrix function $\Theta(z)$ with value I at infinity which satisfies conditions (18.3.18i)–(18.3.18iii) if and only if the Hermitian matrix*

(18.3.19) $$H = \left[\frac{\varphi_i J \varphi_j^*}{z_i + \bar{z}_j} \right]_{1 \leq i,j \leq n}$$

is invertible. In this case the unique such matrix function $\Theta(z)$ is given by

(18.3.20) $$\Theta(z) = I - \sum_{j=1}^{n} (z + \bar{z}_j)^{-1} J \varphi_j^* \gamma_j$$

with

(18.3.21) $$\Theta(z)^{-1} = I - \sum_{i=1}^{n} (z - z_i)^{-1} \chi_i \varphi_i$$

where

$$\begin{bmatrix} \gamma_1 \\ \vdots \\ \gamma_n \end{bmatrix} = H^{-1} \begin{bmatrix} \varphi_1 \\ \vdots \\ \varphi_n \end{bmatrix}$$

and

$$[\chi_1, \ldots, \chi_n] = [-J\varphi_1^*, \ldots, -J\varphi_n^*] \cdot H^{-1}.$$

PROOF. Suppose $\Theta(z)$ is a solution of (18.3.18i)–(18.3.18iii). From (18.3.18i) together with (18.3.18ii) we see that the poles of $\Theta(z)^{-1}$ in $\overline{\Pi^+}$ occur precisely at the points $z_1, z_2, \ldots, z_n$. Moreover, from the identity $\Theta(z)^{-1} = J\Theta(-\bar{z})^* J$ implied by (18.3.18iii), the analyticity of Θ in the closed right half plane implies the analyticity of $\Theta(z)^{-1}$ in the closed left half plane. Hence in fact $z_1, \ldots, z_n$ account for all the poles of $\Theta(z)^{-1}$ in the entire complex plane including infinity. From condition (18.3.18i) together with Lemma 18.3.1 we know that the pole of $\Theta(z)^{-1}$ at z_i is simple with residue of the form $\chi_i \varphi_i$ for some nonzero column vector χ_i $(1 \le i \le n)$. Then by Liouville's Theorem, a solution $\Theta(z)$ of (18.3.18i)–(18.3.18iii) must have the partial fraction form

$$(18.3.22) \qquad \Theta(z)^{-1} = I + \sum_{i=1}^{n} (z - z_i)^{-1} \chi_i \varphi_i.$$

From (18.3.22) we compute

$$J\Theta(-\bar{z})^{*-1} J = I - \sum_{i=1}^{n} (z + \bar{z}_i)^{-1} J \varphi_i^* \gamma_i$$

(where γ_i is the nonzero row vector $\chi_i^* J$). But from (18.3.18iii) $\Theta(z) = J\Theta(-\bar{z})^{*-1} J$. Thus we arrive at

$$(18.3.23) \qquad \Theta(z) = I - \sum_{i=1}^{n} (z + \bar{z}_i)^{-1} J \varphi_i^* \gamma_i$$

where $\gamma_i \neq 0$ for $1 \le i \le n$. Now by Lemma 18.3.1 and the discussion above, we conclude that $\Theta(z)$ is a solution of a problem of the form (18.3.6i)–(18.3.6iii) (with Θ in place of W), where we set $w_i = -\bar{z}_i$ and $\psi_i = -J\varphi_i^*$ for $1 \le i \le n$. Then by Theorem 18.3.3 the matrix H given by

$$H = \left[\frac{\varphi_i J \varphi_j^*}{z_i + \bar{z}_j} \right]_{1 \le i,j \le n}$$

is necessarily invertible if Θ is a solution of (18.3.18i)–(18.3.18iii). Conversely, if the matrix H is invertible, by Theorem 18.3.3 we may construct a solution Θ of (18.3.6i)–(18.3.6iii) with the data having the special form $w_i = -\bar{z}_i$ and $\psi_i = -J\varphi_i^*$; it remains only to check that this function indeed satisfies all the conditions (18.3.18i)–(18.3.18iii).

From (18.3.6i) we see that $\Theta(z)$ satisfies (18.3.18i). From the formulas (18.3.20) and (18.3.21) we see that (18.3.18ii) is satisfied. Finally, to see that $\Theta(z)$

satisfies (18.3.18iii), note that when we compute $J\Theta(-\bar{z})^*J$ from (18.3.20) we recover (18.3.21).   □

To solve the tangential Nevanlinna-Pick interpolation problem, in addition to conditions (18.3.18i)–(18.3.18iii) for $\Theta(z)$ we require

$$(18.3.18iv) \qquad\qquad \text{for } z \in \Pi^+, \qquad \Theta(z)^*J\Theta(z) \le J.$$

Since the solution to (18.3.18i)–(18.3.18iii) is already unique (when it exists), it remains only to find out when this unique solution (18.3.20) also satisfies (18.3.18iv). This is accomplished by the following result. The matrix H is the "associated Hermitian coupling matrix" in the language of Section 6.2; hence the following theorem is a special case of Theorem 6.2.2.

THEOREM 18.3.5.   *There exists a matrix function* $\Theta(z)$ *with value* I *at infinity which satisfies the conditions* (18.3.18i)–(18.3.18iv) *if and only if the Hermitian matrix*

$$H = \left[\frac{\varphi_i J \varphi_j^*}{z_i + \bar{z}_j}\right]_{1 \le i,j \le n}$$

is positive definite. In this case such a $\Theta(z)$ *is unique and is given by* (18.3.20) *with inverse given by* (18.3.21).

PROOF. The proof is a computation as in the proof of Theorem 6.2.2. Write (18.3.20) in the form

$$\Theta(z) = I - JB^*(zI + A^*)^{-1}H^{-1}B$$

where

$$B = \begin{bmatrix} \varphi_1 \\ \vdots \\ \varphi_n \end{bmatrix}, \qquad A = \begin{bmatrix} z_1 & & 0 \\ & \ddots & \\ 0 & & z_n \end{bmatrix}.$$

We now compute

$$\begin{aligned}
\Theta(z)^*J\Theta(z) &= [I - B^*H^{-1}(\bar{z}I + A)^{-1}BJ]J[I - JB^*(zI + A^*)^{-1}H^{-1}B] \\
&= J - B^*H^{-1}(\bar{z}I + A)^{-1}B - B^*(zI + A^*)^{-1}H^{-1}B \\
&\quad + B^*H^{-1}(\bar{z}I + A)^{-1}BJB^*(zI + A^*)^{-1}H^{-1}B \\
&= J - B^*H^{-1}(\bar{z}I + A)^{-1}[H(zI + A^*) + (\bar{z}I + A)H \\
&\quad - BJB^*](zI + A^*)^{-1}H^{-1}B.
\end{aligned}$$

Noting that H satisfies the Lyapunov equation

$$HA^* + AH = BJB^*$$

we get

$$H(zI + A^*) + (\bar{z}I + A)H - BJB^* = (z + \bar{z})H.$$

Thus

$$\Theta(z)^* J \Theta(z) = J - B^* H^{-1} (\bar{z} I + A)^{-1} (z + \bar{z}) H (zI + A^*)^{-1} H^{-1} B$$

so

(18.3.24) $$\frac{J - \Theta(z)^* J \Theta(z)}{z + \bar{z}} = B^* H^{-1} (\bar{z} I + A)^{-1} H (zI + A^*)^{-1} H^{-1} B.$$

From this we see that $\Theta(z)$ is J-contractive in Π^+ if the Hermitian matrix H is positive definite.

Conversely, suppose $\Theta(z)$ is J-contractive for z in Π^+. From (18.3.24) we see that to show that H is positive definite it suffices to show that

(18.3.25) $$\mathrm{Span}\{(zI + A^*)^{-1} H^{-1} B y : z \in \Pi^+, \ y \in \mathbf{C}^p\} = \mathbf{C}^n.$$

By duality (18.3.25) follows if we show that the only vector $x \in \mathbf{C}^n$ which satisfies

(18.3.26) $$x^T (zI + A^*)^{-1} H^{-1} B y = 0$$

for all $z \in \Pi^+$ and $y \in \mathbf{C}^p$ is $x = 0$. Recalling that

$$A = \begin{bmatrix} z_1 & & 0 \\ & \ddots & \\ 0 & & z_n \end{bmatrix}, \qquad H^{-1} B = \begin{bmatrix} \gamma_1 \\ \vdots \\ \gamma_n \end{bmatrix}$$

and writing out the components of x as $x^T = [x_1, \ldots, x_n]$, (18.3.26) can be written out more explicitly as

(18.3.27) $$R(z) := \sum_{i=1}^{n} (z + \bar{z}_i)^{-1} x_i^T \gamma_i y = 0$$

for all $z \in \Pi^+$ and $y \in \mathbf{C}^p$. By analytic continuation (18.3.27) continues to hold for $z \in \mathbf{C} \backslash \{-\bar{z}_1, \ldots, -\bar{z}_n\}$, and hence

(18.3.28) $$0 = \lim_{z \to -\bar{z}_i} (z + \bar{z}_i) R(z) = x_i^T \gamma_i y$$

for $i = 1, \ldots, n$ for all $y \in \mathbf{C}^p$. As in the proof of Theorem 18.3.3, we know that the rows γ_i of $H^{-1} B$ are nonzero. Thus for each i there is a choice of y in $\mathbf{C}^p$ so that $\gamma_i y \neq 0$. Thus (18.3.28) forces the scalar x_i to be 0 for each i, and hence $x^T = [x_1, \ldots, x_n] = 0$ as required. $\square$

We are now in a position to complete the proof of Theorem 18.2.2.

PROOF OF THEOREM 18.2.2. Note first that if a rational matrix function $\Theta(z)$ satisfies $\Theta(z)^* J \Theta(z) = J$ for $z \in i\mathbf{R}$ then since $z = -\bar{z}$ for $z \in i\mathbf{R}$ we have

(18.3.29) $$\Theta(-\bar{z})^* J \Theta(z) = J$$

for $z \in i\mathbf{R}$. As $\Theta(-\bar{z})^*$ is analytic at any z_0 such that $\Theta(z)$ is analytic at $-\bar{z}_0$, (18.3.29) continues to hold at all points z of analyticity of $\Theta(z)$. With this observation we see that the problem of constructing a matrix function $\Theta(z)$ which satisfies (I)–(IV) is equivalent to producing a $\Theta(z)$ satisfying (18.3.18i)–(18.3.18iv) with the special choice of data

$$\varphi_i = [x_i, -y_i], \qquad J = I_M \oplus -I_N.$$

$(p = M + N)$. When this form of the data is plugged into Theorem 18.3.4, the matrix H in (18.3.19) is the same as the matrix Λ in (18.2.9) and Theorem 18.2.2 is immediate from Theorem 18.3.5. □

   This concludes the proof of Theorem 18.2.2 for the case when $\Delta = \Pi^+$. If $\Delta = \mathcal{D}$ (and the points z_i are different from zero), the proof is obtained using analogous arguments. Let us outline briefly these arguments.

   Given n nonzero $1 \times p$ row vectors $\varphi_1, \ldots, \varphi_n$ and given n distinct nonzero points $z_1, \ldots, z_n$ in the open unit disc $\mathcal{D}$. We consider the problem of finding rational $p \times p$ matrix function $\Theta(z)$ which is analytic with invertible value at infinity and such that

(18.3.30i)
$$\text{For } i = 1, \ldots, n, \ \Theta(z) \text{ is analytic at } z_i,$$
$$\det \Theta(z) \text{ has a simple zero at } z_i \text{ and } \operatorname{Ker}_\ell \Theta(z_i) = \operatorname{Span}\{\varphi_i\};$$

(18.3.30ii)
$$\Theta(z) \text{ is analytic in the closed unit disc } \overline{\mathcal{D}}$$
$$\text{and the only zeros of } \det \Theta(z) \text{ in } \overline{\mathcal{D}} \text{ are } z_1, z_2, \ldots, z_n;$$

(18.3.30iii)
$$\text{At all points } z \neq 0 \text{ and } \bar{z}^{-1} \text{ of analyticity}$$
$$\text{of } \Theta(z) \text{ the identity } (\Theta(\bar{z}^{-1}))^* J \Theta(z) = J \text{ holds.}$$
$$\text{Here } J = J^* = J^{-1} \text{ is a fixed } p \times p \text{ signature matrix.}$$

A criterion for existence of such $\Theta(z)$ is given by the following result which is parallel to Theorem 18.3.4; it is a special case of Theorem 7.3.1.

   THEOREM 18.3.6. *There exists a rational matrix function $\Theta(z)$ with invertible value at infinity and which satisfies the conditions* (18.3.30i)–(18.3.30iii) *if and only if the Hermitian matrix*

(18.3.31)
$$H = \left[\frac{\varphi_i J \varphi_j^*}{1 - z_i \bar{z}_j}\right]_{1 \le i,j \le n}$$

is invertible. In this case, one such $\Theta(z)$ is given by the formula

(18.3.32)
$$\Theta(z) = D - \sum_{j=1}^{n} (1 - z\bar{z}_j)^{-1} J \varphi_j^* \gamma_j D,$$

where

$$\begin{bmatrix} \gamma_1 \\ \vdots \\ \gamma_n \end{bmatrix} = H^{-1} \begin{bmatrix} \varphi_1 \\ \vdots \\ \varphi_n \end{bmatrix},$$

and D is a constant matrix. Also,

$$(18.3.33) \qquad \Theta(z)^{-1} = D_1 - \sum_{i=1}^{n} (z - z_i)^{-1} D_1 \widetilde{\chi}_i \varphi_i,$$

where

$$[\widetilde{\chi}_1 \widetilde{\chi}_2 \cdots \widetilde{\chi}_n] = [\overline{z}_1^{-1} J \varphi_1^*, \cdots, \overline{z}_n^{-1} J \varphi_n^*] H^{-1},$$

and D_1 is a constant matrix.

The formula (18.3.32) is a particular case of Theorem 7.3.1 (with

$$A_\zeta = \begin{bmatrix} z_1 & & & 0 \\ & z_2 & & \\ & & \ddots & \\ 0 & & & z_n \end{bmatrix}, \qquad B = \begin{bmatrix} \varphi_1 \\ \varphi_2 \\ \vdots \\ \varphi_n \end{bmatrix},$$

and with C and A_π empty). We give a direct proof of this theorem; it follows generally the pattern of the proof of Theorem 18.3.4.

PROOF OF THEOREM 18.3.6. Assume there is rational $\Theta(z)$ with invertible value at infinity which satisfies (18.3.30i)–(18.3.30ii). The condition (18.3.30iii) allows us to verify that the only zeros of

$$\Theta(z)^{-1} = J\big(\Theta(\overline{z}^{-1})\big)^* J$$

occur at $\overline{z}_1^{-1}, \ldots, \overline{z}_n^{-1}$, and these points are simple zeros of $\det \Theta(z)^{-1}$. Using Lemma 18.3.1 and the Liouville's Theorem, just as in the proof of Theorem 18.3.4, we conclude that

$$(18.3.34) \qquad \Theta(z)^{-1} = D + \sum_{i=1}^{n} (z - z_i)^{-1} \chi_i \varphi_i$$

for some nonzero column vectors χ_i, where $D = \big(\Theta(\infty)\big)^{-1}$. So

$$J\Theta(\overline{z}^{-1})^{*-1} J = J\left(D^* + \sum_{i=1}^{n} (z^{-1} - \overline{z}_i)^{-1} \varphi_i^* \chi_i^* \right) J.$$

On the other hand, $J\Theta(\overline{z}^{-1})^{*-1} J = \Theta(z)$. So

$$\Theta(z) = JD^* J + \sum_{i=1}^{n} (z^{-1} - \overline{z}_i)^{-1} J \varphi_i^* \chi_i^* J$$

$$(18.3.35) \qquad = JD^* J - \sum_{i=1}^{n} \overline{z}_i^{-1} J \varphi_i^* \chi_i^* J - \sum_{i=1}^{n} (z - \overline{z}_i^{-1})^{-1} \overline{z}_i^2 J \varphi_i^* \chi_i^* J$$

$$= D_0 - \sum_{i=1}^{n} (z - \overline{z}_i^{-1})^{-1} \overline{z}_i^2 J \varphi_i^* \chi_i^* J,$$

where $D_0 = \Theta(\infty)$. By Lemma 18.3.1(b) $\det \Theta(z)^{-1}$ has a simple zero at $\bar{z}_i^{-1}$, and

$$\mathrm{Ker}\big(\Theta(z)^{-1}\big)_{z=\bar{z}_i^{-1}} = \mathrm{Span}\{J\varphi_i^*\}.$$

We now use Theorem 18.3.3 applied to $W(z) = \Theta(z)D_0^{-1}$ and conclude that the matrix

(18.3.36)
$$S = \left[\frac{\varphi_i J \varphi_j^*}{\bar{z}_j^{-1} - z_i}\right]_{1 \le i,j \le n}$$

is invertible. Then H (given by (18.3.31)) is invertible as well. The same Theorem 18.3.3 provides also the formulas for $W(z)$ and its inverse:

$$\Theta(z)D_0^{-1} = I + \sum_{j=1}^{n}(z - \bar{z}_j^{-1})^{-1}J\varphi_j^*\gamma_j;$$

$$D_0\Theta(z)^{-1} = I - \sum_{i=1}^{n}(z - z_i)^{-1}\tilde{\chi}_i\varphi_i;$$

where

$$\begin{bmatrix} \gamma_1 \\ \gamma_2 \\ \vdots \\ \gamma_n \end{bmatrix} = S^{-1}\begin{bmatrix} \varphi_1 \\ \vdots \\ \varphi_n \end{bmatrix}; \qquad [\tilde{\chi}_1 \cdots \tilde{\chi}_n] = [J\varphi_1^* \cdots J\varphi_n^*]S^{-1}.$$

These formulas can be easily rewritten in the form (18.3.32) and (18.3.33).

Conversely, assume that H is invertible. Then the matrix S (given by (18.3.36)) is invertible as well. By Theorem 18.3.3 a rational matrix function $\Theta(z)$ exists with the following properties:

(α) The only zeros of $\det \Theta(z)$ in the extended complex plane are $z_1, \ldots, z_n$; these zeros are simple and

$$\mathrm{Ker}_\ell \Theta(z_i) = \mathrm{Span}\{\varphi_i\}; \qquad i = 1, \ldots, n;$$

(β) The only zeros of $\det \Theta(z)^{-1}$ in the extended complex plane are $\bar{z}_1^{-1}, \ldots, \bar{z}_n^{-1}$; these zeros are simple and

$$\mathrm{Ker}\,\Theta(z_i)^{-1}\big|_{z=\bar{z}_i^{-1}} = \mathrm{Span}\{J\varphi_i^*\}.$$

If we consider $\Theta_1(z) = \Theta(z)D^{-1}$ in place of $\Theta(z)$, these properties still hold true, and in addition $\Theta_1(\infty) = I$. Then $\Theta_1(z)$ is given by the formula (which follows from (18.3.16))

(18.3.37)
$$\Theta_1(z) = I + \sum_{j=1}^{n}(z - \bar{z}_j^{-1})^{-1}J\varphi_j^*\gamma_j$$

where

$$
\begin{bmatrix} \gamma_1 \\ \vdots \\ \gamma_n \end{bmatrix} = S^{-1} \begin{bmatrix} \varphi_1 \\ \vdots \\ \varphi_n \end{bmatrix} = \begin{bmatrix} \overline{z}_1^{-1} & & 0 \\ & \ddots & \\ 0 & & \overline{z}_n^{-1} \end{bmatrix} H^{-1} \begin{bmatrix} \varphi_1 \\ \vdots \\ \varphi_n \end{bmatrix}
$$

and the inverse $\Theta_1(z)^{-1}$ is given by

(18.3.38)
$$
\Theta_1(z)^{-1} = I - \sum_{i=1}^{n} (z - z_i)^{-1} \chi_i \varphi_i,
$$

where

$$
[\chi_1 \cdots \chi_n] = [J\varphi_1^* \cdots J\varphi_n^*] S^{-1}
$$
$$
= [J\varphi_1^* \cdots J\varphi_n^*] \operatorname{diag}(\overline{z}_1^{-1}, \ldots, \overline{z}_n^{-1}) H^{-1}.
$$

Using formula (18.3.37) we now have

$$
J\Theta_1(\overline{z}^{-1})^* J = J \left(I + \sum_{j=1}^{n} (z^{-1} - z_j^{-1})^{-1} \gamma_j^* \varphi_j J \right) J
$$

(18.3.39)
$$
= I + \sum_{j=1}^{n} (z^{-1} - z_j^{-1})^{-1} J\gamma_j^* \varphi_j
$$
$$
= \left[I - \sum_{j=1}^{n} z_j J\gamma_j^* \varphi_j \right] - \sum_{j=1}^{n} (z - z_j)^{-1} z_j^2 J\gamma_j^* \varphi_j.
$$

If we introduce the matrices

$$
A_\zeta = \begin{bmatrix} z_1 & & 0 \\ & \ddots & \\ 0 & & z_n \end{bmatrix}, \qquad B_\varphi = \begin{bmatrix} \varphi_1 \\ \vdots \\ \varphi_n \end{bmatrix}
$$

and use the formulas for $\gamma_1, \ldots, \gamma_n$ in (18.3.37) and for $\chi_1, \ldots, \chi_n$ in (18.3.38), then (18.3.38) and (18.3.39) can be given matrix forms

(18.3.40)
$$
\Theta_1(z)^{-1} = I - JB_\varphi^* A_\zeta^{*-1} H^{-1} (zI - A_\zeta)^{-1} B_\varphi
$$

and

$$
J\Theta_1(\overline{z}^{-1})^* J = [I - JB_\varphi^* H^{-1} B_\varphi] - JB_\varphi^* H^{-1} A_\zeta (zI - A_\zeta)^{-1} B_\varphi.
$$

Since none of $z_1, \ldots, z_n$ is zero, we know that $\Theta_1(0)$ is invertible, and hence

$$
J\Theta_1(0)^* J = I - JB_\varphi^* H^{-1} B_\varphi
$$

is invertible. Let us now form the matrix function

(18.3.41)
$$
\Psi(z) := [I - JB_\varphi^* H^{-1} B_\varphi]^{-1} J\Theta_1(\overline{z}^{-1})^* J
$$
$$
= I - [I - JB_\varphi^* H^{-1} B_\varphi]^{-1} JB_\varphi^* H^{-1} A_\zeta (zI - A_\zeta)^{-1} B_\varphi.
$$

From the Stein equation satisfied by H, namely

$$H - A_\zeta H A_\zeta^* = B_\varphi J B_\varphi^*$$

we have

$$[I - J B_\varphi^* H^{-1} B_\varphi] J B_\varphi^* A_\zeta^{*-1} H^{-1}$$
$$= J B_\varphi^* A_\zeta^{*-1} H^{-1} - J B_\varphi^* H^{-1} [H - A_\zeta H A_\zeta^*] A_\zeta^{*-1} H^{-1}$$
$$= J B_\varphi^* H^{-1} A_\zeta,$$

that is,

$$J B_\varphi^* A_\zeta^{*-1} H^{-1} = [I - J B_\varphi^* H^{-1} B_\varphi]^{-1} J B_\varphi^* H^{-1} A_\zeta.$$

Comparison of (18.3.40) with (18.3.41) combined with this identity gives that $\Psi(z) = \Theta_1(z)^{-1}$. From the definition (18.3.41) of $\Psi(z)$, this gives that

$$\Theta_1(z^{-1})^* J \Theta_1(z) = J - J B_\varphi^* H^{-1} B_\varphi.$$

If we now define $\Theta(z)$ by

$$\Theta(z) = \Theta_1(z) \Theta_1(1)^{-1}$$

then $\Theta(z)$ meets all desired requirements. $\square$

We are now ready to complete the proof of Theorem 18.2.1.

PROOF OF THEOREM 18.2.1. We suppose Θ is a matrix function which satisfies (I)–(IV) of Theorem 18.2.1. Let G be an $M \times N$ matrix function which satisfies (i) and (ii) of Theorem 18.2.1. Since Θ satisfies (I) and (II), $\Theta_{22}(z)$ is analytic and invertible on $\overline{\Delta}$ (including at infinity) and

$$\sup_{z \in \Delta} \|\Theta_{22}(z)^{-1} \Theta_{21}(z)\| < 1$$

by Proposition 13.1.2. Thus $\sup_{z \in \Delta} \|\Theta_{22}(z)^{-1} \Theta_{21}(z) G(z)\| < 1$ and hence $\Theta_{21} G + \Theta_{22} = \Theta_{22}(\Theta_{22}^{-1} \Theta_{21} G + I)$ is analytic and invertible on Δ. We conclude that $F = (\Theta_{11} G + \Theta_{12})(\Theta_{21} G + \Theta_{22})^{-1}$ is analytic on $\overline{\Delta}$. Once this is established, the implication (18.2.6) is rigorous, so we have established that F satisfies (i) and (ii) of Theorem 18.2.1. To show that F satisfies also (iii) of Theorem 18.2.1 is equivalent to showing that F satisfies (18.2.7), or equivalently that (18.2.8) holds for $1 \le i \le n$. As X and G are analytic at each z_i this is an immediate consequence of (III).

The converse is somewhat more delicate. Suppose F is a solution of (i), (ii), and (iii) of Theorem 18.2.1. Define rational matrix functions G_1 and G_2 by

(18.3.42)
$$\begin{bmatrix} F \\ I \end{bmatrix} = \Theta \begin{bmatrix} G_1 \\ G_2 \end{bmatrix}.$$

We claim G_1 and G_2 are analytic on Δ, i.e., that $\Theta^{-1} \begin{bmatrix} F \\ I \end{bmatrix}$ is analytic on Δ.

By (IV) the only poles of Θ^{-1} in Δ occur at $z_1, \ldots, z_n$. If z_i is one of these poles, then by (III) and (IV) together with Lemma 18.3.1, $\Theta(z)^{-1}$ has a Laurent expansion near z_i of the form

$$\Theta(z)^{-1} = \sum_{j=-1}^{\infty} \Theta_j^{(i)}(z - z_i)^j$$

where

$$\operatorname{Im}_\ell \Theta_{-1}^{(i)} = \operatorname{Span}\{[x_i, -y_i]\}.$$

From the interpolation condition (iii) of Theorem 18.2.1, we conclude that $\Theta^{-1}\begin{bmatrix} F \\ I \end{bmatrix}$ is analytic at z_i, for each $i = 1, 2, \ldots, n$. Thus $\begin{bmatrix} G_1 \\ G_2 \end{bmatrix} = \Theta^{-1}\begin{bmatrix} F \\ I \end{bmatrix}$ is analytic on Δ as asserted.

We next claim that $\det G_2$ does not vanish on Δ. Since $\Theta(z)$ is J-unitary for $z \in \partial\Delta$, we have

(18.3.43)
$$\begin{aligned}
0 > F(z)^*F(z) - I &= [\, F(z)^* \quad I\,]J\begin{bmatrix} F(z) \\ I \end{bmatrix} \\
&= [\, G_1(z)^* \quad G_2(z)^*\,]\Theta(z)^*J\Theta(z)\begin{bmatrix} G_1(z) \\ G_2(z) \end{bmatrix} \\
&= [\, G_1(z)^* \quad G_2(z)^*\,]J\begin{bmatrix} G_1(z) \\ G_2(z) \end{bmatrix} \\
&= G_1(z)^*G_1(z) - G_2(z)^*G_2(z)
\end{aligned}$$

for $z \in \partial\Delta$. Thus if $G_2(z)x = 0$ and $x \neq 0$ for some $z \in \partial\Delta$ we have

$$0 > \|G_1(z)x\|^2 \geq 0$$

a contradiction. Thus $\det G_2(z) \neq 0$ for $z \in \partial\Delta$, and the above argument implies

$$\|G_1(z)G_2(z)^{-1}\| < 1$$

for $z \in \partial\Delta$. As by Proposition 13.1.2 $\|\Theta_{22}(z)^{-1}\Theta_{21}(z)\| < 1$ for $z \in \partial\Delta$, we obtain that

(18.3.44)
$$I + t\Theta_{22}(z)^{-1}\Theta_{21}(z)G_1(z)G_2(z)^{-1}$$

is invertible for $z \in \partial\Delta$ and $0 \leq t \leq 1$.

At this point we will use some properties of the winding number of a continuous nowhere zero complex valued function $f(z)$ defined on $\partial\Delta$ (notation: wno $f(z)$). By definition, wno $f(z)$ is the net change in the argument of $f(z)$ along $\partial\Delta$. Thus,

$$\operatorname{wno} f(z) = \frac{1}{2\pi}\left[\lim_{w \to \infty} \arg f(iw) - \lim_{w \to -\infty} \arg f(iw)\right], \qquad \Delta = \Pi^+$$

and

$$\text{wno } f(z) = \frac{1}{2\pi}\left[\lim_{\theta \to 2\pi-} \arg f(e^{i\theta}) - \lim_{\theta \to 0+} \arg f(e^{i\theta})\right], \qquad \Delta = \mathcal{D},$$

where in both cases a continuous branch of the argument function is taken. Among the basic properties of the winding number that we need is the logarithmic property:

$$\text{wno}\big(f(z)g(z)\big) = \text{wno } f(z) + \text{wno } g(z).$$

Also, if $f(z)$ admits meromorphic continuation into Δ, then by the Argument Principle

$$\text{wno } f(z) = (\text{number of zeros of } f(z) \text{ in } \Delta)$$
$$- (\text{number of poles of } f(z) \text{ in } \Delta),$$

with zeros and poles counted with multiplicities.

Returning to the functions (18.3.44), observe that the winding number of the function

$$\det\big(I + t\Theta_{22}(z)^{-1}\Theta_{21}(z)G_1(z)G_2(z)^{-1}\big)$$

is a continuous, integer-valued function of $t \in [0, 1]$, and hence is independent of t. Thus,

$$\text{wno } \det\big(I + \Theta_{22}(z)^{-1}\Theta_{21}(z)G_1(z)G_2(z)^{-1}\big) = \text{wno } \det I = 0.$$

From (18.3.42) we have $I = \Theta_{21}G_1 + \Theta_{22}G_2$, so

(18.3.44)
$$\begin{aligned}
0 &= \text{wno } \det I = \text{wno } \det(\Theta_{21}G_1 + \Theta_{22}G_2)\\
&= \text{wno } \det(\Theta_{22}[\Theta_{22}^{-1}\Theta_{21}G_1G_2^{-1} + I]G_2)\\
&= \text{wno } \det \Theta_{22} + \text{wno } \det(\Theta_{22}^{-1}\Theta_{21}G_1G_2^{-1} + I) + \text{wno } \det G_2\\
&= \text{wno } \det G_2.
\end{aligned}$$

Here we used the fact established in the first part of the proof that Θ_{22} is analytic and invertible on Δ, so wno $\det \Theta_{22} = 0$. As we have already established that G_2 is analytic on Δ, again by the Argument Principle (18.3.44) implies that $\det G_2$ has no zeros in Δ as claimed. We now define $G = G_1G_2^{-1}$. As G_1 and G_2^{-1} are each analytic on Δ we see that G satisfies (i). By (18.3.43), $\sup_{z \in \partial\Delta}\|G(z)\| < 1$. Then by the maximum modulus theorem G satisfies (ii). Moreover, from (18.3.42) we see that

$$F = \Theta_{11}G_1 + \Theta_{12}G_2 = (\Theta_{11}G + \Theta_{12})G_2$$

and

$$I = \Theta_{21}G_1 + \Theta_{22}G_2 = (\Theta_{21}G + \Theta_{22})G_2.$$

Combining these two gives that $F = (\Theta_{11}G + \Theta_{12})(\Theta_{21}G + \Theta_{22})^{-1}$. Thus any solution F of (i), (ii), (iii) of Theorem 18.2.1 has the form (18.2.1) with G satisfying (i) and (ii) as asserted. $\square$

18.4 GENERALIZED NEVANLINNA-PICK INTERPOLATION: PROBLEM FORMULATION

In this section we state a far reaching generalization of the matrix Nevanlinna-Pick interpolation problem discussed in Section 18.2.

We again assume either $\Delta = \Pi^+$ or $\Delta = \mathcal{D}$. As everywhere in this book, we denote by $\mathcal{R}_{M \times N}(\Delta)$ the set of all $M \times N$ rational matrix functions without poles in Δ, and by $\mathcal{BR}_{M \times N}(\Delta)$ the set of all $M \times N$ rational matrix functions $F(z) \in \mathcal{R}_{M \times N}(\Delta)$ such that

$$\sup_{z \in \Delta} \|F(z)\| < 1.$$

The general problem is to find a $M \times N$ matrix function $F \in \mathcal{BR}_{M \times N}(\Delta)$ which satisfies interpolation conditions of the form

$$(18.4.1) \qquad \frac{1}{2\pi i} \int_{\gamma} (zI - A_\zeta)^{-1} B_+ F(z) dz = -B_-$$

(in the right hand side we write $-B_-$ rather than B_- as a matter of convenience having in mind subsequent applications)

$$(18.4.2) \qquad \frac{1}{2\pi i} \int_{\gamma} F(z) C_- (zI - A_\pi)^{-1} dz = C_+$$

and

$$(18.4.3) \qquad \frac{1}{2\pi i} \int_{\gamma} (zI - A_\zeta)^{-1} B_+ F(z) C_- (zI - A_\pi)^{-1} dz = S.$$

Here γ is a suitable contour in Δ enclosing $\sigma(A_\pi) \cup \sigma(A_\zeta)$ and the data set $w = (C_+, C_-, A_\pi; A_\zeta, B_+, B_-; S)$ consists of a full range pair (A_ζ, B_+) of matrices of sizes $n_\zeta \times n_\zeta$ and $n_\zeta \times M$ respectively, an $n_\zeta \times N$ matrix B_-, a null kernel pair of matrices (C_-, A_π) of sizes $N \times n_\pi$ and $n_\pi \times n_\pi$ respectively, C_+ is a $M \times n_\pi$ matrix and S is a $n_\zeta \times n_\pi$ matrix. We assume in addition (without loss of generality) that $\sigma(A_\zeta)$ and $\sigma(A_\pi)$ are contained in Δ. Using residues, (18.4.1)–(18.4.3) can be written in an equivalent form which does not involve γ:

$$\sum_{z_0 \in \Delta} \text{Res}_{z=z_0} F(z) C_- (zI - A_\pi)^{-1} = C_+$$

$$\sum_{z_0 \in \Delta} \text{Res}_{z=z_0} (zI - A_\zeta)^{-1} B_+ F(z) = -B_-$$

$$\sum_{z_0 \in \Delta} \text{Res}_{z=z_0} (zI - A_\zeta)^{-1} B_+ F(z) C_- (zI - A_\pi)^{-1} = S.$$

As usual, $\text{Res}_{z=z_0} W(z)$ stands for the residue of the matrix function $W(z)$ at z_0.

The case where C_-, A_π, C_+ and S are all vacuous was considered in Nudelman [1971]. We therefore will call this general problem the *two-sided Nudelman problem*. A set of data w satisfying all the conditions laid out above we shall call an *admissible data set* for a two-sided Nudelman problem over Δ.

Thus, the two-sided Nudelman problem is the TSCII problem (described in Section 16.9) with $\sigma = \Delta$ and with the extra condition that $F \in \mathcal{BR}_{M \times N}(\Delta)$. So Lemma 16.9.1 is applicable, and we do not gain in generality by removing the full range assumption on (A_ζ, B_+), the null kernel assumption on (C_-, A_π), or by dropping the requirement that $\sigma(A_\zeta) \cup \sigma(A_\pi) \subset \Delta$. Also, if there is a solution $F \in \mathcal{BR}_{M \times N}(\Delta)$ to the interpolation problem (18.4.1)–(18.4.3), then necessarily S satisfies the Sylvester equation

$$(18.4.4) \qquad SA_\pi - A_\zeta S = B_- C_- + B_+ C_+ .$$

Recall that if the spectra of A_π and A_ζ are disjoint then the solution S to the Sylvester equation (18.4.4) is unique. With this compatible choice of S, the interpolation condition (18.4.3) follows automatically from condition (18.4.1) and (18.4.2). Thus for the case where A_ζ and A_π have disjoint spectra in Δ, the two-sided Nudelman problem collapses to only the two conditions (18.4.1) and (18.4.2).

For example and illustrations of the interpolation condition (18.4.1)–(18.4.3), we refer the reader to Examples 16.2.1–16.2.4 and to Theorems 16.9.3 and 16.10.1 for the general picture.

We now return to the general two-sided Nudelman problem (18.4.1)–(18.4.3). As in Chapter 16, we say that the set consisting of seven matrices

$$w = (C_+, C_-, A_\pi; A_\zeta, B_+, B_-; S)$$

is Δ-*admissible Sylvester data set* if the following properties are valid:

(i) C_+, C_-, A_π, A_ζ, B_+, B_-, S have sizes $M \times n_\pi$, $N \times n_\pi$, $n_\pi \times n_\pi$, $n_\zeta \times n_\zeta$, $n_\zeta \times M$, $n_\zeta \times N$, $n_\zeta \times n_\pi$, respectively;

(ii) (C_-, A_π) is a null kernel pair and (A_ζ, B_+) is a full range pair;

(iii) $\sigma(A_\pi) \cup \sigma(A_\zeta) \subset \Delta$;

(iv) the matrix S satisfies the Sylvester equation

$$SA_\pi - A_\zeta S = B_+ C_+ + B_- C_- .$$

By Lemma 16.9.1, without loss of generality we can assume that the data set for the two-sided Nudelman problem is Δ-admissible Sylvester. This assumption will be made throughout the rest of this chapter.

We conclude this section with a discussion of the divisor-remainder formulation of the two-sided Nudelman interpolation problem. This amounts to writing the interpolation conditions (18.4.1)–(18.4.3) in the equivalent form presented in Theorem 16.9.3. Given are rational matrix functions K, ψ, φ of respective sizes $M \times N$, $M \times M$ and $N \times N$ having no poles in $\overline{\Delta}$. We also suppose that ψ and φ are regular on $\partial \Delta$ (no pole or zero on $\partial\Delta$). The interpolation problem we consider here is that of finding all $F \in \mathcal{BR}_{M \times N}(\Delta)$ which are in the coset $K + \psi\mathcal{R}_{M \times N}(\Delta)\varphi$. This formulation of the interpolation problems has been studied at length by Ball and Helton [1983] and originally by Sarason [1967] (with nonrational functions) for the scalar case. Let us call this formulation the *Sarason coset interpolation problem*. By Theorem 16.9.3, one can see that this problem (with the above analyticity assumptions on K, ψ, φ) is exactly equivalent to a two-sided Nudelman problem. Indeed, given K, ψ, φ as above, let (A_ζ, B_+) be a left null pair for $\psi(z)$ over Δ, let (C_-, A_π) be a right null pair for $\varphi(z)$ over Δ, and define matrices B_-, C_+ and S by

$$(18.4.5) \qquad -B_- = \frac{1}{2\pi i} \int_{\partial\Delta} (zI - A_\zeta)^{-1}B_+K(z)dz$$

$$(18.4.6) \qquad C_+ = \frac{1}{2\pi i} \int_{\partial\Delta} K(z)C_-(zI - A_\pi)^{-1}dz$$

$$(18.4.7) \qquad S = \frac{1}{2\pi i} \int_{\partial\Delta} (zI - A_\zeta)^{-1}B_+K(z)C_-(zI - A_\pi)^{-1}dz.$$

Then by Lemma 16.9.1 and its proof S satisfies the Sylvester equation (18.4.4) and the collection of data $w = (C_+, C_-, A_\pi; A_\zeta, B_+, B_-; S)$ is a Δ-admissible Sylvester TSCII data set.

Using Theorem 16.9.3 we conclude that the set

$$[K + \psi\mathcal{R}_{M \times N}(\Delta)\varphi] \cap \mathcal{BR}_{M \times N}(\Delta)$$

coincides with the set of solutions of the two-sided Nudelman problem associated with the data set w.

18.5 SOLUTION OF TWO-SIDED NUDELMAN PROBLEM

In this section we present our solution of the two-sided Nudelman interpolation problem over a domain $\Delta = \Pi^+$ or $\Delta = \mathcal{D}$. We are given Δ-*admissible Sylvester TSCII data* set ω, i.e. a collection of matrices of suitable sizes

$$(18.5.1) \qquad \omega = (C_+, C_-, A_\pi; A_\zeta, B_+, B_-; S)$$

where (A_ζ, B_+) is a full range pair with $\sigma(A_\zeta) \subset \Delta$, (C_-, A_π) is a null kernel pair with $\sigma(A_\pi) \subset \Delta$, B_-, C_+ and S are matrices of compatible zeros and S satisfies the Sylvester equation

$$(18.5.2) \qquad SA_\pi - A_\zeta S = B_-C_- + B_+C_+.$$

To state the solution of the two-sided Nudelman problem we need to introduce the *Pick matrix* $\Lambda(\omega)$ for the problem defined by

$$(18.5.3) \qquad \Lambda(\omega) = \begin{bmatrix} S_1 & S^* \\ S & S_2 \end{bmatrix}$$

where S_1 and S_2 are the unique (necessarily Hermitian) solutions of the Lyapunov equations

(18.5.4a)
$$S_1 A_\pi + A_\pi^* S_1 = C_-^* C_- - C_+^* C_+$$

(18.5.4b)
$$S_2 A_\zeta^* + A_\zeta S_2 = B_+ B_+^* - B_- B_-^*$$

for the case $\Delta = \Pi^+$, or where S_1 and S_2 are the unique (necessarily Hermitian) solutions of the Stein equations

(18.5.5a)
$$S_1 - A_\pi^* S_1 A_\pi = C_-^* C_- - C_+^* C_+$$

(18.5.5b)
$$S_2 - A_\zeta S_2 A_\zeta^* = B_+ B_+^* - B_- B_-^*$$

if $\Delta = \mathcal{D}$. For the case where $\Lambda = \Lambda(\omega)$ is invertible we introduce an auxiliary matrix function $\Theta(z)$ constructed completely from the data set ω, defined by

(18.5.6)
$$\Theta(z) = I + \begin{bmatrix} C_+ & -B_+^* \\ C_- & B_-^* \end{bmatrix} \begin{bmatrix} (zI - A_\pi)^{-1} & 0 \\ 0 & (zI + A_\zeta^*)^{-1} \end{bmatrix}$$
$$\times \Lambda(\omega)^{-1} \begin{bmatrix} -C_+^* & C_-^* \\ B_+ & B_- \end{bmatrix}$$

if $\Delta = \Pi^+$, and by

(18.5.7a)
$$\Theta(z) = D + \begin{bmatrix} C_+ & -B_+^* \\ C_- & B_-^* \end{bmatrix} \begin{bmatrix} (zI - A_\pi)^{-1} & 0 \\ 0 & (I - zA_\zeta^*)^{-1} \end{bmatrix}$$
$$\times (\Lambda(\omega))^{-1} \begin{bmatrix} A_\pi^{*-1} C_+^* & -A_\pi^{*-1} C_-^* \\ B_+ & B_- \end{bmatrix} D$$

where, fixing any $\alpha \in \mathbb{C}$ with $|\alpha| = 1$,

(18.5.7b)
$$D = I - \begin{bmatrix} C_+ & B_+^* A_\zeta^{*-1} \\ C_- & -B_-^* A_\zeta^{*-1} \end{bmatrix}$$
$$\times (\Lambda(\omega))^{-1} \begin{bmatrix} (I - \alpha A_\pi)^{-1} & 0 \\ 0 & (\alpha I - A_\zeta)^{-1} \end{bmatrix} \begin{bmatrix} -C_+^* & C_-^* \\ B_+ & B_- \end{bmatrix}$$

if $\Delta = \mathcal{D}$ and both A_π and A_ζ are nonsingular.

In case $\Delta = \mathcal{D}$ and at least one of A_π and A_ζ is singular, we use the more general formula

(18.5.7c)
$$\Theta(z) = I + (z - z_0) \begin{bmatrix} C_+ & -B_+^* \\ C_- & B_-^* \end{bmatrix} \begin{bmatrix} (zI - A_\pi)^{-1} & 0 \\ 0 & (I - zA_\zeta^*)^{-1} \end{bmatrix}$$
$$\times \Lambda(\omega)^{-1} \begin{bmatrix} (I - z_0 A_\pi^*)^{-1} C_+^* & -(I - z_0 A_\pi^*)^{-1} C_-^* \\ (A_\zeta - z_0 I)^{-1} B_+ & (A_\zeta - z_0 I)^{-1} B_- \end{bmatrix},$$

where z_0 is a chosen point on the unit circle.

We state now our main results on the solution of the two-sided Nudelman problem. It will be convenient to have two separate statements for the case $\Delta = \Pi^+$ and the case $\Delta = \mathcal{D}$.

THEOREM 18.5.1. *Let* $\Delta = \Pi^+$, *and let* $\omega = (C_+, C_-, A_\pi; A_\zeta, B_+, B_-; S)$ *be a* Δ-*admissible Sylvester TSCII data set. Let* S_1 *and* S_2 *be the unique solutions of the Lyapunov equations* (18.5.4a)–(18.5.4b)

$$(18.5.8) \qquad S_1 A_\pi + A_\pi^* S_1 = C_-^* C_- - C_+^* C_+$$

$$(18.5.9) \qquad S_2 A_\zeta^* + A_\zeta S_2 = B_+ B_+^* - B_- B_-^*.$$

Let $\Lambda = \Lambda(\omega)$ *be the Hermitian matrix given by* (18.5.3)

$$(18.5.10) \qquad \Lambda = \left[\begin{array}{cc} S_1 & S^* \\ S & S_2 \end{array} \right].$$

Then there exists a rational $M \times N$ *matrix function* F *satisfying the interpolation conditions*

$$(18.5.11a) \qquad \sum_{z_0 \in \Pi^+} \operatorname{Res}_{z=z_0} F(z) C_- (zI - A_\pi)^{-1} = C_+$$

$$(18.5.11b) \qquad \sum_{z_0 \in \Pi^+} \operatorname{Res}_{z=z_0} (zI - A_\zeta)^{-1} B_+ F(z) = -B_-$$

$$(18.5.11c) \qquad \sum_{z_0 \in \Pi^+} \operatorname{Res}_{z=z_0} (zI - A_\zeta)^{-1} B_+ F(z) C_- (zI - A_\pi)^{-1} = S$$

together with the metric constraint $\sup_{z \in \Pi^+} \|F(z)\| < 1$ *if and only if* Λ *is positive definite. In this case, the set of all solutions* F *is given by*

$$(18.5.12) \qquad F = (\Theta_{11} G + \Theta_{12})(\Theta_{21} G + \Theta_{22})^{-1}$$

where G *is an arbitrary* $M \times N$ *matrix function analytic on* Π^+ *with* $\sup_{z \in \Pi^+} \|G(z)\|$, *and*

$$\Theta(z) = \left[\begin{array}{cc} \Theta_{11}(z) & \Theta_{12}(z) \\ \Theta_{21}(z) & \Theta_{22}(z) \end{array} \right]$$

is the matrix function determined from the data set ω *by* (18.5.6).

THEOREM 18.5.2. *Let* $\Delta = \mathcal{D}$, *and let* $\omega = (C_+, C_-, A_\pi; A_\zeta, B_+, B_-; S)$ *be a* Δ-*admissible Sylvester TSCII data set. Let* S_1 *and* S_2 *be the unique solutions of the Stein equations* (18.5.5a)–(18.5.5b)

$$(18.5.13) \qquad S_1 - A_\pi^* S_1 A_\pi = C_-^* C_- - C_+^* C_+$$

and

(18.5.14) $$S_2 - A_\zeta S_2 A_\zeta^* = B_+ B_+^* - B_- B_-^*$$

and let $\Lambda = \Lambda(\omega)$ *be the Hermitian matrix* $\Lambda = \begin{bmatrix} S_1 & S^* \\ S & S_2 \end{bmatrix}$. *Then there exists a rational* $M \times N$ *matrix function* F *satisfying the interpolation conditions*

(18.5.15a) $$\sum_{z_0 \in \mathcal{D}} \operatorname{Res}_{z=z_0} F(z) C_- (zI - A_\pi)^{-1} = C_+$$

(18.5.15b) $$\sum_{z_0 \in \mathcal{D}} \operatorname{Res}_{z=z_0} (zI - A_\zeta)^{-1} B_+ F(z) = -B_-$$

(18.5.15c) $$\sum_{z_0 \in \mathcal{D}} \operatorname{Res}_{z=z_0} (zI - A_\zeta)^{-1} B_+ F(z) C_- (zI - A_\pi)^{-1} = S$$

together with the metric constraints $\sup_{z \in \mathcal{D}} \|F(z)\| < 1$ *if and only if* Λ *is positive definite.*

In this case, the set of all solutions F is given by (18.5.12) where G is an arbitrary rational $M \times N$ matrix function analytic on $\mathcal{D}$ with $\sup_{z \in \mathcal{D}} \|G(z)\| < 1$, and $\Theta(z) = \begin{bmatrix} \Theta_{11}(z) & \Theta_{12}(z) \\ \Theta_{21}(z) & \Theta_{22}(z) \end{bmatrix}$ is the matrix function determined by the data set ω by formulas (18.5.7a) and (18.5.7b) if both A_π and A_ζ are nonsingular, and more generally, by formula (18.5.7c) if at least one of A_π and A_ζ is singular.

Full proofs of Theorems 18.5.1 and 18.5.2 will be given in the next section.

The variation of the two-sided Nudelman interpolation problem — when only the conditions (18.5.11a) and (18.5.11b) (if $\Delta = \Pi^+$) or (18.5.15a) and (18.5.15b) (if $\Delta = \mathcal{D}$) are prescribed — can also be analyzed. In other words, we leave out the interpolation conditions (18.5.11c) (if $\Delta = \Pi^+$) or (18.5.15c) if $\Delta = \mathcal{D}$. For this variation of the Nudelman interpolation problem it is natural to consider the data set ω in the form

(18.5.16) $$\omega = (C_+, C_-, A_\pi; A_\zeta, B_+, B_-),$$

where the matrices C_+, C_-, A_π, A_ζ, B_+ and B_- are of sizes $M \times n_\pi$, $N \times n_\pi$, $n_\pi \times n_\pi$, $n_\zeta \times n_\zeta$, $n_\zeta \times M$, and $n_\zeta \times N$, respectively, and, in addition, the following properties hold:

(i) (C_-, A_π) is a null kernel pair and (A_ζ, B_+) is a full range pair;

(ii) $\sigma(A_\pi) \cup \sigma(A_\zeta) \subset \Delta$.

The set of all solutions to this modified problem in general is not given by a single linear fractional formula and the problem in general cannot be converted to a two-sided divisor-remainder form. The complete description of the set of all solutions to this modified two-sided Nudelman interpolation problem is easily obtained from Theorems 18.5.1 and 18.5.2, as follows.

THEOREM 18.5.3. *Suppose the data set* $\omega = (C_-, C_+, A_\pi; A_\zeta, B_+, B_-)$ *has the above properties* (i) *and* (ii). *Let* S_1 *and* S_2 *be the unique solutions of the Lyapunov equations* (18.5.8), (18.5.9) (*if* $\Delta = \Pi^+$) *or the Stein equations* (18.5.13), (18.5.14) (*if* $\Delta = \mathcal{D}$). *Then there exists a rational* $M \times N$ *matrix function* F *satisfying the interpolation conditions*

(18.5.17a)
$$\sum_{z_0 \in \Delta} \mathrm{Res}_{z=z_0} F(z)C_-(zI - A_\pi)^{-1} = C_+$$

(18.5.17b)
$$\sum_{z_0 \in \Delta} \mathrm{Res}_{z=z_0}(zI - A_\zeta)^{-1}B_+F(z) = -B_-$$

and with

(18.5.18)
$$\sup_{z \in \Delta} \|F(z)\| < 1$$

if and only if for some solution S *of the equation*

(18.5.19)
$$SA_\pi - A_\zeta S = B_+C_+ + B_-C_-$$

the associated matrix $\Lambda = \begin{bmatrix} S_1 & S^* \\ S & S_2 \end{bmatrix}$ *is positive definite. In this case all solutions of the interpolation problem* (18.5.17a), (18.5.17b) *with the metric constraints* (18.5.18) *are given by the formula*

$$F = (\Theta_{11}G + \Theta_{12})(\Theta_{21}G + \Theta_{22})^{-1},$$

where G *is an arbitrary rational* $M \times N$ *matrix function analytic in* Δ *with* $\sup_{z \in \Delta} \|G(z)\| < 1$, *and* $\Theta(z) = \begin{bmatrix} \Theta_{11}(z) & \Theta_{12}(z) \\ \Theta_{21}(z) & \Theta_{22}(z) \end{bmatrix}$ *is the matrix function determined by the data set* $(C_+, C_-, A_\pi; A_\zeta, B_+, B_-; S)$ *by formulas* (18.5.6), *or* (18.5.7a), (18.5.7b), *or* (18.5.7c), *as appropriate, with* S *sweeping through all solutions of* (18.5.19) *for which* $\begin{bmatrix} S_1 & S^* \\ S & S_2 \end{bmatrix}$ *is positive definite.*

PROOF. Suppose a solution F of (18.5.17a), (18.5.17b), (18.5.18) exists. Define S by

(18.5.20)
$$S = \sum_{z_0 \in \Delta} \mathrm{Res}_{z=z_0}(zI - A_\zeta)^{-1}B_+F(z)C_-(zI - A_\pi)^{-1}.$$

Then it follows that equality (18.5.19) is valid (see the proof of Lemma 16.9.1), and by Theorem 18.5.1 or 18.5.2 (as appropriate) the matrix Λ is positive definite. Conversely, if for some solution S of (18.5.19) the matrix Λ is positive definite, then by Theorem 18.5.1 or 18.5.2 there is a solution F to the interpolation problem (18.5.17a), (18.5.17b),

(18.5.18), (18.5.20). The rest of Theorem 18.5.3 follows again from Theorems 18.5.1 and 18.5.2. □

To illustrate the main results of this section, we compute an explicit form for the Pick matrix $\Lambda(\omega)$ for some specific examples.

EXAMPLE 18.5.1. Suppose C_-, A_π, C_+ and S are all vacuous (i.e. $n_\pi = 0$) and B_+, B_- and A_ζ are of the form

$$
B_+ = \begin{bmatrix} x_1 \\ \vdots \\ x_n \end{bmatrix}, \qquad B_- = -\begin{bmatrix} y_1 \\ \vdots \\ y_n \end{bmatrix}, \qquad A_\zeta = \begin{bmatrix} z_1 & & 0 \\ & \ddots & \\ 0 & & z_n \end{bmatrix}
$$

where $x_1, \ldots, x_n$ are nonzero row vectors in $\mathbf{C}^{1 \times M}$, $y_1, \ldots, y_n$ are row vectors in $\mathbf{C}^{1 \times N}$ and $z_1, \ldots, z_n$ are distinct points in Δ. Thus the interpolation conditions (18.5.11a)–(18.5.11c) (or (18.5.15a)–(18.5.15c)) collapse to (18.5.11b) (or (18.5.15b)) only. This in turn assumes the form of the set of left tangential interpolation conditions

$$
x_i F(z_i) = y_i, \qquad 1 \le i \le n
$$

as was discussed in Example 16.2.3.

Since C_-, A_π, C_+ and S are vacuous, the Pick matrix $\Lambda = \Lambda(\omega)$ given by (18.5.3) collapses to $\Lambda = S_2$, where

$$
S_2 A_\zeta^* + A_\zeta S_2 = B_+ B_+^* - B_- B_-^* \quad \text{if} \quad \Delta = \Pi^+
$$

and

$$
S_2 - A_\zeta S_2 A_\zeta^* = B_+ B_+^* - B_- B_-^* \quad \text{if} \quad \Delta = \mathcal{D}.
$$

Using the explicit form of A_ζ, B_+, B_-, and the results of Sections A.1 and A.2 on solutions of Sylvester and Stein equations (or by direct check), we get

$$
\Lambda = \begin{bmatrix} \dfrac{x_i x_j^* - y_i y_j^*}{\rho_\Delta(z_i, z_j)} \end{bmatrix}_{1 \le i,j \le n}
$$

where as usual

$$
\rho_\Delta(z_i, z_j) = z_i + \overline{z}_j \quad \text{if} \quad \Delta = \Pi^+
$$
$$
\rho_\Delta(z_i, z_j) = 1 - z_i \overline{z}_j \quad \text{if} \quad \Delta = \mathcal{D}. \quad \square
$$

EXAMPLE 18.5.2. This time suppose that A_ζ, B_+, B_- and S are all vacuous ($n_\zeta = 0$) and C_-, A_π, C_+ are given by

$$
C_- = [u_1, \ldots, u_m], \qquad C_+ = [v_1, \ldots, v_m], \qquad A_\pi = \begin{bmatrix} w_1 & & 0 \\ & \ddots & \\ 0 & & w_m \end{bmatrix}
$$

where $u_1, \ldots, u_m$ are nonzero vectors in $\mathbf{C}^N$, $v_1, \ldots, v_m$ are vectors in $\mathbf{C}^M$ and $w_1, \ldots, w_m$ are distinct points in Δ. This is just the transposed version of Example

18.5.1. Hence the set of interpolation conditions (18.5.11a)–(18.5.11c) (or (18.5.15a)–(18.5.15c)) collapses to a set of right tangential interpolation conditions

$$F(w_j)u_j = v_j, \qquad 1 \le j \le m.$$

The Pick matrix Λ (18.5.3) in this case collapses to $\Lambda = S_1$ and is given by

$$\Lambda = \left[\frac{u_i^* u_j - v_i^* v_j}{\rho_\Delta(w_j, w_i)} \right]_{1 \le i,j \le m} . \quad \square$$

EXAMPLE 18.5.3. This example combines Examples 18.5.1 and 18.5.2. We assume that A_ζ, B_+ and B_- are as in Example 18.5.1 and C_-, A_π, C_+ are as in Example 18.5.2 and that S is an $n \times m$ matrix. As noted in Example 18.5.1, the interpolation condition (18.5.11b) (or (18.5.15b)) amounts to

$$x_j F(z_j) = y_j, \qquad 1 \le j \le n$$

while by Example 18.5.2 the interpolation condition (18.5.11a) (or (18.5.15a)) amounts to

$$F(w_j)u_j = v_j, \qquad 1 \le j \le m.$$

As discussed in Section 16.8 (see (16.8.1)–(16.8.6)), the third interpolation condition (18.5.11c) (or (18.5.15c)) is automatic if no point z_i is equal to a point w_j for some i and j, as long as S satisfies the necessary consistency condition (18.5.2). In general the interpolation condition (18.5.11c) adds the set of bitangential interpolation conditions on $F'(z)$ given by

$$x_i F'(\xi_{ij})u_j = \rho_{ij} \quad \text{if} \quad z_i = \xi_{ij} = w_j$$

for each pair of indices (i, j) for which $z_i = w_j$, where ρ_{ij} is the (i, j) entry in S. The Pick matrix Λ in this case has the full 2×2 block decomposition

$$\Lambda = \left[\begin{array}{cc} S_1 & S^* \\ S & S_2 \end{array} \right]$$

where

$$S_1 = \left[\frac{u_i^* u_j - v_i^* v_j}{\rho_\Delta(w_j, w_i)} \right]_{1 \le i,j \le m}$$

$$S = [s_{ij}]_{1 \le i \le n, 1 \le j \le m}$$

with

$$s_{ij} = \frac{y_i u_j - x_i v_j}{z_i - w_j} \quad \text{if} \quad z_i \ne w_j$$

and

$$s_{ij} = \rho_i \quad \text{if} \quad z_i = w_j$$

and finally

$$S_2 = \left[\frac{x_i x_j^* - y_i y_j^*}{\rho_\Delta(z_i, z_j)} \right]_{1 \leq i,j \leq n} . \quad \square$$

EXAMPLE 18.5.4. Take C_-, A_π, C_+ and S vacuous ($n_\pi = 0$) while B_+ is the $nM \times M$ matrix

$$B_+ = \begin{bmatrix} I_M \\ \vdots \\ I_M \end{bmatrix}$$

and B_- has the form

$$B_- = - \begin{bmatrix} X_1 \\ \vdots \\ X_n \end{bmatrix}$$

where $X_1, \ldots, X_n$ are given $M \times N$ matrices, and A_ζ is the $nM \times nM$ block diagonal matrix

$$A_\zeta = \begin{bmatrix} z_1 I_M & & 0 \\ & \ddots & \\ 0 & & z_n I_M \end{bmatrix} .$$

Then the interpolation conditions (18.5.11a)–(18.5.11c) collapse to the single condition (18.5.11b) which in turn takes the same form as in Example 16.2.1:

$$F(z_j) = X_j \quad \text{for} \quad j = 1, \ldots, n.$$

As in Example 18.5.1, C_-, A_π, C_+ and S are vacuous, so Λ collapses to S_2. It is easy then to get that Λ has the block matrix form

$$\Lambda = \left[\frac{I - X_i X_j^*}{\rho_\Delta(z_i, z_j)} \right]_{1 \leq i,j \leq n} . \quad \square$$

EXAMPLE 18.5.5. Let us again take C_-, A_π, C_+ and S vacuous ($n_\pi = 0$) but let

$$B_+ = \begin{bmatrix} x_1 \\ \vdots \\ x_n \end{bmatrix}, \qquad B_- = - \begin{bmatrix} y_1 \\ \vdots \\ y_n \end{bmatrix}$$

and let A_ζ be the $n \times n$ lower triangular Jordan block

$$A_\zeta = \begin{bmatrix} z_0 & 0 & & \cdots & & 0 \\ 1 & z_0 & & & & \vdots \\ \vdots & \ddots & & \ddots & & 0 \\ 0 & 0 & & \cdots & 1 & z_0 \end{bmatrix}$$

with eigenvalue z_0 in Δ. The relevant interpolation condition (18.5.11b) takes the form

$$\frac{d^{j-1}}{dz^{j-1}}\{x(z)F(z)\}\Big|_{z=z_0} = \frac{d^{j-1}}{dz^{j-1}}y(z)\Big|_{z=z_0}, \qquad 1 \leq j \leq n$$

where we have set

$$x(z) = \sum_{j=1}^{n} x_j(z - z_0)^{j-1}$$

and

$$y(z) = \sum_{j=1}^{n} y_j(z - z_0)^{j-1},$$

as in Example 16.2.2. Here we have $\Lambda = S_2$, where S_2 is the solution of the Lyapunov equation

$$S_2 A_\zeta^* + A_\zeta S_2 = B_+ B_+^* - B_- B_-^* \quad \text{if} \quad \Delta = \Pi^+$$

or the Stein equation

$$S_2 - A_\zeta S_2 A_\zeta^* = B_+ B_+^* - B_- B_-^* \quad \text{if} \quad \Delta = \mathcal{D}.$$

We refer to the Appendix for the computation of the explicit form of S_2. $\quad\square$

18.6 PROOFS OF THEOREMS 18.5.1 AND 18.5.2

In what follows $\Delta = \Pi^+$ or $\Delta = \mathcal{D}$ as appropriate. Let $\omega = (C_-, C_+, A_\pi; A_\zeta, B_+, B_-; S)$ be a Δ-admissible Sylvester TSCII data set. The problem is to find a rational matrix function $F(z)$ analytic on Δ satisfying the associated set of interpolation conditions

(18.6.1a)
$$\sum_{z_0 \in \Delta} \text{Res}_{z=z_0} F(z)C_-(zI - A_\pi)^{-1} = C_+$$

(18.6.1b)
$$\sum_{z_0 \in \Delta} \text{Res}_{z=z_0}(zI - A_\zeta)^{-1}B_+ F(z) = -B_-$$

(18.6.1c)
$$\sum_{z_0 \in \Delta} \text{Res}_{z=z_0}(zI - A_\zeta)^{-1}B_+ F(z)C_-(zI - A_\pi)^{-1} = S$$

together with the additional metric constraint

(18.6.2)
$$\sup\{\|F(z)\|: z \in \Delta\} < 1.$$

Theorem 16.10.1 gives a description of all matrix functions F analytic on Δ which satisfy only (18.6.1a)–(18.6.1c); the description is in terms of a matrix function $\Theta(z)$ which has

(18.6.3)
$$\widetilde{\omega} = \left(\begin{bmatrix} C_+ \\ C_- \end{bmatrix}, A_\pi; A_\zeta, [B_+, B_-], S\right)$$

as a left null-pole triple over Δ. The idea of the proof of Theorems 18.5.1 and 18.5.2 is to insist that Θ has the additional symmetry property $\Theta(z)^* J \Theta(z) = J$ for $z \in \partial\Delta$ and then analyze which solutions of (18.6.1a)–(18.6.1c) meet the additional metric constraint (18.6.2).

We point out at this stage the origin of the formulas (18.5.6) and (18.5.7a)–(18.5.7c) for $\Theta(z)$ in Theorems 18.5.1 and 18.5.2. Indeed, by Theorem 6.3.1 (if $\Delta = \Pi^+$) or Theorem 7.4.2 (if $\Delta = \mathcal{D}$), the condition that $\Lambda = \Lambda(\omega)$ (18.5.3) be invertible is equivalent to the condition that there exists a rational matrix function regular on $\partial\Delta$ (including at infinity if $\Delta = \Pi^+$) which has $\tilde\omega$ as a null-pole triple over Δ, and in this case, the formulas (18.5.6) and (18.5.7a)–(18.5.7c) give the essentially unique such function $\Theta(z)$ (in case $\Delta = \mathcal{D}$ and A_π and A_ζ are nonsingular, the reader should refer to Theorem 7.3.1 rather than 7.4.2 for the formulas (17.5.7a)–(18.5.7b)). Also, Λ is the associated Hermitian coupling matrix for $\Theta(z)$ discussed in Sections 6.2 and 7.2, so positive definiteness of $\Theta(z)$ corresponds to the associated kernel $\frac{J - \Theta(z)J\Theta(w)^*}{\rho_\Delta(z,w)}$ being positive on Δ.

In general for $G \in \mathcal{R}_{M \times N}(\Delta)$ (i.e. G is an $M \times N$ rational matrix function with no poles in Δ) we denote

$$\|G\|_\infty = \sup\{\|G(z)\| : z \in \Delta\}.$$

The formal proof of Theorems 18.5.1 and 18.5.2 will be organized in three steps.

STEP 1. If a solution $F \in \mathcal{R}_{M \times N}(\Delta)$ satisfying (18.6.1)–(18.6.2) exists and if Λ is invertible, then Λ is in fact positive definite and F has the form $F = (\Theta_{11}G + \Theta_{12})(\Theta_{21}G + \Theta_{22})^{-1}$ for some $G \in \mathcal{R}_{M \times N}(\Delta)$ with $\|G\|_\infty < 1$.

STEP 2. If Λ is positive definite and if $G \in \mathcal{R}_{M \times N}(\Delta)$ has $\|G\|_\infty < 1$, then $F = (\Theta_{11}G + \Theta_{12})(\Theta_{21}G + \Theta_{22})^{-1}$ is a solution of (18.6.1)–(18.6.2).

STEP 3. If a solution F of (18.6.1)–(18.6.2) exists, then Λ is positive definite (and hence in particular invertible).

PROOF OF STEP 1. Let $\tilde\omega$ be given by (18.6.3). Clearly, $\tilde\omega$ is a Δ-admissible Sylvester data set. As Λ is assumed to be invertible, by Theorem 6.3.1 (if $\Delta = \Pi^+$) or by Theorem 7.3.1 (if $\Delta = \mathcal{D}$) there exists a rational matrix function $\Theta(z)$ regular and J-unitary on $\partial\Delta$ for which $\tilde\omega$ is a left null-pole triple over Δ (here $J = I_M \oplus -I_N$). As usual, we partition

$$\Theta(z) = \begin{bmatrix} \Theta_{11}(z) & \Theta_{12}(z) \\ \Theta_{21}(z) & \Theta_{22}(z) \end{bmatrix}.$$

Since $\tilde\omega$ is a Δ-admissible Sylvester data set, in particular $\left(\begin{bmatrix} C_+ \\ C_- \end{bmatrix}, A_\pi\right)$ is a null kernel pair and $(A_\zeta, [B_+, B_-])$ is a full range pair. Thus by (ii) $\Leftrightarrow$ (iv) in Theorem 13.2.3 we see that positive definiteness of Λ is equivalent to the analyticity of $\Theta_{22}^{-1}\Theta_{21}$ in Δ. Thus the proof of Step 1 is complete once we show that the existence of a solution F of (18.6.1)–(18.6.2) leads to the analyticity of $\Theta_{22}^{-1}\Theta_{21}$ on Δ.

By Theorem 16.10.1, since F is an element of $\mathcal{R}_{M \times N}(\overline{\Delta})$ satisfying (18.6.1), F can be represented as

$$(18.6.4) \qquad F = (\Theta_{11}\widetilde{G}_1 + \Theta_{12}\widetilde{G}_2)(\Theta_{21}\widetilde{G}_1 + \Theta_{22}\widetilde{G}_2)^{-1}$$

where $\widetilde{G}_1 \in \mathcal{R}_{M \times N}(\overline{\Delta})$ and $\widetilde{G}_2 \in \mathcal{R}_{N \times N}(\overline{\Delta})$ are such that

$$\varphi(\Theta_{21}\widetilde{G}_1 + \Theta_{22}\widetilde{G}_2) \text{ is analytic and invertible on } \Delta$$

where φ^{-1} is a rational $N \times N$ matrix function having $(C_-, A_\pi; 0, 0; 0)$ as a $\overline{\Delta}$-null-pole triple. Rewrite (18.6.4) in matrix form as

$$(18.6.5) \qquad \begin{aligned} \begin{bmatrix} F \\ I \end{bmatrix}\varphi^{-1} &= \begin{bmatrix} \Theta_{11} & \Theta_{12} \\ \Theta_{21} & \Theta_{22} \end{bmatrix}\begin{bmatrix} \widetilde{G}_1 \\ \widetilde{G}_2 \end{bmatrix}(\Theta_{21}\widetilde{G}_1 + \Theta_{22}\widetilde{G}_2)^{-1}\varphi^{-1} \\ &= \begin{bmatrix} \Theta_{11} & \Theta_{12} \\ \Theta_{21} & \Theta_{22} \end{bmatrix}\begin{bmatrix} G_1 \\ G_2 \end{bmatrix} \end{aligned}$$

where $G_1 = \widetilde{G}_1(\Theta_{21}\widetilde{G}_1 + \Theta_{22}\widetilde{G}_2)^{-1}\varphi^{-1} \in \mathcal{R}_{M \times N}(\Delta)$ and $G_2 = \widetilde{G}_2(\Theta_{21}\widetilde{G}_1 + \Theta_{22}\widetilde{G}_2)^{-1}\varphi^{-1} \in \mathcal{R}_{N \times N}(\Delta)$.

From the bottom component of (18.6.5)

$$(18.6.6) \qquad \varphi^{-1} = \Theta_{21}G_1 + \Theta_{22}G_2.$$

Now we use that $\|F(z)\| < 1$ for $z \in \partial\Delta$ together with (18.6.5). Thus, for $z \in \partial\Delta$,

$$\begin{aligned} 0 &> F(z)^*F(z) - I \\ &= [\, F(z)^* \quad I \,]J\begin{bmatrix} F(z) \\ I \end{bmatrix} \\ &= \varphi(z)^*[\, G_1(z)^* \quad G_2(z)^* \,]\{\Theta(z)^*J\Theta(z)\}\begin{bmatrix} G_1(z) \\ G_2(z) \end{bmatrix}\varphi(z) \\ &= \varphi(z)^*[\, G_1(z)^* \quad G_2(z)^* \,]J\begin{bmatrix} G_1(z) \\ G_2(z) \end{bmatrix}\varphi(z) \end{aligned}$$

and thus

$$(18.6.7) \qquad G_1(z)^*G_1(z) - G_2(z)^*G_2(z) < 0.$$

Thus if $G_2(z)x = 0$ for some vector x, then for some $\delta > 0$,

$$-\delta x^*x \geq x^*[G_1(z)^*G_1(z) - G_2(z)^*G_2(z)]x = \|G_1(z)x\|^2$$

which forces $x = 0$. Thus, since $G_2(z)$ has square size $N \times N$, we see that $G_2(z)$ is invertible for $z \in \partial\Delta$. In particular $\det G_2(z) \not\equiv 0$ and we may define $G(z) = G_1(z)G_2(z)^{-1} \in \mathcal{R}_{M \times N}$. From (18.6.7) we see that $G(z)$ is a strict contraction for $z \in \partial\Delta$.

Recall the notation wno $w(z)$ for the net change in the argument (or winding number) along $\partial\Delta$ of the scalar continuous function $w(z)$ defined and nonzero-valued on $\partial\Delta$. Since $\Theta(z)$ is J-unitary on $\partial\Delta$, by Proposition 13.1.2 $(\Theta_{22}^{-1}\Theta_{21})(z)$ is a strict contraction for $z \in \partial\Delta$. Since G is also contractive on $\partial\Delta$ we have that $t(\Theta_{22}^{-1}\Theta_{21}G) + I$ is invertible on $\partial\Delta$ for $0 \leq t \leq 1$, so $\mathrm{wno}\det(t\Theta_{22}^{-1}\Theta_{21}G + I)$ is defined for $0 \leq t \leq 1$. By a standard homotopy argument

$$(18.6.8) \qquad \mathrm{wno}\det(\Theta_{22}^{-1}\Theta_{21}G + I) = \mathrm{wno}\det I = 0.$$

From (18.6.6)

$$(18.6.9) \qquad \begin{aligned} \mathrm{wno}\det\varphi^{-1} &= \mathrm{wno}\det(\Theta_{21}G_1 + \Theta_{22}G_2) \\ &= \mathrm{wno}\det(\Theta_{22}[\Theta_{22}^{-1}G + I]G_2) \\ &= \mathrm{wno}\det\Theta_{22} + \mathrm{wno}\det(\Theta_{22}^{-1}\Theta_{21}G + I) + \mathrm{wno}\det G_2. \end{aligned}$$

We have just argued that $\mathrm{wno}\det(\Theta_{22}^{-1}\Theta_{21}G + I) = 0$. Since G_2 is analytic on Δ, by the argument principle

$$(18.6.10) \qquad \mathrm{wno}\det G_2 \geq 0.$$

We claim next that

$$(18.6.11) \qquad \mathrm{wno}\det\Theta_{22} \geq \mathrm{wno}\det\varphi^{-1}.$$

To prove (18.6.11) we first note from Theorem 12.3.1 that the subspace $\Theta\mathcal{R}_{M+N}(\Delta)$ is characterized as

$$\Theta\mathcal{R}_{M+N}(\Delta) = \left\{ \begin{bmatrix} C_+ \\ C_- \end{bmatrix} (zI - A_\pi)^{-1}x + \begin{bmatrix} h_1(z) \\ h_2(z) \end{bmatrix} : \right.$$
$$x \in \mathbf{C}^{n_\pi}, \; h_1 \in \mathcal{R}_M(\Delta), \; h_2 \in \mathcal{R}_N(\Delta) \text{ such that}$$
$$\left. \sum_{z_0 \in \Delta} \mathrm{Res}_{z=z_0}(zI - A_\zeta)^{-1}(B_+h_1(z) + B_-h_2(z)) = Sx \right\}.$$

Thus
$$(18.6.12)$$
$$\begin{aligned} [\,\Theta_{21} \quad \Theta_{22}\,]\mathcal{R}_{M+N}(\Delta) &= [\,0 \quad I\,]\Theta\mathcal{R}_{M+N}(\Delta) \\ &= \{C_-(zI - A_\pi)^{-1}x + h_2(z): x \in \mathbf{C}^{n_\pi}, \; h_2 \in \mathcal{R}_N(\Delta) \text{ such that} \\ &\qquad \sum_{z_0 \in \Delta} \mathrm{Res}_{z=z_0}(zI - A_\zeta)^{-1}B_-h_2(z) - Sx \\ &\qquad = -\sum_{z_0 \in \Delta} \mathrm{Res}_{z=z_0}(zI - A_\zeta)^{-1}B_+h_1(z) \text{ for some } h_1 \in \mathcal{R}_N(\Delta)\}. \end{aligned}$$

But since by assumption (A_ζ, B_+) is a full range pair, by Lemma 12.2.2 the map $h \to \sum_{z_0 \in \sigma} \mathrm{Res}_{z=z_0}(zI - A_\zeta)^{-1}B_+h(z)$ maps $\mathcal{R}_M(\Delta)$ onto $\mathbf{C}^{n_\zeta}$. Thus given any $h_2 \in \mathcal{R}_N(\Delta)$ one can always solve for an $h_1 \in \mathcal{R}_M(\Delta)$ so that the condition $\sum_{z_0 \in \Delta} \mathrm{Res}_{z=z_0}(zI -$

$A_\zeta)^{-1} B_- h_2(z) - Sx = -\sum_{z_0 \in \Delta} \operatorname{Res}_{z=z_0}(zI - A_\zeta)^{-1} B_+ h_1(z)$ is satisfied. We conclude that

$$[\,\Theta_{21} \quad \Theta_{22}\,] \mathcal{R}_{M+N}(\Delta) = \{C_-(zI - A_\pi)^{-1} x + h(z) : x \in \mathbf{C}^{n_\pi},\ h \in \mathcal{R}_N(\Delta)\}.$$

Since by construction φ^{-1} has $(C_-, A_\pi; 0, 0; 0)$ as a Δ-null-pole triple, by Theorem 12.3.1 again this in turn is equal to

$$[\,\Theta_{21} \quad \Theta_{22}\,] \mathcal{R}_{M+N}(\Delta) = \varphi^{-1} \mathcal{R}_N(\Delta).$$

Hence

(18.6.13) $$\Theta_{22} \mathcal{R}_N(\Delta) \subset \varphi^{-1} \mathcal{R}_N(\Delta).$$

This implies that the poles of $\det \Theta_{22}$ in Δ form a subset of the poles of $\det \varphi^{-1}$, where we include multiplicities in the count. Since $\det \varphi^{-1}$ has no zeros in Δ, (18.6.11) now follows by the Argument Principle.

When we combine the equality (18.6.9) with the inequalities (18.6.8), (18.6.10) and (18.6.11) we get the term-wise equalities

(18.6.14) $$\operatorname{wno}\,\det \Theta_{22} = \operatorname{wno}\,\det \varphi^{-1}$$

and

(18.6.15) $$\operatorname{wno}\,\det G_2 = 0.$$

From (18.6.14) we deduce that (18.6.13) must hold with equality, so

$$\Theta_{22} \mathcal{R}_N(\Delta) = [\,\Theta_{21} \quad \Theta_{22}\,] \mathcal{R}_{M+N}(\Delta)$$
$$= \Theta_{22}[\,\Theta_{22}^{-1} \Theta_{21} \quad I_N\,] \mathcal{R}_{M+N}(\Delta).$$

This in turn leads to

$$\Theta_{22}^{-1} \Theta_{21} \mathcal{R}_M(\Delta) \subset \mathcal{R}_N(\Delta)$$

i.e., $\Theta_{22}^{-1} \Theta_{21}$ has no poles in Δ, as desired.

If we combine (18.6.15) with (18.6.5) we can go further. Since G_2 is known to be analytic on Δ, $\operatorname{wno}\,\det G_2 = 0$ implies that G_2 has no zeros in Δ, so G_2^{-1} and hence also $G = G_1 G_2^{-1}$ is analytic on Δ. We already established above that $\|G(z)\| < 1$ for $z \in \partial\Delta$. Hence by the maximum modulus theorem $\sup_{z \in \Delta} \|G(z)\| < 1$. Moreover from (18.6.5) we have

(18.6.16) $$\begin{bmatrix} F \\ I \end{bmatrix} = \begin{bmatrix} \Theta_{11} & \Theta_{12} \\ \Theta_{21} & \Theta_{22} \end{bmatrix} \begin{bmatrix} G \\ I \end{bmatrix} G_2 \varphi.$$

From the bottom components,

$$I = (\Theta_{21} G + \Theta_{22}) G_2 \varphi$$

so

$$G_2\varphi = (\Theta_{21}G + \Theta_{22})^{-1}.$$

Then from the top components in (18.6.16) we get

$$F = (\Theta_{11}G + \Theta_{12})G_2\varphi$$
$$= (\Theta_{11}G + \Theta_{12})(\Theta_{21}G + \Theta_{22})^{-1}.$$

Thus any solution F of the interpolation problem has the prescribed form as in Theorems 18.5.1 and 18.5.2.

PROOF OF STEP 2. Conversely, suppose that Λ is positive definite. As Λ is the associated Hermitian matrix for a minimal realization of $\Theta(z)$, by (ii) $\Leftrightarrow$ (iv) in Theorem 13.2.3 again the function $\Theta_{22}^{-1}\Theta_{21}$ has no poles in Δ. By reversing an argument in the first part of the proof, we then have

(18.6.17)
$$\varphi^{-1}\mathcal{R}_N(\Delta) = [\ \Theta_{21} \quad \Theta_{22}\ ]\mathcal{R}_{M+N}(\Delta)$$
$$= \Theta_{22}[\ \Theta_{22}^{-1}\Theta_{21} \quad I\ ]\mathcal{R}_{M+N}(\Delta)$$
$$= \Theta_{22}\mathcal{R}_N(\Delta).$$

Now let G be an arbitrary element of $\mathcal{R}_{M\times N}(\Delta)$ with $\sup_{z\in\Delta}\|G(z)\| < 1$. Then as in the first part of the proof $\|[\Theta_{22}^{-1}\Theta_{21}G](z)\| < 1$ for $z \in \partial\Delta$. As $\Theta_{22}^{-1}\Theta_{21}$ and G are analytic on Δ, by the maximum modulus theorem this continues to hold for z in Δ, so $\Theta_{22}^{-1}\Theta_{21}G + I$ is analytic and invertible on Δ. Thus

$$(\Theta_{21}G + \Theta_{22})\mathcal{R}_N(\Delta) = \Theta_{22}(\Theta_{22}^{-1}\Theta_{21}G + I)\mathcal{R}_N(\Delta)$$
$$= \Theta_{22}\mathcal{R}_N(\Delta) = \varphi^{-1}\mathcal{R}_N(\Delta)$$

where we used (18.6.17) for the last step. From this we get

$$\varphi(\Theta_{21}G + \Theta_{22})\mathcal{R}_N(\Delta) = \mathcal{R}_N(\Delta)$$

so $\varphi(\Theta_{21}G+\Theta_{22})$ is analytic and invertible on Δ. Now by Theorem 16.10.1 (with $Q_1 = G$ and $Q_2 = I$) it follows that $F = (\Theta_{11}G+\Theta_{12})(\Theta_{21}G+\Theta_{22})^{-1}$ satisfies the interpolation conditions (18.6.1). Moreover, since $\begin{bmatrix} F \\ I \end{bmatrix} = \Theta\begin{bmatrix} G \\ I \end{bmatrix} W$ (where $W = \Theta_{21}G+\Theta_{22}$) and since Θ is J-unitary on $\partial\Delta$, we have for $z \in \partial\Delta$

$$F(z)^*F(z) - I = [\ F(z)^* \quad I\ ]J\begin{bmatrix} F(z) \\ I \end{bmatrix}$$
$$= W(z)^*[\ G(z)^* \quad I\ ]\Theta(z)^*J\Theta(z)\begin{bmatrix} G(z) \\ I \end{bmatrix}W(z)$$
$$= W(z)^*(G(z)^*G(z) - I)W(z) < 0$$

and so $\|F(z)\| < 1$ for $z \in \partial\Delta$. As F is analytic on Δ, we conclude by the maximum modulus theorem that $\sup_{z\in\Delta}\|F(z)\| < 1$. Thus we have established that if $\Lambda > 0$

then there are many solutions F of the interpolation problem and in fact any F of the indicated linear fractional form in Theorems 18.5.1 and 18.5.2 is a solution.

PROOF OF STEP 3. Suppose there exists $F \in \mathcal{BR}_{M \times N}(\Delta)$ such that

$$(18.6.18a) \qquad \sum_{z_0 \in \Delta} \operatorname{Res}_{z=z_0} F(z) C_-(zI - A_\pi)^{-1} = C_+;$$

$$(18.6.18b) \qquad \sum_{z_0 \in \Delta} \operatorname{Res}_{z=z_0} (zI - A_\zeta)^{-1} B_+ F(z) = -B_-;$$

$$(18.6.18c) \qquad \sum_{z_0 \in \Delta} \operatorname{Res}_{z=z_0} (zI - A_\zeta)^{-1} B_+ F(z) C_-(zI - A_\pi)^{-1} = S.$$

Consider now the function $r^{-1} F(z)$, where r is a real parameter. We assume $r > 1 - \varepsilon$, where $\varepsilon > 0$ is chosen sufficiently close to zero. Clearly, for such r we have $r^{-1} F \in \mathcal{BR}_{M \times N}(\Delta)$ and $r^{-1} F$ satisfies the equations (18.6.18) with C_- replaced by rC_-, B_+ replaced by rB_+, and S replaced by rS. The corresponding matrix $\Lambda = \Lambda(r)$ given by (18.5.3) is now a function of r; we have

$$(18.6.19) \qquad \Lambda(r) = \begin{bmatrix} S_1(r) & rS^* \\ rS & S_2(r) \end{bmatrix},$$

where

$$(18.6.20a) \qquad S_1(r) A_\pi + A_\pi^* S_1(r) = r^2 C_-^* C_- - C_+^* C_+;$$

$$(18.6.20b) \qquad S_2(r) A_\zeta^* + A_\zeta S_2(r) = r^2 B_+ B_+^* - B_- B_-^*,$$

if $\Delta = \Pi^+$, and

$$(18.6.21a) \qquad S_1(r) - A_\pi^* S_1(r) A_\pi = r^2 C_-^* C_- - C_+^* C_+;$$

$$(18.6.21b) \qquad S_2(r) - A_\zeta S_2(r) A_\zeta^* = r^2 B_+ B_+^* - B_- B_-^*,$$

if $\Delta = \mathcal{D}$.

We can further write

$$(18.6.22) \qquad \Lambda(r) = \begin{bmatrix} r^2 F_{1-} - F_{1+} & rS^* \\ rS & r^2 F_{2+} - F_{2-} \end{bmatrix},$$

where $F_{1\pm}$ and $F_{2\pm}$ satisfy the equations

$$F_{1\pm} A_\pi + A_\pi^* F_{1\pm} = C_\pm^* C_\pm,$$
$$F_{2\pm} A_\zeta^* + A_\zeta F_{2\pm} = B_\pm B_\pm^*,$$

or

$$F_{1\pm} - A_{\pi}^* F_{1\pm} A_{\pi} = C_{\pm}^* C_{\pm},$$
$$F_{2\pm} - A_{\zeta} F_{2\pm} A_{\zeta}^* = B_{\pm} B_{\pm}^*,$$

as appropriate. At this point we make use of Theorems A.3.1 and A.3.3 in the Appendix to conclude that F_{1-} and F_{2+} are positive definite and F_{1+} and F_{2-} are positive semidefinite.

In particular, F_{1-} and F_{2+} are invertible, and therefore the determinant of $\Lambda(r)$ is not identically zero. We conclude that $\Lambda(r)$ is invertible in a punctured neighborhood $0 < |r - 1| < \delta$ of 1 (where $\delta > 0$ is sufficiently small). By the already proved Step 1 the matrix $\Lambda(r)$ is actually positive definite for $0 < |r - 1| < \delta$, and consequently $\Lambda(1)$ is positive semidefinite. We have to prove that in fact $\Lambda(1) > 0$. To this end observe that

$$\Lambda'(1) := \left. \frac{d\Lambda(r)}{dr} \right|_{r=1} = \begin{bmatrix} 2F_{1-} & S^* \\ S & 2F_{2+} \end{bmatrix} = \begin{bmatrix} F_{1-} & 0 \\ 0 & F_{2+} \end{bmatrix} + \Lambda(1) + \begin{bmatrix} F_{1+} & 0 \\ 0 & F_{2-} \end{bmatrix}$$

which is positive definite in view of the positive definiteness of F_{1-}, F_{2+} and positive semidefiniteness of $\Lambda(1)$, F_{1+} and F_{2-}. Now if there existed $x \neq 0$ such that $\Lambda(1)x = 0$, then we would have for r smaller than, but sufficiently close to, 1:

$$x^* \Lambda(r)x = x^* \left[\frac{1}{2}(r - 1)^2 \Lambda''(1) + (r - 1)\Lambda'(1) \right] x$$
$$= (r - 1)\left[x^* \left(\frac{1}{2}(r - 1)\Lambda''(1) + \Lambda'(1) \right) x \right] < 0,$$

a contradiction to the positive definiteness of $\Lambda(r)$. So $\Lambda(1) > 0$, and the proof of Step 3 is completed.

CHAPTER 19

MATRIX NEVANLINNA-PICK-TAKAGI INTERPOLATION

We have seen in the previous chapter that, in order that there exist a (scalar) rational function on the unit disk $\mathcal{D}$ with modulus less than 1 there which in addition satisfies some interpolation conditions

$$(1) \qquad\qquad f(z_i) = w_i, \qquad 1 \le i \le n,$$

it is necessary and sufficient that the associated Pick matrix

$$\Lambda = \left[\frac{1 - \overline{w}_i w_j}{1 - \overline{z}_i z_j} \right]_{1 \le i,j \le n}$$

be positive definite. Thus if Λ has some negative eigenvalues, such a function f cannot exist. A natural question is what can still be said about the class of interpolating functions if Λ has some negative eigenvalues. The solution turns out to be rather elegant. Namely, if κ is the number of negative eigenvalues of Λ, then one can always find a rational function f with κ poles in $\mathcal{D}$ and with modulus on the boundary $\partial \mathcal{D}$ at most 1 which satisfies the interpolation conditions (1) at each z_i which is a point of analyticity. Conversely, Λ has κ negative eigenvalues whenever such an f exists. This type of result was first obtained by Takagi [1924] in the context of the Schur problem where one specifies the first few Taylor coefficients of f at the origin.

In this chapter we obtain the analogue of this result in the full generality of the two-sided Nudelman problem discussed in the previous chapter.

19.1 TAKAGI-NUDELMAN INTERPOLATION PROBLEM

Let us revisit the two-sided Nudelman interpolation problem. As in the previous chapter, we let Δ be either the open unit disc $\mathcal{D}$ or the open right half-plane Π^+. Consider the Nudelman interpolation conditions (same as (18.4.1)–(18.4.3)):

$$(19.1.1) \qquad \sum_{z_0 \in \sigma(A_\zeta)} \mathrm{Res}_{z=z_0} (zI - A_\zeta)^{-1} B_+ F(z) = -B_-;$$

$$(19.1.2) \qquad \sum_{z_0 \in \sigma(A_\pi)} \mathrm{Res}_{z=z_0} F(z) C_- (zI - A_\pi)^{-1} = C_+;$$

$$(19.1.3) \qquad \sum_{z_0 \in \sigma(A_\pi) \cup \sigma(A_\zeta)} \mathrm{Res}_{z=z_0} (zI - A_\zeta)^{-1} B_+ F(z) C_- (zI - A_\pi)^{-1} = S.$$

Here

(19.1.4) $\omega = (C_+, C_-, A_\pi; A_\zeta, B_+, B_-; S)$

is a Δ-*admissible Sylvester* (interpolation) *data* set, i.e. (A_ζ, B_+) is a full range pair, (C_-, A_π) is a null kernel pair, $\sigma(A_\zeta) \cup \sigma(A_\pi) \subset \Delta$, and the Sylvester equation

(19.1.5) $SA_\pi - A_\zeta S = B_- C_- + B_+ C_+$

is satisfied. The sizes of matrices C_+, C_-, A_π, A_ζ, B_+, B_-, S are $M \times n_\pi$, $N \times n_\pi$, $n_\pi \times n_\pi$, $n_\zeta \times n_\zeta$, $n_\zeta \times M$, $n_\zeta \times N$, $n_\zeta \times n_\pi$, respectively. We consider rational $M \times N$ matrix functions $F(z)$ which are solutions of (19.1.1)–(19.1.3) and bounded by 1 on the boundary $\partial\Delta$, but (in contrast with the two-sided Nudelman problem studied in Chapter 18) $F(z)$ may have poles in $\Delta \backslash (\sigma(A_\zeta) \cup \sigma(A_\pi))$. Here is a precise formulation of this interpolation problem.

TAKAGI-NUDELMAN PROBLEM. *Let a* Δ-*admissible Sylvester data set* ω *(19.1.4) be given. Find the smallest nonnegative integer* κ *for which there exists a rational* $M \times N$ *matrix function* $F(z)$ *such that*

1. *F has* κ *poles (counting multiplicities) in* Δ, *none of which occur in* $\sigma(A_\zeta) \cup \sigma(A_\pi)$.

2. $\sup\{\|F(z)\|: z \in \partial\Delta\} < 1$.

3. *F satisfies the interpolation conditions* (19.1.1)–(19.1.3).

Further, for this smallest value of κ *describe the set of all such functions* $F(z)$.

Here and everywhere in this chapter the multiplicity of a pole z_0 of a rational matrix function $F(z)$ is taken to mean the sum of the partial pole multiplicities of $F(z)$ at z_0.

Specific instances of the two-sided Nudelman interpolation conditions are given in Examples 16.2.1–16.2.4. For the Takagi-Nudelman problem one seeks solutions of the explicit interpolation conditions given there which are allowed to have up to κ poles at points in Δ away from interpolation nodes and with norm less than 1 only on $\partial\Delta$.

We now discuss the divisor-remainder formulation of the Takagi-Nudelman problem (analogous to the coset formulation of the Nudelman problem given in Section 18.4). As in Section 18.4, we suppose that we are given rational matrix functions K, ψ, φ of sizes $M \times N$, $M \times M$ and $N \times N$, respectively, having no poles in $\overline{\Delta}$, where moreover ψ and φ have no zeros on $\partial\Delta$. For κ a given nonnegative integer we define $\mathcal{R}_{M \times N; \kappa}(\Delta)$ to be the class of rational $M \times N$ matrix functions having at most κ poles in Δ including multiplicities. The interpolation problem which we consider here (the Takagi version of the divisor-remainder formulation given in Section 18.4) is:

Find all $M \times N$ *rational matrix functions with*

$$\sup\{\|F(z)\|: z \in \partial\Delta\} < 1$$

such that

$$F \in K + \psi \mathcal{R}_{M \times N, \kappa}(\Delta)\varphi.$$

In general, find the smallest nonnegative integer κ for which such a matrix function F exists.

Let us call this the (*two-sided*) *Takagi-Sarason problem*. To relate this formulation to the Takagi-Nudelman problem posed above, proceed as follows (cf. Section 16.11). Let (A_ζ, B_+) be a left null pair for $\psi(z)$ over Δ, let (C_-, A_π) be a right null pair for $\varphi(z)$ over Δ, and define matrices B_-, C_+ and S by

$$(19.1.6) \qquad\qquad B_- = -\frac{1}{2\pi i}\int_\gamma (zI - A_\zeta)^{-1} B_+ K(z) dz$$

$$(19.1.7) \qquad\qquad C_+ = \frac{1}{2\pi i}\int_\gamma K(z) C_- (zI - A_\pi)^{-1} dz$$

and

$$(19.1.8) \qquad\qquad S = \frac{1}{2\pi i}\int_\gamma (zI - A_\zeta)^{-1} B_+ K(z) C_- (zI - A_\pi)^{-1} dz.$$

Here γ can be any contour in Δ enclosing $\sigma(A_\zeta) \cup \sigma(A_\pi)$. Suppose the rational matrix function $F(z)$ has the form

$$F(z) = K(z) + \psi(z) R(z) \varphi(z)$$

where $R(z)$ has at most κ poles in Δ (counting multiplicities). Then if $R(z)$ is analytic on $\sigma(A_\pi) \cup \sigma(A_\zeta)$ (i.e. none of the κ poles occur at a zero of ψ or φ), then by Theorem 16.9.3 $F(z)$ satisfies the interpolation conditions (19.1.1)–(19.1.3). Moreover, $F = K + \psi R \varphi$ has at most κ poles in Δ since K, ψ, φ are all analytic on Δ and R has at most κ poles in Δ. Thus, if also $\sup\{\|F(z)\|: z \in \partial\Delta\} < 1$, then F is a solution of the Takagi-Nudelman problem. Conversely, assume F is a solution of the Takagi-Nudelman problem. Then by the converse direction of Theorem 16.9.3 we know that $F = K + \psi R \varphi$, where R is analytic at least on $\sigma(A_\pi) \cup \sigma(A_\zeta)$. Using that K is analytic on Δ and that ψ^{-1} and φ^{-1} are analytic on Δ away from $\sigma(A_\pi) \cup \sigma(A_\zeta)$, we see that $R = \psi^{-1}(F - K)\varphi^{-1}$ has the same number of poles in $\Delta \backslash (\sigma(A_\pi) \cup \sigma(A_\zeta))$ as does F, namely, at most κ. Hence we conclude that $F \in K + \psi \mathcal{R}_{M \times N, \kappa}(\Delta)\varphi$.

We conclude that the Takagi-Sarason problem is more general than the Takagi-Nudelman problem in that the matrix function R appearing in a solution of the Takagi-Sarason problem is allowed to have poles in $\sigma(A_\zeta) \cup \sigma(A_\pi)$. Roughly, solutions of the Takagi-Sarason problem form the closure of the set of solutions of the Takagi-Nudelman problem; except in some special cases it is difficult to pin down this closure explicitly in terms of interpolation conditions. As we shall see, the Takagi-Sarason problem (where one does not demand analyticity of a solution at the zeros of ψ and φ) is more natural, both mathematically and in the context of applications.

19.2 SOLUTION OF THE TAKAGI-NUDELMAN INTERPOLA-TION PROBLEM: MAIN RESULTS

In this section we present a solution of the two-sided Takagi-Nudelman inter-polation problem over the domain $\Delta = \Pi^+$ or $\Delta = \mathcal{D}$. As we shall see the solution is more satisfactory if we use the Takagi-Sarason formulation. Let us assume we are given a Δ-admissible Sylvester data set

$$(19.2.1) \qquad \omega = (C_+, C_-, A_\pi; A_\zeta, B_+, B_-; S).$$

Recall, by "admissible Sylvester" we mean

(i) (A_ζ, B_+) is a full range pair;

(ii) (C_-, A_π) is a null kernel pair;

(iii) $\sigma(A_\zeta) \cup \sigma(A_\pi) \subset \Delta$;

and

(iv) S satisfies the Sylvester equation

$$SA_\pi - A_\zeta S = B_- C_- + B_+ C_+.$$

From the data set ω, we build the Pick matrix

$$(19.2.2) \qquad \Lambda(\omega) = \begin{bmatrix} S_1 & S^* \\ S & S_2 \end{bmatrix}$$

where for the case $\Delta = \Pi^+$, S_1 and S_2 are the unique solutions of the Lyapunov equations

$$(19.2.3a) \qquad S_1 A_\pi + A_\pi^* S_1 = C_-^* C_- - C_+^* C_+$$

$$(19.2.3b) \qquad S_2 A_\zeta^* + A_\zeta S_2 = B_+ B_+^* - B_- B_-^*$$

while for the case $\Delta = \mathcal{D}$, S_1 and S_2 are determined from the Stein equations

$$(19.2.4a) \qquad S_1 - A_\pi^* S_1 A_\pi = C_-^* C_- - C_+^* C_+$$

$$(19.2.4b) \qquad S_2 - A_\zeta S_2 A_\zeta^* = B_+ B_+^* - B_- B_-^*.$$

For the case where $\Lambda = \Lambda(\omega)$ is invertible we let $\Theta(z) = \begin{bmatrix} \Theta_{11}(z) & \Theta_{12}(z) \\ \Theta_{21}(z) & \Theta_{22}(z) \end{bmatrix}$ be the essentially unique rational matrix function which is regular and J-unitary on $\partial \Delta$ and has $\widetilde{\omega}$ as a left null-pole triple over Δ, where

$$\widetilde{\omega} = \left(\begin{bmatrix} C_+ \\ C_- \end{bmatrix}, A_\pi; A_\zeta, [\, B_+ \quad B_- \,]; S \right)$$

(cf. Section 18.6). Thus $\Theta(z)$ is given explicitly by (18.5.6) for $\Delta = \Pi^+$ and (18.5.7a) (or (18.5.7c)) for $\Delta = \mathcal{D}$. We can now state the solution of the Takagi-Nudelman problem.

Recall that $\mathcal{BR}_{M \times N}(\Delta)$ denotes the class of rational $M \times N$ matrix functions G with no poles in Δ for which

$$\sup\{\|G(z)\|: z \in \Delta\} < 1.$$

THEOREM 19.2.1. *Let ω as in (19.2.1) be a Δ-admissible Sylvester data set for a Takagi-Nudelman interpolation problem over either $\Delta = \Pi^+$ or $\Delta = \mathcal{D}$. Let $\Lambda = \Lambda(\omega)$ be the Pick matrix (19.2.2) defined either by (19.2.3) or (19.2.4). Assume Λ is invertible and let $\Theta(z)$ be the matrix function defined either by (18.5.6) if $\Delta = \Pi^+$ or by (18.5.7a) or (18.5.7c) if $\Delta = \mathcal{D}$ and let $\varphi(z)$ be an $N \times N$ matrix function analytic on $\overline{\Delta}$ having (C_-, A_π) as a right null pair over $\overline{\Delta}$. Then the smallest κ for which the Takagi-Nudelman interpolation problem for the data set ω has a solution is the number of negative eigenvalues of $\Lambda(\omega)$, with multiplicities counted. Moreover for this smallest choice of κ, every solution F has the form*

$$(19.2.5) \qquad F = (\Theta_{11}G + \Theta_{12})(\Theta_{21}G + \Theta_{22})^{-1}$$

for a $G \in \mathcal{BR}_{M \times N}(\Delta)$ such that

$$(19.2.6) \qquad \varphi(\Theta_{21}G + \Theta_{22}) \text{ has no zero in } \sigma(A_\zeta) \cup \sigma(A_\pi).$$

Conversely, if $G \in \mathcal{BR}_{M+N}(\Delta)$ satisfies (19.2.6), then (19.2.5) defines a solution F of the Takagi-Nudelman interpolation problem for the smallest choice of κ.

For the Takagi-Sarason formulation of the problem, the set of all solutions is parametrized via Θ by the whole class $\mathcal{BR}_{M \times N}(\Delta)$, as is explained by the following result.

THEOREM 19.2.2. *Let K, ψ, φ be given rational matrix functions of respective sizes $M \times N$, $M \times N$, $N \times N$ with no poles in $\overline{\Delta}$ such that ψ^{-1} and φ^{-1} have no poles on $\partial\Delta$, where $\Delta = \Pi^+$ or $\Delta = \mathcal{D}$. Let (A_ζ, B_+) be a left null pair for $\psi(z)$ over Δ, let (C_-, A_π) be a right null pair for $\varphi(z)$ over Δ and define matrices B_-, C_+ and S by (19.1.6)–(19.1.8). Let $\Lambda = \Lambda(\omega)$ be the Pick matrix (19.2.2) defined by (19.2.3) or (19.2.4). Assume that Λ is invertible, and let $\Theta(z) = \begin{bmatrix} \Theta_{11}(z) & \Theta_{12}(z) \\ \Theta_{21}(z) & \Theta_{22}(z) \end{bmatrix}$ be the matrix function defined by (18.5.6), (18.5.7a) or (18.5.7c), as appropriate. Then the smallest κ for which the Takagi-Sarason interpolation problem based on data K, ψ, φ has a solution is the number of negative eigenvalues of $\Lambda(\omega)$, counted with multiplicities. For the smallest choice of κ, a matrix function F is a solution if and only if F has the form*

$$F = (\Theta_{11}G + \Theta_{12})(\Theta_{21}G + \Theta_{22})^{-1}$$

where G is an arbitrary matrix function in $\mathcal{BR}_{M \times N}(\Delta)$.

We have already argued in Section 19.1 that solutions of the Takagi-Nudelman problem coincide with solutions $F = K + \psi R\varphi$ of an associated Takagi-Sarason problem for which R is analytic on $\sigma(A_\zeta) \cup \sigma(A_\pi)$. For this reason we shall prove Theorem 19.2.2 in detail and sketch the modifications needed for Theorem 19.2.1.

19.3 PROOFS OF THE MAIN RESULTS

To prove Theorem 19.2.2 we first derive some basic lemmas. Specializing the following development to $\kappa = 0$ gives an alternative approach to the proof of Theorems 18.5.1 and 18.5.2 which words more directly with the Sarason divisor-remainder formulation rather than the Nudelman formulation. The following result is just a restatement of Lemma 16.10.5.

LEMMA 19.3.1. *Let K, ψ, φ be a set of functions satisfying the hypotheses of the Takagi-Sarason interpolation problem. Let (A_ζ, B_+) be a left null pair for $\psi(z)$ over Δ, let (C_-, A_π) be a right null pair for $\varphi(z)$ over Δ and define matrices B_-, C_+ and S by (19.1.6)–(19.1.8). Finally let $L(z)$ be the rational $(M+N) \times (M+N)$ matrix function given by*

$$L(z) = \begin{bmatrix} \psi & K\varphi^{-1} \\ 0 & \varphi^{-1} \end{bmatrix}.$$

Then

$$\widetilde{\omega} = \left(\begin{bmatrix} C_+ \\ C_- \end{bmatrix}, A_\pi; A_\zeta, [B_+, B_-]; S \right)$$

is a null-pole triple for $L(z)$ over Δ.

The next lemma is the needed adaptation of Theorem 16.11.2.

LEMMA 19.3.2. *Let K, ψ, φ be an admissible data set of functions for a Takagi-Sarason interpolation problem. Let (A_ζ, B_+) be a left null pair for $\psi(z)$ over Δ, let (C_-, A_π) be a right null pair for $\varphi(z)$ over Δ, and define matrices B_-, C_+ and S by (19.1.6)–(19.1.8) and then the Pick matrix $\Lambda = \Lambda(\omega)$ (19.2.2) by (19.2.3) (if $\Delta = \Pi^+$) or by (19.2.4) (if $\Delta = \mathcal{D}$). Assume that Λ is invertible and define $\Theta(z)$ by (18.5.6) or (18.5.7a) (or (18.5.7c)), as appropriate. Then a rational $M \times N$ matrix function $F(z)$ has the form*

$$F = K + \psi R\varphi$$

for a rational matrix function $R \in \mathcal{R}_{M \times N, \kappa}(\Delta)$ if and only if, for some $W \in \mathcal{R}_{N \times N, \kappa}(\Delta)$ having precisely κ zeros in Δ (counting multiplicities), the $(M+N) \times N$ matrix function $\begin{bmatrix} F \\ I_N \end{bmatrix} \varphi^{-1} W$ has a factorization

$$(19.3.1) \qquad\qquad \begin{bmatrix} F \\ I_N \end{bmatrix} \varphi^{-1} W = \Theta \begin{bmatrix} G_1 \\ G_2 \end{bmatrix}$$

where the rational matrix functions G_1 and G_2 have no poles in Δ.

PROOF. By Proposition 12.1.1, the factorization (19.3.1) is equivalent to

$$\begin{bmatrix} F \\ I_N \end{bmatrix} \varphi^{-1} W x \in \mathcal{S}_\Delta(\Theta)$$

for each $x \in \mathbb{C}^N$, where $\mathcal{S}_\Delta(\Theta) = \Theta \mathcal{R}_{M+N}(\Delta)$ is the left null-pole subspace for $\Theta(z)$ over Δ. Recall that by construction $\Theta(z)$ has

$$\widetilde{\omega} = \left(\begin{bmatrix} C_+ \\ C_- \end{bmatrix}, A_\pi; A_\zeta, [B_+, B_-]; S \right)$$

as a null-pole triple over Δ. By Lemma 19.3.1 $\tilde{\omega}$ is also a null-pole triple for $L(z) = \begin{bmatrix} \psi(z) & K(z)\varphi^{-1}(z) \\ 0 & \varphi^{-1}(z) \end{bmatrix}$ over Δ, and hence by Corollary 12.3.3 $\mathcal{S}_\Delta(L) = \mathcal{S}_\Delta(\Theta)$. Thus the factorization (19.3.8) is equivalent to a factorization

$$(19.3.2) \qquad \begin{bmatrix} F \\ I_N \end{bmatrix} \varphi^{-1} W = L \begin{bmatrix} G_1 \\ G_2 \end{bmatrix}$$

with G_1 and G_2 having no poles in Δ.

Let us now suppose that $F = K + \psi R \varphi$ where $R \in \mathcal{R}_{M \times N, \kappa}(\Delta)$. Then R has a factorization

$$R = R_0 W^{-1}$$

where R_0 and W have no poles in Δ and where W has precisely κ zeros (counting multiplicities) in Δ; existence of such a factorization follows easily from the Smith form of R (see Lemma 11.1.2). Thus we compute

$$\begin{bmatrix} F \\ I_N \end{bmatrix} \varphi^{-1} W = \begin{bmatrix} K\varphi^{-1}W + \psi(R_0W^{-1})\varphi \cdot \varphi^{-1}W \\ \varphi^{-1}W \end{bmatrix}$$

$$= \begin{bmatrix} K\varphi^{-1}W + \psi R_0 \\ \varphi^{-1}W \end{bmatrix} = L \begin{bmatrix} R_0 \\ W \end{bmatrix}.$$

Thus $\begin{bmatrix} F \\ I_N \end{bmatrix} \varphi^{-1} W$ has the factorization (19.3.2) with $G_1 = R_0$ and $G_2 = W$ having no poles in Δ. Conversely, if the factorization (19.3.2) holds with G_1 and G_2 having no poles in Δ, then

$$F\varphi^{-1}W = \psi G_1 + K\varphi^{-1}G_2; \qquad \varphi^{-1}W = \varphi^{-1}G_2.$$

Thus necessarily $G_2 = W$, and solving the first equality for F we get

$$F = K + \psi(G_1 W^{-1})\varphi,$$

where clearly $R := G_1 W^{-1} \in \mathcal{R}_{M \times N, \kappa}(\Delta)$. $\quad\square$

PROOF OF THEOREM 19.2.2. Suppose that $F = K + \psi R\varphi$ (where $R \in \mathcal{R}_{M \times N, \kappa}(\Delta)$) is a solution of the Takagi-Sarason problem. Then by Lemma 19.3.2 there is a rational matrix function W with no poles and precisely κ zeros in Δ such that

$$(19.3.3) \qquad \begin{bmatrix} F \\ I \end{bmatrix} \varphi^{-1} W = \Theta \begin{bmatrix} G_1 \\ G_2 \end{bmatrix} = \begin{bmatrix} \Theta_{11}G_1 + \Theta_{12}G_2 \\ \Theta_{21}G_1 + \Theta_{22}G_2 \end{bmatrix}$$

where G_1 and G_2 are analytic on Δ. As in the proof of Step 1 in Section 18.6, we see that

$$0 > G_1(z)^* G_1(z) - G_2(z)^* G_2(z), \qquad z \in \partial\Delta$$

and that $G_2(z)$ is invertible on $\partial\Delta$. Therefore $G(z) := G_1(z)G_2(z)^{-1}$ is a well-defined rational matrix function and

$$(19.3.4) \qquad \|G(z)\| < 1, \qquad z \in \partial\Delta.$$

From the bottom block row of (19.3.3) we read off

(19.3.5)
$$\varphi^{-1}W = \Theta_{21}G_1 + \Theta_{22}G_2$$
$$= \Theta_{22}(\Theta_{22}^{-1}\Theta_{21}G + I)G_2.$$

Just as in the proof of Step 1 in Section 18.6, we have

(19.3.6) $\text{wno } \det(\Theta_{22}^{-1}\Theta_{21}G + I) = 0.$

By construction

$$\varphi^{-1}\mathcal{R}_N(\Delta) = [\Theta_{21}, \Theta_{22}]\mathcal{R}_{M+N}(\Delta)$$
$$= \Theta_{22}[\Theta_{22}^{-1}\Theta_{21}, I]\mathcal{R}_{M+N}(\Delta).$$

We conclude that

(19.3.7)
$$\dim[\Theta_{22}^{-1}\Theta_{21}, I]\mathcal{R}_{M+N}(\Delta)/\mathcal{R}_N(\Delta) = \dim \varphi^{-1}\mathcal{R}_N(\Delta)/\Theta_{22}\mathcal{R}_N(\Delta)$$
$$= \dim(\varphi\Theta_{22})^{-1}\mathcal{R}_N(\Delta)/\mathcal{R}_N(\Delta).$$

Here the dimensions are understood as dimensions of vector spaces over **C**. The left hand side in (19.3.14) can be identified with the number of poles (counted with multiplicities) of $\Theta_{22}^{-1}\Theta_{21}$ in Δ, while the right-hand side in (19.3.14) can be identified with wno $\det \Theta_{22} -$ wno $\det \varphi^{-1}$. Thus,

(19.3.8)
$$\text{wno } \det \varphi^{-1} - \text{wno } \det \Theta_{22}$$
$$= -[\text{number of poles of } \Theta_{22}^{-1}\Theta_{21} \text{ in } \Delta \text{ (counting multiplicities)}].$$

On the other hand, from (19.3.5) we see that

(19.3.9)
$$\text{wno } \det \varphi^{-1} - \text{wno } \det \Theta_{22} = \text{wno } \det(\Theta_{22}^{-1}\Theta_{21}G + I) + \text{wno } \det G_2 - \kappa$$
$$\geq -\kappa.$$

We conclude that $\Theta_{22}^{-1}\Theta_{21}$ has at most κ poles in Δ. Now from the implication (ii) $\Rightarrow$ (iv) in Theorem 13.2.3 we conclude that $\Lambda = \Lambda(\omega)$ has at most κ negative eigenvalues.

   Suppose now that Λ has exactly κ negative eigenvalues and that $F = K + \psi R\varphi$ is a solution of the Takagi-Sarason problem as above. Then by (ii) $\Rightarrow$ (iv) in Theorem 13.2.3, $\Theta_{22}^{-1}\Theta_{21}$ has precisely κ poles, so from (19.3.8)

$$\text{wno } \det \varphi^{-1} - \text{wno } \det \Theta_{22} = -\kappa.$$

Then from (19.3.9) we deduce

$$\text{wno } \det G_2 = 0.$$

Thus G_2 has no zeros in Δ, so $G := G_1 G_2^{-1}$ is pole-free on Δ. By the contractivity of $G(z)$ on $\partial\Delta$ (formula (19.3.4)), we conclude that $G \in \mathcal{BR}_{M+N}(\Delta)$. Now from (19.3.3) we get $F = (\Theta_{11}G + \Theta_{12})(\Theta_{12}G + \Theta_{22})^{-1}$ in the desired form.

Conversely, suppose Λ has κ negative eigenvalues and $\sup_{z\in\partial\Delta}\|G(z)\| < 1$. Define the rational matrix functions F_1 and F_2 by

(19.3.10)
$$\begin{bmatrix} F_1 \\ F_2 \end{bmatrix} = \Theta \begin{bmatrix} G \\ I \end{bmatrix}.$$

Again by Theorem 13.2.3 ((ii) $\Rightarrow$ (iv)), we know that $\Theta_{22}^{-1}\Theta_{21}$ has κ poles in Δ, and hence the subspace $\Theta_{22}\mathcal{R}_N(\Delta)$ has codimension κ in $\Theta_{22}[\Theta_{22}^{-1}\Theta_{21}, I]\mathcal{R}_{M+N}(\Delta) = \varphi^{-1}\mathcal{R}_N(\Delta)$. This then implies

$$\text{wno } \det \varphi^{-1} - \text{wno } \det \Theta_{22} = -\kappa.$$

From (19.3.10) we have

$$F_2 = \Theta_{21}G + \Theta_{22} = \Theta_{22}(\Theta_{22}^{-1}\Theta_{21}G + I).$$

Thus, if we set $W = \varphi F_2$, then

(19.3.11)
$$\text{wno } \det W = \text{wno } \det \varphi + \text{wno } \det \Theta_{22} + \text{wno } \det(\Theta_{22}^{-1}\Theta_{21}G + I)$$
$$= \kappa.$$

Also, from (19.3.10) we know that $F_2\mathcal{R}_N(\Delta) \subset \varphi^{-1}\mathcal{R}_N(\Delta)$, so $W = \varphi F_2$ has no poles on Δ. Then (19.3.11) implies that W has precisely κ zeros in Δ. Since Θ is J-unitary on $\partial\Delta$ and $\sup_{z\in\partial\Delta}\|G(z)\| < 1$ the equality (19.3.10) implies

(19.3.12)
$$F_1(z)^* F_1(z) - F_2(z)^* F_2(z) < 0$$

on $\partial\Delta$. By the argument in the proof of Step 1 in Section 18.6, we know that $F_2(z)$ is invertible for $z \in \partial\Delta$, and thus $F := F_1 F_2^{-1}$ is a well defined rational $M \times N$ matrix function. From (19.3.12) we see that $\|F(x)\| < 1$ for $z \in \partial\Delta$ and from (19.3.10) we get

(19.3.13)
$$\begin{bmatrix} F \\ I \end{bmatrix} \varphi^{-1}W = \begin{bmatrix} F_1 \\ F_2 \end{bmatrix} = \Theta \begin{bmatrix} G \\ I \end{bmatrix}.$$

So by Lemma 19.3.2 $F(z)$ has the form $F = K + \psi R\varphi$ for an $R \in \mathcal{R}_{M\times N,\kappa}(\Delta)$. Finally from (19.3.19) we see that $F = (\Theta_{11}G+\Theta_{12})(\Theta_{21}G+\Theta_{22})^{-1}$. Thus each matrix function F of this form is a solution of the Takagi-Sarason interpolation problem with κ equal to the number of negative eigenvalues of $\Lambda = \Lambda(\omega)$. □

PROOF OF THEOREM 19.2.1. To prove Theorem 19.2.1, follow the proof of Theorem 19.2.2 as above but use Lemma 16.11.2 in place of Lemma 19.3.2 as the starting point. The only change in detail is that the matrix function W appearing in the proof of Theorem 19.2.2 still has κ zeros (counting multiplicities) in Δ, but now none of these are permitted to occur in $\sigma(A_\pi) \cup \sigma(A_\zeta)$. □

CHAPTER 20

NEHARI INTERPOLATION PROBLEM

The classical Nehari theorem identifies the distance in the infinity norm of a given scalar function defined on the unit circle to the functions which are bounded and analytic on the unit disc as the norm of a certain Hankel operator. In the spirit of this book, we assume that the given function is matrix valued and rational. In this chapter we reformulate the problem as a more general type of interpolation problem with metric constraint and prove theorems of Nehari type for this class of functions, using the ideas of the previous chapters. Moreover, we shall obtain a linear fractional description of the set of rational matrix functions bounded and analytic in the unit disc for which the distance to the given function is suboptimal.

20.1 THE NEHARI PROBLEM AND THE HANKEL OPERATOR: THE UNIT DISC CASE

The usual formulation of the matrix Nehari problem is as a distance problem. Namely, we are given an $M \times N$ matrix function $K(z)$ uniformly bounded on the unit circle $\partial \mathcal{D}$ and with to compute the L_∞-distance (i.e. the distance with respect to the sup norm on $\partial \Delta$) of $K(z)$ to the H_∞-space of all $M \times N$ matrix valued functions on the unit circle having uniformly bounded analytic continuation to the open unit disk $\mathcal{D}$, and to find the best matrix H_∞ approximants. We shall assume that the original function $K(z)$ is rational; then without loss of generality we may assume that the analytic approximants are rational as well. The distance in question then is

$$\inf\{\|K - R\|_\infty : R \in \mathcal{R}_{M \times N}(\mathcal{D} \cup \partial \mathcal{D})\}$$

where, as usual, we denote by $\mathcal{R}_{M \times N}(\sigma)$ the set of $M \times N$ rational matrix functions with no poles in σ, and $\| \cdot \|_\infty$ is defined by

$$\|F\|_\infty = \sup\{\|F(z)\| : z \in \partial \mathcal{D}\}.$$

In the spirit of this book, we can think of the Nehari distance problem as really being an interpolation problem in the following way. Let $\{w_1, \ldots, w_k\}$ be the poles of $K(z)$ in $\mathcal{D}$ and let

$$K(z) = (z - w_i)^{-k_i} K_{-k_i}^{(i)} + (z - w_i)^{-k_i+1} K_{-k_i+1}^{(i)} + \cdots + (z - w_i)^{-1} K_{-1}^{(i)} + [\text{analytic at } w_i]$$

be the Laurent expansion for $K(z)$ at w_i ($i = 1, \ldots, k$), where $K_{-k_i}^{(i)} \neq 0$. Then it is easily seen that a rational matrix function F has the form $F = K + R$ with R a rational matrix function analytic on $\overline{\mathcal{D}} := \mathcal{D} \cup \partial \mathcal{D}$ if and only if F has the same singular parts in its local Laurent expansions at points in $\overline{\mathcal{D}}$:

$$F \text{ analytic on } \overline{\mathcal{D}} \backslash \{w_1, \ldots, w_k\}$$

and

$$F(z) = (z - w_i)^{-k_i} K_{-k_i}^{(i)} + \cdots + (z - w_i)^{-1} K_{-1}^{(i)} + [\text{analytic at } w_i].$$

In Nevanlinna-Pick interpolation (as well as in Caratheodory-Toeplitz interpolation, see Chapter 22), one specifies the first few Taylor coefficients of the unknown function at various points in the disk; in the Nehari problem, one allows poles and specifies the singular parts of the local Laurent series at various points in $\mathcal{D}$. In both contexts, one is also trying to control the norm of the interpolant on the boundary while maintaining the local interpolation conditions on the interior.

We now formulate the classical solution of the matrix Nehari problem.

Without loss of generality we may suppose that $K(z)$ has no poles outside the unit disc and vanishes at infinity (indeed, if $K(z)$ is not such, we can add to $K(z)$ a suitable rational matrix function analytic in the closure of the unit disc). Thus $K(z)$ has a Laurent series expansion $K(z) = \sum_{i=1}^{\infty} K_i z^{-i}$ valid for $|z| \geq 1$. The *Hankel matrix* induced by $K(z)$ is defined to be

(20.1.1) $$\mathcal{H}_K = [K_{i+j-1}]_{1 \leq i,j < \infty}.$$

We let $\|\mathcal{H}_K\|$ denote the induced operator norm of $\mathcal{H}_K$ as an operator mapping the space ℓ_N^2 of square summable $\mathbb{C}^N$-valued sequences into the space ℓ_M^2 of square summable $\mathbb{C}^M$-valued sequences. The classical theorem of Nehari can now be stated as follows:

THEOREM 20.1.1. *The infinity norm distance of a given rational function $K(z)$ to the space of rational functions analytic and bounded on $\mathcal{D}$ is equal to the norm of the associated Hankel operator $\mathcal{H}_K$:*

$$\|\mathcal{H}_K\| = \inf\{\|K - R\|_\infty : R \in \mathcal{R}_{M \times N}(\mathcal{D} \cup \partial\mathcal{D})\}.$$

We shall give in Section 20.3 a proof of this theorem using the construction of a rational matrix function with given null-pole triple. Moreover, we shall obtain a linear fractional description of the set of all F of the form $K + R$ with $R \in \mathcal{R}_{M \times N}(\mathcal{D})$ and $\|F\|_\infty \leq 1$ for the case where $\|\mathcal{H}_K\| < 1$.

The reader should note that the problem has the same form as the Sarason divisor-remainder interpolation problem (discussed in Chapters 18 and 19); the data set (K, ψ, φ) of functions can be taken as $K = K$, $\psi = I_M$, $\varphi = I_N$. The problem does not fit in the class of Nevanlinna-Pick or Nudelman interpolation problems due to the presence of poles of $K(z)$ in $\mathcal{D}$.

Nevertheless, any two-sided Nudelman (or equivalently Sarason) interpolation problem can be modified to an equivalent Nehari problem. Namely, let the admissible data set (K, ψ, φ) for a Nudelman interpolation problem (in the divisor-remainder formulation) be given. Without loss of generality we may suppose that ψ and φ are inner, i.e., in addition to being analytic on $\overline{\mathcal{D}}$, φ and ψ have unitary boundary values on $\partial\mathcal{D}$. (Indeed, given φ analytic in $\mathcal{D}$ and analytic and invertible on $\partial\mathcal{D}$, let θ be a rational matrix function without poles in $\overline{\mathcal{D}}$ which is unitary on $\partial\mathcal{D}$ and which has the same left null pair over $\mathcal{D}$ as φ has. By Theorem 4.5.8, $\varphi = \theta\varphi_0$ for some φ_0 which is analytic and invertible on $\overline{\mathcal{D}}$; analogously, $\psi = \psi_0\theta$, where θ is inner and ψ_0 is analytic and invertible in $\overline{\mathcal{D}}$.) Then $F = K + \psi R\varphi$ and $\psi^{-1}F\varphi^{-1} = \psi^{-1}K\varphi^{-1} + R$ have the same norm on $\partial\mathcal{D}$. From this we see that F is a solution of the Sarason interpolation problem with data (K, ψ, φ) if and only if $\psi^{-1}F\varphi^{-1}$ is a solution of the Nehari (suboptimal) approximation problem (find $F = \widetilde{K} + R$ with $R \in \mathcal{R}_{M \times N}(\mathcal{D})$ and $\|F\|_\infty < 1$) for $\widetilde{K} = \psi^{-1}K\varphi^{-1}$.

One approach for solving the Nehari problem therefore is to reduce it to an equivalent Sarason problem and use our previous results to solve this problem. However, the explicit linear fractional formulas for the set of all solution obtained this way will be inefficient compared to what can be obtained by deriving the results directly. Also by taking a direct route we shall gain insight into the different mathematical structure of the two problems. For these reasons we shall analyze the Nehari problem directly from the beginning.

We assume that $K(z) = \sum_{j=1}^{\infty} K_j z^{-j}$ is rational of size $M \times N$ with all poles inside the unit disc $\mathcal{D}$ and with value 0 at infinity. Thus $K(z)$ has a minimal realization

$$(20.1.2) \qquad K(z) = C(zI - A)^{-1}B$$

where $\sigma(A) \subset \mathcal{D}$ and A has size $n \times n$. Ultimately we seek explicit formulas for solutions $F = K + R$ of the Nehari problem in terms of the matrices A, B, C. We would also like a finite dimensional test for when solutions exist; as $\mathcal{H}_K$ is an operator on the infinite dimensional space ℓ_N^2, the criterion $\|\mathcal{H}_K\| \le 1$ is unsatisfactory from this point of view.

To this end we introduce the *controllability gramian*

$$(20.1.3) \qquad P = \sum_{k=0}^{\infty} A^k BB^* A^{*k} = \Xi\Xi^*$$

and the *observability gramian*

$$(20.1.4) \qquad Q = \sum_{k=0}^{\infty} A^{*k} C^* C A^k = \Omega^*\Omega$$

where we denote by

$$(20.1.5) \qquad \Xi = [B, AB, A^2B, \ldots]$$

the *controllability operator* from ℓ_N^2 into space $\mathbb{C}^n$, and by

$$(20.1.6) \qquad \Omega = \mathrm{col}(CA^k)_{k=0}^{\infty}$$

the *observablity operator* from $\mathbb{C}^n$ into ℓ_M^2. Note that the operator Ξ is defined and bounded on all of ℓ_N^2 and the operator Ω is bounded and has range inside ℓ_M^2 since by assumption $\sigma(A) \subset \mathcal{D}$. Also we know that Ξ is surjective and Ω is injective since by assumption (20.1.2) is a *minimal realization* for $K(z)$, so (A, B) is a full range pair and (C, A) is a null kernel pair. The controllability and observability gramians alternatively are characterized as the unique solutions of the respective Stein equations

$$(20.1.7) \qquad P - APA^* = BB^*$$

(20.1.8) $$Q - A^*QA = C^*C.$$

We will need (in this Chapter and in Chapter 24) some properties of the Hankel operator, and especially about its norm, which are summarized in the following theorem.

THEOREM 20.1.2. *Let* $K(z) = \sum_{i=1}^{\infty} K_i z^i$ *be a given rational matrix function with minimal realization* (20.1.2). *Let* $\mathcal{H}_K$, P *and* Q *be the associated Hankel operator* (20.1.1), *controllability gramian* (20.1.3) *and observability gramian* (20.1.4) *respectively. Then:*

(i) $\mathcal{H}_K$ *is a finite rank operator and the rank of* $\mathcal{H}_K$ *is equal to the McMillan degree of* K (*i.e. the size of the matrix* A *in the minimal realization* (20.1.2));

(ii) *The nonzero singular values*

$$s_1(\mathcal{H}_k) \geq s_2(\mathcal{H}_k) \geq \cdots \geq s_n(\mathcal{H}_K) > 0$$

of the operator $\mathcal{H}_K$ *are given by*

$$s_k(\mathcal{H}_K) = \sqrt{\lambda_k(PQ)}, \qquad k = 1, \ldots, n,$$

where

$$\lambda_1(PQ) \geq \lambda_2(PQ) \geq \cdots \geq \lambda_n(PQ)$$

are the eigenvalues (necessarily positive) of the $n \times n$ *matrix* PQ. *In particular,*

$$\|\mathcal{H}_K\|^2 = \lambda_{\max}(PQ)$$

where $\lambda_{\max}(PQ)$ *is the largest eigenvalue of the matrix* PQ.

PROOF. From the realization (20.1.2) for $K(z)$ compute the Laurent expansion for $K(z)$ as

$$K(z) = C(zI - A)^{-1}B = \sum_{j=1}^{\infty} CA^{j-1}Bz^{-j} = \sum_{j=1}^{\infty} K_j z^{-j}.$$

Equating coefficients gives

$$K_j = CA^{j-1}B.$$

Thus the (i,j)-th entry K_{i+j-1} of the Hankel matrix $\mathcal{H}_K$ is $CA^{i+j-2}B$. This implies that $\mathcal{H}_K$ factors as

$$\mathcal{H}_K = \Omega \Xi$$

where Ω is the observability operator (20.1.6) and Ξ is the controllability operator (20.1.5). In particular, $\mathcal{H}_K$ is a finite rank operator, and, moreover, since Ω is surjective and Ξ is injective, the rank of $\mathcal{H}_K$ is equal to the size of A. This proves (i).

For the proof of part (ii) write (for $k = 1, \ldots, n$):

$$\left(s_k(\mathcal{H}_K)\right)^2 = \left(s_k(\Omega \Xi)\right)^2 = \lambda_k(\Xi^* \Omega^* \Omega \Xi).$$

Use now the general result that for linear bounded operators $S: \mathcal{X} \to \mathcal{Y}$, $T: \mathcal{Y} \to \mathcal{X}$ acting between Banach spaces $\mathcal{X}$ and $\mathcal{Y}$, the nonzero spectrum of ST coincides with the nonzero spectrum of TS (to verify this fact observe the equality

$$
\begin{bmatrix} I & \lambda S \\ 0 & I \end{bmatrix}^{-1}
\begin{bmatrix} I - \lambda ST & 0 \\ -T & I \end{bmatrix}
\begin{bmatrix} I & \lambda S \\ 0 & I \end{bmatrix} =
\begin{bmatrix} I & 0 \\ -T & I - \lambda TS \end{bmatrix},
$$

where $\lambda \in \mathbb{C}$). We conclude that

$$
\left(s_k(\mathcal{H}_k) \right)^2 = \lambda_k(\Xi \Xi^* \Omega^* \Omega) = \lambda_k(PQ). \quad \square
$$

Finally, we need a norm inequality for the Hankel operator. Given any $M \times N$ rational matrix function $F(z)$ with no poles on the unit circle $\partial \mathcal{D}$, we have the Laurent series expansion

$$
F(z) = \sum_{i=-\infty}^{\infty} F_i z^{-i}
$$

in a neighborhood of the unit circle. This allows us to introduce the Hankel operator $\mathcal{H}_F$ by the same formula as in (20.1.1):

$$
\mathcal{H}_F = [F_{i+j-1}]_{1 \leq i,j < \infty}.
$$

THEOREM 20.1.3. *For any rational $M \times N$ matrix function $F(z)$ with no poles on the unit circle, the inequality*

$$(20.1.9) \qquad\qquad \|\mathcal{H}_F\| \leq \|F\|_\infty$$

holds.

Observe that this is a very particular instance of Theorem 20.1.1 (obtained by taking $R = 0$ in the formula there).

PROOF. Introduce the discrete Fourier transform operators

$$(20.1.10) \qquad\qquad \mathcal{F}_+: \ell_N^2 \to H_N^2; \qquad F_-: \ell_M^2 \to (H_M^2)^\perp.$$

Let us define the Hardy spaces H_N^2 that appear in (20.1.10): H_N^2 is the space of all N-dimensional vector functions $f(z)$ analytic on $\mathcal{D}$ such that

$$
f(z) = \sum_{j=0}^{\infty} f_j z^j, \qquad z \in \mathcal{D}
$$

with $\sum_{j=0}^{\infty} \|f_j\|^2 < \infty$, and the norm of $f(z)$ in H_N^2 is defined by

$$
\|f\| = \left(\sum_{j=0}^{\infty} \|f_j\|^2 \right)^{1/2}.
$$

Often a function $f(z) \in H_N^2$ will be understood (by taking non-tangential limits almost everywhere) as an element of the Hilbert space $L_N^2(\partial \mathcal{D})$ of square summable (with respect to the Lebesgue measure) N-dimensional vector functions on the unit circle $\partial \mathcal{D}$. The orthogonal complement $(H_M^2)^\perp$ to H_M^2 in $L_M^2(\partial \mathcal{D})$ can be identified (again via non-tangential limits almost everywhere) with the space of M-dimensional vector functions $g(z) = \sum_{j=1}^\infty g_j z^{-j}$ analytic in $(\mathbb{C} \cup \{\infty\}) \setminus \overline{\mathcal{D}}$ with $g(\infty) = 0$ and with $\sum_{j=1}^\infty \|g_j\|^2 < \infty$; the norm in $(H_M^2)^\perp$ is given by

$$\|g\| = \left(\sum_{j=1}^\infty \|g_j\|^2 \right)^{1/2}.$$

The operators $\mathcal{F}_+$, $\mathcal{F}_-$ in (20.1.10) are defined by

$$\mathcal{F}_+\{u_n\}_{n \geq 1} = \sum_{n=1}^\infty u_n z^{n-1}, \qquad \mathcal{F}_-\{v_n\}_{n \geq 1} = \sum_{n=1}^\infty v_n z^{-n},$$

where $\{u_n\}_{n \geq 1} \in \ell_N^2$, $\{v_n\}_{n \geq 1} \in \ell_M^2$. Clearly, $\mathcal{F}_-$ and $\mathcal{F}_+$ are isometric isomorphisms. Let

(20.1.11) $$\widehat{\mathcal{H}}_F := \mathcal{F}_- \mathcal{H}_F \mathcal{F}_+^* : H_N^2 \to (H_M^2)^\perp,$$

where F is as in the statement of Theorem 20.1.3. A straightforward verification shows that

(20.1.12) $$\widehat{\mathcal{H}}_F = P_{(H_M^2)^\perp} \mathcal{M}_{F|H_N^2},$$

where $P_{(H_M^2)^\perp}$ is the orthogonal projector on $(H_M^2)^\perp$ and $\mathcal{M}_F : L_N^2(\partial \mathcal{D}) \to L_M^2(\partial \mathcal{D})$ is the multiplication operator induced by $F(z)$:

$$(\mathcal{M}_F f)(z) = F(z) f(z); \qquad f(z) \in L_N^2(\partial \mathcal{D}).$$

Now using (20.1.12) for every $h(z) \in H_N^2$ we obtain

$$\|\widehat{\mathcal{H}}_F h\|^2 = \|P_{(H_M^2)^\perp} \mathcal{M}_F h\|^2 \leq \|\mathcal{M}_F h\|^2$$

$$= \frac{1}{2\pi} \int_0^{2\pi} \|F(e^{i\theta}) h(e^{i\theta})\|^2 \, d\theta$$

$$\leq \max_{0 \leq \theta \leq 2\pi} \|F(e^{i\theta})\|^2 \cdot \frac{1}{2\pi} \int_0^{2\pi} \|h(e^{i\theta})\|^2 \, d\theta$$

$$= \|F\|_\infty^2 \|h\|^2.$$

Thus, $\|\widehat{\mathcal{H}}_F\| \leq \|F\|_\infty$. Taking into account (20.1.11) and the isometry properties of $\mathcal{F}_-$ and $\mathcal{F}_+$, we obtain $\|\mathcal{H}_F\| \leq \|F\|_\infty$, as required. $\square$

20.2 THE NEHARI PROBLEM AND THE HANKEL OPERA-TOR: THE HALF-PLANE CASE

Here we state results analogous to those given in the previous section for the setting where the open left half plane Π^- replaces the unit disc $\mathcal{D}$. For this setting, given a rational matrix function $F(z)$ defined on the imaginary axis we let $\|F\|_\infty$ be the supremum norm over the imaginary axis:

$$\|F\|_\infty = \sup\{\|F(ix)\|: -\infty < x < \infty\}.$$

For a given rational $M \times N$ matrix function $K(z)$ with no poles on the imaginary axis and at infinity, the Nehari problem over Π^- is to compute

$$\inf\{\|K - R\|_\infty: R \in \mathcal{R}_{M \times N}(\overline{\Pi^-})\}$$

where $\overline{\Pi^-} = \Pi^- \cup \{i\mathbf{R}\} \cup \{\infty\}$. As was explained in Section 20.1 for the case of the unit disk, the problem can also be formulated as a problem of interpolation of prescribed local singular Laurent coefficients in the left half plane subject to a norm constraint on the boundary. The solution involves the Hankel integral operator $\mathcal{H}_K$ associated with K and defined as an operator from $L^2_N(0, \infty)$ into $L^2_M(0, \infty)$ ($\mathbf{C}^N$ and $\mathbf{C}^M$-valued L^2 space over $(0, \infty)$ respectively, with respect to the Lebesgue measure). Without loss of generality we may assume that K has all poles in the left half plane and vanishes at infinity. Then $K(z)$ has an inverse Laplace transform $\check{K}(t)$ the matrix entries of which are absolutely integrable over $(0, \infty)$:

$$K(z) = \int_0^\infty e^{-tz} \check{K}(t)dt, \qquad \operatorname{Re} z \geq 0.$$

Actually, $\check{K}(t)$ is exponentially decaying, i.e.

(20.2.1) $$\|\check{K}(t)\| \leq \alpha e^{-\beta t}, \qquad t \geq 0,$$

where $\alpha > 0$ and $\beta > 0$ are constants independent of t (β can be taken any positive number such that all poles of $K(z)$ have real parts smaller than $-\beta$). We define $\mathcal{H}_K: L^2_N(0, \infty) \to L^2_M(0, \infty)$ as the integral operator

(20.2.2) $$(\mathcal{H}_K f)(t) = \int_0^\infty \check{K}(s + t)f(s)ds, \qquad f \in L^2_N(0, \infty).$$

Because of (20.2.1) the linear operator $\mathcal{H}_K$ is bounded, and we will see that $\mathcal{H}_K$ is actually a finite rank operator. Let $\|\mathcal{H}_K\|$ denote the induced operator norm of $\mathcal{H}_K$ as an operator from $L^2_N(0, \infty)$ to $L^2_M(0, \infty)$:

$$\|\mathcal{H}_K\| = \sup\{\|\mathcal{H}_K u\|_2: u \in L^2_N(0, \infty), \|u\|_2 \leq 1\},$$

where $\|\cdot\|_2$ stands for the norm in the Hilbert spaces $L^2_N(0, \infty)$ and $L^2_M(0, \infty)$. Then solution to the Nehari problem for Π^- can be stated as follows. The proof will follow from developments in the next section.

THEOREM 20.2.1. *Let K be a given rational matrix function vanishing at infinity and having all poles in the open left half plane. Let $\mathcal{H}_K$ be the Hankel integral operator* (20.2.2) *from $L^2_N(0, \infty)$ to $L^2_M(0, \infty)$ induced by $K(z)$. Then*

$$\|\mathcal{H}_K\| = \inf\{\|K - R\|_\infty : R \in \mathcal{R}_{M \times N}(\overline{\Pi^-})\}.$$

As for the disk case in the previous section, we assume that we know a minimal realization

$$(20.2.3) \qquad\qquad K(z) = C(zI - A)^{-1}B$$

for $K(z)$, where A is $n \times n$. By assumption $K(z)$ has all poles in the left half plane, so $\sigma(A) \subset \Pi^-$. The *controllability* and *observability gramians* for this case are given by

$$(20.2.4) \qquad\qquad P = \int_0^\infty e^{sA}BB^*e^{sA^*}ds = \Xi\Xi^*$$

and

$$(20.2.5) \qquad\qquad Q = \int_0^\infty e^{sA^*}C^*Ce^{sA}ds = \Omega^*\Omega,$$

respectively, where the controllability operator $\Xi: L^2_N(0, \infty) \to \mathbf{C}^n$ is given by

$$(20.2.6) \qquad\qquad \Xi f = \int_0^\infty e^{sA}Bf(s)ds, \qquad f \in L^2_N(0, \infty)$$

and the observability operator $\Omega: \mathbf{C}^n \to L^2_M(0, \infty)$ is given by

$$(20.2.7) \qquad\qquad (\Omega x)(t) = Ce^{tA}x, \qquad 0 \le t < \infty.$$

Note that all these operators are well-defined since $\sigma(A) \subset \Pi^-$. The controllability and obserability gramians are alternatively characterized as the unique solutions of the Lyapunov equations

$$(20.2.8) \qquad\qquad AP + PA^* = -BB^*$$

$$(20.2.9) \qquad\qquad A^*Q + QA = -C^*C$$

(cf. Section A.3). A standard fact is that Ξ is surjective since (A, B) is a full range pair. Indeed, for a fixed $x \in \mathbf{C}^N$ and fixed $\gamma > 0$ let $c_\gamma \in L^2_N(0, \infty)$ be defined by $c_\gamma(t) = x$ if $0 \le t \le \gamma$, $c_\gamma(t) = 0$ if $t > \gamma$. One verifies that

$$\Xi c_\gamma = (e^{\gamma A} - I)A^{-1}Bx$$

$$= \gamma \sum_{m=0}^\infty \frac{1}{(m+1)!}\gamma^m A^m Bx \in \text{Range}(\Xi).$$

Consequently, all coefficients of this power series in γ are in the range of Ξ; in other words, $A^m Bx \in \text{Range}(\Xi)$ for $m = 0, 1, \ldots$ and every $x \in \mathbf{C}^N$. As (A, B) is a full range pair, the operator Ξ is surjective. Also, Ω is injective since (C, A) is a null kernel pair.

The analogue of Theorem 20.1.2 holds also in the half-plane case:

THEOREM 20.2.2. *Let $K(z)$ be a given rational matrix function with minimal realization of the form (20.2.3) with $\sigma(A) \subset \Pi^-$, and let $\mathcal{H}_K$, P and Q denote the Hankel integral operator (20.2.2), controllability gramian (20.2.4) and observability gramian (20.2.5), respectively. Then:*

(i) *$\mathcal{H}_K$ has finite rank equal to the McMillan degree n of K;*

(ii) *the equalities*

$$s_k(\mathcal{H}_K) = \sqrt{\lambda_k(PQ)}, \qquad k = 1, \ldots, n$$

hold, where $s_k(\mathcal{H}_K)$ are the nonzero singular values of $\mathcal{H}_K$ arranged in the nonincreasing order, and $\lambda_k(PQ)$ are the eigenvalues of PQ (also arranged in the nonincreasing order). In particular,

$$\|\mathcal{H}_K\|^2 = \lambda_{\max}(PQ),$$

the largest eigenvalue of PQ.

PROOF. As in the proof of Theorem 20.1.2, we check first that the factorization

$$\mathcal{H}_K = \Omega \Xi$$

holds. To see this, note that $K(z) = C(zI - A)^{-1}B$ is the Laplace transform of

$$\check{K}(t) = Ce^{tA}B.$$

Thus

$$[\mathcal{H}_K(f)](t) = \int_0^\infty Ce^{(t+s)A}Bf(s)ds$$

$$= Ce^{tA}\int_0^\infty e^{sA}Bf(s)ds$$

$$= \Omega \Xi f$$

from the definitions (20.2.5) and (20.2.7).

Once this factorization is established, the proof proceeds exactly as the proof of Theorem 20.1.2. $\square$

Again, as in the previous section, we need a norm inequality for the Hankel operator. For an $M \times N$ rational matrix function $F(z)$ without poles on the imaginary line (including infinity) write $F = F_1 + F_2$, where F_1 has all poles in the open left half plane and $F_1(\infty) = 0$, while F_2 is analytic in the closed left half plane and at infinity.

Now the Hankel operator $\mathcal{H}_F$ is defined by the formula (20.2.2) applied to F_1:

$$(\mathcal{H}_F f)(t) = \int\limits_0^\infty \check{F}_1(s+t)f(s)ds, \qquad f \in L_N^2(0,\infty).$$

THEOREM 20.2.3. *For any $M \times N$ rational matrix function $F(z)$ without poles on the imaginary line (including infinity) the inequality*

(20.2.10) $$\|\mathcal{H}_F\| \le \|F\|_\infty := \sup\{\|F(z)\| : z \in \partial\Pi^-\}$$

holds, where $\partial\Pi^- = i\mathbf{R} \cup \{\infty\}$ is the boundary of Π^-.

Again, as in the disk case, the inequality (20.2.10) is a very particular instance of the Nehari theorem 20.2.1.

PROOF. The proof is done in a similar way as the proof of Theorem 20.1.2, by using the Laplace transforms

$$\mathcal{F}_+ : L_M^2(0,\infty) \to H_M^2(\Pi^+)$$
$$\mathcal{F}_- : L_N^2(0,\infty) \to H_N^2(\Pi^-)$$

given by

$$(\mathcal{F}_+ h)(z) = \int\limits_0^\infty e^{-tz} h(t)dt, \qquad h \in L_M^2(0,\infty)$$

$$(\mathcal{F}_- h)(z) = \int\limits_0^\infty e^{tz} h(t)dt, \qquad h \in L_N^2(0,\infty).$$

Here $H_M^2(\Pi^+)$ stands for the Hardy space of M-dimensional vector valued square summable functions on the imaginary line that are nontangential limits (almost everywhere) of analytic functions in the open right half plane Π^+. The space $H_N^2(\Pi^-)$ is defined analogously, with Π^+ replaced by the open left half plane Π^-.

One verifies that

$$\mathcal{F}\mathcal{H}_K\mathcal{F}_- = P_{H_M^2(\Pi+)}\mathcal{M}_K|_{H_N^2(\Pi-)},$$

where $\mathcal{M}_K$ is the multiplication operator induced by K and $P_{H_M^2(\Pi+)}$ is the orthogonal projection (in the Hilbert space $L_M^2(i\mathbf{R})$) onto $H_M^2(\Pi^+)$. From this point the proof goes just as the proof of Theorem 20.1.2 using the fact that both $\mathcal{F}_+$ and $\mathcal{F}_-$ are isometric isomorphisms. $\square$

20.3 SOLUTION OF THE NEHARI PROBLEM

Let us suppose $K(z)$ is a given $M \times N$ rational matrix function with minimal realization

$$K(z) = C(zI - A)^{-1}B$$

where $\sigma(A) \subset \Delta$, the size of A is $n \times n$ and either $\Delta = \mathcal{D}$ or $\Delta = \Pi^-$. We would like to characterize rational matrix functions F of the form $F = K + R$ such that $R \in \mathcal{R}_{M \times N}(\Delta)$ and

$$\|F\|_\infty := \sup\{\|F(z)\| : z \in \partial\Delta\} \leq 1.$$

(*Nehari problem*). Define controllability and observability gramians P and Q by either (20.1.3) and (20.1.4) or by (20.2.4) and (20.2.5) depending on whether $\Delta = \mathcal{D}$ or $\Delta = \Pi^-$.

To state the solution of the Nehari problem, we introduce the Sylvester data set

$$\widetilde{\omega} = \left(\begin{bmatrix} C \\ 0 \end{bmatrix}, A; A, [\, 0 \quad B\,]; I \right).$$

We seek a matrix function (for reasons which will be apparent later) $\Theta(z)$ which is regular and J-unitary on $\partial\Delta$ and which has $\widetilde{\omega}$ as a left null-pole triple over Δ. Here $J = \begin{bmatrix} I_n & 0 \\ 0 & -I_n \end{bmatrix}$. Our intention is to use results of Chapter 6 (adapted to Π^- rather than Π^+). In carrying through this adaptation the following easily verified general fact will be handy.

PROPOSITION 20.3.1. $(C, A_\pi; A_\zeta, B; S)$ *is a left null-pole triple of a rational matrix function* $W(z)$ *over the set* $\sigma \subset \mathbb{C}$ *if and only if* $(C, -A_\pi; -A_\zeta, B; -S)$ *is a left null-pole triple for* $W(-z)$ *over the set*

$$-\sigma = \{-z : z \in \sigma\}.$$

By a suitably adapted Theorem 6.3.1 there is $\Theta(z)$ regular and J-unitary on $\partial\Pi^-$ and with a left null-pole triple $\widetilde{\omega}$ over Π^- if and only if the matrix

$$(20.3.1) \qquad\qquad \Lambda = \begin{bmatrix} S_1 & I \\ I & S_2 \end{bmatrix}$$

is invertible, where

$$S_1 A + A^* S_1 = -[\, C^* \quad 0\,] \begin{bmatrix} I & 0 \\ 0 & -I \end{bmatrix} \begin{bmatrix} C \\ 0 \end{bmatrix} = -C^* C$$

and

$$S_2 A^* + A S_2 = [\, 0 \quad B\,] \begin{bmatrix} I & 0 \\ 0 & -I \end{bmatrix} \begin{bmatrix} 0 \\ B^* \end{bmatrix} = -B B^*.$$

Comparing with (20.2.8) and (20.2.9) we conclude that $S_1 = Q$ and $S_2 = P$, so

$$(20.3.2) \qquad\qquad \Lambda = \begin{bmatrix} Q & I \\ I & P \end{bmatrix}, \qquad \Delta = \Pi^-.$$

For $\Delta = \mathcal{D}$, by Theorem 7.3.1 the relevant matrix Λ is of the form (20.3.1) but now S_1 and S_2 arise as solutions of the Stein equations

$$S_1 - A^* S_1 A = -[\, C^* \quad 0\,] \begin{bmatrix} I & 0 \\ 0 & -I \end{bmatrix} \begin{bmatrix} C \\ 0 \end{bmatrix} = -C^* C$$

and

$$S_2 - AS_2 A^* = [\, 0 \;\; B \,] \begin{bmatrix} I & 0 \\ 0 & -I \end{bmatrix} \begin{bmatrix} 0 \\ B^* \end{bmatrix} = -BB^*.$$

Comparing with (20.1.7) and (20.1.8) shows that in this case $S_1 = -Q$ and $S_2 = -P$, so

(20.3.3)
$$\Lambda = \begin{bmatrix} -Q & I \\ I & -P \end{bmatrix}, \qquad \Delta = \mathcal{D}.$$

Let us consider the case $\Delta = \Pi^-$ for the moment. By a standard Schur complement argument we have

(20.3.4)
$$\Lambda = \begin{bmatrix} Q & I \\ I & P \end{bmatrix} = \begin{bmatrix} I & 0 \\ P & I \end{bmatrix} \begin{bmatrix} 0 & I \\ I - PQ & 0 \end{bmatrix} \begin{bmatrix} I & 0 \\ Q & I \end{bmatrix}$$

We see that invertibility of Λ is equivalent to invertibility of $I - PQ$, i.e., to 1 not being an eigenvalue of PQ. When this is the case we define Z by

(20.3.5)
$$Z = (I - PQ)^{-1}.$$

Then Λ^{-1} can be computed explicitly from (20.3.4) as

(20.3.6)
$$\Lambda^{-1} = \begin{bmatrix} -ZP & Z \\ Z^* & -QZ \end{bmatrix}, \qquad \Delta = \Pi^-$$

where we also used that $ZP = PZ^*$. For the case $\Delta = \mathcal{D}$, the algebra is exactly the same, but with $-P$ in place of P and $-Q$ in place of Q. Thus

(20.3.7)
$$\Lambda^{-1} = \begin{bmatrix} ZP & Z \\ Z^* & QZ \end{bmatrix}, \qquad \Delta = \mathcal{D}$$

where again $Z = (I - PQ)^{-1}$. Finally an explicit formula for a rational $(M+N) \times (M+N)$ matrix function which is J-unitary on $\partial \Delta$ and has $\left(\begin{bmatrix} C \\ 0 \end{bmatrix}, A; A, [0, B]; I \right)$ as a left null pole triple over Δ is therefore given by

(20.3.8)
$$\Theta(z) = \begin{bmatrix} I_M & 0 \\ 0 & I_N \end{bmatrix} + \begin{bmatrix} C & 0 \\ 0 & B^* \end{bmatrix} \begin{bmatrix} (zI - A)^{-1} & 0 \\ 0 & (zI + A^*)^{-1} \end{bmatrix}$$
$$\times \begin{bmatrix} -ZP & Z \\ Z^* & -QZ \end{bmatrix} \begin{bmatrix} -C^* & 0 \\ 0 & B \end{bmatrix}$$

if $\Delta = \Pi^-$ and

(20.3.9)
$$\Theta(z) = D + \begin{bmatrix} C & 0 \\ 0 & B^* \end{bmatrix} \begin{bmatrix} (zI - A)^{-1} & 0 \\ 0 & (I - zA^*)^{-1} \end{bmatrix}$$
$$\times \begin{bmatrix} ZP & Z \\ Z^* & QZ \end{bmatrix} \begin{bmatrix} A^{*-1}C^* & 0 \\ 0 & B \end{bmatrix} D,$$

where

$$D = I - \begin{bmatrix} C & 0 \\ 0 & -B^*A^{*-1} \end{bmatrix} \begin{bmatrix} ZP & Z \\ Z^* & QZ \end{bmatrix} \begin{bmatrix} (I - \alpha A)^{-1} & 0 \\ 0 & (\alpha I - A)^{-1} \end{bmatrix} \begin{bmatrix} -C^* & 0 \\ 0 & B \end{bmatrix}$$

where $|\alpha| = 1$, if $\Delta = \mathcal{D}$ and A is invertible.

If it happens that A is not invertible (and $\Delta = \mathcal{D}$) $\Theta(z)$ is given by the general formula (7.4.7) which in our case takes the form

(20.3.10)
$$\Theta(z) = I + (z - z_0) \begin{bmatrix} C & 0 \\ 0 & B^* \end{bmatrix} \begin{bmatrix} (zI - A)^{-1} & 0 \\ 0 & (I - zA^*)^{-1} \end{bmatrix}$$
$$\times \begin{bmatrix} ZP & Z \\ Z^* & QZ \end{bmatrix} \begin{bmatrix} (I - z_0 A^*)^{-1}C^* & 0 \\ 0 & (A - z_0 I)^{-1}B \end{bmatrix}$$

where z_0 is a fixed point on the unit circle.

The complete solution of the Nehari problem for the completely indeterminant case (1 not an eigenvalue of PQ) is then as follows.

THEOREM 20.3.2. *Let $\Delta = \Pi^-$ or $\mathcal{D}$ and let $K(z)$ be a rational $M \times N$ matrix function with minimal realization*

$$K(z) = C(zI - A)^{-1}B$$

where $\sigma(A) \subset \Delta$. Introduce the controllability gramian P and observability gramian Q by (20.1.3) and (20.1.4) if $\Delta = \mathcal{D}$, and by (20.2.4) and (20.2.5) if $\Delta = \Pi^-$. Assume $\lambda = 1$ is not an eigenvalue of PQ and let $\Theta(z) = \begin{bmatrix} \Theta_{11}(z) & \Theta_{12}(z) \\ \Theta_{21}(z) & \Theta_{22}(z) \end{bmatrix}$ be given by (20.3.8) if $\Delta = \Pi^-$ and (20.3.9) or (20.3.10) if $\Delta = \mathcal{D}$. Then the Nehari problem associated with K and Δ has a solution, i.e. there exists $R(z) \in \mathcal{R}_{M \times N}(\Delta)$ such that

$$\sup\{\|K(z) + R(z)\|: z \in \partial\Delta\} \le 1,$$

if and only if $\lambda_{\max}(PQ)$, the maximal eigenvalue of PQ, is less than 1. In this case, solutions $F = K + R$ are characterized as matrix functions of the form

$$F = (\Theta_{11}G + \Theta_{12})(\Theta_{21}G + \Theta_{22})^{-1}$$

where G is an arbitrary rational $M \times N$ matrix function such that

(20.3.11)
$$\sup_{z \in \Delta} \|G(z)\| \le 1.$$

The rest of this chapter will be devoted to the proof of Theorem 20.3.2.

20.4 PROOF OF THEOREM 20.3.2

Before proceeding with the proof of Theorem 20.3.2 we derive some preliminary lemmas.

LEMMA 20.4.1. *Let $K(z)$ be a rational matrix function with minimal realization*

$$K(z) = C(zI - A)^{-1}B$$

with $\sigma(A) \subset \Pi^-$. Let P and Q be the controllability and observability gramians given by the equations

(20.4.1) $$AP + PA^* = -BB^*$$

(20.4.2) $$A^*Q + QA = -C^*C.$$

Assume that 1 is not an eigenvalue of PQ, and set $Z = (I - PQ)^{-1}$. Then the number of eigenvalues of PQ bigger than 1 (counted with multiplicities) equals the number of negative eigenvalues of QZ. In particular, $\lambda_{\max}(PQ) < 1$ if and only if QZ is positive definite.

A similar statement holds with Π^- replaced by $\mathcal{D}$, in which case the controllability and obserability gramians are defined by (20.1.3) and (20.1.4) instead of (20.4.1) and (20.4.2).

PROOF. The full range property of (A, B) and the null kernel property of (C, A) imply that both P and Q are positive definite matrices (Theorem A.3.1). So QZ is invertible, and the equality $QZ = Z^*Q$ implies that QZ is Hermitian. So the second statement in the lemma is indeed a particular case of the first statement.

Next note that

$$\begin{aligned} QZ &= Q(I - PQ)^{-1} \\ &= Q^{1/2}(Q^{-1/2} - PQ^{1/2})^{-1} \\ &= Q^{1/2}(I - Q^{1/2}PQ^{1/2})^{-1}Q^{1/2}. \end{aligned}$$

Thus the number of negative eigenvalues of QZ is the same as the number of negative eigenvalues of $I - Q^{1/2}PQ^{1/2}$, that is, the number of eigenvalues of $Q^{1/2}PQ^{1/2}$ which are larger than 1. By the general fact that XY and YX have the same nonzero eigenvalues, we see that this is the same as the number of eigenvalues of PQ larger than 1. $\square$

LEMMA 20.4.2. *Let $K(z)$ be a given $M \times N$ rational matrix function with minimal realization $C(zI - A)^{-1}B$ with $\sigma(A) \subset \Delta$, where $\Delta = \mathcal{D}$ or $\Delta = \Pi^-$. Let $\Theta(z)$ be a rational matrix function having $\left(\begin{bmatrix} C \\ 0 \end{bmatrix}, A; A, [\, 0 \;\; B \,]; I \right)$ as its left null-pole triple over Δ. Then the rational matrix function F is of the form $F = K + H$ with H having at most κ poles in Δ if and only if there is a regular rational $N \times N$ matrix function b having κ zeros and no poles in Δ such that*

(20.4.3) $$\begin{bmatrix} F \\ I_N \end{bmatrix} b = \Theta \begin{bmatrix} G_1 \\ G_2 \end{bmatrix}$$

where $\begin{bmatrix} G_1 \\ G_2 \end{bmatrix} \in \mathcal{R}_{(M+N) \times N}(\Delta)$. In particular F is of the form $F = K + H$ with $H \in \mathcal{R}_{M \times N}(\Delta)$ if and only if $\begin{bmatrix} F \\ I_N \end{bmatrix}$ itself admits the factorization

$$\begin{bmatrix} F \\ I_N \end{bmatrix} = \Theta \begin{bmatrix} G_1 \\ G_2 \end{bmatrix}$$

with $\begin{bmatrix} G_1 \\ G_2 \end{bmatrix} \in \mathcal{R}_{(M+N)\times N}(\Delta).$

PROOF. Set $L(z) = \begin{bmatrix} I & K(z) \\ 0 & I \end{bmatrix}$. From the minimal realization $K(z) = C(zI - A)^{-1}B$ for $K(z)$ we see that $L(z)$ has the minimal realization

$$L(z) = \begin{bmatrix} I & 0 \\ 0 & I \end{bmatrix} + \begin{bmatrix} C \\ 0 \end{bmatrix}(zI - A)^{-1}[0, B].$$

From this realization use Proposition 4.4.1 to deduce that

$$\widetilde{\omega} = \left(\begin{bmatrix} C \\ 0 \end{bmatrix}, A; A, [0, B]; I \right)$$

is a (left) null-pole triple for $L(z)$ over Δ. But by construction $\widetilde{\omega}$ is a null-pole triple for $\Theta(z)$ over Δ. Thus by Theorem 4.5.8, the factorization (20.4.3) is equivalent to a factorization

(20.4.4) $$\begin{bmatrix} F \\ I \end{bmatrix} b = L \begin{bmatrix} \widetilde{G}_1 \\ \widetilde{G}_2 \end{bmatrix} = \begin{bmatrix} I & K \\ 0 & I \end{bmatrix} \begin{bmatrix} \widetilde{G}_1 \\ \widetilde{G}_2 \end{bmatrix}$$

with $\begin{bmatrix} \widetilde{G}_1 \\ \widetilde{G}_2 \end{bmatrix} \in \mathcal{R}_{(M+N)\times N}(\Delta)$. But (20.4.4) is in turn equivalent to the system of equations

(20.4.5) $$\begin{aligned} Fb &= \widetilde{G}_1 + K\widetilde{G}_2 \\ b &= \widetilde{G}_2 \end{aligned}$$

from which we get immediately

$$F = K + \widetilde{G}_1 b^{-1}.$$

It remains only to note, e.g. by using the theory of coprime factorization in Chapter 11, that a rational $M \times N$ matrix has at most κ poles in Δ if and only if H admits a factorization $H = \widetilde{G}_1 b^{-1}$ with both $\widetilde{G}_1$ and b analytic on Δ and with b regular with at most κ zeros in Δ. $\square$

PROOF OF THEOREM 20.3.2. Suppose a rational $M \times N$ matrix function $F(z)$ exists of the form $F = K + H$ with $H \in \mathcal{R}_{M\times N}(\Delta)$ and $\|F\|_\infty \le 1$ holds. Then $\mathcal{H}_K = \mathcal{H}_F$, so

(20.4.6) $$\|\mathcal{H}_K\| = \|\mathcal{H}_F\| \le \|F\|_\infty \le 1$$

(we have used here Theorems 20.1.3 and 20.2.3) so necessarily $\|\mathcal{H}_K\| \le 1$, or equivalently by Theorem 20.1.2 or 20.2.2, $\lambda_{\max}(PQ) \le 1$. As the matrix PQ does not have eigenvalue 1 by assumption, the first part of Theorem 20.3.1 is proved. We want to proceed further and argue that F has the form

(20.4.7) $$F = (\Theta_{11}G + \Theta_{12})(\Theta_{21}G + \Theta_{22})^{-1}$$

with $G \in \mathcal{R}_{M \times N}(\Delta)$ satisfying (20.3.11). By Lemma 20.4.2 we know that $\begin{bmatrix} F \\ I_N \end{bmatrix}$ has a factorization

$$(20.4.8) \qquad \begin{bmatrix} F \\ I_N \end{bmatrix} = \Theta \begin{bmatrix} G_1 \\ G_2 \end{bmatrix}$$

where $\begin{bmatrix} G_1 \\ G_2 \end{bmatrix}$ has no poles in Δ. Since Θ is J-unitary on $\partial\Delta$ and $\|F\|_\infty \le 1$, we have

$$
\begin{aligned}
0 &\ge F(z)^* F(z) - I \\
(20.4.9) \qquad &= [G_1(z)^* \, G_2(z)^*]\Theta(z)^* J\Theta(z) \begin{bmatrix} G_1(z) \\ G_2(z) \end{bmatrix} \\
&= G_1(z)^* G_1(z) - G_2(z)^* G_2(z)
\end{aligned}
$$

for $z \in \partial\Delta$. If for some $z \in \partial\Delta$, $G_2(z)$ had degenerate rank, then necessarily by (20.4.9) $\begin{bmatrix} G_1(z) \\ G_2(z) \end{bmatrix}$ would also have degenerate rank. But the left side of (20.4.8) is maximal rank at all points of $\partial\Delta$ and Θ is regular on $\partial\Delta$, so $\begin{bmatrix} G_1 \\ G_2 \end{bmatrix}$ necessarily is maximal rank on $\partial\Delta$. Thus $G_2(z)$ is invertible on $\partial\Delta$ and then by (20.4.9), if we set $F = G_1 G_2^{-1}$ we see that

$$(20.4.10) \qquad \|G(z)\| \le 1 \quad \text{for} \quad z \in \partial\Delta.$$

Once we show that G_2^{-1} has no poles in Δ, then we can conclude that G has no poles in Δ, so by (20.4.10) we will have $\|G(z)\| \le 1$ for $z \in \Delta$. Finally the desired form (20.4.7) of F then follows easily from (20.4.8) and the definition $G = G_1 G_2^{-1}$ of G.

We argue that G_2^{-1} has no poles in Δ as follows. From (20.4.8) we get

$$
(20.4.11) \qquad
\begin{aligned}
I_N &= \Theta_{21} G_1 + \Theta_{22} G_2 \\
&\Theta_{22}(\Theta_{22}^{-1}\Theta_{21} G + I)G_2.
\end{aligned}
$$

By Proposition 13.1.2 we know that

$$\|\Theta_{22}^{-1}\Theta_{21}\|_\infty = \sup\{\|\Theta_{22}^{-1}(z)\Theta_{21}(z)\| : z \in \partial\Delta\} < 1$$

and we have already observed that $\|G\|_\infty \le 1$. Thus, by the usual homotopy argument (cf. Section 18.6),

$$(20.4.12) \qquad \mathrm{wno}\det(\Theta_{22}^{-1}\Theta_{21} G + I) = 0.$$

Since $\Theta(z)$ has a right pole pair over Δ of the form $\left(\begin{bmatrix} C \\ 0 \end{bmatrix}, A\right)$, it is clear that Θ_{22} has no poles in Δ. Thus

$$(20.4.13) \qquad \mathrm{wno}\det\Theta_{22} \ge 0.$$

Since we also know that G_2 has no poles in Δ, we also have

(20.4.14)   wno $\det G_2 \geq 0.$

Also by (20.4.11) we get

(20.4.15) $0 = $ wno $\det \Theta_{22} + $ wno $\det(\Theta_{22}^{-1}\Theta_{21}G + I) + $ wno $\det G_2.$

Combining (20.4.12)–(20.4.14) with (20.4.15) forces

(20.4.16)   wno $\det \Theta_{22} = 0,$ wno $\det G_2 = 0.$

We conclude that G_2^{-1} has no poles in Δ as desired.

Conversely, suppose $\lambda_{\max}(PQ) < 1$. In view of Lemma 20.4.1 the matrix QZ (where $Z = (I - PQ)^{-1}$) is positive definite. Let $G \in \mathcal{R}_{M \times N}(\Delta)$ be such that $\sup\{\|G(z)\| : z \in \Delta\} \leq 1$ and define $\begin{bmatrix} F_1 \\ F_2 \end{bmatrix} \in \mathcal{R}_{(M+N) \times N}$ by

(20.4.17) $\begin{bmatrix} F_1 \\ F_2 \end{bmatrix} = \Theta \begin{bmatrix} G \\ I \end{bmatrix}.$

We will show that $F := F_1 F_2^{-1}$ is of the form $F = K + H$ with $H \in \mathcal{R}_{M \times N}(\Delta)$ and $\|F\|_\infty \leq 1$ holds. Since $\Theta(z)$ is J-unitary and $\|G(z)\| \leq 1$ for $z \in \partial\Delta$, we have

(20.4.18)
$$F_1(z)^* F_1(z) - F_2(z)^* F_2(z) \leq [\, G(z)^* \quad I\,]\Theta(z)^* J\Theta(z) \begin{bmatrix} G(z) \\ I \end{bmatrix}$$
$$= G(z)^* G(z) - I \leq 0$$

for $z \in \partial\Delta$. Since Θ is analytic and invertible on $\partial\Delta$, and the matrix $\begin{bmatrix} G \\ I \end{bmatrix}$ is clearly maximal rank on $\partial\Delta$, the left side of (20.4.17) must be also maximal rank on $\partial\Delta$. This combined with (20.4.18) then forces $F_2(z)$ to be invertible and F to have norm at most 1 on $\partial\Delta$. We can now rewrite (20.4.17) as

(20.4.19) $\begin{bmatrix} F \\ I \end{bmatrix} = \Theta \begin{bmatrix} GF_2^{-1} \\ F_2^{-1} \end{bmatrix}.$

If we can show that F_2^{-1} has no poles in Δ, then Lemma 20.4.2 implies that F has the desired form $F = K + H$ with $H \in \mathcal{R}_{M \times N}(\Delta)$.

Thus to complete the proof of Theorem 20.3.2, we need only show that F_2 has no zeros in Δ. From (20.4.17) we have

(20.4.20) $F_2 = \Theta_{21}G + \Theta_{22} = \Theta_{22}(\Theta_{22}^{-1}\Theta_{21}G + I).$

Applying again the standard homotopy argument (as in Section 18.6) we see that

$$\text{wno}\det(\Theta_{22}^{-1}\Theta_{21}G + I) = 0.$$

Therefore

$$\text{wno}\det F_2 = \text{wno}\det \Theta_{22}.$$

It is easy to see that Θ satisfies condition (ii) in Theorem 13.2.4 (since its Hermitian coupling matrix is I). Now in the case $\Delta = \Pi^-$ we have

$$
\frac{J - \Theta(z)J\Theta(w)^*}{-z - \overline{w}} = -\begin{bmatrix} C & 0 \\ 0 & B^* \end{bmatrix}\begin{bmatrix} (zI - A)^{-1} & 0 \\ 0 & (zI + A^*)^{-1} \end{bmatrix}\begin{bmatrix} -ZP & Z \\ Z^* & -QZ \end{bmatrix}
$$
$$
\times \begin{bmatrix} (\overline{w}I - A^*)^{-1} & 0 \\ 0 & (\overline{w}I + A)^{-1} \end{bmatrix}\begin{bmatrix} C^* & 0 \\ 0 & B \end{bmatrix},
$$

so the kernel function

$$
K(z, w) := \begin{bmatrix} 0 & I_N \end{bmatrix}\frac{J - \Theta(z)J\Theta(w)^*}{-z - \overline{w}}\begin{bmatrix} 0 \\ I_N \end{bmatrix}
$$

is equal to

$$(20.4.21) \qquad K(z, w) = B^*(zI + A^*)^{-1}QZ(\overline{w}I + A)^{-1}B$$

and hence is positive definite in view of the positive definiteness of QZ and the full range property of (A, B). For the case $\Delta = \mathcal{D}$ we have (assuming $0 \notin \sigma(A)$)

$$
\frac{J - \Theta(z)J\Theta(w)^*}{\rho_\Delta(z, w)} = \frac{J - \Theta(z)J\Theta(w)^*}{1 - z\overline{w}}
$$
$$
= \begin{bmatrix} C & 0 \\ 0 & -B^* \end{bmatrix}\begin{bmatrix} (zI - A)^{-1} & 0 \\ 0 & (zI - A^{*-1})^{-1} \end{bmatrix}\begin{bmatrix} ZP & Z \\ Z^* & QZ \end{bmatrix}
$$
$$
\times \begin{bmatrix} (\overline{w}I - A^*)^{-1} & 0 \\ 0 & (\overline{w}I - A^{-1})^{-1} \end{bmatrix}\begin{bmatrix} C^* & 0 \\ 0 & -B \end{bmatrix},
$$

and the kernel function

$$
K(z, w) := [0, I_N]\frac{J - \Theta(z)J\Theta(w)^*}{1 - z\overline{w}}\begin{bmatrix} 0 \\ I_N \end{bmatrix}
$$

is equal to

$$
B^*(zI - A^{*-1})^{-1}QZ(\overline{w}I - A^{-1})^{-1}B.
$$

Again, we conclude that $K(z, w)$ is positive definite. (If A is singular, a similar argument works by using formula (20.3.10).) In both cases ($\Delta = \Pi^-$ and $\Delta = \mathcal{D}$) by the implication (ii) $\Rightarrow$ (iv) in Theorem 13.2.4 we conclude that $\Theta_{22}^{-1}\Theta_{21}$ has no poles in Δ. Then by (iii) in Theorem 13.2.4 we see that Θ_{22}^{-1} has no poles in Δ, and thus

$$\text{wno}\det\Theta_{22} = 0.$$

Now by (20.4.20) we get

$$\text{wno}\det F_2 = 0,$$

so F_2 has no zeros in Δ, as desired.

20.5 THE NEHARI-TAKAGI PROBLEM

In this section we study a generalization of the Nehari problem when for a given rational $M \times N$ matrix function $K(z)$ the distance is sought to the set of rational matrix functions with at most κ poles (counted with multiplicities) in Δ. We call this problem the Nehari-Takagi problem. As in the rest of this chapter, $\Delta = \mathcal{D}$ or $\Delta = \Pi^-$ (the open left half-plane). If $\kappa = 0$, we obtain the Nehari problem studied in the previous sections.

We reformulate this problem in the same way as in Section 20.3. Namely, it is assumed that $K(z)$ has a minimal realization of the form

(20.5.1) $$K(z) = C(zI - A)^{-1}B$$

where $\sigma(A) \subset \Delta$, and we will characterize rational matrix functions F of the form $F = K + R$ such that $R(z)$ is an $M \times N$ rational matrix function with at most κ poles in Δ and

$$\|F\|_\infty := \sup\{\|F(z)\| : z \in \partial\Delta\} \leq 1.$$

The solution is given by the following canonical generalization of Theorem 20.3.1.

THEOREM 20.5.1. Let $\Delta = \Pi^-$ or $\Delta = \mathcal{D}$, and let (20.5.1) be a minimal realization for a rational $M \times N$ matrix function $K(z)$ with $\sigma(A) \subset \Delta$. Further, let

$$P = \sum_{k=0}^\infty A^k BB^* A^{*k}$$

$$Q = \sum_{k=0}^\infty A^{*k} C^* C A^k$$

be the controllability and observability gramians (if $\Delta = \mathcal{D}$), and if $\Delta = \Pi^-$, define the gramians by

$$P = \int_0^\infty e^{sA} BB^* e^{sA^*}\, ds$$

$$Q = \int_0^\infty e^{sA^*} C^* C e^{sA}\, ds.$$

Assume 1 is not an eigenvalue of PQ. Then there is a rational matrix function $R(z)$ with at most κ poles (counted with multiplicities) in Δ such that

(20.5.2) $$\|K + R\|_\infty \leq 1$$

if and only if the matrix PQ has at most κ eigenvalues (counted with multiplicities) bigger than 1. Moreover, if κ_0 is the number of eigenvalues of PQ bigger than 1, then the rational matrix functions $F = K + R$ satisfying (20.5.2) and such that R has precisely κ_0 poles in Δ, are given by the formula

(20.5.3) $$F = (\Theta_{11}G + \Theta_{12})(\Theta_{21}G + \Theta_{22})^{-1},$$

where G is an arbitrary rational $M \times N$ matrix function satisfying

$$\sup_{z \in \Delta} \|G(z)\| \leq 1.$$

Here $\Theta(z) = \begin{bmatrix} \Theta_{11}(z) & \Theta_{12}(z) \\ \Theta_{21}(z) & \Theta_{22}(z) \end{bmatrix}$ *is given by (20.3.8) if* $\Delta = \Pi^-$ *and (20.3.9) or* (20.3.10) *if* $\Delta = \mathcal{D}$, *where* $Z = (I - PQ)^{-1}$.

The rest of this section will be devoted to the proof of Theorem 20.5.1. The ideas are the same as in the proof of Theorem 20.3.1, suitably adapted.

Suppose there is a rational $M \times N$ matrix function $F(z)$ of the form $F = K + H$, where the rational matrix function H has at most κ poles in Δ (counted with multiplicities) and the inequality $\|F\|_\infty \leq 1$ holds. This inequality implies, in particular, that H has no poles on the boundary $\partial\Delta$ (including infinity if $\Delta = \Pi^-$). Write $H = H_1 + H_2$, where H_1 is analytic in Δ and H_2 has all its poles in Δ and $H_2(\infty) = 0$. It follows that $H_2(z)$ admits a minimal realization

$$H_2(z) = C(zI - A)^{-1}B,$$

where the size of A is equal to the number of poles of $H_2(z)$, which is less than or equal to κ by assumption. Now we have

$$\mathcal{H}_K + \mathcal{H}_{H_2} = \mathcal{H}_F.$$

So by Theorem 20.1.3 (or 20.2.3)

(20.5.4) $$s_1(\mathcal{H}_F) = \|\mathcal{H}_K + \mathcal{H}_{H_2}\| \leq 1,$$

where we denote by $s_k(V)$ the k-th singular value of the finite rank operator V (the singular values are assumed to be arranged in the nonincreasing order). Using a well-known fact that the $(\kappa + 1)$-st singular value of an operator Z is equal to the distance from Z to the set of all operators whose rank is finite and does not exceed κ (see, e.g., Section 2.2 in Gohberg-Krein [1969]) we obtain from (20.5.3) that $s_{\kappa+1}(\mathcal{H}_K) \leq 1$. Using Theorem 20.2.2 (or 20.1.2) and the hypothesis that $1 \notin \sigma(PQ)$ it follows that actually $s_{\kappa+1}(\mathcal{H}_K) < 1$, and by the same theorem the first part of Theorem 20.5.1 follows.

It remains to show that the set of all solutions F of the Nehari-Takagi problem are given by the linear fractional formula (20.5.3) where the free parameter G is any rational $M \times N$ matrix function with norm at most 1 on Δ. For this we need the following sharpening of Theorem 13.2.4 for the situation where the Sylvester data set $\tau = \left(\begin{bmatrix} C \\ 0 \end{bmatrix}, A; A, [0, B]; I \right)$ has the special form associated with the Nehari-Takagi problem.

LEMMA 20.5.2. *Let the rational matrix function* $K(z) = C(zI - A)^{-1}B$ *and* $\Theta(z) = \begin{bmatrix} \Theta_{11}(z) & \Theta_{12}(z) \\ \Theta_{21}(z) & \Theta_{22}(z) \end{bmatrix}$ *and the matrices* P, Q *be as in Theorem 20.5.1. Then the number of eigenvalues of* PQ *bigger than 1 is equal to the number of poles of* Θ_{22}^{-1} *(counting multiplicities) in* Δ.

PROOF. By construction $\Theta(z)$ is a rational matrix function with J-unitary values on $\partial\Delta$ ($J = I_M \oplus -I_N$) which has $\tau = \left(\begin{bmatrix} C \\ 0 \end{bmatrix}, A; A, [\, 0 \; B \,]; I \right)$ as a null-pole triple over Δ. Since the coupling matrix in τ is I, trivially condition (ii) in Theorem 13.2.4 is satisfied. By (ii) $\Leftrightarrow$ (iv) in Theorem 13.2.4, we see that the number of poles κ_0 of $\Theta_{22}^{-1}\Theta_{21}$ in Δ is the same as the number of negative squares had by the kernel function

$$K(z, w) = [\, 0 \;\; I_N \,]\frac{J - \Theta(z)J\Theta(z)^*}{\rho_\Delta(z, w)}\begin{bmatrix} 0 \\ I_N \end{bmatrix}$$

over Δ, where $\rho_\Delta(z, w) = 1 - z\overline{w}$ if $\Delta = \mathcal{D}$ and $\rho_\Delta(z, w) = -z - \overline{w}$ if $\Delta = \Pi^-$. By (20.4.21) and (20.4.22) and by using the full range property of (A, B), we conclude that the number of negative squares of $K(z, w)$ over Δ is equal to the number of negative eigenvalues of QZ. By Lemma 20.4.1 this in turn is the number of eigenvalues of PQ bigger than 1. We conclude that the number of eigenvalues of PQ bigger than 1 is κ_0.

On the other hand, by (ii) $\Leftrightarrow$ (iii) in Theorem 13.2.4, the pole subspace of Θ_{22}^{-1} is a subspace of the pole subspace of $\Theta_{22}^{-1}\Theta_{21}$. However, since the pole pair for Θ has the form $\left(\begin{bmatrix} C \\ 0 \end{bmatrix}, A \right)$, Θ_{21} is analytic in Δ, so the reverse containment is automatic. Thus the number κ_0 of poles of $\Theta_{22}^{-1}\Theta_{21}$ in Δ is also equal to the number of poles of Θ_{22}^{-1} in Δ, and the lemma follows. $\square$

CONTINUATION OF THE PROOF OF THEOREM 20.5.1. Suppose now that PQ has κ_0 eigenvalues bigger than 1 and F has the form $F = K + H$ with $\|F\|_\infty \leq 1$ where H has κ_0 poles in Δ. By Lemma 20.4.2 there is a regular rational $N \times N$ matrix function b having κ zeros and no poles in Δ such that $\begin{bmatrix} F \\ I_N \end{bmatrix} b$ admits a factorization

$$(20.5.5) \qquad\qquad \begin{bmatrix} F \\ I_N \end{bmatrix} b = \Theta \begin{bmatrix} G_1 \\ G_2 \end{bmatrix}$$

where $\begin{bmatrix} G_1 \\ G_2 \end{bmatrix}$ has no poles in Δ. Since Θ is J-unitary on $\partial\Delta$ and $\|F\|_\infty \leq 1$ we compute as in the proof of Theorem 20.3.2 that

$$
\begin{aligned}
0 &\geq b(z)^*[I - F(z)^*F(z)]b(z) \\
(20.5.6) \qquad &= [G_1(z)^*, G_2(z)^*]\Theta(z)^* J\Theta(z)\begin{bmatrix} G_1(z) \\ G_2(z) \end{bmatrix} \\
&= G_1(z)^*G_1(z) - G_2(z)^*G_2(z)
\end{aligned}
$$

for $z \in \partial\Delta$. Since $\begin{bmatrix} F(z) \\ I \end{bmatrix} b(z)$ has maximal rank for $z \in \partial\Delta$ and $\Theta(z)$ is regular on $\partial\Delta$, necessarily $\begin{bmatrix} G_1 \\ G_2 \end{bmatrix}$ has maximal rank on $\partial\Delta$. Then (20.5.6) implies that $G_2(z)$ is invertible on $\partial\Delta$ and hence $G := G_1 G_2^{-1}$ is a well-defined rational matrix function. From (20.5.6) we see that $\|G\|_\infty \leq 1$ and (20.5.5) implies that F has the desired linear fractional form (20.5.3). It remains to show that the matrix function G arising in this representation for F has no poles in Δ. For this it suffices to show that G_2^{-1} has no poles in Δ.

We argue that G_2^{-1} has no poles in Δ by adapting the proof of the corresponding point in the proof of Theorem 20.3.2. From (20.5.5) we get

$$
\begin{aligned}
b &= \Theta_{21} G_1 + \Theta_{22} G_2 \\
  &= \Theta_{22} (\Theta_{22}^{-1} \Theta_{21} G + I) G_2.
\end{aligned}
$$
(20.5.7)

By Proposition 13.1.2 we know that $\|\Theta_{22}^{-1}\Theta_{21}\|_\infty < 1$ and we have already observed that $\|G\|_\infty \le 1$. Then by the usual homotopy argument from Section 18.6,

$$
\text{wno} \det(\Theta_{22}^{-1}\Theta_{21} G + I) = 0.
$$
(20.5.8)

Since $\Theta(z)$ has a right pole pair over Δ of the form $\left(\begin{bmatrix} C \\ 0 \end{bmatrix}, A \right)$, we see that Θ_{22} has no poles in Δ. But also by Lemma 20.5.2 we know that Θ_{22}^{-1} has precisely κ_0 poles in Δ (counting multiplicities). We conclude that

$$
\text{wno} \det \Theta_{22} = \kappa_0.
$$
(20.5.9)

Moreover, by construction b is analytic with κ_0 zeros and no poles in Δ, so

$$
\text{wno} \det b = \kappa_0.
$$
(20.5.10)

From the identify (20.5.7) we get

$$
\text{wno} \det b = \text{wno} \det \Theta_{22} + \text{wno} \det(\Theta_{22}^{-1}\Theta_{21} G + I) + \text{wno} \det G_2.
$$

Plugging in (20.5.8), (20.5.9) and (20.5.10), we can solve for $\text{wno} \det G_2$ to get

$$
0 = \text{wno} \det G_2.
$$

Since G_2 is known to be analytic on Δ, this equality is enough to imply that $\det G_2$ has no zeros in Δ, and hence $G := G_1 G_2^{-1}$ is analytic on Δ as wanted.

Conversely, suppose that PQ has κ_0 eigenvalues bigger than 1 and G is any rational $M \times N$ matrix function analytic on Δ with $\|G\|_\infty \le 1$. We claim that $F := (\Theta_{11} G + \Theta_{12})(\Theta_{21} G + \Theta_{22})^{-1}$ is well-defined, has the form $F = K + H$ where H has κ_0 poles in Δ, and $\|F\|_\infty \le 1$.

To show this, we first define matrix functions F_1, F_2 by

$$
\begin{bmatrix} F_1 \\ F_2 \end{bmatrix} = \Theta \begin{bmatrix} G \\ I \end{bmatrix}.
$$
(20.5.11)

By Lemma 20.4.2, we need only show that F_2 is a regular matrix function with κ_0 zeros and no poles in Δ; indeed, then $F := F_1 F_2^{-1}$, $b := F_2$ satisfy all the conditions of Lemma

20.4.2 and hence F is a solution of the Nehari-Takagi problem (with $\kappa = \kappa_0$). Also, once regularity of F_2 is known, the linear fractional from $F = (\Theta_{11}G + \Theta_{12})(\Theta_{21}G + \Theta_{22})^{-1}$ is an immediate consequence of the factorization (20.5.11).

From (20.5.11) we get

(20.5.12) $$F_2 = \Theta_{22}(\Theta_{22}^{-1}\Theta_{21}G + I).$$

As in the previous part of the proof, Proposition 13.1.2 and $\|G\|_\infty \leq 1$ together with the standard homotopy argument from Section 18.6 imply that

(20.5.13) $$\text{wno} \det(\Theta_{22}^{-1}\Theta_{21}G + I) = 0.$$

By Lemma 20.5.2

(20.5.14) $$\text{wno} \det \Theta_{22} = \kappa_0.$$

From (20.5.12) we have

$$\text{wno} \det F_2 = \text{wno} \det \Theta_{22} + \text{wno} \det(\Theta_{22}^{-1}\Theta_{21}G + I).$$

Plugging in (20.5.13) and (20.5.14) gives

(20.5.15) $$\text{wno} \det F_2 = \kappa_0.$$

Since Θ has a right null pair of the form $\left(\begin{bmatrix} C \\ 0 \end{bmatrix}, A \right)$ over Δ, the rational matrix function $[\, \Theta_{21} \ \Theta_{22} \,]$ is analytic. But then $F_2 = \Theta_{21}G + \Theta_{22}$ is analytic as well. From (20.5.15) we are now able to conclude that F_2 has precisely κ_0 zeros in Δ, as required. □

CHAPTER 21

BOUNDARY NEVANLINNA-PICK INTERPOLATION

In this chapter we study a variant of the Nevanlinna-Pick interpolation problem where one seeks a function analytic and bounded by 1 on a domain which meets a set of interpolation conditions, some of which may now be prescribed on the boundary of the domain. When the interpolation conditions are of a suitable form, it develops that the set of all solutions is parametrized by a linear fractional map induced by a J-inner function having poles on the boundary, a phenomenon originally observed by Nevanlinna. Matrix extensions of this result provide an area of application for results in Chapters 6 and 7 concerning the structure of such J-inner functions, and form a natural extension of the results of Chapters 18 and 19 to the boundary case. This boundary interpolation is also sometimes called Loewner interpolation.

21.1 INTERPOLATION OF ANGULAR DERIVATIVES: THE SCALAR CASE

In this section we discuss some of the simplest boundary interpolation problems for the scalar case. We take the domain Δ to be the unit disk $\mathcal{D}$ and let $z_1, \ldots, z_k$ be k distinct points on the unit circle $\mathbf{T} = \partial \mathcal{D}$ while $w_1, \ldots, w_k$ are k distinct complex numbers. We then consider the problem of finding a scalar rational function $f(z)$ such that

$$
(21.1.1) \qquad \begin{array}{c} f \text{ is analytic on } \overline{\mathcal{D}} \text{ with} \\ \sup\{|f(z)|: z \in \overline{\mathcal{D}}\} \leq 1 \end{array}
$$

and

$$
(21.1.2) \qquad f(z_i) = w_i \quad \text{for} \quad i = 1, \ldots, k.
$$

Obviously, if $|w_i| > 1$ for some i condition (21.1.2) is a contradiction of condition (21.1.1) so no solution is possible. If on the other hand $|w_i| < 1$ for all i, then for any positive number $r < 1$ and sufficiently close to 1 the matrix

$$
(21.1.3) \qquad \left[\frac{1 - \overline{w}_i w_j}{1 - r^2 \overline{z}_i z_j} \right]_{1 \leq i, j \leq k}
$$

is positive definite (since the diagonal entries can be made arbitrarily large while the off-diagonal entries remain bounded as r tends to 1). Hence by Theorem 18.1.1, there exists a rational function $f_r(z)$ analytic on $\overline{\mathcal{D}}$ with $|f_r(z)| \leq 1$ on $\overline{\mathcal{D}}$ for which

$$
(21.1.4) \qquad f_r(rz_i) = w_i \quad \text{for} \quad i = 1, \ldots, k.
$$

If we next set $f(z) = f_r(rz)$, then f is a solution of (21.1.1) and (21.1.2); in fact f is analytic and bounded by 1 on the larger disk $\{z: |z| < 1/r\}$. Thus the problem (21.1.1)–(21.1.2) always has solution if $|w_i| < 1$ and $|z_i| = 1$ for all i; formally the Pick matrix

(21.1.3) in the limit as $r \nearrow 1$ has diagonal entries equal to positive infinity for all i and so is positive definite regardless of the off diagonal entries.

If $|w_{i_0}| < 1$ for some particular i_0, it can be shown that the full interpolation problem (21.1.1)–(21.1.2) has a solution if and only if the problem (21.1.1)–(21.1.2) with the interpolation condition associated with $i = i_0$ $(f(z_{i_0}) = w_{i_0})$ dropped has a solution.

This leads us to study the case where $|w_i| = 1$ for all i. In this case the limiting values of the diagonal entires of the Pick matrix (21.1.4) involve the so-called angular derivatives of the interpolant, so we next introduce this notion and develop a few of its elementary properties.

The basic result concerning the boundary behavior of rational functions mapping the unit disk into itself is given by the following result.

THEOREM 21.1.1. *Let $f(z)$ be a (scalar) rational function which maps the unit circle ∂D into the closed unit disk $\overline{D}$ and suppose $f(z_0)$ has unit modulus at some point z_0 on ∂D. Then the quantity $\rho_0 = z_0 \overline{f(z_0)} f'(z_0)$ is real. Moreover, if $f(z)$ maps the open unit disk D into itself, then $\rho_0 > 0$.*

PROOF. Since f maps ∂D into $\overline{D}$, certainly f has no poles on ∂D, so $m(t) = |f(e^{it})|^2$ is a smooth function of the real parameter t. Since by assumption $f(z_0)$ has unit modulus, we see that $m(t)$ attains a local maximum at $t = t_0$, where $z_0 = e^{it_0}$, and hence

$$\frac{dm}{dt}(t_0) = 0.$$

A straightforward computation gives that

$$\frac{dm}{dt}(t) = -ie^{-it}\overline{f'(e^{it})}f(e^{it}) + ie^{it}\overline{f(e^{it})}f'(e^{it})$$

$$= i[e^{it}\overline{f(e^{it})}f'(e^{it}) - \{e^{it}\overline{f(e^{it})}f'(e^{it})\}].$$

We conclude from this that $z_0\overline{f(z_0)}f'(z_0)$ is real.

Now suppose that $f(z)$ maps the open unit disk into itself; in this case we have

$$M(r) = \frac{1 - |f(rz_0)|^2}{1 - r^2} > 0$$

for $0 \le r < 1$. We compute the limit as $r \nearrow 1$ by L'Hospital's rule to get

$$0 \le \lim_{r \nearrow 1} M(r) = \lim_{r \nearrow 1} \frac{-\overline{z_0}\,\overline{f'(rz_0)}f(rz_0) - z_0\overline{f(rz_0)}f'(rz_0)}{-2r}$$

$$= \text{Re}\{z_0\overline{f(z_0)}f'(z_0)\}$$

$$= z_0\overline{f(z_0)}f'(z_0)$$

where we used the result of the first part of the proof for the last step. It follows that the real quantity $\rho = z_0\overline{f(z_0)}f'(z_0)$ is nonnegative. That $\rho > 0$ follows from the inequality

$$\frac{1 - |f(z)|^2}{1 - |z|^2} \ge \frac{1 - |f(0)|}{1 + |f(0)|}$$

for $|z| < 1$. To see this inequality, note that the Pick matrix

$$\begin{bmatrix} \frac{1-|f(z)|^2}{1-|z|^2} & 1 - f(z)\overline{f(0)} \\ 1 - \overline{f(z)}f(0) & 1 - |f(0)|^2 \end{bmatrix}$$

must be positive definite for any rational function f mapping $\mathcal{D}$ into $\mathcal{D}$ (see Chapter 18, in particular (18.1.3)) and hence has a positive determinant:

$$\left[\frac{1 - |f(z)|^2}{1 - |z|^2} \right][1 - |f(0)|^2] \geq |1 - f(z)\overline{f(0)}|^2.$$

Noting that

$$|1 - f(z)\overline{f(0)}| \geq 1 - |f(z)\overline{f(0)}| \geq 1 - |f(0)|$$

we get that

$$\frac{1 - |f(z)|^2}{1 - |z|^2} \geq \frac{(1 - |f(0)|)^2}{1 - |f(0)|^2} = \frac{1 - |f(0)|}{1 + |f(0)|} > 0$$

and hence $\rho = \lim_{r \nearrow 1} M(r) > 0$. $\square$

Theorem 21.1.1 is true more generally for holomorphic functions defined on the disk $\mathcal{D}$ which map $\mathcal{D}$ into itself and which have a radial limit of unit modulus at some point on the boundary (see Caratheodory [1954], Nevanlinna [1929, 1970] for function theory proofs and Sarason [1988] for a Hilbert space proof), but we will not need such a level of generality here.

If $f(z)$ is a rational function mapping the unit circle $\partial\mathcal{D}$ into the closed unit disk $\mathcal{D}$ and z_0 is a point on $\partial\mathcal{D}$ where $f(z_0)$ has unit modulus, then we shall refer to $f'(z_0)$ as the angular derivative of f at z_0. If f has no poles in $\mathcal{D}$ and is not a constant of unit modulus, then the phase of $f(z_0)$ is completely determined from the condition that $z_0\overline{f(z_0)}f'(z_0)$ is positive. Since $f(z_0)$ is also assumed to be of unit modulus, we see that the value $f(z_0)$ of f itself at z_0 is completely determined by the value of its angular derivative $f'(z_0)$ according to

$$f(z_0) = z_0 \frac{f'(z_0)}{|f'(z_0)|}.$$

Let us now return to the interpolation problem (21.1.1)–(21.1.2) for the case where both all the points $z_1, \ldots, z_k$ and the numbers $w_1, \ldots, w_k$ have modulus 1. In this case we now see that the limiting value as $r \nearrow 1$ of the i-th diagonal entry of the Pick matrix (21.1.3) should be taken to be the positive quantity $\rho_i = z_i\overline{f(z_i)}f'(z_i) = z_i\overline{w}_i f'(z_i)$ associated with a solution $f(z)$. In particular if no constraint is placed on the value $f'(z_i)$ of the angular derivative of f at z_i, then we may take ρ_i to be arbitrarily large and again we may arrange that the limiting Pick matrix is positive definite regardless of what the limiting off diagonal entries are. Therefore we expect that the problem (21.1.1)–(21.1.2) has a solution for any choice of data $z_1, \ldots, z_k$, $w_1, \ldots, w_k$ if all the numbers $z_1, \ldots, z_k$, $w_1, \ldots, w_k$ are of modulus 1 and if $z_1, \ldots, z_k$ are distinct; as we shall see, this turns out indeed to be the case.

The analysis also suggests that a more canonical interpolation problem is had if one prescribes not only the modulus 1 values $w_1, \ldots, w_k$ of $f(z)$ at $z_1, \ldots, z_k$ but also values $\gamma_1, \ldots, \gamma_k$ for $f'(z)$ at $z_1, \ldots, z_k$. Due to the connections between $f(z)$ and $f'(z)$ at boundary points where $|f(z)| = 1$ if f maps $\mathcal{D}$ into $\mathcal{D}$, we see that the argument of w_i is determined from γ_i, so we need only specify the numbers $\gamma_1, \ldots, \gamma_k$. This leads to the following *angular derivative interpolation* (ADI) problem.

(ADIa)   *Given k distinct points $z_1, \ldots, z_k$ of unit modulus and k complex numbers $\gamma_1, \ldots, \gamma_k$, find a rational matrix function f mapping $\mathcal{D}$ into $\mathcal{D}$ such that for $j = 1, \ldots, k$*

  i) $|f(z_j)| = 1$

and

  ii) $f'(z_j) = \gamma_j$.

Alternatively, note that the phase of the angular derivative $f'(z_0)$ is determined by the value $f(z_0)$ at a point z_0 in $\partial \mathcal{D}$ by the fact that ρ_0 is positive in Theorem 21.1.1. Thus only the modulus ρ_0 of $f'(z_0)$ is not determined by the (unit modulus) value $f(z_0)$ of f at z_0, i.e. we may turn in a complex parameter ($\gamma_0 = f'(z_0)$) for two real parameters (the phase of the modulus 1 number $w_0 = f(z_0)$ and the modulus of the angular derivative $f'(z_0)$). This leads to the following alternative formulation of the *angular derivative interpolation problem.*

(ADIb)   *Given k distinct points $z_1, \ldots, z_k$ in T, k complex numbers $w_1, \ldots, w_k$ of unit modulus and k positive real numbers $\rho_1, \ldots, \rho_k$, find all rational functions $f(z)$ mapping $\mathcal{D}$ into $\mathcal{D}$ such that for $j = 1, \ldots, k$*

  i) $f(z_j) = w_j$

and

  ii) $f'(z_j) = \bar{z}_j w_j \rho_j$.

We now state the solution of problem (ADIb). We shall give a complete proof (with the added assumption that Λ is invertible) in Section 21.4 where a much more general result will be proved; to handle the case where Λ is not necessarily invertible, we refer the reader to other treatments (see Ball-Helton [1986b]).

THEOREM 21.1.2.   *Let $z_1, \ldots, z_k$, $w_1, \ldots, w_k$ and $\rho_1, \ldots, \rho_k$ be as in (ADIb), and define a matrix $\Lambda = [\Lambda_{ij}]_{1 \leq i,j \leq k}$ by*

$$\Lambda_{ij} = \begin{cases} \frac{1 - \overline{w}_i w_j}{1 - \overline{z}_i z_j}, & i \neq j \\ \rho_i, & i = j. \end{cases}$$

Then a necessary condition for the problem (ADIb) to have a solution is that Λ be positive semidefinite and a sufficient condition is that Λ be positive definite. In the latter case the set of all solutions is given by

$$f(z) = \bigl(\Theta_{11}(z) g(z) + \Theta_{12}(z)\bigr)\bigl(\Theta_{21}(z) g(z) + \Theta_{22}(z)\bigr)^{-1}$$

where $g(z)$ is an arbitrary scalar rational function analytic on $\mathcal{D}$ with $\sup\{|g(z)| : z \in \mathcal{D}\} \leq 1$ such that $\Theta_{21}(z) g(z) + \Theta_{22}(z)$ has a simple pole at the points $z_1, \ldots, z_k$. Here

$$\Theta(z) = \begin{bmatrix} \Theta_{11}(z) & \Theta_{12}(z) \\ \Theta_{21}(z) & \Theta_{22}(z) \end{bmatrix} \text{ is given by}$$

$$\Theta(z) = I + (z - z_0)C_0(zI - A_0)^{-1}\Lambda^{-1}(I - z_0 A_0^*)^{-1}C_0^* J$$

where

$$C_0 = \begin{bmatrix} w_1 \cdots w_k \\ 1 \cdots 1 \end{bmatrix}, \qquad A_0 = \begin{bmatrix} z_1 & & 0 \\ & \ddots & \\ 0 & & z_k \end{bmatrix}, \qquad J = \begin{bmatrix} 1 & 0 \\ 0 & -1 \end{bmatrix}$$

and z_0 is any complex number of modulus 1 not equal to any of the numbers $z_1, \ldots, z_k$.

Note that if only the modulus one values $w_1, \ldots, w_k$ are specified at the boundary points $z_1, \ldots, z_k$, then the matrix Λ in Theorem 21.1.1 can be made positive definite by choosing the unspecified quantities $\rho_1, \ldots, \rho_k$ sufficiently large. This leads to the following immediate corollary.

COROLLARY 21.1.3. *Let $2k$ complex numbers of modulus 1*

$$z_1, \ldots, z_k, \ w_1, \ldots, w_k$$

be given, where $z_1, \ldots, z_k$ are distinct. Then there always exist scalar rational functions $f(z)$ analytic on $\mathcal{D}$ with

$$\sup\{|f(z)|\colon z \in \mathcal{D}\} \le 1$$

which satisfy the set of interpolation conditions

$$f(z_i) = w_i \quad \text{for} \quad 1 \le i \le k.$$

In Section 21.4 of this chapter we shall present a more general result which allows for simultaneous interpolation conditions in the interior of $\mathcal{D}$ and for solutions to have a prescribed number of poles in $\mathcal{D}$.

We remark that the matrix function $f(z)$ given in Theorem 21.1.2 is constructed to have J-unitary values at all regular points of the unit circle. Thus if g is chosen to be analytic on $\mathcal{D}$ with modulus 1 values on the unit circle, it is not hard to show that $(\Theta_{11}g + \Theta_{12})(\Theta_{21}g + \Theta_{22})^{-1}$ has modulus 1 values on the unit circle. Hence, if g is chosen to also satisfy the nondegeneracy condition that $\Theta_{21}g + \Theta_{22}$ inherits the simple pole of $[\Theta_{21}, \Theta_{22}]$ at $z_1, \ldots, z_k$, we see that the problem (ADIb) has solutions f having modulus 1 values on the unit circle.

All of the above analysis can be repeated with the right half plane Π^+ in place of the unit disk. We let $z_1, \ldots, z_k$ be k distinct points on the imaginary line, and let $w_1, \ldots, w_k$ be distinct complex numbers. We then consider the problem of finding a scalar rational function $f(z)$ such that

(21.1.5) f is analytic on the closed right half plane
$\overline{\Pi^+}$ and $\sup\{f(z)\colon z \in \overline{\Pi^+}\} \le 1$

and

$$f(z_i) = w_i \quad \text{for} \quad i = 1, \ldots, k.$$

As was the case for the disk, the interesting case is when $|w_i| = 1$ for $i = 1, \ldots, k$. We next develop the basic facts concerning angular derivatives for functions mapping the right half plane into the unit disk analogous to those developed above for the unit disk. The following basic result can be proved in the same manner as Theorem 21.1.1, one should do the same argument but now with $m(t) = |f(it)|^2$ and $M(r) = \frac{1 - |f(it_0 + r)|^2}{2r}$ $(r > 0)$.

THEOREM 21.1.4. *Let* $f(z)$ *be a (scalar) rational function which maps the imaginary line* $i\mathbb{R}$ *into the closed unit disk* $\overline{\mathcal{D}}$ *and suppose* $f(it_0)$ *has unit modulus at some point* t_0 *on* $i\mathbb{R}$. *Then the quantity* $\rho_0 = -\overline{f(it_0)} f'(it_0)$ *is real. Moreover, if* $f(z)$ *maps the open right half plane* Π^+ *into the open unit disk* $\mathcal{D}$, *then* $\rho_0 > 0$.

This suggests the following angular derivative interpolation problem for the right half plane.

(ADIc)   *Given* k *distinct numbers* $z_1, \ldots, z_k$ *on* $i\mathbb{R}$ *and* k *complex numbers* $\gamma_1, \ldots, \gamma_k$, *find a rational function* f *mapping* Π^+ *into* $\mathcal{D}$ *such that*

   i) $|f(z_i)| = 1$

and

   ii) $f'(z_i) = \gamma_i$

for $i = 1, \ldots, n$.

Alternatively we may want to specify the modulus 1 values $w_1, \ldots, w_n$ of $f(z_1), \ldots, f(z_k)$ and then also specify only the moduli $\rho_1, \ldots, \rho_k$ of $f'(z_1), \ldots, f'(z_k)$.

(ADId)   *Given* k *distinct numbers* $z_1, \ldots, z_k$ *on* $i\mathbb{R}$, k *complex numbers* $w_1, \ldots, w_k$ *of unit modulus and* n *positive real numbers* $\rho_1, \ldots, \rho_n$, *find all rational functions* $f(z)$ *mapping* Π^+ *into* $\mathcal{D}$ *such that*

   i) $f(z_i) = w_i$

and

   ii) $f'(z_i) = -w_i \rho_i$

for $i = 1, \ldots, k$.

The solution for the right half plane case parallels that for the unit disk. Again the proof (at least for the case Λ invertible) will follow from a much more general result to be proved in a later section of this chapter.

THEOREM 21.1.5.   *Let* $z_1, \ldots, z_k$, $w_1, \ldots, w_k$ *and* $\rho_1, \ldots, \rho_k$ *be as in* (ADId), *and define a matrix* $\Lambda = [\Lambda_{ij}]_{1 \leq i,j \leq k}$ *by*

$$\Lambda_{ij} = \begin{cases} \frac{1 - \overline{w}_i w_j}{\overline{z}_i + z_j}, & i \neq j \\ \rho_i, & i = j. \end{cases}$$

Then a necessary condition for the problem (ADId) *to have a solution is that* Λ *be positive semidefinite and a sufficient condition is that* Λ *be positive definite. In the latter case the set of all solutions is given by*

$$f(z) = \left(\Theta_{11}(z)g(z) + \Theta_{12}(z)\right)\left(\Theta_{21}(z)g(z) + \Theta_{22}(z)\right)^{-1}$$

where $g(z)$ is an arbitrary scalar rational function analytic on Π^+ with $\sup\{|g(z)|\colon z \in \Pi^+\} \leq 1$ for which $\Theta_{21}(z)g(z) + \Theta_{22}(z)$ has simple poles at the points $z_1, \dots, z_k$. Here $\Theta(z) = \begin{bmatrix} \Theta_{11}(z) & \Theta_{12}(z) \\ \Theta_{21}(z) & \Theta_{22}(z) \end{bmatrix}$ is given by

$$\Theta(z) = I - C_0(zI - A_0)^{-1}\Lambda^{-1}C_0^* J$$

where

$$C_0 = \begin{bmatrix} w_1 \cdots w_k \\ 1 \cdots 1 \end{bmatrix}, \qquad A_0 = \begin{bmatrix} z_1 & & 0 \\ & \ddots & \\ 0 & & z_k \end{bmatrix}, \qquad J = \begin{bmatrix} 1 & 0 \\ 0 & -1 \end{bmatrix}.$$

As in the case for the unit disk we shall in fact prove a much more general version of Theorem 21.1.5 which allows higher order interpolation conditions, simultaneous interpolation at points in the interior of Π^+, as well as the possibility of solutions having some prescribed number of poles in Π^+.

21.2 TANGENTIAL INTERPOLATION OF MATRICIAL ANGULAR DERIVATIVES ON THE UNIT CIRCLE

In this section we introduce matrix interpolation problems which generalize in a natural way the angular derivative interpolation problems considered in the previous section.

Let us first consider a set of right interpolation conditions at points on the unit circle for a rational $n \times n$ matrix function $F(z)$. From the discussion in Chapter 16, we know such a set of interpolation conditions can be written in the compact form

$$\sum_{z_0 \in \sigma(A_0)} \operatorname{Res}_{z=z_0} F(z)C_{0-}(zI - A_0)^{-1} = C_{0+}$$

where C_{0-}, C_{0+}, A_0 are matrices of sizes $N \times n_0$, $N \times n_0$ and $n_0 \times n_0$ such that (C_{0-}, A_0) is a null kernel pair and $\sigma(A_0)$ is contained in the unit circle $\mathbf{T}$. We take the point of view that we wish to consider only interpolation problems for which $F(\bar{z}^{-1})^{*-1}$ is again a solution whenever $F(z)$ is. In this situation the set of all solutions is parametrized by a linear fractional map. This setting is also a natural generalization of the scalar case discussed in Section 21.1, where we have seen that the boundary interpolation problem is interesting only when the boundary interpolation values are 1 in modulus; in the matrix case this is interpreted as unitary boundary values. The result is as follows.

THEOREM 21.2.1. *Suppose (C_{0-}, A_0) is a null kernel pair with $\sigma(A_0)$ contained in the unit circle $\mathbf{T}$ and suppose that $F(z)$ is a rational $N \times N$ matrix function analytic on $\sigma(A_0)$ such that $F(z) = F(\bar{z}^{-1})^{*-1}$ for all $z \in \mathbf{C}\backslash\{0\}$ which are not poles of $F(z)$. Define matrices C_{0+} and H by*

(21.2.1) $$C_{0+} = \sum_{z_0 \in \sigma(A_0)} \operatorname{Res}_{z=z_0} F(z)C_{0-}(zI - A_0)^{-1}$$

and

$$(21.2.2) \qquad H = \sum_{z_0 \in \sigma(A_0)} \operatorname{Res}_{z=z_0} (zI - A_0^{*-1})^{-1} A_0^{*-1} C_{0+}^* F(z) C_{0-}(zI - A_0)^{-1}.$$

Then $(A_0^{*-1}, A_0^{*-1}C_{0+}^*)$ is a full range pair and F satisfies the interpolation condition

$$(21.2.3) \qquad \sum_{z_0 \in \sigma(A_0)} \operatorname{Res}_{z=z_0} (zI - A_0^{*-1})^{-1} A_0^{*-1} C_{0+}^* F(z) = A_0^{*-1} C_{0-}^*.$$

Moreover, H is Hermitian and satisfies the Stein equation

$$(21.2.4) \qquad H - A_0^* H A_0 = C_{0-}^* C_{0-} - C_{0+}^* C_{0+}.$$

PROOF. We first check that $(A_0^{*-1}, A_0^{*-1}C_{0+}^*)$ is a full range pair, or by duality, that $(C_{0+}A_0^{-1}, A_0^{-1})$ is a null kernel pair. Note that since A_0 is invertible, $\bigcap_{j=0}^{n-1}(C_{0+}A_0^{-1})(A_0^{-1})^j = \{0\}$ if and only if $\bigcap_{j=0}^{n-1} C_{0+}A_0^j = \bigcap_{j=0}^{n-1}(C_0 A_0^{-1})(A_0^{-1})^j A_0^n = \{0\}$. Thus the problem is to show that (C_{0+}, A_0) is a null kernel pair. But by Lemma 12.2.2 (C_{0+}, A_0) is a null kernel pair if and only if the map $x \to C_{0+}(zI - A_0)^{-1}x$ is injective. By Lemma 16.9.4 we know that the interpolation condition (21.2.1) is equivalent to the function

$$(21.2.5) \qquad X(z) = F(z)C_{0-}(zI - A_0)^{-1} - C_{0+}(zI - A_0)^{-1}$$

being analytic on $\sigma(A_0)$. If x is a vector in $\mathbf{C}^{n_0}$ for which $C_{0+}(zI - A_0)^{-1}x$ is identically zero, we conclude that $X(z)x = F(z)C_{0-}(zI - A_0)^{-1}x$ is analytic on $\sigma(A_0)$. Since $F(z)^{-1} = F(\bar{z}^{-1})^*$ is also analytic on $\sigma(A_0)$, we then get that $F(z)^{-1}X(z)x = C_{0-}(zI - A_0)^{-1}x$ is analytic on $\sigma(A_0)$, and hence also on all of $\mathbf{C}$. As the value at infinity is 0, we conclude by Liouville's Theorem that $C_{0-}(zI - A_0)^{-1}x$ is zero identically. Since by assumption (C_{0-}, A_0) is a null kernel pair, Lemma 12.2.2 implies that $x = 0$. Hence the map $x \to C_{0+}(zI - A_0)^{-1}x$ is injective and $(A_0^{*-1}, A_0^{*-1}C_{0+}^*)$ is a full range pair as asserted.

To see that $F(z)$ satisfies the interpolation condition (21.2.3) we proceed as follows. As was already mentioned, by Lemma 16.9.4 the interpolation condition (21.2.1) is equivalent to the function $X(z)$ in (21.2.5) being analytic on $\sigma(A_0)$. We now use that $F(z) = F(\bar{z}^{-1})^{*-1}$ to write

$$X(z) = F(\bar{z}^{-1})^{*-1}C_{0-}(zI - A_0)^{-1} - C_{0+}(zI - A_0)^{-1}.$$

As $\sigma(A_0) \subset \mathbf{T}$, $X(z)$ is analytic on $\sigma(A_0)$ if and only if $X(\bar{z}^{-1})^*$ is analytic on $\sigma(A_0)$ where

$$\begin{aligned} X(\bar{z}^{-1})^* &= (z^{-1}I - A_0^*)^{-1}C_{0-}^* F(z)^{-1} - (z^{-1}I - A_0^*)^{-1}C_{0+}^* \\ &= -z(zI - A_0^{*-1})^{-1}A_0^{*-1}C_{0-}^* F(z)^{-1} + z(zI - A_0^{*-1})^{-1}A_0^{*-1}C_{0+}^*. \end{aligned}$$

Since $F(z)$ and $F(z)^{-1}$ are analytic on $\sigma(A_0)$, this is equivalent to $z^{-1}X(\bar{z}^{-1})^*F(z) = (zI - A_0^{*-1})^{-1}A_0^{*-1}C_{0+}^* F(z) - (zI - A_0^{*-1})^{-1}A_0^{*-1}C_{0-}^*$ being analytic on $\sigma(A_0)$. The interpolation condition (21.2.3) is now an immediate consequence of the residue theorem and the Riesz functional calculus (see the easy direction of Lemma 16.9.4).

If H is defined by (21.2.2), then it follows from the interpolation conditions (21.2.1) and (21.2.3) satisfied by $F(z)$ by Lemma 16.9.1 that H satisfies the Sylvester equation

$$HA_0 - A_0^{*-1}H = A_0^{*-1}C_{0+}^*C_{0+} - A_0^{*-1}C_{0-}^*C_{0-}.$$

By simple algebra one sees that this is equivalent to the Stein equation (21.2.4). It remains only to show that H is Hermitian.

From the definition (21.2.2) of H we compute an expression for H^* as follows. By Lemma 16.10.4, we know that

$$\sum_{z_0 \in \sigma(A_0)} \text{Res}_{z=z_0}(zI - A_0^{*-1})^{-1}A_0^{*-1}C_{0+}^*C_{0+}(zI - A_0)^{-1} = 0.$$

Hence we may write a perturbed version of (21.2.2) which still represents H:

$$(21.2.6) \quad H = \sum_{z_0 \in \sigma(A_0)} \text{Res}_{z=z_0}(zI - A_0^{*-1})^{-1}A_0^{*-1}C_{0+}^*[F(z)C_{0-} - C_{0+}](zI - A_0)^{-1}.$$

On the other hand, from the interpolation condition (21.2.1) and Lemma 16.9.4 (as was used above), $A_0^{*-1}C_{0+}^*[F(z)C_{0-} - C_{0+}](zI - A_0)^{-1}$ is analytic on $\sigma(A_0)$. Now by Lemma 16.9.4 again, we conclude from the identity (21.2.6) that the matrix function

$$Y(z) = (zI - A_0^{*-1})^{-1}A_0^{*-1}C_{0+}^*[F(z)C_{0-} - C_{0+}](zI - A_0)^{-1} - (zI - A_0^{*-1})^{-1}H$$

is analytic on $\sigma(A_0)$. As $\sigma(A_0) \subset \mathbf{T}$, then $Y(\bar{z}^{-1})^*$ is also analytic on $\sigma(A_0)$, where

$$\begin{aligned}
Y(\bar{z}^{-1})^* &= (z^{-1}I - A_0^*)^{-1}[C_{0-}^*F(z)^{-1} - C_{0+}^*]C_{0+}A_0^{-1}(z^{-1} - A_0^{-1})^{-1} \\
&\quad - H^*(z^{-1}I - A_0^{-1})^{-1} \\
&= z^2(zI - A_0^{*-1})^{-1}A_0^{*-1}[C_{0-}^*F(z)^{-1} - C_{0+}^*]C_{0+}(zI - A_0)^{-1} \\
&\quad + zH^*(zI - A_0)^{-1}A_0.
\end{aligned}$$

Then also

$$\begin{aligned}
(21.2.7) \quad z^{-2}Y(\bar{z}^{-1})^* &= (zI - A_0^{*-1})^{-1}A_0^{*-1}[C_{0-}^*F(z)^{-1} \\
&\quad - C_{0+}^*]C_{0+}(zI - A_0)^{-1} + z^{-1}H^*(zI - A_0)^{-1}A_0.
\end{aligned}$$

Again by Lemma 16.10.4

$$(21.2.8) \quad \sum_{z_0 \in \sigma(A_0)} \text{Res}_{z=z_0}(zI - A_0^{*-1})^{-1}A_0^{*-1}C_{0+}^*C_{0+}(zI - A_0)^{-1} = 0.$$

From the identity

$$z^{-1}A_0(zI - A_0)^{-1} = -z^{-1}I + (zI - A_0)^{-1}$$

we see that

(21.2.9)
$$\sum_{z_0 \in \sigma(A_0)} \text{Res}_{z=z_0}\, z^{-1} H^*(zI - A_0)^{-1} A_0 = H^*.$$

Combining (21.2.8) and (21.2.9) with (21.2.7) and using that the sum of the residues of $z^{-2} Y(\bar{z}^{-1})^*$ over $\sigma(A_0)$ is zero, we get

(21.2.10) $\quad H^* = - \displaystyle\sum_{z_0 \in \sigma(A_0)} \text{Res}_{z=z_0}(zI - A_0^{*-1})^{-1} A_0^{*-1} C_{0-}^* F(z)^{-1} C_{0+}(zI - A_0)^{-1}.$

To identify this expression with H we need to find the set of interpolation conditions satisfied by $F(z)^{-1}$ implied by $F(z)$ satisfying (21.2.1), (21.2.2) and (21.2.3). This gap is filled by the following lemma; we state it in a somewhat more general form than what is needed to complete the proof of Proposition 21.2.1.

LEMMA 21.2.2. *Let σ be a subset of $\mathbb{C}$ and suppose $\omega = (C_+, C_-, A_\pi; A_\zeta, B_+, B_-; \Gamma)$ is a σ-admissible Sylvester TSCII data set. Suppose that an $n \times n$ rational matrix function $F(z)$ is analytic and invertible on σ and satisfies the associated set of interpolation conditions*

(21.2.11a)
$$\sum_{z_0 \in \sigma} \text{Res}_{z=z_0}\, F(z) C_-(zI - A_\pi)^{-1} = C_+$$

(21.2.11b)
$$\sum_{z_0 \in \sigma} \text{Res}_{z=z_0}(zI - A_\zeta)^{-1} B_+ F(z) = -B_-$$

(21.2.11c)
$$\sum_{z_0 \in \sigma} \text{Res}_{z=z_0}(zI - A_\zeta)^{-1} B_+ F(z) C_-(zI - A_\pi)^{-1} = \Gamma.$$

Then $\omega_1 = (C_-, C_+, A_\pi; A_\zeta, B_-, B_+; \Gamma)$ is also a σ-admissible Sylvester TSCII data set and $F(z)^{-1}$ satisfies the associated set of interpolation conditions

(21.2.12a)
$$\sum_{z_0 \in \sigma} \text{Res}_{z=z_0}\, F(z)^{-1} C_+(zI - A_\pi)^{-1} = C_-$$

(21.2.12b)
$$\sum_{z_0 \in \sigma} \text{Res}_{z=z_0}(zI - A_\zeta)^{-1} B_- F(z)^{-1} = -B_+$$

(21.2.12c)
$$\sum_{z_0 \in \sigma} \text{Res}_{z=z_0}(zI - A_\zeta)^{-1} B_- F(z)^{-1} C_+(zI - A_\pi)^{-1} = \Gamma.$$

PROOF. Recall that ω is σ-admissible Sylvester means that (C_-, A_π) is a null kernel pair with $\sigma(A_\pi) \subset \sigma$, (A_ζ, B_+) is a full range pair with $\sigma(A_\zeta) \subset \sigma$, and that Γ satisfies the Sylvester equation

$$\Gamma A_\pi - A_\zeta \Gamma = B_+ C_+ + B_- C_-.$$

To show that ω_1 is σ-admissible Sylvester, all that needs to be done is to show that (C_+, A_π) is a null kernel pair and that (A_ζ, B_-) is a full range pair. The proof that (C_+, A_π) is a null kernel pair follows exactly as in the first part of the proof of Proposition 21.2.1 that (C_{0+}, A_0) is a null kernel pair; one needs to use only that $F(z)$ satisfies (21.2.11a) and is analytic and invertible on σ. The proof that (A_ζ, B_-) is a full range pair follows by an analogous dual argument.

It remains to show that $F(z)^{-1}$ satisfies the set of interpolation conditions (21.2.12a)–(21.2.12c). By Lemma 16.9.4, (21.2.11a) is equivalent to the analyticity of

$$X(z) = F(z)C_-(zI - A_\pi)^{-1} - C_+(zI - A_\pi)^{-1}$$

on σ. As $F(z)^{-1}$ is analytic on σ by assumption, so is $F(z)^{-1}X(z)$; now (21.2.12a) follows from the converse side of Lemma 16.9.4. The left interpolation condition (21.2.12b) follows from (21.2.11b) in a similar way.

The proof of (21.2.12c) is less elementary. Recall from Theorem 16.10.3 that F satisfies (21.2.11a)–(21.2.11c) if and only if

$$\begin{bmatrix} F(z) \\ I \end{bmatrix} h_-(z) \in \mathcal{S}_{\widetilde\omega}(\sigma)$$

whenever $h_-(z) \in \mathcal{S}_{\widetilde\omega_-}(\sigma)$. Here $\mathcal{S}_{\widetilde\omega}(\sigma)$ is the null-pole subspace

$$(21.2.13) \qquad \mathcal{S}_{\widetilde\omega}(\sigma) = \left\{ \begin{bmatrix} C_+ \\ C_- \end{bmatrix} (zI - A_\pi)^{-1}x + \begin{bmatrix} h_1(z) \\ h_2(z) \end{bmatrix} : \right.$$
$$x \in \mathbf{C}^{n_\pi}, \; h_1, h_2 \in \mathcal{R}_n(\sigma) \text{ such that}$$
$$\left. \sum_{z_0 \in \sigma} \mathrm{Res}_{z_0 \in \sigma}(zI - A_\zeta)^{-1}[B_+, B_-] \begin{bmatrix} h_1(z) \\ h_2(z) \end{bmatrix} = \Gamma x \right\}$$

associated with the σ-admissible Sylvester data set $\widetilde\omega := \left(\begin{bmatrix} C_+ \\ C_- \end{bmatrix}, A_\pi; A_\zeta, [B_+, B_-]; \Gamma \right)$;
while

$$\mathcal{S}_{\widetilde\omega_-}(\sigma) = \{C_-(zI - A_\pi)^{-1}x + k(x) : x \in \mathbf{C}^{n_\pi}, \; k \in \mathcal{R}_n(\sigma)\}$$

is the analogous object associated with $\widetilde\omega_- := (C_-, A_\pi; 0, 0; 0)$. From (21.2.12a) combined with Lemma 16.9.4, we see that $F(z)^{-1}C_+(zI - A_\pi)^{-1}x \in \mathcal{S}_{\widetilde\omega_-}(\sigma)$ for each $x \in \mathbf{C}^{n_\pi}$. Thus

$$\begin{bmatrix} F(z) \\ I \end{bmatrix} F(z)^{-1}C_+(zI - A_\pi)^{-1}x = \begin{bmatrix} I \\ F(z)^{-1} \end{bmatrix} C_+(zI - A_\pi)^{-1}x$$

is in $\mathcal{S}_{\widetilde\omega}(\sigma)$ for each x in $\mathbf{C}^{n_\pi}$. From the characterization (21.2.13) of $\mathcal{S}_{\widetilde\omega}(\sigma)$, this gives

$$(21.2.14) \qquad \sum_{z_0 \in \sigma} \mathrm{Res}_{z_0 \in \sigma}(zI - A_\zeta)^{-1}[B_+, B_-]$$
$$\times \begin{bmatrix} 0 \\ F(z)^{-1}C_+(zI - A_\pi)^{-1}x - C_-(zI - A_\pi)^{-1}x \end{bmatrix} = \Gamma x$$

for each x in $\mathbf{C}^{n_\pi}$. By Lemma 16.10.4 again,

$$\sum_{z_0 \in \sigma} \mathrm{Res}_{z=z_0}(zI - A_\zeta)^{-1}B_-C_-(zI - A_\pi)^{-1} = 0.$$

Hence (21.2.14) collapses to (21.2.12c) as required. $\square$

      COMPLETION OF PROOF OF THEOREM 21.2.1. We apply Lemma 21.2.2 with

$$\omega = (C_{0+}, C_{0-}, A_0; A_0^{*-1}, A_0^{*-1}C_{0+}^*, -A_0^{*-1}C_{0-}^*; H).$$

Then (21.2.12c) gives

$$-\sum_{z_0 \in \sigma} \mathrm{Res}_{z=z_0}(zI - A_0^{*-1})^{-1} A_0^{*-1} C_{0-}^* F(z)^{-1} C_{0+} (zI - A_0)^{-1} = H.$$

Comparison with (21.2.10) now gives that $H = H^*$ as required. $\square$

      Theorem 21.2.1 suggests the appropriate generalization of the angular derivative interpolation problem introduced in the previous section to matrix functions and to interpolation conditions of arbitrarily high multiplicity. Indeed, we have seen that if a rational matrix function $F(z)$ with unitary values on the unit circle satisfies a right tangential interpolation condition (21.2.1), that it automatically also satisfies the reflected left interpolation condition (21.2.3) and there exists a Hermitian solution H of a Stein equation (21.2.4) which is defined by (21.2.2). Moreover, we have already seen in Chapters 16 and 18 that one must prescribe the bitangential interpolation condition (21.2.2) in order for the set of all solutions to have a single linear fractional formula parametrization, whenever one prescribes both left and right tangential interpolation conditions (21.2.1) and (21.2.3). This leads us to the following definition of a *generalized angular derivative interpolation* (GADI) problem for rational matrix functions.

      We say that the collection of matrices $(C_{0-}, C_{0+}, A_0; H)$ of sizes $N \times n_0$, $N \times n_0$, $n_0 \times n_0$ and $n_0 \times n_0$ is a *generalized angular derivative interpolation* (GADI) *data set* with respect to the unit circle $\mathbf{T}$ if

      (i) (C_{0-}, A_0) is a null kernel pair with $\sigma(A_0) \subset \mathbf{T}$

and

      (ii) H is a Hermitian solution of the Stein equation

$$(21.2.15) \qquad H - A_0^* H A_0 = C_{0-}^* C_{0-} - C_{0+}^* C_{0+}.$$

The *generalized angular derivative interpolation* (GADI) *problem* associated with the GADI data set $\omega = (C_{0-}, C_{0+}, A_0; H)$ then is: *find an $n \times n$ rational matrix function $F(z)$ analytic on the unit disk $\mathcal{D}$ such that $\|F(z)\| \le 1$ for $z \in \mathcal{D}$ and*

$$(21.2.16a) \qquad \sum_{z_0 \in \sigma(A_0)} \mathrm{Res}_{z=z_0} F(z) C_{0-}(zI - A_0)^{-1} = C_{0+}$$

$$(21.2.16b) \qquad \sum_{z_0 \in \sigma(A_0)} \mathrm{Res}_{z=z_0}(zI - A_0^{*-1})^{-1} A_0^{*-1} C_{0+}^* F(z) = A_0^{*-1} C_{0-}^*$$

and

(21.2.16c)
$$\sum_{z_0 \in \sigma(A_0)} \text{Res}_{z=z_0}(zI - A_0^{*-1})^{-1}A_0^{*-1}C_{0+}^* F(z)C_{0-}(zI - A_0)^{-1} = H.$$

Note that by Theorem 21.2.1, if $F(z) = F(\bar{z}^{-1})^{*-1}$, then (21.2.16b) is actually a consequence of (21.2.16a) and necessarily $H = H^*$.

We next turn to some examples.

EXAMPLE 21.2.1. Take $N = 1$, $n_0 = k$, $C_{0-} = [1, \ldots, 1]$, $C_{0+} = [w_1, \ldots, w_k]$ where $|w_j| = 1$ for $1 \le j \le k$, take

$$A_0 = \begin{bmatrix} z_1 & & 0 \\ & \ddots & \\ 0 & & z_k \end{bmatrix}$$

where $z_1, \ldots, z_k$ are k distinct complex numbers of modulus 1, and define $H = [H_{ij}]_{1 \le i,j \le k}$ by

$$H_{ij} = \begin{cases} \frac{1 - \bar{w}_i w_j}{1 - \bar{z}_i z_j}, & i \ne j \\ \rho_i, & i = j \end{cases}$$

where $\rho_1, \ldots, \rho_k$ are positive real numbers. Note that the matrix $H - A_0^* H A_0$ then has (i,j)-th entry

$$[H - A_0^* H A_0]_{ij} = \begin{cases} 1 - \bar{w}_i w_j, & i \ne j \\ (1 - |z_i|^2)\rho_i = 0, & i = j \end{cases}$$

which is also equal to the (i,j)-th entry of $C_{0-}^* C_{0-} - C_{0+}^* C_{0+}$ (since $|w_i| = 1$ for all i as well). Thus H satisfies the Stein equation (21.2.15). Trivially (C_{0-}, A_0) is null kernel pair and hence $(C_{0-}, C_{0+}, A_0; H)$ is a T-admissible GADI data set. The interpolation condition (21.2.16a) reduces to

(21.2.17)
$$F(z_j) = w_j, \qquad 1 \le j \le k$$

as does (21.2.16b) since $|w_j| = 1$ for $1 \le j \le k$. For $i \ne j$, the (i,j)-th entry of the interpolation condition (21.2.16c) gives

$$z_i \bar{w}_i F(z_i)(z_i - z_j)^{-1} + (z_j - z_i)^{-1} z_i \bar{w}_i F(z_j) = \frac{1 - \bar{w}_i w_j}{1 - \bar{z}_i z_j}$$

or

$$z_i \frac{\bar{w}_i F(z_i) - \bar{w}_i F(z_j)}{z_i - z_j} = \frac{1 - \bar{w}_i w_j}{1 - \bar{z}_i z_j}.$$

Since $|z_j| = |w_j| = 1$ for all j, we see that this is consistent with (21.2.17). Finally for $i = j$ the i-th diagonal entry of (21.2.16c) gives

$$\text{Res}_{z=z_i}(z - z_i)^{-1} z_i \bar{w}_i F(z)(z - z_i)^{-1} = \rho_i$$

or

(21.2.18) $$z_i \overline{w}_i F'(z_i) = \rho_i.$$

This therefore reduces exactly to the angular derivative interpolation conditions in (ADIb) considered in the previous section. $\square$

EXAMPLE 21.2.2. The simplest matrix extension of the previous example is obtained as follows. We let $z_1, \ldots, z_k$ be k distinct complex numbers of modulus 1 and choose $2k$ column vector $u_1, \ldots, u_k$, $v_1, \ldots, v_k$ in $\mathbf{C}^N$, each of unit length. We then set

$$C_{0-} = [u_1, \ldots, u_k], \qquad C_{0+} = [v_1, \ldots, v_k], \qquad A_0 = \begin{bmatrix} z_1 & & 0 \\ & \ddots & \\ 0 & & z_k \end{bmatrix}$$

and define a $k \times k$ Hermitian matrix $H = [H_{ij}]_{1 \le i,j \le k}$ by

$$H_{ij} = \begin{cases} (\overline{z}_i - z_j)^{-1}(u_i^* u_j - v_i^* v_j), & i \ne j, \\ \rho_i, & i = j \end{cases}$$

where $\rho_1, \ldots, \rho_k$ are k preassigned positive numbers. Then it is easily checked as in Example 21.2.1 that $(C_{0-}, C_{0+}, A_0; H)$ is a T-admissible GADI data set and that the corresponding interpolation conditions on a rational matrix function F are

(21.2.19a) $$F(z_i) u_i = v_i,$$

(21.2.19b) $$v_i^* F(z_i) = u_i^*$$

and

(21.2.20) $$z_i v_i^* F'(z_i) u_i = \rho_i, \qquad 1 \le i \le k$$

for $1 \le i \le k$. The reflection (21.2.19b) of (21.2.19a) is automatically also satisfied by F if $F(z) = F(\overline{z}^{-1})^{*-1}$. Note that (21.2.20) is an interpolation condition on a bitangential angular derivative of F. $\square$

EXAMPLE 21.2.3. In this case we take

$$C_{0-} = [u_1, \ldots, u_k], \qquad C_{0+} = [v_1, \ldots, v_k]$$

with $u_1, \ldots, u_k$, $v_1, \ldots, v_k$ column vectors in $\mathbf{C}^N$ with $u_1 \ne 0$, while we take A_0 to be a $k \times k$ Jordan block

$$A_0 = \begin{bmatrix} z_0 & 1 & & & 0 \\ & z_0 & 1 & & \\ & & \ddots & \ddots & 1 \\ 0 & & & & z_0 \end{bmatrix}$$

with eigenvalue z_0 of modulus 1. We take $H = [H_{ij}]_{1 \leq i,j \leq k}$ to be a $k \times k$ Hermitian matrix which satisfies the Stein equation

$$H - A_0^* H A_0 = C_{0-}^* C_{0-} - C_{0+}^* C_{0+}.$$

Since the eigenvalue z_0 of A_0 has unit modulus, that such a solution H exists is already a nontrivial condition on the vector $u_1, \ldots, u_k, v_1, \ldots, v_k$. Then $(C_{0-}, C_{0+}, A_0; H)$ is a T-admissible GADI data set. The interpolation condition (21.2.16a) (or, more precisely, its dual) for this example has already been discussed in Example 16.2.2; it reduces to

(21.2.21) $$F_i u_1 + F_{i-1} u_2 + \cdots + F_1 u_i = v_i$$

for $1 \leq i \leq k$, where $F_1, \ldots, F_k$ are the first k Taylor coefficients for $F(z)$ at z_0;

(21.2.22) $$F(z) = \sum_{j=1}^{\infty} F_j (z - z_0)^{j-1}, \qquad z \text{ close to } z_0.$$

To work out the interpolation condition (21.2.16b) more explicitly we proceed as follows. Note that $(zI - A_0^{*-1})^{-1} A_0^{*-1} = -(I - zA_0^*)^{-1}$ where

$$I - zA_0^* = (1 - z\bar{z}_0)(I - z(1 - z\bar{z}_0)^{-1} J)$$

with J equal to the $k \times k$ nilpotent matrix

$$J = \begin{bmatrix} 0 & 0 & \cdots & & 0 \\ 1 & 0 & \ddots & & 0 \\ \vdots & & \ddots & & \vdots \\ 0 & 0 & \cdots & 1 & 0 \end{bmatrix}.$$

Thus $(I - zA_0^*)^{-1} = (1 - z\bar{z}_0)^{-1} \sum_{\ell=0}^{k-1} z^\ell (1 - z\bar{z}_0)^{-\ell} J^\ell$ is given by

$$(I - zA_0^*)^{-1} = \begin{bmatrix} \Delta & & & 0 \\ z\Delta^2 & & \ddots & \\ \vdots & & & \ddots \\ z^{k-1}\Delta^k & \cdots & z\Delta^2 & \Delta \end{bmatrix}$$

where we have set $\Delta = \Delta(z) = (1 - z\bar{z}_0)^{-1}$. Therefore the (i,j)-th entry on the left side of (21.2.16c) can be computed to be

(21.2.23) $$\mathrm{Res}_{z=z_0} - \sum_{\alpha=1}^{i} \sum_{\beta=1}^{j} v_{i+1-\alpha}^* F(z) u_{j+1-\beta} z^{\alpha-1} (1 - z\bar{z}_0)^{-\alpha} (z - z_0)^{-\beta}.$$

Use that

$$(1 - z\bar{z}_0)^{-\alpha} = z_0^\alpha (-1)^\alpha (z - z_0)^{-\alpha}$$

and that

$$z^{\alpha-1} = (z_0 + (z - z_0))^{\alpha-1}$$

$$= \sum_{\ell=0}^{\alpha-1} \binom{\alpha-1}{\ell} z_0^{\alpha-1-\ell}(z - z_0)^\ell$$

to see that (21.2.23) is the same as

$$\text{Res}_{z=z_0} - \sum_{\alpha=1}^{i} \sum_{\beta=1}^{j} \sum_{\ell=0}^{\alpha-1} v_{i+1-\alpha}^* F(z) u_{j+1-\beta} (-1)^\alpha \binom{\alpha-1}{\ell} z_0^{2\alpha-1-\ell}(z - z_0)^{\ell-\alpha-\beta}.$$

Hence the interpolation condition (21.2.16c) explicitly is given by

$$(21.2.24) \qquad \sum_{\alpha=1}^{i} \sum_{\beta=1}^{j} \sum_{\ell=0}^{\alpha-1} (-1)^\alpha z_0^{2\alpha-1-\ell} \binom{\alpha-1}{\ell} v_{i+1-\alpha}^* F_{\alpha+\beta-\ell} u_{j+1-\beta} = -H_{ij}$$

for $1 \leq i, j \leq k$, if F is given by (21.2.22).

By a change of variable and a rearrangement of the order of summation one can also write this as
(21.2.25)

$$\sum_{p=2}^{i+j} \left[\sum_{\beta=1}^{\min\{p-1,j\}} \sum_{\alpha=p-\beta}^{i} (-1)^\alpha z_0^{\alpha+p-\beta-1} \binom{\alpha-1}{\alpha+\beta-p} v_{i+1-\alpha}^* F_p u_{j+1-\beta} \right] = -H_{ij}.$$

The reflected interpolation condition (21.2.16b) we have seen is the right tangential interpolation condition (21.2.16a) holding for $F(\overline{z}^{-1})^{*-1}$ in place of $F(z)$. The condition (21.2.16a) for this example, namely (21.2.21), can also be written as

$$F(z)u(z) - v(z) = (z - z_0)^k [\text{analytic at } z_0]$$

where we have set

$$u(z) = \sum_{i=1}^{k} u_i(z - z_0)^{i-1}, \qquad v(z) = \sum_{i=1}^{k} v_i(z - z_0)^{i-1}.$$

This holding for $F(\overline{z})^{*-1}$ in place of $F(z)$ is easily seen to be equivalent to

$$(21.2.26) \qquad v(\overline{z}^{-1})^* F(z) - u(\overline{z}^{-1})^* = (z - z_0)^k [\text{analytic at } z_0].$$

We conclude that (21.2.26) is an equivalent formulation of the interpolation condition (21.2.16b) for this example. Alternatively, one can do a computation similar to that just done above to derive the form (21.2.25) for the bitangential interpolation condition to get that (21.2.16b) for this example can also be expressed directly in terms of the Taylor coefficients $u_1, \ldots, u_k, v_1, \ldots, v_k$ for $u(z)$ and $v(z)$ as

$$\sum_{p=1}^{i} \left[\sum_{\alpha=p}^{i} (-1)^\alpha z_0^{\alpha+p-1} \binom{\alpha-1}{\alpha-p} v_{i+1-\alpha}^* \right] F_p = \sum_{\alpha=1}^{i} (-1)^\alpha \overline{z}_0^\alpha u_\alpha^*.$$

21.3 TANGENTIAL INTERPOLATION OF MATRICIAL ANGULAR DERIVATIVES ON THE IMAGINARY LINE

The plan of this section parallels that of the previous section; here we simply set down the analogue of all of the ideas with the imaginary line in place of the unit circle.

We consider a set of right tangential interpolation conditions at points now on the imaginary line $i\mathbf{R}$ instead of on the unit circle. From Chapter 16 we know that we write such a set of interpolation conditions in the form

$$\sum_{z_0 \in \sigma(A)} \mathrm{Res}_{z=z_0} F(z) C_{0-}(zI - A_0)^{-1} = C_{0+}$$

where (C_{0-}, A_0) is a null kernel pair and now $\sigma(A_0) \subset i\mathbf{R}$. We wish to characterize sets of interpolation conditions which arise from an $i\mathbf{R}$-admissible Sylvester TSCII data set (and so by the results of Chapter 16 the set of all solutions has a single linear fractional formula solution) and for which $F(-\bar{z})^{*-1}$ is a solution whenever $F(z)$ is. This leads us to define a collection of matrices $(C_{0-}, C_{0+}, A_0; H)$ of respective sizes $N \times n_0$, $N \times n_0$, $n_0 \times n_0$ and $n_0 \times n_0$ to be an $i\mathbf{R}$-admissible *generalized angular derivative interpolation* (GADI) data set if

(21.3.1a) (C_{0-}, A_0) *is a null kernel pair with* $\sigma(A_0) \subset i\mathbf{R}$

(21.3.1b)   *H is a Hermitian solution of the Lyapunov equation*
$$HA_0 + A_0^* H = C_{0-}^* C_{0-} - C_{0+}^* C_{0+}.$$

With an $i\mathbf{R}$-admissible GADI data set $\omega = (C_{0-}, C_{0+}, A_0; H)$ we associate the set of interpolation conditions for an $N \times N$ matrix function $F(z)$ analytic on $\sigma(A_0)$

(21.3.2a) $$\sum_{z_0 \in \sigma(A_0)} \mathrm{Res}_{z=z_0} F(z) C_{0-}(zI - A_0)^{-1} = C_{0+}$$

(21.3.2b) $$\sum_{z_0 \in \sigma(A_0)} \mathrm{Res}_{z=z_0} (zI + A_0^*)^{-1} C_{0+}^* F(z) = C_{0-}^*$$

and

(21.3.2c) $$\sum_{z_0 \in \sigma(A_0)} \mathrm{Res}_{z=z_0} -(zI + A_0^*)^{-1} C_{0+}^* F(z) C_{0-}(zI - A_0)^{-1} = H.$$

The problem of finding rational $N \times N$ matrix functions $F(z)$ analytic on Π^+ with $\sup\{\|F(z)\| : z \in \Pi^+\} \leq 1$ satisfying the interpolation conditions (21.3.2a)–(21.3.2c) we refer to as the *generalized angular derivative interpolation* (GADI) problem associated with the GADI data set $\omega = (C_{0-}, C_{0+}, A_0; H)$. The following analogue of Proposition 21.2.1 justifies the terminology.

PROPOSITION 21.3.1. *Suppose* (C_{0-}, A_0) *is a null kernel pair with* $\sigma(A_0)$ *contained in* $i\mathbb{R}$ *and suppose that* $F(z)$ *is a rational* $N \times N$ *matrix function analytic and invertible on* $\sigma(A_0)$ *such that* $F(z) = F(-\bar{z})^{*-1}$. *Define matrices* C_{0+} *and* H *by*

$$(21.3.3) \qquad C_{0+} = \sum_{z_0 \in \sigma(A_0)} \text{Res}_{z=z_0} F(z) C_{0-} (zI - A_0)^{-1}$$

and

$$(21.3.4) \qquad H = \sum_{z_0 \in \sigma(A_0)} \text{Res}_{z=z_0} -(zI + A_0^*)^{-1} C_{0+}^* F(z) C_{0-} (zI - A_0)^{-1}.$$

Then $(-A_0^*, -C_{0+}^*)$ *is a full range pair and* F *satisfies the interpolation condition*

$$(21.3.5) \qquad \sum_{z_0 \in \sigma(A_0)} \text{Res}_{z=z_0} (zI + A_0^*)^{-1} C_{0+}^* F(z) = C_{0-}^*.$$

Moreover, H *is Hermitian and satisfies the Lyapunov equation*

$$(21.3.6) \qquad HA_0 + A_0^* H = C_{0-}^* C_{0-} - C_{0+}^* C_{0+}.$$

PROOF. By simple duality $(-A_0^*, -C_{0+}^*)$ being a full range pair is equivalent to (C_{0+}, A_0) being a null kernel pair. This follows from $F(z)$ being invertible on $\sigma(A_0)$ and satisfying (21.3.3) together with (C_{0-}, A_0) being a null kernel pair as in the proof of Theorem 21.2.1. The identity (21.3.5) follows from (21.3.3) much as the proof of (21.2.3) in Theorem 21.2.1, where now one uses that $F(z) = F(-\bar{z})^{*-1}$ in place of $F(z) = F(\bar{z}^{-1})^{*-1}$. If H is defined by (21.3.4) and F satisfies the right interpolation condition (21.3.3) and the left interpolation condition (21.3.5), then by the consistency condition for the solvability of the TSCII problem (21.3.3)–(21.3.5) (see Lemma 16.9.1 and its proof) the matrix H satisfies the Lyapunov equation (21.3.6).

Finally, to show that H is Hermitian, we first compute an expression for H^* as follows. By using Lemmas 16.10.4 and 16.9.4, we see that (21.3.4) is equivalent to

$$Y(z) = -(zI + A_0^*)^{-1} C_{0+}^* [F(z) C_{0-} - C_{0+}] (zI - A_0)^{-1} - (zI + A_0^*)^{-1} H$$

being analytic on $\sigma(A_0)$. As $\sigma(A_0) \subset i\mathbb{R}$, this is equivalent to $Y(-\bar{z})^*$ being analytic on $\sigma(A_0)$, where

$$Y(-\bar{z})^* = -(-zI - A_0^*)^{-1} [C_{0-}^* F(-\bar{z})^* - C_{0+}^*] C_{0+} (-zI + A_0)^{-1} - H^* (-zI + A_0)^{-1}.$$

Using that $F(-\bar{z})^* = F(z)^{-1}$, this becomes

$$(21.3.7) \quad Y(-\bar{z})^* = -(zI + A_0^*)^{-1} [C_{0-}^* F(z)^{-1} - C_{0+}^*] C_{0+} (zI - A_0)^{-1} + H^* (zI - A_0)^{-1}.$$

By Lemma 16.10.4,

$$\sum_{z_0 \in \sigma(A_0)} \text{Res}_{z=z_0} (zI + A_0^*)^{-1} C_{0-}^* C_{0+} (zI - A_0)^{-1} = 0$$

and by the Riesz functional calculus

$$\sum_{z_0 \in \sigma(A_0)} \text{Res}_{z=z_0} H^* (zI - A_0)^{-1} = H^*.$$

Thus, from the fact that

$$\sum_{z_0 \in \sigma(A_0)} \text{Res}_{z=z_0} Y(-\overline{z})^* = 0$$

since $Y(-\overline{z})^*$ is analytic on $\sigma(A_0)$, we get from (21.3.7) that

$$(21.3.8) \qquad \sum_{z_0 \in \sigma(A_0)} \text{Res}_{z=z_0} (zI + A_0^*)^{-1} C_{0-}^* F(z)^{-1} C_{0+} (zI - A_0)^{-1} = H^*.$$

On the other hand from Lemma 21.2.2 applied to $\omega = (C_{0+}, C_{0-}, A_0; -A_0^*, -C_{0+}^*, C_{0-}^*; H)$, we get that the left side of (21.3.8) also gives H. We conclude that $H = H^*$ as asserted. $\square$

We next illustrate with some examples.

EXAMPLE 21.3.1. Take $N = 1$, $n_0 = k$, $C_{0-} = [1, \ldots, 1]$, $C_{0+} = [w_1, \ldots, w_k]$ where $|w_j| = 1$ for $1 \le j \le k$, and

$$A_0 = \begin{bmatrix} z_1 & & 0 \\ & \ddots & \\ 0 & & z_k \end{bmatrix}$$

where $z_1, \ldots, z_k$ are k distinct points on the imaginary line, and define $H = [H_{ij}]_{1 \le i,j \le k}$ by

$$H_{ij} = \begin{bmatrix} \frac{1 - \overline{w}_i w_j}{\overline{z}_i + z_j}, & i \neq j \\ \rho_i, & i = j \end{bmatrix}$$

where $\rho_1, \ldots, \rho_k$ are positive real numbers. One easily verifies that conditions (21.3.1a)–(21.3.b) are satisfied, and hence $(C_{0-}, C_{0+}, A_0; H)$ is an $i\mathbb{R}$-admissible GADI data set. The interpolation conditions (21.3.2a) and (21.3.2b) for a scalar rational function F analytic in $i\mathbb{R}$ both collapse to

$$(21.3.9) \qquad F(z_i) = w_i, \qquad 1 \le i \le k$$

(since $|w_i| = 1$ and $z_i = -\overline{z}_i$ for all i) while (21.3.2c) becomes

$$(21.3.10a) \qquad -\overline{w}_i F(z_i)(z_i - z_j)^{-1} - (z_j - z_i)^{-1} \overline{w}_i F(z_j) = \frac{1 - \overline{w}_i w_j}{\overline{z}_i + z_j}, \qquad i \neq j$$

(21.3.10b) $$-\overline{w}_i F'(z_i) = \rho_i, \qquad i = j$$

for $1 \leq i, j \leq k$. Note that (21.3.10a) is implied by (21.3.9) and that the form of condition (21.3.10b) is consistent with the angular derivative interpolation constraint for the right half plane discussed in Section 21.1 (see (ADId)). □

EXAMPLE 21.3.2. A simple matrix extension of the previous example is to take again $z_1, \ldots, z_k$ to be k distinct numbers on the imaginary axis $i\mathbf{R}$ and now let $u_1, \ldots, u_k, \ v_1, \ldots, v_k$ be $2k$ column vectors in $\mathbf{C}^N$ of unit length. We then set

$$C_{0-} = [u_1, \ldots, u_k], \qquad C_{0+} = [v_1, \ldots, v_k], \qquad A_0 = \begin{bmatrix} z_1 & & 0 \\ & \ddots & \\ 0 & & z_k \end{bmatrix}$$

and define a $k \times k$ Hermitian matrix $H = [H_{ij}]_{1 \leq i, j \leq k}$ by

$$H_{ij} = \begin{cases} (\overline{z}_i + z_j)^{-1}(u_i^* u_j - v_i^* v_j), & i \neq j \\ \rho_i, & i = j \end{cases}$$

where again $\rho_1, \ldots, \rho_k$ are k positive real numbers. Then one easily checks that $(C_{0-}, C_{0+}, A_0; H)$ is an $i\mathbf{R}$-admissible GADI data set, and that the corresponding interpolation conditions on a rational $n \times n$ matrix function $F(z)$ analytic on $i\mathbf{R}$ are

(21.3.11) $$F(z_i)u_i = v_i$$

(21.3.12) $$v_i^* F(z_i) = u_i^*$$

and

(21.3.13) $$-v_i^* F'(z_i)u_i = \rho_i$$

for $1 \leq i \leq k$. Note that a consequence of (21.3.11) is that

$$v_i^* F(z_i)u_i = v_i^* v_i = 1 = u_i^* u_i.$$

If it is known that $\|F(z_i)\| \leq 1$, this gives equality in Schwarz's inequality

$$|v_i^* F(z_i)u_i| \leq \|v_i^*\| \|F(z_i)u_i\|,$$

and we deduce (21.3.12) as a consequence of (21.3.11). □

EXAMPLE 21.3.3. This example parallels Example 21.2.3. We take

$$C_{0-} = [u_1, \ldots, u_k], \qquad C_{0+} = [v_1, \ldots, v_k]$$

with $u_1, \ldots, u_k, \ v_1, \ldots, v_k$ column vectors in $\mathbf{C}^n$ with $u_1 \neq 0$, while we take A_0 to be a $k \times k$ Jordan block

$$A_0 = \begin{bmatrix} z_0 & 1 & & 0 \\ & z_0 & \ddots & \\ & & \ddots & 1 \\ 0 & & & z_0 \end{bmatrix}$$

with eigenvalue z_0 on the imaginary line $i\mathbf{R}$. Let $H = [H_{ij}]_{1 \leq i,j \leq k}$ be a $k \times k$ Hermitian matrix which satisfies the Lyapunov equation

$$(21.3.14) \qquad\qquad H A_0 + A_0^* H = C_{0-}^* C_{0-} - C_{0+}^* C_{0+}.$$

If we let T be the diagonal $k \times k$ matrix

$$T = \begin{bmatrix} -1 & & & 0 \\ & 1 & & \\ & & \ddots & \\ 0 & & & (-1)^k \end{bmatrix}$$

then $T = T^{-1}$ and

$$T(-A_0^*)T = A_0^T$$

so the above Lyapunov equation (21.3.14) may be rewritten as a Sylvester equation for the unknown TH:

$$(21.3.15) \qquad\qquad (TH)A_0 - A_0^T(TH) = T(C_{0-}^* C_{0-} - C_{0+}^* C_{0+})$$

where the (i,j)-th entry Γ_{ij} of the right hand side $T(C_{0-}^* C_{0-} - C_{0+}^* C_{0+})$ is given by

$$\Gamma_{ij} = (-1)^i(u_i^* u_j - v_i^* v_j).$$

We conclude from Theorem A.1.2 in the Appendix that a necessary and sufficient conditions for a solution TH of (21.3.15) to exist (and hence for a solution H of (21.3.14) to exist) is that

$$(21.3.16) \qquad\qquad \sum_{t=1}^{M}(-1)^{M+1-t}[u_{M+1-t}^* u_t - v_{M+1-t}v_t^*] = 0$$

for $1 \leq M \leq k$. In this case a solution $H = (-1)^{k+1}T(TH)$ of (21.3.14) necessarily has the form (by formulas (A.1.7)–(A.1.9))

$$
\begin{aligned}
(21.3.17) \qquad H_{\alpha\beta} &= -(-1)^\alpha \sum_{\ell=1}^{\beta}(-1)^{\alpha+\beta+1-\ell}[u_{\alpha+\beta+1-\ell}^* u_\ell - v_{\alpha+\beta+1-\ell}^* v_\ell \\
&= (-1)^\alpha \sum_{\ell=\beta+1}^{\alpha+\beta}(-1)^{\alpha+\beta+1-\ell}[u_{\alpha+\beta+1-\ell}^* u_\ell - v_{\alpha+\beta+1-\ell}^* v_\ell]
\end{aligned}
$$

if $\alpha + \beta \leq k$, and

$$(21.3.18) \quad H_{\alpha\beta} = \widetilde{R}_{\alpha+\beta} - (-1)^\alpha \sum_{\ell=\alpha+\beta-k+1}^{\beta}(-1)^{\alpha+\beta+1-\ell}[u_{\alpha+\beta+1-\ell}^* u_\ell - v_{\alpha+\beta+1-\ell}^* v_\ell]$$

where $\widetilde{R}_{\alpha+\beta}$ is arbitrary, if $\alpha+\beta > k$. The consistency condition (21.3.16) guarantees that the two formulas in (21.3.17) are consistent, and also guarantees that the resulting matrix H is Hermitian as long as the numbers $\widetilde{R}_M$ ($k < M \leq 2k$) are chosen appropriately.

As in Example 16.2.2, the interpolation condition (21.3.2a) for this case takes the form

$$(21.3.19) \qquad F_i u_1 + F_{i-1} u_2 + \cdots + F_1 u_i = v_i$$

for $1 \leq i \leq k$, where $F(z) = \sum_{\ell=1}^{\infty} F_\ell (z - z_0)^{\ell-1}$ is the Taylor expansion of $F(z)$ at z_0. The bitangential interpolation condition (21.3.2c) works out to be

$$\sum_{\alpha=1}^{i} \sum_{\beta=1}^{j} (-1)^{\alpha+1} v_{i+1-\alpha}^* F_{\alpha+\beta} u_{j+1-\beta} = H_{ij}$$

or equivalently

$$(21.3.20) \qquad \sum_{p=2}^{i+j} \left[\sum_{\beta=\max\{1,p-i\}}^{j} (-1)^{p-\beta+1} v_{i+1+\beta-p}^* F_p u_{j+1-\beta} \right] = H_{ij},$$

for $1 \leq i, j \leq k$. The reflected left tangential interpolation condition (21.3.2b) reduces to

$$(21.3.21) \qquad \sum_{\alpha=1}^{i} (-1)^\alpha v_{i+1-\alpha}^* F_\alpha = u_i^*$$

for $1 \leq i \leq k$. The condition (21.3.20) is a consequence of H satisfying the Lyapunov equation (21.3.14) and the interpolation conditions (21.3.19) and (21.3.21) for the cases where $i + j \leq k$. The fact that the quantity H_{ij} on the left side of (21.3.20) necessarily satisfies $H_{ij} = \overline{H}_{ji}$ for $i + j > k$ if $F(z) = F(-\bar{z})^{*-1}$ can be viewed as an extension of Theorem 21.1.4 to higher order angular derivatives.

21.4 SOLUTION OF THE GENERALIZED ANGULAR DERIVATIVE INTERPOLATION PROBLEM

In this section we state and prove the results discussed in the previous sections even in the more general context where solutions F of the interpolation problem are allowed to have a prescribed number of poles in the interior of the region Δ (either $\Delta = \mathcal{D}$ or $\Delta = \Pi^+$), and moreover, where additional interpolation constraints may be prescribed inside Δ.

We therefore assume that we are given a $\partial\Delta$-admissible GADI data set $(C_{0-}, C_{0+}, A_0; H)$ together with a Δ-admissible Sylvester TSCII data set $(C_+, C_-, A_\pi; A_\zeta, B_+, B_-; \Gamma)$, where Δ is either the unit disk $\mathcal{D}$ or the right half plane Π^+. The interpolation problem to be considered in this section is: find all rational $N \times N$ matrix functions $F(z)$ such that

$$(21.4.1) \qquad F \text{ is analytic on } \partial\Delta \text{ with } \sup\{\|F(z)\|: z \in \partial\Delta\} \leq 1.$$

F satisfies the GADI *conditions*

$$\sum_{z_0 \in \sigma(A_0)} \mathrm{Res}_{z=z_0} F(z)C_{0-}(zI - A_0)^{-1} = C_{0+}$$

(21.4.2)

$$\sum_{z_0 \in \sigma(A_0)} \mathrm{Res}_{z=z_0}(zI - A_0^{*-1})^{-1}A_0^{*-1}C_{0+}^* F(z) = A_0^{*-1}C_{0-}^*$$

$$\sum_{z_0 \in \sigma(A_0)} \mathrm{Res}_{z=z_0}(zI - A_0^{*-1})^{-1}A_0^{*-1}C_{0+}^* F(z)C_{0-}(zI - A_0)^{-1} = H$$

if $\Delta = \mathcal{D}$, *or*

$$\sum_{z_0 \in \sigma(A_0)} \mathrm{Res}_{z=z_0} F(z)C_{0-}(zI - A_0)^{-1} = C_{0+}$$

$$\sum_{z_0 \in \sigma(A_0)} \mathrm{Res}_{z=z_0}(zI + A_0^*)^{-1}C_{0+}^* F(z) = C_{0-}^*$$

$$\sum_{z_0 \in \sigma(A_0)} \mathrm{Res}_{z=z_0} -(zI + A_0^*)^{-1}C_{0+}^* F(z)C_{0-}(zI - A_0)^{-1} = H$$

if $\Delta = \Pi^+$,

F is analytic on $\sigma(A_\pi) \cup \sigma(A_\zeta)$

and satisfies the TSCII *conditions on the interior of* Δ

$$\sum_{z_0 \in \sigma(A_\pi)} \mathrm{Res}_{z=z_0} F(z)C_-(zI - A_\pi)^{-1} = C_+$$

(21.4.3)

$$\sum_{z_0 \in \sigma(A_\zeta)} \mathrm{Res}_{z=z_0}(zI - A_\zeta)^{-1}B_+ F(z) = -B_-$$

$$\sum_{z_0 \in \sigma(A_\pi) \cup \sigma(A_\zeta)} \mathrm{Res}_{z=z_0}(zI - A_\zeta)^{-1}B_+ F(z)C_-(zI - A_\pi)^{-1} = \Gamma.$$

Note that we no longer require that F be analytic on Δ. However we are interested in keeping track of the number of poles (counting multiplicities) in Δ. As usual the multiplicity of a pole z_0 for a rational matrix function $F(z)$ is taken to be the negative of the sum of the negative partial multiplicities of $F(z)$ in the local Smith-McMillan form for $F(z)$ at z_0. The following result subsumes all the interpolation results discussed in the previous sections of this chapter.

THEOREM 21.4.1. *Suppose* $(C_{0-}, C_{0+}, A_0; H)$ *is a* $\partial\Delta$-*admissible* GADI *data set, and* $(C_+, C_-, A_\pi; A_\zeta, B_+, B_-; \Gamma)$ *is a* Δ-*admissible Sylvester* TSCII *data set, where either* $\Delta = \mathcal{D}$ *or* $\Delta = \Pi^+$. *Associate with these data a matrix* Λ *given by*

(21.4.4)
$$\Lambda = \begin{bmatrix} \Lambda_{11} & \Lambda_{21}^* & \Gamma^* \\ \Lambda_{21} & H & \Lambda_{32}^* \\ \Gamma & \Lambda_{32} & \Lambda_{33} \end{bmatrix}$$

where Λ_{11}, Λ_{21}, Λ_{32} *and* Λ_{33} *are defined as the unique solutions of the respective Sylvester/Stein equations*

(21.4.5a)
$$\Lambda_{11} - A_\pi^* \Lambda_{11} A_\pi = C_-^* C_- - C_+^* C_+$$

(21.4.5b)
$$A_0^* \Lambda_{21} A_\pi - \Lambda_{21} = C_{0+}^* C_+ - C_{0-}^* C_-$$

(21.4.5c)
$$\Lambda_{32} A_0 - A_\zeta \Lambda_{32} = B_+ C_{0+} - B_- C_{0-}$$

(21.4.5d)
$$\Lambda_{33} - A_\zeta \Lambda_{33} A_\zeta^* = B_+ B_+^* - B_- B_-^*$$

if $\Delta = \mathcal{D}$, while Λ_{11}, Λ_{21}, Λ_{32} and Λ_{33} are defined as the unique solutions of the respective Sylvester/Lyapunov equations

(21.4.6a)
$$\Lambda_{11} A_\pi + A_\pi^* \Lambda_{11} = C_+^* C_+ - C_-^* C_-$$

(21.4.6b)
$$\Lambda_{21} A_\pi + A_0^* \Lambda_{21} = C_{0+}^* C_+ - C_{0-}^* C_-$$

(21.4.6c)
$$\Lambda_{32} A_0 - A_\zeta \Lambda_{32} = B_+ C_{0+} - B_- C_{0-}$$

(21.4.6d)
$$\Lambda_{33} A_\zeta^* + A_\zeta \Lambda_{33} = B_- B_-^* - B_+ B_+^*$$

if $\Delta = \Pi^+$. Then the following statements hold true.

(1) *Assume also that Λ is invertible. If there is a rational $N \times N$ matrix function $F(z)$ analytic on $\partial\Delta \cup \sigma(A_\pi) \cup \sigma(A_\zeta)$ satisfying (21.4.1)–(21.4.3) and having κ poles (counting multiplicities) in Δ, then the associated Hermitian Pick matrix Λ given by (21.4.4) has at most κ negative eigenvalues.*

(2) *If Λ is invertible and the number of negative eigenvalues of Λ is κ, then there exist solutions F of (21.4.1)–(21.4.3) having κ poles (counting multiplicities) in Δ. In fact the set of such interpolants F is given by*

(21.4.7)
$$F = (\Theta_{11} G + \Theta_{12})(\Theta_{21} G + \Theta_{22})^{-1}$$

where G is any rational $N \times N$ matrix function satisfying $\sup\{\|G(z)\|: z \in \partial\Delta\} \leq 1$ and having no poles in Δ for which $\det(\Theta_{22}^{-1}\Theta_{21} G + I)$ is nonzero on $\sigma(A_0)$. Here $\Theta(z) = \begin{bmatrix} \Theta_{21}(z) & \Theta_{22}(z) \\ \Theta_{21}(z) & \Theta_{22}(z) \end{bmatrix}$ is given by

(21.4.8)
$$\begin{aligned}
\Theta(z) = I_{2N} + (z - z_0) &\begin{bmatrix} C_+ & C_{0+} & -B_+^* \\ C_- & C_{0-} & -B_-^* \end{bmatrix} \\
\times &\begin{bmatrix} (zI - A_\pi)^{-1} & 0 & 0 \\ 0 & (zI - A_0)^{-1} & 0 \\ 0 & 0 & (I - zA_\zeta^*)^{-1} \end{bmatrix} \Lambda^{-1} \\
\times &\begin{bmatrix} (I - z_0 A_\pi^*)^{-1} C_+^* & -(I - z_0 A_\pi^*)^{-1} C_-^* \\ (I - z_0 A_0^*)^{-1} C_{0+}^* & -(I - z_0 A_0^*)^{-1} C_{0-}^* \\ (A_\zeta - z_0 I)^{-1} B_+ & -(A_\zeta - z_0 I)^{-1} B_- \end{bmatrix}
\end{aligned}$$

where z_0 is some point on $\mathbf{T}$ not in $\sigma(A_0)$ if $\Delta = \mathcal{D}$, while
(21.4.9)

$$\Theta(z) = I_{2N} + \begin{bmatrix} C_+ & C_{0+} & -B_+^* \\ C_- & C_{0-} & -B_-^* \end{bmatrix}$$

$$\times \begin{bmatrix} (zI - A_\pi)^{-1} & 0 & 0 \\ 0 & (zI - A_0)^{-1} & 0 \\ 0 & 0 & (zI + A_\zeta^*)^{-1} \end{bmatrix} \Lambda^{-1} \begin{bmatrix} -C_+^* & C_-^* \\ -C_{0+}^* & C_{0-}^* \\ B_+ & -B_- \end{bmatrix}$$

if $\Delta = \Pi^+$.

We state explicitly as a corollary the specialization of Theorem 21.4.1 to the case where interpolation conditions are prescribed only on $\partial\Delta$.

COROLLARY 21.4.2. *Suppose $(C_{0-}, C_{0+}, A_0; H)$ is a $\partial\Delta$-admissible GADI data set, where either $\Delta = \mathcal{D}$ or $\Delta = \Pi^+$. Then the following statements hold true.*

(1) If there is a rational $N \times N$ matrix function $F(z)$ analytic on $\partial\Delta$ satisfying the GADI conditions (21.4.1)–(21.4.2) and having κ poles (counting multiplicities) in Δ, then the Hermitian matrix H has at most κ negative eigenvalues.

(2) If H is invertible and the number of negative eigenvalues of H is κ, then there exist solutions F of (21.4.1)–(21.4.2) having κ poles (counting multiplicities) in Δ. In fact the set of such interpolants F is given by

$$F = (\Theta_{11} G + \Theta_{12})(\Theta_{21} G + \Theta_{22})^{-1}$$

where G is any rational $N \times N$ matrix function satisfying (21.4.1) and having no poles in Δ for which $\det(\Theta_{22}^{-1}\Theta_{21}G + I)$ is nonzero on $\sigma(A_0)$. Here $\Theta(z) = \begin{bmatrix} \Theta_{11}(z) & \Theta_{12}(z) \\ \Theta_{21}(z) & \Theta_{22}(z) \end{bmatrix}$ is given by

$$(21.4.10) \qquad \Theta(z) = I_{2N} + (z - z_0) \begin{bmatrix} C_{0+} \\ C_{0-} \end{bmatrix} (zI - A_0)^{-1}(I - z_0 A_0^*)^{-1}[C_{0+}^*, -C_{0-}^*]$$

where z_0 is a point on the unit circle not in $\sigma(A_0)$ if $\Delta = \mathcal{D}$, and by

$$(21.4.11) \qquad \Theta(z) = I_{2N} + \begin{bmatrix} C_{0+} \\ C_{0-} \end{bmatrix} (zI - A_0)^{-1} H^{-1}[-C_{0+}^*, C_{0-}^*]$$

if $\Delta = \Pi^+$.

Note that Corollary 21.4.2 is simply a restatement of Theorem 21.4.1 for the case where the Δ-admissible TSCII data set $(C_+, C_-, A_\pi; A_\zeta, B_+, B_-; \Gamma)$ is taken to be vacuous. We remark also that Theorems 21.1.2 and 21.1.5 are immediate consequences of Corollary 21.4.2, at least for the case where $H = \Lambda$ is invertible; one simply specializes the data to the form considered in Examples 21.2.1 and 21.3.1.

We remark also that when Λ is positive definite and we are in the simple situation of Theorems 21.1.2 and 21.1.5, it is possible to characterize the set of rational functions f of the form $f = (\Theta_{11} g + \Theta_{12})(\Theta_{21} g + \Theta_{22})^{-1}$ with g a rational function analytic on Δ with modulus at most 1 (and for which $\Theta_{22}^{-1}\Theta_{21}g + 1$ may have a zero at some interpolation node z_i) as all f analytic and bounded by 1 in modulus on Δ which meet the interpolation conditions on the values of f

$$f(z_i) = w_i, \qquad 1 \le i \le k$$

and satisfy the angular derivative interpolation conditions with inequality

$$z_i \overline{w}_i f'(z_i) \le \rho_i, \qquad 1 \le i \le k \text{ if } \Delta = \mathcal{D}$$

or

$$-\overline{w}_i f'(z_i) \le \rho_i, \qquad 1 \le i \le k \text{ if } \Delta = \Pi^+.$$

However we shall not pursue this point here. For more discussion, see Ball [1983].

PROOF OF THEOREM 21.4.1. Note that Theorem 21.4.1 subsumes Theorem 18.5.3 and Theorem 19.2.1. The idea of the proof of Theorem 21.4.1 is the same as that for Theorems 18.5.3 and 19.2.1 but some details are new and more subtle, so we shall proceed carefully through the whole proof.

We assume that the matrix Λ given by (21.4.4)–(21.4.6) is invertible. Then by Theorem 7.5.2 if $\Delta = \mathcal{D}$ and Theorem 6.4.2 if $\Delta = \Pi^+$, the $2N \times 2N$ matrix function $\Theta(z)$ in Theorem 21.4.1 (given by (21.4.8) for $\Delta = \mathcal{D}$ and (21.4.9) for $\Delta = \Pi^+$) is simply a choice of $2N \times 2N$ matrix function having $\omega_0 = \left(\begin{bmatrix} C_{0+} \\ C_{0-} \end{bmatrix}, A_0; A_0^{*-1}, A_0^{*-1}[C_{0+}^*, -C_{0-}^*], H \right)$ if $\Delta = \mathcal{D}$ (respectively $\left(\begin{bmatrix} C_{0+} \\ C_{0-} \end{bmatrix}, A_0; -A_0^*, [-C_{0+}^*, C_{0-}^*]; H \right)$ if $\Delta = \Pi^+$) as $\partial\Delta$-null-pole triple, having $\omega = \left(\begin{bmatrix} C_+ \\ C_- \end{bmatrix}, A_\pi; A_\zeta, [B_+, B_-]; \Gamma \right)$ as Δ-null-pole triple and having J-unitary values on $\partial\Delta$ (where $J = I_N \oplus -I_N$); when $\Delta = \Pi^+$ we also insist that $\Theta(z)$ is regular at infinity. Note that the GADI conditions (21.4.2) are just the TSCII conditions associated with the data set

$$\omega_0 = (C_{0+}, C_{0-}, A_0; A_0^{*-1}, A_0^{*-1}C_{0+}^*, -A_0^{*-1}C_{0-}^*; H)$$

if $\Delta = \mathcal{D}$, or

$$\omega_0 = (C_{0+}, C_{0-}, A_0; -A_0^*, -C_{0+}^*, C_{0-}^*; H)$$

if $\Delta = \Pi^+$. Then by Theorem 16.11.1 the set of all rational $N \times N$ matrix functions $F(z)$ analytic on $\sigma(A_0) \cup \sigma(A_\pi) \cup \sigma(A_\zeta)$ and having at most κ poles (counting multiplicity) in $\overline{\Delta}$ is given by

$$(21.4.12) \qquad F = (\Theta_{11}G_1 + \Theta_{12}G_2)(\Theta_{21}G_1 + \Theta_{22}G_2)^{-1}$$

where G_1, G_2 are rational $N \times N$ matrix functions such that

$(21.4.13) \qquad \varphi(\Theta_{21}G_1 + \Theta_{22}G_2)$ *is analytic on* $\overline{\Delta}$ *(including infinity if* $\Delta = \Pi^+$*) and* $\det(\varphi[\Theta_{21}G_1 + \Theta_{22}G_2])$ *has at most* κ *zeros (counting multiplicities) in* $\overline{\Delta}$, *none of which occur on* $\sigma(A_0) \cup \sigma(A_\pi) \cup \sigma(A_\zeta)$.

Here we take φ to be any rational $N \times N$ matrix function analytic on $\overline{\Delta}$ (including infinity if appropriate) for which (C_{0-}, A_0) is a right null pair over $\partial\Delta$ and (C_-, A_π) is a right null pair over Δ.

Now suppose that there exists a solution F of (21.4.2) and (21.4.3) with the additional property (21.4.1) and which has κ poles (counting multiplicities) in Δ. By Theorem 16.11.2 combined with Theorem 12.3.1 we see that formula (21.4.12) with the side condition (21.4.13) can be expressed in the more refined form

$$(21.4.14) \qquad \begin{bmatrix} F \\ I \end{bmatrix} \varphi^{-1} W = \begin{bmatrix} \Theta_{11} & \Theta_{12} \\ \Theta_{21} & \Theta_{22} \end{bmatrix} \begin{bmatrix} G_1 \\ G_2 \end{bmatrix}$$

for some rational $N \times N$ matrix functions G_1, G_2, W analytic on $\overline{\Delta}$ for which $\det W$ has at most κ zeros in $\overline{\Delta}$, none of which are in $\sigma(A_\pi) \cup \sigma(A_\zeta) \cup \sigma(A_0)$. Since $\|F(z)\| \leq 1$ for z on $\partial\Delta$ and $\Theta(z)$ is J-unitary at all regular points of $\partial\Delta$, we get for a generic point $z \in \partial\Delta$,

$$0 \geq F(z)^* F(z) - I$$
$$= \varphi(z)^* W(z)^{*-1} [G_1(z)^* \; G_2(z)^*] \Theta(z)^* J \Theta(z) \begin{bmatrix} G_1(z) \\ G_2(z) \end{bmatrix} W(z)^{-1} \varphi(z)$$
$$= \varphi(z)^* W(z)^{*-1} \{G_1(z)^* G_1(z) - G_2(z)^* G_2(z)\} W(z)^{-1} \varphi(z).$$

Thus

$$(21.4.15) \qquad 0 \geq G_1(z)^* G_1(z) - G_2(z)^* G_2(z), \qquad z \in \partial\Delta.$$

Moreover from (21.4.14) we see that the $2N \times N$ matrix function $\begin{bmatrix} G_1(z) \\ G_2(z) \end{bmatrix}$ must have full rank N for all but finitely many points in $\mathbf{C}$. From (21.4.15), $G_1(z)$ has degenerate rank whenever $G_2(z)$ does for $z \in \partial\Delta$; we conclude that $G_2(z)$ has full rank, i.e. $\det G_2(z)$ is nonzero, for all but finitely many points z in $\mathbf{C}$. Thus we may define a rational $N \times N$ matrix $G(z)$ by $G(z) = G_1(z) G_2(z)^{-1}$. Then (21.4.15) gives that

$$(21.4.16) \qquad G(z) \text{ is analytic with } \|G(z)\| \leq 1 \text{ for } z \in \partial\Delta.$$

Now the bottom component of (21.4.14) may be written as

$$(21.4.17) \qquad W = \varphi\Theta_{22}(\Theta_{22}^{-1}\Theta_{21} G + I)G_2.$$

We now recall from Proposition 13.1.2 that

$$(21.4.18) \qquad [\Theta_{22}^{-1}\Theta_{21}](z) \text{ is analytic with } \|[\Theta_{22}^{-1}\Theta_{21}](z)\| \leq 1 \text{ on } \partial\Delta.$$

Also, since $(A_0^{*-1}, A_0^{*-1} C_+^*)$ (for $\Delta = \mathcal{D}$) or $(-A_0^*, C_+^*)$ and (A_ζ, B_+) are full range pairs, one can argue as in the proof of Lemma 16.10.2 (see (16.10.5)) that

$$\varphi^{-1}\mathcal{R}_N(\overline{\Delta}) = [\; \Theta_{21} \quad \Theta_{22} \;]\mathcal{R}_{2N}(\overline{\Delta})$$

and hence in particular $\varphi\Theta_{22}$ is analytic on $\overline{\Delta}$. We have already observed that $\Theta_{22}^{-1}\Theta_{21}$, G and G_2 are analytic on $\partial\Delta$. Moreover, since by (21.4.1) F is analytic on $\partial\Delta$ (including infinity if appropriate) and satisfies the GADI condition (21.4.2) necessarily by Theorem 16.10.1 $W(z)$ is analytic and invertible on $\partial\Delta$ (including infinity). Thus (21.4.17) forces all three factors $\varphi\Theta_{22}$, $\Theta_{22}^{-1}\Theta_{21}G + I$ and G_2 to be analytic and invertible on $\partial\Delta$. Moreover, for $0 \le t < 1$, $\|t[\Theta_{22}^{-1}\Theta_{21}G](z)\| < 1$ for $z \in \partial\Delta$, since $\Theta_{22}^{-1}\Theta_{21}$ and G are each contractions on $\partial\Delta$. Thus $t\Theta_{22}^{-1}\Theta_{21}G + I$ is analytic and invertible on $\partial\Delta$ for $0 \le t \le 1$. By a standard homotopy argument we conclude that

$$\text{wno}\det(\Theta_{22}^{-1}\Theta_{21}G + I) = \text{wno}(I) = 0.$$

Then from (21.4.17) we get

$$(21.4.19) \qquad \text{wno}\det W = \text{wno}\det(\varphi\Theta_{22}) + \text{wno}\det G_2.$$

Recall that W was guaranteed to be analytic on $\overline{\Delta}$ with $\det W$ having at most κ zeros in $\overline{\Delta}$. Thus

$$(21.4.20) \qquad \kappa \ge \text{wno}\det W.$$

Also by the construction G_2 is analytic on $\overline{\Delta}$; hence

$$(21.4.21) \qquad \text{wno}\det G_2 \ge 0.$$

For $\varphi\Theta_{22}$, we note

$$(21.4.22) \qquad \begin{aligned} \varphi^{-1}\mathcal{R}_N(\overline{\Delta}) &= [\Theta_{21}, \Theta_{22}]\mathcal{R}_{2N}(\overline{\Delta}) \\ &= \Theta_{22}[\Theta_{22}^{-1}\Theta_{21}, I]\mathcal{R}_{2N}(\overline{\Delta}). \end{aligned}$$

As we noted above, $\varphi\Theta_{22}$ is analytic in $\overline{\Delta}$, so $\text{wno}\det\varphi\Theta_{22}$ can be identified as follows:

$$\text{wno}\det\varphi\Theta_{22} = \dim\{\mathcal{R}_N(\overline{\Delta})/\varphi\Theta_{22}\mathcal{R}_N(\overline{\Delta})\}.$$

But

$$\begin{aligned} \dim\mathcal{R}_N(\overline{\Delta})/\varphi\Theta_{22}\mathcal{R}_N(\overline{\Delta}) &= \dim\varphi^{-1}\mathcal{R}_N(\overline{\Delta})/\Theta_{22}\mathcal{R}_N(\overline{\Delta}) \\ &= \dim\Theta_{22}[\Theta_{22}^{-1}\Theta_{21}, I]\mathcal{R}_{2N}(\overline{\Delta})/\Theta_{22}\mathcal{R}_N(\overline{\Delta}) \\ &\qquad (\text{where we used (21.4.22)}) \\ &= \dim[\Theta_{22}^{-1}\Theta_{21}, I]\mathcal{R}_{2N}(\overline{\Delta})/\mathcal{R}_N(\overline{\Delta}) \\ &= \dim P_{\mathcal{R}_N^0(\overline{\Delta}^c)}\{\Theta_{22}^{-1}\Theta_{21}\mathcal{R}_N(\overline{\Delta})\}. \end{aligned}$$

Now by (ii) $\Rightarrow$ (iv) in Theorem 13.2.3 we conclude that this last quantity is the number κ_Λ of negative eigenvalues of Λ. Thus we get

$$(21.4.23) \qquad \text{wno}\det\varphi\Theta_{22} = \kappa_\Lambda.$$

Combining (21.4.19) with (21.4.20), (21.4.21) and (21.4.23) gives $\kappa \ge \kappa_\Lambda$, as asserted. Moreover, in the case $\kappa = \kappa_\Lambda$, then (21.4.19) forces $\text{wno}\det G_2 = 0$ in which case $\det G_2$

can have no zeros on Δ, so $G = G_1 G_2^{-1}$ is analytic on $\overline{\Delta}$ (including infinity if appropriate). As $\|G(z)\| \leq 1$ for $z \in \partial\Delta$, the maximum modulus theorem implies that $\|G(z)\| \leq 1$ for all z in $\overline{\Delta}$.

Conversely, suppose Λ is invertible and has κ negative eigenvalues. Let G be any rational $N \times N$ matrix function with no poles in $\overline{\Delta}$ and with $\|G(z)\| \leq 1$ for z in $\overline{\Delta}$ for which $\det(\varphi[\Theta_{21}G + \Theta_{22}])$ does not vanish on $\sigma(A_0) \cup \sigma(A_\pi) \cup \sigma(A_\zeta)$. (We observe that the set of G's with these properties is generic. Indeed, $\varphi\Theta_{22}$ is analytic on $\overline{\Delta}$; so the condition that $\det(\varphi[\Theta_{21}G + \Theta_{22}])$ does not vanish on $\sigma(A_0) \cup \sigma(A_\pi) \cup \sigma(A_\zeta)$ can be expressed in the form

$$\mathrm{Im} \left[\begin{array}{c} G(z) \\ I \end{array} \right] \cap \mathrm{Ker}\left[\varphi(z)[\, \Theta_{21}(z) \quad \Theta_{22}(z) \,]\right] = \{0\}$$

for $z \in \sigma(A_0) \cup \sigma(A_\pi) \cup \sigma(A_\zeta)$. This condition can be in turn achieved by letting $G(z)$ be a solution of a certain interpolation problem over $\sigma(A_0) \cup \sigma(A_\pi) \cup \sigma(A_\zeta)$, and, moreover, the norm $\sup_{z \in \overline{\Delta}} \|G(z)\|$ can be made as small as we wish.) We have already observed (cf. the proof of Lemma 16.10.2) that $\varphi\Theta_{22}$ is analytic on $\overline{\Delta}$ and (see Proposition 13.1.2) that $\Theta_{22}^{-1}\Theta_{21}$ is analytic with norm at most 1 on $\partial\Delta$. We wish now to make the finer point that $\|[\Theta_{22}^{-1}\Theta_{21}](z)\| < 1$ for $z \in \partial\Delta \backslash \sigma(A_0)$; indeed this follows from the fact that $\Theta_{22}(z)$ is analytic and invertible on $\partial\Delta \backslash \sigma(A_0)$ together with the identity (for $z \in \partial\Delta \backslash \sigma(A_0)$)

$$\Theta_{22}(z)^{-1}\Theta_{21}(z)\Theta_{21}(z)^*\Theta_{22}(z)^{*-1} = I - \Theta_{22}(z)^{-1}\Theta_{22}(z)^{*-1},$$

a consequence of the identity (13.1.5). As $\|G(z)\| \leq 1$ for $z \in \partial\Delta$, we conclude $\|[\Theta_{22}^{-1}\Theta_{21}G](z)\| < 1$ for $z \in \partial\Delta \backslash \sigma(A_0)$, so $\Theta_{22}^{-1}\Theta_{21}G + I$ is invertible at least on $\partial\Delta \backslash \sigma(A_0)$. But we also know that $\varphi(\Theta_{21}G + \Theta_{22})$ is constrained to be analytic and invertible on $\sigma(A_0)$, where we have the factorization

$$\varphi(\Theta_{21}G + \Theta_{22}) = (\varphi\Theta_{22})(\Theta_{22}^{-1}\Theta_{21}G + I).$$

As each factor is analytic on $\partial\Delta$ and the product is analytic and invertible on $\sigma(A_0)$, we conclude that each factor in fact is also invertible on $\sigma(A_0)$. We have now established that $\Theta_{22}^{-1}\Theta_{21}G + I$ is invertible on all of $\partial\Delta$. Since both $\Theta_{22}^{-1}\Theta_{21}$ and G have norm at most 1 on $\partial\Delta$, then certainly $t\Theta_{22}^{-1}\Theta_{21}G + I$ is analytic and invertible on all of $\partial\Delta$ for $0 \leq t < 1$. Hence, by the usual homotopy argument,

$$\mathrm{wno}\det(\Theta_{22}^{-1}\Theta_{21}G + I) = \mathrm{wno}\det(I) = 0.$$

We now define $N \times N$ matrix functions F_1 and F_2 by

(21.4.24)
$$\left[\begin{array}{c} F_1 \\ F_2 \end{array} \right] = \Theta \left[\begin{array}{c} G \\ I \end{array} \right] = \left[\begin{array}{c} \Theta_{11}G + \Theta_{12} \\ \Theta_{21}G + \Theta_{22} \end{array} \right].$$

Since $\|G(z)\| \leq 1$ on $\overline{\Delta}$ and $\Theta(z)$ is J-unitary at all regular points of $\partial\Delta$, we see that

(21.4.25)
$$F_1(z)^* F_1(z) - F_2(z)^* F_2(z) = [\, G(z)^* \quad I \,]\Theta(z)^* J\Theta(z) \left[\begin{array}{c} G(z) \\ I \end{array} \right]$$

$$= G(z)^* G(z) - I \leq 0, \qquad z \in \partial\Delta.$$

From (21.4.24) we see that $\begin{bmatrix} F_1(z) \\ F_2(z) \end{bmatrix}$ has full rank N for all but finitely many $z \in \mathbb{C}$; (21.4.25) then forces $F_2(z)$ to have full rank for all but finitely many z, so $F = F_1 F_2^{-1}$ is well defined and has norm at most 1 on $\partial \Delta$. We already established that $\varphi \Theta_{22}$ is analytic and invertible on $\sigma(A_0)$. On $\partial \Delta \backslash \sigma(A_0)$, φ is analytic and invertible by construction, and Θ_{22} is analytic and invertible as a consequence of the identity (13.1.5) since $\Theta(z)$ is analytic with J-unitary values on $\partial \Delta \backslash \sigma(A_0)$. Thus $\varphi F_2 = \varphi \Theta_{22}(\Theta_{22}^{-1} \Theta_{21} G + I)$ is analytic and invertible on $\partial \Delta$ and therefore $\det(\varphi F_2)$ has a well defined winding number. From the factorization

$$\varphi F_2 = \varphi \Theta_{22}(\Theta_{22}^{-1} \Theta_{21} G + I)$$

we get

$$\text{wno} \det(\varphi F_2) = \text{wno} \det \varphi \Theta_{22} + \text{wno} \det(\Theta_{22}^{-1} \Theta_{21} G + I)$$
$$= \text{wno} \det \varphi \Theta_{22}.$$

In the first part of the proof we established that $\text{wno} \det \varphi \Theta_{22}$ is the number κ of negative eigenvalues of Λ. As F_2 is analytic we conclude that $\det(\varphi F_2) = \det \varphi (\Theta_{21} G + \Theta_{22})$ has κ zeros in $\overline{\Delta}$. From the representation

$$F = (\Theta_{11} G + \Theta_{12})(\Theta_{21} G + \Theta_{22})^{-1}$$

and Theorem 16.11.1, it follows that F satisfies the interpolation conditions (21.4.2)–(21.4.3) and has at most κ poles (counting multiplicities) in $\overline{\Delta}$. $\square$

CHAPTER 22

CARATHEODORY-TOEPLITZ INTERPOLATION

In this brief chapter we indicate how the techniques of the previous chapters used for generalized Nevanlinna-Pick interpolation problems can also be used to handle the classical problem of Caratheodory-Toeplitz and its generalizations. The problem here is to find all functions (scalar or matrix valued) analytic on the right half plane or on the unit disk with positive definite real part there which meet some additional interpolation conditions.

22.1 THE CLASSICAL SCALAR CASE

In the classical Caratheodory-Toeplitz problem, we are given $n + 1$ complex numbers $a_0, a_1, \ldots, a_n$, and seek a complex-valued function $f(z)$ analytic on the unit disk of the form

$$f(z) = a_0 + a_1 z + \cdots + a_n z^n + O(|z|^{n+1})$$

such that $f(z)$ has positive real part on the unit disk $\mathcal{D}$

$$\operatorname{Re} f(z) \geq 0 \quad \text{for} \quad |z| < 1.$$

The classical solution is that such a function exists if and only if the Hermitian Toeplitz matrix

$$T(a_0, a_1, \ldots, a_n) = \begin{bmatrix} a_0 + \overline{a}_0 & \overline{a}_1 & \cdots & \overline{a}_n \\ a_1 & a_0 + \overline{a}_0 & \cdots & \overline{a}_{n-1} \\ a_2 & a_1 & & \vdots \\ \vdots & \vdots & \ddots & \\ a_n & a_{n-1} \cdots & & a_0 + \overline{a}_0 \end{bmatrix}$$

is positive definite. Moreover one can obtain a linear fractional description of all such functions.

More generally, one could demand that f satisfy interpolation conditions at various other points $z_1, \ldots, z_k$ in the unit disk, e.g.

$$\frac{1}{j!} f^{[j]}(z_i) = c_{ij}, \qquad 1 \leq i \leq k, \ 0 \leq j \leq m_i$$

and look for such f which are analytic and have positive real part values on the unit disk.

We state our solution of the problem formally as follows. Matrix generalizations will be considered in the following section.

THEOREM 22.1.1. *Let $n + 1$ complex numbers $a_0, a_1, \ldots, a_n$ be given. In order that there exists a scalar rational function $f(z)$ analytic on the closed unit disk such that*

(i) $\operatorname{Re} f(z) > 0$ *for* $|z| < 1$

and

(ii) $\frac{1}{j!} f^{[j]}(0) = a_j$ *for* $j = 0, 1, \ldots n$

it is necessary and sufficient that the Hermitian Toeplitz matrix

$$(22.1.1) \qquad \Lambda = \begin{bmatrix} a_0 + \overline{a}_0 & a_1 & \cdots & a_n \\ \overline{a}_1 & & & \\ \vdots & \ddots & \ddots & \vdots \\ & & & a_1 \\ \overline{a}_n & \cdots & \overline{a}_1 & a_0 + \overline{a}_0 \end{bmatrix}$$

be positive definite. Moreover, in this case the set of all such $f(z)$ consists of those functions $f(z)$ with a representation of the form

$$f(z) = \big(\Theta_{11}(z)g(z) + \Theta_{12}(z)\big)\big(\Theta_{21}(z)g(z) + \Theta_{22}(z)\big)^{-1}$$

where g is any rational function analytic on the closed unit disk with positive real part there. Here $\Theta(z) = \begin{bmatrix} \Theta_{11}(z) & \Theta_{12}(z) \\ \Theta_{21}(z) & \Theta_{22}(z) \end{bmatrix}$ is given by

$$(22.1.2)$$

$$\Theta(z) = \begin{bmatrix} 1 & 0 \\ 0 & 1 \end{bmatrix} - (z-1) \begin{bmatrix} a_0 & a_1 & \cdots & a_n \\ 1 & 0 & \cdots & 0 \end{bmatrix} \begin{bmatrix} z^{-1} & z^{-2} & \cdots & z^{-(n+1)} \\ 0 & z^{-1} & \ddots & \\ \vdots & \vdots & \ddots & \vdots \\ & & & z^{-2} \\ 0 & 0 & \cdots & z^{-1} \end{bmatrix}$$

$$\times \Lambda^{-1} \begin{bmatrix} 1 & \overline{a}_0 \\ 1 & \overline{a}_0 + \overline{a}_1 \\ \vdots & \vdots \\ 0 & \overline{a}_0 + \overline{a}_1 + \cdots + \overline{a}_n \end{bmatrix}.$$

A similar solution can be obtained for the problem involving first order interpolation conditions at a collection of points in the disk.

THEOREM 22.1.2. *Suppose k distinct points $z_1, \ldots, z_k$ in the unit disk and k complex numbers $w_1, \ldots, w_k$ are specified. In order that there exist a scalar rational function $f(z)$ which is analytic with positive real part on the closed unit disk and which satisfies the interpolation conditions*

$$f(z_j) = w_j \quad \text{for} \quad j = 1, \ldots, k$$

it is necessary and sufficient that the associated Caratheodory-Pick matrix

$$(22.1.3) \qquad \Lambda = \left[\frac{\overline{w}_i + w_j}{1 - \overline{z}_i z_j} \right]_{1 \le i, j \le k}$$

be positive definite. Moreover in this case the set of all such functions consists of all f having a representation

$$f = (\Theta_{11}g + \Theta_{12})(\Theta_{21}g + \Theta_{22})^{-1}$$

with g an arbitrary rational function analytic with positive real part on the closed unit disk. Here $\Theta(z) = \begin{bmatrix} \Theta_{11}(z) & \Theta_{12}(z) \\ \Theta_{21}(z) & \Theta_{22}(z) \end{bmatrix}$ is the 2×2 matrix function given by

(22.1.4)

$$\Theta(z) = \begin{bmatrix} 1 & 0 \\ 0 & 1 \end{bmatrix} + (z-1)^{-1} \begin{bmatrix} w_1 \cdots w_k \\ 1 \cdots 1 \end{bmatrix} \begin{bmatrix} (z-z_1)^{-1} & & 0 \\ & \ddots & \\ 0 & & (z-z_k)^{-1} \end{bmatrix}$$

$$\times \Lambda^{-1} \begin{bmatrix} (\bar{z}_1-1)^{-1} & (\bar{z}_1-1)^{-1}\overline{w}_1 \\ (\bar{z}_2-1)^{-1} & (\bar{z}_2-1)^{-1}\overline{w}_2 \\ \vdots & \vdots \\ (\bar{z}_k-1)^{-1} & (\bar{z}_k-1)^{-1}\overline{w}_k \end{bmatrix}$$

A result completely analogous to Theorem 22.1.2 holds if the underlying domain is taken to be the right half plane rather than the unit disk.

THEOREM 22.1.3. *Suppose k distinct points $z_1, \ldots, z_k$ in the right half plane and k complex numbers $w_1, \ldots, w_k$ are specified. In order that there exists a scalar rational function $f(z)$ which is analytic with positive real part on the closed right half plane and which satisfies the interpolation conditions*

$$f(z_j) = w_j \quad \text{for} \quad j = 1, \ldots, k$$

it is necessary and sufficient that the associated Caratheodory-Pick matrix

(22.1.5)
$$\Lambda = \left[\frac{\overline{w}_i + w_j}{\overline{z}_i + z_j} \right]$$

be positive definite. Moreover in this case the set of all such functions consists of all f having representations

$$f = (\Theta_{11}g + \Theta_{12})(\Theta_{21}g + \Theta_{22})^{-1}$$

with g an arbitrary rational function analytic with positive real part on the closed right half plane. Here $\Theta(z) = \begin{bmatrix} \Theta_{11}(z) & \Theta_{12}(z) \\ \Theta_{21}(z) & \Theta_{22}(z) \end{bmatrix}$ is the 2×2 matrix function given by

$$\Theta(z) = \begin{bmatrix} 1 & 0 \\ 0 & 1 \end{bmatrix} + \begin{bmatrix} w_1 \cdots w_k \\ 1 \cdots 1 \end{bmatrix} \begin{bmatrix} (z-z_1)^{-1} & & 0 \\ & \ddots & \\ 0 & & (z-z_k)^{-1} \end{bmatrix}$$

(22.1.6)
$$\times \Lambda^{-1} \begin{bmatrix} 1 & \overline{w}_1 \\ 1 & \overline{w}_2 \\ \vdots & \vdots \\ 1 & \overline{w}_k \end{bmatrix}.$$

22.2   MATRIX CARATHEODORY-TOEPLITZ INTERPOLATION PROBLEMS

In this section we consider the problem of constructing a $N \times N$ matrix function $F(z)$ defined and analytic on the region $\overline{\Delta}$ such that $\operatorname{Re} F(z) = \frac{1}{2}[F(z) + F(z)^*]$ is positive definite on $\overline{\Delta}$ and F satisfies a TSCII problem on $\overline{\Delta}$. The formal result for Δ equal to the unit disk $\mathcal{D}$ is as follows.

THEOREM 22.2.1. *Suppose that a $\mathcal{D}$-admissible Sylvester TSCII data set $\omega = (C_+, C_-, A_\pi; A_\zeta, B_+, B_-; \Gamma)$ is given where C_- and C_+ have N rows and B_+ and B_- have N columns. Define matrices Λ_1 and Λ_2 as the unique solutions of the Stein equations*

$$(22.2.1) \qquad \Lambda_1 - A_\pi^* \Lambda_1 A_\pi = C_+^* C_- + C_-^* C_+$$

and

$$\Lambda_2 - A_\zeta \Lambda_2 A_\zeta^* = -B_+ B_-^* - B_- B_+^*$$

and let Λ be the 2×2 block matrix

$$(22.2.2) \qquad \Lambda = \begin{bmatrix} \Lambda_1 & \Gamma^* \\ \Gamma & \Lambda_2 \end{bmatrix}$$

Then there exists a rational $N \times N$ matrix function $F(z)$ analytic with positive definite real part on $\mathcal{D}$ which satisfies the associated TSCII conditions

$$(22.2.3) \qquad \sum_{z_0 \in \mathcal{D}} \operatorname{Res}_{z=z_0} F(z) C_- (zI - A_\pi)^{-1} = C_+$$

$$(22.2.4) \qquad \sum_{z_0 \in \mathcal{D}} \operatorname{Res}_{z=z_0} (zI - A_\zeta)^{-1} B_+ F(z) = -B_-$$

and

$$(22.2.5) \qquad \sum_{z_0 \in \mathcal{D}} \operatorname{Res}_{z=z_0} (zI - A_\zeta)^{-1} B_+ F(z) C_- (zI - A_\pi)^{-1} = \Gamma$$

if and only if the matrix Λ given by (22.2.2) is positive definite. Moreover, in this case the set of all such matrix function $F(z)$ is given by

$$(22.2.6) \qquad F = (\Theta_{11} G + \Theta_{12})(\Theta_{21} G + \Theta_{22})^{-1}$$

where G is an arbitrary rational $N \times N$ matrix function analytic and with positive definite real part on $\overline{\mathcal{D}}$. Here $\Theta(z) = \begin{bmatrix} \Theta_{11}(z) & \Theta_{12}(z) \\ \Theta_{21}(z) & \Theta_{22}(z) \end{bmatrix}$ is given by

$$(22.2.7) \qquad \begin{aligned} \Theta(z) = {} & I_{2N} + (z-1) \begin{bmatrix} C_+ & B_-^* \\ C_- & B_+^* \end{bmatrix} \begin{bmatrix} (zI - A_\pi)^{-1} & 0 \\ 0 & (I - zA_\zeta^*)^{-1} \end{bmatrix} \\ & \times \Lambda^{-1} \begin{bmatrix} (A_\pi^* - I)^{-1} C_-^* & (A_\pi^* - I)^{-1} C_+^* \\ (A_\zeta - I)^{-1} B_+ & (A_\zeta - I)^{-1} B_- \end{bmatrix}. \end{aligned}$$

Before commencing with the proof of Theorem 22.2.1, we consider some simple examples.

EXAMPLE 22.2.1. Take $N = 1$, A_ζ, B_+, B_- and Γ are vacuous, and

(22.2.8a)
$$C_- = [1, 0, \ldots, 0], \qquad C_+ = [a_0, \ldots, a_n]$$

(22.2.8b)
$$A_\pi = \begin{bmatrix} 0 & 1 & & \\ & 0 & \ddots & \\ & & \ddots & 1 \\ & & & 0 \end{bmatrix}$$

(where A_π is $(n+1) \times (n+1)$). Then the interpolation conditions (22.2.3)–(22.2.5) reduce to

$$\frac{1}{j!} f^{[j]}(0) = a_j, \qquad j = 0, 1, \ldots, n.$$

The matrix Λ collapses to Λ_1 and so is given by the Stein equation (22.2.1). For C_- and C_+ as in (22.2.8a) we see that $\Omega := C_+^* C_- + C_-^* C_+$ has the form

$$\Omega = \begin{bmatrix} \bar{a}_0 + a_0 & a_1 & \cdots & a_n \\ \bar{a}_1 & 0 & \cdots & 0 \\ \vdots & \vdots & & \vdots \\ \bar{a}_n & 0 & \cdots & 0 \end{bmatrix}.$$

From formula (A.2.8) or by direct verification we see that the solution $\Lambda = \Lambda_1$ of (22.2.1) for this example is

$$\Lambda = \begin{bmatrix} \bar{a}_0 + a_0 & a_1 & \cdots & a_n \\ \bar{a}_1 & \bar{a}_0 + a_0 & & \\ \vdots & & \ddots & \\ \bar{a}_n & \cdots & & \bar{a}_0 + a_0 \end{bmatrix},$$

in agreement with formula (22.1.1) in Theorem 22.1.1. We leave it for the reader to check that the formula (22.1.2) for $\Theta(z)$ is a direct specialization of the general formula (22.2.7). □

EXAMPLE 22.2.2. Again take $N = 1$ and A_ζ, B_+, B_-, Γ vacuous, but in this case take

$$C_- = [1, 1, \ldots, 1], \qquad C_+ = [w_1, w_2, \ldots, w_k]$$

$$A_\pi = \begin{bmatrix} z_1 & & 0 \\ & \ddots & \\ 0 & & z_k \end{bmatrix}$$

with $z_1, \ldots, z_k$ in $\mathcal{D}$. Then the interpolation conditions (22.2.3)–(22.2.5) collapse to

$$f(z_j) = w_j \quad \text{for} \quad j = 1, \ldots, k$$

and Λ again collapses to the solution $\Lambda = \Lambda_1$ of the Stein equation (22.2.1). It is straightforward to see that the solution is the Caratheodory-Pick matrix (22.1.3) given in Theorem 22.1.2 and that the formula (22.1.4) for $\Theta(z)$ in Theorem 22.1.2 results upon specializing the general formula (22.2.7) in Theorem 22.2.1. □

The result parallel to Theorem 22.2.1 for the case where the underlying domain Δ is taken to be the right half plane Π^+ rather than the unit disk $\mathcal{D}$ is as follows.

THEOREM 22.2.2. *Suppose that a Π^+-admissible Sylvester TSCII data set $\omega = (C_+, C_-, A_\pi; A_\zeta, B_+, B_-; \Gamma)$ is given (where C_- and C_+ have N rows, B_+ and B_- have N columns). Define matrices Λ_1 and Λ_2 as the unique solutions of the respective Lyapunov equations*

(22.2.9) $$\Lambda_1 A_\pi + A_\pi^* \Lambda_1 = C_+^* C_- + C_-^* C_+$$

and

(22.2.10) $$\Lambda_2 A_\zeta^* + A_\zeta \Lambda_2 = -B_+ B_-^* - B_- B_+^*$$

and set Λ equal to

(22.2.11) $$\Lambda = \begin{bmatrix} \Lambda_1 & \Gamma^* \\ \Gamma & \Lambda_2 \end{bmatrix}.$$

Then there exists a rational $N \times N$ matrix function $F(z)$ analytic with positive definite real part on $\overline{\Pi^+}$ which satisfies the associated TSCII conditions

(22.2.12) $$\sum_{z_0 \in \Pi^+} \mathrm{Res}_{z=z_0} F(z) C_- (zI - A_\pi)^{-1} = C_+$$

(22.2.13) $$\sum_{z_0 \in \Pi^+} \mathrm{Res}_{z=z_0} (zI - A_\zeta)^{-1} B_+ F(z) = -B_-$$

(22.2.14) $$\sum_{z_0 \in \Pi^+} \mathrm{Res}_{z=z_0} (zI - A_\zeta)^{-1} B_+ F(z) C_- (zI - A_\pi)^{-1} = \Gamma$$

if and only if the matrix Λ given by (22.2.11) is positive definite. Moreover, in this case the set of all such matrix functions $F(z)$ is given by

(22.2.15) $$F = (\Theta_{11} G + \Theta_{12})(\Theta_{21} G + \Theta_{22})^{-1}$$

where G is an arbitrary rational $N \times N$ matrix function analytic and with positive definite real part on $\overline{\Pi^+}$. Here $\Theta(z) = \begin{bmatrix} \Theta_{11}(z) & \Theta_{12}(z) \\ \Theta_{21}(z) & \Theta_{22}(z) \end{bmatrix}$ is given by

(22.2.16)
$$\Theta(z) = I_{2N} + \begin{bmatrix} C_+ & B_-^* \\ C_- & B_+^* \end{bmatrix} \begin{bmatrix} (zI - A_\pi)^{-1} & 0 \\ 0 & (zI + A_\zeta^*)^{-1} \end{bmatrix} \Lambda^{-1} \begin{bmatrix} C_-^* & C_+^* \\ B_+ & B_- \end{bmatrix}.$$

We next illustrate Theorem 22.2.2 with a simple example.

EXAMPLE 22.2.3. We take $N = 1$, A_ζ, B_+, B_-, Γ vacuous, and

$$C_- = [1, 1, \ldots, 1], \qquad C_+ = [w_1, w_2, \ldots, w_k]$$

and

$$A_\pi = \begin{bmatrix} z_1 & & 0 \\ & \ddots & \\ 0 & & z_k \end{bmatrix}$$

as in Example 22.2.2, but now with the points $z_1, \ldots, z_k$ in the right half plane. The interpolation conditions (22.2.12)–(22.2.14) collapse to

$$f(z_j) = w_j \quad \text{for} \quad j = 1, \ldots, k$$

and Λ as in (22.2.11) collapses to Λ_1 given by the Lyapunov equation (22.2.9). The solution is easily seen to be the Caratheodory-Pick matrix Λ given by (22.1.5) in Theorem 22.1.3. The general formula (22.2.16) for $\Theta(z)$ in Theorem 22.2.2 collapses to the formula (22.1.6) in Theorem 22.1.3 as a special case of Theorem 22.2.2. □

22.3 PROOFS OF THE MAIN RESULTS

In this section we give complete proofs of Theorems 22.2.1 and 22.2.2. The idea is to study an equivalent two-sided Nudelman interpolation problem for $S = (F - I)(F + I)^{-1}$ in place of F and simply quote the results concerning this problem already derived in Chapter 18.

It is an elementary fact that if X is a matrix for which $X + X^*$ is positive definite, then $X + I$ is invertible and $Y = (X - I)(X + I)^{-1}$ has $\|Y\| < 1$; indeed this can be seen from

$$I - Y^*Y = (X^* + I)^{-1}[(X^* + I)(X + I) - (X^* - I)(X - I)](X + I)^{-1}$$
$$= 2(X^* + I)^{-1}[X^* + X](X + I)^{-1}.$$

Conversely, if Y is a strict contraction, then $X = (Y + I)(-Y + I)^{-1}$ has positive definite real part and we recover Y as $Y = (X - I)(X + I)^{-1}$. We wish to apply this transformation with $X = F(z)$ taken to be a rational matrix function with values having positive definite real part over a region Δ. The following lemma indicates how TSCII conditions transform under this mapping.

LEMMA 22.3.1. *Suppose $(C_+, C_-, A_\pi; A_\zeta, B_+, B_-; \Gamma)$ is a Δ-admissible Sylvester TSCII data set and F is a rational $N \times N$ matrix function analytic on Δ for which $\det(I + F(z))$ has no zeros in Δ. Then F satisfies the interpolation conditions*

(22.3.1a)
$$\sum_{z_0 \in \Delta} \text{Res}_{z=z_0} F(z)C_-(zI - A_\pi)^{-1} = C_+$$

(22.3.1b)
$$\sum_{z_0 \in \Delta} \text{Res}_{z=z_0}(zI - A_\zeta)^{-1}B_+F(z) = -B_-$$

$$(22.3.1c) \qquad \sum_{z_0 \in \Delta} \text{Res}_{z=z_0}(zI - A_\zeta)^{-1}B_+F(z)C_-(zI - A_\pi)^{-1} = \Gamma$$

if and only if $S(z) = (F(z) - I)(F(z) + I)^{-1}$ *satisfies the interpolation conditions*

$$(22.3.2a) \qquad \sum_{z_0 \in \Delta} \text{Res}_{z=z_0} S(z)(C_+ + C_-)(zI - A_\pi)^{-1} = -C_+ + C_-$$

$$(22.3.2b) \qquad \sum_{z_0 \in \Delta} \text{Res}_{z=z_0}(zI - A_\zeta)^{-1}(B_+ - B_-)S(z) = -B_+ - B_-$$

$$(22.3.2c) \quad \sum_{z_0 \in \Delta} \text{Res}_{z=z_0}(zI - A_\zeta)^{-1}(B_+ - B_-)S(z)(C_+ + C_-)(zI - A_\pi)^{-1} = 2\Gamma.$$

PROOF. By Lemma 16.9.4 the interpolation condition (22.3.1a) being satisfied by a rational matrix function $F(z)$ analytic on Δ is equivalent to

$$T(z) = F(z)C_-(zI - A_\pi)^{-1} - C_+(zI - A_\pi)^{-1}$$

being analytic on Δ. Trivially, we may write $T(z)$ in the form

$$T(z) = (F(z) + I)C_-(zI - A_\pi)^{-1} - (C_+ + C_-)(zI - A_\pi)^{-1}.$$

By assumption $F(z) + I$ is analytic and invertible on Δ. Hence T is analytic on Δ if and only if the function

$$-(F(z) + I)^{-1}T(z) = (F(z) + I)^{-1}(C_+ + C_-)(zI - A_\pi)^{-1} - C_-(zI - A_\pi)^{-1}$$

has no poles on Δ. Since $F(z) - I$ is analytic on Δ, in this case we also have

$$(F(z) - I)(F(z) + I)^{-1}T(z) = -S(z)(C_+ + C_-)(zI - A_\pi)^{-1} + (I - F(z))C_-(zI - A_\pi)^{-1}$$

is analytic on Δ. Now use that $F(z)C_-(zI - A_\pi)^{-1} - C_+(zI - A_\pi)^{-1}$ is analytic on Δ to conclude that

$$S(z)(C_+ + C_-)(zI - A_\pi)^{-1} + (C_+ - C_-)(zI - A_\pi)^{-1}$$

is analytic on Δ. This last condition is equivalent to the interpolation condition (22.3.2a). We have thus shown that (22.3.1a) implies (22.3.2a). Conversely, one can show that (22.3.2a) leads to (22.3.1a) in a similar way by using that $F(z) = (S(z) + I)(-S(z) + I)^{-1}$.

The equivalence between (22.3.1b) and (22.3.2b) follows in a similar way; in this case one uses that the interpolation condition (22.3.1b) is equivalent to the analyticity of $\widetilde{T}(z) = (zI - A_\zeta)^{-1}B_+F(z) + (zI - A_\zeta)^{-1}B_-$ on Δ.

To deduce (22.3.2c) from (22.3.1a)–(22.3.1c) we proceed as follows. By Lemma 16.10.4, $\sum_{z_0 \in \Delta} \text{Res}_{z=z_0}(zI - A_\zeta)^{-1}B_-C_-(zI - A_\pi)^{-1} = 0$ and hence (22.3.1c) is equivalent to

$$(22.3.3) \qquad \sum_{z_0 \in \Delta} \text{Res}_{z=z_0}(zI - A_\zeta)^{-1}[B_+F(z) + B_-]C_-(zI - A_\pi)^{-1} = \Gamma.$$

Again by Lemma 16.9.4 the interpolation condition (22.3.1b) is equivalent to the analyticity of the matrix function

$$(22.3.4) \qquad W(z) := (zI - A_\zeta)^{-1}[B_+F(z) + B_-]$$

on Δ. A reapplication of the same lemma to (22.3.3) then gives the equivalence of (22.3.3) to the analyticity of

$$(22.3.5) \qquad V(z) := (zI - A_\zeta)^{-1}[B_+F(z) + B_-]C_-(zI - A_\pi)^{-1} - \Gamma(zI - A_\pi)^{-1}$$

on Δ. As we observed above the interpolation condition (22.3.1a) is equivalent to the analyticity of $U(z) = F(z)C_-(zI - A_\pi)^{-1} - C_+(zI - A_\pi)^{-1}$ on Δ. Rewriting $U(z)$ as

$$U(z) := \big(F(z) + I\big)C_-(zI - A_\pi)^{-1} - (C_+ + C_-)(zI - A_\pi)^{-1}$$

and using the analyticity of $\big(F(z) + I\big)^{-1}$ on Δ, we get that

$$(22.3.6) \quad \big(F(z) + I\big)^{-1}U(z) = C_-(zI - A_\pi)^{-1} - \big(F(z) + I\big)^{-1}(C_+ + C_-)(zI - A_\pi)^{-1}$$

is analytic on Δ. From (22.3.4) and (22.3.5) we see that

$$\begin{aligned} X(z) &:= V(z) - W(z)\big(F(z) + I\big)^{-1}U(z) \\ &= (zI - A_\zeta)^{-1}[B_+F(z) + B_-]\big(F(z) + I\big)^{-1} \\ &\quad \times (C_+ + C_-)(zI - A_\pi)^{-1} - \Gamma(zI - A_\pi)^{-1} \end{aligned}$$

is analytic on Δ. Next use the identity

$$2[B_+F(z) + B_-] = (B_+ - B_-)\big(F(z) - I\big) + (B_+ + B_-)\big(F(z) + I\big)$$

to see that

$$(22.3.7) \quad \begin{aligned} 2X(z) &= (zI - A_\zeta)^{-1}\big[(B_+ - B_-)\big(F(z) - I\big) + (B_+ + B_-)\big(F(z) + I\big)\big] \\ &\quad \times \big(F(z) + I\big)^{-1}(C_+ + C_-)(zI - A_\pi)^{-1} - 2\Gamma(zI - A_\pi)^{-1} \\ &= (zI - A_\zeta)^{-1}[(B_+ - B_-)S(z) + (B_+ + B_-)] \\ &\quad \times (C_+ + C_-)(zI - A_\pi)^{-1} - 2\Gamma(zI - A_\pi)^{-1}. \end{aligned}$$

The analyticity of $X(z)$ over Δ gives that

$$\sum_{z_0 \in \Delta} \text{Res}_{z=z_0} 2X(z) = 0.$$

But also by Lemma 16.10.4 again,

$$\sum_{z_0 \in \Delta} \text{Res}_{z=z_0}(zI - A_\zeta)^{-1}(B_+ + B_-)(C_+ + C_-)(zI - A_\pi)^{-1} = 0$$

and by the Riesz functional calculus

$$\sum_{z_0 \in \Delta} \text{Res}_{z=z_0} 2\Gamma(zI - A_\pi)^{-1} = 2\Gamma.$$

These facts combined with the formula (22.3.7) for $X(z)$ lead to the interpolation condition (22.3.2c) being satisfied by $S(z)$ as claimed. The converse direction (that (22.3.2a)–(22.3.2c) implies (22.3.1c)) follows in a similar way from the connection $F(z) = (S(z) + I)(-S(z) + I)^{-1}$. $\square$

In our procedure for reducing nonhomogeneous interpolation conditions to homogeneous problems as discussed already in Chapter 16, we consider the homogeneous interpolation conditions on the $2N \times N$ matrix function $\begin{bmatrix} F(z) \\ I_N \end{bmatrix}$ rather than nonhomogeneous interpolation conditions on the $N \times N$ matrix function $F(z)$ itself. As we are now interested in interpolants F with positive real part, it is now appropriate to note that the positive real part condition on F can be expressed in terms of $\begin{bmatrix} F \\ I \end{bmatrix}$ as

$$0 > -F^* - F = \begin{bmatrix} F^* & I \end{bmatrix} \begin{bmatrix} 0 & -I \\ -I & 0 \end{bmatrix} \begin{bmatrix} F \\ I \end{bmatrix}.$$

If we follow the idea of the proof of Theorems 18.5.1 and 18.5.2, we expect the linear fractional parameterizer Θ for the set of all solutions in Theorems 22.2.1 and 22.2.2 to be a rational $2N \times 2N$ matrix function satisfying the following conditions:

(22.3.8) Θ has $\left(\begin{bmatrix} C_+ \\ C_- \end{bmatrix}, A_\pi; A_\zeta, [B_+, B_-]; \Gamma \right)$ as a null-pole triple over $\overline{\Delta}$

(22.3.9) $\Theta(z) J_1 \Theta(z) = J_1$ for $z \in \partial \Delta$

(22.3.10) $\Theta(z) J_1 \Theta(z) \leq J_1$ for $z \in \Delta$

where we have set

$$J_1 = \begin{bmatrix} 0 & -I_N \\ -I_N & 0 \end{bmatrix}.$$

That this does give the correct result is essentially the content of Theorems 22.2.1 and 22.2.2. Indeed the formula (22.2.7) for $\Theta(z)$ in Theorem 22.2.1 is just the formula (7.4.7) for a $\Theta(z)$ in Theorem 7.4.2 which satisfies (22.3.8) and (22.3.9) with $\Delta = \mathcal{D}$ and $J_1 = \begin{bmatrix} 0 & -I_N \\ -I_N & 0 \end{bmatrix}$. The condition that Λ as in (22.2.2) be invertible guarantees that such a $\Theta(z)$ exists and the positive definiteness of Λ is equivalent to the validity of (22.3.10) (by Theorem 7.4.3). Similarly the formula (22.2.16) for $\Theta(z)$ in Theorem 22.2.2 arises in an analogous way from the formula (6.3.4) in Theorem 6.3.1 for a $\Theta(z)$ satisfying (22.3.8) and (22.3.9), but now with $\Delta = \Pi^+$. Again invertibility of Λ as in (22.2.11) is equivalent to the existence of such a $\Theta(z)$ and positive definiteness of Λ is equivalent to the validity of (22.3.10). Thus Theorems 22.2.1 and 22.2.2 will be completely proved once we show that the existence of solutions F of (22.3.1a)–(22.3.1c) with positive definite real part over $\overline{\Delta}$ is equivalent to the existence of a rational $2N \times 2N$ matrix function $\Theta(z)$ satisfying (22.3.8)–(22.3.10) and the verification that such a $\Theta(z)$ (when it exists) parametrizes the set of all such $F(z)$.

To reduce the proof of Theorems 22.2.1 and 22.2.2 in an efficient manner to the results already presented in Chapter 18, we simply need to establish the connection between J_1-inner matrix functions and J-inner matrix functions, where $J_1 = \begin{bmatrix} 0 & -I_N \\ -I_N & 0 \end{bmatrix}$ and $J = \begin{bmatrix} I_N & 0 \\ 0 & -I_N \end{bmatrix}$, including the correspondence between their Δ-null-pole triples. Here we say that Θ is a J_1-*inner* function (with respect to Δ equal to either Π^+ or $\mathcal{D}$) if Θ satisfies (22.3.9) and (22.3.10), and an analogous definition applies for the J-inner functions. The following lemma establishes the needed connections.

LEMMA 22.3.2. *The $2N \times 2N$ matrix function $\Theta(z)$ is J_1-inner if and only if the matrix function $T^*\Theta(z)T$ is J-inner, where T is the unitary matrix*

$$T = \frac{1}{\sqrt{2}} \begin{bmatrix} I_N & I_N \\ -I_N & I_N \end{bmatrix}.$$

PROOF. This is an easy consequence of the congruence

$$J_1 = TJT^*$$

between J_1 and J. □

LEMMA 22.3.3. *Let T be an invertible matrix. The Δ-admissible Sylvester data set $\omega = (C, A_\pi; A_\zeta, B; \Gamma)$ is a Δ-null-pole triple for the rational matrix function $\Theta(z)$ if and only if*

$$\widetilde{\omega} = (T^{-1}C, A_\pi; A_\zeta, BT; \Gamma)$$

is a Δ-null-pole triple for $\Psi(z) = T^{-1}\Theta(z)T$.

PROOF. The proof is straightforward if one uses the characterization of Δ-null-pole triple for a given function $\Theta(z)$ in terms of null-pole subspaces given by Theorem 12.3.1. Thus, $\omega = (C, A_\pi; A_\zeta, B; \Gamma)$ is a Δ-null-pole triple for $\Theta(z)$ means that the null-pole subspace over Δ (i.e. the subspace $\Theta \mathcal{R}_M(\Delta)$ where $M \times M$ is the size of Θ) is expressible in the form

(22.3.11)
$$\Theta \mathcal{R}_M(\Delta) = \{C(zI - A_\pi)^{-1}x + h(z) : x \in \mathbb{C}^{n_\pi}, \ h \in \mathcal{R}_M(\Delta) \text{ such that} \\ \sum_{z_0 \in \Delta} \operatorname{Res}_{z=z_0}(zI - A_\zeta)^{-1}Bh(z) = \Gamma x\}.$$

If $\Psi(z) = T^{-1}\Theta(z)T$, then it is clear that

$$\Psi\mathcal{R}_M(\Delta) = T^{-1}\Theta\mathcal{R}_M(\Delta).$$

On the other hand, applying T^{-1} to both sides of (22.3.11) and setting $\tilde{h} = T^{-1}h$, we get

$$T^{-1}\Theta\mathcal{R}_M(\Delta) = \{T^{-1}C(zI - A_\pi)^{-1}x + \tilde{h}(z): \ x \in \mathbf{C}^{n_\pi} \text{ and } \tilde{h} \in \mathcal{R}_M(\Delta) \text{ such that}$$
$$\sum_{z_0 \in \Delta} \text{Res}_{z=z_0}(zI - A_\zeta)^{-1}BT\tilde{h}(z) = \Gamma x\}.$$

Thus $\Psi\mathcal{R}_M(\Delta) = T^{-1}\Theta\mathcal{R}_M(\Delta)$ is the null-pole subspace associated with the Δ-admissible Sylvester data set $\tilde{\omega} = (T^{-1}C, A_\pi; A_\zeta, BT; \Gamma)$. Combining Theorems 12.2.4 and 12.3.1 gives that $\tilde{\omega}$ is similar to (and hence is equal to) a null pole triple over Δ for $\Psi(z)$. The converse assertion follows by interchanging the roles of Ψ and Θ. $\square$

We are now ready to complete the proofs of Theorems 22.3.1 and 22.3.2. By Lemma 22.3.1, there exists a rational $N \times N$ matrix function $F(z)$ with positive definite imaginary part on $\overline{\Delta}$ which satisfies the interpolation conditions (22.3.1a)–(22.3.1c) on Δ if and only if there exists an $N \times N$ matrix function $S(z)$ with norm strictly less than 1 on $\overline{\Delta}$ which satisfies the interpolation conditions (22.3.2a)–(22.3.2c) over $\overline{\Delta}$. By Theorems 18.5.1 and 18.5.2 such a matrix function $S(z)$ exists if and only if there exists a rational $2N \times 2N$ matrix function $\Psi(z)$ such that

$$(22.3.12) \qquad \tau = \left(\begin{bmatrix} -C_+ + C_- \\ C_+ + C_- \end{bmatrix}, A_\pi; A_\zeta; [B_+ - B_-, B_+ + B_-]; 2\Gamma \right)$$
$$\text{is a } \Delta\text{-null-pole triple for } \Psi(z)$$

and

$$(22.3.13) \qquad\qquad \Psi(z) \text{ is } J\text{-inner.}$$

One easily sees that the Sylvester data set τ appearing in (22.3.12) is similar to the Sylvester data set

$$(22.3.14) \qquad \tilde{\omega} = \left(T^{-1} \begin{bmatrix} -C_+ \\ C_- \end{bmatrix}, A_\pi; A_\zeta, [B_+, B_-]T; \Gamma \right)$$

where $T = \frac{1}{\sqrt{2}} \begin{bmatrix} I_N & I_N \\ -I_N & I_N \end{bmatrix} = T^{*-1}$ is as in Lemma 22.3.2. Thus Lemmas 22.3.2 and 22.3.3 combine to say that the existence of a matrix function $\Psi(z)$ satisfying (22.3.12) and (22.3.13) is equivalent to the existence of a $\Theta(z)$ which satisfies (22.3.8), (22.3.9) and (22.3.10). By the discussion immediately preceding Lemma 22.3.2, it remains only to show that such a matrix function satisfying (22.3.8)–(22.3.10) parametrizes the set of all solutions of (22.3.1a)–(22.3.1c) which have positive definite real part in $\overline{\Delta}$.

This last issue is settled as follows. From the results of Chapter 18 (see Theorems 18.5.1 and 18.5.2) we know that the strictly contractive solutions $S(z)$ of

(22.3.2a)–(22.3.2c) are parametrized as $S = (\Psi_{11}U + \Psi_{12})(\Psi_{21}U + \Psi_{22})^{-1}$ where U is analytic and strictly contractive on $\overline{\Delta}$, and we have set $\Psi = \begin{bmatrix} \Psi_{11} & \Psi_{12} \\ \Psi_{21} & \Psi_{22} \end{bmatrix}$. This relation can be written in a more linear (in fact projective) form as

$$\begin{bmatrix} S \\ I \end{bmatrix} X = \Psi \begin{bmatrix} U \\ I \end{bmatrix}$$

where we have set $X = \Psi_{21}U + \Psi_{22}$. Using that $\Psi = T^{-1}\Theta T$ (where $T = \frac{1}{\sqrt{2}}\begin{bmatrix} I_N & I_N \\ -I_N & I_N \end{bmatrix}$) this may be written as

$$T\begin{bmatrix} S \\ I \end{bmatrix} X = \Theta T \begin{bmatrix} U \\ I \end{bmatrix}.$$

Using the explicit form of T and multiplying out gives

$$\begin{bmatrix} S+I \\ -S+I \end{bmatrix} X = \Theta \begin{bmatrix} U+I \\ -U+I \end{bmatrix}.$$

If we set $F = (S+I)(-S+I)^{-1}$ and $G = (U+I)(-U+I)^{-1}$, this can be written as

$$\begin{bmatrix} F \\ I \end{bmatrix}(-S+I)X = \Theta \begin{bmatrix} G \\ I \end{bmatrix}(-U+I)$$

from which we get

$$F = (\Theta_{11}G + \Theta_{12})(\Theta_{21}G + \Theta_{22})^{-1}.$$

By Lemma 22.3.1 we know that the formulas

$$F = (S+I)(-S+I)^{-1}, \qquad S = (F-I)(F+I)^{-1}$$

establish a one-to-one correspondence between solutions F of (22.3.1a)–(22.3.1c) with positive definite real part and strictly contractive solutions S of (22.3.2a)–(22.3.2c) and similarly the formulas

$$G = (U+I)(-U+I), \qquad U = (G-I)(G+I)^{-1}$$

give a one-to-one correspondence between free parameter matrix functions G with positive definite real part and free parameter strictly contractive matrix functions U analytic on $\overline{\Delta}$. In this way we see that the parametrization of all strictly contractive solutions $S(z)$ of (22.3.2a)–(22.3.2c) via $\Psi(z)$ in terms of strictly contractive parameter functions $U(z)$ is equivalent to the parametrization of all solutions $F(z)$ of (22.3.1a)–(22.3.1c) with positive definite real part via $\Theta(z)$ in terms of free parameter functions $G(z)$ having positive definite real part on $\overline{\Delta}$. This completes the proofs of Theorems 22.2.1 and 22.2.2.

NOTES FOR PART V

In the notes for this part, we describe only developments for the matrix valued case of the various problems; for the scalar case we refer the reader to the following books: Donoghue [1974], Walsh [1960], Garnett [1981] (see also Akhiezer [1965], Akhiezer-Krein [1962], Krein-Nudelman [1977], Rosenblum-Rovnyak [1985]).

CHAPTER 18   Theorem 18.1.1 in case of interpolation of full matrix values probably appears for the first time in Kovalishina-Potapov [1988]; there it was proved by a different method. Theorems 18.5.1, 18.5.2 and 18.5.3 are generalizations of results of Nudelman [1977]; these theorems and their proofs were taken from Ball-Gohberg-Rodman [1989a]. Here in this chapter we are following the method developed in Ball-Gohberg-Rodman [1988, 1989c]. These papers in turn are based on ideas on interpolation from Helton [1980], Ball-Helton [1983] and papers on rational matrix functions (Gohberg-Kaashoek-Lerer-Rodman [1984], Ball-Ran [1987a], Alpay-Gohberg [1988]). For other expository treatments of the Ball-Helton approach, see Ball [1982, 1988] and Sarason [1985].

There are now a variety of different methods for solving Nevanlinna-Pick type problems for rational matrix functions, each with its own advantages and disadvantages. We now list some of them here. The first matrix interpolation problem was solved by Sz.-Nagy-Koranyi [1956, 1958] where the interpolation condition was formulated for the full matrix values. Their method was based on extension results for operators defined on reproducing kernel Hilbert spaces. The tangential Nevanlinna-Pick problem (simple multiplicity case) was first solved by Fedčina [1972] via a reduction to the matrix Nehari problem and then an application of work of Adamian-Arov-Krein [1971]. The term "tangential" for this class of problems was suggested by M.G. Krein. In 1974 for the case of simple multiplicity with interpolation of full matrix values, Kovalishina-Potapov [1988] solved the Nevanlinna-Pick interpolation problem and obtained for the first time explicit formulas for the linear fractional map. This paper was based on ideas of Potapov [1955] on splitting off elementary factors and on the use of some fundamental matrix inequalities. The problem was solved by a Schur-type algorithm where the interpolation conditions were picked off one at a time. In later work (Kovalishina-Potapov [1982]), the method was improved so as to rely on a direct factorization of a matrix valued function rather than on a recursive procedure.

In a second paper, Fedčina [1975] solved the higher multiplicity tangential Nevanlinna-Pick problem by a multi-dimensional analogue of the Schur-Nevanlinna recurrence method. The generalization of the Nevanlinna-Pick problem where the interpolation conditions are given in contour integral form was introduced and solved by Nudelman [1977]. In a short note without proofs, he sketched a method based on solving a Lyapunov or Stein equation and obtained the formula for the linear fractional parametrizer in realization form. Later Delsarte-Genin-Kamp [1979] gave a solution of the matrix Nevanlinna-Pick problem (with simple multiplicities and interpolation of full matrix values) via a method also based on the Schur-Nevanlinna algorithm, and gave other properties.

Sarason [1967] introduced a new operator theoretic method now known as the commutant lifting approach and applied it to scalar interpolation problems. Sz.-Nagy-Foias [1968] (see also Sz.-Nagy-Foias [1970]) extended Sarason's result considerably; this created a strong method for proving existence of solutions to interpolation problems. In the 1970's Foias and his school made the next important step of obtaining a description of all solutions of a lifting problem in a more explicit form (see Arsene-Ceauşescu-Foias [1980], Ceauşescu-Foias [1978, 1979]). An interesting monograph (Foias-Frazho [1989]) dedicated to the commutant lifting approach for interpolation problems will appear soon. Later Helton [1980] applied the commutant lifting theorem to handle tangential Nevanlinna-Pick interpolation problems. Rosenblum and Rovnyak [1980] invented another method using the commutant lifting theorem to solve Nudelman's problem. The commutant lifting approach was extended by Abrahamse [1979] (for the scalar case) and Ball [1979] (for the matrix valued case) to handle Nevanlinna-Pick type interpolation problems for functions on finitely connected domains. There now exists also a "time-dependent" version of the commutant lifting theorem which for the finite dimensional case appeared in Ball-Gohberg [1985] and was generalized to the infinite dimensional case by Paulsen-Power [1988] (see also Paulsen-Power-Ward [1988]); roughly, in this generalization the algebra of bounded analytic functions of the unit disk is replaced by the algebra of upper triangular matrices. Also appearing recently is a nonlinear generalization of the commutant lifting theorem, where the algebra of bounded analytic functions is replaced by the set of causal, stable, nonlinear operators (see Ball-Foias-Helton-Tannenbaum [1987]).

Since the discovery of the engineering applications of Nevanlinna-Pick interpolation, there have appeared a number of proposals for solution procedures for such problems. For example, Chang-Pearson [1984] embed a bitangential Nevanlinna-Pick problem in a problem of interpolation of full matrix values and then use the adaptation of the Schur algorithm worked out by Delsarte-Genin-Kamp [1979]. More direct adaptations of the Schur algorithm to the bitangential Nevanlinna-Pick problem were elucidated by Kimura [1987] and Limebeer-Anderson [1988]; these authors were motivated by applications in H^∞-control theory (see Part VI). Later Kimura [1989] developed a nonrecursive "J-conjugation" method which overlaps our inverse spectral method for constructing the J-inner matrix function Θ. Dewilde-Dym [1981, 1984], motivated by different applications, also worked out the Schur-Nevanlinna algorithm where the tangential Nevanlinna-Pick interpolation was given in the context of the lossless inverse scattering problem. A special case of this work, when all interpolation conditions are given at the point 0, goes back to the classical work of Levinson. Recent results can be found in Lev-Ari-Kailath [1984]. For a collection of articles on the Schur algorithm in general and its applications, see the volume edited by Gohberg [1986].

In 1983 Ball [1983] adapted the method of Sz.-Nagy-Koranyi [1956] to handle more general tangential Nevanlinna-Pick interpolation problems (including the generalizations of Takagi and the boundary case discussed in Chapters 19 and 21), but did not obtain the linear fractional description of all solutions. Dym [1989] developed a new approach to interpolation using reproducing kernel Hilbert spaces and incorporating ideas of de Branges and Rovnyak to obtain the linear fractional description of all solutions.

Dym-Gohberg [1986] analyzed the Nevanlinna-Pick problem with interpolation of full matrix values by using the standard method to reduce to the Nehari problem (see Section 20.1). In this way it is possible to define a special extension which is called the maximum entropy extension and has the property that it is maximizing a certain integral.

Interpolation problems where the solution is required to satisfy additional symmetries were considered in Ball-Helton [1986a] and Dyukarev-Katsnelson [1986]. Other variations involving conditions on the singular values of the matrix values of the interpolant have been studied in Helton [1980], Ball-Helton [1982] and Young [1986].

Recently Bercovici-Foias-Tannenbaum [1989] introduced and solved the problem of finding a matrix function analytic on the unit disk assuming prescribed values at certain points whose matrix values have spectral radius at most 1 on the unit disk.

In our discussion of Nevanlinna-Pick interpolation in the text we consider only the formulation where a solution is required to have norm strictly less than 1 and the Pick matrix is positive definite. In many of the approaches mentioned above, the solution is allowed to have norm less than or equal to 1 and a parametrization of the set of all solutions is obtained in situations where the Pick matrix is singular. A similar remark applies to the problems discussed in Chapters 19–22 as well.

A theory of nonlinear interpolation has been initiated in Ball-Helton [1988b], using the same fusion of ideas on generalized Beurling-Lax theory and inverse spectral problems as was done here for the linear case.

Finally we mention that there is a lot of deep classical work on interpolation of scalar functions over infinitely many points which we do not touch here. Beside the Nevanlinna-Pick problem for infinitely many points which was already solved by Nevanlinna, we mention the theory of interpolating sequences. By definition a sequence of points $z_1, z_2, \ldots$ in the unit disk is an interpolating sequence if, given any bounded sequence of numbers $w_1, w_2, \ldots$, there is a bounded function f analytic on the unit disk for which $f(z_j) = w_j$. Such sequences were characterized by Carleson [1958]; for a summary of the theory, see Garnett [1981]. For latest extensions and generalizations, see Nikolskii [1986] and Nikolskii-Khrushchev [1988]. A number of interesting results on interpolation was recently obtained also by Katsnelson [1987].

CHAPTER 19   The theme of this chapter originates with the work of Takagi [1924]. Theorems 19.2.1, 19.2.2 are generalizations of results announced without proofs by Nudelman [1981]. This chapter provides details of the proofs of the results sketched for the simple multiplicity case in Ball-Gohberg-Rodman [1988]. This theme was also incorporated in the work of Ball [1983] using the approach of Sz.-Nagy-Koranyi, and of Ball-Helton [1983] using their approach through generalized Beurling-Lax representations. Similar results were given by Golinsky [1983] for the case of interpolation of full matrix values.

CHAPTER 20   The approach used here is a natural continuation of the results of the previous chapter and in principle has the same sources as in the previous chapters.

A landmark paper in the theory of the matrix Nehari problem is the paper of Adamian-Arov-Krein [1978]. Here the one-step extension method was introduced and studied, and the description of all solutions of the matrix Nehari problem via a linear fractional formula was obtained; this work became influential in the engineering community (where it is known simply as AAK) and led to many attempts to make the solution more explicit and computationally practical. The existence of solutions of the matrix Nehari problem based on the Sz.-Nagy-Foias lifting theorem was obtained by Page [1970]. The matrix Nehari-Takagi problem was solved (without state space formulas) by Ball-Helton [1983] and Kung-Lin [1983]. The latter succeeded in adapting ideas of Adamian-Arov-Krein in the Nehari problem. The main results for rational matrix valued functions for the half plane version, including state space formulas for the linear fractional map, were first obtained by Glover [1984] via a different approach motivated by engineering ideas. Later Ball-Ran [1986, 1987c] provided another derivation for the circle and the line versions by combining the ideas of Ball-Helton [1983] with the state space approach to Wiener-Hopf factorization from Bart-Gohberg-Kaashoek [1979].

The Nehari problem was analyzed by Dym-Gohberg [1986] in the framework of a general band extension method developed by the same authors (Dym-Gohberg [1982/83]). The band extension method was developed further with many applications in the papers of Gohberg-Kaashoek-Woerdeman [1989a, 1989b] which also include the linear fractional description of the set of all solutions. A maximum entropy principle for contractive interpolants is discussed in Arov-Krein [1983] and Dym-Gohberg [1986]. A general approach to entropy problems in the framework of band extensions is proposed by Gohberg-Kaashoek-Woerdeman [1989c]. The analysis of unitary interpolants and their factorization indices was done in the three papers Dym-Gohberg [1983a, 1983b], Ball [1984]. The papers Dym-Gohberg [1983a, 1986] were used by Gohberg-Kaashoek-van Schagen [1982] to produce state space formulas. Another theory of generalized Nehari problems is studied in a series of papers of Arocena [1989a, 1989b], Cotlar-Sadosky [1985, 1988].

CHAPTER 21   The approach used here again is a natural continuation of the results of the previous chapters with the same sources as before. The adaptation of the Ball-Helton approach to the boundary case was worked out in Ball-Helton [1986b].

Dewilde-Dym [1984] worked out the Schur algorithm for boundary tangential Nevanlinna-Pick interpolation in the context of the lossless inverse scattering problem and, as already mentioned, Ball [1983] handled the boundary case by adapting the Sz.-Nagy-Koranyi method. Unlike the problems considered in earlier chapters of this part, boundary Nevanlinna-Pick interpolation so far has not been amenable to the commutant lifting approach.

There are other boundary interpolation problems associated with the name of Loewner, for example, that of finding a function interpolating a given function on a whole arc of the boundary. For a discussion and further references and generalizations on this type of boundary interpolation, see Rosenblum-Rovnyak [1984].

CHAPTER 22   The method used to obtain Theorem 22.1.1 is simply a bookkeeping device, namely, a linear fractional change of variable, to reduce to the setting already studied in Chapter 18. By the same method we obtain generalizations of the usual Caratheodory-Toeplitz theorem (see Theorems. 22.1.2, 22.1.3 and 22.2.1). In this way it also possible to prove a theorem of Caratheodory-Toeplitz-Takagi type, where the Caratheodory kernel function $\frac{F(z)+F(w)^*}{1-z\overline{w}}$ or $\frac{F(z)+F(w)^*}{z+\overline{w}}$ is allowed to have a prescribed number of negative squares over the unit disk (or the right half plane).

Sz.-Nagy-Koranyi [1958] solved the full matrix valued Caratheodory-Toeplitz problem with simple multiplicity over the upper half plane but did not obtain the linear fractional description of all solutions. Alpay-Bruinsma-Dijksma-de Snoo [1989] have refined the same approach to handle the general multiplicity tangential problem, and also incorporated ideas of Krein and Krein-Langer on generalized resolvents to obtain the linear fractional description of all solutions.

We have insisted that our solutions be positive definite in the right half plane. If one works with positive semi-definite valued solutions, then it is natural to expand the class of solutions to include some which intuitively take on the value infinity identically on a subspace; this is done rigorously also in Alpay-Bruinsma-Dijksma-de Snoo [1989].

Problems of the Caratheodory-Toeplitz type naturally can be included in the framework of band extensions which were developed by Dym-Gohberg [1979, 1980, 1982/83]. These papers led to problems of interpolation for finite matrices which now form an independent topic in the literature (Dym-Gohberg [1981], Ball-Gohberg [1985, 1986a, 1986b, 1986c], Grone-Johnson-de Sá-Wolkowicz [1984], Agler-Helton-McCullough-Rodman [1988], Paulsen-Power-Smith [1989], Paulsen-Rodman [1989], Ellis-Gohberg-Lay [1986], Ben-Artzi-Ellis-Gohberg-Lay [1987]) and also led to the notions of maximum entropy for finite matrices. Recently some indefinite cases have also been studied by Johnson-Rodman [1984] and Ellis-Gohberg-Lay [1987]. The general band extension method initiated by Dym-Gohberg was developed further and supplemented with linear fractional description of all solutions by Gohberg-Kaashoek-Woerdeman [1989a,b,c,d].

A method of solving Caratheodory-Toeplitz problems by a one-step extensions and parametrizing the solutions by choice sequences is developed in work by Arsene, Ceauşescu, Foias, Constantinescu, and Frazho (see Ceauşescu-Foias [1978, 1979], Foias [1978], Arsene-Ceauşescu-Foias [1980], Constantinescu [1986], Foias-Frazho [1986, 1989], Frazho [1986]). In a recent series of papers Fritzsche and Kirstein are also analyzing different aspects of the Caratheodory-Toeplitz problem (see Fritzsche-Kirstein [1987a, 1987b, 1988a, 1988b]).

PART VI

APPLICATIONS TO CONTROL AND SYSTEMS THEORY

In this part we show how the theory of interpolation developed in the earlier parts of the book can be used to solve some sample problems in control and systems theory. All three chapters of this part deal with engineering problems associated with uncertainties, either due to approximations introduced by using the mathematical model of the physical system of interest or due to outside disturbances or measurement errors affecting the signals entering and leaving the system. The sensitivity minimization problem dealt with in Chapter 23 concerns choosing an appropriate compensator in a feedback configuration to maintain system performance in the face of unknown outside disturbances affecting the input signal. The model reduction problem in Chapter 24 is concerned with how to choose a less complex model for a given system without changing the associated input-output map appreciably. Finally Chapter 25 deals with the robust stabilization problem, namely, the problem of designing a compensator in a feedback configuration in such a way that the system is guaranteed to be internally stable even if the original plant is known only approximately. All three topics are central problems in the area of engineering now called H^∞-control theory which are developed in the past decade and a half or so.

CHAPTER 23

SENSITIVITY MINIMIZATION

In this chapter we discuss how the sensitivity minimization problem from H^∞-control theory can be solved in a simple way by using the framework of this book.

23.1 ENGINEERING MOTIVATION

By a *system* we mean of system of differential equations of the form

(23.1.1)
$$\frac{dx(t)}{dt} = Ax(t) + Bu(t); \qquad x(0) = 0$$
$$y(t) = Cx(t) + Du(t)$$

where A, B, C and D are constant (complex) matrices of appropriate sizes (note that A must be square). Here $x(t)$, $u(t)$ and $y(t)$ are continuous functions of $t \geq 0$ whose values are column vectors of appropriate dimensions with generally complex coordinates; $x(t)$ describes the *state* of the system at time t, $u(t)$ is an *input* signal and $y(t)$ is the corresponding *output* signal. It is possible to solve (23.1.1) explicitly for $x(t)$ and $y(t)$ in terms of $u(t)$; the result is

(23.1.2)
$$x(t) = \int_0^t e^{(t-s)A} Bu(s)ds, \qquad t \geq 0$$

$$y(t) = \int_0^t Ce^{(t-s)A} Bu(s)ds + Du(t), \qquad t \geq 0.$$

Then input-output map $u(t) \to y(t)$ in (23.1.2) assumes a simpler form if we use Laplace transform. Denote by the capital Roman letter the Laplace transform of the (scalar or vector) function designated by the corresponding small letter; thus

$$Z(\lambda) = \int_0^\infty e^{-\lambda s} z(s)ds.$$

It is assumed here that for $t \geq 0$, $z(t)$ is a continuous function (generally, complex valued) of t such that $|z(t)| \leq Ke^{\mu t}$ for some positive constants K and μ. This ensures that $Z(\lambda)$ is well-defined for all complex λ with $\mathrm{Re}\,\lambda > \mu$. The system (23.1.1) then takes the form

$$\lambda X(\lambda) = AX(\lambda) + BU(\lambda)$$
$$Y(\lambda) = CX(\lambda) + DU(\lambda).$$

Solving the first equation of $X(\lambda)$ and substituting into the second equation we get

$$Y(\lambda) = [D + C(\lambda I - A)^{-1}B]U(\lambda).$$

So multiplication by the function $W(\lambda) = D + C(\lambda I - A)^{-1}B$ implements the input-output map of the system (23.1.1) when the input and output signals are expressed via their Laplace transforms; for this reason $W(\lambda)$ is said to be the *transfer function* of the system and (A, B, C, D) is said to *realize* $W(\lambda)$ as a transfer function. We saw in Chapter 4 that a rational matrix function $W(\lambda)$ has such a realization if and only if $W(\lambda)$ is *proper* (i.e. analytic at infinity).

As a side remark we note that the associated inverse Laplace transform $\check{W}(t) = Ce^{tA}B + D\delta(t)$ (where $\delta(t)$ is the delta function) relates outputs to inputs in the original time domain

$$y(t) = \int_0^t \check{W}(t-s)u(s)ds$$

(see formula (23.1.2)).

Of crucial importance for us will be the notion of *stability* for a system (A, B, C, D). More precisely, we shall say that the system (A, B, C, D) is *stable* if bounded inputs $u(t)$, $0 \le t < \infty$ lead to bounded outputs $y(t)$, $0 \le t < \infty$ in (23.1.1). (This notion is sometimes called *bounded-input-bounded-output* (BIBO) *stability* in the literature to distinguish it from other notions of stability.)

The following is a useful characterization of BIBO stability.

PROPOSITION 23.1.1. *The system (23.1.1) is stable if and only if the matrix function $Ce^{tA}B$ is integrable over $[0, \infty)$.*

PROOF. Assume first that $Ce^{tA}B$ is integrable over $[0, \infty)$. Then, by formula (23.1.2) for bounded $u(t)$ we have

$$\|y(t)\| \le M \int_0^t \|Ce^{(t-s)A}B\|ds$$

$$= M \int_0^t \|Ce^{sA}B\|ds$$

$$\le M \int_0^\infty \|Ce^{sA}B\|ds$$

where $M = \sup_{t \ge 0} \|u(t)\|$ and hence $y(t)$ is bounded. Conversely, if $Ce^{tA}B$ is not integrable, then some matrix entry $h(t) = [Ce^{tA}B]_{ij}$ of $Ce^{tA}B$ is not integrable. Choose the vector function $u(t)$ to be of the form

$$u(t) = \text{col}(\delta_{kj}u_0(t))_{k=1}^N, \qquad t \ge 0,$$

where $u_0(t)$ is a scalar function, and $\delta_{kj} = 1$ if $k = j$, $\delta_{kj} = 0$ if $k \ne j$. Then the i-th component of $y(t)$ has the form

(21.1.3)

$$y_i(t) = \int_0^t h(t-s)u_0(s)ds + D_{ij}u_0(t)$$

$$= \int_0^t h(s)u_0(t-s)ds + D_{ij}u_0(t).$$

If $h(t) \geq 0$ then it is clear that $u_0(t) \equiv 1$ results in an unbounded $y_i(t)$, since h is by assumption not integrable. In general, by considering separately the real and imaginary parts of $h(t)$, we can assume that $h(t)$ is real-valued. Since h is not integrable, either $h^+(t) = \sup\{h(t), 0\}$ or $h^-(t) = \sup\{-h(t), 0\}$ is not integrable. In the former case, we can choose $u_0(t)$ to be piecewise-constant with values 1 and 0 in such a way to arrange that $y_i(t)$ given by (23.1.3) is unbounded. The case where $h^-(t)$ is not integrable is handled similarly. $\square$

The next result gives a characterization of stability of the system (23.1.1) completely in terms of the transfer function $W(\lambda) = D + C(\lambda I - A)^{-1}B$ for the system.

PROPOSITION 23.1.2. *The system* (23.1.1) *is stable, i.e. if* $u(t)$ *is continuous and* $\sup_{t\geq 0}\|u(t)\| < \infty$ *then also* $\sup_{t\geq 0}\|y(t)\| < \infty$, *if and only if all poles of the transfer function* $W(\lambda) = D + C(\lambda I - A)^{-1}B$ *are in the open left half plane.*

PROOF. Let the size of A be $m \times m$. By Theorem 4.1.2 we know that $\mathbb{C}^n$ can be decomposed as a direct sum $\mathbb{C}^m = \mathcal{L} \dotplus \mathcal{M} \dotplus \mathcal{N}$ such that, A, B, C when expressed conformally with respect to this decomposition have the form

$$A = \begin{bmatrix} A_{11} & A_{12} & A_{13} \\ 0 & A_0 & A_{23} \\ 0 & 0 & A_{33} \end{bmatrix}, \qquad B = \begin{bmatrix} B_1 \\ B_0 \\ 0 \end{bmatrix}, \qquad C = [\, 0 \quad C_0 \quad C_1 \,]$$

where (A_0, B_0) is a full range pair and (C_0, A_0) is a null kernel pair. From this decomposition it is easily seen that the formula for $y(t)$ in (23.1.2) collapses to

$$y(t) = \int_0^t C_0 e^{(t-s)A_0} B_0 u(s)ds + Du(t);$$

hence (A_0, B_0, C_0, D) is BIBO stable. We also already observed in Chapter 4 that (A, B, C, D) and (A_0, B_0, C_0, D) have the same transfer functions. Hence we may assume without loss of generality that (A, B, C, D) is a *minimal* realization (i.e. that (A, B) is a full range pair and (C, A) is a null kernel pair) for $W(\lambda)$ in the first place.

Given that the realization

$$W(\lambda) = D + C(\lambda I - A)^{-1}B$$

for $W(\lambda)$ is minimal, it is possible to show that the eigenvalues of A are precisely the poles of $W(\lambda)$. Indeed, arguing by contradiction, suppose that $\lambda_0 \in \sigma(A)$ is not a pole of

$W(\lambda)$. By applying a similarity transformation $(A, B, C) \to (S^{-1}AS, S^{-1}B, CS)$ which has no effect on the transfer function $W(\lambda)$, we may assume without loss of generality that

$$A = \begin{bmatrix} A_0 & 0 \\ 0 & A_1 \end{bmatrix},$$

where λ_0 is the only eigenvalue of A_0 but λ_0 is not an eigenvalue of A_1. Partition C and B accordingly

$$B = \begin{bmatrix} B_0 \\ B_1 \end{bmatrix}, \qquad C = [\, C_0 \quad C_1 \,]$$

and write

$$C_0(\lambda I - A_0)^{-1}B_0 = W(\lambda) - C_1(\lambda I - A_1)^{-1}B_1.$$

The right-hand side of this equality by our assumption and construction is analytic at λ_0. The left-hand side is analytic in $\mathbb{C}\backslash\{\lambda_0\}$ and takes value 0 at infinity. By the Liouville's theorem, $C_0(\lambda I - A_0)^{-1}B_0 \equiv 0$, and we obtain

$$W(\lambda) = D + C_1(\lambda I - A_1)^{-1}B_1,$$

a contradiction with the minimality of the realization A, B, C, D.

By Proposition 23.1.1 it remains to show that, under the assumption of minimality of the realization (A, B, C, D) of $W(\lambda)$, the matrix function $Ce^{tA}B$ is integrable over $[0, \infty)$ if and only if all eigenvalues of A have negative real parts. Again without loss of generality we may assume that A is in Jordan form $A = \bigoplus_{k=1}^{m} J(\lambda_k)$, where $J(\lambda_k)$ is the Jordan block with eigenvalue λ_k of size (say) $m_k \times m_k$. Then one easily verifies that $e^{tJ(\lambda_k)}$ has the form

$$e^{tJ(\lambda_k)} = \begin{bmatrix} e^{\lambda_k t} & te^{\lambda_k t} & \cdots & \frac{t^{m_k-1}}{(m_k-1)!}e^{\lambda_k t} \\ 0 & \ddots & \ddots & \\ \vdots & & & \vdots \\ & & & te^{\lambda_k t} \\ 0 & 0 & \cdots & e^{\lambda_k t} \end{bmatrix}.$$

From this it follows easily that $Ce^{tA}B$ is integrable if all eigenvalues of A are in the open left half plane Π^-. Conversely, suppose some eigenvalue of A, say λ_1, is not in the open left half plane. By the argument above we see that $Ce^{tA}B$ has the form

(23.1.4) $$Ce^{tA}B = \sum_{k=1}^{s} \sum_{j=0}^{p_k-1} t^j e^{\lambda_k t} X_{kj}$$

where $\lambda_1, \ldots, \lambda_s$ are the distinct eigenvalues of A, p_k is the largest size of Jordan blocks with eigenvalue λ_k in A, and X_{kj} are some matrix coefficients. Taking Laplace transforms of both sides of (23.1.4) yields

(23.1.5) $$C(\lambda I - A)^{-1}B = \sum_{k=1}^{s} \sum_{j=0}^{p_k-1} (\lambda - \lambda_k)^{-j-1} X_{kj}.$$

If $X_{1j} = 0$ for $j = 0, \ldots, p_1 - 1$, from (23.1.5) we see that λ_1 is not a pole of $W(\lambda)$ after all, a contradiction. Thus $X_{1j} \neq 0$ from some j and from (23.1.4) we see that $\operatorname{Re} \lambda_1 \geq 0$ forces $Ce^{tA}B$ not to be integrable. □

It is also important in applications to have a measure of the relative size of the output signal $y(t)$ as compared to the input signal $u(t)$. In many contexts it is appropriate to use the L^2-norm

$$\|w\|_2^2 = \int\limits_0^\infty \|w(t)\|^2 dt$$

as the measure of energy or strength of a vector valued signal $w(t)$. A measure of the size of an input-output map $\mathcal{M}_W \colon u(t) \to y(t)$ given by (23.1.2) is the induced operator norm

$$\|\mathcal{M}_W\|_{op} = \sup\{\|y(t)\| \colon y = \mathcal{M}_W[u] \text{ is related to } u \text{ as in } (23.1.2),\ \|u\|_2 \leq 1\}.$$

Let $H_n^2(\Pi^+)$ be the standard Hardy space of square summable $\mathbf{C}^n$-valued functions on the imaginary axis whose values are the almost everywhere defined nontangential limiting values of a $\mathbf{C}^n$-valued function analytic on the open right half plane (see, e.g., Rosenblum-Rovnyak [1985] for more information on this Hardy space).

It is elementary to check that after Laplace transformation $\mathcal{M}_W$ becomes the multiplication operator M_W from $H_N^2(\Pi^+) \to H_M^2(\Pi^+)$ given by

$$U(\lambda) \to Y(\lambda) = [M_W U](\lambda) = W(\lambda)U(\lambda);$$

here N (resp. M) is the number of dimensions of $u(t)$ (resp. $y(t)$). Moreover,

$$\|\mathcal{M}_W\|_{op} = \|W\|_\infty,$$

where in general

$$\|W\|_\infty = \sup_{-\infty < w < \infty} \{\|W(i\omega)\|\}.$$

Thus, the infinity norm $\|\cdot\|_\infty$ appears naturally as a mathematical expression (after the Laplace transform) of the measure of the size of an input-output map.

In the sequel it will be convenient to describe systems (referred to as *plants*) in terms of their transfer functions.

One of the main problems in systems theory (and practice) is the problem of stabilization. That is, given a plant which is not stable, one wishes to design a controller that will stabilize the system. A standard configuration is the following:

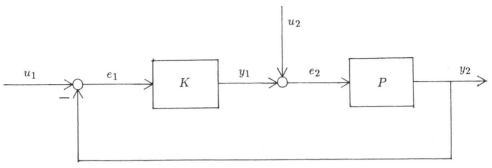

Figure 23.1.1

Here we take P and K to be systems given by their transfer functions. For physical reasons, the transfer functions are assumed to be proper (i.e. analytic at infinity) rational matrix functions. The plant P is the *nominal plant* (considered as given and known), and K is the *controller* (to be designed). The feedback system in Figure 1 is called *internally stable* if the four transfer functions from u_1, u_2 to e_1, e_2 (or equivalently, the four transfer functions from u_1, u_2 to y_1, y_2) are stable, i.e. have all their poles in the open left half plane. Equivalently, if a bounded input $u_1(t)$ is put into the system and $u_2(t)$ is taken to be zero, then not only $y(t)$ but also all the internal signals e_1 and e_2 will be bounded. Representing P and K by their respective transfer functions (which are proper rational matrix functions) we have

$$U_1 = E_1 + PE_2$$
$$U_2 = -KE_1 + E_2$$

or in matrix form

$$\left[\begin{array}{c} U_1 \\ U_2 \end{array} \right] = \left[\begin{array}{cc} I & P \\ -K & I \end{array} \right] \left[\begin{array}{c} E_1 \\ E_2 \end{array} \right].$$

Thus, the feedback system is internally stable if and only if the matrix function $\left[\begin{array}{cc} I & P \\ -K & I \end{array} \right]$ is invertible with inverse having no poles in the closed right half-plane (including infinity). By a Schur complement argument, one can see that $\left[\begin{array}{cc} I & P \\ -K & I \end{array} \right]$ is invertible if and only if $(I + PK)^{-1}$ is invertible, and then the inverse can be computed explicitly as

$$\left[\begin{array}{cc} I & P \\ -K & I \end{array} \right]^{-1} = \left[\begin{array}{cc} (I+PK)^{-1} & -P(I+KP)^{-1} \\ (I+KP)^{-1}K & (I+KP)^{-1} \end{array} \right].$$

Thus *internal stability* means: the four rational matrix functions $(I+PK)^{-1}$, $-P(I+KP)^{-1}$, $(I+KP)^{-1}K$ and $(I+KP)^{-1}$ should be proper and have all poles in the open left half-plane.

Of particular interest is $S = (I+PK)^{-1}$, the transfer function from u_1 to e_1. This measures the effect of the input signal u_1 on the signal e_1 which is fed into the controller K, and is called the *sensitivity*; for physical reasons one wants to minimize the effect of u_1 on the signal e_1 fed into K. Another transfer function of interest is that from u_1 to y_2 which is given by $PK(I + PK)^{-1} = I - S$. This is often called the *complementary sensitivity* and minimizing its effect over certain operating ranges is important for so-called tracking problems. The transfer function $K(I + PK)^{-1} = KS$ from u_1 to y_1 is important for the study of *robust stability*, and will be discussed in Chapter 25.

In this chapter we consider only the problem of sensitivity minimization: Find

$$\min_{K \text{ stabilizes } P} \|(I + PK)^{-1}\|_\infty.$$

The physical significance of the choice of the infinity norm is its identification with the induced operator norm on L^2 in the time domain (as was discussed above). "K stabilizes P" is understood in the sense of internal stability of the feedback system in Figure 23.1.1. In engineering practice one chooses weight functions W_1 and W_2 (usually taken to be rational with no poles or zeros in the closed right half plane) and seeks to minimize the weighted sensitivity norm

$$\min_{K \text{ stabilizes } P} \|W_1(I + PK)^{-1}W_2\|_\infty.$$

In the subsequent sections we analyze and solve the sensitivity minimization problem using the techniques and results on interpolation of rational matrix functions described in earlier chapters.

23.2   INTERPOLATION CHARACTERIZATION OF STABLE SENSITIVITIES: THE SIMPLE CASE

The problem we consider in this section is that of characterizing all sensitivity functions $S = (I + PK)^{-1}$ associated with stabilizing compensators K. More precisely, the mathematical problem to be solved in this section is: *given the rational matrix function P* (with no poles and zeros on the imaginary axis including ∞) *describe all rational matrix functions S of the form*

$$S = (I + PK)^{-1}$$

where K is any rational matrix function such that the four functions

$$S = (I + PK)^{-1}, \ K(I + PK)^{-1}, \ (I + PK)^{-1}P, \ (I + KP)^{-1}$$

are all stable (i.e. have all poles in the open left half plane and no pole at ∞). Note that here and in the sequel for mathematical convenience we drop the constraint that K be proper; this constraint can always reinstated later. Our techniques in general apply to the case where the plant P is a rational square (say $n \times n$) matrix function with $\det P$ not vanishing identically, and such that both P and P^{-1} have no poles on the imaginary axis (including infinity). We will consider this general problem in Section 23.3. Here we keep the discussion elementary and assume that all poles and zeros of P in the right half plane are simple; by this we mean that in the local Smith form for $P(z)$ at a point z_0 in the right-half plane

$$P(z) = E(z) \cdot \text{diag}((z - z_0)^{\mu_1}, \ldots, (z - z_0)^{\mu_n}) F(z)$$

(where E and F are analytic and invertible at z_0), at most one index μ_j is positive in which case it is 1, and at most one index μ_k is negative in which case it is -1. Thus all but at most two indices are nonzero at each point z_0 in the right half plane. Let $z_1, \ldots, z_p$ be the zeros of $P(z)$ in the open right half plane Π^+ (i.e. points z_0 where some index is 1) and let $w_1, \ldots, w_q$ be the poles of $P(z)$ in Π^+ (i.e. points z_0 where some index is -1); note that we allow some points ξ_{ij} to be both a zero and a pole ($z_i = \xi_{ij} = w_j$).

Intuitively, as is explained in detail in Chapters 1 and 3, a consequence of our assumptions is that the zeros and poles of P occur only in certain directions. This we make precise in the following way.

Recall that if z_i is a zero but not a pole of $P(z)$, we say that the row vector $x_i \neq 0$ is a left null vector for $P(z)$ at z_i if

$$x_i P(z_i) = 0.$$

If z_i is also a pole, then in general we need a *left null function* $x_i(z)$, i.e. a row vector function analytic at z_i with $x_i(z_i) \neq 0$ such that the product $x_i(z)P(z)$ has analytic continuation to z_i with value at z_i equal to zero. If the Taylor expansion of $x_i(z)$ is given by

$$(23.2.1) \qquad\qquad x_i(z) = \sum_{j=0}^{\infty} x_{ij}(z - z_i)^j,$$

then the constant term x_{i0} is said to be a *left null vector* for $P(z)$ at z_i. A column vector u_j is said to be a *right pole vector* for $P(z)$ at the pole w_j if there exist a column vector function $\phi(z)$ analytic at w_j such that $P(z)\phi(z)$ has the form

$$P(z)\phi(z) = (z - w_j)^{-1} u_j + [\text{analytic at } w_j].$$

If the point $\xi_{ij} = z_i = w_j$ is both a zero and a pole for $P(z)$, the only piece of information regarding the vector x_{i1} in the Taylor series (23.2.1) at ξ_{ij} which we shall need is the scalar quantity

$$\rho_{ij} = -x_{i1} u_j$$

called the *null-pole coupling number* for $P(z)$ at ξ_{ij}. The following result shows how the null and pole vector together with the associated coupling number can be used to solve a factorization problem.

PROPOSITION 23.2.1. *Suppose the rational matrix function $P(z)$ has a simple zero and a simple pole at the point $\xi_{ij} = z_i = w_j$ with left null vector x_i, right pole vector u_j and associated coupling number ρ_{ij}. Then a rational vector function $f(z)$ factors as*

$$f(z) = P(z)g(z)$$

where g is a rational vector function which is analytic at ξ_{ij} if and only if there is a scalar constant α for which

$$f(z) = \alpha(z - \xi_{ij})^{-1} u_j + h(z)$$

where h is analytic at ξ_{ij} and satisfies

$$x_i h(\xi_{ij}) = \rho_{ij}\alpha.$$

This proposition is a particular case of the general Theorem 4.5.8, but can be also easily proved directly. We can now state our characterization of sensitivities $S = (I + PK)^{-1}$ formed from stabilizing compensators for the case of a plant P having only simple zeros and poles in the open right half plane. We denote by $\mathcal{S}$ the set of all *stable* rational matrix functions, i.e. those that are analytic in the closed right half-plane and at infinity.

THEOREM 23.2.2. *Let $P(z)$ be a regular rational $n \times n$ matrix function with no poles and no zeros on the imaginary axis (including infinity) and having only simple zeros $z_1, \ldots, z_p$ and simple poles $w_1, \ldots, w_q$, in the open right half plane. Let x_i be a left null vector for $P(z)$ at z_i, u_j a right pole vector for $P(z)$ at w_j and ρ_{ij} the associated null-pole coupling number for $P(z)$ at ξ_{ij} if $\xi_{ij} = z_i = w_j$, for $1 \leq i \leq p$, $1 \leq j \leq q$. Then the regular rational matrix function $S(z)$ has the form*

$$S(z) = \big(I + P(z)K(z)\big)^{-1}$$

for a stabilizing compensator $K(z)$ if and only if

(i) $S \in \mathcal{S}$

(ii) $x_i S(z_i) = x_i$ for $1 \leq i \leq p$

(iii) $S(w_j)u_j = 0$ for $1 \leq j \leq q$

and

(iv) $x_i S'(\xi_{ij})u_j = -\rho_{ij}$ *whenever $\xi_{ij} = z_i = w_j$.*

PROOF. We have seen that K is a stabilizing compensator if and only if the four functions $(I + PK)^{-1} = S$, $(I + PK)^{-1}P = SP$, $-K(I + PK)^{-1} = P^{-1}(S - I)$ and $(I + KP)^{-1} = P^{-1}SP$ are all stable. Conversely if S is a stable regular rational matrix function such that the three auxiliary matrix functions SP, $P^{-1}(S - I)$ and $P^{-1}SP$ are all stable, then $K = P^{-1}(S^{-1} - I)$ is a stabilizing compensator and $S = (I + PK)^{-1}$. Thus the problem is to characterize the regular stable functions S for which the three auxiliary functions SP, $P^{-1}(S - I)$ and $P^{-1}S$ are all stable.

We first analyze SP. Since S is stable, the only possible poles of SP in Π^+ are at the poles $w_1, \ldots, w_q$ of P in Π^+. The poles of P in Π^+ occurs at $w_1, \ldots, w_q$ and by assumption are all simple with associated pole vector u_j at w_j. This means that $P(z)$ has Laurent expansion at w_j of the form

$$P(z) = (z - w_j)^{-1}X_j + [\text{analytic at } w_j]$$

where

$$\operatorname{Im} X_j = \operatorname{Span}\{u_j\}.$$

From this one easily sees that $S \in \mathcal{S}$ satisfying the interpolation condition (iii) is equivalent to $S \cdot P$ being stable.

Next consider $P^{-1}(S - I)$. By assumption the zeros of P in Π^+ occur at $z_1, \ldots, z_p$ and are all simple with null vector x_i at the zero z_i. As $S - I$ is stable, the only possible poles of $P^{-1}(S - I)$ in Π^+ are at the poles of P^{-1} in Π^+; namely, at the zeros $z_1, \ldots, z_p$ of P in Π^+. If $x_i(z) = x_i + (z - z_i)x_{i1}$ is a left null function associated with left null vector x_i for $P(z)$ at z_i, then

$$x_i(z)P(z) = (z - z_i)f(z)$$

where f is a row vector function analytic at z_i. Multiplying both sides by $P(z)^{-1}(z - z_i)^{-1}$ gives

$$x_i(z)(z - z_i)^{-1} = f(z)P(z)^{-1}.$$

This shows that the row vector x_i is a *left pole vector* for $P(z)^{-1}$ at z_i. As the pole z_i of $P(z)^{-1}$ is simple (since the zero z_i of $P(z)$ is simple), we conclude that $P(z)^{-1}$ has Laurent expansion at z_i

$$P(z)^{-1} = (z - z_i)^{-1}Y_i + [\text{analytic at } z_i]$$

where the left image $\text{Im}_\ell Y_i = \{yY_i : y \text{ is an } 1 \times n \text{ row vector}\}$ is given by

$$\text{Im}_\ell Y_i = \text{Span}\{x_i\}.$$

From this we easily see that $S \in \mathcal{S}$ satisfying the interpolation condition (ii) is equivalent to $P^{-1}(S - I)$ being stable.

Finally we consider $P^{-1}SP = P^{-1}(S - I)P + I$. For the analysis of this condition we assume that $S \in \mathcal{S}$ and the interpolation conditions (ii) and (iii) hold. Clearly $P^{-1}SP \in \mathcal{S}$ is equivalent to $P^{-1}(S - I)P \in \mathcal{S}$. If $S \in \mathcal{S}$ then the only possible poles of $P^{-1}(S - I)P$ in Π^+ are at the poles of P^{-1} and P in Π^+ (i.e., $z_1, \ldots, z_p$, $w_1, \ldots, w_q$). If the zero z_i of P is not also pole of P, then P is analytic at z_i and we saw above that $P^{-1}(S - I)$ is analytic exactly when the interpolation condition (ii) holds; hence in this case the product $P^{-1}(S - I) \cdot P$ is analytic at z_i. On the other hand, if the pole w_j of P is not also a zero, then P^{-1} is analytic at w_j and we saw above that SP is analytic at w_j exactly when the interpolation condition (iii) holds; thus in this case $P^{-1}(S - I)P = P^{-1} \cdot SP - I$ is analytic at w_j.

Finally if $\xi_{ij} = z_i = w_j$ is both a zero and a pole, we use the interpolation condition (iv) and Proposition 23.2.1. Note that $P^{-1}(S - I)P$ is analytic at ξ_{ij} if and only if

(23.2.2) $$(S - I)P\phi = P \cdot [\text{analytic at } \xi_{ij}]$$

for every rational column vector function ϕ analytic at ξ_{ij}. By Proposition 23.2.1 we know that $P\phi$ has the form

(23.2.3) $$P\phi = \alpha(z - \xi_{ij})^{-1}u_j + h(z)$$

where h is analytic at ξ_{ij} with

(23.2.4) $$x_i h(\xi_{ij}) = \rho_{ij}\alpha.$$

Then $(S - I)P\phi$ assumes the form

(23.2.5) $(S - I)P\phi = -\alpha(z - \xi_{ij})^{-1}u_j + [\alpha(z - \xi_{ij})^{-1}S(z)u_j + S(z)h(z) - h(z)].$

From the interpolation condition (iii), $(z - \xi_{ij})^{-1}S(z)u_j$ is analytic at ξ_{ij} and hence

(23.2.6) $(z - \xi_{ij})^{-1}S(z)u_j|_{z=\xi_{ij}} = S'(\xi_{ij})u_j.$

Now by (23.2.6) and Proposition 23.2.1 we see that $(S - I)P\phi = P \cdot$ [analytic at ξ_{ij}] if and only if

(23.2.7) $x_i[\alpha S'(\xi_{ij})u_j + S(\xi_{ij})h(\xi_{ij}) - h(\xi_{ij})] = -\rho_{ij}\alpha.$

By the interpolation condition (ii) $x_i S(\xi_{ij})h(\xi_{ij}) = x_i h(\xi_{ij})$. Thus (23.2.7) collapses to

$$\alpha x_i S'(\xi_{ij})u_j = -\alpha\rho_{ij}.$$

This holding true for all scalars α is just the interpolation condition (iv). We conclude that, given that (i), (ii) and (iii) hold, (iv) is equivalent to $P^{-1}SP \in \mathcal{S}$ as needed. □

REMARK. For the case where the zeros $z_1, \ldots, z_p$ of the plant $P(z)$ in Π^+ are disjoint from the poles $w_1, \ldots, w_q$, it follows from the equivalences in the proof of Theorem 23.2.2 that the fourth function $(I + KP)^{-1}$ is stable once the other three $(I + PK)^{-1}$, $(I + PK)^{-1}P$ and $-K(I + PK)^{-1}$ are all stable. In particular this applies in the scalar case (when the functions P and K are scalar valued); of course one also sees this in the scalar case immediately from the identity $(I + KP)^{-1} = (I + PK)^{-1}$. For the general matrix case where zeros and poles of P may intersect, we see that the stability of $(I + KP)^{-1}$ does not necessarily follow from the stability of $(I + PK)^{-1}$, $(I + PK)^{-1}P$ and $K(I + KP)^{-1}$.

Thus, the sensitivities $S = (I + PK)^{-1}$ which correspond to stabilizing compensators K are described as solutions to a certain two-sided Lagrange-Sylvester interpolation problem. In the next section we will see that this is the case in the general situation as well. So we can (and will) use the results of Chapters 16 and 18 to solve the sensitivity minimization problem.

23.3   INTERPOLATION CHARACTERIZATION OF STABLE SENSITIVITIES: THE GENERAL CASE

As in Section 23.2, we consider here the characterization of all sensitivity functions $S = (I + PK)^{-1}$ associated with stabilizing compensators K. The rational matrix function P is assumed to be of square size $(n \times n)$ and without zeros and poles on the imaginary axis (including infinity). However, here we drop the assumption that zeros and poles of P are simple, and allow the most general structure of zeros and poles of P. Thus, given P as above, we describe all rational matrix functions of the form $S = (I + PK)^{-1}$ where K is any rational matrix function such that the functions S, KS, SP and $(I + KP)^{-1}$ are all stable as all solutions of a two sided contour integral interpolation problem as discussed in Chapter 16. Recall that we denote by $\text{Res}_{z=z_0} f(z)$ the residue of a rational matrix or vector function $f(z)$ at z_0 (here $z_0 \in \mathbb{C} \cup \{\infty\}$).

THEOREM 23.3.1. *Let P be an $n \times n$ rational matrix function without poles and zeros on the imaginary axis (including infinity), and with left null-pole triple $(C, A_\pi; A_\zeta, B; \Gamma)$ corresponding to the right half-plane.*

A stable rational matrix function S has the form $S = (I + PK)^{-1}$ for some rational matrix K such that SP, KS and $(I + KP)^{-1}$ are also stable if and only if $\det S \not\equiv 0$ and the following equations are satisfied:

$$\text{(23.3.1)} \qquad \sum_{z_0 \in \Pi^+} \operatorname{Res}_{z=z_0} S(z)C(zI - A_\pi)^{-1} = 0;$$

$$\text{(23.3.2)} \qquad \sum_{z_0 \in \Pi^+} \operatorname{Res}_{z=z_0} (zI - A_\zeta)^{-1} BS(z) = B;$$

$$\text{(23.3.3)} \qquad \sum_{z_0 \in \Pi^+} \operatorname{Res}_{z=z_0} (zI - A_\zeta)^{-1} BS(z)C(zI - A_\pi)^{-1} = -\Gamma.$$

PROOF. Suppose $S(z)$ is of the form $\bigl(I + P(z)K(z)\bigr)^{-1}$ and assume all four functions

$$S(z), \quad S(z)P(z), \quad -K(z)S(z), \quad \bigl(I + K(z)P(z)\bigr)^{-1}$$

are analytic in $\overline{\Pi^+}$ (here and elsewhere we denote by $\overline{\Pi^+}$ the closure of Π^+ in the extended complex plane; thus, $\overline{\Pi^+} = \{z \colon \operatorname{Re} z \geq 0\} \cup \{\infty\}$). Since S is the inverse of a rational matrix function, certainty $\det S \not\equiv 0$. It will be convenient for our analysis to rewrite the conditions (23.3.1)–(23.3.3) in the equivalent contour integral form:

$$\text{(23.3.1')} \qquad \frac{1}{2\pi i} \int_\gamma S(z)C(zI - A_\pi)^{-1} dz = 0;$$

$$\text{(23.3.2')} \qquad \frac{1}{2\pi i} \int_\gamma (zI - A_\zeta)^{-1} BS(z) dz = B;$$

$$\text{(23.3.3')} \qquad \frac{1}{2\pi i} \int_\gamma (zI - A_\zeta)^{-1} BS(z)C(zI - A_\pi)^{-1} dz = -\Gamma.$$

Here γ is simple rectifiable contour in the right half-plane such that all eigenvalues of A_ζ and A_π are inside γ.

By Corollary 3.3.3 we may find matrices $\widetilde{C}$ and $\widetilde{B}$ such that the differences

$$V(z) = P(z)^{-1} - \widetilde{C}(zI - A_\zeta)^{-1} B$$

and

$$U(z) = P(z) - C(zI - A_\pi)^{-1}\widetilde{B}$$

are analytic in $\overline{\Pi^+}$, where $(\widetilde{C}, A_\zeta)$ is a null kernel pair and $(A_\pi, \widetilde{B})$ is a full range pair. As $S(z)P(z)$ is analytic in $\overline{\Pi^+}$, it follows that so is also

$$S(z)C(zI - A_\pi)^{-1}\widetilde{B}.$$

Now for a fixed integer $k \geq 0$ we have

(23.3.4)

$$\frac{1}{2\pi i}\int_\gamma S(z)C(zI - A_\pi)^{-1}A_\pi^k\widetilde{B}dz$$

$$= \frac{1}{2\pi i}\int_\gamma S(z)C(zI - A_\pi)^{-1}(A_\pi - zI)A_\pi^{k-1}\widetilde{B}dz$$

$$+ \frac{1}{2\pi i}\int_\gamma S(z)C(zI - A_\pi)^{-1}zA_\pi^{k-1}\widetilde{B}dz$$

$$= \frac{1}{2\pi i}\int_\gamma S(z)C(zI - A_\pi)^{-1}zA_\pi^{k-1}\widetilde{B}dz,$$

and continuing in this manner we find that

$$\frac{1}{2\pi i}\int_\gamma S(z)C(zI - A_\pi)^{-1}A_\pi^k\widetilde{B}dz = \frac{1}{2\pi i}\int_\gamma S(z)C(zI - A_\pi)^{-1}z^k\widetilde{B}dz = 0.$$

So

$$\frac{1}{2\pi i}\int_\gamma S(z)C(zI - A_\pi)^{-1}dz \cdot [\widetilde{B}, A_\pi\widetilde{B}, \ldots, A_\pi^k\widetilde{B}] = 0; \qquad k = 0, 1, \ldots,$$

and since the pair $(A_\pi, \widetilde{B})$ is a full range pair, the equality (23.3.1') follows.

Next, we verify (23.3.2'). We have

(23.3.5) $-KS = -K(I + PK)^{-1} = P^{-1}(I - (I + PK))(I + PK)^{-1} = P^{-1}(S - I),$

and hence the function $P^{-1}(S - I)$ is analytic in $\overline{\Pi^+}$. So

$$\widetilde{C}(zI - A_\zeta)^{-1}B(S(z) - I)$$

is analytic in $\overline{\Pi^+}$ as well. Arguing as in the verification of (23.3.1') we conclude that

$$\frac{1}{2\pi i}\int_\gamma (zI - A_\zeta)^{-1}B(S(z) - I)dz = 0,$$

which implies (23.3.2').

To verify (23.3.3'), recall that we denote by $\mathcal{R}_{n \times n}(\overline{\Pi^+})$ the set of all $n \times n$ rational matrix functions which are analytic in $\overline{\Pi^+}$. Write

$$(I + KP)^{-1} = P^{-1}(I + PK)^{-1}P = P^{-1}SP.$$

As $(I + KP)^{-1}$ is analytic in $\overline{\Pi^+}$, it follows that $P^{-1}(S - I)P$ is analytic in $\overline{\Pi^+}$ as well, and

$$-P + SP = (S - I)P \in P\mathcal{R}_{n \times n}(\overline{\Pi^+}).$$

Now

$$-C(zI - A_\pi)^{-1}\widetilde{B} + S(z)C(zI - A_\pi)^{-1}\widetilde{B} = -P(z) + U(z) + S(z)\big(P(z) - U(z)\big),$$

where $U(z) \in \mathcal{R}_{n \times n}(\overline{\Pi^+})$. Thus

$$-C(zI - A_\pi)^{-1}\widetilde{B} + S(z)C(zI - A_\pi)^{-1}\widetilde{B} = \big(-P(z) + S(z)P(z)\big) + \big(I - S(z)\big)U(z),$$

and both summands on the right-hand side belong to $P\mathcal{R}_{n \times n}(\overline{\Pi^+})$ (for the second summand we use the equality (23.3.5) that implies $S - I \in P\mathcal{R}_{n \times n}(\overline{\Pi^+})$). So $-C(zI - A_\pi)^{-1}\widetilde{B} + S(z)C(zI - A_\pi)^{-1}\widetilde{B} \in P\mathcal{R}_{n \times n}(\overline{\Pi^+})$. Using the full range property of $(A_\pi, \widetilde{B})$ it is not difficult to deduce that for every $x \in \mathbb{C}^n$ we have

$$(23.3.6) \qquad -C(zI - A_\pi)^{-1}x + S(z)C(zI - A_\pi)^{-1}x \in P\mathcal{R}_{n \times n}(\overline{\Pi^+}).$$

The subspace $P\mathcal{R}_{n \times n}(\overline{\Pi^+})$ is the left null-pole subspace of P over $\overline{\Pi^+}$. It follows from Theorem 12.3.1 that (23.3.6) is precisely equivalent to (23.3.3').

Conversely, assume S is stable with $\det S \not\equiv 0$ and (23.3.1')–(23.3.3') are satisfied. If we define K by

$$K = P^{-1}(S^{-1} - I)$$

then S has the form $S = (I + PK)^{-1}$. Using the calculation (23.3.4) we find that

$$\frac{1}{2\pi i} \int_\gamma z^k S(z)C(zI - A_\pi)^{-1}\widetilde{B}dz = 0, \qquad k = 0, 1, \ldots,$$

and hence the function $S(z)C(zI - A_\pi)^{-1}\widetilde{B}$ is analytic in $\overline{\Pi^+}$. As the difference

$$P(z) - C(zI - A_\pi)^{-1}\widetilde{B}$$

is analytic in $\overline{\Pi^+}$ as well, we conclude that SP is stable. An analogous argument implies that $P^{-1}(S - I)$ is stable, and hence in view of (23.3.5) so is KS. Finally (23.3.3') is equivalent to (23.3.6) which in turn implies (using the stability of $P^{-1}(S - I)$) that $-P + SP \in P\mathcal{R}_{n \times n}(\overline{\Pi^+})$. Consequently, $P^{-1}(I - S)P$ is analytic in $\overline{\Pi^+}$, and hence so is $(I + KP)^{-1} = P^{-1}SP$. $\square$

We remark that the proof of Theorem 23.3.1 reveals that given the stability of S, the function SP is stable if and only if (23.3.1) holds, and the function KS is stable if and only if (23.3.2) holds.

The problem (23.3.1)–(23.3.3) (or, equivalently, (23.3.1′)–(23.3.3′)) is precisely the two-sided Lagrange-Sylvester interpolation problem studied in Chapter 16. Combining Theorems 23.3.1 and 16.10.1 we obtain the following result.

THEOREM 23.3.2. *Let P be as in Theorem 23.3.1 and assume that S has the form $S = (I + PK)^{-1}$. Then S is stable and is associated with a stabilizing compensator K if and only if S has the form*

$$S = (\Theta_{11}Q_1 + \Theta_{12}Q_2)(\Theta_{21}Q_1 + \Theta_{22}Q_2)^{-1}.$$

Here

$$\Theta = \left[\begin{array}{cc} \Theta_{11} & \Theta_{12} \\ \Theta_{21} & \Theta_{22} \end{array} \right]$$

is a fixed rational matrix function analytic and invertible on the imaginary axis and at infinity and having the set

$$\left(\left[\begin{array}{c} 0 \\ C \end{array} \right], A_\pi; A_\zeta, [B, -B], -\Gamma \right)$$

as a null-pole triple over $\overline{\Pi^+}$, ϕ^{-1} is a fixed rational matrix function analytic and invertible on the imaginary axis and at infinity having $(C, A_\pi; 0, 0; 0)$ as its null-pole triple over Π^+, and Q_1, Q_2 are any stable rational matrix functions (of suitable sizes) such that $\phi(\Theta_{21}Q_1 + \Theta_{22}Q_2)$ has no zeros in $\overline{\Pi^+}$.

Our next step will be to take into account the norm inequality $\|S\|_\infty \leq \mu$ (where μ is given positive number) as well. This will be done in the next two sections by using solutions to the Nudelman problem given in Chapter 18.

23.4   SOLUTION TO THE SENSITIVITY MINIMIZATION PROBLEM

We return now to the sensitivity minimization problem formulated in Section 23.1:

(23.4.1) $$\min_{K \text{ stabilizes } P} \|(I + PK)^{-1}\|_\infty.$$

First, we find out the minimum in (23.4.1):

THEOREM 23.4.1. *Let $P(z)$ be a rational $n \times n$ matrix function with no poles or zeros on the imaginary axis (including infinity) and let $\tau = (C, A_\pi; A_\zeta, B; \Gamma)$ be a left null-pole triple for $P(z)$ over the open right half-plane Π^+. Then the number $\mu_{\inf}$ defined by $\mu_{\inf} = \{\mu : \|S(z)\|_\infty \leq \mu$ for some S of the form $S = (I + PK)^{-1}$, where K is a stabilizing compensator for $P\}$ is given by*

(23.4.2) $$\mu_{\inf} = \|I + \Gamma_2^{-1/2}\Gamma\Gamma_1^{-1}\Gamma^*\Gamma_2^{-1/2}\|^{1/2}$$

where Γ_1 and Γ_2 are given by

(23.4.3)
$$\Gamma_1 A_\pi + A_\pi^* \Gamma_1 = C^* C;$$

(23.4.4)
$$\Gamma_2 A_\zeta^* + A_\zeta \Gamma_2 = BB^*.$$

Using the inertia results of the Appendix, it follows that the unique solutions Γ_1 and Γ_2 to (23.4.3) and (23.4.4) are positive definite. In particular Γ_1 and Γ_2 are invertible, and formula (23.4.2) makes sense.

PROOF. We use Theorem 18.5.1. Note that S satisfies (23.3.1)–(23.3.3) together with $\|S\|_\infty < \mu$ if and only if $\mu^{-1}S$ satisfies (23.3.1) together with

(23.4.5)
$$\sum_{z_0 \in \Pi+} \text{Res}_{z=z_0}(zI - A_\zeta)^{-1}B(\mu^{-1}S)(z)dz = \mu^{-1}B;$$

(23.4.6)
$$\sum_{z_0 \in \Pi+} \text{Res}_{z=z_0}(zI - A_\zeta)^{-1}B(\mu^{-1}S)(z)C(zI - A_\pi)^{-1}dz = -\mu^{-1}\Gamma,$$

and $\|\mu^{-1}S\|_\infty < 1$. This is the type of problem considered in Theorem 18.5.1. Applying this theorem for $F = \mu^{-1}S$ we obtain that there exists a stable solution $\mu^{-1}S$ to (23.3.1), (23.4.5), (23.4.6) satisfying $\|\mu^{-1}S\|_\infty < 1$ if and only if the matrix

(23.4.7)
$$\Lambda_\mu = \begin{bmatrix} \Gamma_1 & -\mu^{-1}\Gamma^* \\ -\mu^{-1}\Gamma & (1-\mu^{-2})\Gamma_2 \end{bmatrix}$$

is positive definite (here Γ_1 and Γ_2 are solutions of (23.4.3) and (23.4.4) respectively, and it is assumed in advance that μ is such that Λ_μ is invertible). Hence

$$\mu_{\text{inf}} = \inf\{\mu > 0 \colon \Lambda_\mu \text{ it positive definite}\}.$$

Use the Schur complements:

$$\Lambda_\mu = X \begin{bmatrix} \Gamma_1 & 0 \\ 0 & \Gamma_2 - \mu^{-2}(\Gamma_2 + \Gamma\Gamma_1^{-1}\Gamma^*) \end{bmatrix} X^*,$$

where $X = \begin{bmatrix} I & 0 \\ -\mu^{-1}\Gamma\Gamma_1^{-1} & I \end{bmatrix}$. Thus Λ_μ is positive definite if and only if $\Gamma_2 - \mu^{-2}(\Gamma_2 + \Gamma\Gamma_1^{-1}\Gamma^*)$ is positive definite, that is to say if and only if

$$I + \Gamma_2^{-1/2}\Gamma\Gamma_1^{-1}\Gamma^*\Gamma_2^{-1/2} < \mu^2 I.$$

From this criterion the theorem follows.   □

For the case where $\mu > \mu_{\text{inf}}$ we see from the proof of Theorem 23.4.1 that the matrix Λ_μ given by (23.4.7) is positive definite (and in particular invertible). In this situation we define a rational $(2n \times 2n)$ matrix function $\Theta_\mu(z) = \begin{bmatrix} \Theta_{\mu 11}(z) & \Theta_{\mu 12}(z) \\ \Theta_{\mu 21}(z) & \Theta_{\mu 22}(z) \end{bmatrix}$ by

(23.4.8)
$$\Theta_\mu(z) = \begin{bmatrix} I & 0 \\ 0 & I \end{bmatrix} + \begin{bmatrix} 0 & \mu^{-1}B^* \\ C & B^* \end{bmatrix} \begin{bmatrix} (zI - A_\pi)^{-1} & 0 \\ 0 & (zI - A_\zeta^*)^{-1} \end{bmatrix}$$
$$\times \Lambda_\mu^{-1} \begin{bmatrix} 0 & C^* \\ B & -\mu^{-1}B \end{bmatrix}.$$

This function arises in the parametrization of all suboptional sensitivities as shown in the next theorem. Once one has explicit formulas for S, the stabilizing compensator K itself can be found as $K = P^{-1}(S^{-1} - I)$.

THEOREM 23.4.2. *Let* $P(z)$, $\tau = (C, A_\pi; A_\zeta, B; \Gamma)$ *and* $\mu_{\inf}$ *be as in Theorem 23.4.1 and suppose that* $\mu > \mu_{\inf}$. *Then a rational matrix function* S *satisfies the two conditions*

(i) $\|S\|_\infty < \mu$

and

(ii) $S = (I + PK)^{-1}$ *for some stabilizing compensator* K *for* P *if and only if* $\det S \not\equiv 0$ *and*

$$S = \mu(\Theta_{\mu 11} G + \Theta_{\mu 12})(\Theta_{\mu 21} G + \Theta_{\mu 22})^{-1}$$

where $\Theta_{\mu ij}$ $(i,j = 1,2)$ *are given by* (23.4.8) *where* G *is an arbitrary rational* $n \times n$ *matrix function with no poles in* $\overline{\Pi^+}$ *and such that* $\|G\|_\infty < 1$.

The proof of Theorem 23.4.2 follows immediately by combining Theorem 18.5.1 and 23.3.1.

23.5 SENSITIVITY MINIMIZATION WITH WEIGHTS

In this section we extend the results of the previous section to the problem of weighted sensitivity minimization. For the weighted sensitivity minimization problem, besides the plant $P(z)$ and a null-pole triple $\tau = (C, A_\pi; A_\zeta, B; \Gamma)$ for $P(z)$ over Π^+, we assume that we are given also rational $n \times n$ matrix weight functions $W_1(z)$ and $W_2(z)$ (in practice generally taken to be scalar) along with minimal realizations

(23.5.1) $W_j(z) = I + C_j(zI - A_j)^{-1}B_j$

for $j = 1, 2$. We also assume that both $W_1(z)$ and $W_2(z)$ have no zeros and no poles in $\overline{\Pi^+}$. The problem then is to minimize the norm of the weighted sensitivity $W_1(z)(I + P(z)K(z))^{-1}W_2(z)$ where K ranges over all stabilizing compensators for P.

We shall state and prove two main results: One is the characterization of the minimal weighted sensitivity norm; the other is parametrization of all suboptimal solutions in terms of a linear fractional map.

To describe the results we need to introduce more matrices. Given the null-pole triple τ over Π^+ for $P(z)$ and minimal realizations for $W_j(z)$ as in (23.5.1) for

$j = 1, 2$, define matrices Q_2, E_1 and E_2 by

$$(23.5.2) \qquad\qquad Q_2 A_\pi - (A_2 - B_2 C_2) Q_2 = B_2 C$$

$$(23.5.3) \qquad\qquad E_1(A_1 - B_1 C_1) - A_\zeta E_1 = BC_1$$

$$(23.5.4) \qquad\qquad E_2 A_2 - A_\zeta E_2 = BC_2.$$

Note that by assumption the spectra of A_π and A_ζ are in the open right half plane while the spectra of $A_2 - B_2 C_2$, $A_1 - B_1 C_1$ and A_2 are in the open left half plane, so the Sylvester equations (23.5.2), (23.5.3) and (23.5.4) have unique solutions Q_2, E_1 and E_2. Similarly we may define matrices Γ_1, Γ_{2a} and Γ_{2b} as the unique solutions of the Lyapunov equations

$$(23.5.5) \qquad \Gamma_1 A_\pi + A_\pi^* \Gamma_1 = (C^* - Q_2^* C_2^*)(C - C_2 Q_2)$$

$$(23.5.6) \qquad \Gamma_{2a} A_\zeta^* + A_\zeta \Gamma_{2a} = (E_1 B_1 + B)(B_1^* E_1^* + B^*)$$

$$(23.5.7) \qquad \Gamma_{2b} A_\zeta^* + A_\zeta \Gamma_{2b} = (E_2 B_2 - B)(B_2^* E_2^* - B^*).$$

It is not difficult to verify that Γ_1 is positive definite. Indeed, in view of Theorem A.3.1 we have only to check that $(C - C_2 Q_2, A_\pi)$ is a null kernel pair. Let $\mathcal{M} = \bigcap_{i \geq 0} \mathrm{Ker}(C - C_2 Q_2) A_\pi^i$. The subspace $\mathcal{M}$ is clearly A_π-invariant, and equality (23.5.2) gives

$$(23.5.8) \qquad\qquad Q_2 A_\pi x = A_2 Q_2 x, \qquad x \in \mathcal{M}.$$

As $\sigma(A_\pi|_{\mathcal{M}}) \cap \sigma(A_2) = \emptyset$, we obtain from (23.5.8) that $Q_2 x = 0$ for all $x \in \mathcal{M}$. Now using (23.5.2) it is not difficult to see that $C_2 Q_2 A_\pi^i x$, where $x \in \mathcal{M}$, has the form

$$C_2 Q_2 A_\pi^i x = \sum_{j=0}^{i-1} X_{ji} C A_\pi^j x + Y_i Q_2 x \qquad (i \geq 1)$$

for some matrices X_{ji} and Y_i. So for $x \in \mathcal{M}$ we have

$$\begin{aligned}
C A_\pi^i x &= (C - C_2 Q_2) A_\pi^i x + C_2 Q_2 A_\pi^i x \\
&= C_2 Q_2 A_\pi^i x = \sum_{j=0}^{i-1} X_{ji} C A_\pi^i x + Y_i Q_2 x \\
&= \sum_{j=0}^{i-1} X_{ji} C A_\pi^i x, \qquad i = 1, 2, \ldots .
\end{aligned}$$

By induction on i it is now evident that every $x \in \mathcal{M}$ belongs to $\bigcap_{i \geq 0} \text{Ker}(CA_\pi^i)$. Using the null kernel property of (C, A_π) we conclude that $\mathcal{M} = \{0\}$, as required. We can now give a computation of the minimal weighted sensitivity norm.

THEOREM 23.5.1. *Let $P(z)$ be a rational $n \times n$ matrix function having $\tau = (C, A_\pi; A_\zeta, B; \Gamma)$ as a null-pole triple over the open right half plane Π^+ and having no zeros or poles on the imaginary line and at infinity. Let W_1 and W_2 be rational $n \times n$ matrix weighting functions having no poles or zeros in $\overline{\Pi^+}$ and with minimal realizations (23.5.1). Then the infimum μ_{inf} of the set of positive real numbers μ for which there exists a matrix function S such that*

(i) *$S = (I + PK)^{-1}$ for a stabilizing compensator K for P*

and

(ii) *$\|W_1 S W_2\|_\infty \leq \mu$ is given by*

$$\mu_{\text{inf}} = \inf\{\mu > 0 : \Gamma_{2a} - \mu^{-2}\Gamma_{2b} - \mu^{-2}(\Gamma - E_2 Q_2)\Gamma_1^{-1}(\Gamma - E_2 Q_2)^*$$
is positive definite}.

For $\mu > \mu_{\text{inf}}$ we can also obtain a linear fractional parametrization of the set of all weighted sensitivities associated with stabilizing compensators and having norm at most μ. To do this we introduce the Hermitian matrix Λ_μ by

(23.5.9)
$$\Lambda_\mu = \begin{bmatrix} \Gamma_1 & -\mu^{-1}Q_2^* E_2^* - \mu^{-1}\Gamma^* \\ -\mu^{-1}E_2 Q_2 - \mu^{-1}\Gamma & \Gamma_{2a} - \mu^{-2}\Gamma_{2b} \end{bmatrix}$$

where $\Gamma_1, \Gamma_{2a}, \Gamma_{2b}, Q_2, E_2$ are defined by (23.5.2)–(23.5.7). If $\mu > \mu_{\text{inf}}$, then the matrix Λ_μ is positive definite. For these values of μ we introduce a rational $2n \times 2n$ matrix function $\Theta_\mu(z)$ by
(23.5.10)
$$\begin{aligned}
\Theta_\mu(z) &= \begin{bmatrix} \Theta_{\mu 11}(z) & \Theta_{\mu 12}(z) \\ \Theta_{\mu 21}(z) & \Theta_{\mu 22}(z) \end{bmatrix} \\
&= \begin{bmatrix} I & 0 \\ 0 & I \end{bmatrix} + \begin{bmatrix} 0 & -B_1^* E_1^* - B^* \\ -C_2 Q_2 + C & \mu^{-1}B_2^* E_2^* - \mu^{-1}B^* \end{bmatrix} \\
&\quad \times \begin{bmatrix} (zI - A_\pi)^{-1} & 0 \\ 0 & (zI + A_\zeta^*)^{-1} \end{bmatrix} \Lambda_\mu^{-1} \begin{bmatrix} 0 & -Q_2^* C_2^* + C^* \\ E_1 B_1 + B & \mu^{-1}E_2 B_2 - \mu^{-1}B \end{bmatrix}.
\end{aligned}$$

Here is our second main result, on the linear fractional parametrization of suboptimal solutions.

THEOREM 23.5.2. *Let $P(z)$, $\tau = (C, A_\pi; A_\zeta, B; \Gamma)$, W_j ($j = 1, 2$) and μ_{inf} be as in Theorem 23.5.1 and let the matrices $Q_2, E_1, E_2, \Gamma_1, \Gamma_{2a}, \Gamma_{2b}$ and Λ_μ be given by (23.5.2)–(23.5.7) and (23.5.9) where $\mu > \mu_{\text{inf}}$. Then a matrix function $\widetilde{S}$ satisfies the two conditions*

(i) *$\|\widetilde{S}\|_\infty < \mu$*

and

(ii) *$\widetilde{S} = W_1(I + PK)^{-1}W_2$*

for some stabilizing compensator K for P if and only if $\det \widetilde{S} \not\equiv 0$ and

$$\widetilde{S} = \mu(\Theta_{\mu 11}G + \Theta_{\mu 12})(\Theta_{\mu 21}G + \Theta_{\mu 22})^{-1}$$

where $\Theta_{\mu ij}$ $(i,j = 1,2)$ are given by (23.5.10) and where G is an arbitrary $n \times n$ rational matrix function with no poles in $\overline{\Pi^+}$ and with $\|G\|_\infty < 1$.

The rest of this section will be devoted to the proofs of Theorems 23.5.1 and 23.5.2.

We start with a preliminary lemma which is also independently interesting.

LEMMA 23.5.3. *Let $P(z)$ be $n \times n$ rational matrix function without poles and zeros on the imaginary axis and at infinity. Suppose P has $(C_\pi, A_\pi; A_\zeta, B_\zeta; \Gamma)$ as a (left) null-pole triple over Π^+ and $W(z)$ is rational $n \times n$ matrix function with no poles and no zeros in $\overline{\Pi^+}$ having the minimal realization*

$$(23.5.11) \qquad W(z) = I + C(zI - A)^{-1}B.$$

Define matrices Q and L as the unique solutions of the Sylvester equation

$$(23.5.12) \qquad QA_\pi - AQ = BC_\pi$$

and

$$(23.5.13) \qquad L(A - BC) - A_\zeta L = B_\zeta C.$$

Then a (left) null-pole triple for the product $W \cdot P$ over Π^+ is given by

$$(23.5.14) \qquad (CQ + C_\pi, A_\pi; A_\zeta, LB + B_\zeta; LQ + \Gamma).$$

Note that (under the hypotheses of Lemma 23.5.3) each equation (23.5.12) and (23.5.13) has a unique solution. Indeed, $\sigma(A_\pi) \cup \sigma(A_\zeta) \subset \Pi^+$ by the definition of a null-pole triple of P over Π^+. On the other hand, the minimality of realization (23.5.11) ensures that the eigenvalues of A coincide with the poles of W, so $\sigma(A)$ is contained in the open left half-plane Π^-. Also, as a consequence of (23.5.11) one has a minimal realization for W^{-1} as well:

$$W(z)^{-1} = I - C\big(zI - (A - BC)\big)^{-1}B,$$

hence the zeros of $W(z)$ (= poles of $W(z)^{-1}$) coincide with the eigenvalues of $A - BC$, and in particular $\sigma(A - BC) \subset \Pi^-$.

PROOF. Without loss of generality we shall assume $P(\infty) = I$. Next, we show that we can also assume that the zeros and poles of P in the left half-plane are disjoint from the zeros and poles of W. Indeed by Theorem 4.5.8 the (left)null-pole triple (over Π^+) of a rational matrix function does not change if the function is postmultiplied by a rational matrix function without poles and zeros in $\overline{\Pi^+}$. Thus, one can replace P by a product $P\widetilde{P}$, where $\widetilde{P}$ is any rational matrix function with all its poles and zeros in the open left half-plane. It remains to show that $\widetilde{P}$ can be chosen in such a way that $P\widetilde{P}$ has all its poles and zeros away from the poles and zeros of W. To this end use Theorem 4.5.1 to construct a rational matrix function $R(z)$ having null-pole triple over Π^+ equal to $(C_\pi, A_\pi; A_\zeta, B_\zeta; \Gamma)$, having no poles or zeros on the imaginary axis (including infinity), and having its poles and zeros in Π^- disjoint from those of $W(z)$. By Theorem 4.5.8 R has the form $R = P\widetilde{P}$ where $\widetilde{P}$ has all its poles and zeros in the left half plane. Thus without loss of generality, we may assume that the poles and zeros of P are disjoint from the poles and zeros of W.

We now return to the rational matrix function P. Let $(C_{\pi c}, A_{\pi c}; A_{\zeta c}, B_{\zeta c}; \Gamma_c)$ be a left null-pole triple for P with respect to the open left half-plane Π^-. It follows from Theorem 4.5.5 that P can be written in the form

$$(23.5.15) \qquad P(z) = I + C_{\pi e}(zI - A_{\pi e})^{-1}\Gamma_e^{-1}B_{\zeta e},$$

where

$$C_{\pi e} = [C_\pi, C_{\pi c}]; \qquad A_{\pi e} = \begin{bmatrix} A_\pi & 0 \\ 0 & A_{\pi c} \end{bmatrix}; \qquad B_{\zeta e} = \begin{bmatrix} B_\zeta \\ B_{\zeta c} \end{bmatrix};$$

and

$$\Gamma_e = \begin{bmatrix} \Gamma & \Gamma_1 \\ \Gamma_2 & \Gamma_3 \end{bmatrix}$$

with Γ_1 and Γ_2 equal to the unique solutions of the Sylvester equations

$$(23.5.16) \qquad \Gamma_1 A_{\pi c} - A_\zeta \Gamma_1 = B_\zeta C_{\pi c}$$

$$(23.5.17) \qquad \Gamma_2 A_\pi - A_{\zeta c}\Gamma_2 = B_{\zeta c}C_\pi.$$

(Recall that the matrices C_π, A_π, A_ζ, B_ζ and Γ are taken from the null-pole triple of P over Π^+ given in the statement of Lemma 23.5.3.) The matrix Γ_e is invertible by Theorem 4.5.5, so (23.5.15) makes sense. Moreover, denoting

$$A_{\zeta e} = \begin{bmatrix} A_\zeta & 0 \\ 0 & A_{\zeta c} \end{bmatrix},$$

the Sylvester equations holds:

$$(23.5.18) \qquad \Gamma_e A_{\pi e} - A_{\zeta e}\Gamma_e = B_{\zeta e}C_{\pi e}.$$

Since we have assumed that W and P have no common poles and no common zeros, by Corollary 4.1.7 the product WP has a minimal realization

$$(23.5.19) \qquad W(z)P(z) = I + \underline{C}(zI - \underline{A})^{-1}\underline{B},$$

where

$$\underline{A} = \begin{bmatrix} A & BC_{\pi e} \\ 0 & A_{\pi e} \end{bmatrix}; \qquad \underline{B} = \begin{bmatrix} B \\ \Gamma_e^{-1}B_{\zeta e} \end{bmatrix}; \qquad \underline{C} = [C, C_{\pi e}].$$

We calculate that $\underline{A}^\times := \underline{A} - \underline{BC}$ is given by

$$\underline{A}^\times = \left[\begin{array}{cc} A - BC & 0 \\ -\Gamma_e^{-1}B_\zeta eC & A_{\pi e} - \Gamma_e^{-1}B_\zeta eC_{\pi e} \end{array} \right]$$

and thus $(WP)^{-1}$ has a minimal realization

(23.5.20) $$\left(W(z)P(z)\right)^{-1} = I - \underline{C}(zI - \underline{A}^\times)^{-1}\underline{B}.$$

Writing in more detail, we have

$$\underline{C} = [C, C_\pi, C_{\pi c}]; \quad \underline{A} = \left[\begin{array}{ccc} A & BC_\pi & BC_{\pi c} \\ 0 & A_\pi & 0 \\ 0 & 0 & A_{\pi c} \end{array} \right]; \quad \left[\begin{array}{ccc} I & 0 & 0 \\ 0 & \Gamma & \Gamma_1 \\ 0 & \Gamma_2 & \Gamma_c \end{array} \right] \underline{B} = \left[\begin{array}{c} B \\ B_\zeta \\ B_{\zeta c} \end{array} \right];$$

$$\left[\begin{array}{ccc} I & 0 & 0 \\ 0 & \Gamma & \Gamma_1 \\ 0 & \Gamma_2 & \Gamma_c \end{array} \right] \underline{A}^\times = \left[\begin{array}{ccc} A - BC & 0 & 0 \\ -B_\zeta C & A_\zeta & 0 \\ -B_{\zeta c}C & 0 & A_{\zeta c} \end{array} \right] \left[\begin{array}{ccc} I & 0 & 0 \\ 0 & \Gamma & \Gamma_1 \\ 0 & \Gamma_2 & \Gamma_c \end{array} \right].$$

We now construct a null-pole triple for WP over Π^+ using Corollary 4.5.9. To this end we need the spectral $\underline{A}$-invariant subspace $\mathcal{M}$ corresponding to the eigenvalues of $\underline{A}$ in the open right half-plane, and we need also the Riesz projection Z onto the spectral $\underline{A}^\times$-invariant subspace corresponding to the eigenvalues of $\underline{A}^\times$ lying in Π^+.

We claim that

$$\mathcal{M} = \text{Im} \left[\begin{array}{c} Q \\ I \\ 0 \end{array} \right]$$

where Q is determined by (23.5.12). Indeed,

$$\underline{A} \left[\begin{array}{c} Q \\ I \\ 0 \end{array} \right] = \left[\begin{array}{c} Q \\ I \\ 0 \end{array} \right] A_\pi,$$

Hence $\mathcal{M}$ is $\underline{A}$-invariant and $\underline{A}|_\mathcal{M}$ is similar to A_π. Since both $\sigma(A)$ and $\sigma(A_{\pi c})$ lie in the open left half plane, the dimensional considerations show that $\mathcal{M}$ is indeed the spectral $\underline{A}$-invariant subspace corresponding to the eigenvalues in Π^+.

Next, we will verify that the projection Z is given by

(23.5.21) $$Z = \left[\begin{array}{ccc} 0 & 0 & 0 \\ \Delta_1 L & \Delta_1 \Gamma & \Delta_1 \Gamma_1 \\ \Delta_3 L & \Delta_3 \Gamma & \Delta_3 \Gamma_1 \end{array} \right],$$

where L is determined by (23.5.13) and

$$\left[\begin{array}{cc} \Delta_1 & \Delta_2 \\ \Delta_3 & \Delta_4 \end{array} \right] = \left[\begin{array}{cc} \Gamma & \Gamma_1 \\ \Gamma_2 & \Gamma_c \end{array} \right]^{-1}.$$

It is convenient to rewrite (23.5.21) in the form

$$
Z = \begin{bmatrix} I & 0 & 0 \\ 0 & \Gamma & \Gamma_1 \\ 0 & \Gamma_2 & \Gamma_c \end{bmatrix}^{-1} \begin{bmatrix} 0 & 0 & 0 \\ L & I & 0 \\ 0 & 0 & 0 \end{bmatrix} \begin{bmatrix} I & 0 & 0 \\ 0 & \Gamma & \Gamma_1 \\ 0 & \Gamma_2 & \Gamma_c \end{bmatrix}.
$$

Using formula (23.5.20) we reduce the verification of (23.5.21) to the proof that

$$
Z' := \begin{bmatrix} 0 & 0 & 0 \\ L & I & 0 \\ 0 & 0 & 0 \end{bmatrix}
$$

is the Riesz projection on the spectral invariant subspace for the matrix

$$
A' := \begin{bmatrix} A - BC & 0 & 0 \\ -B_\zeta C & A_\zeta & 0 \\ -B_{\zeta c} C & 0 & A_{\zeta c} \end{bmatrix}
$$

corresponding to the eigenvalues in Π^+. Now

$$
\operatorname{Im} Z' = \operatorname{Im} \begin{bmatrix} 0 \\ I \\ 0 \end{bmatrix}; \qquad \operatorname{Ker} Z' = \operatorname{Im} \begin{bmatrix} I & 0 \\ -L & 0 \\ 0 & I \end{bmatrix}
$$

and

$$
(23.5.22) \qquad A' \begin{bmatrix} 0 \\ I \\ 0 \end{bmatrix} = \begin{bmatrix} 0 \\ I \\ 0 \end{bmatrix} A_\zeta;
$$

$$
(23.5.23) \qquad A' \begin{bmatrix} I & 0 \\ -L & 0 \\ 0 & I \end{bmatrix} = \begin{bmatrix} I & 0 \\ -L & 0 \\ 0 & I \end{bmatrix} \begin{bmatrix} A - BC & 0 \\ -B_\zeta C & A_{\zeta c} \end{bmatrix}.
$$

The equality (23.5.22) show that $\operatorname{Im} Z'$ is A'-invariant and the restriction of A' to $\operatorname{Im} Z'$ is similar to A_ζ, in particular, this restriction has all its eigenvalues in Π^+. Analogously, (23.5.23) shows that $\operatorname{Ker} Z'$ is A'-invariant and the restriction of A' to $\operatorname{Ker} Z'$ has all its eigenvalues in the open left half-plane. This proves our claim.

Finally, we are ready to finish the proof of Lemma 23.5.3. By definition,

$$
(23.5.24) \qquad (\underline{C}|_{\mathcal{M}}, \underline{A}|_{\mathcal{M}}; \underline{A}^\times|_{\operatorname{Im} Z}, Z\underline{B}; Z|_{\mathcal{M}})
$$

is a null-pole triple for WP with respect to Π^+. Choose a basis in $\mathcal{M}$ in the form

$$
\left\{ \begin{bmatrix} Q e_i \\ e_i \\ 0 \end{bmatrix} \,\middle|\, e_1, \ldots, e_{n_\pi} \text{ is the standard basis in } \mathbb{C}^{n_\pi} \right\}
$$

and a basis in $\operatorname{Im} Z$ in the form

$$\left\{ \begin{bmatrix} 0 \\ \Delta_1 f_j \\ \Delta_3 f_j \end{bmatrix} \, \middle| f_1, \ldots, f_{n_\zeta} \text{ is the standard basis in } \mathbb{C}^{n_\zeta} \right\}.$$

With this choice of bases, the above formulas for $\underline{C}$, $\underline{A}$, $\underline{A}^\times$, $\underline{B}$, A, and $\mathcal{M}$ show that (23.5.24) actually coincides with (23.5.14). $\square$

We now return to the proofs of Theorems 23.5.1 and 23.5.2. The idea of the proofs is to reduce weighted sensitivity optimization problem to the non-weighted one which was treated in the previous section. To make this idea work we have to calculate how the various components of the proofs of Theorems 23.4.1 and 23.4.2 behave under the transformation $S \to W_1 S W_2$. We start with the two-sided Lagrange-Sylvester interpolation problem.

Let there be given a set of interpolation data

(23.5.26) $$\omega = (C_-, C_+, A_\pi; A_\zeta, B_+, B_-; \Gamma)$$

where the matrices C_-, C_+, A_π, A_ζ, B_+, B_-, Γ are of sizes $N \times n_\pi$, $M \times n_\pi$, $n_\pi \times n_\pi$, $n_\zeta \times n_\zeta$, $n_\zeta \times M$, $n_\zeta \times N$, $n_\zeta \times n_\pi$, respectively, with the following properties:

(α) all eigenvalues of A_π and A_ζ are in Π^+;

(β) the pair (C_-, A_π) is a null kernel pair, and the pair (A_ζ, B_+) is a full range pair;

(γ) the Sylvester equation

$$\Gamma A_\pi - A_\zeta \Gamma = B_+ C_+ + B_- C_-$$

is satisfied.

(In other words, ω is a Π^+-admissible Sylvester TSCII data set.) Under these circumstances by Theorem 4.5.1 there exists an $(M + N) \times (M + N)$ rational matrix function $L(z)$ which is analytic and invertible on the imaginary axis and at infinity and having

(23.5.27) $$\left(\begin{bmatrix} C_+ \\ C_- \end{bmatrix}, A_\pi; A_\zeta, [\, B_+ \quad B_- \,]; \Gamma \right)$$

as its null-pole triple with respect to Π^+; note that such $L(z)$ is not unique. We say that $L(z)$ is *associated* with the interpolation data (23.5.26).

LEMMA 23.5.4. *Let $L(z)$ be a rational matrix function associated with (23.5.26). Given two rational matrix functions W_1 and W_2 of sizes $M \times M$ and $N \times N$, respectively, without poles and zeros in $\overline{\Pi^+}$, let*

(23.5.28) $$\left(\begin{bmatrix} \widetilde{C}_+ \\ \widetilde{C}_- \end{bmatrix}, \widetilde{A}_\pi; \widetilde{A}_\zeta, [\, \widetilde{B}_+ \quad \widetilde{B}_- \,]; \widetilde{\Gamma} \right)$$

be a null-pole triple over Π^+ for the product $\begin{bmatrix} W_1(z) & 0 \\ 0 & W_2^{-1}(z) \end{bmatrix} L(z)$. *Then a stable $M \times N$ rational matrix $F(z)$ satisfies the interpolation conditions*

(i) $\frac{1}{2\pi i} \int\limits_{\gamma} F(z)C_-(zI - A_\pi)^{-1}dz = C_+$

(ii) $\frac{1}{2\pi i} \int\limits_{\gamma} (zI - A_\zeta)^{-1}B_+F(z)dz = -B_-$

and

(iii) $\frac{1}{2\pi i} \int\limits_{\gamma} (zI - A_\zeta)^{-1}B_+F(z)C_-(zI - A_\pi)^{-1}dz = \Gamma$

if and only if $\widetilde{F} := W_1FW_2$ satisfies

(iv) $\frac{1}{2\pi i} \int\limits_{\gamma} \widetilde{F}(z)\widetilde{C}_-(zI - \widetilde{A}_\pi)^{-1}dz = \widetilde{C}_+$

(v) $\frac{1}{2\pi i} \int\limits_{\gamma} (zI - \widetilde{A}_\zeta)^{-1}\widetilde{B}_+\widetilde{F}(z)dz = -\widetilde{B}_-$

(vi) $\frac{1}{2\pi i} \int\limits_{\gamma} (zI - \widetilde{A}_\zeta)^{-1}\widetilde{B}_+\widetilde{F}(z)\widetilde{C}_-(zI - \widetilde{A}_\pi)^{-1}dz = \widetilde{\Gamma}.$

Here γ is a simple closed contour in the right half-plane such that A_π, A_ζ, $\widetilde{A}_\pi$ and $\widetilde{A}_\zeta$ all have spectra inside γ.

PROOF. Let ψ and ϕ be rational matrix functions without poles in $\overline{\Pi^+}$ of sizes $M \times M$ and $N \times N$, respectively, such that $(C_-, A_\pi; 0, 0; 0)$ is a null-pole triple of ϕ^{-1} over Π^+ and $(0, 0; A_\zeta, B_+; 0)$ is a null-pole triple of ψ over Π^+ (it is assumed also that ϕ and ψ have no zeros on the imaginary axis and at infinity).

Let K be any stable interpolating rational matrix function satisfying (i), (ii) and (iii). Then by Theorem 16.9.3 the interpolation conditions (i), (ii) and (iii) for stable F are equivalent to F having the form

(23.5.29) $$F = K + \psi Q \phi$$

for some stable Q. Introduce the function

$$L = \begin{bmatrix} \psi & K\phi^{-1} \\ 0 & \phi^{-1} \end{bmatrix}.$$

It follows from Lemma 16.10.5 that (23.5.27) is a null-pole triple over Π^+ for L. Next note that stable F satisfies (23.5.29) if and only if $\widetilde{F} = W_1FW_2$ has the form

$$\widetilde{F} = \widetilde{K} + \widetilde{\psi} Q \widetilde{\phi}$$

for a stable Q, where

$$\widetilde{K} = W_1KW_2; \qquad \widetilde{\psi} = W_1\psi; \qquad \widetilde{\phi} = \phi W_2.$$

When we form the associated matrix function $\widetilde{L} = \begin{bmatrix} \widetilde{\psi} & \widetilde{K}\widetilde{\phi}^{-1} \\ 0 & \widetilde{\phi}^{-1} \end{bmatrix}$ from this transformed data, the result is

$$\widetilde{L} = \begin{bmatrix} W_1 & 0 \\ 0 & W_2^{-1} \end{bmatrix} L.$$

It remains to apply Theorem 16.9.3 again. □

We are now in a position to prove Theorems 23.5.1 and 23.5.2.

PROOF OF THEOREM 23.5.1. By Theorem 23.3.1 a rational $n \times n$ matrix function S has the form $S = (I + PK)^{-1}$ for a stabilizing compensator K for P if and only if S is stable and satisfies the interpolation conditions (23.3.1)–(23.3.3).

We intend to use Lemma 23.5.4 to construct the interpolation conditions for the function $W_1(\mu^{-1}S)W_2$, here $\mu > 0$ is a parameter. So we need to compute a null-pole triple for $\widetilde{L} := \begin{bmatrix} W_1 & 0 \\ 0 & W_2^{-1} \end{bmatrix} \cdot L(z)$, where $L(z)$ has

$$\left(\begin{bmatrix} 0 \\ C \end{bmatrix}, A_\pi; A_\zeta, [B, -\mu^{-1}B]; -\mu^{-1}\Gamma \right)$$

as a null-pole triple over Π^+. As the realizations (23.5.1) for W_1 and W_2 are minimal, the realization

$$\begin{bmatrix} W_1(z) & 0 \\ 0 & W_2^{-1}(z) \end{bmatrix} = I + \begin{bmatrix} C_1 & 0 \\ 0 & -C_2 \end{bmatrix} \left(zI - \begin{bmatrix} A_1 & 0 \\ 0 & A_2 - B_2C_2 \end{bmatrix} \right)^{-1} \begin{bmatrix} B_1 & 0 \\ 0 & B_2 \end{bmatrix}$$

is minimal as well. We apply Lemma 23.5.3 to find a null-pole triple for $\widetilde{L}$ over Π^+; so define matrices Q and E which are solutions to the equations

(23.5.30)
$$QA_\pi - \begin{bmatrix} A_1 & 0 \\ 0 & A_2 - B_2C_2 \end{bmatrix} Q = \begin{bmatrix} B_1 & 0 \\ 0 & B_2 \end{bmatrix} \begin{bmatrix} 0 \\ C \end{bmatrix}$$

and

(23.5.31)
$$E \begin{bmatrix} A_1 - B_1C_1 & 0 \\ 0 & A_2 \end{bmatrix} - A_\zeta E = [B, -\mu^{-1}B] \begin{bmatrix} C_1 & 0 \\ 0 & -C_2 \end{bmatrix}.$$

Write Q as $\begin{bmatrix} Q_1 \\ Q_2 \end{bmatrix}$ and E as $[E_{1\mu}, E_{2\mu}]$. We deduce from (23.5.30) that $Q_1 = 0$ and Q_2 is given by (23.5.2), and from (23.5.31) we deduce that $E_{1\mu} = E_1$, where E_1 is given by (23.5.3), and $E_{2\mu} = \mu^{-1}E_2$, where E_2 is given by (23.5.4). By Lemma 23.5.3 we conclude that a null-pole triple for $\widetilde{L}$ over Π^+ is given by

$$\left(\begin{bmatrix} C_1 & 0 \\ 0 & -C_2 \end{bmatrix} \begin{bmatrix} 0 \\ Q_2 \end{bmatrix} + \begin{bmatrix} 0 \\ C \end{bmatrix}, A_\pi; \right.$$
$$A_\zeta, [E_1, \mu^{-1}E_2] \begin{bmatrix} B_1 & 0 \\ 0 & B_2 \end{bmatrix} + [B, -\mu^{-1}B]; [E_1, \mu^{-1}E_2] \begin{bmatrix} 0 \\ Q_2 \end{bmatrix} - \mu^{-1}\Gamma \right)$$
$$= \left(\begin{bmatrix} 0 \\ -C_2Q_2 + C \end{bmatrix}, A_\pi; A_\zeta, [E_1B_1 + B, \mu^{-1}E_2B_2 - \mu^{-1}B]; \right.$$
$$\left. -\mu^{-1}E_2Q_2 - \mu^{-1}\Gamma \right).$$

We conclude by Lemma 23.5.4 that S is a stable sensitivity for P if and only if the function $F = \mu^{-1}W_1 S W_2$ is stable and satisfies the interpolation conditions

$$\sum_{z_0 \in \Pi^+} \text{Res}_{z=z_0} F(z)(-C_2 Q_2 + C)(zI - A_\pi)^{-1} = 0;$$

$$\sum_{z_0 \in \Pi^+} \text{Res}_{z=z_0} (zI - A_\zeta)^{-1}(E_1 B_1 + B)F(z) = -\mu^{-1}E_2 B_2 + \mu^{-1}B;$$

$$\sum_{z_0 \in \Pi^+} \text{Res}_{z=z_0} (zI - A_\zeta)^{-1}(E_1 B_1 + B)F(z)(-C_2 Q_2 + C)(zI - A_\pi)^{-1} =$$
$$\qquad - \mu^{-1}E_2 Q_2 - \mu^{-1}\Gamma.$$

Next, plug these data into Theorem 18.5.1 to compute the matrix (18.5.3) for this interpolation problem; the result is

$$(23.5.32) \qquad \Lambda_\mu = \begin{bmatrix} \Gamma_{1\mu} & -\mu^{-1}(Q_2^* E_2^* E_2 + \Gamma^*) \\ -\mu^{-1}(E_2 Q_2 + \Gamma) & \Gamma_{2\mu} \end{bmatrix}$$

where

$$\Gamma_{1\mu} A_\pi + A_\pi^* \Gamma_{1\mu} = (C^* - Q_2^* C_2^*)(C - C_2 Q_2)$$
$$\Gamma_{2\mu} A_\zeta^* + A_\zeta \Gamma_{2\mu} = (E_1 B_2 + B)(B_1^* E_1^* + B^*)$$
$$\qquad - (\mu^{-1}E_2 B_2 - \mu^{-1}B)(\mu^{-1}B_2^* E_2^* - \mu^{-1}B^*).$$

Thus $\Gamma_{1\mu} = \Gamma_1$ is independent of μ and is given by (23.5.5), and $\Gamma_{2\mu}$ can be written as

$$\Gamma_{2\mu} = \Gamma_{2a} - \mu^{-1}\Gamma_{2b}$$

where Γ_{2a} and Γ_{2b} are given by (23.5.6) and (23.5.7), respectively. Plugging these expressions into (23.5.32) we see that Λ_μ is also as in (23.5.9). To test when Λ_μ is positive definite, since Γ_1 is positive definite it suffices to check when the Schur complement

$$\Gamma_{2a} - \mu^{-2}\Gamma_{2b} - \mu^{-2}(\Gamma - P_2 Q_2)\Gamma_1^{-1}(\Gamma - P_2 Q_2)^*$$

of Γ_1 is positive definite. By Theorem 18.5.1 this yields the formula for μ_{inf} in Theorem 23.5.1 and completes the proof of Theorem 23.5.1. $\quad\square$

PROOF OF THEOREM 23.5.2. This is just a continuation of the proof of Theorem 23.5.1. Observe that $\Theta_\mu(z)$ given by (23.5.10) is J-unitary on the imaginary axis having

$$\left(\begin{bmatrix} 0 \\ -C_2 Q_2 + C \end{bmatrix}, A_\pi; A_\zeta, [E_1 B_1 + B, \mu^{-1}E_2 B_2 - \mu^{-1}B]; -\mu^{-1}E_2 Q_2 - \mu^{-1}\Gamma \right)$$

as its null-pole triple over Π^+ (see formula (18.5.7)). Now Theorem 23.5.2 follows by using Theorem 18.5.1. $\quad\square$

CHAPTER 24
MODEL REDUCTION

In many engineering applications the choice of a model for a physical process of interest is often complicated by competing desirable goals. On the one hand one would like a model which reflects the characteristics and predicts the behavior of the actual physical device with a useful amount of accuracy. On the other hand one would like the model to be sufficiency simple so as to make various numerical procedures and computational algorithms practical and easy. Unfortunately making the model more realistic and accurate usually leads to increasing the number of parameters needed to describe it which in turn leads to greater and greater complexity for computational manipulation. If some of the parameters in fact have practically very little influence on the model, it makes good physical sense to discard them. More generally, the problem of *model reduction* is concerned with approximating (in some appropriate sense) a given model by a model which is less complex and thereby easier to work with.

In the engineering literature there are many approaches to model reduction. Here we discuss only one, optimal Hankel norm model reduction, which fits well with the techniques and ideas of this book.

24.1 INTRODUCTION

As was described in Chapter 23, a system (continuous-time model) is often thought of as a black box which manifests itself in an input-output behavior, say $u(t) \rightarrow y(t)$, where $u(t)$ is a vector-valued input function and $y(t)$ is a vector-valued output function and the variable t represents time. Discrete-time models of systems will be discussed later in this chapter. We assume that the system is linear, time-invariant, causal and finite-dimensional; then, as described in Chapter 23, the system can be modelled by a first order system of linear differential equations

(24.1.1)
$$\frac{dx}{dt}(t) = Ax(t) + Bu(t); \qquad x(0) = 0;$$
$$y(t) = Cx(t) + Du(t)$$

where A, B, C, D are constant matrices of appropriate size, and $x(t)$ is the state variable taking values in the state space, say $\mathbb{C}^n$. One measure of the complexity of such a system is the number of parameters required to describe the state vector $x(t)$, i.e., the dimension n of the state space $\mathbb{C}^n$.

As was pointed out in Chapter 23, after Laplace transformation the input-output map $U(\lambda) \rightarrow Y(\lambda)$ of the system (24.1.1) is given by a matrix function multiplication

$$Y(\lambda) = W(\lambda)U(\lambda)$$

where $W(\lambda) = D + C(\lambda I - A)^{-1}B$ is the *transfer function*. If (C, A) is a null kernel and (A, B) is a full range pair, then the size n of the state space coincides with the McMillan degree of the rational matrix function $W(\lambda)$, and is already the smallest possible size for the state space among all systems of the form (24.1.1) which give rise to the same input-output behavior. If this minimal size is still larger than one would like for practical purposes, one may want to achieve only a reasonable approximation of the exact input-output behavior in order to achieve a smaller state space size or McMillan degree. This then leads to the problem of model reduction. It remains to discuss what measure of approximation is reasonable and practical to use. We shall return to this issue in Section 24.3 after we introduce the Hankel norm in the next section.

24.2 THE HANKEL NORM

In this section we wish to discuss various sorts of norms and operators associated with a stable matrix transfer function $W(z)$ and the connections and correspondences among them. This background information will be useful for our discussion of the model reduction problem in the next section, and in particular for motivation of the Hankel norm as the measure of closeness in the model reduction problem.

Let us suppose that $W(z)$ is a *strictly proper* (i.e. analytic at infinity, with $W(\infty) = 0$) rational $M \times N$ matrix function with all poles in the open left half plane. Such rational matrix functions will be called *strictly stable* in this chapter. Thus $W(z)$ has a minimal realization of the form

$$W(z) = C(zI - A)^{-1}B$$

where the spectrum $\sigma(A)$ of A is contained in the open left half plane Π^-. As discussed in Section 23.1, the inverse Laplace transform $\check{K}(t)$ satisfies

(24.2.1)
$$W(z) = \int_0^\infty e^{-tz}\check{W}(t)dt, \qquad \mathrm{Re}\,z \geq 0,$$

in realization form is given by

$$\check{W}(t) = Ce^{tA}B$$

and is exponentially decaying on the real line, i.e.

$$\|\check{W}(t)\| \leq \alpha e^{-\beta t}, \qquad t \geq 0$$

where $\alpha > 0$ and where $\beta > 0$ is such that all poles of $W(z)$ have real parts less than $-\beta$.

Associated with $W(z)$ are several operators of importance. The *input-output operator* $\mathcal{M}_K$ associated with the system

$$\frac{dx}{dt}(t) = Ax(t) + Bu(t); \qquad x(0) = 0$$
$$y(t) = Cx(t)$$

is the convolution operator defined from $L_N^2(0, \infty)$ into $L_M^2(0, \infty)$ given by

(24.2.2)
$$\mathcal{M}_W[u](t) = \int_0^t \check{W}(t-s)u(s)ds$$
$$= \int_0^t Ce^{(t-s)A}Bu(s)ds, \qquad t \geq 0$$

and was discussed in Section 23.1. There we saw that the induced operator norm

$$\|\mathcal{M}_W\|_{op} := \sup\{\|\mathcal{M}_W[u]\|_2 : u \in L_N^2(0, \infty), \ \|u\|_2 \leq 1\}$$

where by $\| \cdot \|_2$ we denote the norm in $L_N^2(0, \infty)$ and in $L_M^2(0, \infty)$, is the same as the infinity norm of W:

$$\|\mathcal{M}_W\| = \|W\|_\infty := \sup_{-\infty < w < \infty} \|W(iw)\|.$$

The Hankel operator, introduced in Section 20.2 (see (20.2.2)) is defined also from $L_N^2(0, \infty)$ to $L_M^2(0, \infty)$ by

(24.2.3)
$$\mathcal{H}_W[u](t) = \int_0^\infty \check{W}(s+t)u(s)ds, \qquad u \in L_N^2(0, \infty),$$

with the induced operator norm $\|\mathcal{H}_W\|$. By Nehari's theorem (Theorem 20.2.1), we also know that $\|\mathcal{H}_W\|$ can be identified with the distance of W to the set of antistable rational matrix functions (i.e. functions with all poles in the right half plane) in the infinity norm, i.e.

(24.2.4)
$$\|\mathcal{H}_W\| = \inf\{\|W - R\|_\infty : R \in \mathcal{R}_{M \times N}(\overline{\Pi^-})\}.$$

We saw in Theorem 20.2.2 that for W a stable rational matrix function $\mathcal{H}_W$ is a finite rank operator with rank equal to the McMillan degree n of W. In particular $\mathcal{H}_W$ has only finitely many nonzero singular values

$$s_1(\mathcal{H}_W) \geq s_2(\mathcal{H}_W) \geq \cdots \geq s_n(\mathcal{H}_W) > 0$$

and hence has finite trace norm

(24.2.5)
$$\mathrm{Tr}(|\mathcal{H}_W|) = \sum_{j=1}^\infty s_j(\mathcal{H}_W)$$
$$= s_1(\mathcal{H}_W) + \cdots + s_n(\mathcal{H}_W)$$
$$\leq ns_1(\mathcal{H}_W) = n\|\mathcal{H}_W\|.$$

The following result gives another useful connection between the Hankel norm and the infinity norm.

THEOREM 24.2.1. *Let $W(z)$ be a strict proper rational $M \times N$ matrix function with all poles in the open left half plane and with McMillan degree equal to n. Then*

$$\|\mathcal{H}_W\| \le \|W\|_\infty \le 2n\|\mathcal{H}_W\|.$$

For the proof of Theorem 24.2.1 we first require the following general lemma.

LEMMA 24.2.2. *Suppose $K(t,s)$ is a continuous norm integrable $M \times N$ matrix function defined for $0 \le t, s < \infty$ for which the associated integral operator $\mathcal{K}: L_N^2(0,\infty) \to L_M^2(0,\infty)$ defined by*

$$\mathcal{K}[f](t) = \int_0^\infty K(t,s)f(s)ds$$

has finite rank, and let $s_1(\mathcal{K}) \ge s_2(\mathcal{K}) \ge \cdots \ge s_n(\mathcal{K})$ be the nonzero singular values of $\mathcal{K}$. Then

$$\int_0^\infty \|K(t,t)\|dt \le \sum_{j=1}^n s_j(\mathcal{K}).$$

PROOF. Since $\mathcal{K}$ has finite rank, there exists an orthonormal set of column vector functions $\varphi_1(t), \ldots, \varphi_n(t)$ in $L_N^2(0,\infty)$ and $\psi_1(t), \ldots, \psi_n(t)$ in $L_M^2(0,\infty)$ such that K has the separable form (Schmidt expansion)

$$K(t,s) = \sum_{j=1}^n s_j \varphi_j(t)\psi_j(s)^*$$

where $s_j = s_j(\mathcal{K})$ (see, e.g., Section 2.2 in Gohberg-Krein [1969]). We set $t = s$ and estimate

$$\int_0^\infty \|K(t,t)\|dt \le \sum_{j=1}^n s_j \int_0^\infty \|\varphi_j(t)\psi_j(t)^*\|dt$$

$$\le \sum_{j=1}^n s_j \int_0^\infty \|\varphi_j(t)\|\|\psi_j(t)\|dt$$

$$\le \sum_{j=1}^n s_j\|\varphi_j\|_2\|\psi_j\|_2 = \sum_{j=1}^n s_j. \quad \square$$

PROOF OF THEOREM 24.2.1. The first inequality $\|\mathcal{H}_W\| \le \|W\|_\infty$ is an immediate consequence of (24.2.4). For the second inequality $\|W\|_\infty \le 2n\|\mathcal{H}_W\|$, we first use the connection (24.2.1) between W and $\check{W}$ to estimate

$$\|W\|_\infty \le \int_0^\infty \|\check{W}(t)\|dt =: \|\check{W}\|_1.$$

On the other hand, if we apply Lemma 24.2.2 with $K(t,s) = \check{W}(s+t)$ we get

$$\int\limits_0^\infty \|\check{W}(2t)\| dt \leq \sum_{j=1}^n s_j$$

where $s_j = s_j(\mathcal{H}_W)$. Since by a simple change of variable we have

$$\int\limits_0^\infty \|\check{W}(2t)\| dt = \frac{1}{2} \int\limits_0^\infty \|\check{W}(t)\| dt = \frac{1}{2}\|\check{W}\|_1,$$

we thus get

$$\|W\|_\infty \leq \|\check{W}\|_1 \leq 2 \sum_{j=1}^n s_j \leq 2ns_1 = 2n\|\mathcal{H}_W\|$$

as needed. $\square$

24.3 OPTIMAL HANKEL NORM MODEL REDUCTION

The problem of model reduction more precisely can be described as follows. We are given a strictly stable (i.e. all poles in the open left half plane and zero at infinity) rational matrix function $P(z)$ with a minimal realization

$$P(z) = C(zI - A)^{-1}B$$

and of McMillan degree $\delta(P) = n$. Thus the spectrum $\sigma(A)$ of A is contained in the left half plane, and the size of A is $n \times n$. We are also given a desired tolerance level $\mu > 0$. The problem then is to find

$$\kappa_0 = \min\{\kappa: \text{there exists strictly stable rational}$$
$$\widetilde{P} \text{ with } \delta(\widetilde{P}) \leq \kappa, \ \|P - \widetilde{P}\|_* < \mu\}$$

where $\|\cdot\|_*$ is some appropriate choice of norm, and then to describe the set of all such degree κ_0 approximants:

$$\{\widetilde{P}(z): \widetilde{P} \text{ strictly stable}, \ \delta(\widetilde{P}) = \kappa_0, \ \|P - \widetilde{P}\|_* < \mu\}.$$

A physically appropriate choice of norm is the infinity norm $\|W\|_\infty = \|\mathcal{M}_W\|_{op}$, as this is the induced operator norm for the input-output map $\mathcal{M}_W$ associated with the error system $W = P - \widetilde{P}$ acting between vector valued L^2-spaces. However the mathematical problem is difficult to solve in this formulation. A reasonable compromise is to use the Hankel norm $\|W\|_* = \|\mathcal{H}_W\|$. Indeed, by Theorem 24.2.1 the Hankel norm and the infinity norm are equivalent as long as we restrict ourselves to transfer functions of bounded McMillan degree. Once this choice has been made, the model reduction problem reduces immediately to the Nehari-Takagi problem. More precisely we have the following result.

THEOREM 24.3.1. *Let P be a strictly stable rational $M \times N$ matrix function and suppose $\mu > 0$. Then*

$$\kappa_0 = \min\{\kappa: \text{ there exists strictly stable rational } \widetilde{P}$$

$$\text{with } \delta(\widetilde{P}) = \kappa \text{ and } \|\mathcal{H}_{P-\widetilde{P}}\| < \mu\}$$

is given by

$$\kappa_0 = \min\{\kappa: s_{\kappa+1}(\mathcal{H}_P) \leq \mu\}$$

where $s_1(\mathcal{H}_P) \geq \cdots \geq s_n(\mathcal{H}_P) > 0$ are the singular values of the Hankel operator associated with P.

PROOF. By Nehari's theorem (Theorem 20.2.1) we know that

$$\|\mathcal{H}_{P-\widetilde{P}}\| = \inf\{\|P - \widetilde{P} - R\|_\infty : R \in \mathcal{R}_{M \times N}(\overline{\Pi^-})\}.$$

Thus we have

$$\inf\{\|\mathcal{H}_{P-\widetilde{P}}\|: \widetilde{P} \text{ strictly stable, } \delta(\widetilde{P}) \leq \kappa\}$$

$$= \inf\{\|P - [\widetilde{P} + R]\|_\infty : \widetilde{P} \text{ strictly stable with } \delta(\widetilde{P}) \leq \kappa, \ R \in \mathcal{R}_{M \times N}(\overline{\Pi^-})\}.$$

By the Nehari-Takagi theorem (Theorem 20.5.1) and also by taking into account Theorem 20.2.2(ii), this last infimum is simply $s_{\kappa+1}(\mathcal{H}_P)$. The theorem follows. $\square$

We now recall some notions from Section 20.2 which enable us to express $\mathcal{H}_P$ is state space terms. Specifically, if $P(z)$ has a minimal realization of the form

$$P(z) = C(zI - A)^{-1}B,$$

we may define the controllability operator $\Xi: L^2_N(0, \infty) \to \mathbb{C}^n$ and the observability operator $\Omega: \mathbb{C}^n \to L^2_M(0, \infty)$ by

$$\Xi f = \int_0^\infty e^{sA} B f(s) ds, \qquad f \in L^2_N(0, \infty)$$

and

$$(\Omega x)(t) = Ce^{tA} x \quad (0 \leq t < \infty), \qquad x \in \mathbb{C}^n,$$

and then controllability and observability gramians Π and Q respectively by

$$\Pi = \Xi \Xi^*; \qquad Q = \Omega^* \Omega.$$

As was also mentioned in Section 20.2, the finite matrices Π and Q can also be found as solutions of Lyapunov equations

(24.3.1) $$A\Pi + \Pi A^* = -BB^*$$

(24.3.2) $$A^*Q + QA = -C^*C.$$

The following corollary then gives a restatement of Theorem 24.3.1 in state space terms.

COROLLARY 24.3.2. *Let P be a strictly stable rational $M \times N$ matrix function with minimal realization*

$$P(z) = C(zI - A)^{-1}B$$

and let the controllability and observability gramians Π and Q for this realization be given by (24.3.1) and (24.3.2). Then, for $\mu > 0$, we have

$$\min\{\kappa: \text{ there exists strictly stable rational } \widetilde{P} \text{ with } \delta(\widetilde{P}) = \kappa \text{ and } \|\mathcal{H}_{P-\widetilde{P}}\| \le \mu\}$$
$$= \min\{\kappa: \lambda_{\kappa+1}(\Pi Q) \le \mu^2\}.$$

For the proof of this corollary use Theorem 24.3.1 in conjunction with Theorem 20.2.2(ii).

We are now in a position to use the detailed state space results of Chapter 20 to obtain a linear fractional parametrization of the set of all solutions of the model reduction problem for a given tolerance μ, at least up to a free antistable (i.e. analytic in $\overline{\Pi^-}$, including infinity) additive term. More precisely, the result is as follows.

THEOREM 24.3.3. *Let $P(z)$ be a strictly stable proper rational $M \times N$ matrix function of McMillan degree n having minimal realization*

$$P(z) = C(zI - A)^{-1}B$$

with controllability and observability gramians Π and Q given by (24.3.1) and (24.3.2) respectively. Choose $\mu > 0$ so that

$$\lambda_{\kappa+1}(\Pi Q) < \mu^2 < \lambda_\kappa(\Pi Q)$$

and define a rational $(M + N) \times (M + N)$ matrix function $\Theta(z) = \begin{bmatrix} \Theta_{11}(z) & \Theta_{12}(z) \\ \Theta_{21}(z) & \Theta_{22}(z) \end{bmatrix}$
by

$$\Theta(z) = \begin{bmatrix} I_M & 0 \\ 0 & I_N \end{bmatrix} + \begin{bmatrix} \mu^{-1}C & 0 \\ 0 & B^* \end{bmatrix} \begin{bmatrix} (zI - A)^{-1} & 0 \\ 0 & (zI + A^*)^{-1} \end{bmatrix}$$
$$\times \begin{bmatrix} -Z\Pi & Z \\ Z^* & -\mu^{-2}QZ \end{bmatrix} \begin{bmatrix} -\mu^{-1}C^* & 0 \\ 0 & B \end{bmatrix}$$

where

$$Z = (I - \mu^{-2}\Pi Q)^{-1}.$$

Then a rational matrix function $\widetilde{P}(z)$ has the properties

(i) $\|\mathcal{H}_{P-\widetilde{P}}\| < \mu$

and

(ii) $\widetilde{P}$ is strictly stable with $\delta(\widetilde{P}) = \kappa$

if and only if

$$P - \widetilde{P} = \mathrm{st}[(\Theta_{11}G + \Theta_{12})(\Theta_{21}G + \Theta_{22})^{-1}]$$

for an antistable rational $M \times N$ matrix function G with $\|G\|_\infty < 1$. (Here in general if W is a rational matrix function $\mathrm{st}[W]$ stands for the strictly stable part of W, i.e. the sum of all terms in the partial fraction expansion for W with poles in the left half plane.)

PROOF. This is an immediate consequence of Theorem 20.5.1 with $K = \mu^{-1}P$. □

24.4 MODEL REDUCTION FOR DISCRETE TIME SYSTEMS

In this section we describe the analogues of the results given in the previous sections of this chapter on model reduction for discrete time systems.

For the discrete time system, the time variable (now denoted by k rather than t), is assumed to take integer values rather than continuous real values; if we assume that the process starts at time $k = 0$, then input and output signals are vector functions defined on the nonnegative integers (i.e. vector-valued sequences) $k \to u_k$ and $k \to y_k$. The system manifests itself through an input-output map $\{u_k\}_{k \geq 0} \to \{y_k\}_{k \geq 0}$. If we assume that the system is *linear time-invariant* and *finite-dimensional*, then we may assume that the system is driven by a first order difference equation

(24.4.1)
$$\begin{aligned} x_{k+1} &= Ax_k + Bu_k, \qquad x_0 = 0 \\ y_k &= Cx_k + Du_k. \end{aligned}$$

Here A, B, C, D are constant matrices of compatible sizes (with A being square), and the vector x_k describes the *state* of the system at time k. One can solve the system (24.4.1) for $\{y_k\}$ explicitly in terms of $\{u_k\}$:

(24.4.2)
$$y_k = Du_k + \sum_{j=0}^{k-1} CA^{k-j-1}Bu_j, \qquad k \geq 0.$$

In place of the Laplace transform, we use the Z-transform defined by

$$\{v_k\}_{k \geq 0} \to V(z) = \sum_{j=0}^{\infty} v_j z^{-j}.$$

If the sequence $\{v_k\}$ has a geometrical decay at infinity, then $V(z)$ represents a vector-valued function analytic in some neighborhood of infinity. If we apply the Z-transform to (24.4.1) we get the system

$$\begin{aligned} zX(z) &= AX(z) + BU(z) \\ Y(z) &= CX(z) + DU(z). \end{aligned}$$

We can then solve for $Y(z)$ in terms of $U(z)$ and get

$$Y(z) = W(z)U(z)$$

where

(24.4.3) $$W(z) = D + C(zI - A)^{-1}B$$

is the *transfer function* of the discrete time system (24.4.1). Stability is defined as in the continuous time case (see Section 23.1): bounded input sequences $\{u_k\}_{k \geq 0}$ must map to bounded output sequences $\{y_k\}_{k \geq 0}$. It is not difficult to show that this is equivalent to $W(z)$ having all its poles inside the unit disk $\mathcal{D} = \{z : |z| < 1\}$. For completeness, we prove this statement. Assume all the poles of $W(z)$ are in $\mathcal{D}$; then we have

$$W(z) = \sum_{j=0}^{\infty} W_j z^{-j}$$

and the series converges for all $z \in \mathbb{C} \cup \{\infty\}$ with $|z| > r$, where $0 < r < 1$. So

$$K := \sum_{j=0}^{\infty} \|W_j\| < \infty.$$

Now for a bounded input sequence $\{u_k\}_{k=0}^{\infty}$ and the corresponding output sequence $\{y_k\}_{k=0}^{\infty}$ we have

$$\|y_j\| = \left\| \sum_{k=0}^{j} W_{j-k} u_k \right\|$$

$$\leq \sup\{\|u_k\| : k = 0, 1, \ldots\} \sum_{k=0}^{j} \|W_{j-k}\|$$

$$\leq K \sum \{\|u_k\| : k = 0, 1, \ldots\}.$$

Conversely, let z_0 be a pole of $W(z)$ of maximal absolute value, and assume $|z_0| \geq 1$. We have to eshibit a bounded input sequence $\{u_k\}_{k=0}^{\infty}$ for which the corresponding output sequence $\{y_k\}_{k=0}^{\infty}$ is unbounded. As $W(z)$ is a rational function, the series

$$\sum_{j=0}^{\infty} W_j z_0^{-j}$$

diverges, where W_j are the coefficients of the Laurent series (centered at infinity) for $W(z)$. Now for the bounded input sequence $u_k = z_0^{-k}$, $k = 0, 1, \ldots$, we have

$$y_j = \sum_{k=0}^{j} W_{j-k} u_k = z_0^j \sum_{k=0}^{j} W_k z_0^{-k},$$

which is unbounded.

If the realization (A, B, C, D) is minimal (i.e. the size of A is minimal among all realizations of $W(z)$), then by results in Section 4.2 it follows that bounded input sequences must map to bounded output sequences if and only if $\sigma(A) \subset \mathcal{D}$.

The model reduction problem for the discrete time case can now be stated as for the continuous time case. We are given a stable system (24.4.1) for which the size $n \times n$ of the matrix A is larger than is desirable. Without loss of generality we may assume that (C, A) is a null kernel pair and (A, B) is a full range pair, so it is impossible to reduce the size of A without changing the input-output behavior. The problem then is to reduce the size of A while changing the input-output behavior as little as possible.

For the discrete-time case we define the infinity norm $\|W\|_\infty$ of the transfer function W by

(24.4.4) $$\|W\|_\infty = \sup\{\|W(z)\|\colon |z| = 1\}.$$

Analogously to the continuous time case, it turns out that $\|W\|_\infty$ is the induced operator norm of the input-output map $\{u_k\}_{k\geq 0} \to \{y_k\}_{k\geq 0}$ for the system (24.4.1), if one uses the ℓ^2-norm on spaces of vector-valued sequences

$$\|\{v_k\}_{k\geq 0}\|_2^2 = \sum_{k=0}^\infty \|v_k\|^2.$$

For a matrix function $W(z)$ analytic in a neighborhood of infinity with value at infinity equal to 0 with power series representation

$$W(z) = \sum_{j=1}^\infty \breve{W}_j z^{-j} \qquad (|z| \text{ large}),$$

the associated Hankel matrix $\mathcal{H}_W$ was introduced in Section 20.1; it is the infinite matrix with rows and columns indexed by the positive integer numbers with (i,j)-th entry $[\mathcal{H}_W]_{i,j}$ given by

(24.4.5) $$[\mathcal{H}_W]_{i,j} = W_{i+j-1}.$$

If $W(z)$ is rational with all poles inside the unit disk and has $M \times N$ matrices (say) as values, then the matrix $\mathcal{H}_W$ defines a bounded operator form ℓ_N^2 into ℓ_M^2 (where in general ℓ_r^2 is the space of $\mathbf{C}^r$-valued sequences which are square summable in norm). Let $\|\mathcal{H}_W\|$ be the induced operator norm of $\mathcal{H}_W$ as an operator from ℓ_N^2 into ℓ_M^2. By Theorem 20.1.1 (Nehari's theorem), $\|\mathcal{H}_W\|$ expresses the distance from W to the space of antistable rational matrix functions (i.e. functions with all poles outside the closed unit disk) in the infinity norm

(24.4.6) $$\|\mathcal{H}_W\| = \inf\{\|W - R\|_\infty \colon R \in \mathcal{R}_{M\times N}(\mathcal{D} \cup \partial\mathcal{D})\}.$$

As is the case for continuous time systems, the Hankel matrix $\mathcal{H}_W$ has a factorization

(24.4.7) $$\mathcal{H}_W = \Omega\Xi$$

where Ω is the observability operator

$$\Omega = \mathrm{col}(CA^k)_{0 \le k < \infty} : \mathbb{C}^n \to \ell_M^2$$

and Ξ is the controllability operator

$$\Xi = [B, AB, A^2 B, \ldots] : \ell_N^2 \to \mathbb{C}^n$$

(see the proof of Theorem 20.1.2). In particular, $\mathcal{H}_W$ has finite rank n if W is stable with McMillan degree n. Define controllability and observability gramians Π and Q by

(24.4.8)
$$\Pi = \sum_{k=0}^{\infty} A^k BB^* A^{*k} = \Xi\Xi^*$$

and

(24.4.9)
$$Q = \sum_{k=0}^{\infty} A^{*k} C^* C A^k = \Omega^* \Omega.$$

Then an equivalent form of Nehari's theorem (Theorem 20.1.1) asserts that (24.4.6) is equal to the largest eigenvalue of the finite matrix ΠQ:

$$\|\mathcal{H}_W\| = \lambda_{\max}(\Pi Q).$$

Moreover Π and Q are directly computable as solutions of the Stein equations

(24.4.10)
$$\Pi - A\Pi A^* = BB^*$$

(24.4.11)
$$Q - A^* Q A = C^* C.$$

The following analogue of Theorem 24.2.1 gives the connection between the infinity norm and the Hankel norm for the discrete time case.

THEOREM 24.4.1. *Let $W(z)$ be a proper rational $M \times N$ matrix function of McMillan degree n with all poles in the unit disk $\mathcal{D}$, and define $\|W\|_\infty$ and $\mathcal{H}_W$ by (24.4.4) and (24.4.5). Then*

$$\|\mathcal{H}_W\| \le \|W\|_\infty \le 2n\|\mathcal{H}_W\|.$$

The proof relies on the following discrete analogue of Lemma 24.2.2.

LEMMA 24.4.2. *Suppose $[K_{ij}]_{0 \le i,j < \infty}$ is a block matrix (with blocks of size $M \times N$) which defines an operator of finite rank n, denoted by $\mathcal{K}$, from ℓ_N^2 into ℓ_M^2, and let $s_1(\mathcal{K}) \ge \cdots \ge s_n(\mathcal{K})$ denote the nonzero singular values of $\mathcal{K}$. Then*

$$\sum_{j=0}^{\infty} \|K_{jj}\| \le \sum_{j=1}^{n} s_j(\mathcal{K}).$$

PROOF. The proof parallels exactly the proof of Lemma 24.2.2. Since $\mathcal{K}$ is of finite rank n, there exists two orthonormal sets $\{\{\varphi_i^{(k)}\}_{i\geq 0}: 1 \leq k \leq n\}$ and $\{\{\psi_i^{(k)}\}_{i\geq 0}: 1 \leq k \leq n\}$ in ℓ_M^2 and ℓ_N^2 respectively such that

$$K_{ij} = \sum_{k=1}^{n} s_k \varphi_i^{(k)} [\psi_j^{(k)}]^*, \qquad i, j \geq 0$$

where $s_k = s_k(\mathcal{K})$. Thus

$$\sum_{i=0}^{\infty} \|K_{ii}\| \leq \sum_{k=1}^{n} s_k \left[\sum_{i=0}^{\infty} \|\varphi_i^{(k)}\| \|\psi_i^{(k)}\| \right]$$

$$\leq \sum_{k=1}^{n} s_k \left\{ \sum_{i=0}^{\infty} \|\varphi_i^{(k)}\|^2 \right\}^{1/2} \left\{ \sum_{i=0}^{\infty} \|\psi_i^{(k)}\|^2 \right\}^{1/2}$$

$$= \sum_{k=1}^{n} s_k$$

as claimed. □

PROOF OF THEOREM 24.4.1. The inequality $\|\mathcal{H}_W\| \leq \|W\|_\infty$ follows from Nehari's theorem (24.4.6). For the second inequality $\|W\|_\infty \leq 2n\|\mathcal{H}_W\|$, note first from the expansion $W(z) = \sum_{j=0}^{\infty} \breve{W}_j z^{-j}$ valid on the unit circle that

$$(24.4.12) \qquad \|W\|_\infty = \sup_{|z|=1} \|W(z)\| \leq \sum_{j=1}^{\infty} \|\breve{W}_j\|.$$

Also, if we apply Lemma 24.4.2 to the Hankel matrix $\mathcal{H}_W$ ($[\mathcal{H}_W]_{ij} = \breve{W}_{i+j-1}$), we get

$$(24.4.13) \qquad \sum_{j=1}^{\infty} \|\breve{W}_{2j-1}\| \leq \sum_{j=1}^{n} s_j(\mathcal{H}_W) \leq n s_1(\mathcal{H}_W) = n\|\mathcal{H}_W\|.$$

If we consider instead the shifted Hankel matrix $S^*\mathcal{H}_W$ with matrix entries $[S^*\mathcal{H}_W]_{ij} = \breve{W}_{i+j}$, we get

$$(24.4.14) \qquad \begin{aligned} \sum_{j=1}^{\infty} \|\breve{W}_{2j}\| &\leq \sum_{j=1}^{n} s_j(S^*\mathcal{H}_W) \\ &\leq n s_1(S^*\mathcal{H}_W) \leq n\|\mathcal{H}_W\|. \end{aligned}$$

We conclude by putting together (24.4.12)–(24.4.14) that

$$\|W\|_\infty \leq \sum_{j=1}^{\infty} \|\breve{W}_j\| \leq 2n\|\mathcal{H}_W\|. \qquad □$$

Motivation for use of the Hankel norm (i.e. the norm of the Hankel operator $\mathcal{H}_W$) for the discrete time model reduction problems can be given in the same way as was given for the continuous time version. The infinity norm is motivated physically as the induced operator norm of the input-output map on ℓ^2, but Theorem 24.4.1 implies that the Hankel norm is an equivalent norm on transfer functions of bounded McMillan degree and so is a reasonable substitute.

The problem of model reduction for a discrete time system we now state formally as follows. Given is a strictly proper rational $M \times N$ matrix function $P(z)$ of McMillan degree n. We also assume that $P(z)$ is *stable*, which in this section only will mean that all poles of P are in the unit disk $\mathcal{D}$. We are also given a tolerance level $\mu > 0$. The problem then is to find the integer

$$\kappa_0 = \min\{\kappa\colon \text{there exists stable rational } \widetilde{P} \text{ with } \delta(\widetilde{P}) \leq \kappa, \; \|\mathcal{H}_{P-\widetilde{P}}\| < \mu\}.$$

The solution is essentially given by the Nehari-Takagi theorem for the unit disk (see Theorem 20.5.1). The proof of the following follows in the same way as Theorem 24.3.1.

THEOREM 24.4.3. *Let P be a stable (in the discrete time sense) strictly proper rational $M \times N$ matrix function with minimal realization*

$$P(z) = C(zI - A)^{-1}B.$$

Let Π and Q be the controllability and observability gramians associated with this realization given by (24.4.8) and (24.4.9), and let $\mu > 0$. Then

$$\kappa_0 := \min\{\kappa\colon \text{there exists stable rational } \widetilde{P}$$
$$\text{with } \delta(\widetilde{P}) = \kappa \text{ and } \|\mathcal{H}_{P-\widetilde{P}}\| < \mu\}$$

is given by

$$\kappa_0 = \min\{\kappa\colon s_{\kappa+1}(\mathcal{H}_P) < \mu\}$$

where $s_1(\mathcal{H}_P) \geq \cdots \geq s_n(\mathcal{H}_P)$ are the non-zero singular values of the Hankel operator $\mathcal{H}_P$, or equivalently by

$$\kappa_0 = \min\{\kappa\colon \lambda_{\kappa+1}(\Pi Q) \leq \mu^2\}$$

where $\lambda_1(\Pi Q) \geq \cdots \geq \lambda_n(\Pi Q)$ are the eigenvalues of the finite matrix ΠQ.

As in Theorem 24.3.3 for the continuous time case, we may use the results of Chapter 20 to parametrize all solutions of the model matching problem for a given prescribed level of performance μ, at least up to an antistable additive term. For the discrete time case, for $W(z)$ a given rational matrix function we define the stable part st$[W]$ of W to be the sum of all terms in the partial fraction expansion for $W(z)$ associated with poles in the open unit disk. The following is the discrete time analogue of Theorem 24.3.3 and is a direct consequence of Theorem 20.5.1.

THEOREM 24.4.4. *Let $P(z)$ be a strictly proper, stable (in the discrete time sense) rational $M \times N$ matrix function of McMillan degree n having minimal realization*

$$P(z) = C(zI - A)^{-1}B$$

with controllability and observability gramians given by (24.4.8) *and* (24.4.9). *Choose* $\mu > 0$ *so that*

$$\lambda_{\kappa+1}(\Pi Q) < \mu^2 < \lambda_\kappa(\Pi Q)$$

and define a rational $(M + N) \times (M + N)$ *matrix function* $\Theta(z)$ *by*

(24.4.15)
$$\Theta(z) = I - (z - z_0) \left[\begin{array}{cc} \mu^{-1}C & 0 \\ 0 & B^* \end{array} \right] \left[\begin{array}{cc} (zI - A)^{-1} & 0 \\ 0 & (I - zA^*)^{-1} \end{array} \right]$$
$$\times \left[\begin{array}{cc} Z\Pi & Z \\ Z^* & QZ \end{array} \right] \left[\begin{array}{cc} (I - z_0 A^*)^{-1}\mu^{-1}C^* & 0 \\ 0 & (A - z_0 I)^{-1}B \end{array} \right],$$

where z_0 *is a fixed point on the unit circle and* $Z = (I - \Pi Q)^{-1}$. *Then a rational matrix function* $\widetilde{P}(z)$ *has the properties*

(i) $\|\mathcal{H}_{P - \widetilde{P}}\| < \mu$

and

(ii) $\widetilde{P}$ *is stable* (*in the discrete time sense*) *with* $\delta(\widetilde{P}) = \kappa$

if and only if

$$P - \widetilde{P} = \text{st}[(\Theta_{11}G + \Theta_{12})(\Theta_{21}G + \Theta_{22})^{-1}]$$

where G *is an* $M \times N$ *rational matrix function with all poles outside the unit disk* $\mathcal{D}$ *for which* $\sup\{\|G(z)\| : |z| = 1\} < 1$.

If it happens that the matrix A is invertible, then (under the hypotheses of Theorem 24.4.4) $\Theta(z)$ can be chosen alternatively in the form

$$\Theta(z) = D + \left[\begin{array}{cc} \mu^{-1}C & 0 \\ 0 & B \end{array} \right] \left[\begin{array}{cc} (zI - A)^{-1} & 0 \\ 0 & (I - zA^*)^{-1} \end{array} \right]$$
$$\times \left[\begin{array}{cc} Z\Pi & Z \\ Z^* & QZ \end{array} \right] \left[\begin{array}{cc} A^{*-1}\mu^{-1}C & 0 \\ 0 & B \end{array} \right] D,$$

where

$$D = I - \left[\begin{array}{cc} \mu^{-1}C & 0 \\ 0 & -B^*A^{*-1} \end{array} \right] \left[\begin{array}{cc} Z\Pi & Z \\ Z^* & QZ \end{array} \right]$$
$$\times \left[\begin{array}{cc} (I - \alpha A)^{-1} & 0 \\ 0 & (\alpha I - A)^{-1} \end{array} \right] \left[\begin{array}{cc} -\mu^{-1}C^* & 0 \\ 0 & B \end{array} \right]$$

and α is any fixed point on the unit circle.

CHAPTER 25

ROBUST STABILIZATION

In many engineering problems the plant is known only approximately. It is important to know therefore that the system will perform adequately if the true plant is only approximately equal to the model or nominal plant around which the system is designed. This is particularly crucial with regard to the issue of stability. In this chapter we show how one particular form of this robust stability issue reduces to an interpolation problem which in turn can be solved explicitly via the formalism developed in Chapter 18.

25.1 INTRODUCTION

Consider the system in Figure 25.1.1. As was explained in Chapter 23, internal stability of the closed loop system $\Sigma(P, K)$ corresponds to stability of the input-output maps from (u_1, u_2) to (e_1, e_2), or equivalently, from (u_1, u_2) to (e_1, e_2). After Laplace transformation the input-output maps associated with the plant P and compensator K become multiplication by their respective transfer functions $P = P(z)$ and $K = K(z)$. Then, as we have seen in Chapter 23, internal stability is equivalent to the four transfer functions $(I + PK)^{-1}$, $(I + PK)^{-1}P$, $K(I + PK)^{-1}$ and $(I + KP)^{-1}$ being proper and having all poles in the open left half plane.

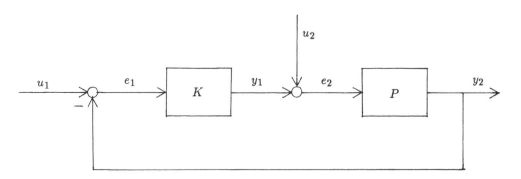

Figure 25.1.1

However, as has often been pointed out in the engineering literature (see the notes after this chapter), the plant is not usually known exactly. In fact, one assumes that one has a nominal plant P and the true plant is some slightly perturbed plant $\tilde{P} = P + \Delta$ of the nominal plant. It is usually assumed that the size of the perturbation Δ is known, e.g. one might specify that

$$\|W_1 \Delta W_2\|_\infty \leq \varepsilon$$

where W_1 and W_2 are known rational matrix weight functions for which W_1, W_1^{-1}, W_2, W_2^{-1} are all proper and stable and $\varepsilon > 0$, although in practice one may want to specify some more structured form of uncertainty (see Notes). We shall also restrict consideration to perturbed plants $\widetilde{P} = P + \Delta$ which have the same number of right half plane poles (counting multiplicities) as the nominal plant P, since ultimately this condition turns out to be satisfied when robust stability holds (see Notes). The precise definition then is: *the system $\Sigma(P, K)$ in Figure 25.1.1 is robustly stable (with respect to the matrix weights W_1, W_2 and positive number ε) if and only if the system $\Sigma(P+\Delta, K)$ is internally stable for all rational $N \times N$ matrix valued perturbations Δ such that*

(25.1.1) Δ *is proper with no poles on the imaginary line*

(25.1.2) $\begin{array}{l} P \text{ and } P + \Delta \text{ have the same number of poles} \\ \text{(counting multiplicities) in the right half plane} \end{array}$

and

(25.1.3) $\|W_1 \Delta W_2\|_\infty \le \varepsilon.$

In this definition W_1 and W_2 are rational $N \times N$ matrix functions such that

(25.1.4) $W_1^{\pm 1}, W_2^{\pm 1}$ *are proper and stable (all poles in the open left half plane Π^-).*

The following result is the first step in reducing the *robust stabilization problem* to the generalized Nevanlinna-Pick interpolation problem already discussed in depth in Chapter 18.

THEOREM 25.1.1. *Let P be a proper rational $N \times N$ matrix function with no poles on the imaginary line, let $\varepsilon > 0$ and let W_1, W_2 be rational $N \times N$ matrix functions satisfying (25.1.4). Then K is a rational matrix function for which $\bigl(\Sigma(P, K), W_1, W_2, \varepsilon\bigr)$ is robustly stable if and only if*

(25.1.5) $\|W_2^{-1} K(I + PK)^{-1} W_1^{-1}\|_\infty < \varepsilon^{-1}.$

The proof relies on the following winding number characterization of internal stability. As in earlier chapters we denote by $\operatorname{wno} f$ the change in argument of the continuous scalar function $f(z)$ along the imaginary axis if $f(z)$ has no zeros on the axis. If $f(z)$ has finitely many zeros and/or poles on the axis, we define $\operatorname{wno} f$ to be the change in argument of f along the contour consisting of the imaginary axis with slight indendations into the left half plane around the zeros and poles f on the axis.

LEMMA 25.1.2. *The system in Figure 25.1.1 is internally stable if and only if $I + PK$ is proper and*

(25.1.6) $\operatorname{wno} \det(I + PK) = -(n_P + n_K)$

where

$n_P = $ *number of poles (counting multiplicities)*

of P in the closed right half plane $\overline{\Pi^+}$ (including infinity)

and

$$n_K = number\ of\ poles\ (counting\ multiplicities)$$
$$of\ K\ in\ \overline{\Pi^+}\ (including\ infinity).$$

PROOF. For this argument we work with a left coprime factorization

$$(25.1.7) \qquad\qquad P = \widetilde{D}_P^{-1} \widetilde{N}_P$$

of the plant P, and a right coprime factorization

$$(25.1.8) \qquad\qquad K = N_K D_K^{-1}$$

of the compensator K, localized to $\overline{\Pi^+}$. By this we mean that $\widetilde{N}_P$ and $\widetilde{D}_P$ in the representation (25.1.7) for P are *left coprime* over $\overline{\Pi^+}$, i.e., there exist stable rational $N \times N$ matrix functions $\widetilde{X}$, $\widetilde{Y}$ for which

$$(25.1.9) \qquad\qquad \widetilde{N}_P \widetilde{X} + \widetilde{D}_P \widetilde{Y} = I_N.$$

Similarly, N_K and D_K in the representation (25.1.8) for K are *right coprime* over $\overline{\Pi^+}$ in the sense that there exist stable rational matrix functions X and Y for which

$$(25.1.10) \qquad\qquad X N_K + Y D_K = I_N.$$

Existence of such representations for P and K follows from Theorem 11.1.5 after a linear fractional change of variable is applied to move infinity to a point in the left half plane.

From the representation (25.1.7) for P and (25.1.8) for K we get

$$(25.1.11) \qquad\qquad I + PK = \widetilde{D}_P^{-1} \Delta D_K^{-1}$$

where

$$(25.1.12) \qquad\qquad \Delta = \widetilde{D}_P D_K + \widetilde{N}_P N_K,$$

and consequently

$$\text{wno} \det(I + PK) = -\text{wno} \det D_P + \text{wno} \det \Delta - \text{wno} \det D_K.$$

By the argument principle, wno $\det D_P$ is the total number of zeros of $\det D_P$ in $\overline{\Pi^+}$. By Corollary 1.1.7, this in turn is the sum of all the zero partial multiplicities for $D_P(z)$ over all zeros in $\overline{\Pi^+}$. By Theorem 11.1.6, this in turn is exactly n_P, the sum of all the pole partial multiplicities for $P(z)$ over poles in $\overline{\Pi^+}$. Similarly wno $\det D_K = n_K$. Thus the condition (25.1.6) of the lemma is equivalent to

$$\text{wno} \det \Delta = 0.$$

As Δ is analytic on $\overline{\Pi^+}$ by construction, this is equivalent in turn to Δ^{-1} being stable. Thus the lemma is proved once we show that internal stability of the system $\Sigma(P, K)$ is equivalent to stability of the single function Δ^{-1} where Δ is given by (25.1.12). (Note that the well-posedness assumption on $\Sigma(P, K)$, i.e. the condition $\det(I + PK) \neq 0$, guarantees that $\det \Delta \not\equiv 0$, so Δ^{-1} exists as a rational matrix function.)

Internal stability, as explained in Chapter 23, is equivalent to stability of the four matrix functions $(I+PK)^{-1}$, $(I+PK)^{-1}P$, $K(I+PK)^{-1}$ and $(I+KP)^{-1}$. From (25.1.11) we can express these as

(25.1.13)
$$(I + PK)^{-1} = D_K \Delta^{-1} \widetilde{D}_P$$

(25.1.14)
$$(I + PK)^{-1}P = D_K \Delta^{-1} \widetilde{N}_P$$

(25.1.15)
$$K(I + PK)^{-1} = N_K \Delta^{-1} \widetilde{D}_P$$

and

(25.1.16)
$$K(I + PK)^{-1}P = N_K \Delta^{-1} \widetilde{N}_P.$$

If Δ^{-1} is stable then it is clear that each of the functions (25.1.13)–(25.1.16) is stable. Conversely, suppose each of (25.1.13)–(25.1.16) is stable. Multiply (25.1.13) on the right by $\widetilde{Y}$ and (25.1.14) on the right by $\widetilde{X}$ and then add to get (by (25.1.9)) that $D_K \Delta^{-1}$ is stable. By a similar argument applied to (25.1.15) and (25.1.16) we get that $N_K \Delta^{-1}$ is stable. Finally use (25.1.10) to get that

$$X \cdot N_K \Delta^{-1} + Y \cdot D_K \Delta^{-1} = (XN_K + YD_K)\Delta^{-1} = \Delta^{-1}$$

is stable. We conclude that internal stability is equivalent to stability of Δ^{-1} as needed.
$\square$

PROOF OF THEOREM 25.1.1. We consider the system $\Sigma(P, K)$ as in Figure 25.1.1 and also the system $\Sigma(\widetilde{P}, K)$ with the same compensator K but perturbed plant $\widetilde{P} = P + \Delta$. By assumption the number of right half plane poles $n_{\widetilde{P}}$ for $\widetilde{P}$ is the same as the number n_P for P. To apply Lemma 25.1.2, we note the identity

(25.1.17) $I + \widetilde{P}K = I + PK + \Delta K = (I + \Delta K(I + PK)^{-1})(I + PK).$

Thus

$$\text{wno} \det(I + \widetilde{P}K) = \text{wno} \det(I + \Delta K(I + PK)^{-1}) + \text{wno} \det(I + PK).$$

By Lemma 25.1.2, since $\Sigma(P, K)$ is internally stable we have

$$\text{wno} \det(I + PK) = -n_P - n_K$$

and internal stability of $\Sigma(\widetilde{P}, K)$ is equivalent to

$$\text{wno} \det(I + \widetilde{P}K) = -n_{\widetilde{P}} - n_K.$$

Since $n_{\widetilde{P}} = n_P$ this in turn is equivalent to

$$\operatorname{wno}\det(I + \widetilde{P}K) = \operatorname{wno}\det(I + PK).$$

From (25.1.16) this comes down to

(25.1.18) $\operatorname{wno}\det\left(I + \Delta K(I + PK)^{-1}\right) = 0.$

Thus robust stability we now see is equivalent to (25.1.18) holding for all Δ with

(25.1.19) $\|W_1 \Delta W_2\|_\infty < \varepsilon$

subject to $P + \Delta$ and P having the same number of right half plane poles. Suppose first that

(25.1.20) $\|W_2^{-1} K(I + PK)^{-1} W_1^{-1}\|_\infty < \varepsilon^{-1}$

and Δ satisfies (25.1.19). Then we claim that $[I + t\Delta K(I + PK)^{-1}](ix)$ is invertible for all real x (including at infinity) and for all t, $0 \le t \le 1$. In general, we denote by $\rho(X)$ the spectral radius of a matrix X

$$\rho(X) = \max\{|\lambda| \colon \lambda \text{ is an eigenvalue of } X\}.$$

Then, since the spectrum is invariant under similarity, we have

(25.1.21) $\rho\big([t\Delta K(I + PK)^{-1}](ix)\big) = \rho\big(t[(W_1 \Delta W_2) \cdot W_2^{-1} K(I + PK)^{-1} W_1^{-1}](ix)\big)$

where

$$\|t[(W_1 \Delta W_2) \cdot W_2^{-1} K(I + PK)^{-1} W_1^{-1}](ix)\|$$
$$\le \|W_1 \Delta W_2\|_\infty \|W_2^{-1} K(I + PK)^{-1} W_1^{-1}\|_\infty < 1.$$

Therefore

$$\rho\big(t[W_1 \Delta W_2 \cdot W_2^{-1} K(I - PK)^{-1} W_1^{-1}](ix)\big) < 1,$$

and by taking into account (25.1.21) we obtain the invertibility of $[I + t\Delta K(I + PK)^{-1}](ix)$ for all $x \in \mathbb{R}$ (including ∞) and for all t, $0 \le t \le 1$. Now by a standard homotopy argument, we conclude that

$$\operatorname{wno}\det\big(I + \Delta K(I + PK)^{-1}\big) = \operatorname{wno}\det I = 0$$

as needed.

For the converse we need to show that if $\|W_2^{-1} K(I + PK)^{-1} W_1^{-1}\|_\infty \ge \varepsilon^{-1}$, then there is a perturbation Δ satisfying (25.1.3) and with $\widetilde{P} = P + \Delta$ having the same number of right half plane poles as P but for which $\Sigma(\widetilde{P}, \Delta)$ is not internally stable. Set $Q = K(I + PK)^{-1}$. By assumption there is a point ix_0 on the imaginary axis (possibly equal to infinity) for which $\|[W_2^{-1} Q W_1^{-1}](ix_0)\| \ge \varepsilon^{-1}$. Let

$$[W_2^{-1} Q W_1^{-1}](ix_0) = U\Sigma V$$

be the singular value decomposition of $[W_2^{-1}QW_1^{-1}](ix_0)$ (see Gohberg-Krein [1969], or advanced books on matrices: Horn-Johnson [1985], Lancaster-Tismenetsky [1985], Golub-Van Loan [1989]); thus U and V are unitary matrices and Σ is a diagonal matrix with diagonal entries $\sigma_1 \geq \sigma_2 \geq \cdots \geq \sigma_N \geq 0$. Thus $\|[W_2^{-1}QW_1^{-1}](ix_0)\| = \sigma_1 \geq \varepsilon^{-1}$. Define a matrix function Δ by

$$\Delta(z) = W_1(z)^{-1}V^* \begin{bmatrix} -\sigma_1^{-1} & & & \\ & 0 & & \\ & & \ddots & \\ & & & 0 \end{bmatrix} U^*W_2(z)^{-1}.$$

Thus

$$\|W_1\Delta W_2\|_\infty = \|V^* \operatorname{diag}(-\sigma_1^{-1}, 0, \ldots, 0)U^*\| = \sigma_1^{-1} \leq \varepsilon$$

and $\widetilde{P} = P + \Delta$ has the same number of right half plane poles as P since Δ is stable. Thus Δ is an admissible perturbation. Moreover,

$$I + \Delta K(I + PK)^{-1} = W_1^{-1}[I + W_1\Delta W_2 \cdot W_2^{-1}QW_1^{-1}]W_1$$

where

$$\begin{aligned}[I + W_1\Delta W_2 \cdot W_2^{-1}QW_1^{-1}](ix_0) &= I + V^* \operatorname{diag}(-\sigma_1^{-1}, 0, \ldots, 0)\Sigma V \\ &= V^* \operatorname{diag}(0, 1, \ldots, 1)V\end{aligned}$$

is not invertible. From the factorization (25.1.17), namely,

$$I + \widetilde{P}K = \big(I + \Delta K(I + PK)^{-1}\big)(I + PK),$$

we conclude that $(I + \widetilde{P}K)^{-1}$ has a pole at ix_0, and hence the system $\Sigma(\widetilde{P}, K)$ cannot be stable. This completes the proof of Theorem 25.1.1. $\square$

25.2   INTERPOLATION CHARACTERIZATION OF THE Q-PARAMETER

We consider the system configuration shown in Figure 25.1.1 with nominal plant P and compensator K. As explained in Section 23.1, internal stability of the system is equivalent to stability of the four transfer functions $(I + PK)^{-1}$, $-P(I + KP)^{-1}$, $(I + KP)^{-1}K$, $(I + KP)^{-1}$. In Chapter 23 we obtained a characterization of internal stability in terms of interpolation conditions on the sensitivity $S = (I + PK)^{-1}$. In this section we obtain an analogous characterization of internal stability in terms of interpolation conditions on the transfer function $Q = K(I + PK)^{-1}$ (sometimes called the "Q-parameter" in the literature) rather than on the sensitivity S.

While the interpolation conditions on S equivalent to internal stability were determined by a left null-pole triple for the plant P over the right half plane, the interpolation conditions on Q are determined by a pole triple for P over Π^+ (as introduced in Section 3.3). In particular, if the plant P is stable, the pole triple over Π^+ is trivial and stability of $Q = K(I + PK)^{-1}$ alone implies stability of the three other transfer functions $(I + PK)^{-1}$, $P(I + KP)^{-1}$ and $(I + KP)^{-1}$ as well.

We first illustrate the result for a simple case. Suppose that the poles of $P(z)$ in Π^+ occur at $w_1, \ldots, w_q$ and all these poles are simple with residues of rank 1. Thus at each such pole w_j $(1 \leq j \leq q)$ $P(z)$ has a Laurent expansion of the form

$$(25.2.1) \qquad P(z) = (z - w_j)^{-1} u_j x_j + [\text{analytic at } w_j]$$

where u_j is a column vector and x_j is a row vector. In this situation internal stability can be characterized as follows. Here again for mathematical convenience we drop the requirement that the compensator K be stable.

THEOREM 25.2.1. *Suppose that $P(z)$ is a rational $N \times N$ matrix function regular (i.e. with no poles or zeros) on the $i\mathbb{R}$-axis (including at infinity) with Laurent expansions at its poles $w_1, \ldots, w_q$ in Π^+ of the form (25.2.1) for some column vectors $u_1, \ldots, u_q$ and row vectors $x_1, \ldots, x_q$. Then a stable rational matrix function $Q(z)$ has the form*

$$(25.2.2) \qquad Q = K(I + PK)^{-1}$$

for a rational matrix function K for which the system $\Sigma(P, K)$ in Figure 25.1.1 is internally stable if and only if

$$(25.2.3) \qquad Q(w_j)u_j = 0$$

$$(25.2.4) \qquad x_j Q(w_j) = 0$$

and

$$(25.2.5) \qquad x_j Q'(w_j) u_j = 1$$

for $1 \leq j \leq q$, and $\det(I - PQ) \not\equiv 0$.

PROOF. If $\det(I - PQ) \not\equiv 0$, then Q has the desired form $Q = K(I + PK)^{-1}$ with

$$K = P^{-1}[(I - PQ)^{-1} - I].$$

Next we note the identities

$$(25.2.6) \qquad (I + PK)^{-1} = I - PQ$$

$$(25.2.7) \qquad P(I + KP)^{-1} = P - PQP$$

and

$$(25.2.8) \qquad (I + KP)^{-1} = I - QP$$

if $Q = K(I + PK)^{-1}$. Thus internal stability is equivalent to the stability of the three transfer functions PQ, $P - PQP$ and QP in addition to that of Q. In particular, if P is itself stable, then stability of Q alone guarantees internal stability of the system $\Sigma(P, K)$.

Now suppose that the poles $w_1, \ldots, w_q$ of P in the right half plane have the simple form as described in (25.2.1). Then the residue of $P(z)Q(z)$ at w_j is equal to $u_j x_j Q(w_j)$. Since u_j is a nonzero column vector, analyticity of PQ at w_j is equivalent to Q satisfying the interpolation condition (25.2.3). Similarly analyticity of QP at w_j is equivalent to the interpolation condition (25.2.4). Finally we note that near w_j
(25.2.9)
$$
\begin{aligned}
P(z) - P(z)Q(z)P(z) = {}& (z - w_j)^{-1} u_j x_j - (z - w_j)^{-2} u_j x_j Q(z) u_j x_j \\
& + (z - w_j)^{-1} u_j x_j Q(z) [\text{analytic at } w_j] \\
& + (z - w_j)^{-1} [\text{analytic at } w_j] Q(z) u_j x_j + [\text{analytic at } w_j].
\end{aligned}
$$

If $Q(z)$ satisfies the interpolation conditions (25.2.3) and (25.2.4), then it is clear that the pole at w_j in (25.2.9) is simple with residue equal to

$$
\begin{aligned}
\operatorname{Res}_{z=w_j} [P(z) - P(z)Q(z)P(z)] &= u_j x_j - u_j x_j Q'(w_j) u_j x_j \\
&= u_j [1 - x_j Q'(w_j) u_j] x_j.
\end{aligned}
$$

Since u_j is a nonzero column vector and x_j is a nonzero row vector, this residue being equal to zero is equivalent to the interpolation condition (23.2.5). $\square$

For the general case internal stability is characterized in terms of a two-sided contour integral interpolation problem to be satisfied by the Q-parameter. The data for the interpolation problem comes from a pole triple (Z, T, Y) for the plant over the closed right half plane $\overline{\Pi^+}$. Recall from Section 3.3 that the triple of matrices (Z, T, Y) is a pole triple for the rational matrix function $P(z)$ over $\overline{\Pi^+}$ if

(i) (Z, T) is a right pole pair for $P(z)$ over $\overline{\Pi^+}$;

(ii) (T, Y) is a left pole pair for $P(z)$ over Π^+;

and

(iii) $P(z) - Z(zI - T)^{-1}Y$ is analytic on $\overline{\Pi^+}$.

The precise characterization of internal stability of a feedback system $\Sigma(P, K)$ in terms of the Q-parameter $Q = K(I + PK)^{-1}$ is as follows.

THEOREM 25.2.2. *Suppose that $P(z)$ is a rational $N \times N$ matrix function regular on the $i\mathbb{R}$-axis (including at infinity) with pole triple (Z, T, Y) over the closed right half plane Π^+. Then a stable rational matrix function $Q(z)$ has the form*

$$
Q = K(I + PK)^{-1}
$$

for a rational matrix function K for which the system $\Sigma(P, K)$ in Figure 25.1.1 is internally stable if and only if

(25.2.10)
$$
\sum_{z_0 \in \Pi^+} \operatorname{Res}_{z=z_0} Q(z) Z(zI - T)^{-1} = 0
$$

(26.2.11)
$$
\sum_{z_0 \in \Pi^+} \operatorname{Res}_{z=z_0} (zI - T)^{-1} Y Q(z) = 0
$$

(25.2.12) $$\sum_{z_0 \in \Pi^+} \text{Res}_{z=z_0}(zI - T)^{-1}YQ(z)Z(zI - T)^{-1} = I$$

and

$$\det(I - PQ) \not\equiv 0.$$

REMARK. In the terminology of Chapter 16, Theorem 25.2.2 states that the Q-parameter for an internally stable system $\Sigma(P, K)$ with given plant P must satisfy a TSCII problem with interpolation data set $\omega = (0, Z, T; T, Y, 0; I)$. In general, for the interpolation problem associated with data set $(C_+, C_-, A_\pi; A_\zeta, B_+, B_-, \Gamma)$ to have solutions, we know from Chapter 16 that the necessary consistency condition is that Γ satisfies the Sylvester equation

$$\Gamma A_\pi - A_\zeta \Gamma = B_+ C_+ + B_- C_-.$$

For the data set of the form ω given above, it is easy to check directly that this consistency condition holds

$$I \cdot T - T \cdot I = Y \cdot 0 + 0 \cdot Z.$$

PROOF. The condition $\det(I - PQ) \not\equiv 0$ guarantees that Q has the desired form $Q = K(I + PK)^{-1}$ for the matrix function $K = P^{-1}[(I - PQ)^{-1} - I]$ just as was the case in the proof of Theorem 25.2.1. Also as there, given that Q itself is stable, internal stability of $\Sigma(P, K)$ amounts to the stability of the three additional functions PQ, $P - PQP$ and QP. To analyze this issue in terms of the pole triple for $P(z)$, we need the following general lemma.

LEMMA 25.2.3. *Let $\sigma \subset \mathbb{C}$ and suppose (T, Y) is a full range pair with $\sigma(T) \subset \sigma$ and that $X(z)$ is a rational matrix function which is analytic on σ. Then $X(z)(zI - T)^{-1}$ is analytic on σ if and only if $X(z)(zI - T)^{-1}Y$ is analytic on σ.*

PROOF. This result was actually derived in the course of the proof of Theorem 23.3.1 but we isolate it here again for the sake of completeness. We prove the non-trivial "if" part of the lemma. By assumption $X(z)$ and $X(z)(zI - T)^{-1}Y$ are analytic on σ. From the identity

$$X(z)(zI - T)^{-1}TY = zX(z)(zI - T)^{-1}Y - X(z)Y$$

we get that then $X(z)(zI - T)^{-1}TY$ is also analytic on σ. By induction $X(z)(zI - T)^{-1}T^kY$ is analytic on σ for arbitrarily large k, and hence so also

(25.2.13) $$X(z)(zI - T)^{-1}[Y, TY, \ldots, T^kY].$$

Since (T, Y) is a full range pair, $[Y, TY, \ldots, T^kY]$ has a right inverse once k is large enough. Multiplication of (25.2.13) by this right inverse now gives that $X(z)(zI - T)^{-1}$ is analytic on σ as claimed. $\square$

To complete the proof of Theorem 25.2.2 we let (Z, T, Y) be a pole triple for $P(z)$ over $\overline{\Pi^+}$ (which is the same as a pole triple of $P(z)$ over Π^+) and we wish to give interpolation conditions on the matrix function $Q(z)$ (assumed to be analytic on $\overline{\Pi^+}$) which guarantee analyticity of PQ, $P - PQP$ and QP on Π^+. Since (Z, T, Y) is a pole triple for $P(z)$ over Π^+, we have by definition that $P(z)$ has the form

$$(25.2.14) \qquad\qquad P(z) = Z(zI - T)^{-1}Y + R(z)$$

where R is analytic on Π^+. Thus

$$P(z)Q(z) = Z(zI - T)^{-1}YQ(z) + R(z)Q(z)$$

is analytic on Π^+ if and only if the first term $Z(zI - T)^{-1}YQ(z)$ is analytic on Π^+. As (Z, T) is a null kernel pair and $Q(z)$ is analytic on Π^+, the transposed version of Lemma 25.2.3 gives that this in turn is equivalent to the analyticity of $(zI - T)^{-1}YQ(z)$ on Π^+. Now by Lemma 16.9.4 this is equivalent to the interpolation condition (25.2.11). In a similar way, we see that the analyticity of QP on Π^+ is equivalent to the interpolation condition (25.2.12).

Finally, to analyze $P - PQP$ we substitute in (25.2.14) to get

$$P(z) - P(z)Q(z)P(z)$$
$$= Z(zI - T)^{-1}Y + R(z) - Z(zI - T)^{-1}YQ(z)Z(zI - T)^{-1}Y$$
$$- R(z)Q(z)Z(zI - T)^{-1}Y - Z(zI - T)^{-1}YQ(z)R(z) - R(z)Q(z)R(z).$$

If $Q(z)$ satisfies the interpolation conditions (25.2.10) and (25.2.11), then again by Lemma 16.9.4 $Q(z)Z(zI - T)^{-1}$ and $(zI - T)^{-1}YQ(z)$ are analytic on Π^+. Thus, given that (25.2.10) and (25.2.11) hold, $P - PQP$ is analytic on Π^+ if and only if so also is
(25.2.15)
$$Z(zI-T)^{-1}Y - Z(zI-T)^{-1}YQ(z)Z(zI-T)^{-1}Y = Z(zI-T)^{-1}[Y - YQ(z)Z(zI-T)^{-1}Y].$$

Since the term in brackets is analytic on Π^+ and (Z, T) is a null kernel pair, again by the transposed version of Lemma 25.2.3 analyticity of (25.2.15) on Π^+ is equivalent to analyticity on Π^+ of

$$(25.2.16)\ \ (zI - T)^{-1}[Y - YQ(z)Z(zI - T)^{-1}Y] = [I - (zI - T)^{-1}YQ(z)Z](zI - T)^{-1}Y.$$

Since the term in brackets on the right side of the equality in (25.2.16) is analytic on Π^+ and (T, Y) is a full range pair, another application of Lemma 25.2.3 gives that analyticity of the expression in (25.2.16) on Π^+ is equivalent to analyticity on Π^+ of

$$(25.2.17) \qquad\qquad [I - (zI - T)^{-1}YQ(z)Z](zI - T)^{-1}.$$

Finally, Lemma 16.9.4 gives that analyticity of (25.2.17) on Π^+ is exactly equivalent to the interpolation condition (25.2.12), and Theorem 25.2.2 now follows. $\square$

25.3 SOLUTION OF THE ROBUST STABILIZATION PROBLEM

We are now in position to apply the interpolation results from Chapter 18 (together with some technical extensions developed in Chapter 23) to obtain a complete solution (in state space form) of the robust stabilization problem as stated in Section 25.3.1.

We first recall the problem. We are given a rational $N \times N$ matrix function $P(z)$ and a pole triple (Z, T, Y) for P over the open right half plane Π^+. (We assume that $P(z)$ is regular, i.e. analytic and invertible, on the $i\mathbb{R}$-axis, including at infinity.) We are also given weight functions W_1 and W_2 with minimal realizations

$$(25.3.1) \qquad W_j(z) = I + C_j(zI - A_j)^{-1}B_j$$

($j = 1, 2$). The functions W_1 and W_2 are assumed to be analytic and invertible on the closed right half plane, including at infinity; without loss of generality, we take the values at infinity to be I. We are also given a number $\varepsilon > 0$. A rational $N \times N$ matrix function Δ is said to be an *admissible perturbation* if conditions (25.1.1)–(25.1.3) are satisfied. The problem is to find a single compensator K so that the system $\Sigma(\widetilde{P} = P + \Delta, K)$ is internally stable for all choices of admissible perturbations Δ. If such a K exists, we say that the system $\Sigma(P, W_1, W_2, \varepsilon)$ is *robustly stabilizable*.

To state the solution of the robust stabilization problem, we first need to introduce a few more matrices. Define matrices R and L as the unique solutions of the Sylvester equations

$$(25.3.2) \qquad RT - A_2 R = B_2 Z$$

and

$$(25.3.3) \qquad LA_1 - TL_1 = -YC_1.$$

By assumption T has spectrum in the right half plane while A_1 and A_2 have spectra in the left half plane, so by the results of Section A.1, (25.3.2) and (25.3.3) have unique solutions T and L. Similarly we define Hermitian matrices S_1 and S_2 as the unique solutions of the Lyapunov equations

$$(25.3.4) \qquad S_1 T + T^* S_1 = (R^* C_2^* + Z^*)(C_2 R + Z)$$

$$(25.3.5) \qquad S_2 T^* + T S_2 = (LB_1 + Y)(B_1^* L^* + Y^*).$$

Finally, we define a Hermitian matrix Λ_ε by

$$(25.3.6) \qquad \Lambda_\varepsilon = \begin{bmatrix} S_1 & \varepsilon I \\ \varepsilon I & S_2 \end{bmatrix}.$$

The solution of the robust stabilization problem can now be stated as follows.

THEOREM 25.3.1. *Let $P(z)$ be a rational matrix function regular on $i\mathbb{R} \cup \{\infty\}$ and having (Z, T, Y) as a pole triple over Π^+, let $W_1(z)$ and $W_2(z)$ be rational matrix functions with no poles or zeros in $\overline{\Pi^+}$ with minimal realizations given by (25.3.1), and let $\varepsilon > 0$. Define matrices R, L, S_1, S_2 and Λ_ε by (25.3.2)–(25.3.6). Then the system $\Sigma(P, W_1, W_2, \varepsilon)$ is robustly stabilizable if and only if Λ_ε is positive definite.*

In the case where robustly stabilizing compensators K exist, it is straightforward to parametrize the set of all associated weighted Q-parameters

$$\widetilde{Q} = W_2^{-1} K (I + PK)^{-1} W_1^{-1}.$$

Note that once $\widetilde{Q}$ is known, one can backsolve for K. To do this we define a rational $(2N \times 2N)$-matrix function $\Theta_\varepsilon(z)$ by

$$
\begin{aligned}
(25.3.7) \quad \Theta_\varepsilon(z) &= \begin{bmatrix} \Theta_{\varepsilon 11}(z) & \Theta_{\varepsilon 12}(z) \\ \Theta_{\varepsilon 21}(z) & \Theta_{\varepsilon 22}(z) \end{bmatrix} . \\
&= \begin{bmatrix} I & 0 \\ 0 & I \end{bmatrix} + \begin{bmatrix} 0 & -(B_1^* L^* + Y^*) \\ C_2 R + Z & 0 \end{bmatrix} \\
&\quad \times \begin{bmatrix} (zI - T)^{-1} & 0 \\ 0 & (zI + T^*)^{-1} \end{bmatrix} \Lambda_\varepsilon^{-1} \begin{bmatrix} 0 & R^* C_2^* + Z^* \\ LB_1 + Y & 0 \end{bmatrix}
\end{aligned}
$$

The parametrization result is as follows.

THEOREM 25.3.2. *Let the data P, W_1, W_2, ε for the robust stabilization problem be as in Theorem 25.3.1. Define matrices R, L, S_1, S_2, Λ_ε by (25.3.2)–(25.3.6). Suppose Λ_ε is positive definite and define rational $N \times N$ matrix functions $\Theta_{\varepsilon ij}$ ($1 \leq i, j \leq 2$) by (25.3.7). Then the set of all weighted Q-parameters*

$$\widetilde{Q} = W_2^{-1} K (I + PK)^{-1} W_1^{-1}$$

associated with robustly stabilizing compensators K is given by

$$\widetilde{Q} = \varepsilon^{-1}(\Theta_{\varepsilon 11} G + \Theta_{\varepsilon 12})(\Theta_{\varepsilon 21} G + \Theta_{\varepsilon 22})^{-1}$$

where G is an arbitrary rational $N \times N$ matrix function with no poles in $\overline{\Pi^+}$ and with $\|G\|_\infty < 1$, subject to $\det(I - PW_2 \widetilde{Q} W_1) \not\equiv 0$.

PROOF. The proof of Theorems 25.3.1 and 25.3.2 amounts to a bookkeeping exercise of combining Theorems 25.1.1 and 25.2.2 with the results of generalized Nevanlinna-Pick interpolation from Chapter 18. Theorem 25.1.1 characterizes robustly stabilizing compensators as stabilizing compensators for which the associated Q-parameter $Q = K(I + PK)^{-1}$ satisfies

$$(25.3.8) \qquad \|W_2^{-1} Q W_1^{-1}\| < \varepsilon^{-1}.$$

On the other hand, Theorem 25.2.2 characterizes functions Q arising as Q-parameters for stabilizing compensators K as those stable functions which satisfy the interpolation conditions associated with the Π^+-admissible Sylvester TSCII data set $(0, Z, T; T, Y, 0; I)$, and for which $\det(I - PQ) \not\equiv 0$. If we consider εQ in place of Q, then (25.3.8) becomes

$$(25.3.9) \qquad \|W_2^{-1}(\varepsilon Q) W_1^{-1}\| < 1$$

and the relevant interpolation data set is

(25.3.10) $\omega = (0, Z, T; T, Y, 0; \varepsilon I).$

A construction in Chapter 23 (see Lemma 23.5.4) determines the interpolation conditions satisfied by $W_2^{-1}(\varepsilon Q)W_1^{-1}$ in terms of minimal realizations of the stable and stably invertible functions W_1 and W_2 and the interpolation conditions satisfied by εQ; a straightforward computation then shows that the relevant interpolation conditions for $W_2^{-1}(\varepsilon Q)W_1^{-1}$ are those determined by the Π^+-admissible Sylvester TSCII data set

$$\widetilde{\omega} = (0, C_2 R + Z, T; T, LB_1 + Y, 0; \varepsilon I)$$

where R and L are given by (25.3.2) and (25.3.3), respectively. The condition for stable solutions $\varepsilon \widetilde{Q} = W_2^{-1}(\varepsilon Q)W_1^{-1}$ of the interpolation conditions associated with $\widetilde{\omega}$ to exist which also satisfy (25.3.8) can be determined from Theorem 18.5.1; it is simply that Λ_ε given by (25.3.6) is positive definite. The parametrization of all weighted Q-parameters as in Theorem 25.3.2 is a straightforward specialization of the parametrization in the same Theorem 18.5.1. $\square$

NOTES FOR PART VI

Part VI is a self-contained introduction to some of the modern problems in systems theory. We choose these problems because of the important role played by the theory of interpolation for rational matrix functions in their solutions. This is an introduction only; we do not pretend that the results or literature cited here are complete.

The problems discussed here form a part of the recent direction in systems theory now known as H^∞-control. For more detail on this see Francis [1987]. Applications of matrix Nevanlinna-Pick problems, including some exotic variations in the problem statements, have also arisen in connection with circuit theory. We mention here Youla-Saitoh [1967], Helton [1980], Ball-Helton [1982], Delsarte-Genin-Kamp [1981].

CHAPTER 23 The presentation in this chapter follows Ball-Gohberg-Rodman [1989b]. The sensitivity minimization problem and engineering motivation behind it was introduced by Zames [1981] and then pursued in Zames-Francis [1983], Francis-Zames [1984] and Francis-Helton-Zames [1984]. This corresponds only to the simplest case of the standard H^∞-problem (see Francis [1987]); the general 4-block problem introduced by Doyle [1983] we do not treat here. A particularly important step in the state space solution of the problem was the presentation of results by Doyle at the Honeywell workshop (Doyle [1984]). This method of solution goes through several auxiliary steps (see Francis [1987]), before arriving at a Nevanlinna-Pick or Nehari problem. Clean, direct formulas for the solution analogous to those obtained here for the case of the weighted sensitivity minimization problem have recently been obtained for the general 4-block problem by Doyle-Francis-Glover-Khargonekar [1989].

CHAPTER 24 The main ideas of the exposition here on the model reduction problem come from Glover [1984]. The first step consists in the reduction to the Nehari-Takagi problem and the second is the solution of the latter, already given in Chapter 20. The proof is done here in a direct elementary way different from the method used in Glover [1984]. The same remark applies also to the connection between the Hankel and infinity norm.

CHAPTER 25 Theorem 25.1.1 is due to Doyle-Stein [1981] and Chen-Desoer [1982]. The winding number criterion for internal stability is well known and is called the Nyquist criterion in the engineering literature (see Vidyasagar [1985]). Theorems 25.2.1 and 25.2.2 appear here in this form for the first time; a related result but in terms of coprime factors is due to Bhaya-Desoer [1986]. After the results from Sections 25.1 and 25.2, it remains only to use results from chapter 18 to obtain the solution of the problem as presented in Section 25.3.

Kimura [1984] and Vidyasagar-Kimura [1986] reduced the robust stabilization problem to an interpolation problem. Glover [1986] (see also Curtain-Glover [1986]) obtained a reduction of the problem to one in a Nehari-Takagi form. Verma [1989] analyzed the situation where the hypothesis that P and $\widetilde{P}$ have the same number of right-half-plane poles is removed.

Variations in the form of the perturbation (e.g. multiplicative perturbations or perturbations of coprime factors of a given nominal plant) have also been considered (see e.g. Vidyasagar [1985] and Glover-McFarlane [1989]). Also receiving a lot of attention are "structured perturbations" of the plant rather than the unstructured plant perturbations considered here (see Doyle [1981]).

The robust stabilization problem, like the sensitivity minimization problem (both with or without weights) fits into the general scheme of the standard H^∞-problem (see Francis [1987]) and hence any general scheme used to solve this general problem (such as Doyle-Glover-Khargonekar-Francis [1989]) also applies directly to the robust stabilization problem.

APPENDIX
SYLVESTER, LYAPUNOV AND STEIN EQUATIONS

In this appendix we study the matrix equations associated with the names of Sylvester, Lyapunov and Stein (which appear in a fundamental way throughout the book) as algebraic equations removed from the context of any applications.

A.1 SYLVESTER EQUATIONS

Here we study the Sylvester equation

$$(A.1.1) \qquad SA_\pi - A_\zeta S = \Gamma$$

(where A_π, A_ζ and Γ are given matrices and S is unknown matrix). Observe that this equation is linear in S. It turns out that if A_π and A_ζ are given in Jordan canonical form, then the solution can be given explicitly. Note that if $\widetilde{S}$ satisfies

$$(A.1.2) \qquad \widetilde{S}T_1^{-1}A_\pi T_1 - T_2^{-1}A_\zeta T_2\widetilde{S} = T_2^{-1}\Gamma T_1$$

then $S := T_2\widetilde{S}T_1^{-1}$ satisfies (A.1.1). Thus the case where A_π and A_ζ are in Jordan canonical form gives the structure of the general case.

We proceed by considering some elementary special cases and then putting them together to handle the general case. Let J_r denote the $r \times r$ Jordan block with 0 eigenvalue

$$J_r = \begin{bmatrix} 0 & 1 & & & \\ & 0 & 1 & & \\ & & & \ddots & \\ & & & \ddots & 1 \\ & & & & 0 \end{bmatrix}$$

and I_r the $r \times r$ identity matrix.

THEOREM A.1.1. *Suppose A_π is the $r \times r$ Jordan block $A_\pi = wI_r + J_r$, A_ζ is the matrix $A_\zeta = zI_m + J_m^T$, $\Gamma = [\Gamma_{\alpha\beta}]_{\substack{1 \le \alpha \le m \\ 1 \le \beta \le r}}$ is an $m \times r$ matrix and $z \ne w$. Then the Sylvester matrix equation*

$$(A.1.2) \qquad SA_\pi - A_\zeta S = \Gamma$$

has the unique solution $S = [S_{\alpha\beta}]_{\substack{1 \le \alpha \le m \\ 1 \le \beta \le r}}$ given by

$$(A.1.3) \qquad S_{\alpha\beta} = \sum_{t=0}^{\alpha-1}\sum_{\nu=0}^{\beta-1} \Gamma_{\alpha-t,\beta-\nu}(-1)^\nu \binom{\nu+t}{\nu}(w-z)^{-\nu-t-1}.$$

PROOF. If $S = [S_{\alpha\beta}]_{\substack{1 \leq \alpha \leq m \\ 1 \leq \beta \leq r}}$ is any $m \times r$ matrix and A_π and A_ζ have the form given in the lemma, then

$$[SA_\pi]_{ij} = wS_{ij} + S_{i,j-1}$$
$$[A_\zeta S]_{ij} = zS_{ij} + S_{i-1,j}$$

for $1 \leq i \leq m$, $1 \leq \beta \leq r$, where we set $S_{i,0} = S_{0,j} = 0$. Thus the Sylvester equation (A.1.1) amounts to the system of equations

$$(w - z)S_{ij} + S_{i,j-1} - S_{i-1,j} = \Gamma_{ij}$$

for all i and j. Since $w \neq z$ we can solve for S_{ij} to get

(A.1.4) $$S_{ij} = (w - z)^{-1}\{\Gamma_{ij} + S_{i-1,j} - S_{i,j-1}\}.$$

This gives a recurrence relation for S_{ij} with the initialization $S_{0,j} = S_{i,0} = 0$. By an inductive argument, we see that $[\Gamma_{ij}]$ $(1 \leq i \leq m, 1 \leq \beta \leq r)$ uniquely determines S. It remains only to show that the formula (A.1.3) for $S_{\alpha\beta}$ satisfies the recurrence relation (A.1.4).

Let us now define $S_{\alpha\beta}$ by (A.1.3). We then have

$$S_{ij} = (w - z)^{-1}\Gamma_{ij} + \sum_{t=1}^{i-1}\sum_{\nu=0}^{j-1}\Gamma_{i-t,j-\nu}(-1)^\nu\binom{\nu+t}{\nu}(w - z)^{-\nu-t-1}$$
$$+ \sum_{\nu=1}^{j-1}\Gamma_{i,j-\nu}(-1)^\nu(w - z)^{-\nu-1}.$$

Thus the relation (A.1.4) amounts to

(A.1.5) $$\sum_{t=1}^{i-1}\sum_{\nu=0}^{j-1}\Gamma_{i-t,j-\nu}(-1)^\nu\binom{\nu+t}{\nu}(w - z)^{-\nu-t}$$
$$+ \sum_{\nu=1}^{j-1}\Gamma_{i,j-\nu}(-1)^\nu(w - z)^{-\nu} - S_{i-1,j} + S_{i,j-1} = 0.$$

By shifting the summation index t in (A.1.5) we may write $S_{i-1,j}$ as

$$S_{i-1,j} = \sum_{t=1}^{i-1}\sum_{\nu=0}^{j-1}\Gamma_{i-t,j-\nu}(-1)^\nu\binom{\nu+t-1}{\nu}(w - z)^{-\nu-t}.$$

Similarly, by a shift of the index ν we may write $S_{i,j-1}$ as

$$S_{i,j-1} = \sum_{t=0}^{i-1}\sum_{\nu=1}^{j-1}\Gamma_{i-t,j-\nu}(-1)^{\nu-1}\binom{\nu+t-1}{\nu}(w - z)^{-\nu-t}.$$

Collecting terms we may rewrite (A.1.5) as

$$\sum_{t=1}^{i-1}\sum_{\nu=1}^{j-1}\Gamma_{i-t,j-\nu}\left\{(-1)^{\nu}\binom{\nu+t}{\nu}\right.$$

$$\left.-(-1)^{\nu}\binom{\nu+t-1}{\nu}+(-1)^{\nu-1}\binom{\nu+t-1}{\nu-1}\right\}(w-z)^{-\nu+t}$$

$$+\sum_{t=1}^{i-1}\Gamma_{i-t,j}\{(-1)^{\nu}-(-1)^{\nu}\}(w-z)^{-t}$$

$$+\sum_{\nu=1}^{j-1}\Gamma_{i,j-\nu}\{(-1)^{\nu}+(-1)^{\nu-1}\}(w-z)^{-\nu}=0.$$

To verify this equality it is sufficient to check the identity

$$\binom{\nu+t}{\nu}-\binom{\nu+t-1}{\nu}-\binom{\nu+t-1}{\nu-1}=0.$$

This is easily verified. $\square$

THEOREM A.1.2. *Suppose A_{π} is the $r\times r$ matrix $A_{\pi}=wI_r+J_r$, A_{ζ} is the $m\times m$ matrix $A_{\zeta}=wI_m+J_m^T$ and $\Gamma=[\Gamma_{\alpha\beta}]$ is a $m\times r$ matrix. Then the Sylvester equation*

$$(A.1.1) \qquad\qquad\qquad SA_{\pi}-A_{\zeta}S=\Gamma$$

has a solution S if and only if

$$(A.1.6)\qquad\qquad \sum_{t=1}^{k}\Gamma_{k+1-t,t}=0 \quad if \quad 1\le k\le\min\{m,r\}.$$

In this case the solutions $S=[S_{\alpha\beta}]$ are given by

$$(A.1.7)\qquad\qquad S_{\alpha\beta}=-\sum_{k=1}^{\beta}\Gamma_{\alpha+\beta+1-k,k} \quad if \quad \alpha+\beta\le m$$

$$(A.1.8)\qquad\qquad S_{\alpha\beta}=\sum_{k=\beta+1}^{\alpha+\beta}\Gamma_{\alpha+\beta+1-k,k} \quad if \quad \alpha+\beta\le r$$

$$(A.1.9)\qquad S_{\alpha\beta}=-R_{\alpha+\beta}-\sum_{k=\alpha+\beta-m+1}^{\beta}\Gamma_{\alpha+\beta+1-k,k} \quad if \quad \max\{m,r\}<\alpha+\beta$$

where R_K (max$\{m,r\} < K \leq m+r$) is arbitrary.

PROOF. For the case here where $\sigma(A_\pi) = \sigma(A_\zeta) = \{w\}$, the Sylvester equation (A.1.1) is equivalent to the system of equations

$$(A.1.10) \qquad S_{\alpha,\beta-1} - S_{\alpha-1,\beta} = \Gamma_{\alpha\beta}$$

for $1 \leq \alpha \leq m$, $1 \leq \beta \leq r$, where we have set $S_{\alpha,0} = 0 = S_{0,\beta}$.

Thus, if (A.1.1) has a solution $[S_{\alpha\beta}]$, from (A.1.10) if $1 \leq k \leq \min\{m,r\}$ we get

$$\sum_{t=1}^{k} \Gamma_{k+1-t,t} = \sum_{t=1}^{k} [S_{k+1-t,t-1} - S_{k-t,t}]$$
$$= S_{k,0} + [S_{k-1,1} + S_{k-1,1}] + \cdots + [-S_{1,k-1} + S_{1,k-1}] - S_{0,k}$$
$$= S_{k,0} - S_{0,k} = 0.$$

This shows that (A.1.6) is necessary for solutions to exist.

Conversely, suppose (A.1.6) holds. We may then define $S_{\alpha\beta}$ by (A.1.7), (A.1.8) and (A.1.9). By (A.1.6) formulas (A.1.7) and (A.1.8) are consistent where they overlap ($\alpha + \beta \leq \min\{m,r\}$). We must check the validity of (A.1.10) with this choice of $S_{\alpha\beta}$. When $\alpha = \beta = 1$, (A.1.10) assumes the form $0 = \Gamma_{11}$. This is true by (A.1.6) with $k = 1$. When $\alpha = 1$ and $\beta > 1$, (A.1.10) has the form

$$S_{1,\beta-1} = \Gamma_{1,\beta}.$$

Since $1 + (\beta - 1) \leq r$, this holds by the definition (A.1.8) of $S_{1,\beta-1}$. Similarly, if $\alpha > 1$ and $\beta = 1$, (A.1.10) has the form

$$-S_{\alpha-1,1} = \Gamma_{\alpha 1}.$$

Since $(\alpha-1)+1 \leq m$, this holds by the definition (A.1.9) of $S_{\alpha-1,1}$. For $1 < \min\{\alpha,\beta\} \leq \alpha+\beta \leq m+1$, we may use (A.1.9) as the definition of $S_{\alpha,\beta-1}$ and $S_{\alpha-1,\beta}$; then equation (A.1.10) reduces to the identity

$$S_{\alpha,\beta-1} - S_{\alpha-1,\beta} = -\sum_{k=1}^{\beta-1} \Gamma_{\alpha+\beta-k,k} + \sum_{k=1}^{\beta} \Gamma_{\alpha+\beta-k,k}$$
$$= \Gamma_{\alpha\beta}.$$

If $1 < \min\{\alpha,\beta\} \leq \alpha + \beta \leq r+1$, we use (A.1.8) to define $S_{\alpha,\beta-1}$ and $S_{\alpha-1,\beta}$; then plugging into (A.1.10) gives

$$S_{\alpha,\beta-1} - S_{\alpha-1,\beta} = \sum_{k=\beta}^{\alpha+\beta-1} \Gamma_{\alpha+\beta-k,k} - \sum_{k=\beta+1}^{\alpha+\beta-1} \Gamma_{\alpha+\beta-k,k}$$
$$= \Gamma_{\alpha\beta}.$$

Finally, when $\max\{m+1, r+1\} < \alpha + \beta$, we use (A.1.9) to get for this case

$$S_{\alpha,\beta-1} - S_{\alpha-1,\beta} = -R_{\alpha+\beta-1} - \sum_{k=\alpha+\beta-m}^{\beta-1} \Gamma_{\alpha+\beta-k,k} + R_{\alpha+\beta-1} + \sum_{k=\alpha+\beta-m}^{\beta} \Gamma_{\alpha+\beta-k,k}$$

$$= \Gamma_{\alpha\beta}.$$

We have thus verified that when (A.1.6) holds then (A.1.7), (A.1.8) and (A.1.9) define a solution.

To show that every solution is of this form, we analyze the homogeneous equation (A.1.10) with $\Gamma_{\alpha\beta} = 0$ (for all α, β). By (A.1.10) and a simple inductive argument, in this case we have

$$S_{\alpha,\beta-1} = S_{\alpha-1,\beta} = \cdots = S_{\alpha-k,\beta+k-1}$$

for $1 \leq k \leq \min\{\alpha-1, r-\beta+1\}$, and

$$S_{\alpha-1,\beta} = S_{\alpha,\beta-1} = \cdots = S_{\alpha+k-1,\beta-k}$$

for $1 \leq k \leq \min\{m-\alpha+1, \beta-1\}$. These conditions force $S_{\alpha\beta}$ to have the form

$$S_{\alpha\beta} = R_{\alpha+\beta}$$

for some sequence of numbers $\{R_K: 2 \leq K \leq m+r\}$. But in addition from (A.1.8) we have boundary conditions

$$S_{1\beta} = 0 \quad \text{for} \quad 1 \leq \beta \leq r-1$$
$$S_{\alpha 1} = 0 \quad \text{for} \quad 1 \leq \alpha \leq m-1$$

and thus

$$R_K = 0 \quad \text{for} \quad 2 \leq K \leq \max\{m, r\}.$$

Thus the amount of arbitrariness in the solution (A.1.7), (A.1.8), (A.1.9) matches exactly with the general solution of the homogeneous equation, so Theorem A.1.2 follows. $\square$

THEOREM A.1.3. *Suppose the matrices A_π and A_ζ^T are in Jordan canonical form*

$$A_\pi = \bigoplus_{j=1}^{p} \bigoplus_{\gamma=1}^{e_j} [w_j I_{r_{j\gamma}} + J_{r_{j\gamma}}]$$

$$A_\zeta = \bigoplus_{i=1}^{\ell} \bigoplus_{\mu=1}^{d_i} [z_i I_{m_{i\mu}} + J_{m_{i\mu}}^T].$$

Suppose Γ is a matrix with columns partitioned conformally with A_π and rows conformally with A_ζ:

$$\Gamma = [\Gamma_{ij}]_{1 \leq i \leq \ell, 1 \leq j \leq p}$$

with

$$\Gamma_{ij} = [\Gamma_{ij,\mu\gamma}]_{1\leq\mu\leq d_i, 1\leq\gamma\leq e_j}.$$

Then the Sylvester equation

$$SA_\pi - A_\zeta S = \Gamma$$

has a solution

$$S = [S_{ij}]_{1\leq i\leq\ell, 1\leq j\leq p}$$

with

$$S_{ij} = [S_{ij,\mu\gamma}]_{1\leq\mu\leq d_i, 1\leq\gamma\leq e_i}$$

if and only if each elementary Sylvester equation

$$SA_{\pi,j\gamma} - A_{\zeta,i\mu}S = \Gamma_{ij,\mu\gamma}$$

has a solution $S = S_{ij,\mu\gamma}$, *where*

$$A_{\pi,j\gamma} = w_j I_{r_{j\gamma}} + J_{r_{j\gamma}}$$
$$A_{\zeta,i\mu} = z_i I_{m_{i\mu}} + J_{m_{i\mu}}^T,$$

for $1 \leq i \leq \ell$, $1 \leq j \leq p$, $1 \leq \mu \leq d_i$, $1 \leq \gamma \leq e_j$.

PROOF. This follows immediately from the block diagonal structure of A_π and A_ζ. $\square$

Using Theorems A.1.1–A.1.3, one can easily count the dimension of the set of solutions to the homogeneous equation $SA_\pi - A_\zeta S = 0$, or, what is the same, the number of independent parameters describing the solutions of $SA_\pi - A_\zeta S = \Gamma$ (provided this equation has a solution).

COROLLARY A.1.4. *In the notation introduced in Theorem A.1.3, the dimension of the set of matrices* S *satisfying* $SA_\pi - A_\zeta S = 0$ *is equal to*

$$\sum_{\gamma=1}^{e_j} \sum_{\mu=1}^{d_i} \min(r_{j\gamma}, m_{i\mu}),$$

where the double sum is taken over all pairs of indices (i,j) *such that* $z_i = w_j$. *In particular, the equation* $SA_\pi - A_\zeta S = \Gamma$ *has unique solution if and only if* $\sigma(A_\pi) \cap \sigma(A_\zeta) = \emptyset$.

We note also that in general if $\sigma(A_\pi) \cap \sigma(A_\zeta) = \emptyset$ then the unique solution S of the Sylvester equation (A.1.1) is given by the contour integral formula

$$(A.1.11) \qquad S = \frac{1}{2\pi i} \int_\gamma (\lambda I - A_\zeta)^{-1}\Gamma(\lambda I - A_\pi)^{-1} d\lambda$$

where γ is a simple closed contour in the complex plane such that $\sigma(A_\pi)$ is inside γ and $\sigma(A_\zeta)$ is outside γ. To verify this, plug S given by (A.1.11) into the left side of the Sylvester equation (A.1.1) to get

$$SA_\pi - A_\zeta S = \frac{1}{2\pi i} \int_\gamma [(\lambda I - A_\zeta)^{-1}\Gamma(\lambda I - A_\pi)^{-1}A_\pi - A_\zeta(\lambda I - A_\zeta)^{-1}\Gamma(\lambda I - A_\pi)^{-1}]d\lambda.$$

Add and subtract $\lambda(\lambda I - A_\zeta)^{-1}\Gamma(\lambda I - A_\pi)^{-1}$ in the integrand to get next

$$
\begin{aligned}
SA_\pi - A_\zeta S &= \frac{1}{2\pi i}\int_\lambda [(\lambda I - A_\zeta)^{-1}\Gamma(\lambda I - A_\pi)^{-1}(A_\pi - \lambda I) \\
&\quad + (\lambda I - A_\zeta)(\lambda I - A_\zeta)^{-1}\Gamma(\lambda I - A_\pi)^{-1}]d\lambda \\
&= -\frac{1}{2\pi i}\int_\gamma (\lambda I - A_\zeta)^{-1}\Gamma dz + \frac{1}{2\pi i}\int_\gamma \Gamma(\lambda I - A_\pi)^{-1}d\lambda.
\end{aligned}
$$

Since $\sigma(A_\zeta)$ is outside γ, the first term is 0 by Cauchy's theorem, while the second term is Γ by the Riesz functional calculus for A_π since $\sigma(A_\pi)$ is inside γ, and thus S is the solution of (A.1.1) as asserted. When A_ζ and A_π are in Jordan form, $(\lambda I - A_\zeta)^{-1}$ and $(\lambda I - A_\pi)^{-1}$ can be computed explicitly, and then the right side of (A.1.11) can be calculated by using the theory of residues. This gives an alternative derivation of Theorem A.1.1.

Finally, we consider a special case of the Sylvester equation when the solution can be found explicitly.

LEMMA A.1.5. *Let*

$$
C_\pi = [\,1 \quad 1 \quad \cdots \quad 1\,], \qquad
B_\zeta = \begin{bmatrix} 1 \\ 1 \\ \vdots \\ 1 \end{bmatrix}
$$

(C_π is $1 \times m$ and B_ζ is $n \times 1$) and let A_π, A_ζ be $m \times m$ and $n \times n$ diagonal matrices

$$
A_\pi = \mathrm{diag}(w_1,\ldots,w_m), \qquad A_\zeta = \mathrm{diag}(z_1,\ldots,z_n).
$$

Then the Sylvester equation

$$(A.1.12) \qquad\qquad SA_\pi - A_\zeta S = B_\zeta C_\pi$$

has a solution S if and only if no number z_i coincides with any number w_j, and in this case the solution is unique and given by

$$
S = \left[\frac{1}{w_j - z_i}\right]_{1\le i\le n,\, 1\le j\le m}.
$$

Moreover, S is invertible if and only if $n = m$, and in this case the inverse is given by

$$(A.1.13) \qquad\qquad S^{-1} = [t_{\alpha\beta}]_{1\le\alpha,\beta\le n}$$

where

$$(A.1.14) \qquad
t_{\alpha\beta} = \frac{\prod_{\substack{1\le j\le n \\ j\ne\alpha}}(w_j - z_\beta)\cdot \prod_{1\le k\le n}(w_\alpha - z_k)}{\prod_{\substack{1\le j\le n \\ j\ne\beta}}(z_\beta - z_j)\cdot \prod_{\substack{1\le k\le n \\ k\ne\alpha}}(w_k - w_\alpha)}.
$$

PROOF. Using the special form of C_π, A_π, A_ζ, B_ζ we may compute the (i,j)-th entry of both sides of (A.1.12) to see that $S = [s_{ij}]_{1 \le i \le n; 1 \le j \le m}$ satisfies the Sylvester equation if and only if

$$s_{ij}w_j - z_i s_{ij} = 1$$

for all (i,j). Thus a solution exists if and only if $w_j \ne z_i$ for all (i,j) and then $s_{ij} = \frac{1}{w_j - z_i}$ is the unique solution.

Certainly $n = m$ is a necessary condition for S to be invertible. For $n = m$, the determinant of S can be found (after a minor change of notation) in Problem #3 (p. 92) of Polya-Szego [1972], where it is attributed to Cauchy in the year 1841. The derivation involves elementary row and column operations to obtain a recurrence relation for the determinant with $2n$ points in terms of the determinant with $2(n-1)$ points z_i and w_j. The result is

$$(A.1.15) \qquad \det S = \frac{\prod_{1 \le k < j \le n}\{(z_k - z_j)(w_j - w_k)\}}{\prod_{1 \le j,k \le n}(w_j - z_k)}.$$

From this formula we see that $\det S \ne 0$ (since we are assuming that the z_i's and w_j's are distinct from each other). Next we use the adjoint formula

$$S^{-1} = \frac{1}{\det S} adj\, S$$

to compute the inverse of S. The $(n-1) \times (n-1)$ submatrices of S have the same form as S itself and thus all the cofactors of S can be computed from the formula (A.1.15). The result is the expression (A.1.14) for the entries of S^{-1}. $\square$

A.2 STEIN EQUATIONS

In this section we study a matrix equation related to the Sylvester equation which appears in a fundamental way in Chapters 5 and 7. Suppose A_π, A_ζ and Γ are given matrices of sizes $r \times r$, $m \times m$ and $m \times r$ respectively. The equation

$$(A.2.1) \qquad S - A_\zeta S A_\pi = \Gamma$$

for the unknown $m \times r$ matrix S is called a *Stein equation*. Again, this is a linear equation in S. If A_ζ is invertible, the Stein equation (A.2.1) can be converted to the Sylvester equation

$$(A.2.2) \qquad S A_\pi - A_\zeta^{-1} S = -A_\zeta^{-1} \Gamma.$$

Similarly, if A_π is invertible (A.2.1) can be converted to the Sylvester equation

$$(A.2.3) \qquad S A_\pi^{-1} - A_\zeta S = \Gamma A_\pi^{-1}.$$

From Theorem A.1.1 we know that a unique solution of (A.2.2) exists for any choice of Γ as long as

$$(A.2.4) \qquad \sigma(A_\pi) \cap \sigma(A_\zeta^{-1}) = \emptyset.$$

Similarly (A.2.3) has a unique solution if

$$(A.2.5) \qquad\qquad\qquad \sigma(A_\pi^{-1}) \cap \sigma(A_\zeta) = \emptyset.$$

Note that (A.2.4) and (A.2.5) are both equivalent by

$$(A.2.6) \qquad\qquad w \in \sigma(A_\pi), \qquad z \in \sigma(A_\zeta) \Longrightarrow zw \neq 1.$$

Condition (A.2.6) turns out to be the criterion for the existence of a unique solution in general.

THEOREM A.2.1. *The Stein equation (A.2.1) has a unique solution S if and only if $zw \neq 1$ for each $w \in \sigma(A_\pi)$ and $z \in \sigma(A_\zeta)$. In this case the unique solution is given by*

$$(A.2.7) \qquad\qquad S = \frac{1}{2\pi i} \int_\gamma (\lambda I - A_\zeta)^{-1} \Gamma (I - \lambda A_\pi)^{-1} d\lambda$$

where γ is a simple closed contour in $\mathbf{C}$ such that $\sigma(A_\zeta)$ is inside γ and, for each nonzero $w \in \sigma(A_\pi)$, w^{-1} is outside γ.

PROOF. One can verify that (A.2.7) is a solution of (A.2.1) by direct computation. Indeed, if S is given by (A.2.7), then

$$S - A_\zeta S A_\pi = \frac{1}{2\pi i} \int_\gamma [(\lambda I - A_\zeta)^{-1} \Gamma (I - \lambda A_\pi)^{-1} - A_\zeta (\lambda I - A_\zeta)^{-1} \Gamma (I - \lambda A_\pi)^{-1} A_\pi] d\lambda.$$

Add and subtract $\lambda(\lambda I - A_\zeta)^{-1} \Gamma (I - \lambda A_\pi)^{-1} A_\pi$ to the integrand to get the integrand equal to

$$(\lambda I - A_\zeta)^{-1} \Gamma (I - \lambda A_\pi)^{-1} + (\lambda I - A_\zeta)(\lambda I - A_\zeta)^{-1} \Gamma (I - \lambda A_\pi)^{-1} A_\pi$$
$$+ (\lambda I - A_\zeta)^{-1} \Gamma (I - \lambda A_\pi)^{-1} (-\lambda A_\pi).$$

Note that the first and the third term collapse to $(\lambda I - A_\zeta)^{-1} \Gamma$. Hence

$$S - A_\zeta S A_\pi = \frac{1}{2\pi i} \int_\gamma (\lambda I - A_\zeta)^{-1} \Gamma d\lambda + \frac{1}{2\pi i} \int_\gamma \Gamma (I - \lambda A_\pi)^{-1} A_\pi d\lambda.$$

The first term equals Γ since $\sigma(A_\zeta)$ is inside γ; the integrand in the second term is analytic on and inside γ by the assumptions on γ and hence the second term is 0 by Cauchy's Theorem. Thus (A.2.7) gives a solution S of the Stein equation (A.2.1) as asserted.

To prove uniqueness, note that S is a solution of (A.2.1) if and only if PSQ satisfies

$$(PSQ) - PA_\zeta (PSQ) A_\pi Q = P\Gamma Q$$

for every Riesz, or spectral, projection P for A_ζ and Riesz, or spectral, projection Q for A_π. From this one sees that it is sufficient to consider the case where the spectra of

A_ζ and A_π consist of single points $\{z\}$ and $\{w\}$ respectively. If $z \neq 0$, then the Stein equation can be converted to the Sylvester equation (A.2.2), and uniqueness is equivalent to the condition $zw \neq 0$ by Corollary A.1.4. If $z = 0$, then A_ζ is nilpotent. If S is a solution of the homogeneous Stein equation

$$S - A_\zeta S A_\pi = 0,$$

then

$$S = A_\zeta S A_\pi = A_\zeta (A_\zeta S A_\pi) A_\pi$$
$$= A_\zeta^2 S A_\pi^2 = \cdots = A_\zeta^N S A_\pi^N.$$

For N large enough, $A_\zeta^N = 0$ so $S = 0$ and the homogeneous equation has only the trivial solution. Thus the solution of the nonhomogeneous equation (A.2.1) is unique if $\sigma(A_\zeta) = \{0\}$. $\square$

Explicit formulas for solutions of the Stein equation turn out to be more complicated than for the Sylvester equation. Of course if A_ζ is invertible and bases are chosen so that A_ζ^{-1} and A_π are in Jordan form, then the results of the previous section (applied to the Sylvester equation (A.2.2)) give an explicit representation for S (be it unique or not). A similar statement holds if A_π is invertible and one converts to the Sylvester equation (A.2.3); in this case we should choose bases to represent A_ζ and A_π^{-1} in Jordan form. From the point of view of the interpolation problems discussed in Part V, it would be of interest to have explicit formulas for the case where A_ζ and A_π are both presented in Jordan canonical form. We first consider the case of uniqueness by applying the contour integral formula from Theorem A.2.1.

THEOREM A.2.2. *Suppose that A_π and A_ζ are the Jordan block and the transposed Jordan block*

$$A_\pi = wI_r + J_r, \qquad A_\zeta = zI_m + J_m^T$$

of sizes $r \times r$ and $m \times m$ respectively, where z, w are complex numbers such that $wz \neq 1$. Then the unique solution of the Stein equation

(A.2.1) $$S - A_\zeta S A_\pi = \Gamma,$$

for a given $m \times r$ matrix $\Gamma = [\Gamma_{ij}]$, is given by

$$S = [S_{\alpha\beta}]_{1 \leq \alpha \leq m; 1 \leq \beta \leq r}$$

where

(A.2.8) $$S_{\alpha\beta} = \sum_{k=1}^{\alpha} \sum_{\ell=1}^{\beta} \Gamma_{\alpha+1-k,\beta+1-\ell} T_{k\ell}$$

and where

(A.2.9) $$T_{k\ell} = \sum_{j=0}^{\min\{\ell-1,k-1\}} \binom{\ell-1}{j}\binom{\ell+k-j-2}{k-j-1} z^{\ell-j-1} w^{k-1-j}(1-zw)^{j+1-k-\ell}$$

for $1 \leq k \leq m,\ 1 \leq \ell \leq r.$

PROOF. We use the contour integral formula (A.2.7) for the unique solution S given by Theorem A.2.1. We first need to evaluate the inverses $(\lambda I - A_\zeta)^{-1}$ and $(I - \lambda A_\pi)^{-1}$ for A_ζ and A_π of the special form given in Theorem A.2.2. First we note

$$(\lambda I - A_\zeta)^{-1} = [(\lambda - z)I_m - J_m^T]^{-1}$$

$$= \sum_{j=1}^{m} (\lambda - z)^{-j}(J_m^T)^{j-1}$$

$$= \begin{bmatrix} (\lambda - z)^{-1} & & & 0 \\ (\lambda - z)^{-2} & (\lambda - z)^{-1} & & \\ \vdots & & \ddots & & \ddots \\ (\lambda - z)^{-m} & \cdots & & (\lambda - z)^{-2} & (\lambda - z)^{-1} \end{bmatrix}.$$

Similarly,

$$(I - \lambda A_\pi)^{-1} = [(1 - \lambda w)I_r - \lambda J_r]^{-1}$$

$$= \sum_{j=1}^{r} (1 - \lambda w)^{-j}\lambda^{j-1} J_r^{j-1}$$

$$= \begin{bmatrix} (1 - \lambda w)^{-1} & \lambda(1 - \lambda w)^{-2} & \cdots & \lambda^{r-1}(1 - \lambda w)^{-r} \\ & (1 - \lambda w)^{-1} & \ddots & \vdots \\ & & \ddots & \\ & & & \lambda(1 - \lambda w)^{-2} \\ 0 & & & (1 - \lambda w)^{-1} \end{bmatrix}.$$

Then we compute

$$[\Gamma(I - \lambda A_\pi)^{-1}]_{k\beta} = \sum_{\ell=1}^{\beta} \Gamma_{k\ell}\lambda^{\beta-\ell}(1 - \lambda w)^{-(\beta+1-\ell)}$$

$$= \sum_{\ell=1}^{\beta} \Gamma_{k,\beta+1-\ell}\lambda^{\ell-1}(1 - \lambda w)^{-\ell}.$$

Hence

$$[(\lambda I - A_\zeta)^{-1} \cdot \Gamma(I - \lambda A_\pi)^{-1}]_{\alpha\beta}$$

$$= \sum_{k=1}^{\alpha} (\lambda - z)^{\alpha+1-k} \sum_{\ell=1}^{\beta} \Gamma_{k,\beta+1-\ell}\lambda^{\ell-1}(1 - \lambda w)^{-\ell}$$

$$= \sum_{k=1}^{\alpha} (\lambda - z)^{-k} \sum_{\ell=1}^{\beta} \Gamma_{\alpha+1-k,\beta+1-\ell}\lambda^{\ell-1}(1 - \lambda w)^{-\ell}$$

$$= \sum_{k=1}^{\alpha}\sum_{\ell=1}^{\beta} (\lambda - z)^{-k}\Gamma_{\alpha+1-k,\beta+1-\ell}\lambda^{\ell-1}(1 - \lambda w)^{-\ell}.$$

The only pole of this function of λ inside γ is at $\lambda = z$; to compute the residue at z we need the Taylor expansion of the function $\lambda^{\ell-1}(1 - \lambda w)^{-\ell}$ at z. Start with

$$\lambda^{\ell-1} = [z + (\lambda - z)]^{\ell-1} = \sum_{j=0}^{\ell-1} \binom{\ell-1}{j} z^{\ell-j-1}(\lambda - z)^j$$

while

$$\frac{1}{i!}\frac{d^i}{d\lambda^i}(1 - \lambda w)^{-\ell} = \frac{\ell(\ell+1)\cdots(\ell+i-1)}{i!}(1 - \lambda w)^{-\ell-i}w^i$$

$$= \binom{\ell+i-1}{i}w^i(1 - \lambda w)^{-\ell-i}.$$

Evaluation at $\lambda = z$ gives

$$\frac{1}{i!}\frac{d^i}{d\lambda^i}(1 - \lambda w)^{-\ell}\,|_{\lambda=z} = \binom{\ell+i-1}{i}w^i(1 - zw)^{-\ell-i}.$$

Therefore the function $\lambda^{\ell-1}(1 - \lambda w)^{-\ell}$ has h-th Taylor coefficients at z given by

$$\sum_{j=0}^{\min\{\ell-1,h\}} \binom{\ell-1}{j} z^{\ell-j-1} \binom{\ell+h-j-1}{h-j} w^{h-j}(1 - zw)^{-h+j-\ell}$$

$$= \sum_{j=0}^{\min\{\ell-1,h\}} \binom{\ell-1}{j}\binom{\ell+h-j-1}{h-j} z^{\ell-j-1}w^{h-j}(1 - zw)^{j-h-\ell}.$$

Therefore the residue of $[(\lambda I - A_\zeta)^{-1}\Gamma(I - \lambda A_\pi)^{-1}]_{\alpha\beta}$, the (α, β)-entry of $(\lambda I - A_\zeta)^{-1}\Gamma(I - \lambda A_\pi)^{-1}$, at $\lambda = z$ is given by

$$\mathrm{Res}_{\lambda=z}[(\lambda I - A_\zeta)^{-1}\Gamma(I - \lambda A_\pi)^{-1}]_{\alpha\beta} = \sum_{k=1}^{\alpha}\sum_{\ell=1}^{\beta}\Gamma_{\alpha+1-k,\beta+1-\ell}$$

$$\times \left\{ \sum_{j=0}^{\min\{\ell-1,k-1\}} \binom{\ell-1}{j}\binom{\ell+k-j-2}{k-j-1} z^{\ell-j-1}w^{k-1-j}(1 - zw)^{-k+1+j-\ell} \right\}$$

$$= \sum_{k=1}^{\alpha}\sum_{\ell=1}^{\beta}\Gamma_{\alpha+1-k,\beta+1-\ell}T_{k\ell}.$$

Theorem A.2.2 now follows from Theorem A.2.1 and the Residue Theorem. $\square$

Once the formula (A.2.8)–(A.2.9) is known for the solution S it should be possible to verify it directly by plugging it back into the equation (A.2.1). We did this successfully for the solution to the Sylvester equation in the previous section (Theorem A.2.1); the verification reduced to the well known combinatorial identity

$$\binom{\nu+t}{\nu} = \binom{\nu+t-1}{\nu} + \binom{\nu+t-1}{\nu-1}.$$

A similar verification of formula (A.2.8)–(A.2.9) for the solution of the Stein equation leads to the more complicated combinatorial identity

$$T_{k\ell} = (1 - zw)^{-1}wT_{k-1,\ell} + (1 - zw)^{-1}zT_{k,\ell-1} + (1 - zw)^{-1}T_{k-1,\ell-1}$$

where in general $T_{k\ell}$ is given by (A.2.9).

The case where $zw = 1$ (so the solution is not unique if it exists) can be handled by using a similarity transformation to bring A_π^{-1} to the Jordan canonical form and then applying the analysis of the Sylvester equation in the previous section. In the following we shall use the convention for binomial coefficients that $\binom{-1}{-1} = 1$ and $\binom{p}{q} = 0$ for $q > p$ or $q = -1$ and $p > -1$. The result is as follows.

THEOREM A.2.3. *Suppose that w is a nonzero complex number and A_π and A_ζ are the Jordan block and the transpose of the Jordan block*

$$A_\pi = wI_r + J_r, \qquad A_\zeta = w^{-1}I_m + J_m^T$$

of sizes $r \times r$ and $m \times m$ respectively, and let $\Gamma = [\Gamma_{ij}]_{1\le i\le m,1\le j\le r}$ be a given $m \times r$ matrix. Then the Stein equation

$$(A.2.1) \qquad\qquad\qquad S - A_\zeta S A_\pi = \Gamma$$

has a solution if and only if

$$\sum_{t=1}^{k}\sum_{\ell=t}^{r}\Gamma_{k+1-t,\ell}(-1)^t\left[\binom{t-3}{\ell-2} - \binom{t-2}{\ell-2}\right]w_0^{\ell+t} = 0$$

for $1 \le k \le \max\{m,r\}$. In this case an $m \times r$ matrix S is a solution of (A.2.1) if and only if S has the form

$$S = VM^{-1}$$

where M is the $r \times r$ invertible upper triangular matrix

$$M = \left[(-1)^{j-1}\binom{j-2}{i-2}w_0^{i+j-2}\right]_{1\le i,j\le r}$$

and $V = [V_{\alpha\beta}]_{1\le\alpha\le m,1\le\beta\le r}$ is given by

$$V_{\alpha\beta} = \sum_{k=1}^{\beta}\sum_{\ell=k}^{r}\Gamma_{\alpha+\beta+1-k,\ell}\left[\binom{k-3}{\ell-2} - \binom{k-2}{\ell-2}\right](-1)^{k+1}w_0^{\ell+k-3}$$

$$\text{if } \alpha + \beta \le m$$

$$V_{\alpha\beta} = \sum_{k=\beta+1}^{\alpha+\beta}\sum_{\ell=k}^{r}\Gamma_{\alpha+\beta+1-k,\ell}\left[\binom{k-3}{\ell-2} - \binom{k-2}{\ell-2}\right](-1)^{k}w_0^{\ell+k-3}$$

$$\text{if } \alpha + \beta \le r$$

and

$$V_{\alpha\beta} = R_{\alpha+\beta} - \sum_{k=\alpha+\beta-m+1}^{\beta} \sum_{\ell=k}^{r} \Gamma_{\alpha+\beta+1-k,k} \left[\binom{k-3}{\ell-2} - \binom{k-2}{\ell-2} \right] (-1)^k w_0^{\ell+k-3}$$

if $\max\{m,r\} < \alpha + \beta$

where R_K $(\max\{m,r\} < K \leq m+r)$ *are arbitrary.*

PROOF. Define an $r \times r$ matrix M by

$$M = \left[(-1)^{j+1} \binom{j-2}{i-2} w_0^{i+j-2} \right]_{1 \leq i,j \leq r}.$$

Then M is upper triangular with no zero entries on the diagonal, and so is invertible. By direct multiplication one can verify $A_\pi M \widetilde{A}_\pi = M$, where $\widetilde{A}_\pi = w^{-1} I_r + J_r$, and hence

$(A.2.10)$ $$A_\pi^{-1} = M \widetilde{A}_\pi M^{-1}.$$

(This identity was also used in the proof of Theorem 2.5.3.) Rewrite the Stein equation $(A.2.1)$ as a Sylvester equation

$$S A_\pi^{-1} - A_\zeta S = \Gamma A_\pi^{-1}.$$

Plug in $(A.2.10)$ for A_π^{-1} to rewrite this as a Sylvester equation for SM:

$$(SM)\widetilde{A}_\pi - A_\zeta(SM) = \Gamma M \widetilde{A}_\pi.$$

As $\widetilde{A}_\pi$ and A_ζ are Jordan blocks with the common eigenvalue w^{-1}, one can get the existence criterion and an explicit formula for the solution by applying Theorem A.1.2. When this is done, one gets the result as stated in the theorem. $\square$

One can easily state a result for the Stein equation analogous to Theorem A.1.3. We leave this statement to the reader and only formulate a corollary analogous to Corollary A.1.4.

COROLLARY A.2.4. *Let* A_π *and* A_ζ *be as in Theorem A.1.3. Then the dimension of the set of matrices* S *satisfying* $S - A_\zeta S A_\pi = 0$ *is equal to*

$$\sum_{\gamma=1}^{e_j} \sum_{\mu=1}^{d_i} \min(r_{j\gamma}, m_{i\mu}),$$

where the double sum is taken over all pairs (i,j) *such that* $z_i w_j = 1$.

A.3 LYAPUNOV AND SYMMETRIC STEIN EQUATIONS

In this section we study the Lyapunov and symmetric Stein equations which appeared in a fundamental way in various places in the book (for example the solution of the two-sided Nudelman problem).

Consider the Lyapunov equation

(A.3.1) $$SA^* + AS = W,$$

where A and W are given $n \times n$ matrices and W is Hermitian. We focus on Hermitian solutions S of (A.3.1) (it is easy to see that S solves (A.3.1) if and only if its Hermitian part $\frac{1}{2}(S + S^*)$ solves (A.3.1) and its skew Hermitian part $\frac{1}{2i}(S - S^*)$ solves the homogeneous equation

$$TA^* + AT = 0).$$

It follows from Theorem A.1.1 that (A.3.1) has a unique solution (necessarily Hermitian) provided A has no pairs of eigenvalues symmetric relative to the imaginary axis (i.e. $z_0 \in \sigma(A)$ implies $-\bar{z}_0 \notin \sigma(A)$). In particular, the solution is unique if $\sigma(A) \subset \Pi^+$, and for such A the solution is given by

(A.3.2) $$S = \int_{-\infty}^{0} e^{At} W e^{A^* t} dt$$

(the integral converges because all eigenvalues of A and A^* have positive real parts). Indeed, substituting the right-hand side of (A.3.2) in (A.3.1) we have

$$SA^* + AS = \int_{-\infty}^{0} (e^{At} W e^{A^* t} A^* + A e^{At} W e^{A^* t}) dt$$

$$= \int_{-\infty}^{0} (e^{At} W e^{A^* t})' dt = (e^{At} W e^{A^* t})|_{t=-\infty}^{t=0} = W.$$

Analogously, the solution S of (A.3.1) is unique if $\sigma(A) \subset \Pi^-$ (the open left half-plane), and S is given by the formula

$$S = - \int_{0}^{\infty} e^{At} W e^{A^* t} dt.$$

The following well-known result was needed in the proof of Theorems 18.5.1 and 18.5.2, as well as in other places in the book.

THEOREM A.3.1. *Assume* $\sigma(A) \subset \Pi^+$.

(a) *If* W *is positive semidefinite, then the unique solution* S *of* (A.3.1) *is positive semidefinite.*

(b) *If, moreover,* $W = BB^*$, *where the (not necessarily square) matrix* B *is such that* (A, B) *is a full range pair, then* S *is positive definite.*

PROOF. Part (a) follows from the formula (A.3.2). Assume now the hypotheses of (b), and let $Sx = 0$ for some $x \in \mathbb{C}^n$. Then

$$x^* BB^* x = x^*(SA^* + AS)x = 0,$$

so $x^*B = 0$ and $x^*AS = x^*BB^* - x^*SA^* = 0$. Furthermore,

$$x^*ABB^*A^*x = x^*A(SA^* + AS)A^*x$$
$$= (x^*AS)A^{*2}x + x^*A^2(SA^*x) = 0.$$

Hence $x^*AB = 0$. Repeating this procedure, we obtain $x^*A^jB = 0$ for $j = 0, 1, \ldots$. By the full range property of (A, B) the vector x must be 0. $\square$

We now give formulas for the general Hermitian solution S of (A.3.1) (if such exists) under the assumption that A^* is in the Jordan normal form. Observe that by a suitable similarity transformation every equation of type (A.3.1) is reduced to this situation. So let $A^* = \bigoplus_{j=1}^r K_j$, where $K_1, \ldots, K_r$ are Jordan blocks. Partition accordingly $S = [S_{ij}]_{1 \le i,j \le r}$ and $W = [W_{ij}]_{1 \le i,j \le r}$ (note that $W_{ij} = W_{ji}^*$). Then the equation (A.3.1) reduces to r^2 equations

$$(A.3.3) \qquad\qquad S_{ij}K_j + K_j^*S_{ij} = W_{ij}.$$

Theorems A.1.1 and A.1.2 give the general solution of (A.3.3) for $i > j$; in this connection observe the formula (we denote by w_i the eigenvalue of $K_i = w_iI + J$)

$$(w_iI + J)^* = \text{diag}\big(-1, 1, \ldots, (-1)^{m_i}\big)\big(\overline{w}_iI - J^T\big)\text{diag}\big(-1, 1, \ldots, (-1)^{m_i}\big),$$

where m_i is the size of K_i. Given the general solution of (A.3.3) for $i > j$ the blocks S_{ij} for $i < j$ are obtained by putting $S_{ij} = S_{ji}^*$ ($i < j$). It remains only to consider the equation (A.3.3) with $i = j$, i.e. the case of one Jordan block:

THEOREM A.3.2. *Let* $A^* = wI + J$ *be a Jordan block of size* n *with the eigenvalue* w, *and let* $W = [W_{ij}]_{1 \le i,j \le n}$ *be a Hermitian matrix. If* w *is not pure imaginary, then* (A.3.1) *has unique solution* $S = [S_{\alpha\beta}]_{1 \le \alpha, \beta \le n}$ *given by*

$$S_{\alpha\beta} = \sum_{t=0}^{\alpha-1}\sum_{\nu=0}^{\beta-1} W_{\alpha-t,\beta-\nu}(-1)^{t+\nu}\binom{\nu+t}{\nu}(w+\overline{w})^{-\nu-t-1}.$$

If $w = i\alpha$, $\alpha \in \mathbf{R}$, *then* (A.3.1) *has a solution if and only if*

$$\sum_{t=1}^k (-1)^{k+1-t}W_{k+1-t,t} = 0 \quad \text{for} \quad 1 \le k \le n.$$

In this case the Hermitian solutions $S = [S_{\alpha\beta}]$ *are given by the formulas*

$$S_{\alpha\beta} = -\sum_{k=1}^\beta (-1)^{\beta+1-k}W_{\alpha+\beta+1-k,k} \quad \text{if} \quad \alpha + \beta \le n,$$

and

$$S_{\alpha\beta} = -R_{\alpha+\beta} - \sum_{k=\alpha+\beta-n+1}^\beta (-1)^{\beta+1-k}W_{\alpha+\beta+1-k,k} \quad \text{if} \quad \alpha + \beta > n,$$

where R_γ $(n < \gamma \leq 2n)$ is a complex number whose imaginary part is equal to

$$\frac{1}{2}\left[\sum_{k=\alpha+\beta-n+1}^{\alpha} (-1)^{\alpha+1-k}\overline{W}_{\alpha+\beta+1-k,k} - \sum_{k=\alpha+\beta-n+1}^{\beta} (-1)^{\beta+1-k}W_{\alpha+\beta+1-k,k}\right]$$

but otherwise R_γ is arbitrary.

Theorem A.3.2 follows easily by adapting the formulas given in Theorems A.1.1 and A.1.2.

COROLLARY A.3.3. *Let $A^* = \bigoplus_{j=1}^r K_j$, where K_j is the $m_j \times m_j$ Jordan block with eigenvalue w_j. Then the dimension of the set of Hermitian solutions S (considered as a real vector space) of the homogeneous equation $SA + A^*S = 0$ is equal to*

$$\sum_{i,j} \min(m_i, m_j),$$

where the double sum is over all pairs (i,j) $(1 \leq i, j \leq r)$ such that $w_i = -\overline{w}_j$.

Consider now the symmetric Stein equation

$$(A.3.4) \qquad\qquad S - A^*SA = V,$$

where A and V are given $n \times n$ matrices and V is Hermitian. By Theorem A.2.1 the equation (A.3.4) has unique solution S (necessarily Hermitian) provided $z\overline{w} \neq 1$ for every pair of eigenvalues z, w of A. In particular, this is the case when $\sigma(A) \subset \mathcal{D}$, and then the unique solution is given by

$$(A.3.5) \qquad\qquad S = \sum_{j=0}^\infty A^{*j}VA^j$$

(the series converges because $|z_0| < 1$ for every $z_0 \in \sigma(A)$).

We need an analogue of Theorem A.3.1.

THEOREM A.3.4. *Assume $\sigma(A) \subset \mathcal{D}$.*

(a) *If V is positive semidefinite, then the unique solution S of (A.3.4) is positive semidefinite.*

(b) *If $V = C^*C$, where the pair (C, A) is a null kernel pair, then S is actually positive definite.*

PROOF. Part (a) follows from (A.3.5). Assume the hypotheses of (b), and let $x^*Sx = 0$ for some $x \in \mathbb{C}^n$. As S is given by (A.3.5) it follows that $x^*A^{*j}VA^jx = 0$ for $j = 0, 1, \dots$. Thus, $CA^jx = 0$ for $j = 0, 1, \dots$, and because of the null kernel property of (C, A) the vector x is 0. $\square$

A general Hermitian solution S of (A.3.4) (if such exists) can be described as follows. Let K be a Jordan form of A, so $A = T^{-1}KT$ for some invertible T. Then $\widehat{S} := T^{*-1}ST^{-1}$ solves the equation

$$(A.3.6) \qquad\qquad \widehat{S} - K^*\widehat{S}K = \widehat{V},$$

where $\widehat{V} = T^{*-1}VT^{-1}$. Write $K = \bigoplus_{j=1}^{r} K_j$, where K_j are Jordan blocks, and partition $\widehat{S} = [S_{ij}]_{1 \leq i,j \leq r}$, $\widehat{V} = [V_{ij}]_{1 \leq i,j \leq r}$ accordingly. The equation (A.3.6) decouples:

$$(A.3.7) \qquad\qquad S_{ij} - K_i^* S_{ij} K_j = V_{ij}, \qquad 1 \leq i,j \leq r.$$

Use the formulas in Theorems A.2.2 and A.2.3 to determine the general solution S_{ij} of (A.3.7) for $i > j$, and put $S_{ij} = S_{ji}^*$ for $i < j$. It remains to find general Hermitian solution to (A.3.7) for $i = j$, i.e. in the case of one Jordan block. If the eigenvalue of K_i is not unimodular, then the solution of

$$S_{ii} - K_i^* S_{ii} K_i = V_{ii}$$

is unique and is found by the formulas in Theorem A.2.2. The remaining case of K_i with unimodular eigenvalue requires special attention. It can be dealt with using the formulas of Theorem A.2.3.

Finally, observe that the dimension of the set of Hermitian solutions $\widehat{S}$ of (A.3.6) (with $\widehat{V} = 0$), considered as a real vector space, is

$$\sum_{i,j} \min(m_i, m_j),$$

where m_j is the size of K_j and where the double sum is over all pairs (i,j) $(1 \leq i,j \leq r)$ such that $w_i \overline{w}_j = 1$.

NOTES FOR APPENDIX

Theorems A.1.1 (apart from formula (A.1.3)), A.2.1, A.3.1(a) and A.3.4 are well-known in the literature (see, e.g., Gantmakher [1959], Lancaster-Tismenetsky [1985], Daleckii-Krein [1974]). Theorem A.3.1(b) is a particular case of the results in Snyders-Zakai [1970], as well as of the general inertia theorems due to Chen [1973], Wimmer [1974].

One of the elements for solving interpolation problems is to find an invertible solution of a Sylvester or Lyapunov equation. In this Appendix we give explicit solutions of Sylvester/Stein equations. It is of special interest to know when the solution is invertible, as is evident, for example, in Chapter 4. For further discussion of this point and other applications, see Hearon [1977], Datta-Datta [1986].

REFERENCES

M.B. Abrahamse [1979], The Pick interpolation theorem for finitely connected domains, *Mich. Math. J.* **26**, 195–203.

V.M. Adamjan, D.Z. Arov and M.G. Krein [1978], Infinite block Hankel matrices and their connection with the interpolation problem, *Amer. Math. Soc. Transl.* (2) **111**, 133–156. (Russian original, 1971).

J. Agler, J.W. Helton, J. McCullough and L. Rodman [1988], Positive semidefinite matrices with given sparsity pattern, *Linear Algebra and its Applications* **107**, 101–149.

N.I. Akhiezer [1965], *The Classical Moment Problem*, Oliver & Boyd, Edinburgh and London.

N.I. Akhiezer and M. Krein [1962], *Some Questions in the Theory of Moments*, Amer. Math. Soc. Transl. Math. Monographs, Vol. 2, Providence, RI.

D. Alpay, P. Bruinsma, A. Dijksma and H.S.V. de Snoo [1989], Interpolation problems, extensions of symmetric operators and reproducing kernel spaces I, preprint.

D. Alpay and I. Gohberg [1988], Unitary rational matrix functions, in *Operator Theory: Advances and Applications*, OT 33, Birkhäuser-Verlag, Basel, pp. 175–222.

B.D.O. Anderson and A.C. Antoulas [1989], Rational interpolation and state variable realization, preprint.

B.M. Anderson, F.M. Brasch, Jr. and P.V. Lopresti [1975], The sequential construction of minimal partial realizations from finite input-output data, *SIAM J. Control* **13**, 552–571.

A.C. Antoulas [1986], On recursiveness and related topics in linear systems, *IEEE Trans. Auto. Control* **AC-31**, 1121–1135.

A.C. Antoulas and B.D.O. Anderson [1986], On the scalar rational interpolation problem, *IMA J. Math. Control and Information* **3**, 61–88.

A.C. Antoulas, J.A. Ball, J. Kang and J.C. Willems [1989], On the solution of the minimal rational interpolation problem, preprint.

R. Arocena [1989a], Schur analysis of a class of translation invariant forms, in Collection dedicated to M. Cotlar on occasion of his 75-th birthday, Marcel-Dekker, in press.

R. Arocena [1989b], On the extension problem for a class of translation invariant forms, *J. Operator Theory*, in press.

D.Z. Arov and M.G. Krein [1983], On computation of entropy functionals and their minimums, *Acta Sci. Math.* **45**, 51–66.

Gr. Arsene, Z. Ceauşescu and C. Foias [1980], On intertwining dilations VIII, *J. Operator Theory* **4**, 55–91.

G.A. Baker, Jr. and P. Graves-Morris [1981], *Padé Approximants*, Addison-Wesley, Reading, MA.

J.A. Ball [1979], A lifting theorem for operator models of finite rank on multiply connected domains, *J. Operator Theory* **1**, 3–25.

J.A. Ball [1982], A noneuclidean Lax-Beurling theorem with applications to matricial Nevanlinna-Pick interpolation, in *Operator Theory: Advances and Applications*, OT 4, Birkhäuser-Verlag, Basel, pp. 67–84.

J.A. Ball [1983], Interpolation problems of Pick-Nevanlinna and Loewner types for meromorphic matrix functions, *Integral Equations and Operator Theory* **6**, 804–840.

J.A. Ball [1984], Invariant subspace representations, unitary interpolants and factorization indices, in *Operator Theory: Advances and Applications*, OT 12, Birkhäuser-Verlag, Basel, pp. 11–38.

J.A. Ball [1988], Nevanlinna-Pick interpolation: generalizations and applications, in *Surveys of Some Recent Results in Operator Theory I* (J.B. Conway and B.B. Morrell, eds.), Pitman, 1988, pp. 51–94.

J.A. Ball, N. Cohen and A.C.M. Ran [1988], Inverse spectral problems for regular improper rational matrix functions, in *Operator Theory: Advances and Applications*, OT 33, Birkhäuser-Verlag, Basel, pp. 123–173.

J.A. Ball, N. Cohen and L. Rodman [1989], Zero data and interpolation problems for rectangular matrix polynomials, preprint.

J.A. Ball, C. Foias, J.W. Helton and A. Tannenbaum [1987], On a local nonlinear commutant lifting theorem, *Ind. Univ. Math. J.* **36**, 693–709.

J.A. Ball and I. Gohberg [1985], A commutant lifting theorem for triangular matrices with diverse applications, *Integral Equations and Operator Theory* **8**, 205–267.

J.A. Ball and I. Gohberg [1986a], Classification of shift invariant subspaces of matrices with Hermitian form and completion of matrices, in *Operator Theory: Advances and Applications*, OT 19, Birkhäuser-Verlag, Basel, pp. 23–85.

J.A. Ball and I. Gohberg [1986b], Pairs of shift invariant subspaces of matrices and noncanonical factorization, *Linear and Multilinear Algebra* **20**, 27–61.

J.A. Ball and I. Gohberg [1986c], Shift invariant subspaces, factorization, and interpolation for matrices, I. The canonical case, *Linear Algegra and its Applications* **74**, 87–150.

J.A. Ball, I. Gohberg and L. Rodman [1987], Minimal factorization of meromorphic matrix functions in terms of local data, *Integral Equations and Operator Theory* **10**, 437–465.

J.A. Ball, I. Gohberg and L. Rodman [1988], Realization and interpolation of rational matrix functions, *Operator Theory: Advances and Applications*, OT 33, Birkhäuser-Verlag, Basel, pp. 1–72.

J.A. Ball, I. Gohberg and L. Rodman [1989a], Two-sided Nudelman interpolation problem for rational matrix functions, in *Collection dedicated to M. Cotlar on occasion of his 75th birthday*, Marcel-Dekker, in press.

J.A. Ball, I. Gohberg and L. Rodman [1989b], Sensitivity minimization and bitangential Nevanlinna-Pick interpolation in contour integral form, *Proc. IMA Workshop on Signal Processing*, Springer-Verlag, in press.

J.A. Ball, I. Gohberg and L. Rodman [1989c], Tangential interpolation problems for rational matrix functions, in *Proc. Symposia in Applied Mathematics*, Amer. Math. Soc., Providence, RI, in press.

J.A. Ball, I. Gohberg and L. Rodman [1989d], Two-sided Lagrange-Sylvester interpolation problems for rational matrix functions, in *Proc. Symposia in Pure Mathematics*, Providence, RI, in press.

J.A. Ball, I. Gohberg and L. Rodman [1989e], Common minimal divisors and multiples for rational matrix functions, *Linear Algebra and its Applications*, in press.

J.A. Ball, I. Gohberg and L. Rodman [1989f], Simultaneous interpolation problems for rational matrix functions, preprint.

J.A. Ball and J.W. Helton [1982], Lie groups over the field of rational functions, signed spectral factorization, signed interpolation, and amplifier design, *J. Operator Theory* **8**, 19–64.

J.A. Ball and J.W. Helton [1983], A Beurling-Lax theorem for the Lie group $U(m, n)$ which contains most classical interpolation, *J. Operator Theory* **9**, 107–142.

J.A. Ball and J.W. Helton [1984], Beurling-Lax representations using classical Lie groups with many applications, II. $GL(n, C)$ and Wiener-Hopf factorization, *Integral Equations and Operator Theory* **7**, 291–309.

J.A. Ball and J.W. Helton [1986a], Beurling-Lax representations using classical Lie groups with many applications, III. Groups preserving forms, *Amer. J. Math.* **108**, 95–174.

J.A. Ball and J.W. Helton [1986b], Interpolation problems of Pick-Nevanlinna and Loewner types for meromorphic matrix functions: parametrization of the set of all solutions, *Integral Equations and Operator Theory* **9**, 155–203.

J.A. Ball and J.W. Helton [1986c], Beurling-Lax representations using classical Lie groups with many applications, IV. $GL(n, R)$, $U^*(2n)$, $SL(n, C)$ and a solvable group, *J. Functional Analysis* **69**, 178–206.

J.A. Ball and J.W. Helton [1988a], Shift invariant subspaces, passivity, reproducing kernels and H-infinity-optimization, in *Contributions to Operator Theory and its Applications*, OT 35, Birkhäuser-Verlag, Basel, pp. 265–310.

J.A. Ball and J.W. Helton [1988b], Shift invariant manifolds and nonlinear analytic function theory, *Integral Equations and Operator Theory* **11**, 615–725.

J.A. Ball and J. Kang [1989], Matrix polynomial solutions of tangential Lagrange-Sylvester interpolation conditions of low McMillan degree, preprint.

J.A. Ball and A.C.M. Ran [1986], Hankel norm approximation of a rational matrix function in terms of its realization, in *Modelling, Identification and Robust Control*, Elsevier, pp. 285–296.

J.A. Ball and A.C.M. Ran [1987a], Local inverse spectral problems for rational matrix functions, *Integral Equations and Operator Theory* **10**, 349–415.

J.A. Ball and A.C.M. Ran [1987b], Global inverse spectral problems for rational matrix functions, *Linear Algebra and its Applications* **86**, 237–382.

J.A. Ball and A.C.M. Ran [1987c], Optimal Hankel norm model reduction and Wiener-Hopf factorization, I. The canonical case, *SIAM J. Control Optim.* **25**, 362–382.

H. Bart, I. Gohberg and M.A. Kaashoek [1979], *Minimal Factorization of Matrix and Operator Functions*, Birkhäuser-Verlag, Basel.

H. Bart, I. Gohberg and M.A. Kaashoek [1986], Wiener-Hopf equations with symbols analytic in a strip, in *Operator Theory: Advances and Applications*, OT 21, Birkhäuser-Verlag, Basel, pp. 39–74.

V. Belevitch [1970], Interpolation matrices, *Phillips Res. Reports* **25**, 337–369.

A. Ben-Artzi, R.L. Ellis, I. Gohberg and D.C. Lay [1987], The maximum distance problem and band sequences, *Linear Algebra and its Applications* **87**, 93–112.

H. Bercovici, C. Foias and A. Tannenbaum [1989], A spectral commutant lifting theorem, *Trans. Amer. Math. Soc.*, in press.

A. Beurling [1949], On two problems concerning linear transformations in Hilbert space, *Acta Math.* **81**, 239–255.

A. Bhaya and C.A. Desoer [1986], Necessary and sufficient conditions on Q $(= C(I + PC)^{-1})$ for stabilization of the linear feedback system $S(P,C)$, *Systems and Control Letters* **7**, 35–38.

Y. Bistritz [1983], Nested bases of invariants for minimal realizations of finite matrix sequences, *SIAM J. Control and Optim.* **21**, 804–821.

O.H. Bosgra [1983], On parametrizations for the minimal partial realization problem, *Systems and Control Letters* **3**, 181–187.

M.S. Brodskii [1971], *Triangular and Jordan Representations of Linear Operators and Intermediate Systems*, Transl. Math. Monographs, Vol. 32, Amer. Math. Soc., Providence, RI. (Russian original, 1969).

M.S. Brodskii and M.S. Livsic [1958], Spectral analysis of non-selfadjoint operator and intermediate systems, *Uspehi Mat. Nauk* **13**, 3–85; English translation: *Amer. Math. Soc. Transl.* **13** (1960), 265–346.

V.M. Brodskii, I. Gohberg and M.G. Krein [1970], The definition and basic properties of the characteristic function of a knot, *Funkt. Anal. i Prilozh.* **1**, 88–90.

A. Bultheel [1983] Applications of Padé approximants and continuous fractions in systems theory, in *Mathematical Theory of Networks and System* (P.A. Fuhrmann, ed.), Lecture Notes in Control and Information Sciences, Vol. 58, Springer-Verlag, Berlin, pp. 130–148.

A. Bultheel [1987], *Laurent Series and their Padé Approximations*, Birkhäuser, Boston.

C. Caratheodory [1954], *Theory of Functions of a Complex Variable*, Vol. 2, Chelsea, New York.

L. Carleson [1958], An interpolation for bounded analytic functions, *Amer. J. Math.* **80**, 921–930.

Z. Ceauşescu and C. Foias [1978], On intertwining dilations, V, *Acta Sci. Math.* **40**, 9–32.

Z. Ceauşescu and C. Foias [1979], On intertwining dilations, V (Letter to the Editor), *Acta Sci. Math.* **41**, 457–459.

B.C. Chang and J.B. Pearson [1984], Optimal disturbance reduction in linear multivariable system, *IEEE Trans. Auto. Control* **AC-29**, 880–888.

C.T. Chen [1973], A generalization of the inertia theorem, *SIAM J. Appl. Math.* **25**, 158–161.

M.T. Chen and C.A. Desoer [1982], Necessary and safficient conditions for robust stability of linear distributed feedback systems, *Int. J. Control* **35**, 255–267.

T. Constantinescu [1986], Schur analysis of positive block-matrices, in *Operator Theory: Advances and Applications*, OT 18, Birkhäuser-Verlag, Basel, pp. 191–206.

M. Cotlar and C. Sadosky [1985], Generalized Toeplitz kernels, stationarity and harmonizability, *J. d'Analyse Math.* **44**, 114–133.

M. Cotlar and C. Sadosky [1988], Toeplitz liftings of Hankel forms, in *Lecture Notes in Math.* **1032**, Springer-Verlag, pp. 22–43.

R. Curtain and K. Glover [1986], Robust stabilization of infinite dimensional systems by finite dimensional controllers, *Systems and Control Letters* **7**, 41–47.

Ju.L. Daleckii and M.G. Krein [1974], *Stability of Solutions of Differential Equations in Banach Space*, Amer. Math. Soc. Transl. Math. Monographs, Vol. 43, Providence, RI.

B.N. Datta and K. Datta [1986], Theoretical and computational aspects of some linear algebra problems in control theory, in *Computational and Combinatorial Aspects in Control Theory* (C.I. Byrnes and A. Lindquist, eds.), Elsevier, pp. 201–212.

P.J. Davis [1975], *Interpolation and Approximation*, Dover, New York.

Ph. Delsarte, Y. Genin and Y. Kamp [1979], The Nevanlinna-Pick problem for matrix-valued functions, *SIAM J. Appl. Math.* **36**, 47–61.

Ph. Delsarte, Y. Genin and Y. Kamp [1981], On the role of the Nevanlinna-Pick problem in circuit and system theory, *Circuit Theory and Applications* **9**, 177–187.

P. Dewilde and H. Dym [1981], Lossless chain scattering matrices and optimum linear prediction: the vector case, *Circuit Theory and Applications* **9**, 135–175.

P. Dewilde and H. Dym [1984], Lossless inverse scattering for digital filters, *IEEE Trans. Information Theory* **30**, 644–662.

B.W. Dickinson, M. Morf and T. Kailath [1974], A minimal realization algorithm for matrix sequences, *IEEE Trans. Auto. Control*, **AC-19**, 31–38.

W.F. Donoghue [1974], *Monotone Matrix Functions*, Springer-Verlag, New York-Heidelberg-Berlin.

J. Doyle [1981], Analysis of feedback systems with structured uncertainty, *IEEE Proc.* **129D**, 242–250.

J. Doyle [1983], Synthesis of robust controllers and filters, in *Proc. 22nd Conference on Decision and Control*, San Antonio, TX.

J. Doyle [1984], *Lecture Notes for ONR*, Honeywell Workshop on Advances in Multivariable Control, Minneapolis, MN.

J.C. Doyle, K. Glover, P.P. Khargonekar and B.A. Francis [1989], State space solutions to standard H_2 and H_∞ control problems, *IEEE Trans. Auto. Control* **AC-34**, 831–847.

J. Doyle and G. Stein [1981], Multivariable feedback design: concepts for a classical/modern synthesis, *IEEE Trans. Auto. Control* **AC-26**, 4–16.

H. Dym [1989], *J Contractive Matrix Functions, Reproducing Kernel Hilbert Spaces and Interpolation*, CBMS, No. 71, Amer. Math. Soc., Providence, RI.

H. Dym and I. Gohberg [1981], Extensions of band matrices with band inverses, *Linear Algebra and its Applications* **36**, 1–14.

H. Dym and I. Gohberg [1982/83], Extension of kernels of Fredholm operators, *J. d'Analyse Math.* **42**, 51–97.

H. Dym and I. Gohberg [1983a], Unitary interpolants, factorization indices and infinite Hankel block matrices, *J. Functional Analysis* **54**, 229–289.

H. Dym and I. Gohberg [1983b], Hankel integral operators and isometric interpolants on the line, *J. Functional Analysis* **54**, 290–307.

H. Dym and I. Gohberg [1986], A maximum entropy principle for contractive interpolants, *J. Functional Analysis* **65**, 83–125.

Yu.M. Dyukarev and V.E. Katsnel'son [1986], Multiplicative and additive classes of Stieltjes analytic matrix-valued functions, and interpolation problems associated with them, I, *Amer. Math. Soc. Transl.* (2) **131**, 55–70. (Russian original, 1981).

R.L. Ellis, I. Gohberg and D.C. Lay [1986], Band extensions, maximum entropy and permanence principle, in *Maximum Entropy and Bayesian Methods in Applied Statistics* (J. Justice, ed.), Cambridge University Press, Cambridge.

R.L. Ellis, I. Gohberg and D.C. Lay [1987], Invertible selfadjoint extensions of band matrices and their entropy, *SIAM J. Algebraic and Discrete Methods* **8**, 483–500.

I.P. Fedchina [1972], A criterion for the solvability of the Nevanlinna-Pick tangent problem, *Mat. Issled* **7**, 213–227 (in Russian).

I.P. Fedchina [1975], Tangential Nevanlinna-Pick problem with multiple points, *Doklady Akad. Nauk Arm. SSR* **61**, 214–218 (in Russian).

C. Foias [1978], Contractive intertwining dilations and waves in layered media, in *Proc. International Congress of Mathematicians*, Vol. 2, Helsinki, pp. 605–613.

C. Foias and A.E. Frazho [1986], On the Schur representation in the commutant lifting theorem, I, in *Operator Theory: Advances and Applications*, OT 18, Birkhäuser-Verlag, Basel, pp. 207–217.

C. Foias and A.E. Frazho [1989], *The Commutant Lifting Approach to Interpolation Problems*, Birkhäuser Verlag, to appear.

B.A. Francis [1987], *A Course in H_∞ Control Theory*, Springer-Verlag, Berlin-Heidelberg-New York.

B.A. Francis, J.W. Helton and G. Zames [1984], H_∞-optimal feedback controllers for linear multivariate systems, *IEEE Trans. Auto. Control* **AC-29**, 888–900.

B.A. Francis and G. Zames [1984], On H_∞-optimal sensitivity theory for siso feedback systems, *IEEE Trans. Auto. Control* **AC-29**, 9–16.

A.E. Frazho [1986], Three inverse scattering algorithms for the lifting theorem, in *Operator Theory: Advances and Applications*, OT 18, Birkhäuser-Verlag, Basel, pp. 219–248.

B. Fritzsche and B. Kirstein [1987a, 1988a], An extension problem for nonnegative Hermitian block Toeplitz matrices, I–IV, *Math. Nachrichten* **130**, 121–135; **131**, 287–297; **135**, 319–341.

B. Fritzsche and B. Kirstein [1987b, 1988b], A Schur type matrix extension problem, I–IV, *Math. Nachrichten* **134**, 257–271; **138**, 195–216.

G. Frobenius [1878], *Jour. Reine Angew. Math.* (Crelle) **86**, 146–208.

P.A. Fuhrmann [1976], Algebraic system theory: an analyst's point of view, *J. Franklin Inst.* **30**, 521–540.

P.A. Fuhrmann [1981], *Linear Systems and Operators in Hilbert Space*, McGraw-Hill, New York.

P.A. Fuhrmann [1983], A matrix Euclidean algorithm and matrix continued fraction expansion, *Systems Control Letters* **3**, 263–271.

F.R. Gantmakher [1959], *The Theory of Matrices*, Vols. I and II, Chelsea, New York.

J.B. Garnett [1981], *Bounded Analytic Functions*, Academic Press, New York.

J. Gilewicz [1978], *Approximants de Padé*, Springer-Verlag, Berlin.

K. Glover [1984], All optimal Hankel-norm approximations of linear multivariable systems and their L^∞ error bounds, *Int. J. Control* **39**, 1115–1193.

K. Glover [1986], Robust stabilization of linear multivariable systems: relations to approximation, *Int. J. Control* **43**, 741–766.

K. Glover and D. McFarlane [1989], Robust stabilization of normalized coprime factor plant descriptions with H_∞-bounded uncertainty, *IEEE Trans. Auto. Control* **34**, 821–830.

I. Gohberg [1971], On some questions of spectral theory of finite-meromorphic operator functions, *Izv. Armyan Akad. Nauk* **6**, 160–181 (Russian).

I. Gohberg [1972], The correction to the paper "On some questions of spectral theory of finite-meromorphic operator functions", *Izv. Armyan Akad. Nauk* **7**, 52 (Russian).

I. Gohberg (ed.) [1986], *I. Schur Methods in Operator Theory and Signal Processing*, Operator Theory: Advances and Applications, OT 18, Birkhäuser-Verlag, Basel.

I. Gohberg and M.A. Kaashoek [1987], An inverse spectral problem for rational matrix functions and minimal divisibility, *Integral Equaitons and Operator Theory* 10, 437–465.

I. Gohberg and M.A. Kaashoek [1988], Regular rational matrix functions with prescribed pole and zero structure, in *Operator Theory: Advances and Applications*, OT 33, Birkhäuser-Verlag, Basel, pp. 109–122.

I. Gohberg, M.A. Kaashoek and L. Lerer [1987], On minimality in the partial realization problem, *Systems and Control Letters* 9, 97–104.

I. Gohberg, M.A. Kaashoek, L. Lerer and L. Rodman [1981], Common multiples and common divisors of matrix polynomials, I. Spectral method, *Indiana J. Math.* 30, 321–356.

I. Gohberg, M.A. Kaashoek, L. Lerer and L. Rodman [1982], Common multiples and common divisors of matrix polynomials, II. Vandermonde and resultant matrices, *Linear and Multilinear Algebra* 12, 159–203.

I. Gohberg, M.A. Kaashoek, L. Lerer and L. Rodman [1984], Minimal divisors of rational matrix functions with prescribed zero and pole structure, in *Operator Theory: Advances and Applications*, OT 12, Birkhäuser-Verlag, Basel, pp. 241–275.

I. Gohberg, M.A. Kaashoek and A.C.M. Ran [1988], Interpolation problems for rational matrix functions with incomplete data and Wiener-Hopf factorization, in *Operator Theory: Advances and Applications*, OT 33, Birkhäuser-Verlag, Basel, pp. 73–108.

I. Gohberg, M.A. Kaashoek and A.C.M. Ran [1989a], Regular rational matrix functions with prescribed null and pole data except at infinity, *Linear Algebra and Applications*, in press.

I. Gohberg, M.A. Kaashoek and A.C.M. Ran [1989b], Matrix polynomials with prescribed zero structure in the finite complex plane, preprint.

I. Gohberg, M.A. Kaashoek and L. Rodman [1978a], Spectral analysis of families of operator polynomials and a generalized Vandermonde matrix, I. The finite-dimensional case, in *Topics in Functional Analysis, Advances in Math. Suppl. Studies* 3, 91–128.

I. Gohberg, M.A. Kaashoek and L. Rodman [1978b], Spectral analysis of families of operator polynomials and a generalized Vandermonde matrix, II. The infinite dimensional case, *J. Functional Analysis* 30, 359–389.

I. Gohberg, M.A. Kaashoek and H. Woerdeman [1989a], The band method for positive and contractive extension problems, *J. Operator Theory*, to appear.

I. Gohberg, M.A. Kaashoek and H. Woerdeman [1989b], The band method for positive strictly contractive extension problems: an alternative version and new applications, *Integral Equations and Operator Theory* 12, 343–382.

I. Gohberg, M.A. Kaashoek and H. Woerdeman [1989c], A maximum entropy principle in the general framework of the band method, preprint.

I. Gohberg, M.A. Kaashoek and H. Woerdeman [1989d], The band method for extension problems and maximum entropy, preprint.

I. Gohberg, M.A. Kaashoek and F. van Schagen [1982], Rational matrix and operator functions with prescribed singularities, *Integral Equations and Operator Theory* **5**, 673–717.

I. Gohberg and M.G. Krein [1969], *Introduction to the Theory of Linear Nonselfadjoint Operators*, Amer. Math. Soc., Providence, RI.

I. Gohberg, P. Lancaster and L. Rodman [1978a], Spectral analysis of matrix polynomials, I. Canonical forms and divisors, *Linear Algebra and its Applications* **20**, 1-44.

I. Gohberg, P. Lancaster and L. Rodman [1978b], Spectral analysis of matrix polynomials, II. The resolvent form and spectral divisors, *Linear Algebra and its Applications* **21**, 65-88.

I. Gohberg, P. Lancaster and L. Rodman [1978c], Representation and divisibility of operator polynomials, *Canadian Math. J.*, **30**, 1045–1069.

I. Gohberg, P. Lancaster and L. Rodman [1982], *Matrix Polynomials*, Academic Press, New York.

I. Gohberg, P. Lancaster and L. Rodman [1983], *Matrices and Indefinite Scalar Products*, Operator Theory: Advances and Applications, Vol. 8, Birkhäuser-Verlag, Basel.

I. Gohberg, P. Lancaster and L. Rodman [1986], *Invariant Subspaces of Matrices with Applications*, J. Wiley & Sons, New York.

I. Gohberg, L. Lerer and L. Rodman [1980], On factorization, indices and completely decomposable matrix polynomials, Technical Report 80–47, Dept. of Mathematical Sciences, Tel-Aviv University.

I. Gohberg and A.S. Marcus [1955], On a characteristic property of the kernel of a linear operator, *Doklady Akad. Nauk SSSR* **101**, 893–896 (Russian); **MR 17**, 769 (1956).

I. Gohberg and L. Rodman [1978], On spectral analysis of non-monic matrix and operator polynomials, I. Reduction to monic polynomials, *Israel J. Math.* **30**, 133–151.

I. Gohberg and L. Rodman [1979], On the spectral structure of monic matrix polynomials and the extension problem, *Linear Algebra and its Applications* **24**, 157–172.

I. Gohberg and L. Rodman [1981], Analytic matrix functions with prescribed local data, *J. d'Analyse Math.* **40**, 90–128.

I. Gohberg and L. Rodman [1983], Analytic operator valued functions with prescribed local data, *Acta Math. (Szeged)* **45**, 189–200.

I. Gohberg and L. Rodman [1986], Interpolation and local data for meromorphic matrix and operator functions, *Integral Equations and Operator Theory* **9**, 60–94.

I. Gohberg and S. Rubinstein [1987], Cascade decompositions of rational matrix functions and their stability, *Int. J. Control* **46**, 603–629.

I.C. Gohberg and E.I. Sigal [1971], On operator generalizations of the logarithmic residue theorem and the theorem of Rouché, *Math. USSR Sb* **13**, 603–625.

L.B. Golinskii [1983], On one generalization of the matrix Nevanlinna-Pick problem, *Izv. Akad. Nauk Arm. SSR Math.* **18**, 187–205.

W.B. Gragg and A. Lindquist [1983], On the partial realization problem, *Linear Algebra and its Applications* **50**, 277–319.

G.H. Golub and C.F. Van Loan [1989], *Matrix Computations*, The Johns Hopkins University Press, Baltimore and London.

R. Grone, C.R. Johnson, E. de Sá and H. Wolkowicz [1984], Positive definite completions of partial Hermitian matrices, *Linear Algebra and its Applications* **58**, 109–124.

P.R. Halmos [1961], Shifts on Hilbert space, *J. für die Reine und Angew. Math.* **208**, 102–112.

J.Z. Hearon [1977], Nonsingular solutions of $TA - BT = C$, *Linear Algebra and its Applications* **16**, 57–65.

M. Heins [1982], A bibliography of Pick-Nevanlinna interpolation and cognate questions, private communication.

H. Helson [1964], *Lectures on Invariant Subspaces*, Academic Press, New York.

J.W. Helton [1980], The distance of a function to H^∞ in the Poincaré metric, electrical power transfer, *J. Functional Analysis* **38**, 273–314.

J.W. Helton [1987], *Operator Theory, Analytic Functions, Matrices and Electrical Engineering*, CBMS, No. 68, Amer. Math. Soc., Providence, RI.

J.W. Helton and J.A. Ball [1982], The cascade decomposition of a given system vs. the linear fractional decomposition of its transfer function, *Integral Equations and Operator Theory* **5**, 341–385.

R.A. Horn and C.R. Johnson [1985], *Matrix Analysis*, Cambridge University Press, Cambridge.

C.R. Johnson and L. Rodman [1984], Inertial possibilities for completions of partial Hermitian matrices, *Linear and Multilinear Algebra* **16**, 179–195.

M.A. Kaashoek, C.V.M. van der Mee and L. Rodman [1982], Analytic operator functions with compact spectrum, II. Spectral pairs and factorization, *Integral Equations and Operator Theory* **5**, 791–827.

M.A. Kaashoek, C.V.M. van der Mee and L. Rodman [1983], Analytic operator functions with compact spectrum, III. Hilbert space case: inverse problems and applications, *J. Operator Theory* **10**, 219–250.

T. Kailath [1980], *Linear Systems*, Prentice Hall, Englewood Cliffs, NJ.

R.E. Kalman [1971], On minimal partial realization of a linear input/output map, in *Aspects of Networks and System Theory* (R.E. Kalman and N. De Claris, eds.), Holt, Rinehart and Winston, New York, pp. 385–407.

R.E. Kalman [1979], On partial realization, transfer functions, and canonical forms, *Acta Polytechnica Scandinavica* **31**, 9–32.

R.E. Kalman [1982], Realization of covariance sequences, *Operator Theory: Advances and Applications*, Vol. 4, Birkhäser-Verlag, Basel, pp. 331–342.

R.E. Kalman, P. Falb and M. Arbib [1969], *Topics in Mathematical System Theory*, McGraw-Hill, New York.

V.E. Katsnel'son [1987], Methods of J theory in continuous problems of analysis, I–IV, *Amer. Math. Soc. Transl.* **136**, 49–108. (Russian originals 1981, 1982, 1983, 1983).

M.V. Keldysh [1951], On eigenvalues and eigenfunctions of some classes of nonselfadjoint equations, *DAN SSSR* **77**(1), 11–14 (in Russian).

H. Kimura [1984], Robust stabilizability for a class of transfer functions, *IEEE Trans. Auto. Control* **AC-29**, 788–793.

H. Kimura [1987], Directional interpolation approach to H^∞-optimization and robust stabilization, *IEEE Trans. Auto. Control* **AC-32**, 1085–1093.

H. Kimura [1989], Conjugation, interpolation and model-matching in H^∞, *Int. J. Control* **49**, 269–307.

I.V. Kovalishina and V.P. Potapov [1982], Integral representations of Hermitian positive functions, Private translation by T. Ando, Sapporo, Japan. (Russian original, 1982).

I.V. Kovalishina and V.P. Potapov [1988], Indefinite metric in the Nevanlinna-Pick problem, *Amer. Math. Soc. Transl.* **138**(2), 15–19. (Russian original, 1974).

M.G. Krein and H. Langer [1970], Über die verallgemeinerten Resolventen und die charackteristische Funktion eines isometrisches Operators in Raume π_k, *Colloquia Mathematica Societatis Janos Bolyai* 5, *Hilbert Space Operators*, Tihany, Hungary, pp. 353–399.

M.G. Krein and A.A. Nudelman [1977], *Markov Moment Problem and Extremal Problems*, Amer. Math. Soc. Transl. Math. Monographs, Vol. 50, Providence, RI.

S.G. Krein and V.P. Trofimov [1969], Holomorphic operator-valued functions of several complex variables, *Functional Analysis Appl.* **3**, 330–331.

S.Y. Kung and D.W. Lin [1981], Optimal Hankel-norm model reductions: multivariable systems, *IEEE Trans. Auto. Control* **26**, 832–852.

P. Lancaster and M. Tismenetsky [1985], *Theory of Matrices with Applications*, 2nd ed., Academic press, Orlando.

S. Lang [1965], *Algebra*, Addison-Wesley, Reading, MA.

P.D. Lax [1959], Translation invariant subspaces, *Acta Math.* **101**, 163–178.

H. Lev-Ari and T. Kailath [1984], Lattice filter parametrization and modeling of nonstationary processes, *IEEE Trans. Information Theory* **30**, 2–16.

D.J.N. Limebeer and B.D.O. Anderson [1988], An interpolation theory approach to H^∞ controller degree bounds, *Linear Algebra and Application* **98**, 347–386.

C.C. MacDuffee [1956], *The Theory of Matrices*, Chelsea, New York.

A.S. Marcus [1958], On holomorphic operator functions, *Doklady Akad. Nauk SSSR* **119**, 1099–1102.

A.S. Marcus and I.V. Mereutsa [1973], On complete sets of roots of the operator equation corresponding to an operator bundle, *Izvestiya AN SSSR, Seriya Mathem.* **37**, 1108–1131 (Russian).

R. Nevanlinna [1929], Über beschränkte analytische Funktionen, *Ann. Acad. Sci. Fenn.* **A32**(7).

R. Nevanlinna [1970], *Analytic Functions*, Springer, Berlin-Heidelberg-New York.

N.K. Nikolskii [1986], *Treatise on the Shift Operator*, Springer-Verlag, Berlin.

N.K. Nikolskii and S.V. Khrushchev [1988], A function model and some problems in the spectral theory of functions, *Proc. Steklov Institute of Mathematics* **176**, 101–214. (Russian original, 1987).

A.A. Nudelman [1977], On a new problem of moment type, *Soviet Math. Doklady* **18**, 507–510.

A.A. Nudelman [1981], A generalization of classical interpolation problems, *Soviet Math. Doklady* **23**, 125–128.

L. Page [1970], Bounded and compact vectorial Hankel operators, *Transl. Amer. Math. Soc.* **150**, 529–539.

Ju.A. Palant [1970], On a method of testing for the multiple completeness of a system of eigenvectors and associated vectors of a polynomial operator bundle, *Vestnik Har'kov. Gos. Univ.* **34**(4), 1–13 (in Russian).

V.I. Paulsen and S.C. Power [1988], Lifting theorems for nest algebras, *J. Operator Theory* **20**, 311–327.

V.I. Paulsen, S.C. Power and R.R. Smith [1989], Schur products and matrix completions, *J. Functional Analysis* **85**, 151–178. .

V.I. Paulsen, S.C. Power and J.D. Ward [1988], Semi-discreteness and dilation theory for nest algebras, *J. Functional Analysis* **80**, 76–87.

V.I. Paulsen and L. Rodman [1989], Positive completions of matrices over C^*-algebras, preprint.

G. Polya and G. Szegö [1972], *Problems and Theorems in Analysis*, Springer-Verlag, Berlin, New York.

V.P. Potapov [1960], Multiplicative structures of J-expansive matrix functions, *Amer. Math. Soc. Transl.* **15**(2), 131–244. (Russian original, 1955).

M. Rakowski [1989], *Zero-pole Interpolation for Nonregular Rational Matrix Functions*, Ph.D. Dissertation, Virginia Tech, Blacksburg, Virginia.

L. Rodman [1978], *Spectral Theory of Analytic Matrix Functions*, Ph.D. Thesis, Tel-Aviv University.

L. Rodman [1989], *An Introduction to Operator Polynomials*, in Operator Theory: Advances and Applications, Vol. 38, Birkhäser-Verlag, Basel.

M. Rosenblum and J. Rovnyak [1985], *Hardy Classes and Operator Theory*, Oxford University Press, New York.

J. Rovnyak [1963], *Some Hilbert Spaces of Analytic Functions*, Dissertation, Yale University.

L.A. Sakhnovich [1986], Factorization problems and operator identities, *Russian Math. Surveys* **41**, 1-64.

D. Sarason [1967], Generalized interpolation in H^∞, *Trans. Amer. Math. Soc.* **127**, 179-203.

D. Sarason [1985], Operator theoretic aspects of the Nevanlinna-Pick interpolation problems, in *Operators and Function Theory* (S.C. Power, ed.), Reidel, Dordrecht, pp. 279-314.

D. Sarason [1988], Angular derivatives via Hilbert space, *Complex Variables Theory and Applications* **10**, 1-10.

H.J.S. Smith [1861], *Phil. Trans. Roy. Soc. London* **151**, 293-326.

J. Snyders and M. Zakai [1970], On nonnegative solutions of the equation $AD + DA' = -C^*$, *SIAM J. Appl. Math.* **18**, 704-714.

B. Sz.-Nagy and C. Foias [1968], Dilatation des commutants d'operateurs, *C.R. Acad. Sci. Paris* **A266**, 493-495.

B. Sz.-Nagy and C. Foias [1970], *Harmonic Analysis of Operators on Hilvert Space*, American Elsevier, New York.

B. Sz.-Nagy and A. Koranyi [1956], Relations d'un problème de Nevanlinna et Pick avec la théorie des opérateurs de l'espace hilbertien, *Acta Math. Sci. Hungar.* **7**, 295-302.

B. Sz.-Nagy and A. Koranyi [1958], Operator theoretische Behandlung und Veralgemeinerung eines Problemkreises in der komplexen Funktionentheorie, *Acta Math.* **100**, 171-202.

T. Takagi [1924], On an algebraic problem related to an analytic theorem of Caratheodory and Fejer, *Japan J. Math.* **1**, 83-93.

M. van Barel and A. Bultheel [1989], A canonical matrix continued fraction solution of the minimal (partial) realization problem, *Linear Algebra and its Applications* **122/123/124**, 973-1002.

M.S. Verma [1989], Robust stabilization of linear time-invariant systems, *IEEE Trans. Auto. Control* **34**, 870-875.

M. Vidyasagar [1985], *Control System Synthesis: a Factorization Approach*, The MIT Press, Cambridge, MA.

M. Vidyasagar and H. Kimura [1986], Robust controllers for uncertain linear multivariable systems, *Automatica* **22**, 85-94.

J.L. Walsh [1960], *Interpolation and Approximation by Rational Functions*, Amer. Math. Soc., Providence, RI.

H.K. Wimmer [1974], Inertia theorems for matrices, controllability and linear vibrations, *Linear Algebra and its Applications* **8**, 337-343.

W.M. Wonham [1985], *Linear Multivariable Control*, Springer-Verlag, New York.

B.F. Wyman and M.K. Sain [1981], The zero module and essential inverse systems, *IEEE Trans. Circuits and Systems* **CAS-28**, 112–126.

B.F. Wyman and M.K. Sain [1987], Module theoretic zero structures for system matrices, *SIAM J. Control and Optim.* **25**, 86–99.

B.F. Wyman, M.K. Sain, G. Conte and A.M. Perdon [1989], On the zeros and poles of a transfer function, *Linear Algebra and its Application* **122/123/124**, 123–144.

D.C. Youla and M. Saitoh [1967], Interpolation with positive real functions, *J. Franklin Institute* **284**, 77–108.

N. Young [1986], Nevanlinna-Pick problem for the matrix-valued functions, *J. Operator Theory* **15**, 239–265.

G. Zames [1982], Feedback and optimal sensitivity: model reference transformations, multiplicative seminorms, and approximate inverses, *IEEE Trans. Auto. Control* **AC-23**, 301–320.

G. Zames and B.A. Francis [1983], Feedback, minimax sensitivity, and optimal robustness, *IEEE Trans. Auto. Control* **AC-28**, 585–601.

NOTATIONS AND CONVENTIONS

General

☐   end of proof, end of example

∅   empty set

$S \subset T, T \supset S$ inclusion (not necessarily proper) between sets T and S

$|S|$ number of elements in a finite set S

$\#\{\cdots\}$ number of elements in a finite set $\{\cdots\}$

$A := B$, $A \stackrel{\text{def}}{=} B$ the left hand side A is defined by its equality to the right hand side B

$A =: B$ the right hand side B is defined by its equality to the left hand side A

$\begin{pmatrix} q \\ p \end{pmatrix} = \frac{q!}{(q-p)!p!}$ if $0 \le p \le q$

$\begin{pmatrix} q \\ p \end{pmatrix} = 1$ if $p = q = -1$

$\begin{pmatrix} q \\ p \end{pmatrix} = 0$ for all other integer values of p and q

$\delta_{ij} = \begin{cases} 1 & \text{if } i = j \\ 0 & \text{if } i \ne j \end{cases}$ Kronecker index

Spaces and sets

R real numbers

$i\mathsf{R}$ imaginary axis

C complex numbers

$\mathsf{C}^* = \mathsf{C} \cup \{\infty\}$

$\operatorname{Re} z, \operatorname{Im} z$ real and imaginary parts of the complex number z

$\Pi^+ = \{z \in \mathsf{C} : \operatorname{Re} z > 0\}$ open right half plane

$\Pi^- = \{z \in \mathsf{C} : \operatorname{Re} z < 0\}$ open left half plane

$\mathsf{T} = \{z \in \mathsf{C} : |z| = 1\}$ unit circle

$\mathcal{D} = \{z \in \mathsf{C} : |z| < 1\}$ open unit disk

$\mathcal{D}_e = \{z \in \mathsf{C} : |z| > 1\} \cup \{\infty\}$ exterior of the open unit disk

$\overline{\mathcal{D}} = \{z \in \mathsf{C} : |z| \le 1\}$ closure of $\mathcal{D}$

$\partial \mathcal{D} = \mathsf{T}$ boundary of $\mathcal{D}$

$\partial \Pi^\pm = i\mathbf{R} \cup \{\infty\}$ boundary of Π^+

$\overline{\Pi^\pm} = \Pi^\pm \cup i\mathbf{R} \cup \{\infty\}$ closure of $\Pi^\pm$

For a set $\sigma \subset \mathbf{C}$, we denote by σ^c the complement of σ in $\mathbf{C} \cup \{\infty\}$

$\mathbf{C}^n$ the vector space of n-dimensional column vectors over $\mathbf{C}$

$\mathbf{R}^n$ the vector space of n-dimensional column vectors over $\mathbf{R}$

$\|x\|$ Euclidean norm of a vector in $\mathbf{C}^n$ or $\mathbf{R}^n$

$e_1, \ldots, e_n$ standard basis in $\mathbf{C}^n$ (or $\mathbf{R}^n$): e_j has its j-th component 1, all other
          components zeros (the dimension of e_j is clear from context)

$\mathbf{C}^{p \times q}$ the vector space of $p \times q$ matrices over $\mathbf{C}$

$\{0\}$ the zero subspace

$\mathrm{Span}\{x_1, \ldots, x_k\}$ the subspace spanned by $x_1, \ldots, x_k$

$\dim \mathcal{K}$ complex dimension of a subspace $\mathcal{K}$ (over $\mathbf{C}$)

$\mathcal{K}^\perp$ orthogonal complement to a subspace $\mathcal{K}$

$\mathcal{M} \dotplus \mathcal{N}$ direct sum of subspaces $\mathcal{M}$ and $\mathcal{N}$

Matrices

When convenient an $m \times n$ matrix (with complex entries) is thought of as a linear transformation $\mathbf{C}^n \to \mathbf{C}^m$ represented by this matrix with respect to the chosen bases in $\mathbf{C}^n$ and $\mathbf{C}^m$ (usually the standard bases: $e_1, \ldots, e_n$ and $e_1, \ldots, e_m$)

A^T transpose of a matrix A

A^* conjugate transpose of A

$\mathrm{rank}\, A$ the rank of a matrix A

$\sigma(A)$ spectrum of A (the set of all distinct eigenvalues)

$\|A\|$ spectral, or operator norm, of the matrix A: the largest singular value of A

$\mathrm{col}(Z_j)_{j=r}^s,\ \mathrm{col}(Z_r, Z_{r+1}, \ldots, Z_s)$ block column matrix $\begin{bmatrix} Z_r \\ Z_{r+1} \\ \vdots \\ Z_s \end{bmatrix}$

$\mathrm{row}(Z_j)_{j=r}^s$ block row matrix $[Z_r, Z_{r+1}, \ldots, Z_s]$

$\mathrm{diag}(Z_j)_{j=r}^s,\ \mathrm{diag}(Z_r, Z_{r+1}, \ldots, Z_s),\ Z_r \oplus \cdots \oplus Z_s,\ \bigoplus_{i=r}^s Z_i$

block diagonal matrix $\begin{bmatrix} Z_r & & & 0 \\ & Z_{r+1} & & \\ & & \ddots & \\ 0 & & & Z_s \end{bmatrix}$

I_N, I $N \times N$ identity matrix

αI where $\alpha \in \mathbf{C}$, is often shorened to α

C_L companion matrix of a matrix polynomial $L(z)$

J_α often stands for the $\alpha \times \alpha$ upper triangular nilpotent Jordan block

$\operatorname{Ker} A = \{x \in \mathbf{C}^n : Ax = 0\}$ kernel of an $m \times n$ matrix A

$\operatorname{Im} A = \{Ax \in \mathbf{C}^n : x \in \mathbf{C}^n\}$ image of A

$\operatorname{Ker}_\ell A = \{$all m-dimensional rows x such that $xA = 0\}$ left kernel of an $m \times n$
        matrix A

$\operatorname{Im}_\ell A = \{xA : x$ is an m-dimensional row$\}$ left image of A

$A|\mathcal{M}, A|_{\mathcal{M}}$ restriction of a matrix A (thought of as a linear transformation) to
        a subspace $\mathcal{M}$; often $\mathcal{M}$ is A-invariant

Partial order for hermitian matrices A and B is understood in the sense of Loewner order:

$A \geq B, B \leq A$ means that $A - B$ is positive semidefinite

$A > B, B < A$ means that $A - B$ is positive definite

We use in the book the *Riesz calculus* (the terminology is not common in matrix theory, but in operator theory it is standard, and we will use this terminology in the finite dimensional case as well). For a given $m \times m$ matrix A the Riesz calculus is the algebra homomorphism Ξ which maps the algebra of complex valued functions analytic on $\sigma(A)$ into the algebra of $m \times m$ matrices defined by

$$\Xi(f) = \frac{1}{2\pi i} \int_\gamma (zI - A)^{-1} f(z) dz.$$

Here $f(z)$ is analytic on $\sigma(A)$ and γ is a closed rectifiable contour without selfintersections that lies (together with its interior) in the domain of analyticity of $f(z)$ and surrounds the spectrum of A. This formula allows to define $f(A)$, the value of the function $f(z)$ on A, by

$$f(A) = \frac{1}{2\pi i} \int_\gamma (zI - A)^{-1} f(z) dz.$$

We frequently use the property (spectral mapping theorem) that

$$\sigma(f(A)) = \{f(z_0) : z_0 \in \sigma(A)\}.$$

Of particular interest are the Riesz projections. Given a set σ of eigenvalues of A, define the corresponding *Riesz projection* by

$$P = P(\sigma; A) = \frac{1}{2\pi i} \int_\gamma (zI - A)^{-1} p(z) dz,$$

where $p(z) \equiv 1$ (resp. $p(z) \equiv 0$) in a neighborhood of an eigenvalue z_0 of A which does (resp. does not) belong to σ. The homomorphism properties of Ξ ensure that P is indeed a projection. The kernel (resp. image) of P is the sum of the root subspaces of A corresponding to the eigenvalues off (resp. in) σ.

Resolvent set of a matrix A consists, by definition, of all the complex numbers which are not eigenvalues of A (again, this terminology is borrowed from operator theory).

Functions

$A^{[i]}(z)$ the i-th derivative of a (scalar or matrix) function $A(z)$

wno $f(z)$ the winding number of a scalar function $f(z)$

$O(|z|^k)$ class of (scalar or matrix) functions $f(z)$ with the property that $\|f(z)\| \leq M|z|^k$ for all $|z| < \varepsilon$, where M, ε are positive constants independent of z

$o(|z|^k)$ class of (scalar or matrix) functions $f(z)$ with the property that $\lim_{z \to 0}\left(\|f(z)\| \cdot |z|^{-k}\right) = 0$

$\mathrm{Res}_{z=z_0} f(z)$ residue of a (scalar or matrix) meromorphic function $f(z)$ at z_0

Analytic and meromorphic matrix functions

$\mathrm{Ker}(A; z_0)$ the right eigenspace of $A(z)$ at z_0

$SP\big(A(z_0)\big)$ the singular part of $A(z)$ at z_0

$\sigma\big(A(z)\big)$ spectrum of an analytic matrix function $A(z)$

Rational matrix functions

$\mathcal{R}$ field of scalar rational functions

$\mathcal{R}_n$ the vector space (over $\mathbf{C}$) of n-dimensional column vectors whose components are (scalar) rational functions

$\mathcal{R}_{m \times n}$ the vector space (over $\mathbf{C}$) of $m \times n$ matrices whose entries are rational functions

$\mathcal{R}(\sigma)$, $\mathcal{R}_n(\sigma)$, $\mathcal{R}_{m \times n}(\sigma)$ the sets of scalar, $\mathbf{C}^n$ and $\mathbf{C}^{m \times n}$ valued rational functions, respectively, with no poles in the set $\sigma \subset \mathbf{C}$

$\mathcal{R}^0(\sigma^c)$, $\mathcal{R}_n^0(\sigma^c)$, $\mathcal{R}_{m \times n}^0(\sigma^c)$ the sets of scalar, $\mathbf{C}^n$ and $\mathbf{C}^{m \times n}$ valued rational functions, respectively, with no poles in the set $\sigma^c = (\mathbf{C} \cup \{\infty\})\backslash\sigma$ and value zero at infinity; here $\sigma \subset \mathbf{C}$

$\mathcal{R}_{m \times n, x}(\sigma)$ the set of $m \times n$ rational matrix functions with at most x poles in σ (counted with multiplicities)

P_σ projection onto $\mathcal{R}_n(\sigma)$ along $\mathcal{R}_n^0(\sigma^c)$

$P_{\sigma^c}^0 = I - P_\sigma$ projection onto $\mathcal{R}_n^0(\sigma^c)$ along $\mathcal{R}_n(\sigma)$

$\delta(M)$ McMillan degree of a rational matrix function $M(z)$

$\|W\|_\infty = \sup\{\|W(ix)\| \colon x \in \mathbf{R}\}$ infinity norm in the half plane case

$\|W\|_\infty = \sup\{\|W(z)\| \colon |z| = 1\}$ infinity norm in the disk case

$\mathcal{S}_\sigma(W) = \{Wp \colon p \in \mathcal{R}_n(\sigma)\}$ left null-pole subspace over σ for an $m \times n$ rational matrix
   function $W(z)$

$\mathcal{S}^0_{\sigma^c}(W) = \{Wp \colon p \in \mathcal{R}^0_n(\sigma^c)\}$ complemented left null subspace for an $m \times n$ rational
   matrix function $W(z)$

$\mathcal{N}_{\{z_0\}}(W)$ coefficient form of the local null-pole subspace of $W(z)$ at z_0

$\mathcal{BR}(\Delta)$, $\mathcal{BR}_n(\Delta)$, $\mathcal{BR}_{m \times n}(\Delta)$ the sets of scalar, $\mathbf{C}^n$ or $\mathbf{C}^{m \times n}$ valued rational
   functions $F(z)$, respectively, with the property that $\sup_{z \in \Delta} \|F(z)\| < 1$ (in,
   particular $F(z)$ has no poles in Δ)

$I(w)$ the set of rational functions satisfying the interpolation condition w

$\mathcal{S}$ the set of stable rational matrix functions

Kernel functions

$\rho_\Delta(z, w) = z + \overline{w}$ if $\Delta = \Pi^+$

$\rho_\Delta(z, w) = 1 - z\overline{w}$ if $\Delta = \mathcal{D}$

$K_U(z, w) = \dfrac{J - U(z)JU(w)^*}{\rho_\Delta(z, w)}$

$\widehat{K}_U(z, w) = \dfrac{J - U(w)^* JU(z)}{\rho_\Delta(z, w)}$

Function spaces and operators

$L^2_N(0, \infty)$ the Lebesgue space of square integrable $\mathbf{C}^N$-valued functions on $(0, \infty)$

ℓ^2_N the space of square summable sequences $\{f_i\}_{i=0}^\infty$ with $f_i \in \mathbf{C}^N$

H^2_N the Hardy space of $\mathbf{C}^N$-valued functions $f(z) = \sum_{j=0}^\infty f_j z^j$ which are analytic
   in the unit disk and satisfy $\sum_{j=0}^\infty \|f_j\|^2 < \infty$

$L^2_N(\partial \mathcal{D})$ the Lebesgue space of square integrable $\mathbf{C}^N$-valued functions on the unit
   circle

$H^2_N(\Pi^\pm)$ the Hardy space of $\mathbf{C}^N$-valued square summable functions on the imaginary
   line that are nontangential limits (almost everywhere) of analytic
   functions in $\Pi^\pm$

$\mathcal{H}_F$ Hankel operator induced by a matrix function $F(z)$

$\mathcal{M}_F$, M_F multiplication operator by (scalar or matrix) function $F(z)$

INDEX